Wiley Series of Practical Construction Guides

M.D. Morris, P.E., Editor

William R. Park
CONSTRUCTION BIDDING FOR PROFIT

J. Stewart Stein
CONSTRUCTION GLOSSARY: AN ENCYCLOPEDIC REFERENCE AND MANUAL

Harvey V. Debo and Leo Diamant
CONSTRUCTION SUPERINTENDENT'S JOB GUIDE, Second Edition

Robert M. Koerner and Joseph P. Welsh
CONSTRUCTION AND GEOTECHNICAL ENGINEERING USING SYNTHETIC FABRICS

J. Patrick Powers
CONSTRUCTION DEWATERING: A GUIDE TO THEORY AND PRACTICE

Walter Podolny, Jr. and Jean M. Müller
CONSTRUCTION AND DESIGN OF PRESTRESSED CONCRETE SEGMENTAL BRIDGES

Ben C. Gerwick, Jr. and John C. Woolery
CONSTRUCTION AND ENGINEERING MARKETING FOR MAJOR PROJECT SERVICES

James E. Clyde
CONSTRUCTION INSPECTION: A FIELD GUIDE TO PRACTICE, Second Edition

Julian R. Panek and John Philip Cook
CONSTRUCTION SEALANTS AND ADHESIVES, Second Edition

Courtland A. Collier and Don A. Halperin
CONSTRUCTION FUNDING: WHERE THE MONEY COMES FROM, Second Edition

Harold J. Rosen
CONSTRUCTION MATERIALS FOR ARCHITECTURE

William B. Kays
CONSTRUCTION OF LININGS FOR RESERVOIRS, TANKS, AND POLLUTION CONTROL FACILITIES, Second Edition

Walter Podolny and John B. Scalzi
CONSTRUCTION OF CABLE-STAYED BRIDGES, Second Edition

Edward J. Monahan
CONSTRUCTION OF AND ON COMPACTED FILLS

Ben C. Gerwick, Jr.
CONSTRUCTION OF OFFSHORE STRUCTURES

David M. Greer and William S. Gardner
CONSTRUCTION OF DRILLED PIER FOUNDATIONS

James E. Clyde
CONSTRUCTION FOREMAN'S JOB GUIDE

Leo Diamant
CONSTRUCTION ESTIMATING FOR GENERAL CONTRACTORS

Richard G. Ahlvin and Vernon Allen Smoots
CONSTRUCTION GUIDE FOR SOILS AND FOUNDATIONS, Second Edition

B. Austin Barry
CONSTRUCTION MEASUREMENTS, Second Edition

John E. Traister and Paul Rosenberg
CONSTRUCTION ELECTRICAL CONTRACTING, Second Edition

Harold J. Rosen and Tom Heineman
CONSTRUCTION SPECIFICATIONS WRITING: PRINCIPLES AND PROCEDURES, Third Edition

Leo Diamant and C. R. Tumblin
CONSTRUCTION COST ESTIMATES, Second Edition

Thomas C. Schleifer
CONSTRUCTION CONTRACTORS' SURVIVAL GUIDE

A. C. Houlsby
CONSTRUCTION AND DESIGN OF CEMENT GROUTING: A GUIDE TO GROUTING IN ROCK FOUNDATIONS

CONSTRUCTION AND DESIGN OF CEMENT GROUTING

CONSTRUCTION AND DESIGN OF CEMENT GROUTING

A Guide to Grouting in Rock Foundations

A. C. HOULSBY, *A.S.T.C., F.I.E. Aust., F. ASCE*

A WILEY-INTERSCIENCE PUBLICATION
John Wiley & Sons, Inc.
NEW YORK / CHICHESTER / BRISBANE / TORONTO / SINGAPORE

Library of Congress Cataloging in Publication Data:

Houlsby, A. C. (A. Clive)
Construction and design of cement grouting: a guide to grouting in rock foundations / A.C. Houlsby.
p. cm.—(Wiley series of practical construction guides, ISSN 0271-6011)
"A Wiley-Interscience publication."
Includes bibliographical references.
ISBN 0-471-51629-5
1. Grouting. 2. Grout (Mortar) 3. Foundations—Design and construction. I. Title. II. Series.
TA755.H68 1990
690'.11—dc20 89-27981
CIP

10 9 8

FOREWORD

As a student of grouting for nearly forty years, I have had the opportunity to meet and discuss the subject with many individuals involved therein, as well as review virtually every written word on the subject. It is in this regard that I first became acquainted with Clive Houlsby. He had submitted written discussion to a published paper which I had authored. His discussion exhibited his own wide knowledge of the subject as well as an intense desire to stay totally informed, and perhaps, most significantly, complete objectivity and honesty. This is a quality often sadly lacking where grouting is concerned and many self-proclaimed "magicians" pervade. It is not unusual for such "experts" to present their ideas with a certain amount of mysticism, refusing to discuss them in any detail or offer any documentation for their substantiation, let alone share in any detail their own experiences. Another sad reality to the grouting field is a tendency to "sweep under the rug" those projects where the results have been less than satisfactory. Such is not the case with Clive Houlsby, who is always anxious to share his knowledge and experience (both good and otherwise) and objectively discuss any matter pertaining to the subject.

From my first introduction to this unusual man, a long-term relationship has developed, and I have come to recognize him as the world's most knowledgeable authority on the cement grouting of rock foundations. Through the years, I have grown to know Houlsby well and have, among many other professional activities, been privileged to work with him in presentation of the annual Short Course on the Fundamentals of Grouting sponsored by the University of Missouri-Rolla and held each year since 1979.

Clive Houlsby is a rare individual who quietly goes about his work, humble, curious beyond description, meticulous and a near perfectionist, who attends to every detail, as well as one who assumes the validity of nothing, unless supported by valid documentation. Whereas the grouting field is filled with many "experts," it is seldom that the opinions espoused are backed up by either long-term monitoring or other valid documentation. Such is not the case with Houlsby, whose ideas, although they at first might seem revolutionary, are always validated by thorough documentation, whether it be through research, field observation, long-term monitoring of actual field applications, or, most often, all of the above.

As Engineer of Grouting for the Water Resources Commission of New South Wales, Australia, Houlsby authored and produced what is, I believe, the first ever and probably only "how to do it" Grouting Manual. That publication, although intended only for in-house use, has since its introduction in 1972 undergone four editions and seven printings, has been produced in three languages, and is in use throughout the entire world.

Houlsby entered the grouting field in 1954 as a young engineer working in the field on a grouting crew. Following many years of work with the crews in the field, he progressively rose to more important duties, always, however, maintaining close contact with grouting matters until his final position of Leader of Dam Safety and Monitoring for the Water Resources Commission. In order to have the time to write this volume and to follow other creative pursuits, he took early retirement from the Commission in 1986.

Although he warns the reader in the first chapter that this volume is to be used as a "guide only," what follows is a complete treatise on the subject of cement grouting in rock. Despite the fact that it forms only a small part of the overall volume, all of the material found in the original Grouting Manual, although updated and greatly expanded, is included. This book provides complete coverage of the subject including sections on site investigation, geological and permeability assessments, reasonable improvements to be expected from grouting along with limitations therewith, and the formulation of a rational grouting program when considering all of the factors inherent with the particular site being considered. A complete discussion of materials and constituents and their formulation into optimal mixtures for the individual site condition is included. Several sections are provided on the actual execution of the work in the field with easy-to-follow instructions, and there are many examples of factors that can be expected to occur and the appropriate actions to take in remediation.

The entire text is filled with simple sketches and drawings to aid the reader in understanding the text, and many tables are included which simplify field work and assure good quality. There is included, in fact, so much useful data for job performance that every grouting project site should be provided with a copy for routine reference. Finally, many case histories, which provide the results of various practice (both good and poor), and examples of actual grouting application are included. Of particular interest will be the many examples of field grouting projects, including follow-up inspections to evaluate the long-term effectiveness. In this regard, actual practices used are discussed in a factual manner so that the reader may learn from the extensive prior experience of others. As is typical of Houlsby, every detail of such prior work is provided, including tables of grout takes, order of placement, and the benefits derived. Houlsby's fondness for the field grouting operations, as well as his high regard for the men that actually make it happen, is amply illustrated throughout by his simple, easy-to-follow presentation.

It is thus, indeed, gratifying to introduce this first ever, complete treatise on Cement Grouting in Rock Foundations.

JAMES WARNER

Mariposa, California

SERIES PREFACE

Congratulations! You've just bought a profit-making tool that is inexpensive and requires no maintenance, no overhead, and no amortization. Actually, it will increase in value for you each time you use this volume in the Wiley Series of Practical Construction Guides. This book should contribute toward getting your project done under budget, ahead of schedule, and out of court.

For nearly quarter of a century, over 50 books on various aspects of construction and contracting have appeared in this series. If one is still valid, it is "updated" to stay on the cutting edge. If it ceases to serve, it goes out of print. Thus you get the most advanced construction practice and technology information available from experts who use it on the job.

The Associated General Contractors of America (AGC) statistician advises that the construction industry now represents close to 10% of the gross national product (GNP), some 410 billion dollars worth per year. Therefore, simple, off-the-shelf books won't work. The construction industry is unique in that it is the only one where the factory goes out to the buyer at the point of sale. The constructor takes more than the normal risk in operating a needed service business.

Until the advent of the series, various single books (many by professors), magazine articles, and vendors' literature constituted the total source of information for builders. To fill this need, this series has provided solid usable information and data for and by working constructors. This has increased the contractors' earning capacity while giving the owner a better product. Profit is not a dirty word. The Wiley Series of Practical Construction Guides is dedicated to that cause.

M. D. Morris, P.E.
Series Editor

Ithaca, New York
November 1989

PREFACE

Maybe you are one of the majority who think that grouting is just something done in the bathroom between the tiles. If so, this book will introduce you to other sorts of grouting, and in particular to the process of filling open cracks or spaces in rock foundations with grout made of cement and water. This is injected through drill holes and penetrates through the cracking systems or voids. Afterwards the cement sets to provide a permanent filling that reduces seepage flow through the foundation and strengthens it.

Seems easy, doesn't it? But good grouting calls for a lot of skill, and this isn't quickly gained. This book aims to help in developing it. However, every grouting site has different peculiarities, so you can't just transpose techniques from one to another; an element of art comes into successful grouting and cannot be put into words. I have taken the guidance here as far as is possible—to the brink of the art phase—but remember that this is not a recipe book to be followed precisely. It is a guide.

Foundation grouting is mostly used for reducing permeability under dams, but it also strengthens weak foundations and helps in mining, tunnel driving, and shaft sinking.

The book is intended for anyone who has involvement in grouting: grout operators, inspectors, supervisors, engineers, geologists, academics, and students. It is a hands-on guide covering the subject fully. It explains how to do grouting, describes the best equipment, shows how to handle problems, discusses exploring a new site and the designing of grouting, suggests details for contract specifications, gives a history of grouting, and many other things.

Writing this book has been a retirement project. After 32 years of active involvement in every aspect of this sort of grouting (and other sorts too), I have enjoyed writing down this accumulated experience. A lot of it has never been put on paper before. Some has not even been presented at the grouting courses I regularly give. So you have here the fullest treatment of the subject yet compiled by anyone. Hopefully, it will improve the quality of grouting generally; such improvement is needed on some of the jobs I see and hear about, regrettably. These days it is no longer convincing for members of the "black magic" fraternity to try to excuse shoddy grouting with a cloak of mystery. All is revealed!

I hope you find the book helpful.

Clive Houlsby

ACKNOWLEDGMENTS

My thanks to Jim Warner, Norbert Schmidt, Wally Baker, Don Bruce, Ray Cope, Pete Aberle, Bill Clarke, and other friends in the American grouting fraternity, for their encouragement in the writing of this book. And in England, thanks go to Stuart Littlejohn.

Back home in Australia, Ulf Krause deserves particular mention for his many years of detailed plotting of grouting results, and his meticulous statistical analyses of them. Then there is the work of the engineers, inspectors, geologists and grouting crews who participated in the various grouting jobs over many years.

CONTENTS

CHAPTER 1

Introduction

Grouting is the process of injecting fluids that set into cracks or voids. The most commonly used grout in rock foundations is cement and water; this book is a guide on how to grout with these.

WHAT FOUNDATION GROUTING IS If there is need to fill (or sometimes only partly fill) open cracks, joints, holes, cavities, and so forth in a foundation, this is done by:

1. Drilling holes into the foundation to intercept them. Figure 1.0-0A shows an example.
2. Then pumping grout into the holes. It enters the cracks, joints, or whatever and flows through them and into others, as Figure 1.0-0B shows.

The grout is pushed through the cracking system by pressure from the hole into those openings within reach. It stops moving when it can't be pushed further. When its cement component sets (takes at least 12 hours usually), a hard filling results if the grouting has been done properly.

SCOPE OF BOOK In this book you will find guidance, advice, information, comments, and a lot of other things about grouting, with the emphasis on construction. But site investigation and design are equally as important for success, so these are also covered, although more briefly.

The book is arranged as a reference manual. You don't have to read through it from beginning to end (unless, of course, you are a beggar for punishment). Once the principles of grouting are understood, you can delve into the book anywhere your interest takes you; there you will also find cross references to matters associated with the point of interest.

A word about the format: because most readers will be more interested in getting on with the job than in going into all the fine details, many sections are arranged with the main, essential information given in a generalized manner at first; explanations and fuller details follow for those who want them.

A CAUTION

A casual reading may give the impression that grouting is easy—just a matter of following rules. It is nothing of the sort. Although it can nowadays be classed as an engineered process, it has the peculiarity that there is more art to it than in most engineering work. At times it can be difficult and complex, calling for much experience and skill, and these instances cannot readily be described in words. They are given here as far as practicable, but it must be understood that there is more to good grouting than is written in this or any other book on the subject. Please note well that

**THIS BOOK IS ONLY A GUIDE.
IT IS NOT A RECIPE BOOK AND IS
NOT TO BE FOLLOWED IMPLICITLY.**

Generalized guidelines are given (often in the form of comments or discussion), but adaptation to the particular site conditions of your job is always necessary. There is a strong chance that poor grouting will result if you just follow the guidelines dogmatically, without intelligent thought. Because this has been known to happen, regrettably, by people who have unthinkingly followed suggestions in some of my previous writings and lectures, the warning is stressed throughout this

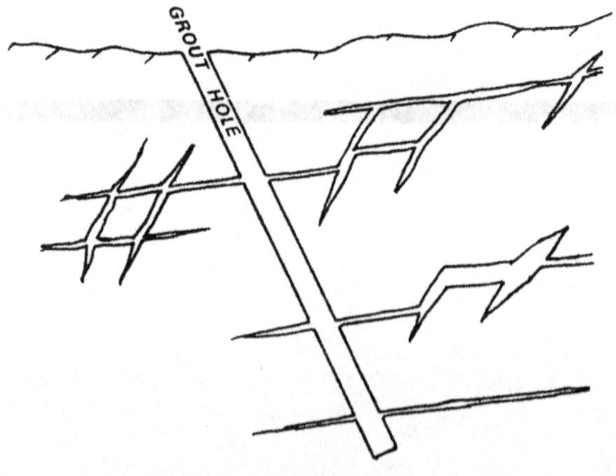

Figure 1.0-0A. *Grout hole intercepting cracks.*

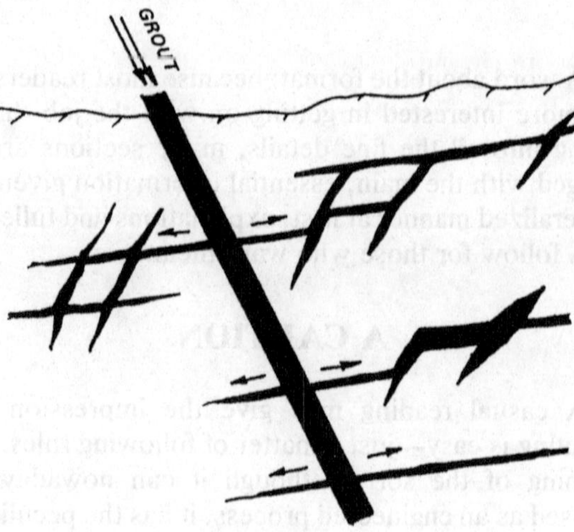

Figure 1.0-0B. *Grout traveling down hole and filling cracks.*

book perhaps even ad nauseam—to curb them and solace me.

While every effort has been made to provide sound and accurate guidance, descriptions, advice, and whatever, no undertaking is given or implied as to their appropriateness or correctness. My opinions and experiences are scattered throughout the book; note that not every grouting practitioner feels the same about some things. Where wide differences of opinion may exist, both sides of the question are explained as far as possible.

While most other types of engineering can safely follow a recipe or rules, grouting cannot and never will be able to. If you can't see why, have a look at some of the vagaries of nature to be contended with discussed in Section 8.1. One of the aims of this book is to advance the science and art of grouting so that standards of work are improved; it will be retrograde if, instead, some readers use this as a recipe book. Please don't!

If you would like a crash course on the subject of grouting, try the following sections:

1.2-1 to 1.2-5	Types of grouting and purposes.
2.1	Layout of gear and functions.
4.1-1	Water : cement ratios.
6.1-1 and 6.1-2	How grout flows in cracks.
5.1-1 and 5.1-2	Standpipes, grout caps, and surface preparation.
7.2-1	How to grout a hole.
7.3-1 and 7.3-6	What to do about leaks and connections.

These should give you a general idea of the subject. If you would like to continue the guided crash course, try the following sections:

5.4-1 to 5.4-6	Stage grouting and methods.
5.5-1 to 5.5-3	Closure grouting.
5.3-1 and 5.3-2	Water testing.
6.2-1	Pressures for grouting.
4.5-1 to 4.5-3	Durability.
8.2-3	The lugeon unit.

1.1 AN OUTLINE OF CEMENT GROUTING

1.1-1 Grouting from Beginning to End

This section is a rundown on all the main steps in cement grouting, from investigation of the site to completion of the job. As well as putting each step into perspective, it provides a quick index as to where to find these steps explained.

(*a*) *Investigation* If no grouting has been previously done on the site, a campaign investigating the geology and the permeability situation lets the grouting design be tailored properly to the conditions. This not only saves money in the long run on many sites by avoiding the need for the hit-or-miss or try-it-and-see sort of waste but properly engineered grouting should result.

Investigation methods for the geology are discussed in Section 8.1 and for the permeability in Section 8.2.

(*b*) *Deciding Whether Grouting Is Necessary* When the details of the site are known, a decision is then made as to whether grouting is really needed. Of course, in some locations the need for grouting is so obvious that this step is hardly necessary; there is, however, probably a need to decide just how extensive the grouting should be in such places.

For decisions about the need for grouting and how much of it to do, see Section 8.3.

(*c*) *Design* If it is decided that grouting is necessary, its extent and intensity are decided on, and the hole pattern, stage arrangements, grouting method, surface preparations, and other matters are then determined.

For guidance on design and hole layouts see Section 8.4, for stage arrangements and grouting methods see Section 5.4, and for surface preparations go to Section 5.1.

(*d*) *Contract Arrangements* If the work is to be done by contract, a specification and drawings are prepared, the contract is awarded, and experienced (hopefully) engineering supervision and Inspectors obtained. Contractural arrangements are discussed in Chapter 9.

(*e*) *Site Preparation* The surface at the grout hole positions is prepared appropriately. This can variously involve cleanup, a grout cap, standpipes, perhaps slush grouting, and other activities. See Section 5.1 for details.

(*f*) *Equipment* The grouting machinery and other equipment is set up ready for grouting. The general layout and the functions of the main items are discussed in Section 2.1. Separately, the main items are commented on in the following sections:

Mixers	Section 2.2
Agitators	Section 2.3
Pumps	Section 2.4
Circulation lines	Section 2.7
Pressure gauges	Section 2.5
Valves	Section 2.5
Packers	Section 2.5

(*g*) *Drilling* When the surface arrangements are ready, grout holes are drilled in preparation for grouting and then washed clean ready to receive the grout. Drills and drilling are discussed in Section 5.2

(*h*) *Water Testing* Water should be pumped into the hole through a meter for about 15 minutes before starting the grouting of each stage. This measures the permeability, which needs to be known when assessing the progress of grouting later. The test also rehearses the grouting somewhat and can give early warning of leaks, rock movements, connections, and other possible troubles for which suitable preparations can then be arranged. This step is described in Section 5.3.

(*i*) *Hook-up* This is the name given to the operation of preparing everything ready to start grouting of a hole. It means getting all the gear ready, with the circulation line hooked on to the hole, making sure adequate supplies are at hand, and having grout operators in position ready and raring to go. This process is included in Section 7.1.

(*j*) *Commencing Grouting* A commencing water : cement ratio appropriate to the hole is selected, the grout is mixed, sent on to an agitator, and then pumped around the circulation line past the grout hole. When a return has been obtained to the agitator, the valve on the hole is opened cautiously and grout is pumped into the hole. Consult Section 7.2 for details.

(*k*) *Looking for Trouble* A careful watch is kept for troubles such as leaks, connections underground to other grout holes, movement of rock, loose standpipes, and so on. The water test usually gives early warning of such problems, but not always. If any of these occur, appropriate action is necessary. Treatment of such problems is discussed in Section 7.3.

(*l*) *Reviewing the Mix* After grouting of a hole has been in progress sufficiently long to indicate trends (often about 15 minutes), the suitability of the water : cement ratio is reviewed, and changes are made if the grouting will be improved by doing so. Further reviews and adjustments are made as necessary during the course of the grouting. See Section 7.4.

(*m*) *Completion* When the hole will not take any more grout (i.e., refusal), the grouting is almost finished. Good practice is to continue applying pressure for a further 15 minutes after refusal. This is also discussed in Section 7.2.

(*n*) *Assessment* Afterwards, a postmortem of each grout application should be made to see whether it was done in the best way. This constructive analysis enables improvements to be made on the next holes and is part of the process of tailoring grouting to suit the site. In addition, overall assessment of areas where the grouting is approaching completion is necessary to decide whether more holes are needed, before regarding the grouting as finished.

Assessment of each hole is described in Section 7.4 and overall assessments in Section 7.5.

(*o*) *Long-Time Performance* If grouting is not done properly, it can be washed away by seepage. Placement of instrumentation and regular surveillance of seepage flows, pressures, and so on to pick up any nondurable grouting are the finale to a thorough grouting job, particularly at dams. Nondurable grouting is deplored in Section 4.5.

1.1-2 The Concealed Nature of Grouting

One of the facts of life about grouting is that you can't see what is happening. An experienced person can deduce how the grout is flowing and how well the grouting is proceeding, but there is no direct way of knowing this. A variety of methods have been tried, of course, such as putting radioactive tracers in the grout (and trying to pick up progress from sensors in nearby holes), coloring the grout, scanning the rock with rays of various sorts to pick up changes as the grout enters the scanned field, and other ingenious methods. But none of these are really direct, and most are no improvement on an experienced person.

This discourages some types of people who like to see the fruits of their work. These are the people who like to build bridges, buildings, roads, and so on, where they get the satisfaction of making something that is visible.

Perhaps it requires a person of a particular temperament to handle grouting properly; unsympathetic and ribald descriptions aside, could this be one who has patience, persistence, and perseverance with the unexciting, invisible work? A sense of humor is a great asset too. And do you think that somebody from Down Under has a natural advantage in being able to look up under the grouting done from On Top?

1.2 TYPES AND PURPOSES OF GROUTING

The following outline is intended for those who may want to get an overall idea of the scope of engineering grouting. It includes all forms, not just cement grouting.

1.2-1 Some Purposes of Grouting

A. To reduce the flow of seepage through a foundation. Typical examples are shown in Figures 1.2-1A and 1.2-1B. Under concrete gravity dams, most of the foundation seepage is stopped by the grout curtain. Drain holes pick up the small quantity of seepage passing the grouting as shown in Figure 1.2-1A.

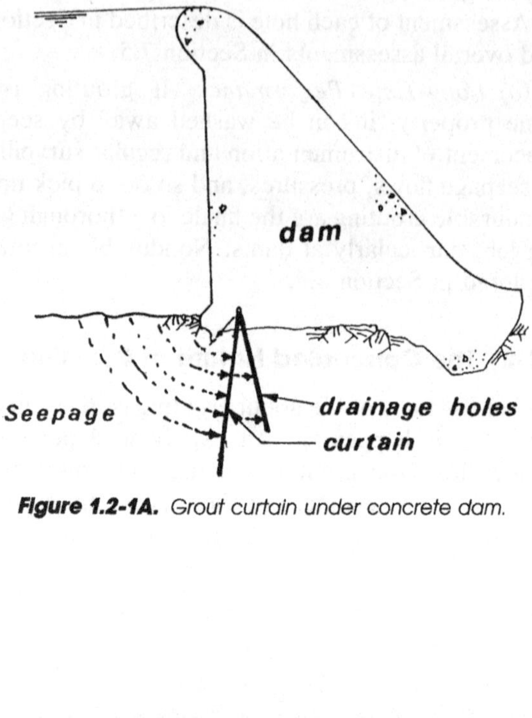

Figure 1.2-1A. *Grout curtain under concrete dam.*

At fill dams, the grout curtain also has the purpose of stopping much of the foundation seepage (Figure 1.2-1B).

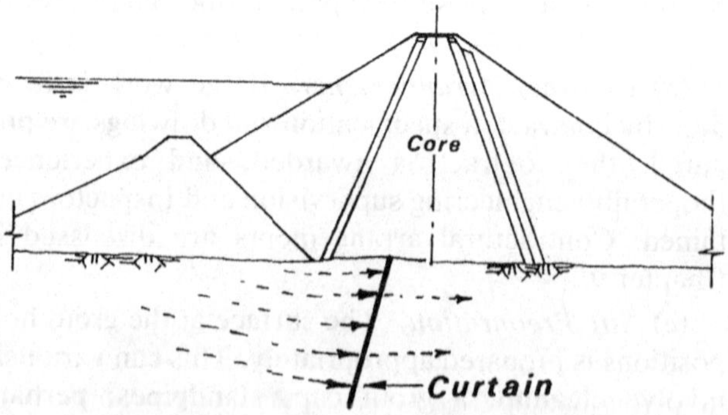

Figure 1.2-1B. *Grout curtain under earth dam.*

B. To strengthen a foundation. Some weak rock foundations can be strengthened by grouting to take a structural load (Figure 1.2-1C). This is known as blanket or consolidation grouting.

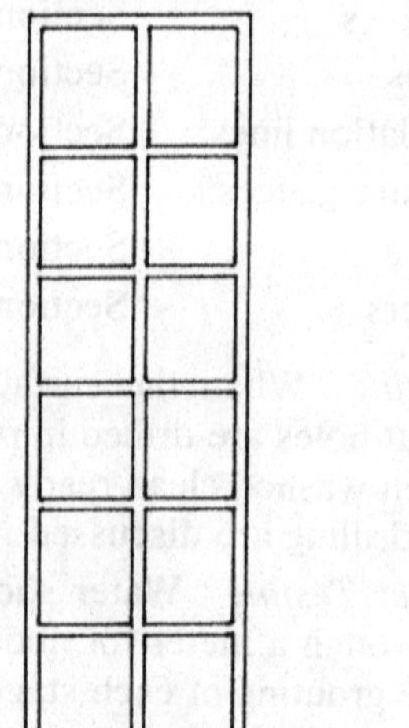

Figure 1.2-1C. *Weak foundation under heavy building strengthened by grouting.*

C. To assist tunneling or mining by reducing water inflow or strengthening the foundation ahead of a drive. An example is in Figure 1.2-1D, where a ring of grout holes is used for grouting the area about to be excavated.

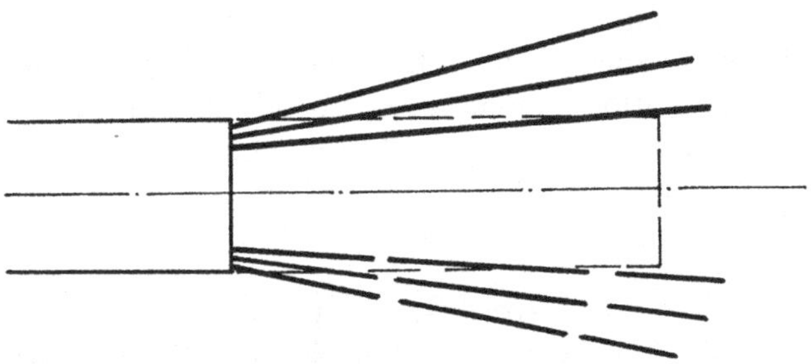

Figure 1.2-1D. *Grouting ahead of a drive.*

D. To assist shaft sinking. Grouting can be used to reduce inflow and strengthen the foundation for the next stage of a shaft sinking operation as indicated in Figure 1.2-1E.

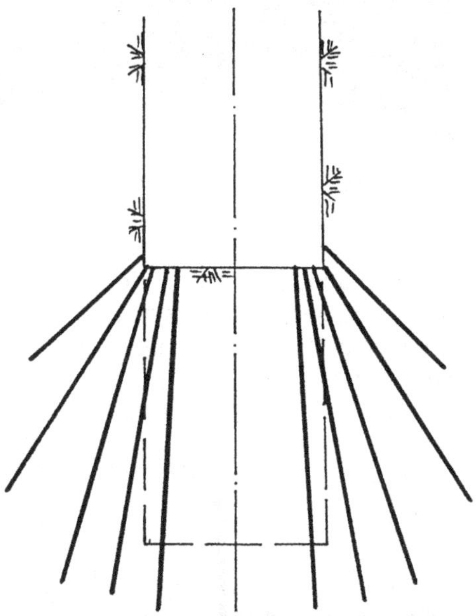

Figure 1.2-1E. *Grouting below a shaft.*

E. To fill voids where concrete cannot. Concrete pumped in a tunnel to make a lining will frequently slump away from the roof of the tunnel and will not rise up into high spots of excavation. Grouting fills the resulting overbreak void using holes that are either formed in the concrete or drilled after (Figure 1.2-1F). This type of grouting is also known as backpack or contact grouting.

When concrete is placed around a steel liner, access for vibration of the concrete is often difficult, leaving small voids under the penstock. These are grouted by means of holes through the liner using a vacuum to exhaust the void and then letting grout in to the evacuated space (Figure 1.2-1G). This process is sometimes called interface grouting.

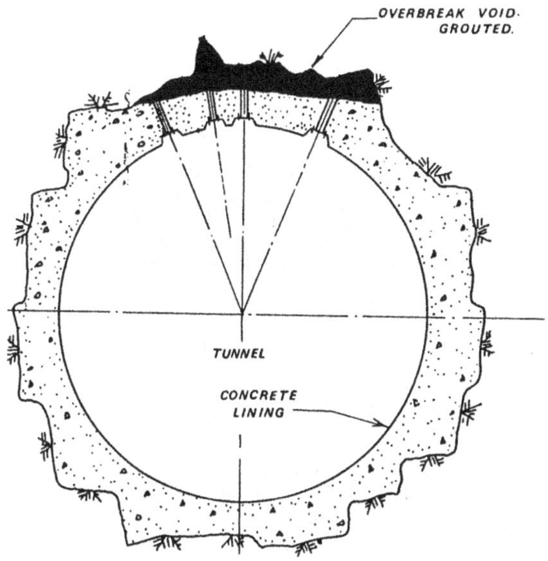

Figure 1.2-1F. *Overbreak void filled by grouting.*

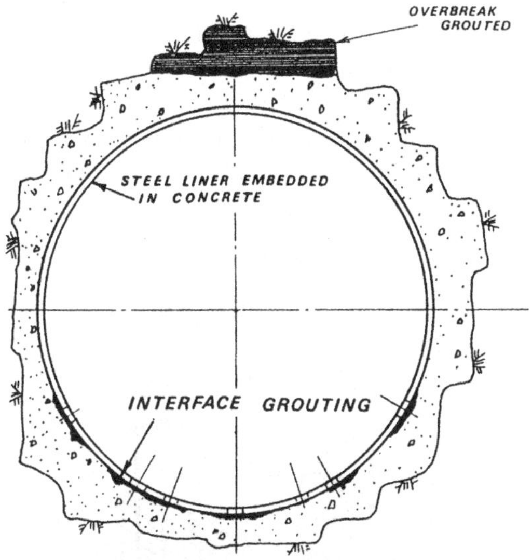

Figure 1.2-1G. *Void under steel liner filled by vacuum grouting.*

F. To fill contraction joints between large blocks of concrete. Contraction joints in large masses of concrete such as dams open up shortly after placement as the result of contraction of the concrete following dissipation of the heat of hydration and other shrinkage processes. The continuity of the arch action is restored across these joints by filling them with grout. Figure 1.2-1H shows an example at an arch dam.

G. To repair the effects of subsidence under structures. Figure 1.2-1I shows compaction grouting used to repair effects of subsidence. In this process a very stiff sand–cement grout is pumped in at high pressure

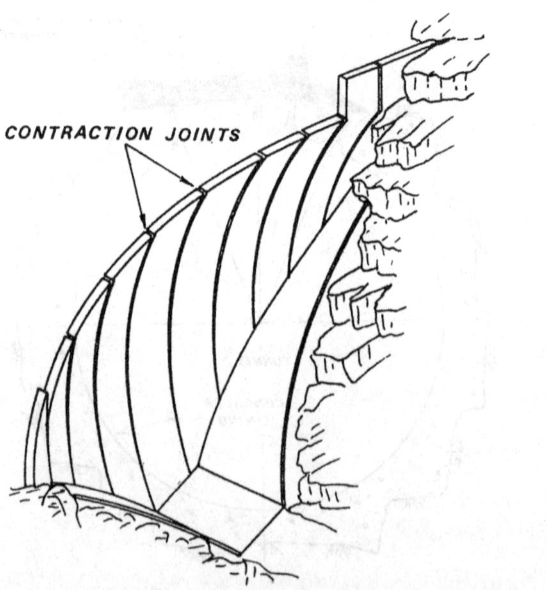

Figure 1.2-1H. *Contraction joints to be grouted at an arch dam.*

to form bulbs or lenses of grout. These compress any soft materials in the foundation and may also be used to jack up structures if required.

H. To make concrete. In this unusual way for making concrete, the coarse aggregate is placed in the forms and then sand–cement grout is piped into it to fill the voids. It is a convenient method for concreting in awkward positions where nonshrink concrete is needed, perhaps underwater. Fragile components can be concreted without damage, or heavy-weight concrete made easily. Crushed pieces of old concrete may be used instead of coarse aggregate ("recycled concrete"). Figure 1.2-1J is an imaginative sketch of various situations. Grout is shown emerging from injection tubes. Various names given to this process include Colcrete, Prepakt, Grouted Concrete, and Pre-placed Aggregate Concrete.

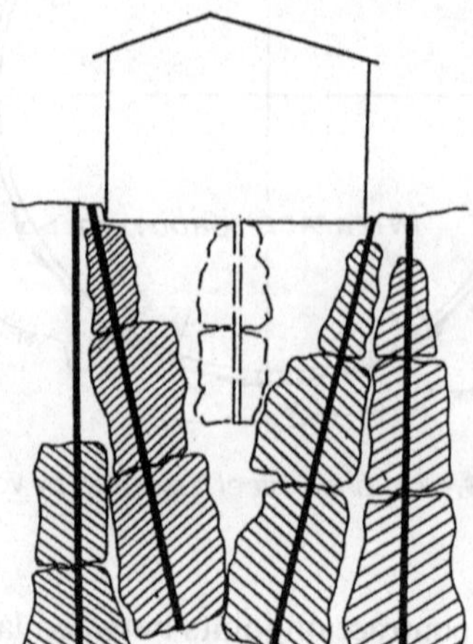

Figure 1.2-1I. *Structure leveled by grouting underneath.*

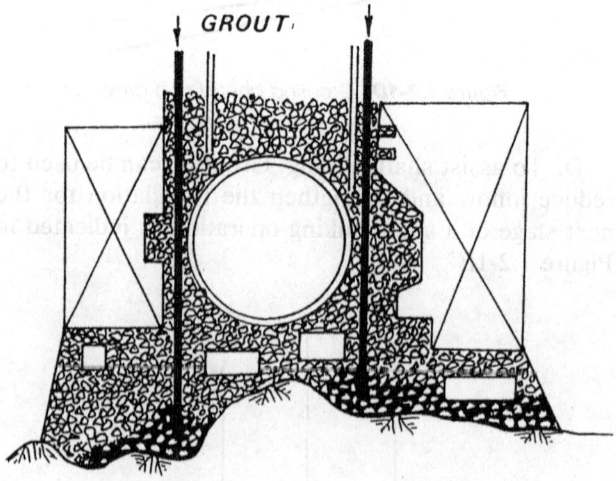

Figure 1.2-1J. *Grouted concrete.*

I. To fill abandoned mines. If access within a mine is dangerous because of a collapsing roof, for example, grouting may be carried out through holes from above (see Figure 1.2-1K).

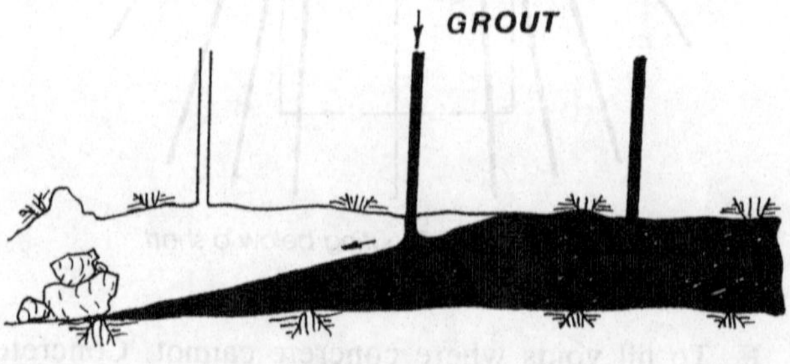

Figure 1.2-1K. *Underground spaces filled with grout containing cheap fillers such as fly ash or tailings.*

1.2-2 Grouting of Gravel and Sand

There is a difference between grouting of cracks in rock and grouting of voids in gravel or sand. Cement grout is only useful in coarse gravel free of finer material, where it can penetrate reasonably well (see Figure 1.2-2A).

The cement grout can only travel in the larger voids. In the finer-sized materials, voids are too small to accept cement grout. However, better penetration might be achieved to some other pockets of coarse

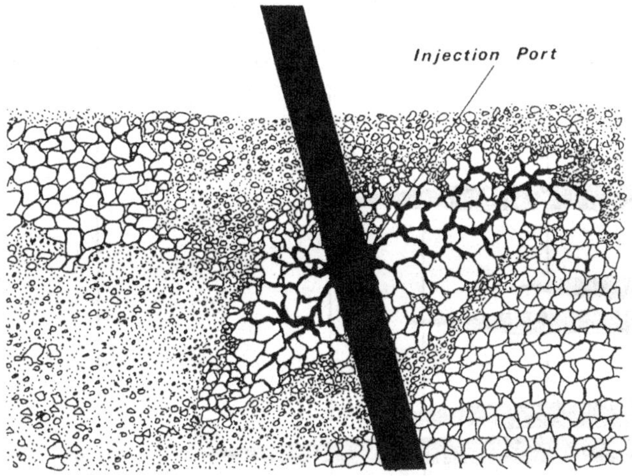

Figure 1.2-2A. *Cement grout, shown black, penetrating out from grout hole in voids between large pieces of gravel but not in voids in finer materials.*

gravel by using excess pressure to fracture paths through the foundation as shown in Figure 1.2-2B. This fracturing can cause damage at some sites and is not always permissible.

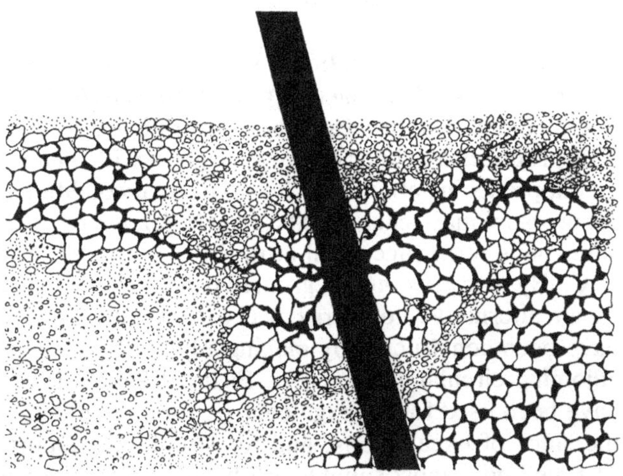

Figure 1.2-2B. *Penetration improved in alluvium through fracture paths.*

Because of the unsuitability of cement grout in sands and fine gravel, different grouts are used such as the following.

CHEMICAL GROUTS These grouts do not have any particles to restrict their flow through fine voids. Some formulations are nearly as fluid as water, others are more viscous; low viscosity gives best penetration usually. Several chemicals are mixed just before injection of the grout. These react after predetermined periods to form a gel when the grout is in place. The periods can be varied from several minutes to an hour or two.

Chemical grouts are relatively expensive and require expertise for effective use. They may not always be permanent, and it is possible to extrude some out of their placed positions. In conditions where dehydration of thick fillings can occur, loss of water from the grout can reduce the volume of gel to only that of the chemicals in it—sometimes as low as 20 or 30% of the original volume.

CLAY GROUTS These are made of clay and water, with perhaps an additive. Bentonite is the usual clay, but some local clays have been successfully used after fine grinding. Extensive testing is always necessary before use.

1.2-3 Injection into Gravel and Sands

Because grout holes in these materials will rarely stay open against internal collapse, some form of casing is necessary to line the holes. Casing is placed with the aid of sacrificed spear points, Odex eccentric bits, "overburden" drilling, or one of a number of proprietary placing methods. However, the most popular injection method is the "sleeve pipe" or "tube-a-manchette" casing, which has holes through it every so often, covered with soft rubber sleeves that open under pressure to let grout out into the job. The grout is supplied to the desired holes by means of a loose internal tube carrying grout from the surface: this tube has two packers that are seated above and below the selected holes and thus isolate them from others. This method has been adapted for injection into rock foundations (known as MPSP); see Section 5.4-12 for details.

1.2-4 Grouting of Silts and Clays

Apart from the coarser silts, these are difficult to effectively grout because their fine voids are too small for even low-viscosity chemical grouts to penetrate. If a barrier against seepage is required, jet grouting (see Section 1.2-6) can provide this in some cases. This method can also provide strengthening of the walls of an excavation or the roof of a tunnel.

1.2-5 Comparison of Grouts

Figure 1.2-5A compares the suitability of various grouts in different foundation materials. This figure is only for the purpose of giving an idea of the uses for

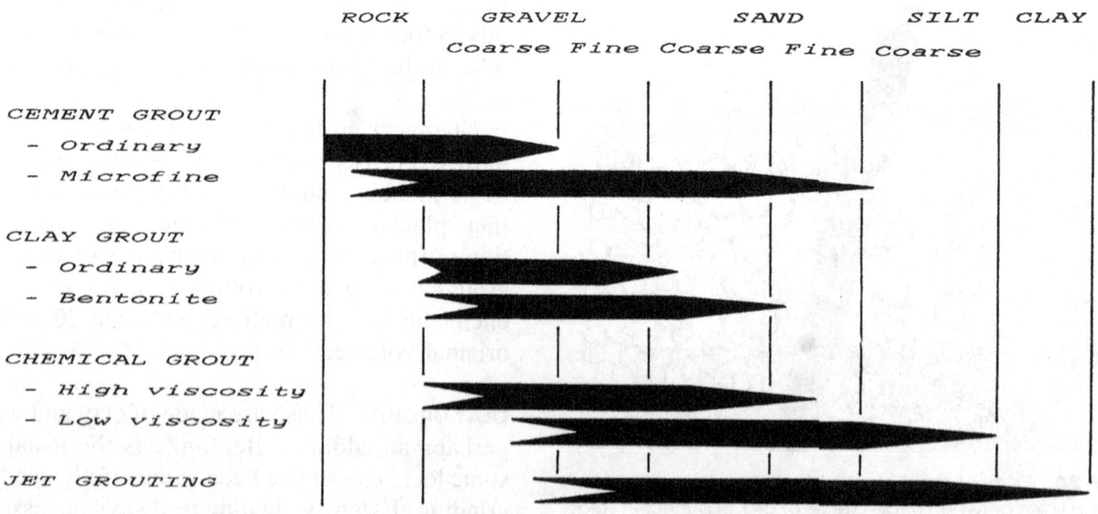

Figure 1.2-5A. *Comparison of grouts.*

the various grouts. It is deliberately vague because anything more precise would be a deception.

Clay grouts are listed separately, but in fact they are usually used with chemical grouts and sometimes with cement.

Clay and chemical grouts are normally too expensive to use in rock foundations. Cement gives a more economical and more durable job. However, there are instances where multiple fine jointing will not accept ordinary cement (see Section 6.1-4), and it is necessary to turn to clay and chemical grouts, if the higher cost can be justified. Where the possible lack of permanency of these cannot be tolerated, microfine cements (see Section 3.1-6) have been successful, but at greater cost still.

Microfine cements can also be used in fine gravels and sands. Their grain sizes are usually sufficiently small for penetration of these materials as alternatives to clay and chemical grouts.

Although this book is about cement grouts in rock foundations, some of the equipment and methods described are similar to those used for the other grouts. The References for this chapter list books for further reading that include references to these other grouts.

1.2-6 Jet Grouting

This process is mentioned for completeness. As Figure 1.2-5A shows, jet grouting is an alternative to chemical and clay grouts in the finer materials. The method erodes soft foundations by means of very high pressure jets and then fills the resulting space with cement grout. Several variations are in use:

A. In the simplest, the grout itself is the high-pressure cutting medium; fine material is flushed out the top of the hole, and the remaining coarser fraction mixes with the grout to form a type of mixed-in-place filling. This works best in very soft foundations and makes a "pile" about 1.5 to 2 ft ⟨0.5 to 0.7 m⟩* diameter.

B. For firmer foundations, where greater effort is needed to erode the space, air and water are the cutting medium. The process is shown in Figure 1.2-6A. This produces a cylindrical mass of grout with a diameter approaching about 5 ft ⟨1.5 m⟩ in good conditions. Jet pressure may reach 9,000 psi ⟨60 MPa⟩; lesser pressures have been successful at times.

C. By using two holes, grouted panels can be made as an alternative to slurry walls. Continuous walls have been constructed by making adjoining panels or adjoining circular masses (see Figure 1.2-6B). These have been used for seepage control, for strengthening the sides of excavations, and for structural support. The method is most suitable in silts and clays, but can also find use in sands and some fine gravels as an alternative and more positive treatment than chemical grouting.

Jet grouting was initially a development by British engineers working in Pakistan (Nicholson, 1963) and later taken up in Japan and elsewhere.

* Metric equivalents are given in angle brackets throughout the text.

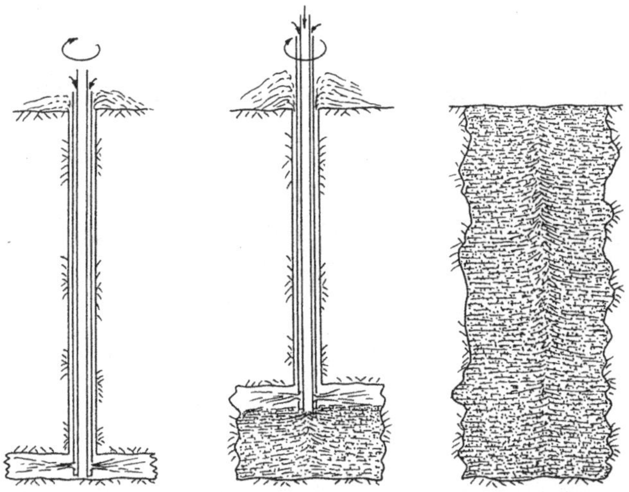

a b c

Figure 1.2-6A. Jet grouting using air and water jets. (a) Hole drilled to full depth of treatment. Tube inserted. Air and water jets turned on and start cutting space around hole. Tube rotated slowly. Fine material flushed out at top. (b) Tube lifted slowly and grout injected through it into the space made by jets. Coarser material falls into grout. (c) Completed grouting. A mixture of coarser material and grout.

1.2-7 Demolition Grout

To round off this discussion of the various types of grout, and passing from constructive grouts to the opposite, a brief mention is made of demolition grout. It is used for excavation of concrete or rock where the shock from explosives must be avoided and perhaps should not really be classed as a grout, although loosely called such.

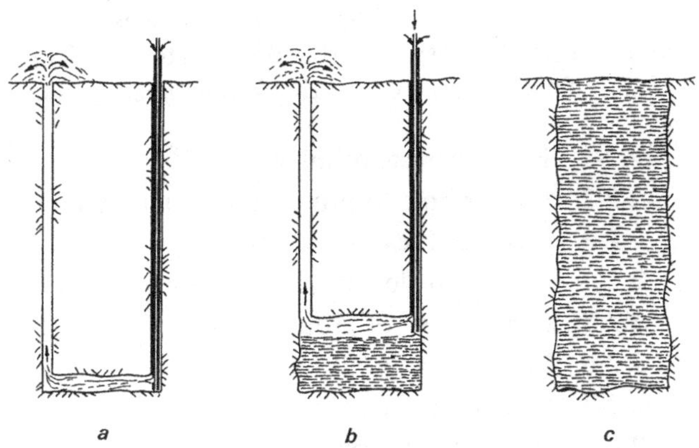

a b c

Figure 1.2-6B. Wall construction using jet grouting. (a) High-pressure jets in one hole force eroded materials out another. (b) Grout injected below jets fills eroded space. (c) Grouted panel completed.

With one type of grout, drill holes are half filled with water, and granules are then poured into them slowly. The granules are gently compacted with a rod until the hole is full. They expand with considerable force over the next 5 to 60 minutes and can crack concrete or rock, enabling its subsequent removal in hunks. Low temperatures produce the slowest reaction. After the first swelling, the grout can be made to swell further by wetting it. Considerable heat is generated. The grout is toxic and protective clothing is necessary. It consists of silicate and an organic compound.

Another type is an inorganic lime compound that generates the expansion when it hydrates. Powder is mixed with water to form the grout and then poured into holes. Expansion occurs over the next 10 to 20 hours.

REFERENCES AND FURTHER READING

Baker, W. H. (1982). "Planning and performing structural chemical grouting." *Proceedings of Conference on Grouting in Geotechnical Engineering*. ASCE, New Orleans, 515–539.

Ewert, F.-K. (1984). *Rock Grouting, with Emphasis on Dam Sites*. Springer-Verlag, Berlin.

Grouts and Drilling Muds in Engineering Practice (1963). Butterworths, London.

Houlsby, A. C. (1982). "Cement grouting for dams." *Proceedings of Conference on Grouting in Geotechnical Engineering*. ASCE, New Orleans, 1–34.

International Congress on Large Dams (1985). Question 58. International Commission on Large Dams, Lausanne.

Karol, R. H. (1983). *Chemical Grouting*. Marcel Dekker, New York.

Mongilardi, E., and Tornaghi, R. (1986). "Construction of large underground openings and use of grouts." Presented in Beijing, September 1986. Rodio S.p.A., Milano, Italy.

Nicholson, A. J. (1963). *Grouts and Drilling Muds in Engineering Practice,* in discussion. Butterworths, London, 108.

Shibazaki, M., and Ohta, S. (1982). "A unique underpinning of soil solidification utilizing super-high pressure liquid jet." *Proceedings of Conference on Grouting in Geotechnical Engineering*. ASCE, New Orleans, 680–693.

Warner, J. (1982). "Compaction grouting—the first thirty years." *Proceedings of Conference on Grouting in Geotechnical Engineering*. ASCE, New Orleans, 694–707.

CHAPTER 2

Grouting Equipment

The grout is mixed, stored for a short while until needed, and then pumped into the grout holes. That's a brief description of the process, but as you might guess, there's a lot more to it than just that; this chapter explains in detail.

The mixers are key machines: woeful mixers produce woeful grouting—and a disturbingly large number of mixers fall in this latter category, I've noticed. Considerable discussion on this follows in Section 2.2.

This chapter is divided into the following sections:

2.1 Layout of equipment and functions of the items.
2.2 Grout mixers.
2.3 Agitators.
2.4 Grout pumps.
2.5 Valves, pressure gauges, washout gear, and packers.
2.6 Fittings on the grout holes.
2.7 Circulation lines and gear setup.
2.8 Water testing equipment.

2.1 GEAR LAYOUT AND FUNCTIONS OF MAIN ITEMS

The basic arrangement of gear for cement grouting is shown on Figure 2.1-0A. Arrows show the direction of grout flow, and the circled numbers indicate the following steps:

(1) The *MIXER* mixes the grout in batches.
(2) Then it sends the grout to the *AGITATOR* for storage.
(3) The *PUMP* draws from the agitator and pumps grout around the circulation line past the grout hole.
(4) *STANDPIPE FITTINGS* on the grout hole control the amount of grout going down the hole.
(5) Surplus grout, not used down the hole, circulates back to the agitator for later use.

2.2 GROUT MIXERS

These mix the grout ingredients, usually in batches. Cement, water, and any other ingredients are measured into the mixer, which then does the mixing and discharges the resulting batch of grout into the agitator or wherever it has to go next.

The technology of mixing has progressed a long way from the days of simple stirring devices. The history of this progress is outlined separately in Section 10.2-2. This section covers the following subjects:

2.2-1 Effectiveness of mixers.
2.2-2 Mixers able to produce high-quality grout.
2.2-3 Mixers of lesser efficiency.
2.2-4 Mixers for lower quality grout.
2.2-5 Other types of grout mixer.
2.2-6 Laboratory grout mixers.
2.2-7 How to mix grout.
2.2-8 Hints about mixing.
2.2-9 Testing for grout quality.
2.2-10 Properties of high-quality grout.
2.2-11 Improving lower quality grout by using additives.
2.2-12 Why grout cannot be a true colloidal suspension.

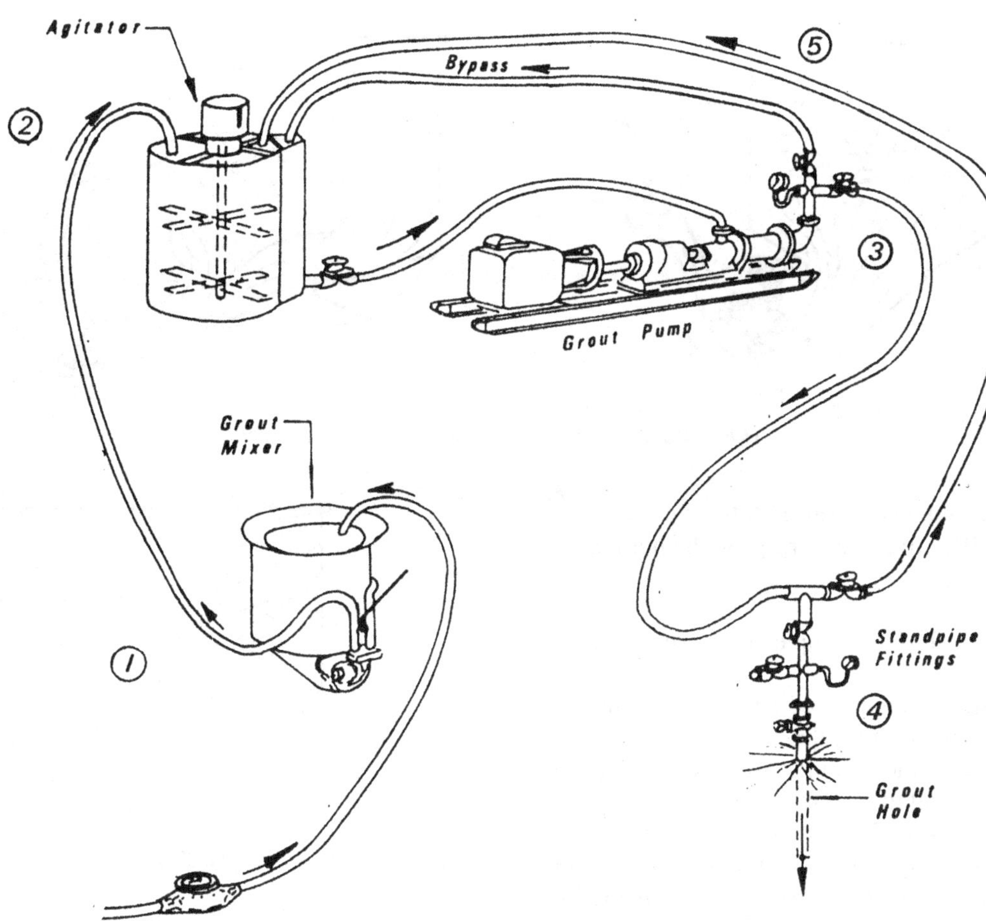

Figure 2.1-0A. Gear layout.

2.2-1 Effectiveness of Mixers

The quality of grouting depends greatly on the effectiveness of the mixer. See Section 2.2-10 for the reasons and for a description of high-quality grout. See also Section 6.1-5 for the effects on penetration in fine cracks and general comments throughout the book.

For comparative purposes here, mixers may be grouped into:

- Those able to produce high-quality grout
- Mixers of lesser efficiency
- Mixers for lower quality work

Opinions about this grouping differ greatly among grouting people, so please understand that the classifications shown and discussed here reflect personal opinions and experiences of this author. The classification of various brands of machines into the groupings is based only on technical considerations; economics does not come into this.

2.2-2 Mixers Able to Produce High-Quality Grout

GENERAL NAME "Semi-colloidal," "High-speed mixer," "High-shear mixer," or combinations of these names are variously used. They are also loosely called "Colloidal," which is not strictly correct; although grout produced in these machines approaches a colloidal condition, it remains essentially a mechanical suspension (see Section 2.2-12).

FEATURES

High-speed rotor running at more than 1,500 revolutions per minute (rpm) in a small chamber.

Vortex drum feeding the rotor (in some types).

Mixing time = 15 seconds (minimum) in the most efficient of the mixers and up to a minute or more in others.

After mixing, the rotor can be used to pump the

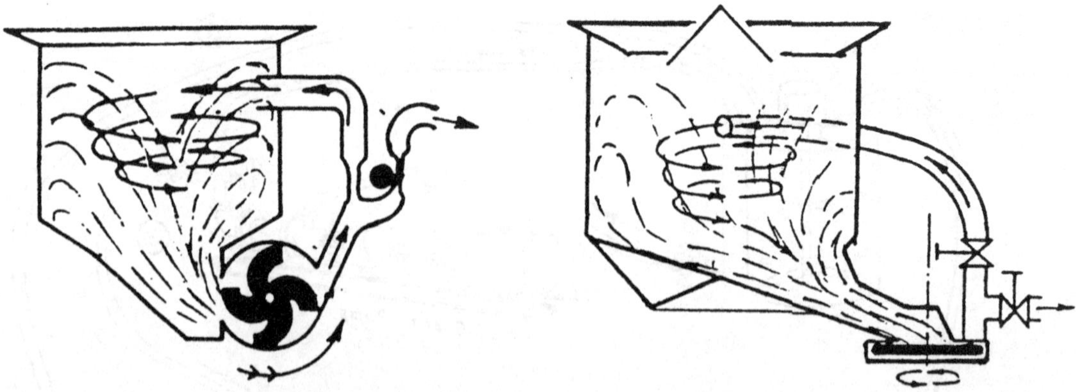

Figure 2.2-2A. Colcrete (left) and Cemix (right) mixers.

batch of grout away to an agitator. Pumping pressure is sufficient to raise the grout higher than the mixer.

Figure 2.2-2A shows examples of these mixers.

MODE OF MIXING A high-speed rotor (see Figure 2.2-2B) produces violent turbulence and high shearing action. This gives rapid and thorough mixing. The high shearing can be created by housing the rotor in a close-fitting chamber, as in Colcrete and Cemix mixers, or by placing it in a specially shaped free-fit chamber where great turbulence does the shearing, as in Häny mixers.

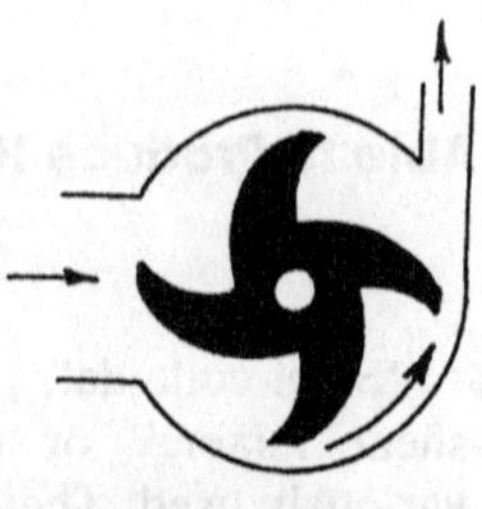

Figure 2.2-2B. High-speed rotor.

FUNCTION OF THE VORTEX DRUM The vortex drum (see Figure 2.2-2C) is a conveniently large receptacle in which to dump the ingredients. These are then drawn into the rotor chamber in a continuous stream for mixing. The vortex acts as a centrifugal separator. The lighter portions, such as water and partly mixed grout, move inwards to the throat and heavy, thick grout is spun to the outside. Well-designed machines feed each particle of cement through the rotor chamber several times during the mixing cycle. These feed the rotor with the relatively watery grout from the throat of the vortex. This then discharges into the outer part of the

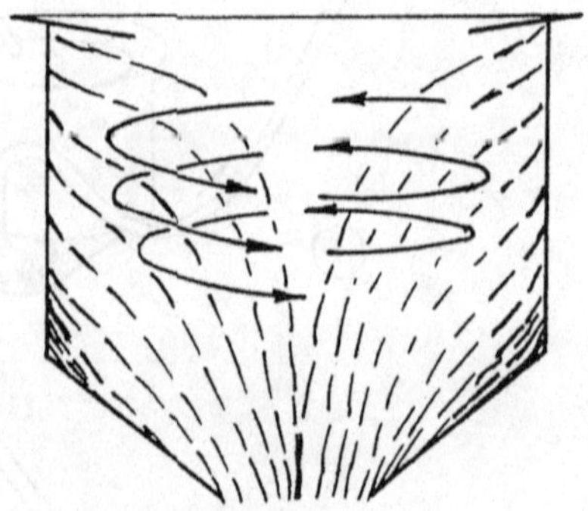

Figure 2.2-2C. Vortex drum.

vortex, amongst the relatively thicker grout there and then migrates inwards to the vortex throat under the influence of the strong centrifugal separation process. The multiple passes through the rotor produce a thicker grout on each pass, and this continues until the grout becomes so uniform that the centrifugal action can no longer separate different densities of grout. The surface of the vortex then has a smooth, uniform appearance.

The vortex also helps to provide rapid assimilation of the ingredients when they are first dumped into the water.

In order to permit formation of a smooth vortex, there should be no baffles in the drum. At the top, a wide rim should project well in from the sides to stop the vortex from spinning up and out, and to minimize splashing.

Note that not all manufacturers make use of these vortex properties, as mentioned later.

FEED TO ROTOR The arrangements for connecting the bottom of the drum to the rotor must have sufficiently ample proportions to be able to pass large lumps of unmixed cement at the start of a mixing cycle (see Figure 2.2-2D). These lumps can be sticky on their

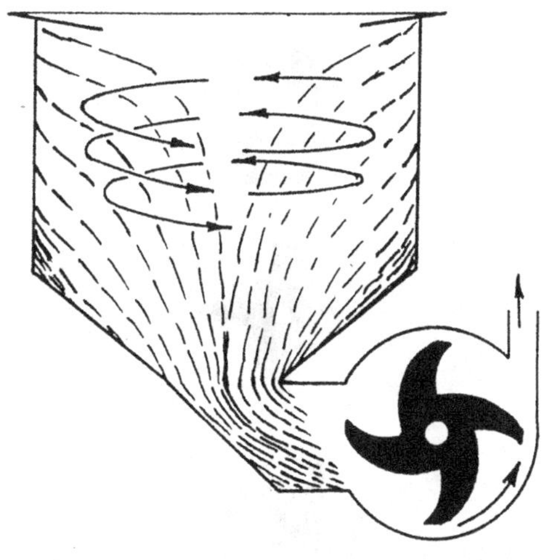

Figure 2.2-2D. Feed chute connecting drum to rotor.

first pass to the rotor before being broken up and can present problems in those mixers where the rotor is not hard up against the drum.

DISCHARGE VALVES During the mixing cycle, discharge from the rotor is sent back into the drum and enters it tangentially (in most machines) to make the vortex. When mixing of the batch is complete, a valve is used to close off this flow and at the same time another is opened to let the grout discharge to an agitator or wherever it has to go (see Figure 2.2-2E). On some mixers, these two valves take the form of a rubber tube on each pipe, one of which is squashed closed by means of a roller on a long handle. Moving the handle then opens that valve and closes the other. This is fool-proof against accidental closure of both valves at once, which can happen if separately operated valves are used and which can cause mechanical damage. One brand of mixer uses a three-way valve.

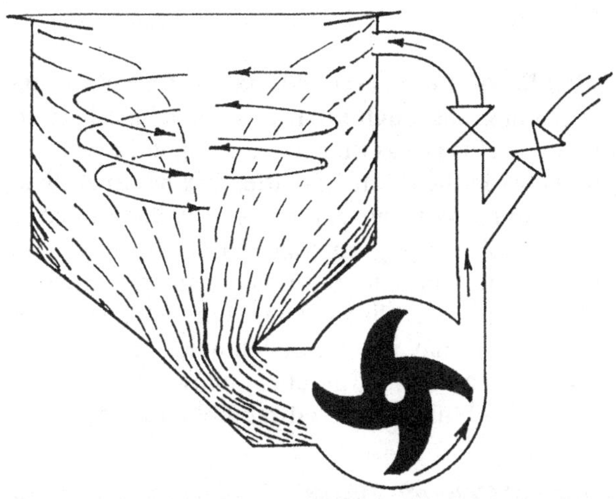

Figure 2.2-2E. Valves control rotor discharge.

MOTORS FOR THESE MIXERS Mixers used for rock grouting usually have motors of between 15 and 25 hp. Drive speed is constant so ordinary electric, air, diesel, internal combustion, or hydraulic motors are suitable: air motors are usually the most reliable in the typically wet conditions. Fuel motors can give problems unless protected from cement spray and the frequent washdowns of the mixers. If electric motors and switchgear are used, they must be sealed against water and cement dust.

MATERIALS ABLE TO BE MIXED In addition to cement, these mixers can usually effectively mix sand, bentonite, fly ash, silica fume, and most other grout ingredients. These materials are described in Chapter 3.

EXAMPLES OF MIXERS Well-known brands of mixers available ready to run include: Colcrete, Cemix, Chemgrout, Häny, and others. Not all of the mixers produced by these makers are in this present category.

Some users make their own by using a centrifugal pump for the rotor or by fabricating rotors and chambers in imitation of the proprietary machines. The vortex drum can be easily made up from sheet steel.

Selected details of some of the commercially available mixers follow, for comparative purposes. This is believed to be a fair rundown of good and not-so-good points of each brand and is not intended as advocacy of any particular brand over any other. The comments about Colcrete and Cemix mixers come from extensive personal use of these machines, whereas those about Chemgrout and Häny are from observing the work of other people.

COLCRETE MIXER Figures 2.2-2F and 2.2-2G are examples of Colcrete mixers.

Features of Rotor

- Operating speed is approx 2,000 rpm.
- Rotor shaft is mounted horizontally.
- Side of rotor is a close fit in its housing.
- Housing is cast steel and can be quickly opened up at a bolted flange located in the plane of the rotor shaft.
- Rotor and housing are a simple form of colloidal mill.
- When rotors became worn, they can be replaced easily, but if this is not done the quality of the mixing can regress to a very poor product.

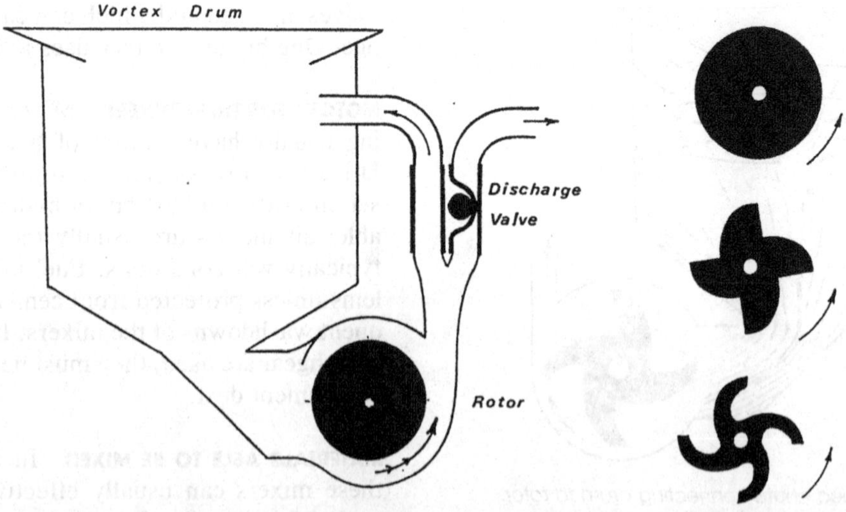

Figure 2.2-2F. Colcrete mixer. Rotor shapes: (Upper) Solid disc for cement mixing only. (Middle) Discar for cement and sand. (Lower) For sand (in double drum model).

Vortex Drum

1. Drum is relatively squat.

2. Top has a wide flange projecting both inside and outside the drum to accommodate the length of a bag of cement without overhang when placed on a central bag splitter and also to restrain the vortex from overflowing the drum when fairly full.

Figure 2.2-2G. A veteran Colcrete mixer. Still in use after 20 years. Mounted with air motor (at back) on steel frame. Conical bag splitter removable. Note rim well whacked from thousands of cement bags. Rotor is in shade under right-hand side of vortex drum with squeezed-tube discharge valves above and vertical discharge tube going off to a distant agitator.

3. The opening at the bottom of the drum is concentric and leads into a short chute sloping down to the rotor entrance. The concentric opening feeds from the throat of the vortex, and thus the rotor is preferentially processing the thinner, more watery, part of the drum contents. This gives efficient, rapid mixing as described earlier. The short chute is of adequate width to avoid problems from blockages by lumps of unmixed cement at the start of mixing.

4. Small batches, where the batching water fills little more than the conical base of the drum, can be mixed effectively because of the use of a concentric drawoff at the base. Grout discharged into the drum after passing through the rotor is able to create the requisite vortex with this arrangement.

5. Grouts as thick as 0.5 : 1 by volume are easily mixed.

Mixing Time Fifteen seconds mixing time is adequate for obtaining thorough mixing in a machine in proper condition. This is less than it takes to put the water and ingredients in the drum, and after mixing, to discharge the contents. With an experienced, willing operator, the cycle of charging, mixing and discharging can be kept down to 1 minute, but this is physically demanding if cement is supplied in bags. For fast holes, where many batches may be required in quick succession, $1\frac{1}{2}$ minutes per batch is a practical allowance unless the cement is in bulk, measured rapidly into the mixer.

Models of Colcrete Mixers *Single-* and *double-drum* mixers were made for many years, but the double-drum version is now rarely used. The second drum was intended for sand–cement grouts; it was for mixing the sand into cement grout that had been mixed in the first drum. However, a single drum can adequately mix both sand and cement, so there is no need for the second drum.

Single-drum mixers are made in several sizes. The size commonly used is Model SD4: its batch capacity is 4.25 cu ft ⟨120 liters⟩. For unusually large jobs or for filling extensive voids, such as mines, larger machines may be warranted such as the double-rotor model (see Figure 2.2-2H).

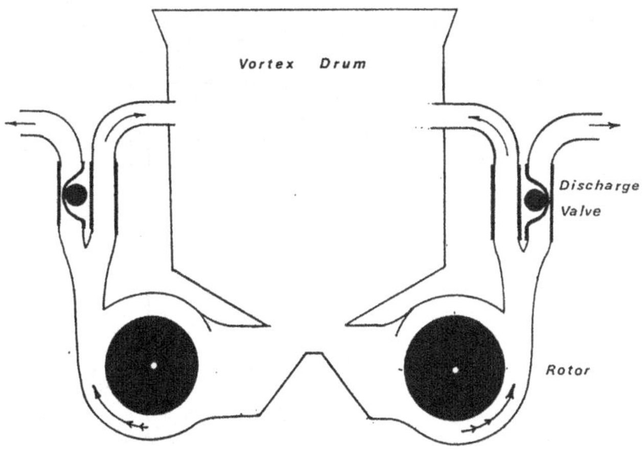

Figure 2.2-2H. *Double-rotor Colcrete mixer.*

The 1989 cost of a Model SD4 mixer, without motor, is £6,500, including freight to the United States. Colcrete is a registered trademark. Colcrete mixers are made by GKN Colcrete Ltd, Thorp Arch Trading Estate, Wetherby, West Yorkshire LS23 7BJ, England. Success breeds imitation; identical machines are made by other people in various countries.

CEMIX MIXER Also known as Craelius and Atlas Copco, Cemix and Craelius are registered trademarks (see Figures 2.2-2I, 2.2-2J, and 2.2-2K).

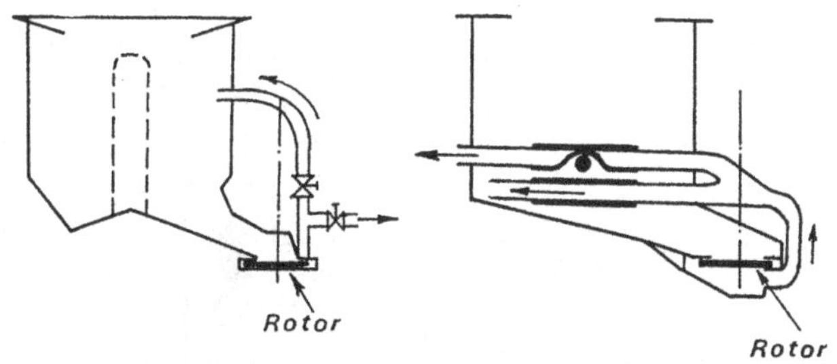

Figure 2.2-2I. *Cemix mixers. Left: Model 175. Right: Model 200.*

Figure 2.2-2J. *Rotor shape.*

Features of Cemix Rotor (See Figure 2.2-2J.)

- Operating speed is approx 1,500 rpm.
- Rotor shaft is vertical.
- Outer ends of blades are a fairly close side fit in housing (seen in Figure 2.2-2K).
- Housing is cast steel and can be opened up at a bolted flange to give access to the rotor (as in Figure 2.2-2K).
- Rotor and housing are a simple form of centrifugal pump.

Figure 2.2-2K. *Rotor of a Cemix mixer. Rotor chamber unbolted and opened up. Air motor directly coupled to rotor; is lying down behind.*

Features of Drum

1. Drum is relatively squat.
2. Top has a wide flange projecting both inside and outside the drum. This accommodates the length of a bag of cement when dumped on a central bag splitter and restrains the vortex from spinning out.
3. A difficulty found by some operators using Model 175 machines (the original model) is the ten-

dency for stiff, nonmobile cement accumulations to build up in the base of the drum at times. This cannot be sucked away by the rotor: dislodgment by a stick is necessary. To minimize this, the manufacturers placed a central pole, about 6 in ⟨150 mm⟩ diameter, in the drum of some machines. This occupies the space that would otherwise be the throat of the vortex; accumulations still build up, however.

Figure 2.2-2L. Cemix grout mixer model 175. Conical bag splitter sits on top of vortex drum. Water meter fixed (by owner) to top of air motor to give an efficient operating position. Rotor chamber at bottom of picture, below motor, with discharge pipe returning at top of drum to make vortex. Note cement in handy position and slat floor to give fast working by one operator.

4. The mixing of small batches that only fill the base of the drum can be troublesome in Model 175 mixers because the discharge pipe from the rotor comes into the drum too high up to form a vortex; the incoming grout merely splashes on the sluggishly moving grout in the base. This has been overcome in more recent models, by bringing the discharge pipe in at a lower level.

5. In Model 175 the opening to the rotor was at one side (Figure 2.2-2I). This was improved in Model 200 (also shown on Figure 2.2-2I) with a concentric feed chute and has been further improved in latest Model 201 (not pictured) with a tapered approach. Grouts as thick as 0.5 : 1 by volume can be mixed in these latest machines.

Mixing Time Thirty seconds or more is necessary for adequate mixing. Cycle time of charging, mixing, and discharging is about $1\frac{1}{2}$ minutes. Two minutes should be allowed for a long run of batches, such as is sometimes necessary in fast holes, unless bulk cement is being used with rapid measuring arrangements.

Models of Cemix Mixers All machines are single drum. The original model was Cemix 175. Drum capacity of this is 6 cu ft ⟨175 liters⟩; working batch size is about 75% of this. Subsequent models are 200 and 201 which have a capacity of 7 cu ft ⟨200 liters⟩, and Model 401 with twice this capacity. The cost of Model 201 is $8,123 (1989). Cemix mixers are made by Diamant Boart Craelius AB, Bristagatan 13, Märsta, Sweden.

CHEMGROUT MIXER This mixer (see Figure 2.2-2M) has undergone substantial modifications and is now (1989) being replaced by the design shown in Figure 2.2-2N). The company makes several other types of mixers.

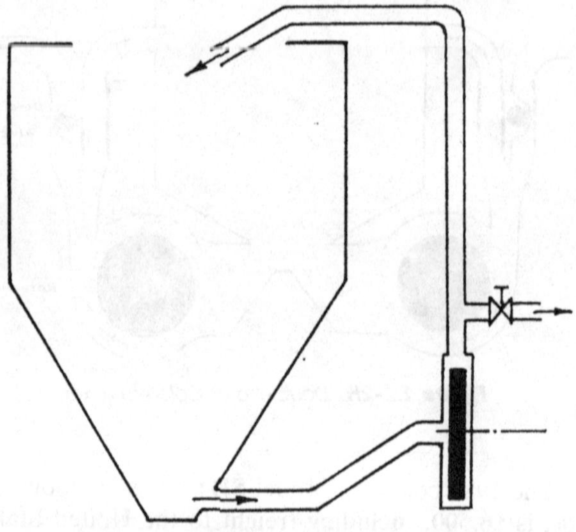

Figure 2.2-2M. The original Chemgrout "colloidal" mixer.

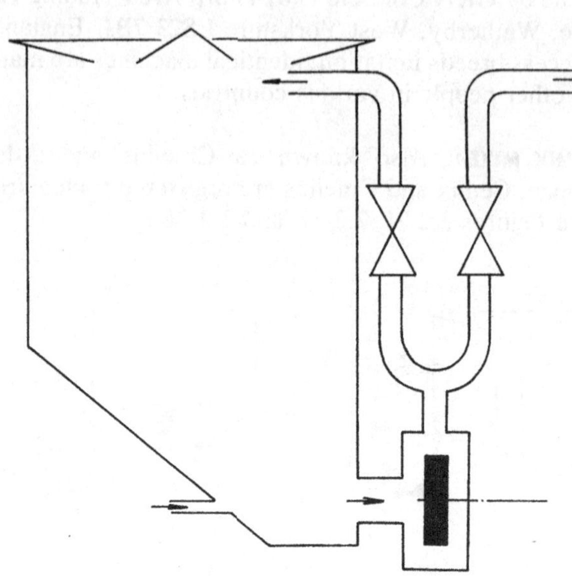

Figure 2.2-2N. Sketch of latest Chemgrout mixer.

Features of Original "Colloidal" Mixer

- Used a modified centrifugal pump described by the maker as "high-pressure, diffuser type centrifugal with resistant impeller and case and adjustable wear plates." This was well separated from the drum.
- Pump shaft horizontal.
- Had a pronounced cone forming the lower half of the drum.
- Drum relatively tall.
- Wide rim at top extended inward only.
- The opening at the bottom leading off to the rotor was at one side and had no tapering entry. This, and the long distance that the partly mixed grout had to travel in a relatively small diameter pipe before reaching the rotor, could be troublesome with the thicker mixes, as explained earlier.
- Discharge from the rotor went vertically into the top of the drum, not the side as in other brands, and hence contributed little to vortex generation, which this maker regarded as nonessential.

Features of Latest "Colloidal" Mixer

- Has a colloidal mixing mill instead of a modified centrifugal pump.
- Rotation speed 1,500 to 2,000 rpm.
- Shorter distance between drum and rotor and larger diameter connection.
- Cone not concentric with drum.
- Discharge from rotor enters top of drum at a tangent to assist vortex action.
- Wider rim on drum with conical bag splitter.
- Thickest mix able to be handled is only 0.8 : 1 in manufacturer's claims. Probably able to do better than this though.

Models Sizes range from small to very large. This manufacturer usually mounts the mixer on a trailer or base together with the agitator and pump as in Figure 2.2-2O. A typical unit of moderate size is CG-625; tank capacity is 8.4 cu ft ⟨237 liters⟩. Some of the larger models are fed from bulk cement silos and have built-in bentonite mixing facilities. Current (1989) price of a complete CG-625 system, including motor, is $22,500. To give an idea of mixer cost alone, CG-620 with price of motor deducted (although all mixers are supplied with motors) is $11,800. These mixers are made by Chemgrout, P.O. Box 1140, La Grange Park, IL 60525, USA (312–354–7112).

Figure 2.2-2O. *Latest Chemgrout mixer is on left. Agitator is on right. Motors and pumps are on floor of trailer. (Courtesy Chemgrout Inc.)*

HÄNY MIXER

Features of Rotor (See Figure 2.2-2P.)

- Operating speed is 1,500 to 1,750 rpm.
- The entire drum contents pass through the rotor 3 to 5 times per minute of mixing.
- Rotor shaft is vertical, with motor placed alongside the drum.
- Rotor is not a close fit in its housing but instead is located at the top of a free-fit chamber. Rotor has deep vanes projecting down and creates great turbulence in this chamber. The maker describes this as a "vortex pump" and considers that a separate vortex in the drum is not necessary.
- Can quickly be disassembled for maintenance.

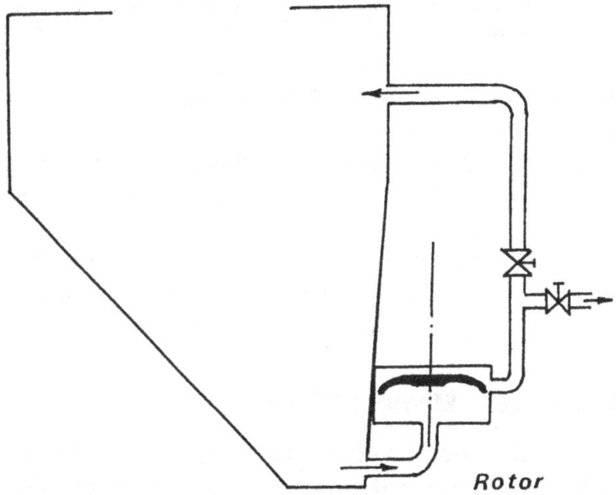

Figure 2.2-2P. *Hāny mixer.*

- Coarse sand is easily handled, and particles up to about an inch ⟨30 mm⟩ do not block or damage the impeller.
- The maker claims that rotor can run dry for several minutes without damage to the shaft seal.

The Drum

- This is made lopsided, one reason being to inhibit vortex formation in it.
- Discharge from rotor is fed square into the drum, not tangentially, to produce two sets of swirling motion in the drum.
- Drum is relatively wide at the top and has a large inward rim. It can comfortably accommodate a bag of cement dumped over its central bag splitter.
- The pipe leading from the drum to rotor is relatively small, and there can be the threat of blockages by unmixed blobs of cement, as explained earlier, although manufacturer has not heard of any problems in use.
- Top of drum set relatively low to facilitate handling cement bags.
- Bag splitter and drum rim removable for easy cleaning.
- Grouts as thick as 0.5 : 1 by volume (0.3 : 1 by weight approximately) can be mixed, the maker claims.

Mixing Time Thirty seconds mixing time is possible. Batch cycling time of about 1.5 minutes is about the minimum possible.

MODELS OF HÄNY MIXER Sizes range from HCM 200 (Figure 2.2-2Q), 6 cu ft ⟨170 liters⟩, to HCM 1250, 42 cu ft ⟨1200 liters⟩. Häny is a registered trade name of Häny & Cie. AG, Bergstrasse 103, CH-8706 Meilen, Switzerland.

OTHER MIXERS OF THIS TYPE The machines detailed above are not the only ones available operating on these general principles. Two others, for instance, are Swibo (made by Proceq, Zurich, Switzerland) and Reader (made by Reader Equipment, Eastman Way, Hemel Hempstead, Hertfordshire HP2 7HB, England). Figure 2.2-2S, the Reader mixer is yet another slight variation on the general theme.

2.2-3 Less Efficient Mixers

This section deals with machines that are less efficient than those described in Section 2.2-2. They are, however, usually cheaper and more compact. It should be remembered, though, that savings in initial cost may be offset by slower output and the need for more machines to give the same output as fewer of the more efficient ones.

Some general names for these mixers are *combined*

Figure 2.2-2Q. *Häny mixer HCM 200 with electric motor mounted above rotor. (Courtesy Häny & Cie.AG.)*

Figure 2.2-2R. *Häny mixer HCM 500. (Courtesy Häny & Cie. AG.)*

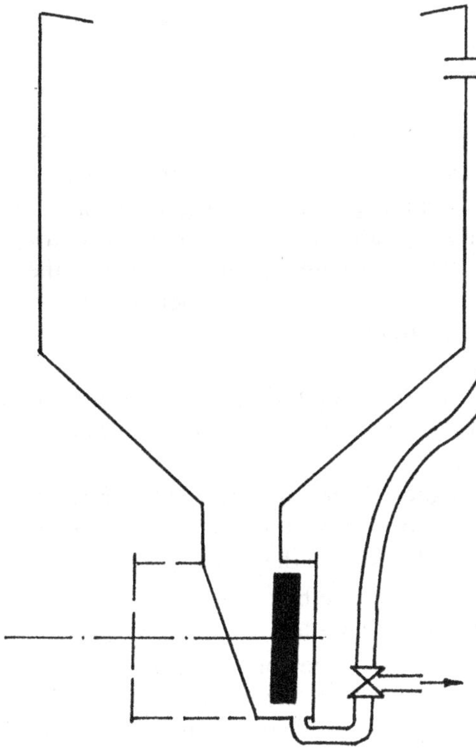

Figure 2.2-2S. Reader grout mixer.

mixer, a name used to a limited extent for some machines in this grouping, but there is no generally accepted generic name, and *roller mixer*.

FEATURES OF COMBINED MIXERS (see Figure 2.2-3A.)

- Rotor operates at about 700 to 1,000 rpm in the base of a drum.
- A small grout pump, usually of the helical rotor type, is mounted near the bottom of the drum.
- Pump circulates back to drum while mixing.
- Pump is sometimes driven by the same drive shaft as rotor.

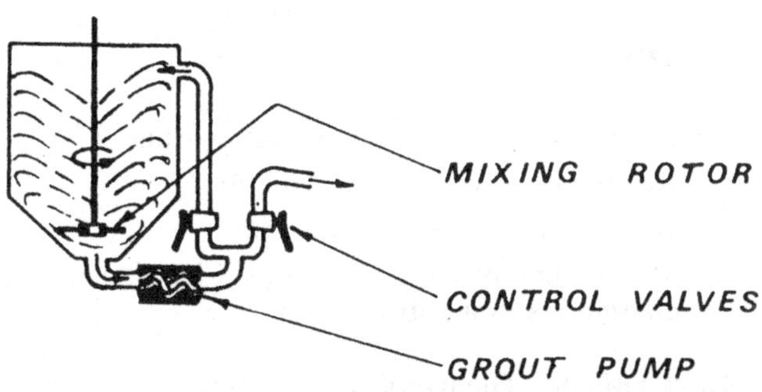

Figure 2.2-3A. Combined mixer.

- Mixing time is about 3 minutes. Charging and discharge is in addition to this.
- After mixing, the pump can discharge to an agitator, or on small jobs it can circulate the grout to the job directly.
- Figure 2.2-3B shows alternative pump positions.

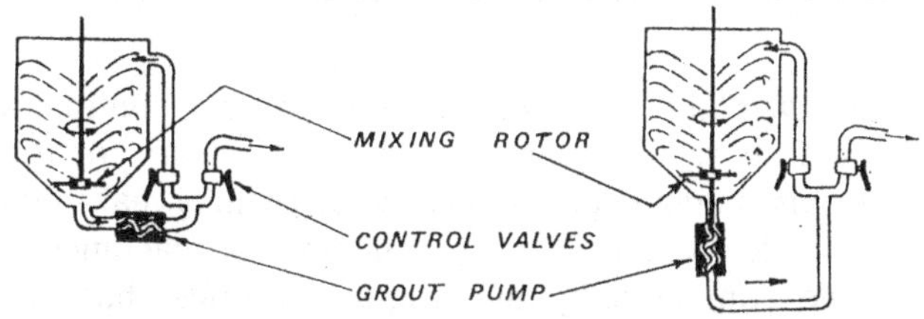

Figure 2.2-3B. Two pump arrangements. Left: On separate drive. Right: On rotor drive.

Mode of Mixing The rotor rapidly stirs the contents of the lower part of the drum. The degree of success in also treating upper parts depends on the shape of the rotor and the drum, and also on the thickness of the mix.

The pump draws off some of the flow from below the rotor and gives some further mixing before discharging tangentially back into the drum. A weak vortex is formed by this discharge, but there is insufficient centrifugal action in most machines to separate various thicknesses of grout.

There is no certainty that inadequately mixed grout will get to the rotor in some machines. The problem is minimized in others by use of propeller-shaped rotor blades to draw grout from above and expel some of it downwards into the pump.

Relatively long mixing times are necessary to give sufficient opportunity for most of the grout to pass the rotor several times.

Well-designed machines can produce a uniformly mixed grout. A significant deficiency of all mixers of this type is their lack of high shear. Therefore, they cannot be relied on to give thorough separation and wetting of cement grains (refer to Section 2.2-10 for comments on this).

Pump The helical rotor-type pump (described in Section 2.4-2) commonly fitted to these mixers has less "wriggles" and is shorter than similar pumps used for general circulation line work (Section 2.7-3). This is not critical because pressures involved during the mixing process are slight. The power needed to drive the pump is also low, and hence it is practical to run the pump off the same shaft as the rotor.

Motors Power requirements are about 3 hp typically, with some machines requiring about 5 hp. Small electric, air, or fuel motors are used. The units are small enough to be manhandled.

Materials Able to Be Mixed Combined mixers are intended for cement mainly. They can handle fine material such as fly ash but not sand.

Manufacture of These Mixers Some contractors make appreciably large numbers of mixers for their own use and do not necessarily sell them or make them available to others. A small number of machines are made commercially in various countries, but none have sufficiently wide distribution to warrant detailing here.

FEATURES OF ROLLER MIXERS The rollers can be either vertical or horizontal. Vertical roller mixers are for ordinary grouts, while the horizontal roller type is for special grouts. They are separately described.

Vertical Roller Mixer (See Figure 2.2-3C.)

- Vertical rollers revolve toward each other in a drum.
- No pump.
- After mixing, the grout is discharged by gravity. The mixers therefore need to be mounted on a platform higher than the agitators.

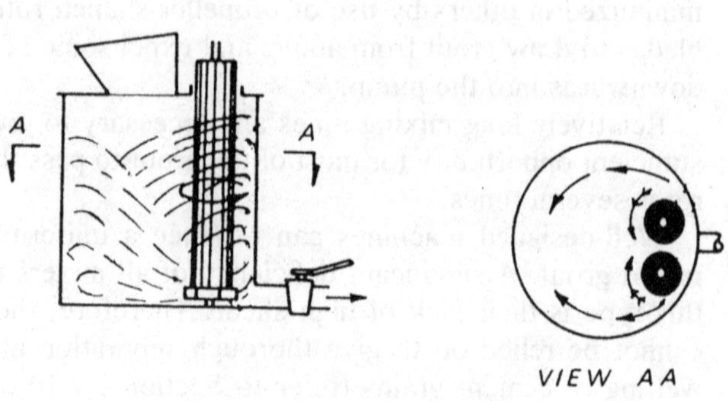

Figure 2.2-3C. *Roller mixer.*

Mode of Mixing The rollers produce circulatory currents in the drum, which draw grout through into the roller contact face. The pressure between the rollers here has the purpose of breaking up lumps of cement. Cement should be put into the mixer slowly to avoid lumps forming; there is insufficient agitation usually for rapid absorption of cement into the mix.

Motors Power requirements are in the range of 5 to 10 hp typically. Motors are electric, air, or internal combustion.

Horizontal Roller Mixers (See Figure 2.2-3D.) These mixers are used for sludgy grouts or unusual types such as foamed grout, lightweight and heavyweight grout, and for mortar, plaster, and hard wearing floor surfaces. They vary among manufacturers, the one in Figure 2.2-3D being made by Colcrete (England), and operating as follows:

- Material is forced through a fine gap at the top of the roller at great velocity in the form of a cascade or jet.
- This material then passes under the roller to the right-hand side of the tank, where it is deflected and fed back over the roller.
- After mixing of a batch, the tank is tipped to discharge the contents.
- The gap above the roller can be adjusted from a fine setting (for cement grout) to a wide setting (sand grouts and those containing fine aggregates).

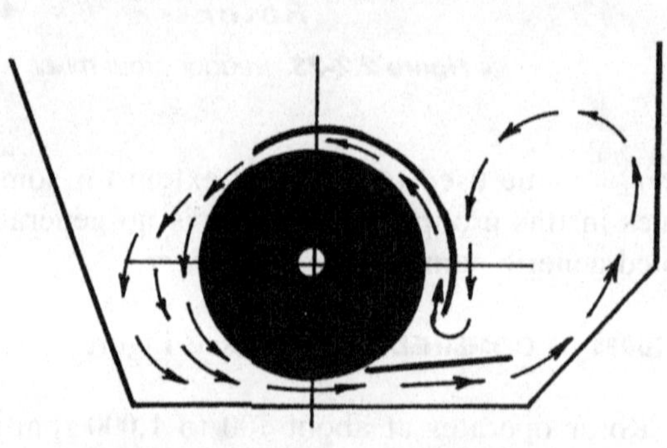

Figure 2.2-3D. *Horizontal roller mixer.*

2.2-4 Mixers for Lower Quality Grout

The general names for mixers used for lower quality grout are paddle, propeller, or an unspecified generality such as grout mixer.

FEATURES

- Paddles or propeller revolve in a drum.
- Relatively slow rate of revolution.
- Low-shear, low-energy mixing.
- Mixing time is about 5 minutes. Charging and discharging is in addition to this.

Paddle Mixer The paddles sweep through part of the volume of the drum (see Figure 2.2-4A). Grout stirred

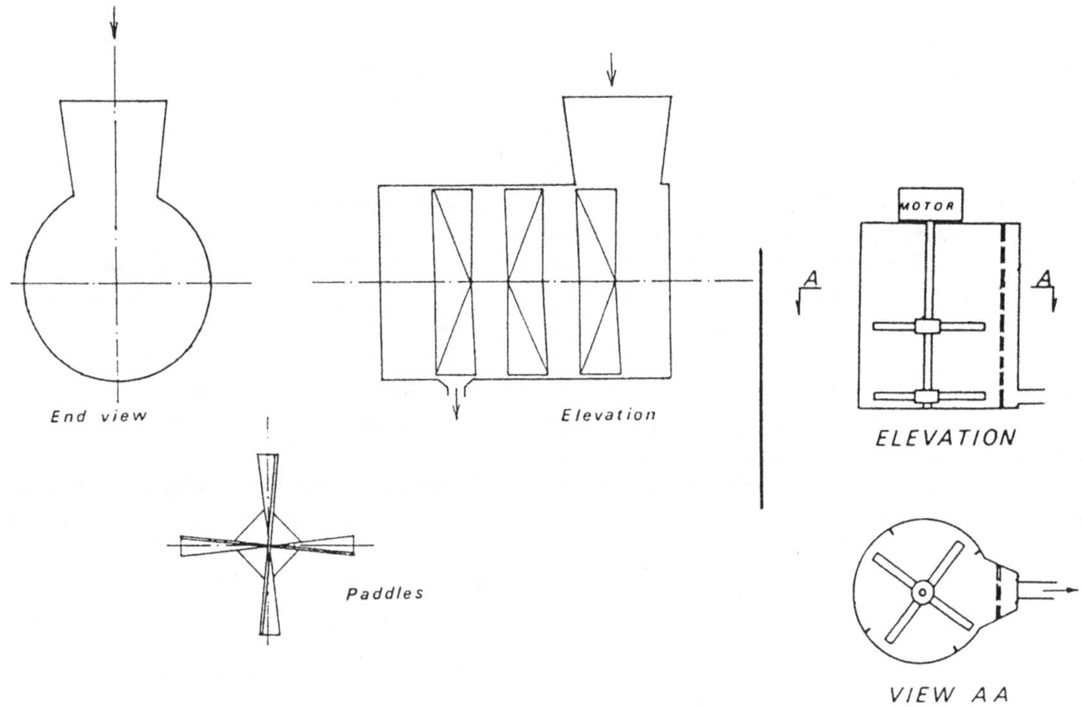

Figure 2.2-4A. *Left: Horizontal paddle mixer. Right: Vertical paddle mixer.*

by the passage of the paddles intermingles with unstirred grout nearby to give a general movement of grout throughout the drum. This movement is more effective when mixing thin grouts than with thick. Speed of revolution ranges from about 100 to 700 rpm. These mixers are similar to agitators (see Section 2.3).

When the cement is dropped in, there is a tendency for it to fall through the water and settle on the bottom of the drum. Well-designed mixers, therefore, have paddles arranged to sweep close to the bottom. In the vertical type it is best to avoid a bearing in grout at the bottom for the reasons mentioned in Section 2.3-3 (agitators).

Paddle mixers are so simple that there is no real scope for much variation in the design. Compare the oldtimer ones pictured in Section 10.2-4 with Figure 2.2-4A. The current model from Chemgrout is shown in Figure 2.2-4B.

Propeller Mixer The propeller or mixing blade (Figure 2.2-4C) provides a strong mixing action in its im-

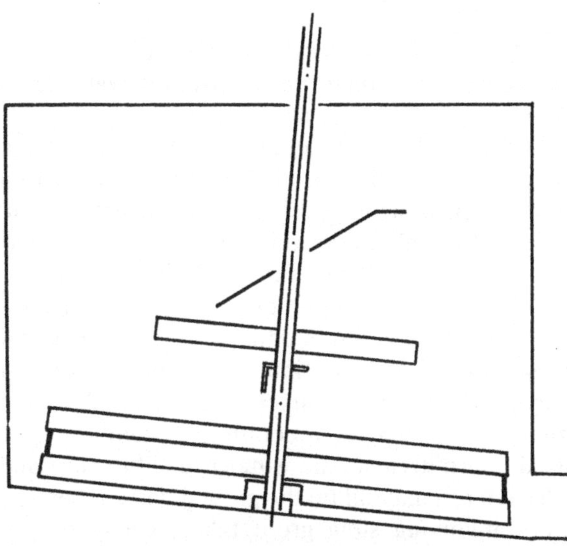

Figure 2.2-4B. *Chemgrout paddle mixer.*

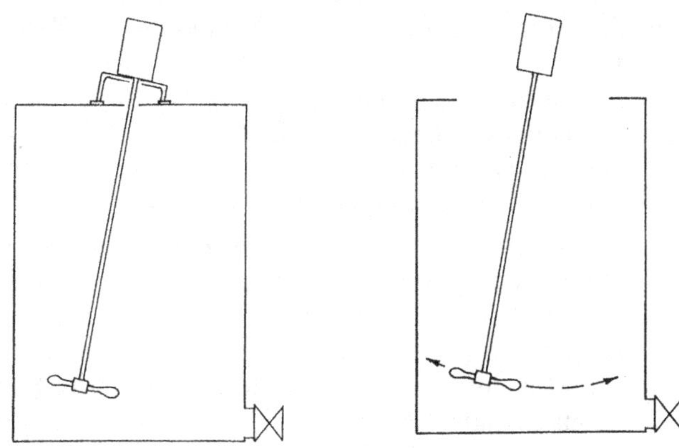

Figure 2.2-4C. *Left: Propeller fixed in drum. Right: Portable propeller.*

mediate vicinity, and this rapid motion spreads outwards, its effect diminishing as distance from the propeller increases. The motion should reach out as far as the side of the drum, otherwise mixing can be very uneven. This is overcome with the portable type of propeller, which is supposed to be moved (by the operator) into all parts of the drum.

The fixed propeller type is usually unable to deal with cement that has settled around the edge of the drum bottom when mixing the thicker grouts. The portable type is a convenient way to improvise if the job is too small to warrant obtaining a proper mixer. Such improvised mixers can be made with a paint stirrer, suitably strengthened, held in the chuck of a portable drill (see Figure 2.2-4D). An improved version of this is made by Chemgrout, among other manufacturers; it has a double-bladed arrangement on a shaft.

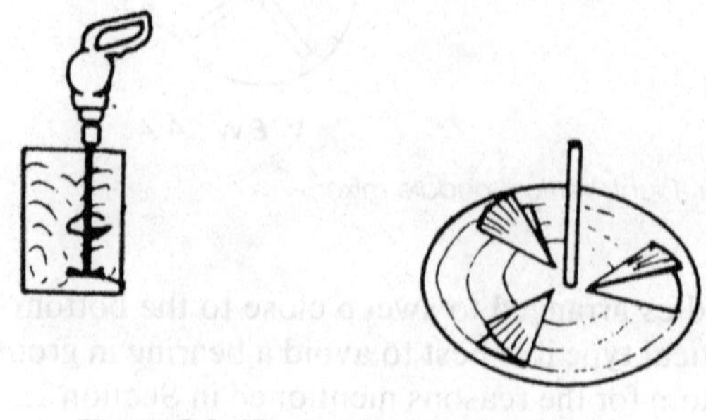

Figure 2.2-4D. *An improvised grout mixer.*

OTHER FEATURES OF PADDLE AND PROPELLER MIXERS After mixing of a batch is complete, the grout is drained rather than pumped out of the mixer. This is relatively slow compared with discharge times obtained from mixers having a pump or rotor.

The inferior mixing action of these machines is often compensated for by the use of additives, as discussed in Section 2.2-11, and there is also some further mixing of the grout as it passes around the circulation system.

- If the mixer has not broken up all the lumps of cement, blockages in equipment can occur, and cracks can become blocked against grout entry. Screens are commonly used to minimize this.
- Particles of cement are rarely individually separated and surrounded with a water layer, as they are in efficient high-shear mixers. Hence fine cracks cannot be grouted effectively, and the grout is not as strong or as durable. Refer to Sections 6.1 and 4.5.
- Mixes thicker than about 0.8 : 1 are usually too thick for these machines to handle.

Motors Typically motors are air or electric, with internal combustion motors confined to the larger plants. Electric motors range from 10 hp down to half horsepower (for the improvised type). Hand-cranked paddle mixers are commercially available.

Materials Able to Be Mixed All of the various ingredients used in grout have been mixed in paddle or propeller machines at various times and with varying degrees of success. As recounted in Section 10.2-2 (history of mixer development), deficiencies and difficulties with these types of mixers have led to the development of better ones, such as the high-speed, high-shear mixers (Section 2.2-2).

Manufacture of These Mixers The majority of these machines are made by contractors for their own use. Some are made commercially, particularly the propeller type. With little difference between the various makes there is no point in comparing them here.

Output The long mixing time and slow charge and discharge rates make outputs relatively low, and hence several mixers are usually needed, working simultaneously, to keep up the supply of grout needed for a fast hole in routine rock grouting work.

For other types of grouting, if very large quantities of grout are required, possibly of medium to low quality, paddle mixers of much larger size than used for rock grouting are available.

2.2-5 Other Types of Grout Mixers

In addition, there are also jet mixers, pump mixers, and others, some quite ingenious, that have been developed in the past and have not come into general use.

There is potential for future development of even better machines than those currently available.

JET MIXERS These mixers (see Figure 2.2-5A) have no working parts. They use the venturi principle to make a vacuum: a strong jet of water discharging into an enlarged chamber does this. The cement is sucked into the vacuum and then mixes with the water jet.

Mixing is not as thorough as in machine mixers, and control of w : c ratios is difficult because the water has to be applied as a continuous jet and cannot be measured in batches. Jet mixers are not suitable for thick mixes. They provide a continuous supply of grout: if used intermittently, the water needed to start and finish each run may dilute the grout greatly. Therefore, in work involving slow grout takes, where only small quantities of grout are taken at wide intervals of time,

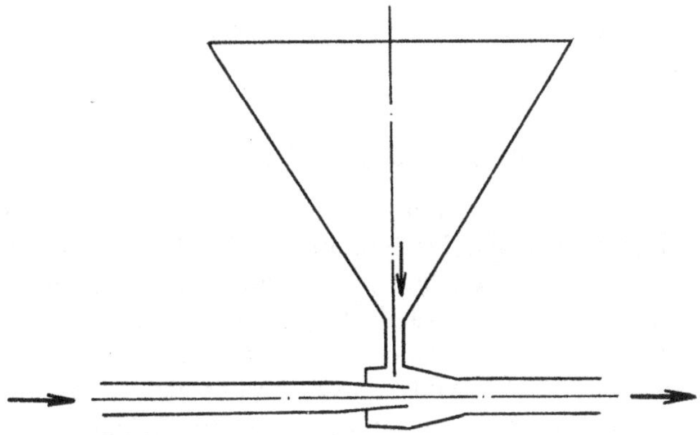

Figure 2.2-5A. *Section through a jet mixer.*

jet mixers are not really suitable. In fairly sophisticated setups, the w : c ratio is controlled by varying the water flow in response to readings from a nuclear density meter mounted on the discharge line.

The principle of these mixers is widely used in industry for entraining powders in fluids and in domestic usages such as applying fertilizer to water in a garden hose. The principle is also used in vacuum grouting (refer to Section 1.2-1).

Jet pumps were introduced in oil well drilling in 1920 for mixing the grout when cementing wells (Section 10.2-2). Use in rock grouting work has been limited for the reasons indicated. However, on jobs where several may be available on drilling rigs, and the required quality of the work is not high, economy may dictate their use rather than bringing in machine mixers.

PUMP MIXERS These use the turbulence produced in centrifugal pumps to do the mixing. The pumps resemble those used with some of the mixers described in Section 2.2-2, but they do not have the vortex drum arrangement. Instead, they take cement and water from a hopper or agitator and pump it directly into the circulation line hoping that the grout is adequately mixed by the time it reaches the hole, but there can be considerable doubt about this. Grout not taken down the hole is returned to the hopper or agitator.

The method is relatively crude and is only tolerated on very low standard, cheap grouting.

2.2-6 Laboratory Grout Mixers

When conducting laboratory tests and research into grouting, mixing of small batches is sometimes necessary. It is, of course, preferable to use full-size mixers for such testing if conditions permit, even though most of the grout mixed may be wasted after the small quantities used in the tests are taken off.

However, in circumstances where use of a small laboratory mixer cannot be avoided, care is needed to ensure that the mixer used is able to produce grout that resembles the grout used in the full-size job to which the tests relate. Not all laboratory mixers are suitable. Furthermore, grouts on actual jobs vary depending on the efficiency of the mixers used, cement type, effect of additives (if used), and so on.

As a general rule, the following features assist in giving compatability between laboratory and job mixers:

1. Laboratory mixers should be of the type used for preparing the fine fraction of soils in classification testing or of the ''milkshake mixer'' type, having one or more rotors mounted on a vertical spindle.
2. The rotors should have the shape of a low-pitch propeller or a waffle plate, of a diameter suited to ensure that the mixing action is effective, but does not propel the grout out the top of the vessel.
3. The speed of rotation of the rotors should be at least 8,000 rpm under no load; 12,000 rpm under no load is better if the mixer is intended to simulate the product of a high-speed, high-shear mixer (Section 2.2-2) operating at 2,000 rpm.

Examples of mixers are shown in American Petroleum Institute's ''Specification for Materials and Testing for Well Cements'' Spec 10, Section 5 (1984).

METHOD OF MIXING The procedure for mixing grout in this type of mixer is as follows:

1. Place the full quantity of water in the mixing vessel.
2. Run the mixer up to full speed.
3. Slowly add any additive to be used in the test.
4. Slowly pour the cement into the water, being careful to ensure that it hits the water in the highly turbulent region of the surface. The surest way to do this is to pour it over the rotor(s).
5. If any sand is to be added, pour it in slowly in the same manner as for the cement.
6. After all ingredients have been added, mix for 1 minute further. This is to simulate the 15-second mixing time of a full-size, high-speed, high-shear mixer.

COMPARATIVE DATA Because not all lab mixers of the type described give satisfactory simulation of job grouts, it is necessary to verify that laboratory mixers are suitable by making comparisons of specific gravity, flow cone times, and possibly viscosity and other somewhat elusive properties.

For details of these tests and comments on them see Section 4.6.

2.2-7 How to Mix Grout

Returning now to full-size mixers, let's consider proper mixing:

1. Put the quantity of water required for the batch into the mixer. Preferably measure this water through a water meter for accuracy and speed.
2. Start the mixer if it is not already running. Get it up to full speed.
3. If additives are to be used, they are usually put in the mixer at this stage.
4. Put the required quantity of cement in the mixer. Use whole bags of cement if batching is by bags. Do not try to use part of a bag.
5. If other solids are required, such as sand, add them now.
6. Mix the grout for the specified time.

2.2-8 Hints about Mixing

Cleanliness pays. If there is a pause of more than a few minutes between mixing of batches, wash the mixer out (unless it is a "combined" type) and leave it full of water until the next batch. Hose any spilt grout off the outside of the mixer. These precautions can lead to lower maintenance costs and greater reliability.

Water meters are better for batching water into mixers than the alternative of measuring in graduated drums or containers. The meters should have a large digital display or a large sweep hand that is still visible when the face is partly obscured by the covering of smeared grout and cement dust it will inevitably acquire. The meter should be set to zero at the start of each batch. Periodical checking for accuracy is necessary.

For opening cement bags, a removable steel cone or sawtooth splitter should be mounted centrally on the mixer drum. Where the design of the mixer permits (some are shown in Section 2.2-2), bags can then be emptied in a second or two by dumping them onto this device and then lifting the two ends of the broken bag. The cement spills into the mixer rapidly. Well-designed mixers can handle this sudden inrush, poorer ones cannot and have to be fed more slowly.

When cement and other ingredients are supplied in bulk, equipment for accurately measuring the amount for each batch should be quick-operating and mounted over the mixer.

2.2-9 Testing for Grout Quality

A simple test is to pour some of the mixer output into a shallow tray to a depth of about 1 in ⟨25 mm⟩. Leave it undisturbed until the next day and then break open a vertical section through the hard grout. If the section shows bands of various shades, then the grout is low quality. Best quality grout is completely uniform across a vertical section and is relatively lightweight.

Other tests for quality are shown in Section 4.6.

2.2-10 Properties of High-Quality Grout

High-quality grout is regarded as having the following properties, which are not necessarily found in the poorer grouts:

- Cement grains are separate from each other. There are no flocs or clumps of grains.
- Each cement grain is thoroughly wetted and surrounded by a film of water. As well as improving penetration in fine cracks (see Section 6.1-5), this chemically activates each particle, giving the full hydration necessary for strength and durability (Section 4.4-3).
- The grout is uniform throughout.
- It has less settlement of the cement when stationary than lower quality grout and has less bleed.

These properties are usually obtained by use of properly designed and maintained high-speed, high-shear mixers. The following summary is condensed from Kravetz (1959) on the subject:

1. Cement grains, when mixed with water, tend to aggregate and form clumps. This slows the wetting process, as does air attached to the grains. The effect of high-speed shearing or laminating, plus the centrifugal effect, is to thoroughly break up the clumps and to separate air bubbles. As a result, each individual grain is rapidly and thoroughly wetted and put into suspension.

2. During cement hydration, needlelike or stringlike elements of hydrates form on the superficial layer of each wetted grain of cement. In a high-speed mixer, the laminating effect and high-speed rotation keeps breaking these hydrates away from the grain of cement, thus exposing new areas to the water and consequently bringing the formation of new elements. These hydrate elements are of colloidal size, and as the amount of these elements in the mixture increases, the grout becomes colloidal in character.

3. Based on this somewhat simplified description of the process in a mixer, it can be said that:
a. High-speed mixing activates the hydration of the cement, or more correctly, the first phase of the hydration of the cement (cement is not fully hydrated even after one year). This activated hydration has, however, no effect on the setting time of the grout.
b. High-speed mixing changes the granular composition of the cement. While this statement may be more a matter of semantics than physics, there is no doubt that the size of a cement grain continuously changes from the moment it is wetted to the very end of the hydration process. At first, its size increases as the hydrates form; then as hydrates are being broken away from the grain and new hydrates develop, the size of the grain starts to decrease. Also, because the rates of hydration of the different chemical components of the cement are different during the first hours of the process, it is not impossible that some of the cement particles may split or break during that time, depending on their chemical components.

4. The gradual formation of hydrates of colloidal size explains why, for an ordinary cement grout, a longer mixing time in a high-speed mixer improves both stability and capacity to carry sand (p. 113).

An indication of the change made by high-speed mixing is given by Mayer (1959):

> The French Cement Research Center has made very thorough studies on high velocity mixing. The abrasive action . . . can be felt on the grains under 20 micron size. The percentage of grains 5 micron in size is twice as large with high speed mixing than with ordinary mixers. The same is true with all colloidal elements which accounts for the fact that the suspensions obtained with high speed mixed cement are much more stable than others (p. 41).

2.2-11 Improving Lower Quality Grout by Using Additives

When economics or predisposition dictates, bentonite, lignosulfanates, water-reducing agents, and various proprietary products are in use to make lower quality grouts more "stable" and to give them improved penetrating ability. This is discussed at length and somewhat decried in Section 3.3, particularly in relation to possible adverse affects on durability and quality.

Use of these additives in conjunction with the cheaper mixers can be more economical than purchasing the more expensive high-shear machines. Note, though, as already mentioned, that the rapid output of the high-speed machines means that fewer of them are necessary.

2.2-12 Why Grout Cannot Be a True Colloidal Suspension

Some of the high-speed, high-shear mixers (Section 2.2-2) are claimed to produce a "colloidal suspension." Sounds good, but is not strictly correct as far as cement grout is concerned.

"Semicolloidal" or "near-colloidal" are more accurate descriptions. Briefly, this is because most cement grains are too large for genuine colloidal suspension.

Probably the best way to understand this is to first think about true chemical solutions where the liquid dissolves the solids added to it. Salt in water is a good example—the salt loses its original properties and reacts with the water. Then compare this with a mixture where the solids are not dissolved such as, say, iron filings in water. However finely ground up, the iron filings always tend to drop out of suspension and will be caught on filter paper if the mixture is filtered, even though the water causes a slight chemical reaction in the iron. Now think of muddy water, smoke, aerosols—these are true colloidal solutions. The solids are fine enough to pass through filter paper and are too fine to be seen. When we come to cement grout, the solids can certainly be filtered out, though perhaps not all of them if the cement is Microfine (Section 3.1-6), and grains can indeed be seen. So cement grout misses out on being a true colloidal suspension.

However, as explained in Section 2.2-10, the effect of high-shear mixing is to break down the sizes of grains, and to make cement hydrates form, of colloidal size. Although being only a small proportion of the grout, these have a considerable influence on fluidic properties, tending toward beneficial colloidal behavior.

2.3 AGITATORS

2.3-1 Purposes

Agitators store the grout and keep it stirred up until it is needed by the pump, and they receive grout returned through a circulation line. Additionally, they are a convenient place to take dipstick measurements of grout consumption. They provide an open surface where an experienced person can see the grout and can watch how fast it is taking (very useful—see Section 7.2-6) and can also check that it is the correct w : c ratio (the ability to do this comes with considerable experience), and that it is uniform.

Agitators consist of a simple tank, with some sort of stirring device, as shown in Figure 2.3-1A.

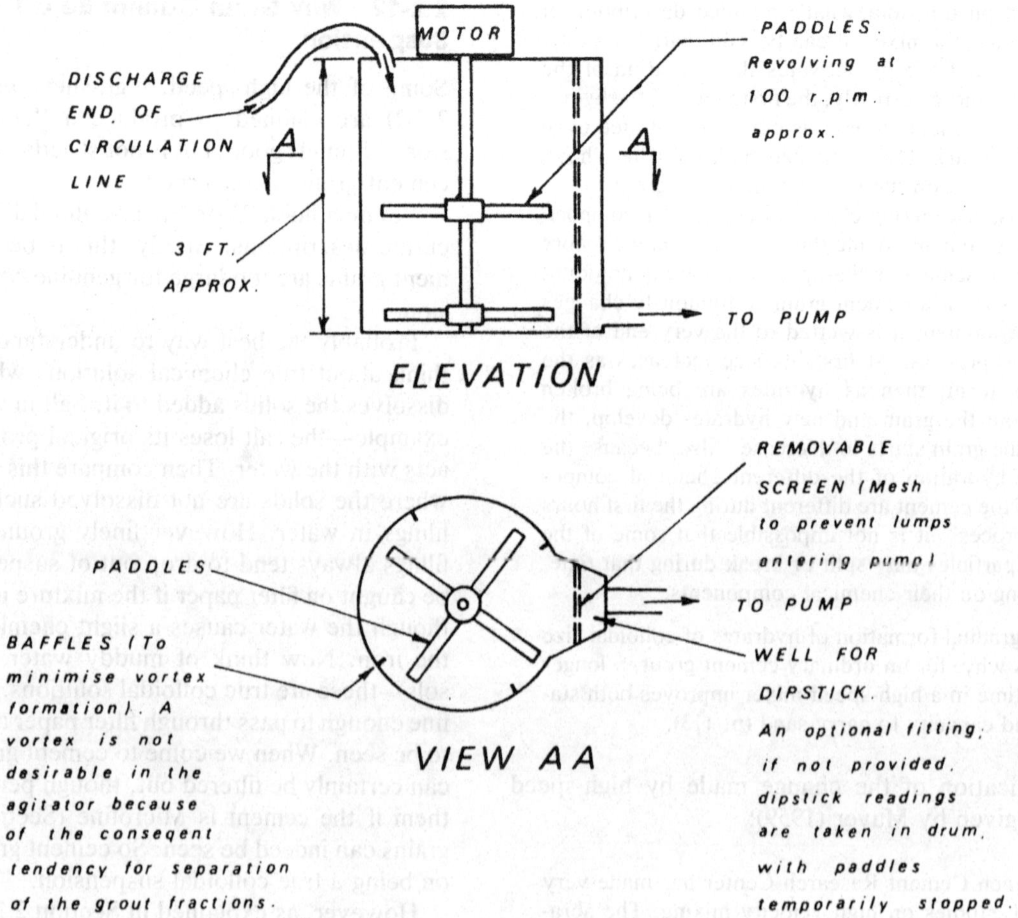

Figure 2.3-1A. *Typical agitator.*

The smallest size of drum is usually about 50 U.S. gallons (200 liters). Agitators are sometimes improvised from fuel drums of this size.

The paddles are usually on a vertical shaft. An operating speed of about 100 rpm is usually adequate.

The lowest paddle should almost scrape the bottom of the drum to minimize build up of stale, settled grout. Because of this almost inevitable buildup, the offtake pipe should be quite large: at least 2 in (50 mm) diameter. To help with washout, a Y-piece is sometimes fitted to it just after it leaves the agitator drum, and a quick-coupling arrangement is fitted so that it can be easily unhooked from the pump for washout.

Agitator drums and paddles must be washed thoroughly to prevent accumulation of layers of old grout. If hard flakes of this fall off, they can produce problems by blocking fitting and hoses.

When used with paddle or propeller mixers, a screen is desirable over the pump offtake to catch lumps. The screen must be easily removable for cleaning. Provision of a well for the screen provides a convenient place to take dipstick measurements without stopping rotation. This arrangement is shown in Figure 2.3-1A. However, when high-speed, high-shear mixers in good condition are in use, lumps should be too rare to need the screen. The layout is then as shown in Figure 2.3-1B.

2.3-2 Taking Measurements in Agitators

Dipstick measurements can be made in agitators to measure quantities of grout taken by the hole, as described in Section 4.1-10. Alternatively, the grout can be measured through flow meters; see Section 7.7-10. To take measurements, dipsticks are either seated on the floor of the agitators or hung from the rim, as shown in Figure 2.3-2A.

Dipstick markings can be on the stick, as shown, or the stick itself may be unmarked and read by placing it alongside a calibrated stick fastened to a wall or convenient place nearby. The grout level on the dipstick is then eyed across to the marked stick to obtain the reading.

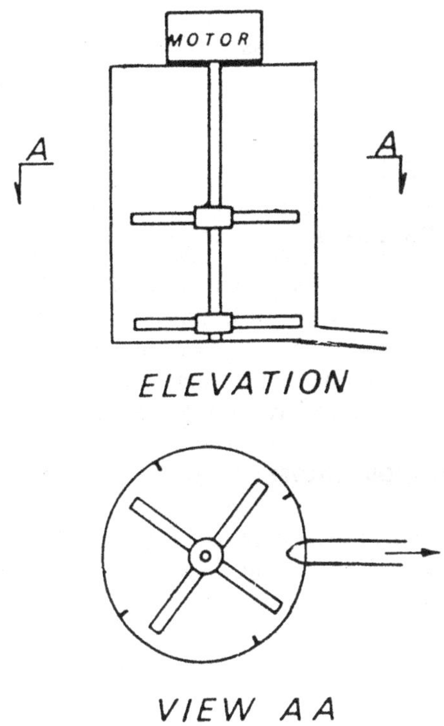

Figure 2.3-1B. *Agitator for use with high-shear mixer.*

The dipsticks that are marked are made of steel with markings etched into them or of timber with saw cuts or marks. They must be wiped after each use and tend to become rather grouty and difficult to read after a number of wipes. The unmarked type has advantages in this regard.

When grout is thickened (Section 4.1-8) during an application to a hole, or thinned (Section 4.1-9) preparatory to going on to the next hole, the makeup mix or the extra water, as appropriate, is added to the existing mix in the agitator. Dipstick readings taken before and after this addition provide a check on the amount added for comparison with other measurements, and they provide data used in the summing up and balancing of cement quantities.

2.3-3 Motors for Agitators

Very little effort is needed to turn the usual size of agitator if there is no sand in the mix. Just a few horsepower will do even when the mix is as thick as 0.5 : 1. Small motors can be used, often mounted on top of the vertical drive shaft and directly coupled to it. Motors used like this are air (the type used to drive power augers for timber boring is useful for this), electric, or hydraulic. Alternatively, bevel gearing can provide transmission from a motor mounted at the side of the agitator. It is rarely worthwhile using internal combustion motors—they are too cumbersome compared with the alternatives. Slightly stronger motors are needed for sand–cement grouts.

Whatever sort of motor is used, it must be able to stand frequent wetting and mechanical abuse. Things are rugged at the top of an agitator: pipe wrenches can get used on the drive shaft to coax it to turn over or to get it restarted under load, and these are not gentle tools.

If the paddle shaft has a bearing at the bottom of the drum, the life of this bearing is short, even with the best seals. It operates in grout—abrasive when liquid and strong when set. A better arrangement is to avoid having bearings in the grout by using them at the top only and suspending the shaft from there. This may let the shaft wobble somewhat, though, and means that paddles cannot be allowed as near the base of the drum. With this arrangement, the shaft, paddles, and motor can be removed for exchange with another set when servicing is needed or the motor gives trouble. This works best when the motor is direct coupled to the shaft and when the assembly sits in forks or has some other means of quick detachment from the drum (see Figure 2.3-3A).

2.3-4 Agitator Placement

Efficient practice is to have several agitators fed by one high-shear, high-speed mixer (Section 2.2-2), with

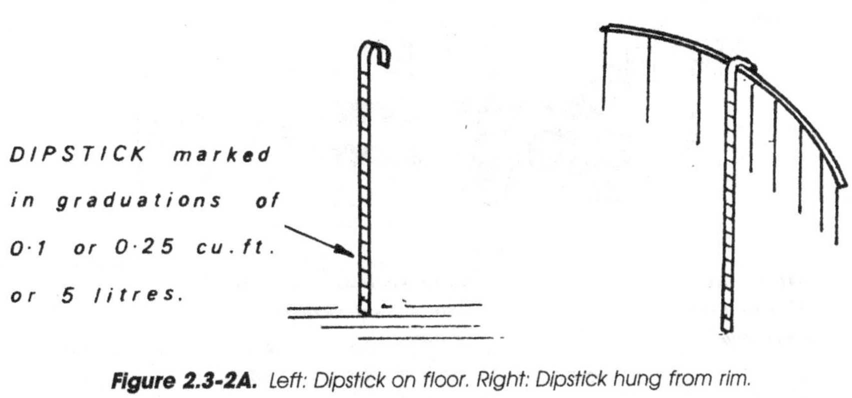

Figure 2.3-2A. *Left: Dipstick on floor. Right: Dipstick hung from rim.*

Figure 2.3-3A. *Three agitators in simultaneous use fed by one Cemix mixer. Agitators made by contractor each with air motor and coupled paddle shaft removable for exchange.*

each agitator supplying one hole. Several holes can easily be grouted simultaneously this way if crews are well trained. Each agitator then has its own pump and circulation line. For a discussion on pumps see Section 2.4 and for a discussion on circulation lines see Section 2.7.

Compare this efficient use of separate machines with the alternative of prepackaged complete grouting units having mixer, agitator, and pump combined (see Section 2.7-10). These can be useful, but they only grout one hole at a time, so each hole then ties up a mixer. This might be tolerated when the mixer is one of the cheaper ones (Sections 2.2-3 and 2.2-4) but is not good economics when the mixer is an expensive, high-efficiency unit (Section 2.2-2).

When space is at a premium, agitators are sometimes mounted one above the other, double-decker fashion.

In some difficult locations it is not sensible to place the agitators at the mixers. Section 2.7-7 goes into arrangements that should then be used, such as using agitators for break-pressure devices and having agitators at various levels.

2.4 PUMPS

2.4-1 Purposes

The pump takes grout from an agitator and pumps it to the hole(s) through a line of hoses or pipes (see Section 2.7 for circulation lines). Several types of pump are used for grouting. They can be divided into two groups: those that have valves and those that don't.

PUMPS WITHOUT VALVES

- Helical rotor, progressive cavity type (Mono or Moyno is one brand).
- Centrifugal.

PUMPS WITH VALVES

- Piston. Usually two cylinder; sometimes four or six.
- Ram.
- Diaphragm.
- Hand pumps.
- Compressed air pans.

2.4-2 Helical Rotor Pumps

Helical rotor pumps use a steel rotor rotating inside a softer stator (see Figure 2.4-2A). Spaces containing the grout advance along the rotor in a positive screwing motion resulting from the single-helix shape of the rotor and the double-helix configuration of the stator. The rotor has a constant circular section, with the centers of sections forming a helix that is eccentric to the

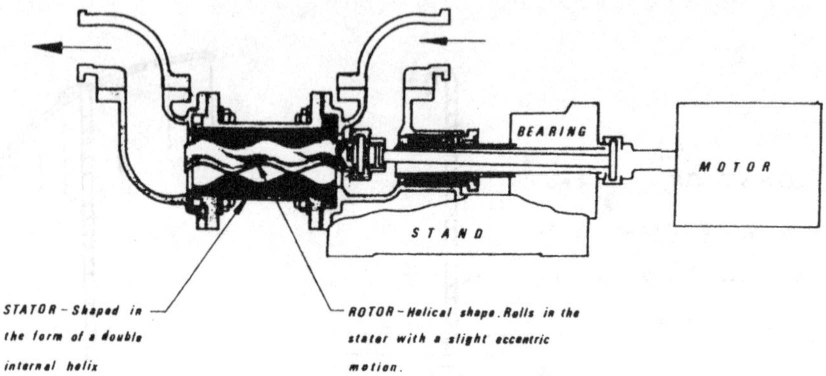

Figure 2.4-2A. *Helical rotor pump (Mono).*

rotor axis. Stator pitch is twice that of the rotor, and engagement between the two is such that the rotor traverses the stator aperture. A positive seal results along the stator, up to certain limiting pressures.

Priming is not necessary. There are no valves, and the only working parts in contact with the grout are the rotor and its drive shaft. For grout service, the rotor revolves relatively slowly at about 300 rpm and is hard chrome plated. The plating can be replated eventually, preferably before the parent metal shows through. The stator is made of suitably wear-resistant rubber or nitrile.

The life of stators varies and depends a lot on how well the pump is operated and washed. Some good grouting crews can get a year or so out of an ordinary stator, and longer from the special ones. Fitting a new stator takes only a few minutes: a flange or housing is opened up and gives immediate access to the stator (see Figure 2.4-2B).

Figure 2.4-2B. *Single-stage Mono pump with stator removed (standing on its end in flanged housing). Rotor visible. Driving end is at left.*

The pressure output is fairly steady when pumps are in good condition, a beneficial feature in my experience (discussed in Section 2.4-12).

This type of pump is now the most popular for grouting in many countries.

MOTORS FOR HELICAL ROTOR PUMPS Power required for the sizes of pump used for grouting work ranges from about 7 to 15 hp.

Air, hydraulic, diesel or internal combustion motors are suitable, and it is usually convenient to use the same type motor as the mixers. However, it is not advisable to use electric motors on these pumps for grouting work, unless a variable-speed drive is fitted. This is because the pumps must be able to operate at any required output pressure and volume; electric motors without variable-speed facilities give too limited a range of output. While this can be overcome with adroit manipulation of a bypass, it provides an undesirable complication in the grouting. Nevertheless, electric motors are often used if everything else is electric.

Motors must be able to start the pumps under full load. Some do not always kick over, particularly rotary air motors if they have stopped in certain phases of their rotation. The rough type of operator overcomes this with a heave on a pipe wrench briefly applied to the drive shaft.

A stuffing box or gland is necessary where the drive shaft enters the pump. This can be a source of annoyance if not well designed. The rotor has an eccentric motion in the stator. This means that some flexible arrangement has to be used to connect the rotor to the drive shaft, and this arrangement has to be located in flowing grout—not a good environment for the conventional flexible coupling. Some makers do use them, however, but better designs have either solid rubber flexible sections in the shaft or have a flexible steel shaft.

OUTPUT These pumps come in a wide range of outputs, and it is not much use suggesting firm output requirements here; they vary too greatly depending on the way the gear is set up. By way of general guidance, for a fairly level site, with the arrangements shown in Sections 2.1 and 2.7-5, suitable helical rotor pumps operating to 150 psi ⟨10 bars⟩ can handle 1,000 gph ⟨about 4,000 liters per hour⟩.

PUMP MANUFACTURERS The helical rotor pump, invented by M. Rene Moineau, was pioneered by the Mono group in Europe. Consequently, their name has become loosely adopted as the generic name for the pump. In the United States the pumps have been known as "Moyno" from the use of this name by the licencees, Robbins and Myers, Inc. (Springfield, Ohio). Probably the majority of pumps used for grouting are of "Mono" or "Moyno" manufacture.

Pumps of the helical rotor type are made for a variety of uses. Those suitable for grout are the "D" and "L" series. The D series is now superseded by Series L (see Figure 2.4-2C), which has an improved drive arrangement and stator housing. Series D had a flexible coupling in the grout as shown in Figure 2.4-2A, and stators were a push fit in a housing pipe.

The Series L pumps replace the flexible coupling with a long flexible steel shaft, fixed to the rotor. The

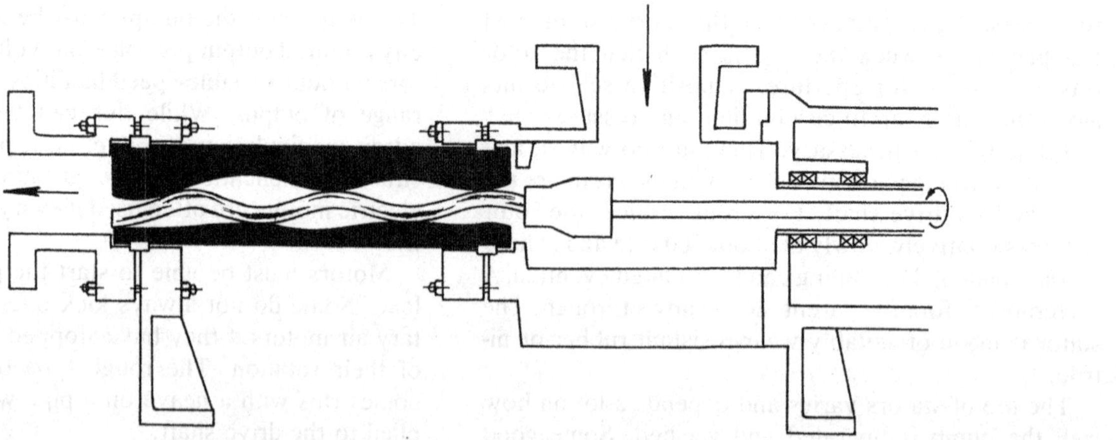

Figure 2.4-2C. *Series L Mono pump, single stage.*

stator is bonded inside a steel tube so replacement is a very simple matter of undoing a few bolts.

In Mono pumps, those suitable for grout develop up to about 75 psi ⟨5 bars⟩ in each stage. Two stages are the minimum practicable for grouting (see Figure 2.4-2D). These can readily provide the 150 psi ⟨10 bars⟩ needed for most rock grouting jobs of any size. Further stages are sometimes used, to a reasonable limit of about four, which gives pressures of about 300 psi ⟨20 bars⟩.

There are other manufacturers such as Bornemann (Germany), Midland (United States), Abeco (New Zealand) and others, some only newly on the market.

Figure 2.4-2D. *Two-stage Mono pumps with diesel motors. Each pump draws from its own agitator and supplies a separate circulation line. Note the gravel bedding to facilitate underfoot safety.*

2.4-3 Centrifugal Pumps

Centrifugal pumps used for grouting are usually adapted from mud or slurry handling duties. Wear of the impeller and casing can be a problem when used with grout and may involve frequent and expensive replacements.

These pumps are not usually the first choice for grouting service. In addition to wear, some may have difficulty in handling the wide range of pressure output often required. Although similar pumps are used in some grout mixers (Section 2.2-2), they run at more or less constant speed.

Limited usage makes it not worthwhile to go into further detail here.

2.4-4 Piston Pumps

These pumps have pistons reciprocating in close-fitting cylinders. Those used for grout are usually double acting: while a piston moves in one direction, expelling grout, suction forms on the other side, pulling grout in, ready for the next expulsion stroke. Nonreturn valves are necessary to close off the outflow port while the inflow is taking place, and vice versa. The basic arrangement is shown in Figure 2.4-4A.

The flow of grout to the pump is coming from underneath at left; the pumped discharge goes out at the top to the right. In Figure 2.4-4A, the piston stroke is toward the right; the arrows indicate the flow. Ball valves are shown (see later for other types of valves). The upper right one has lifted off its seat under pressure from the piston via the grout, and the lower left one has also opened, in this case due to the suction from the back side of the piston. This lets grout into the cylinder ready for the return stroke. The other two valves are closed for the moment; they open for the return stroke.

Piston pumps used for grouting usually have two cylinders, called Duplex. Four- and six-cylinder pumps have been used at times for particularly large jobs; these can be massive pieces of machinery only

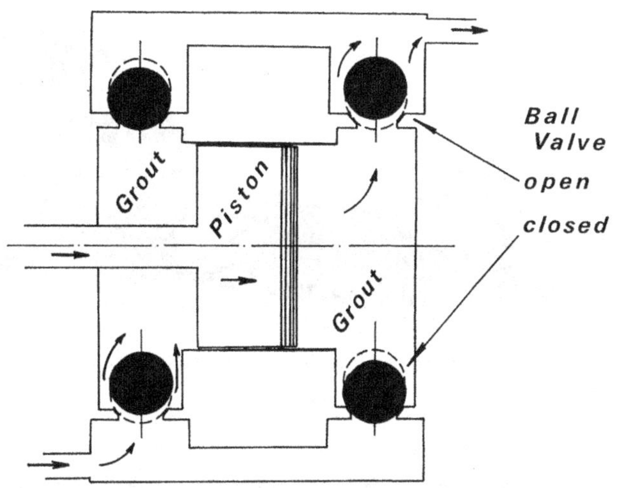

Figure 2.4-4A. *Schematic layout of a piston pump; single cylinder, double acting.*

suited to a large pumping station. The modern much lighter type of two-cylinder pumps mostly have the pistons at each end of one shaft; the shaft is powered in between them, as shown in Figure 2.4-4B.

In this particular arrangement the grout has access only to the outer end of the two pistons—the space between is occupied by the driving fluid, which can be air or hydraulic. Similar to Figure 2.4-4A, this pump shows the piston moving to the right, and grout flows are shown by arrows. These pumps are usually more compact and more portable than those with an external motor.

Pumps with more than two cylinders tend to have the cylinders side by side, similar to a car motor, with the pistons powered by a crankshaft. When working hard most of them vibrate and have to be well anchored to an engine bed or some other substantial base.

Motors for piston pumps can be reciprocating, placed between the pistons as shown, or can be separate, operating the piston directly or through a mechanism. Air, hydraulic, electric, or internal combustion motors can be used. However, in order to get the variable outputs needed, motors that give a constant speed, such as electric and some internal combustion motors, must have variable-speed drives between them and the pump. These drives are usually hydraulic, housed in a large heavy box, which makes the pumps hard to move around a site. They, therefore, tend to be mostly used in large pumping stations (see Section 2.7-12) rather than being moved near to the grout holes.

The power needed for piston pumps ranges from just a few horsepower to hundreds, depending, obviously, on the capacity and output pressure required. Available outputs likewise range from small to very large. Too many makes and variants are on the market for listing here.

Chrome plating or hardening is advisable on the parts in contact with the grout. Even then wear can be rapid, especially when handling a sand–cement grout.

2.4-5 Ram Pumps

Ram pumps are also reciprocating pumps, but the rams move through cylinders adjoining a chamber, so that the pumping comes from the displacement caused by the volume of the pistons or rams, rather than that caused by sweeping a cylinder clean. Valves are necessary.

The older styles have two pistons in line, horizontally opposed across a central valve chamber. The drive force comes from crossarms and push rods moved by a separate motor. The basic arrangement is shown in Figure 2.4-5A.

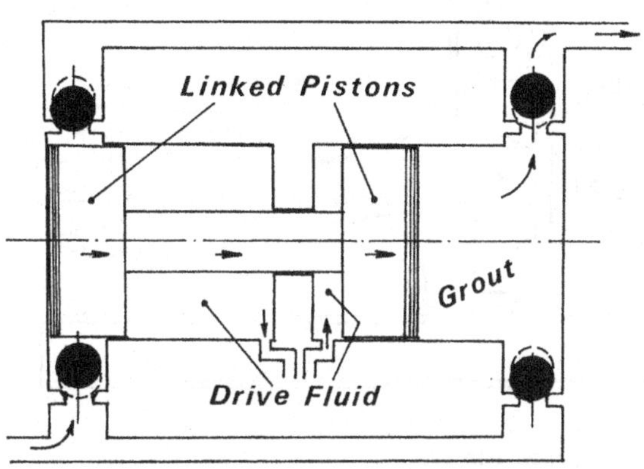

Figure 2.4-4B. *Modern Duplex piston pump.*

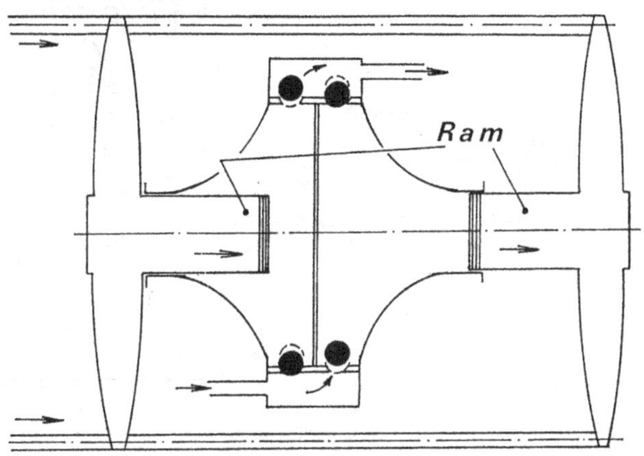

Figure 2.4-5A. *Older style of Duplex ram pump.*

The rams are connected through crossarms and push rods. The grout chamber is situated between them; the valves are located in this chamber. The arrows show direction of grout movement when the rams are moving to the right.

Some pumps are mighty monsters, needing a crane to move them. A popular brand was known by the Italian name of Clivio (see Figure 2.4-5B). However, although this name is almost the same as mine (which is Clive) I do not endorse their use in any way.

Another common brand of ram pump is the Evans (see Figure 2.4-5C). These are still in manufacture.

The big cumbersome types are diminishing in use, although still seen on some jobs done by European contractors. Parry-Davies (1982) regards them as the best pumps for high pressure work. Compact modern versions are made by at least one European manufacturer (Häny) (see Figure 2.4-5D).

Figure 2.4-5B. "Clivio" ram pump.

Figure 2.4-5D. *Modern double-acting ram pump. (Courtesy Häny & Cie AG.)*

2.4-6 Diaphragm Pumps

Diaphragm pumps have single-acting diaphragms, usually in pairs, working in chambers with valves. As the diaphragms flex in and out under the stimulus of pistons or a crank, they suck in grout and then expel it in turn.

The example in Figure 2.4-6A of a single-diaphragm unit has the grout chamber on the right-hand side of the diaphragm. The valves are shown closed. Grout is pushed upwards and out as the diaphragm is flexed to

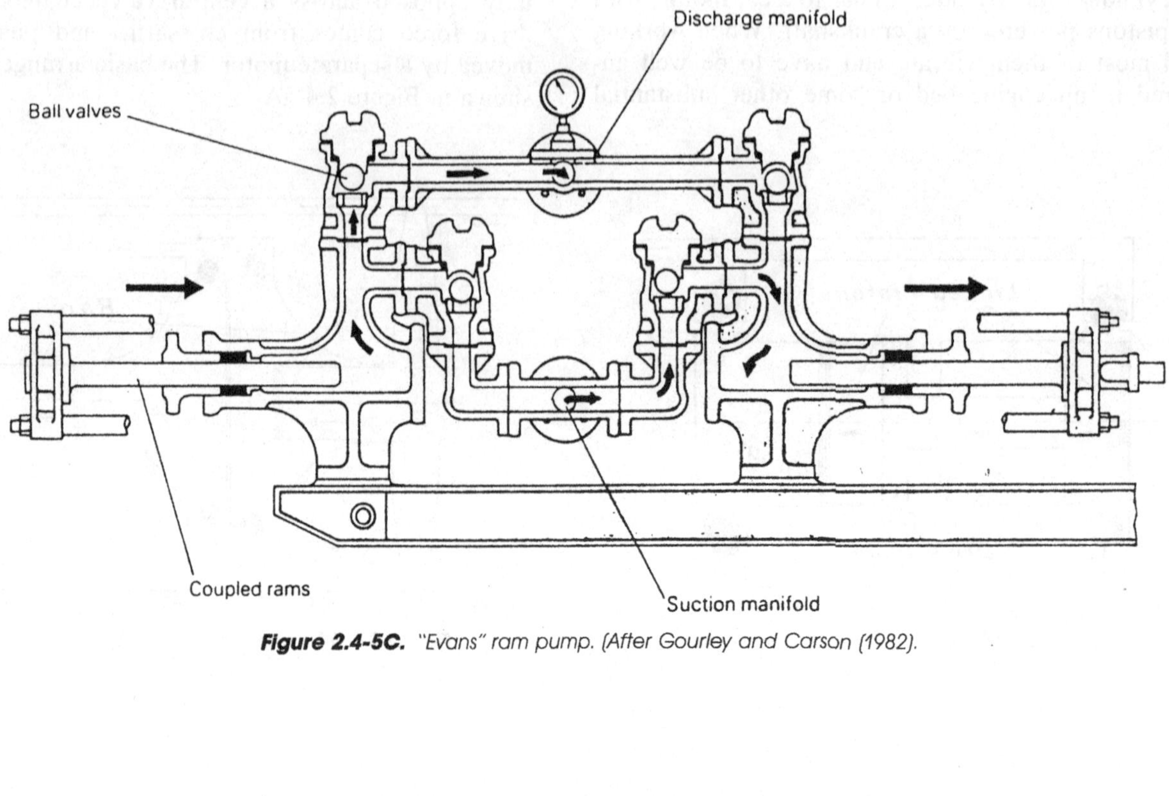

Figure 2.4-5C. *"Evans" ram pump. (After Gourley and Carson (1982).*

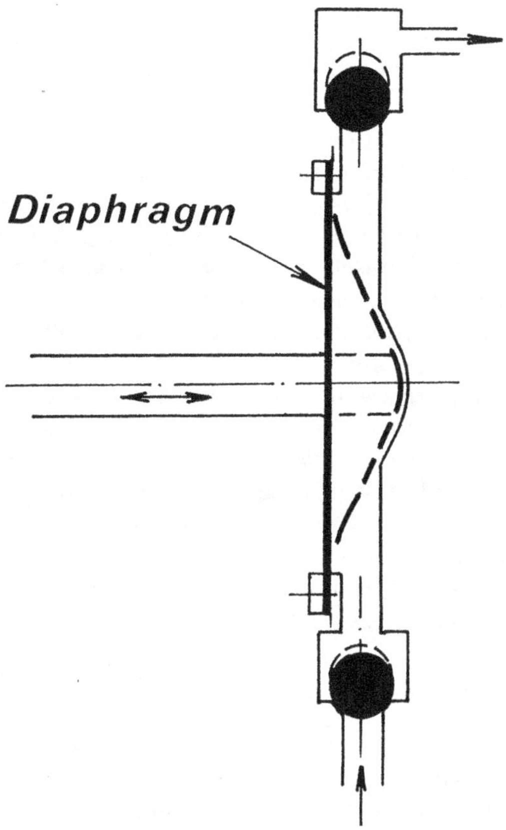

Figure 2.4-6A. *Single-diaphragm pump.*

the right and is sucked in from below on the return stroke.

Modern units can be compact and able to handle substantial pressures. It must be possible to quickly change diaphragms to deal with bursting, tearing, and wear. Pumping sanded grouts with them can be especially severe on the diaphragms.

Output is small relative to other pumps when pumping at high pressure. Typical maximum pressure (Reader brand) is 800 psi ⟨55 bars⟩.

2.4-7 Hand Pumps

Hand pumps are for jobs too small to warrant a powered pump. A long handle operated by a person provides the reciprocating motion for a small piston or diaphragm pump, usually single acting.

2.4-8 Compressed Air Pans

These are history! Yes indeed. They formed an important link between the very earliest primitive methods and the powered pump. Toward the end of last century they were *the* grout injection method. Some are still made and used, but their limitations in comparison to modern pumps discourage use. They use compressed air to expel a charge of grout from a strong, air-tight container. Grouting is intermittent: the pan has to be opened to let each new charge of grout in. The compressed air sometimes intrudes into the grout, altering its properties adversely. Another disadvantage is that if compressed air gets underground, it can be dangerous and can spoil the grouting.

2.4-9 Are Valves Worth the Trouble?

Figure 2.4-9A shows some examples of grout valves, but there are others. Considerable ingenuity has gone

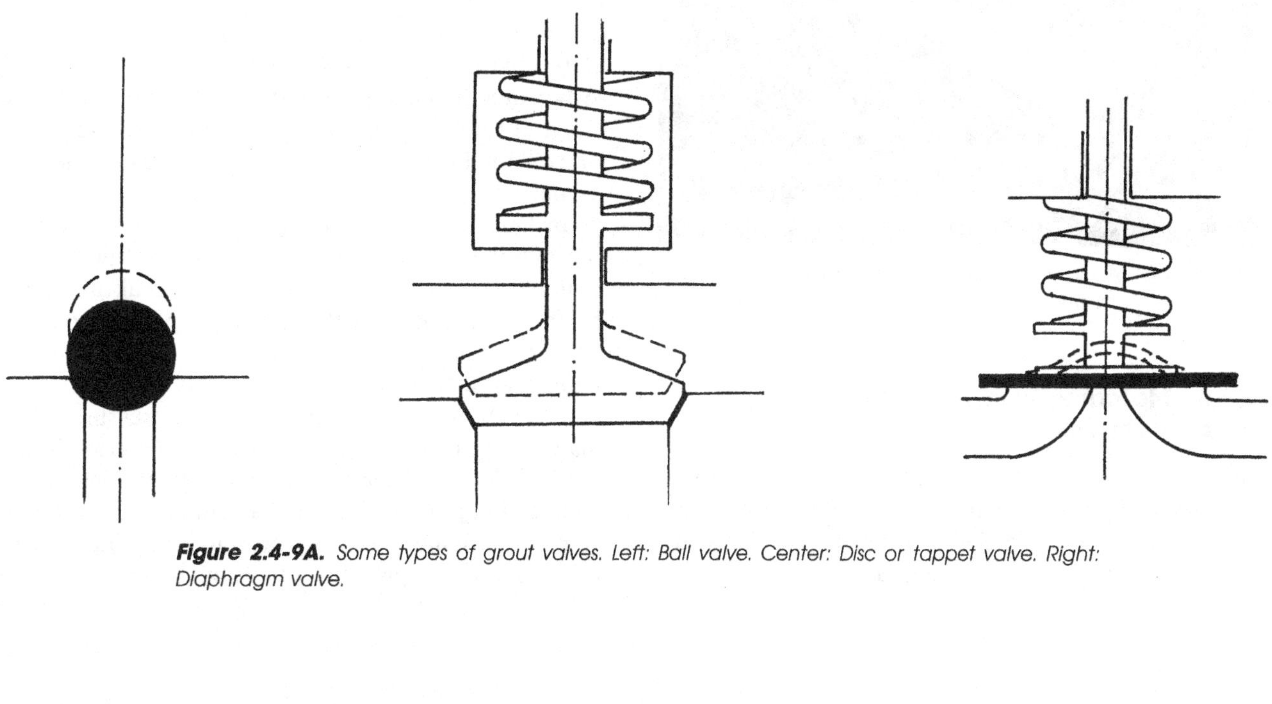

Figure 2.4-9A. *Some types of grout valves. Left: Ball valve. Center: Disc or tappet valve. Right: Diaphragm valve.*

into developing valves that will work reliably with grout. This is not easy to do.

Valves are the "Trouble Center" of a grout pump, so any pump that doesn't have them has distinct advantages over one that has. Maintenance problems come not only from wear but also from blockage by hard grout—if washing is not frequent and thorough. Manufacturers try to extol their own particular creations in terms of superiority and ease of cleaning, but even the best designs are thwarted if there is delay or forgetfullness in washing out the grout chambers. Perhaps the crew may even be called away to some emergency, and the pump has to be stopped—when they return you can guess what they find. These things do happen. The crew must then chip the hard grout out of awkward recesses in the chambers (Figure 2.4-9B); this can cost real money and time. Even having to strip down valves to free them can put a pump out of service for several hours. And then sudden blockage can be a serious problem if it happens during a grouting operation and if a standby pump cannot be quickly brought into action. Problems such as these are some of the reasons why the helical rotor type of pump (no valves!) has become the principal one for grouting when pressures are moderate.

However, for high-pressure work piston pumps are virtually the only option, so valves can't be avoided.

Figure 2.4-9B. A sad case. Someone forgot to wash the pump out. Pneumatic chipper removing hard grout from valve chamber.

2.4-10 Matching Pumps to Circulation Line Alternatives

If the usual type of return circulation line is in use in which unused grout goes back to the agitator (Section 2.1-3), the pump is not called on to operate under near-stalling conditions, and it is not used as a pressure-limiting device. Accordingly, it can be a relatively simple machine running at a comfortable speed without struggling. All types of pumps are suitable for this type of usage.

However, with the nonreturn type of delivery line, which goes from the pump to the grout hole without return (Section 2.7-6), the pump has to be able to cope with decreasing discharge as the grouting of each hole approaches refusal. Then, when refusal is met, the pump has to be able to stand zero-discharge conditions without distress. Piston pumps are usually used in these circumstances, equipped with a variable-speed hydraulic drive unit. The ability of these drives to automatically govern pressures is usually implemented. Units having a suitable capacity for such work tend to be large and are usually set up in central mixing plants. These may also have distribution drawbacks as discussed in Section 2.7-12. However, although portable pumps of this type are marketed, their output is rather low for most rock grouting jobs.

2.4-11 Portability of Pumps

Most helical rotor pumps are small and light enough to be manhandled into locations where access may be difficult. None of the other types are as portable or as compact although some are nearly so. On many sites it is desirable to have the pumps at suitable elevations in relation to grout holes to keep circulation line pressures matched to grouting pressures (Section 2.7-7). The desirable and requisite moving of the pumps every few days to match progress of the work up or downhill is fairly easy when pumps are portable.

2.4-12 Steady Pressure or Cyclic Pressure?

There is some difference of opinion about whether pressures in grout holes should be steady or pulsing. Advocates of steady pressure contend that this keeps the grout moving uniformly and minimizes settlement of cement in cracks and slows the thixotropic stiffening.

In Britain, where pulsing pressure has been in use for many years by some grouting practitioners, the concept is that the pressure cycles break up cement bridges in cracks and thus improve penetration. This concept cuts across my experience that the opposite happens—to the extent that one way to block leaking cracks is to purposely pulse the grout pressure. I have made use of this when dealing with grout leaks through fine cracks in concrete structures (leaks leading off contraction joints). The mechanism appears to be that cement settlement and thixotropic stiffening (Section 4.2-2) in the low-pressure phase of each cycle inhibits penetration. Hence my preference as well as many others' for steady pressures rather than cyclic.

Cyclic pressures are produced by piston, ram, diaphragm, and hand pumps in varying degrees with each stroke of the pistons or diaphragms. The extent can be softened by using pumps with several cylinders or by putting an air dome (called an accumulator sometimes) in the line. When double-acting pumps are hydraulically operated, it is possible to vary the hydraulic pressure during the stroke so as to produce an almost constant output pressure. Doing this with multicylinder pumps can iron out pulsations considerably.

Admittedly, by the time grout travels to the hole in long circulation lines, much of the pulsing is ironed out, particularly if the line is made of hoses; flexing of their walls produces some damping of pressure peaks.

By contrast, the helical rotor pumps produce steady pressure output when in good condition, another reason for their popularity.

2.4-13 Comparison of Maximum Pressures

As a guide to deciding which pump to put on a job, the following comparison indicates maximum pressures. Pump types are listed in preferred order:

- Up to 150 psi ⟨10 bars⟩—Helical rotor and most other pumps.
- From 150 to 300 psi ⟨10 to 20 bars⟩—Piston, ram, diaphragm, and some helical rotor.
- Above 300 psi ⟨20 bars⟩—Ram, piston, and some diaphragm.

This is not a strict classification.

For most cement grouting jobs at surface rock, there is no need for the pumps to be able to exceed 150 psi ⟨10 bars⟩ (see Section 6.2-1 about this) if they are of the light, portable sort; if perhaps greater pressures are needed, these can often be obtained just by setting the pumps up on some high spot, and the additional hydrostatic pressure thus gained in the lines produces the greater pressures at the grout hole (Section 2.7-7). At fixed pumping stations and underground, though, higher pressure pumps are usually needed.

2.5 VALVES, PRESSURE GAUGES, WASHOUT GEAR, AND PACKERS

2.5-1 Control Valves for Grout

Valves used for controlling grout need to have some means for keeping the grout out of any working parts such as screw mechanisms. Otherwise, an early demise is predictable. Ordinary valves used for water such as gate, globe, or disk are no good for grout.

Valves are used for the following operations:

- At the grout hole for controlling the grout going into the hole (Section 2.6-1).
- On the circulation line after it leaves the hole (as explained in Section 2.7-4), and only, of course, if it has a return (see use discussed in Section 2.7-6).
- On the hole at the surface so that it can be closed off after grouting is finished (Section 7.2-9).
- Between the agitator and pump so the hose to the pump can be changed without draining the agitator.
- To divert flow into a bypass if one is used (Section 2.7-3).
- For a number of other minor purposes such as bleeding the hole (Section 4.3-3).

At most of these positions the valve is called on to operate at any setting; at others it is only used in the fully open or fully closed position.

2.5-2 Suitable Valves

Valves successfully used for grout include:

DIAPHRAGM Valves
BALL Valves of certain types
PLUG COCKS of certain types.

2.5-3 Diaphragm Valves

These valves have a flexible diaphragm separating the grout from the working parts, as shown in Figure 2.5-3A and 2.5-3B.

Turning the handwheel at the top causes a threaded spindle to operate the diaphragm toward or away from the weir-shaped base. The valves can usually be operated for several hours without washing and are usually left in position and washed simultaneously with the line or equipment they are on. Cleaning is easy. If

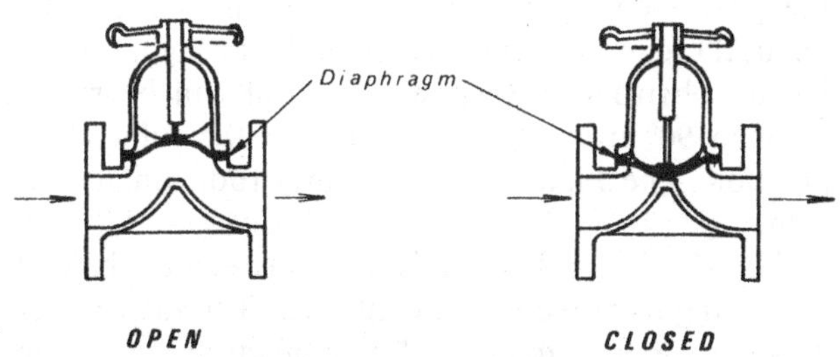

Figure 2.5-3A. "Saunders" valve.

Figure 2.5-3B. *Parts of diaphragm valve. Upper left: Bonnet with handwheel and diaphragm. Lower left: Cast steel base (this one shows wear on the weir). Upper right: A diaphragm. Lower right: An unworn base.*

necessary, the bonnet can be readily unbolted and the diaphragm lifted out for replacement or checking. Provided the correct type of diaphragm is used, they can last for many months of continuous service. Wear is sometimes faster in the cast steel weir.

A particularly useful feature of these is the way the flexing of the diaphragm can be used to free a blockage such as that occurring when the valve is throttled back to almost no flow. By exercising the spindle a few turns, the diaphragm flexes and breaks the grout plug. Flow then starts again—but be ready for the surge!

These valves are almost obligatory when handling a tricky situation where volumes and pressures have to be controlled very precisely and varied rapidly. An example of this is the grouting of a hole in rock liable to move (Section 7.6). No other type of valve is anywhere near as satisfactory for this.

The type having a handwheel is better for grouting than the lever type (which is better suited for fully off or fully open situations).

2.5-4 Ball Valves

These have a rotating steel ball with a central hole through it of the same diameter as the tube carrying the grout (see Figure 2.5-4A). When the valve is fully open, there is an opening completely through, presenting no obstruction to flow. The ball can be rotated through 90° from the fully open to the fully closed position. The Figure shows a fairly rudimentary version.

These valves rely on a tight fit between the ball and its seating to prevent grout entering. The rate of wear depends on the quality of manufacture, and some valves may only last through a few days' use. They

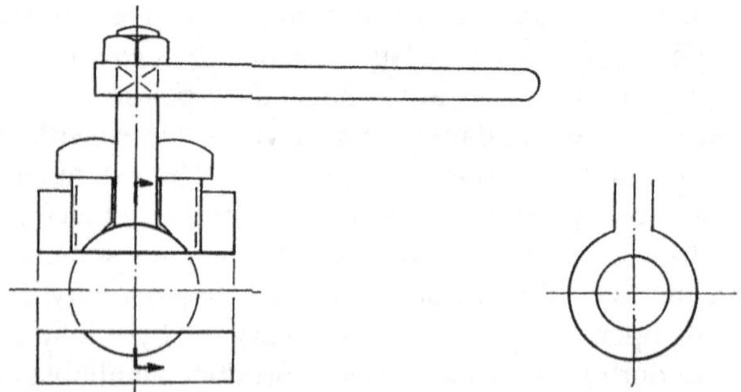

Figure 2.5-4A. *Left: Section through ball valve. Right: Cross section through the ball showing the central hole.*

cannot be serviced and so are discarded when worn too much; they are, however, relatively cheap.

A wide range is available, some made of tough plastic. Most can handle moderate grout pressures such as 150 psi ⟨10 bars⟩. Some can take higher pressures.

They tend to be in the "cheap and nasty" category as far as grouting is concerned because they can easily block at small openings, particularly with thick grouts. When this happens, the only way to free them is to stop grouting, open up the lines, and poke wire through them—altogether undesirable interruptions. This is one reason they should not be permitted on high-quality grouting. The other reason is the difficulty of accurately setting them to precise flows and pressures. A few contractors have a preference for them, though, probably on the score of cost rather than of operational facility.

2.5-5 Plug Cocks

Plug cocks resemble ball valves but have a tapered plug instead of a ball, as shown in Figure 2.5-5A.

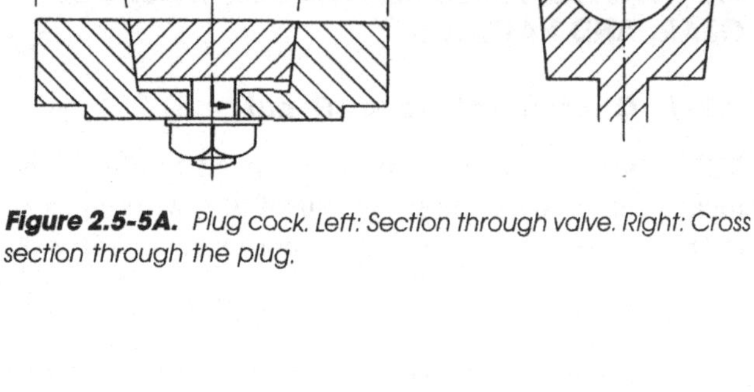

Figure 2.5-5A. *Plug cock. Left: Section through valve. Right: Cross section through the plug.*

The handle at the top turns the plug. In simple versions, such as the one shown, it is difficult to keep the tapered plug sufficiently tight in its seating to prevent entry of grout there while at the same time having it free enough to turn. This is overcome in more expensive versions by using lubricating material to line the seating, and possibly also by passing the spindle through glands.

The simple types should only be used either fully open or fully closed. Any setting in between can result in blockages and interruptions in the same way as with ball valves.

However, well-made plug cocks are the best valves available for handling high pressures. A drawback is the difficulty in making slight changes in flows; the lever control is not sensitive enough, but there is nothing better yet. It would be great if diaphragm valves could be used for really high pressures. Here is an opportunity for someone inventive.

2.5-6 Handles

A small point, but a very vital one: the handwheel or lever of a valve should always be left on. Never take it off and cart it around separately with the idea of slipping it onto the squared operating shaft when a valve has to be turned. And worse still is the practice of using a wrench or stilsen to turn valves. Apart from the inevitable mangling of squared shafts, these practices mean that valves cannot be operated immediately when needed, say to turn off the grout quickly if something has gone wrong. There is no time then to fit a handle or whatever. Valves must be ready for instant operation at all times.

2.5-7 Automatic and Remotely Controlled Valves

Diaphragm valves are sometimes fitted with control equipment permitting remote or automatic operation, generally as part of a computer-controlled operation. See Section 7.7-5 for further discussion.

2.5-8 Pressure-Limiting Valves

On some jobs, these valves are used on the grout lines to ostensibly limit the amount of pressure leaving the pump or going into the grout hole. Note that these are not those sometimes used to limit pump output by restricting air or hydraulic drive pressure (see Section 2.4 for pumps).

There are some experienced grouting people who aver that there is no such thing as a reliable pressure-limiting device for grout. I'm one of them.

A fairly obvious arrangement is to spring load a diaphragm valve to open up if pressure exceeds its setting, and thus dissipate the excess pressure to waste. This sort of thing is used successfully with water, gas, air, and so on, but grout is quite a different matter. The trouble is that it sits there, closed, for hour after hour, with stationary grout in it. Just the right prescription for a blockage. And, of course, thixotropic stiffening (refer to Section 4.2-2) means that the pressure is not being transmitted to it anyway, so it can be a fallacy to expect it to exert control.

Perhaps a fairly sophisticated pressure sensor mounted in the circulation line and connected electrically to a remotely controlled valve is an answer, but this means more things to go wrong, and that, once more, makes reliability questionable. This is not to say that you can't buy systems claimed to be absolutely reliable—but prudent purchasers beware.

It is unfortunate that some contracts make the use of pressure-limiting valves mandatory, and some supervisors have so little confidence in their crews that they install these valves as a precaution against them doing the wrong thing. If the crews are so poor that they cannot be relied on to manually control pressures properly, then they shouldn't be let loose on grouting. I am well aware that some awful crews find their way on to this work.

2.5-9 Pressure Gauges

In grouting, pressure gauges are used to show the following values:

- The pressure in grout holes
- Circulation line pressures
- Pump pressures

Gauges of the Bourdon type are usual.

For grouting there has to be some means of keeping grout out of the Bourdon tube. Devices for this include diaphragms of various types, U-tubes, air tubes, and others.

2.5-10 Diaphragm-Type Gauge Protection

The type of gauge protection shown in Figure 2.5-10A comprises a semiflat rubber diaphragm mounted in a fitting separately from the gauge. The Bourdon tube inside the gauge is filled with oil or grease or glycerine and transmits the pressure from the diaphragm. Grout pushes against the other side of the diaphragm.

The material used in the diaphragm must be sufficiently soft to transmit pressures without loss but at the same time must be strong enough not to tear. In

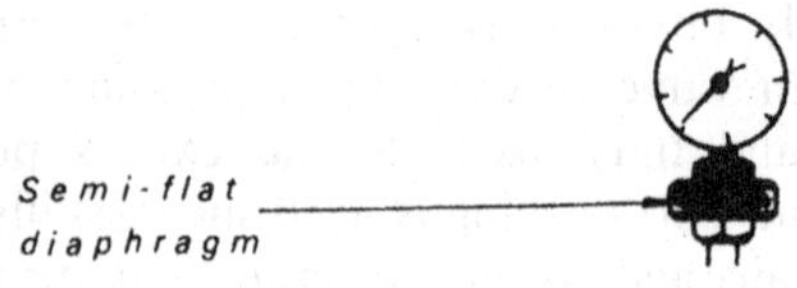

Figure 2.5-10A. *Flat diaphragm gauge saver.*

addition, clamping around the edge must be carefully done so that the diaphragm is not stretched and therefore unable to transfer pressures correctly. Some contractors make their own, but it is usually better to buy diaphragm assemblies ready made. The best have half-round clamps that can be opened up to give access to the diaphragm for changing or adjustment (see Figure 2.5-10B).

Figure 2.5-10B. *Top-quality pressure gauge. Flat diaphragm accessible by undoing clamp, as shown. (Dobbie gauge as described in Section 2.5-13.)*

A different type of protection is the rubber cup which transmits pressure from grout within it to oil outside (see Figure 2.5-10C). These are factory made units used for other things besides grouting.

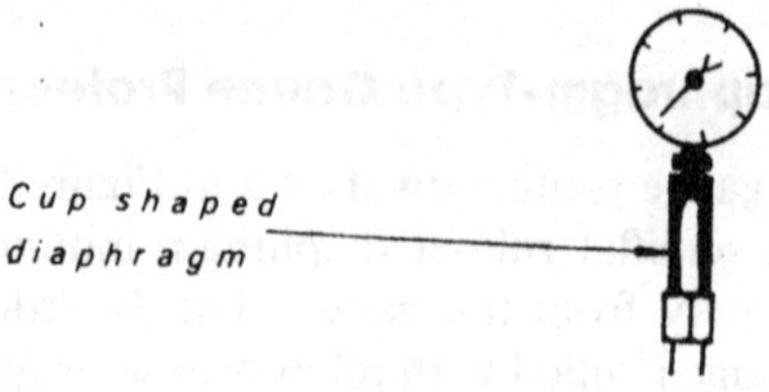

Figure 2.5-10C. *Cup-shaped gauge saver.*

Then there is the tube type of gauge saver shown in Figure 2.5-10D, where the grout passes through the inside of a rubber tube. Oil outside it carries the pressure to the gauge. These are used by some European grouting contractors. However, testing on a number of occasions has found inaccurate transmition of pressure; caution is advised.

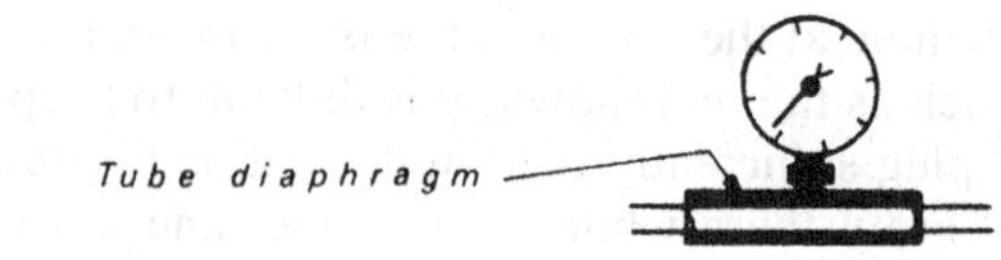

Figure 2.5-10D. *Tube-shaped gauge saver.*

2.5-11 U-Tube Gauge Saver

In the U-tube gauge saver shown in Figure 2.5-11A, there is nothing between the grout and the oil. The grout does not penetrate into the oil in a tube of this

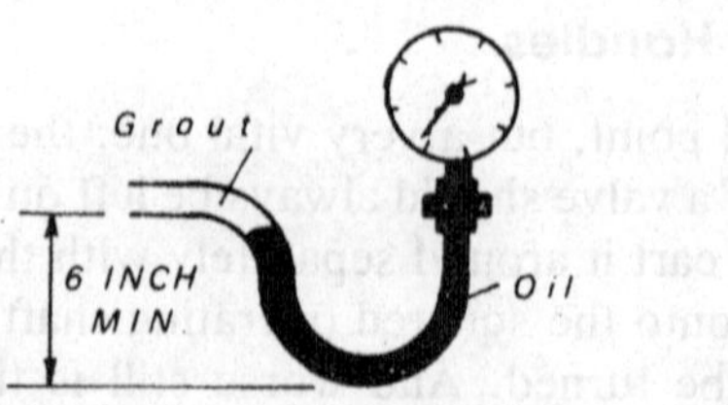

Figure 2.5-11A. *Use of U-tube to keep grout out of gauge.*

shape if it is about 0.25 or 0.50 in ⟨7 or 15 mm⟩ diameter. Copper tubing is best because it can be bent more or less to shape by hand and is stiff enough to retain shape and stand rough handling. The oil should be about SAE 40 viscosity. The method works well, is a simple arrangement, and can be improvised on site. Care is needed to make sure that it is not inadvertently inverted; the oil might then drain out unnoticed, so at the next use the pressure gauge is ruined. Sometimes it is wise to protect pressure gauges from careless handling, and a wire protective cage as shown in Figure 2.5-11B can be worthwhile.

2.5-12 Excessive Protection

Some people always manage to overdo things! Particularly in this matter of gauge protection. One gauge-saving device is quite enough—two is one too many. On various jobs I have seen two diaphragms on one gauge, one after the other, and a diaphragm after a U-tube. Doubling up like this is very likely to dampen pressure transferal and make the gauge read untrue. Not good!

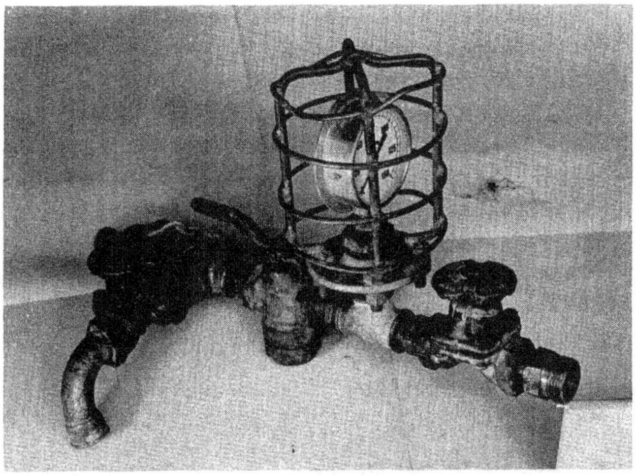

Figure 2.5-11B. *Examples of grout fittings. From the right: Saunders diaphragm valve; pressure gauge, with flat diaphragm saver and protective cage; Audco lubricated plug cock (valve); Saunders diaphragm valve on bleeder offtake.*

2.5-13 Comments on Gauges

Reliability and accuracy are essential. Gauges must be able to function well for several days without attention. Their face must be large enough to be seen clearly, even when smeared with grouty wipings after being in the target zone of the inevitable stray spray. The usual "clean" rags used to wipe them often aren't. Reduced visibility can also occur in poor weather conditions and on night work. The recommended size of face is 4 in ⟨100 mm⟩ diameter. I make the strong recommendation that anything smaller than 3 in ⟨75 mm⟩ should be kept off the site. Nevertheless, plenty of jobs do indeed go in for smaller (and cheaper, of course) gauges, regrettably.

The range of the gauges should be about twice the maximum pressure used. On most jobs this means that several ranges are needed, some for the low-pressure work, and others for higher pressures. Markings must be clear, with large figures.

Glass faces are common, and so is breakage. Better quality gauges nowdays use transparent plastic faces that don't break and can be unscrewed to get to the hand for adjustment.

Should pressure gauges be of the last-for-a-long-time variety or of the throw-away type? Grouting is inherently tough on relatively delicate things like gauges, and some people prefer to buy cheap gauges, not adjustable, and just discard them when they go out of adjustment or are damaged (the "K-Mart" attitude as a leading grouting engineer at the USBR calls them). Other people like to use the very best gauges available (and there are some beauties now) and to look after them carefully, readjusting them whenever needed. These people are usually small contractors doing the grouting personally, or where a large organization maintains its own grouting crews (doesn't let the work out to contractors).

For those who want the best gauges, features now regularly available include tough, thick plastic case able to stand quite a lot of abuse; tough transparent plastic face (called "bezel"), unscrewed from the front by several turns on a coarse pitched thread; nonstick internal works made also of plastic; microadjustment of the hand by means of a screwdriver-operated gear reached by unscrewing the bezel. One manufacturer is Dobbie Instruments, Melbourne, Australia.

It is essential to regularly check the gauges against a carefully kept master gauge at least weekly, and any inaccurate gauges adjusted or discarded, depending on philosophy. Watch for gauges that don't show zero when they should. It probably means that they have gone out of adjustment since the last check.

A somewhat ambivalent attitude to accuracy was noticed on one job where a very inexperienced contractor used two pressure gauges on each hole, apparently on the premise that if one happened to be out of adjustment, the other wasn't. I don't need to explain the fallacy of this, I hope.

Thorough washout after grouting includes unscrewing each gauge below its protective device and washing the tube to there.

2.5-14 Remote Readout Pressure Gauges

Remote sensors are available that transmit the pressure reading by cable or radio to reading arrangements elsewhere. These are used in conjunction with computers usually (Section 7.7-4).

Included in this category are downhole sensors. These may have readout remote from the hole or at it. Sometimes incorporated in packers to measure pressure down in the hole, their use is rare, largely because of inevitable high casuality rates and considerable cost. However, they find some use in investigation water testing (Sections 8.2-6 and 8.2-7), where the chances of survival are better.

2.5-15 Packers

Packers are used for isolating part of a grout hole so that grout only goes into the foundation in that part and for holding the fittings at the top of a grout hole instead of a standpipe doing this. (See Section 5.1-1 for discussion on standpipes).

Essentially, they are a device that can be put freely down a hole in the collapsed or deflated condition, and then when in position, can be expanded or inflated to provide the requisite sealing or isolation.

Packers are of two main types: (a) mechanical and (b) inflated (see Figure 2.5-15A).

The mechanical type has a short sleeve of soft rubber expanded against the hole wall to provide the sealing. This expansion results from squashing the rubber lengthwise by means of a screwing device operated from the surface.

The inflated type uses a rubber tube that is inflated from within by compressed air, nitrogen, or water under pressure, supplied through a small tube.

Which is the best sort of packer? Arguments rage on this. Let me suggest that the mechanical type is generally best in rock where holes are fairly circular and even, while the inflated type is better if holes tend to be rough and uneven.

The mechanical ones are the more robust of the two

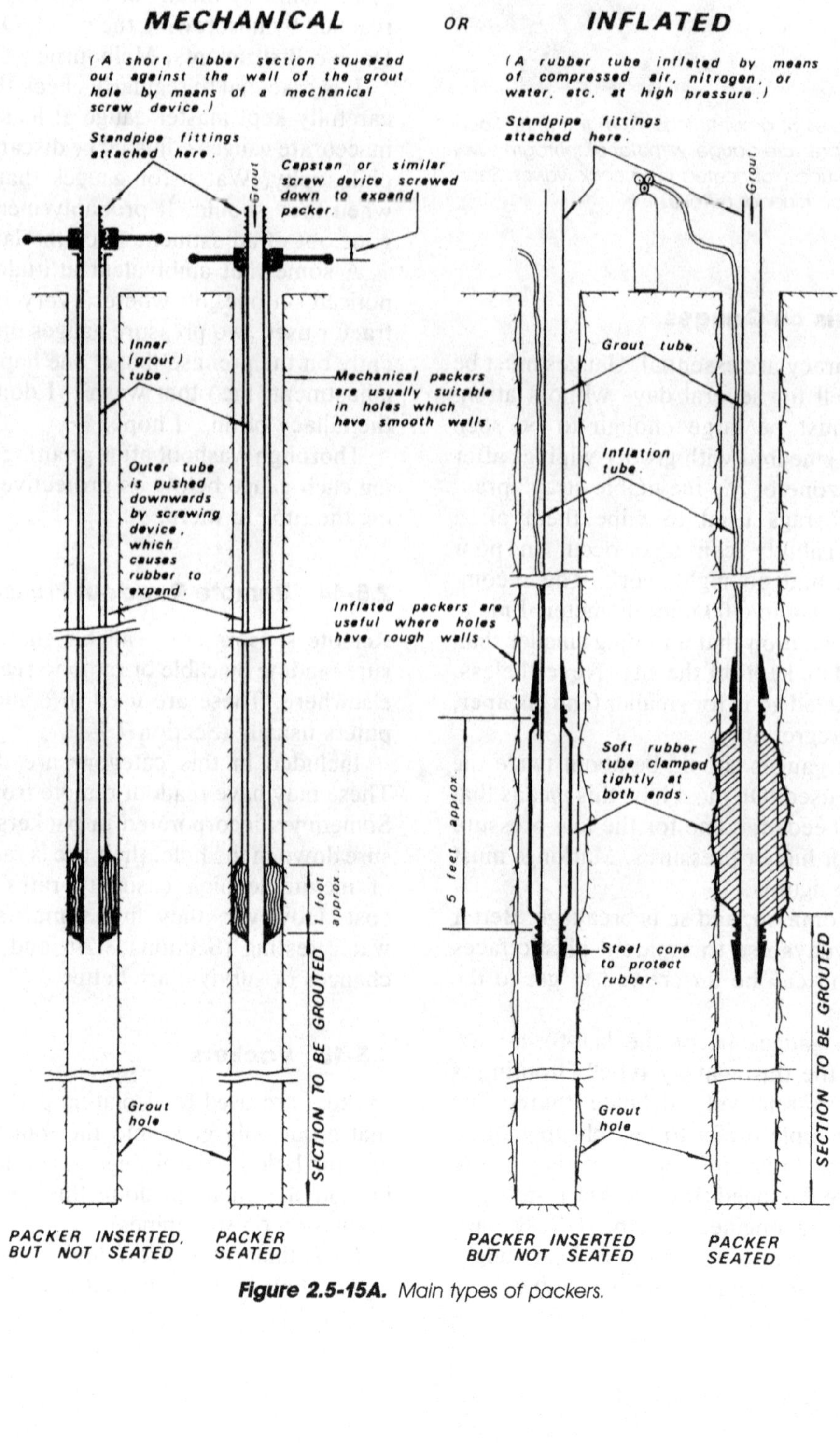

Figure 2.5-15A. *Main types of packers.*

but are much heavier in deep holes, where a crane or A-frame may be necessary just to deal with their weight. Fifty feet ⟨say 15 m⟩ is about the limit for reasonable use.

However, the inflated type can have problems too. It is vulnerable to punctures or tearing on rough projections while being moved in the hole, and there is a limit to the amount of irregularity able to be accommodated when inflated.

One of the problems with packers is that the tube carrying grout through them is relatively small in diameter. This means that bleeding of the hole cannot be properly carried out, and this is a serious disadvantage (the need for bleeding and its effect on the quality of the work is explained in Section 4.3) (see Figure 2.5-15B). This can happen with both downhole and surface packers.

Another problem with downhole packers is the possibility that grout can bypass them through cracks and enter above without the knowledge of the operator; packers can become firmly grouted in when this happens (see Figure 2.5-15C).

If this happens, the packer may have to be drilled out or the hole may have to be replaced alongside, both at considerable expense.

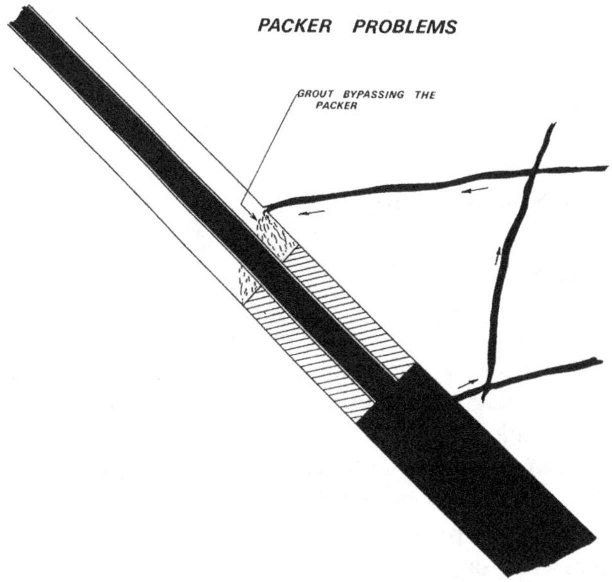

Figure 2.5-15C. *Packer grouted in by grout going around it.*

2.5-16 Why Use Packers?

Having been caught too many times with packer troubles, and never having been convinced in the first

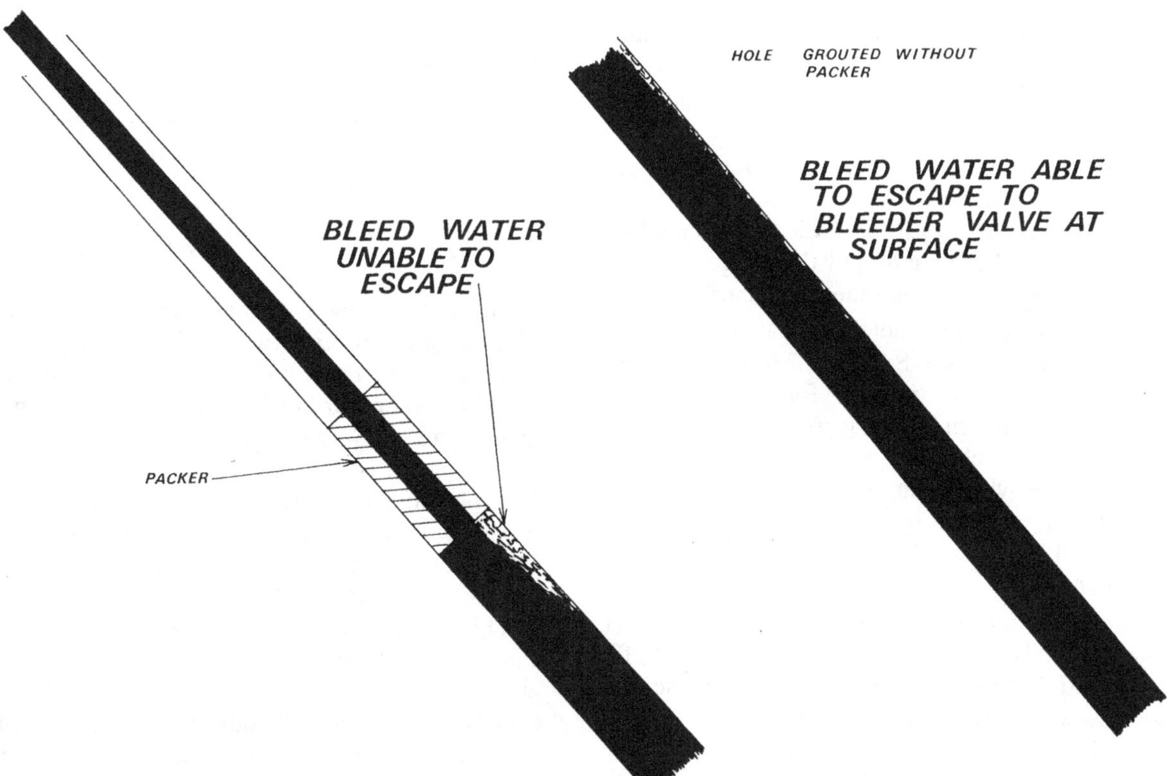

Figure 2.5-15B. *Left: Bleedwater trapped under a packer. Right: Better grouting without a packer.*

place about the arguments advanced for their use, my advice, for what it is worth, is

AVOID PACKERS

unless in dire straits.

DIGRESSING ON THIS

1. In some of the stage grouting methods (Section 5.4-3) packers are used to separate the stages. The point is made in Section 5.4-8 that use of packers can be technically disadvantageous and the intended protection against surface heaving and the intended isolation can be just an illusion.

2. Using packers at the surface instead of standpipes can save money if the rock is sufficiently sound (refer to Section 5.1-2) but can inhibit proper bleeding of the holes. They are not as rigid as standpipes and have been known to lose seating and slide or fall out of the hole during the grouting. This is poor.

AND WHAT ARE THE DIRE STRAITS WHEN MIGHT THEY SAVE THE DAY?

1. Severe surface leaks from shallow cracks where other means of stopping them have failed or are unsuitable (refer to Section 7.3) and the top of the packer is visible down the hole so that any grout entering above it can be seen.

2. Handling some types of rock movement where a single (horizontal usually) crack has to be treated separately from others (Section 7.6-4 discusses handling rock movements). Possible loss of the packer is a minor consideration among the other problems.

3. Grout connections to holes that haven't yet been fitted with standpipes (see Section 7.3-6 for treatment of connections and Section 5.1 for installation of standpipes). Even if quick-setting grout is used for putting standpipes in such holes, a matter of half an hour or so will elapse while doing it, and often this is too long. Hence the solution is to put surface packers in the connected holes so that the connections can be properly handled.

4. When grouting under structures where uplift is undesirable. Packers are sometimes used, but their efficacy is doubtful as always. For pressures in these circumstances see Section 6.2-7.

5. There are other remote happenings that are too rare to warrant space here.

2.5-17 More about Packers

In addition to the two types shown earlier, there are cup packers and other less used types. They have their advocates, but not many.

Double packers have been used for grouting but are liable to give even more trouble than single ones; the chances of the lower packer not seating properly and thus unknowingly ruining its isolation role tend to discourage use.

Nonrecoverable packers are available from several manufacturers. These can be cheapest in difficult holes. Some are made of soft material that can be drilled through later, if necessary. The MPSP method (see Section 5.4-12) uses nonrecoverable tubing as well as grout-filled permanent packers having fabric walls.

Rubber for inflated packers should be reinforced, both to help against bursting and to stiffen it against tearing. However, it must still be pliable enough to expand into crevices and hollows. It has to be clamped securely to the grout tube at one end. In short packers it can be also firmly clamped at the other end, but in long ones (about 5 ft ⟨1.5 m⟩ or more), provision needs to be made for the lengthwise shortening of the sleeve when it is expanded. The sleeve should be able to withstand inflation in the air, out of the hole, to about the intended grout pressure.

2.5-18 Washout Gear

It is no use trying to get grout to travel outwards from grout holes if their walls are covered with mess such as drill cuttings, grout from a previous application, weathered materials from the foundation, or whatever. Thorough washing is the way to get holes nice and clean.

Grout holes have to be thoroughly washed after drilling, after grouting (in certain of the stage grouting methods; refer to Sections 5.4-4 and 5.4-5), and after grouting when further applications are going to be given (refer to Section 7.2-10).

Equipment is the same for each sort of washout. A long plastic hose is the most convenient, but where washout after grouting has been delayed unduly and grout has firmed, some form of cutting teeth on steel pipe is necessary, as shown in Figure 2.5-18A.

This ferocious looking gadget for removing grout can be made up on site from a pipe reducer or a coupling, but there is often no need to go to this extent. Careful job planning should provide that washing out is done before the grout has hardened too much for a plastic hose. In deep holes there can be problems,

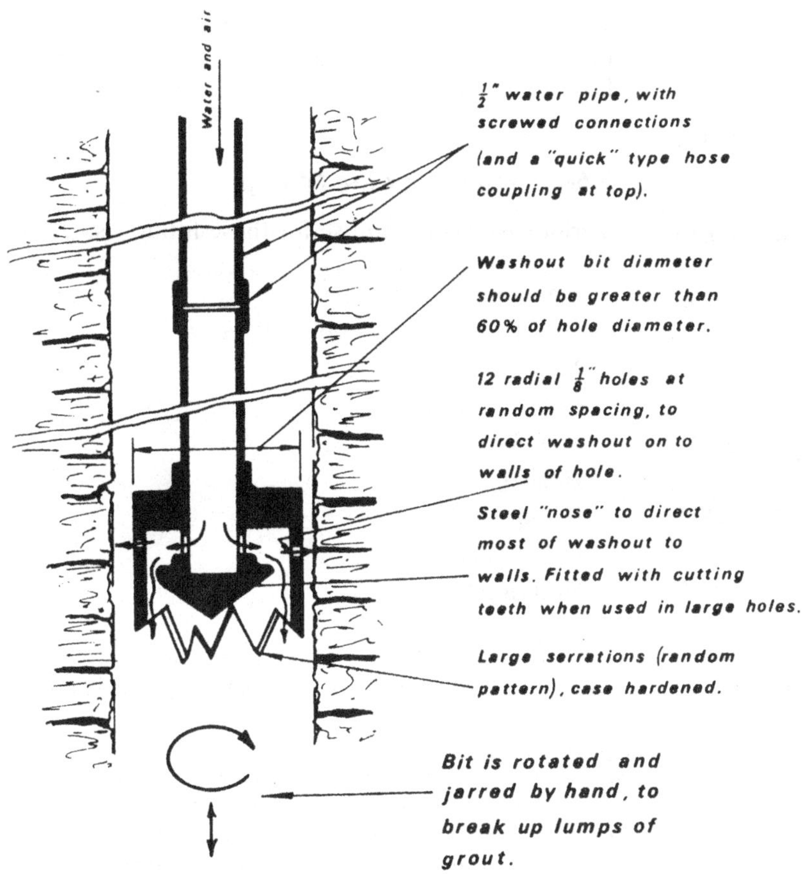

Figure 2.5-18A. *Washout gear for firm (but not set) grout.*

even so, as discussed from contractural aspects in Section 9.2-8.

Washout commences at the top of the hole, with water passing down the tube or hose, and then coming up outside it to emerge at the surface (Figure 2.5-18B).

Figure 2.5-18B. *Washing out a grout hole using a plastic hose. Protective cap is sitting upside down behind standpipe.*

Initially the return water will be dirty or grouty, so washing at that depth is continued until it runs clean. Then the pipe or hose is lowered a little and the process is repeated and so on until the bottom of the hole is reached.

Usually water alone is sufficient, but some foundations with a high proportion of weathered material or dispersive clay tend to pug and need compressed air with the water to provide sufficient agitation to get a return. Use of air can pose a hazard though, if it gets into the cracking to any extent. It can collect in pockets to form cyclic surges and surface eruptions, possibly to the detriment of foundation integrity. Only use air in wash water if absolutely necessary.

As mentioned in Section 5.2-14, the driller should give a start to the washing out process by lifting the rods slightly off the bottom of a completed stage and then giving the hole a burst of drill wash water.

Washing produces mess in the return water, and it sometimes sprays everywhere and splatters gear (and people) within range. This dirty water sometimes cannot be allowed to just run anywhere over the foundation, and sandbag settling ponds and channels may be needed to restrain it.

2.6 FITTINGS ON GROUT HOLES

2.6-1 Functions of Fittings

Fittings are needed on grout holes for the following functions:

- Controlling the quantity of grout going in (refer to Section 7.2).
- Controlling the pressure applied (refer to Section 6.2 for pressures).
- Carrying out regular bleeding-off to remove water and thin grout (see Section 4.3-3).
- Closing down the hole once grouting is complete (refer to Section 7.2-9).
- Attachment of flow meters or other special gear (Section 7.7-4).

2.6-2 Arrangement of Fittings

The arrangement of standpipe fittings and comments on their use are shown in Figure 2.6-2A. The bleeder valve can also be used for checking for blocked fittings, for blowing them out, and for connection of a flow meter (Section 7.7-4) as well as for its main purpose of regular bleeding of the hole (Section 4.3-3).

2.6-3 Names for the Fittings

Various names are in use for these fittings, such as:

Standpipe fittings
Christmas tree (as in oilwell work)
X-tree
Control stand
Manifold
Header
Banjo
Collar gear
Grouting tee

Take your pick. I like "standpipe fittings" and that is what you'll find used in this book, but there's noth-

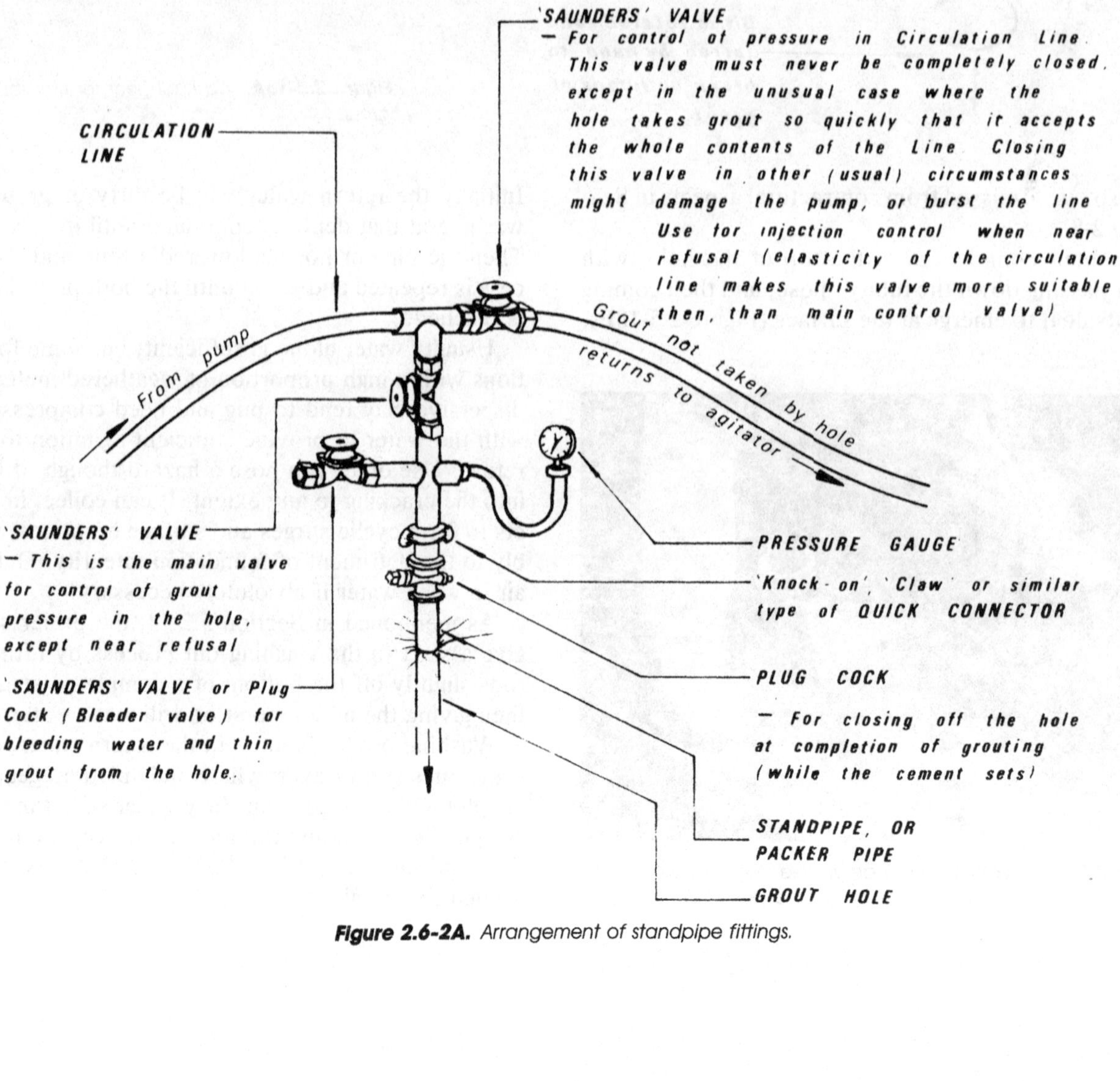

Figure 2.6-2A. *Arrangement of standpipe fittings.*

ing magic about that name. Another is every bit as good.

2.6-4 Quick Setup

By providing quick-connecting type couplings where the fittings are attached to the standpipe (see Section 5.1-1) or surface packer (see Section 2.5-15), it is possible to have grout continuing to circulate in the line while moving the fittings from a hole that has just been completed to one that is about to be started. This arrangement is shown in Figure 2.6-4A.

In a well-organized job with reasonable access, the time taken to do this transferring is typically only a few minutes.

2.6-5 Position of the Bleeder Valve

On sloping holes, the bleeder valve must be placed on the high side of the fittings to let the water and thin grout out; they tend to float on the denser grout. Section 4.3 explains this.

Standpipe fittings must be placed directly on the standpipe or on the packer tube if packers are used. A short hose should not be used between fittings and the hole for the following reasons:

- The hose is more than likely to have a low spot before it reaches the fittings, and this means that attempts to bleed the hole are hopeless.
- When grouting is approaching refusal and flow through such a hose almost stops, thixotropic stiffening (Section 4.2-2) of the grout in the hose will mean that pressure does not transmit through to the hole and hence the "refusal" indicated by take measurements will be false. The hole is then not properly grouted.

2.6-6 Size of Fittings

The pipe fittings and valves should be 1 in ⟨25 mm⟩ in size in most cases. Smaller sizes than this tend to block prematurely. Larger sizes sometimes develop dead spots where grout collects in low-velocity regions. An exception, outside the scope of this book, is vacuum grouting through a penstock (Section 1.2-1), where the grouting is virtually instantaneous and $\frac{1}{2}$–in ⟨12-mm⟩ fittings can be used.

Standpipes on grout holes are usually larger than 1 in ⟨25 mm⟩, and the necessary bushing down to this size should be done gradually to avoid dead recesses.

2.7 CIRCULATION LINES AND GEAR SETUP

2.7-1 The Importance of Careful Setup

Money can be saved by carefully setting up the grouting equipment to suit the nature of the site. On the other hand money can be lost by overdoing this: using unnecessarily large and complex mixing and pumping arrangements; using centralized pumping and long lines when portable pumping and short lines are more appropriate, or perhaps by going too high-tech (but not always a bad thing—see discussion in Section 2.7-12).

2.7-2 Examples of Setups

Stressing once more that suggestions in this book are for guidance, the following is a very generalized outline of what sort of setup to use in a few common cases. The general principles can be extended to comparable jobs not specifically mentioned. The first few cases are when working from the surface; mining and underground grouting come later.

SMALL JOBS Perhaps small jobs are best defined as jobs where only one hole will be grouted at a time (because there is sufficient time for this and/or because the extent of the work is relatively limited).

Portable gear arranged as shown in Section 2.1 is usually suitable. Alternatively, if the job is too small for this, use a combined mixer (Section 2.2-3), but be prepared to interrupt grouting every time a new batch of grout has to be mixed.

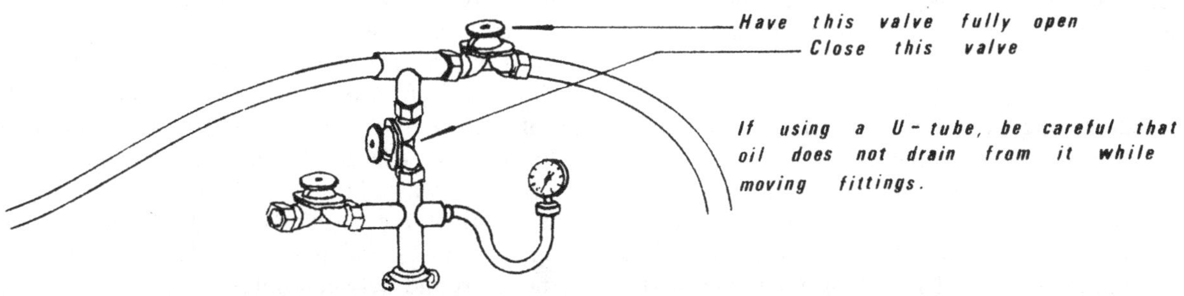

Figure 2.6-4A. *Changing to a new hole with the grout still circulating.*

Portability is improved if the gear is put on a small trailer and used from there. The trailer is moved as close to a group of holes as access permits; then the circulation line hoses are connected up and away goes the grouting. Be careful to leave room on the trailer for lifting bags of cement onto the mixer.

MEDIUM-SIZED JOBS It is difficult to explain the distinctions between the various sizes of jobs—the boundaries are so blurred and depend on so many things—so we might vaguely regard this present group as those jobs where the number of holes to be grouted at one time is between 2 and about 10. Among other things, this covers all but the largest dam sites.

Once again, portable equipment is best from the functional and economical aspects, provided the terrain is suitable for readily moving it around. This portable gear can be set up with one mixer supplying two or three agitators and hence enabling that number of holes to be grouted at the same time from that set of gear (see Sections 2.3-4 and 2.7-8 for comments on this type of setup). It is advisable to have at least two sets of gear like this on the job, with more if faster grouting is wanted. Sometimes the site is divided up into sections, with one set of gear allocated to each section, and a standby set for backup. Wheeled trailers are good for moving the gear if the site is suitable for them. Alternatively, in areas too rough for this, the gear has often been mounted on a platform on skids, towed around by a tractor when it has to be moved; or the individual machines have been handled by crane or tractor and placed down on the foundation in a group near the holes. See Section 2.7-9 for suggestions about mobile setups.

If, on the other hand, the site is too steep or rugged for the use of portable gear, it is usual to place the gear as close as possible (refer to Section 2.7-7) and use long circulation lines to get to the holes. Some sites have a mixture of easy and tough terrain, and so portable gear can be used for both areas, with hoses extending to the tough parts from the nearest easy position.

Occasionally, sites are too hard even for this. An example is when the walls of canyons are to be grouted. Here the gear is set up at a fixed location above the job, and the lines go down to the holes (refer to Section 2.7-7).

REALLY LARGE JOBS These are rare. The general principles suggested for medium-sized jobs apply, extended somewhat. Some very large jobs have been done with several sets of gear mounted on trailers (long levees, jobs in flat country); others have had mighty mixing stations. The use of fixed central stations is discussed in Section 2.7-12.

MINING AND UNDERGROUND GROUTING Access is often the controlling factor here. For small jobs in confined conditions, individual portable units are the only answer. Sometimes there is room to mount them on a trailer, but often not, particularly in mine drives.

In mines where a lot of grouting is done, mixing and pumping plants are set up at the surface or in some large underground cavity, and fixed circulation pipes extend along the drives as permanent features, alongside the air and water mains (see Section 2.7-7.)

EXCEPTIONS There are many exceptions to the suggestions presented here, of course. The issues affecting some are dealt with separately later, such as the effect bulk supply of cement has on the setup (see Section 2.7-11), the effect use of computers has (see Section 2.7-12), and the effect of having no return on the circulation lines (see Section 2.7-6).

In keeping with the general practice in this book, the more complex aspects of setup are left until later. The relatively simple one of circulation line practices is dealt with next.

2.7-3 Circulation Lines

Circulation lines carry the grout from the pump to the grout holes. As shown earlier, and repeated again here, the typical layout is as shown in Figure 2.7-3A.

Circulation lines are made up of hoses and/or pipes with a *maximum* internal diameter of 1 in ⟨25 mm⟩. Anything bigger than this is courting problems. The idea is to keep grout velocity high. Three-quarter-inch ⟨20 mm⟩ pipes have worked even better on very long lines, such as one that was 2 miles ⟨3 km⟩ long. Sizes larger than 1 in ⟨25 mm⟩ give velocities that are too slow usually and hence more blockages happen. Some people like to use such sizes because of a misconception that pumping effort is less, but if the grout is thick enough to be thixotropic (Section 4.2-2), this is not usually the case: high velocities keep it moving as a fluid in the range below the Bingham yield point (easiest pumping), whereas slowly moving grout is above this point, in the paste range (hardest pumping), and is very prone to block lines and fittings.

Another reason for using 1-in ⟨25 mm⟩ or smaller lines is that the minimum volume of grout is held in them, and grout tends to get to the hole quicker than with larger lines. The fresher the grout the better. If lines require a large volume of grout to fill, then a lot of cement is wasted (usually at the contractor's expense) when they are regularly washed out. In addition, smaller hoses and pipes cost less and are easier to move around between holes.

If hoses are used for circulation lines, they should

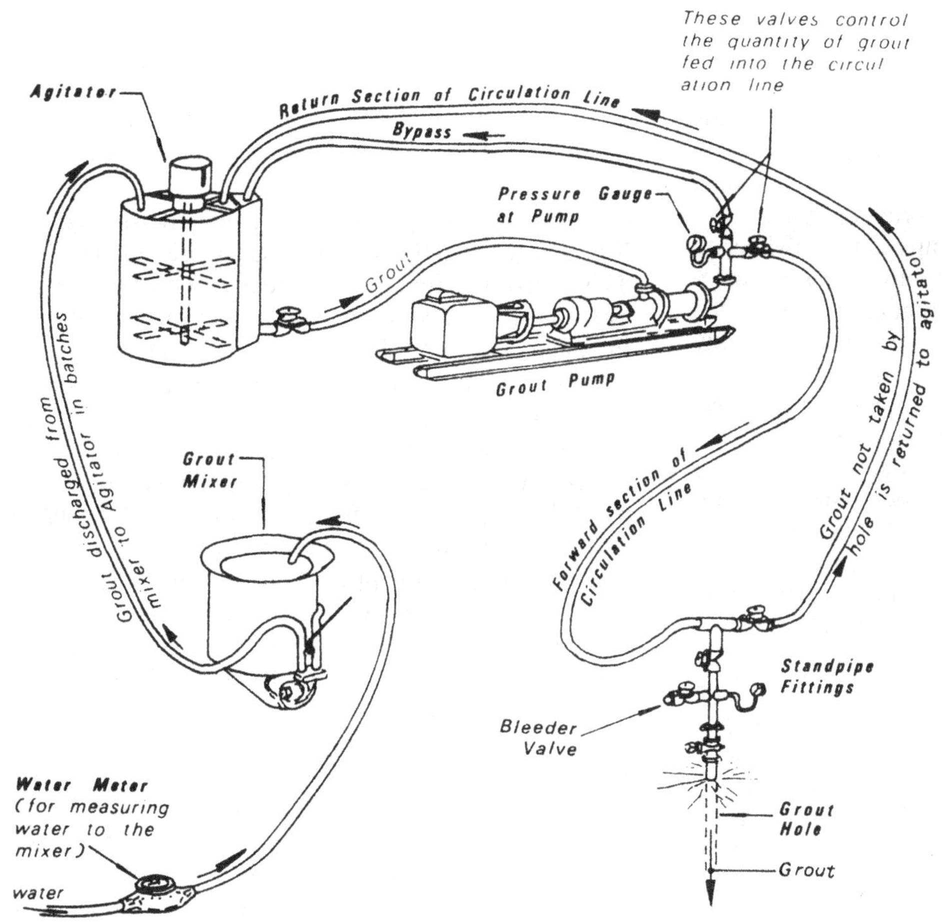

Figure 2.7-3A. Circulation line arrangement.

have snap couplings or else some type of coupling that can be quickly taken apart for clearing blockages. The interior of couplings should be flush with the inside of the hoses so as to avoid projections on which grout might build up.

When pipes are used, unions should be located at top and bottom of each near-vertical run, and also every 200 ft ⟨60 m⟩ or so in other parts of the line, for cleaning out and checking purposes. Lines should be washed out every 8 hours or so, and if used for long periods, pipework should be dismantled every few months and beaten with hammers to dislodge hard flakes of grout from the walls.

In underground work, if lines descend vertically for

Figure 2.7-3B. Example of circulation lines. The pump is in the mobile shed at back. Grout hole with fittings, at right.

more than about 50 ft ⟨15 m say⟩, hoses are too liable to burst, so pipes are better, and they must be securely supported. Piping should also be considered along drives where machinery is moving and liable to hit the lines. Bends in piping should not be sharper than about 4 ft ⟨1.2 m⟩ radius. When first assembled, lines should be tested with water under an overload pressure.

Fitting a bypass on grout pumps is worthwhile, to take some or all of the pump output back to the agitator, as shown in the layout in Figure 2.7-3A. If the circulation line blocks or has to be turned off quickly for some reason, the bypass can be opened up to take the flow without having to stop the pump. Another use of the bypass is to take some of the output when the pump cannot be comfortably throttled back on those occasions when a low pressure is required in the line. Constant output pumps such as those with direct drive electric motors (Section 2.4-2) need this treatment.

Even though these various precautions are taken, blockages will still occur in the best regulated jobs. They can be minimized by cleanliness and thorough washout every few hours. Washout points should be located near the pump in long lines for attaching a water hose; this speeds up washout: while the line is being washed from this, the pump and agitator can be separately washed. In short lines this can be dispensed with, and washing can be done by pumping water from the agitator around the line.

In hot weather on exposed sites, covering the lines may be necessary to keep the temperature of the grout down as much as possible. Conversely, insulation may be needed in cold situations. Refer to Section 4.4-4 for permitted grout temperatures.

2.7-4 Using a Circulation Line

The arrangement shown in Figure 2.7-3A has a return on the circulation line. Not all jobs do it this way, as discussed in Section 2.7-6. This arrangement is, however, used predominantly. It provides a flexible arrangement for the grouting and gives the best chances of dealing with the inevitable troubles that occur. It is operated in the following manner:

1. At the start of grouting a hole, the line is filled with grout (if it was empty), and when grout returns to the agitator out the end of the line, grout is turned into the hole. If the line contains a different mix to the one required for the hole, it is thickened or thinned as appropriate (refer to Sections 4.1-8 and 4.1-9) for the methods of doing this). The idea is to get a return of the right mix before turning into the hole.

2. During the first few minutes after turning into the hole, it is good practice to let grout in at low pressure (see Section 6.2-5) while checks are made for leaks, connections, rock movements, and so on. The circulation line permits control of this to be done at the grout hole by means of the valve on the return part of the line just past the hole (refer to Section 2.6 for standpipe fittings).

3. Pressure in the line at the grout hole should be a little greater than the maximum grout pressure needed. This pressure is reduced as it goes into the hole by use of the control valve in the standpipe fittings, operated in conjunction with the valve on the return line.

4. For the periodical bleeding of the hole, it is necessary to turn the hole off for a short period. This is done by closing the valve on the hole and opening the return line valve fully, as described in Section 4.3-3. The full pump output can then return without having to make any change to the pump setting.

5. As the grouting of the hole proceeds, the rate of take of grout should slow down (see Section 7.2-7) and eventually reach refusal. These changing conditions can be easily accommodated by manipulation of the valves at the holes.

6. Once the pump speed has been set for a hole, there is only occasional need to change it. This means that only two people are needed to grout a hole—one at the hole and one operating the mixer. The mixer operator shares time with the mixing for other holes.

7. In the unusual case of such a fast rate of grout take that the full output of the pump is taken by the hole without appreciable buildup of pressure, the return half of the line can be temporarily disconnected (and washed out) after the valve on it at the standpipe has been closed. Later, when take diminishes somewhat, the return half is connected again and the grouting is taken to refusal in the usual way.

A suggestion about the hoses: it is wise to have couplings on the grout hoses that are different from those on air and water lines. Some canny contractors have learned the wisdom of this after having had hoses purloined for other purposes. This can happen when individuals who have nothing to do with grouting want a hose for other parts of the construction work. Perhaps, in wandering around looking for one (always in short supply), they happen to notice a fine string of them (the grout circulation line) and decide that they'll have one if nobody is in the vicinity. If the line is under pressure, full of grout, when they open up a coupling, they very quickly learn their lesson: if the spray of grout doesn't dampen their ardor, the rain of ribald language from the grouting crews is likely to. Interruptions to the grouting like this can lose a hole, possibly leading to regrouting at the contractor's expense.

Once started, grouting should not be interrupted (Section 7.1-1). Hence the wisdom of using couplings on grout hoses that are visually different from others on the site.

Good communication between the operators at the hole and at the mixer/pump are essential. Experienced crews sometimes have a sign language for this—things are too noisy for shouting to be any use—others use radio. In underground work, telephones are suitable.

2.7-5 Group Grouting Arrangement

Figure 2.7-5A shows the circulation line layout when more than one hole is being grouted at a time. Group

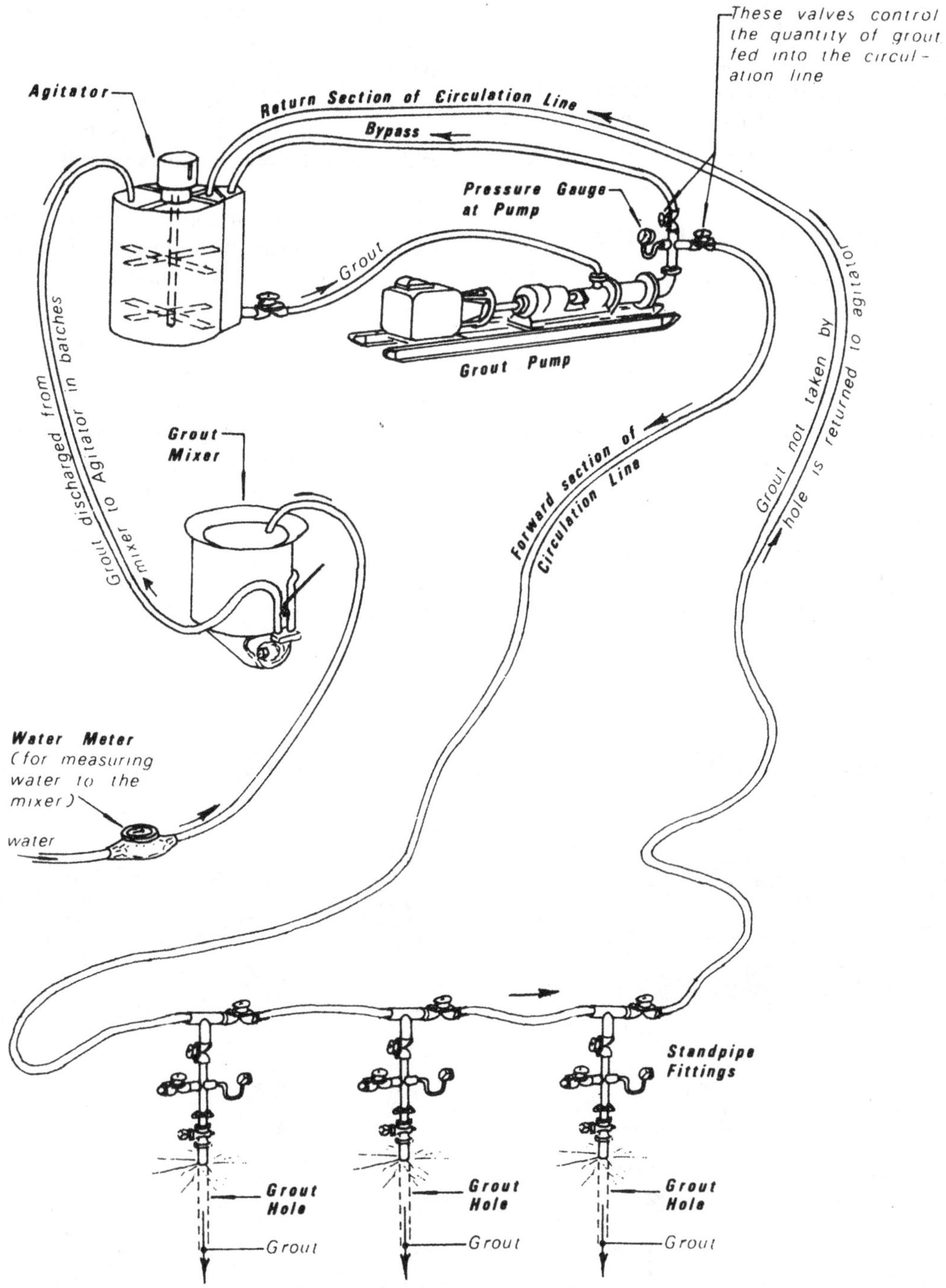

Figure 2.7-5A. *Several holes hooked onto a circulation line.*

grouting like this is only suitable for lower quality work. It is often used with blanket or consolidation grouting. However, it has the disadvantage that if one of the holes takes grout much more rapidly than the others, there is difficulty in knowing which it is and where additional later holes should therefore be placed. A further disadvantage is that a fast hole will take all the grout and leave slower holes starved. This does not make for good grouting.

Note, however, that this aspect can be overcome by putting a flow meter on each hole (Section 7.7-10 Häny unit). A remaining disadvantage though is that w:c ratios cannot be adjusted to suit the requirements of each hole (Sections 4.1-7 and 4.1-8). All the holes necessarily share the same w:c ratio.

2.7-6 Should the Line Have a Return?

The recommended arrangements shown earlier in Figure 2.7-3A have a return section in the circulation line beyond the hole so that unused grout can return to the agitator. Not everyone grouts in this way and so a comparison is warranted. Calling the type of line which returns "circulation" and the type which does not "direct," comparisons are made on Figures 2.7-6A to 2.7-6D.

Figure 2.7-6A shows these alternative systems and Figure 2.7-6B compares their pump requirements, while Figure 2.7-6C comments on the situations relating to control of the grout injections. And in Figure 2.7-6D are comparisons on their relative facility when handling thick grouts and near-refusal conditions.

From these comparisons, the reasons for my preference for the circulation type are fairly evident.

2.7-7 Difficult Sites

If grout holes are in difficult positions, such as in a steep face, or in an area where there is no access for bringing in the cement, the gear is placed somewhere else where access is available, and circulation lines used to take the grout to the holes. Figure 2.7-7A shows the general idea.

Whenever possible, it is best to have the gear above rather than below the holes because:

(a) When the circulation line goes downwards, the hydrostatic head of the grout in it adds to the pump pressure, so that the pump might not have to work very hard at all.

(b) When the circulation line goes uphill, the pump has to work much harder to offset the drop in hydrostatic pressure up to the hole. Stronger pumps may have to be used. A greater consideration, however, can be the way negative pressures may develop in the return part of the circulation line. These come from the grout as it rapidly descends back to the agitator and can cause hoses to flatten and then block at bends. Hence circulation stops. Some control over this can be exerted by putting a valve at the bottom end of the

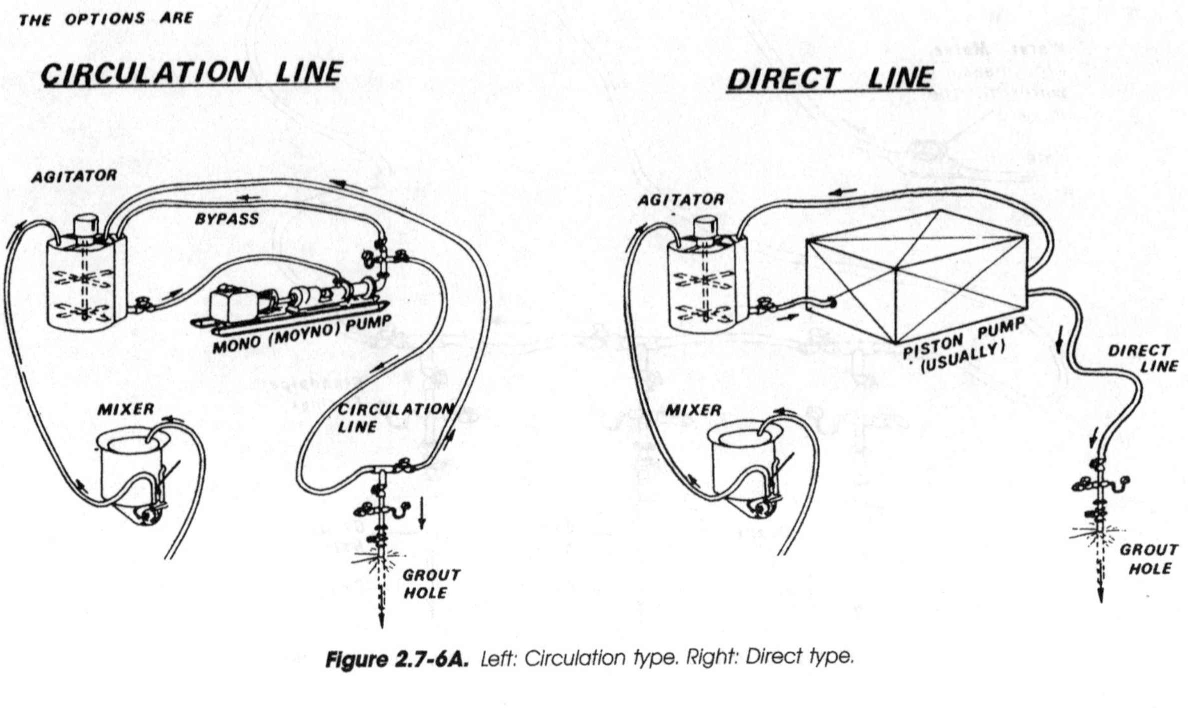

Figure 2.7-6A. *Left: Circulation type. Right: Direct type.*

PUMP REQUIREMENTS FOR

A CIRCULATION LINE	A DIRECT LINE

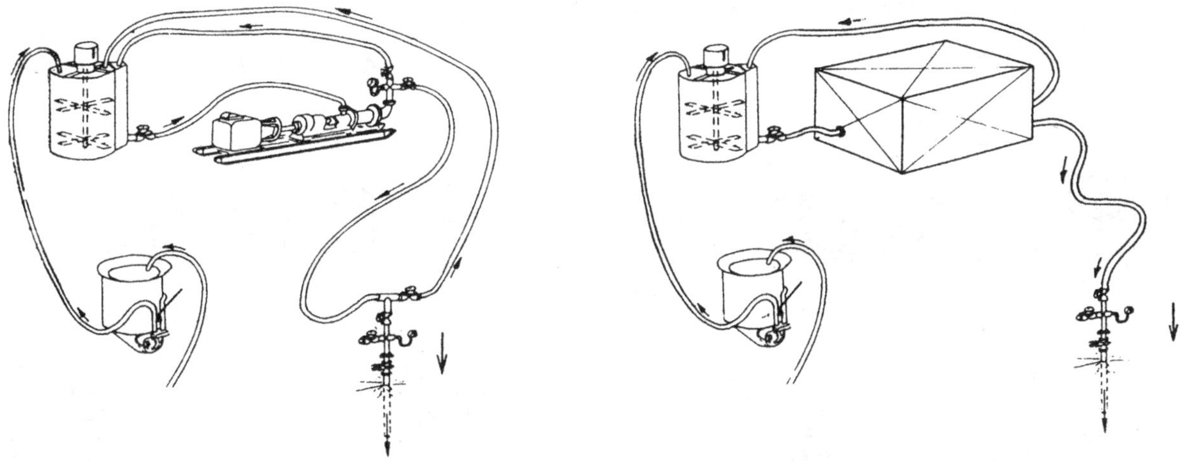

A CIRCULATION LINE	A DIRECT LINE
- PUMP RUNS AT ABOUT SAME SPEED & PRESSURE THROUGHOUT APPLICATION	- PUMP RUNS AT VARYING SPEEDS & PRESSURES DEPENDING ON HOLE REQUIREMENTS
- UNATTENDED MOST OF THE TIME	- IS CONSTANTLY ATTENDED
- PUMP IS LIGHT & PORTABLE. EASILY MOVED AROUND SITE	- PUMP IS HEAVY. USUALLY AT A FIXED LOCATION
- USUALLY A MONO (MOYNO) TYPE	- USUALLY A PISTON TYPE PUMP, MAYBE WITH VARIABLE SPEED DRIVE

Figure 2.7-6B. Comparison of pumping requirements.

WITH

A CIRCULATION LINE	A DIRECT LINE

CONTROL OF HOLE PRESSURE IS AT:–

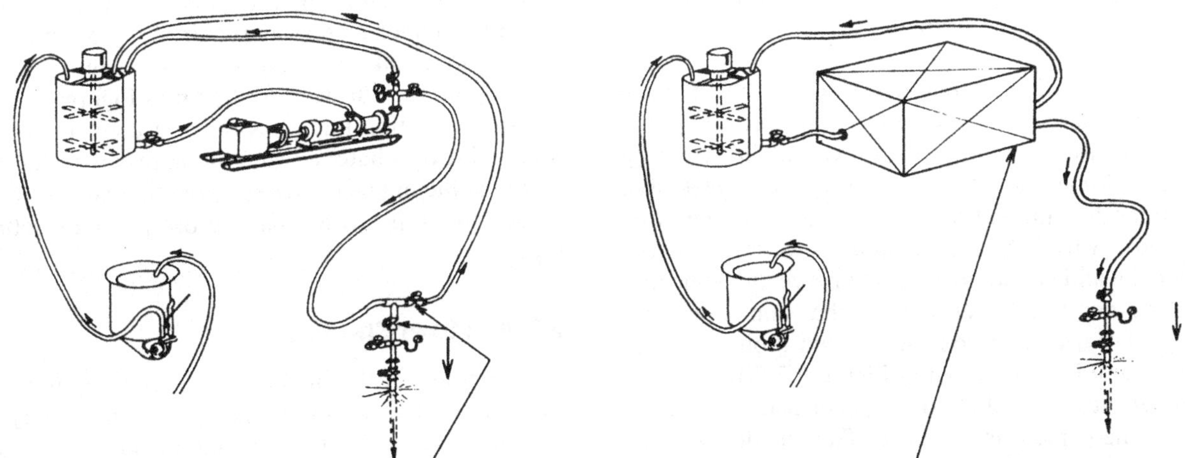

A CIRCULATION LINE	A DIRECT LINE
CONTROL IS AT THE HOLE	CONTROL IS AT THE PUMP - & IS THEREFORE REMOTE FROM THE HOLE - A MAN IS ALSO NEEDED AT THE HOLE FOR PERIODIC BLEEDING
THE PUMP RUNS AT A SUITABLE SPEED TO GIVE LINE PRESSURES A LITTLE GREATER THAN THE HOLE NEEDS	THE LINE PRESSURE IS THE SAME AS IN THE HOLE.

Figure 2.7-6C. Comparison of control arrangements.

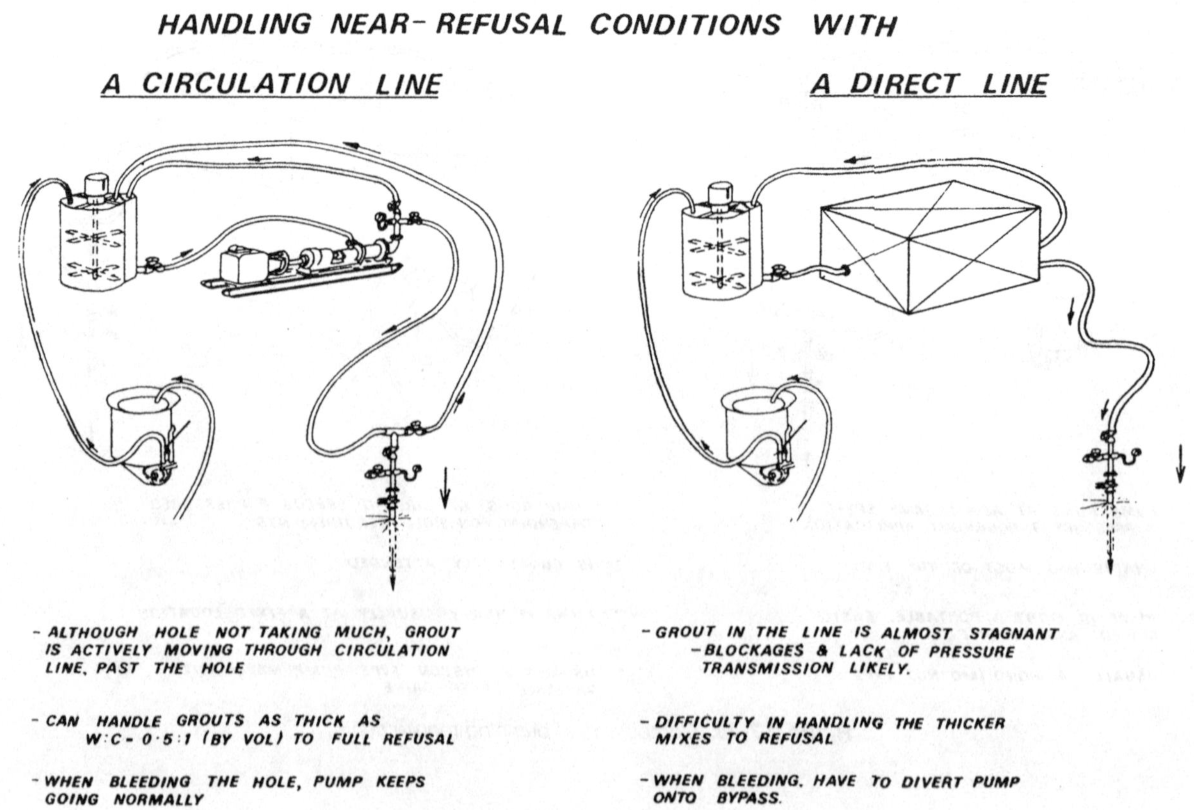

Figure 2.7-6D. Comparison of handling when near refusal.

return and having it partly closed. With judicious adjustment, part closure of this valve slows the returning grout and reduces the problem.

However, with lines below the pump, the situation can occur where the hole is so far below the pump that the hydrostatic pressure greatly exceeds the grouting pressure. Sometimes the high line pressures just have to be tolerated and extra care exerted when operating the control valve. A better solution, where warranted technically and economically, is to move the pump and agitator down to a lower level, perhaps on some temporary platform, and send the mixed grout down to them by gravity, as shown in Figure 2.7-7B.

The grout is mixed at some higher place where access for the cement is available. The line down to the pump can be very long, and in some jobs has had its own pump where it has to traverse ups and downs before going downward.

This general concept is handy for all sorts of difficult and remote situations, not just the steep hillside type of situation shown. Figure 2.7-7C shows an example in mining work.

The underground gear is compact enough to fit in the side of a drive, although if several units are in use, it would be better to put them in a chamber or recess. Thick grout can be mixed at the surface and sent down to the underground agitator where it can be broken down to thinner mixes, as required, by adding water.

Of course, all the machinery can be kept out of the way, up on the surface, as shown in Figure 2.7-7D. A return circulation line goes down a convenient shaft, hole, adit, or whatever, and along drives to the grout holes. Some mines have extensive permanent lines arranged like this, with a large grout pumping setup on top.

2.7-8 Mixer Sharing

If mixers are of the high-shear, high-speed type (Section 2.2-2), their rapid mixing cycles can usually keep up with three grout holes simultaneously, unless fast holes are encountered. These are relatively unusual, most holes being in the moderate to slow taking category. Anyway, if a sudden fast take occurs at the same time as, say, two slower holes are being grouted from the one mixer, the fast hole can be throttled back on the control valve (but not stopped altogether) for a while until the others slow down. Once they approach refusal and are taking little grout, the mixer can give its full output to the fast one.

Even though one mixer is allocated to the three

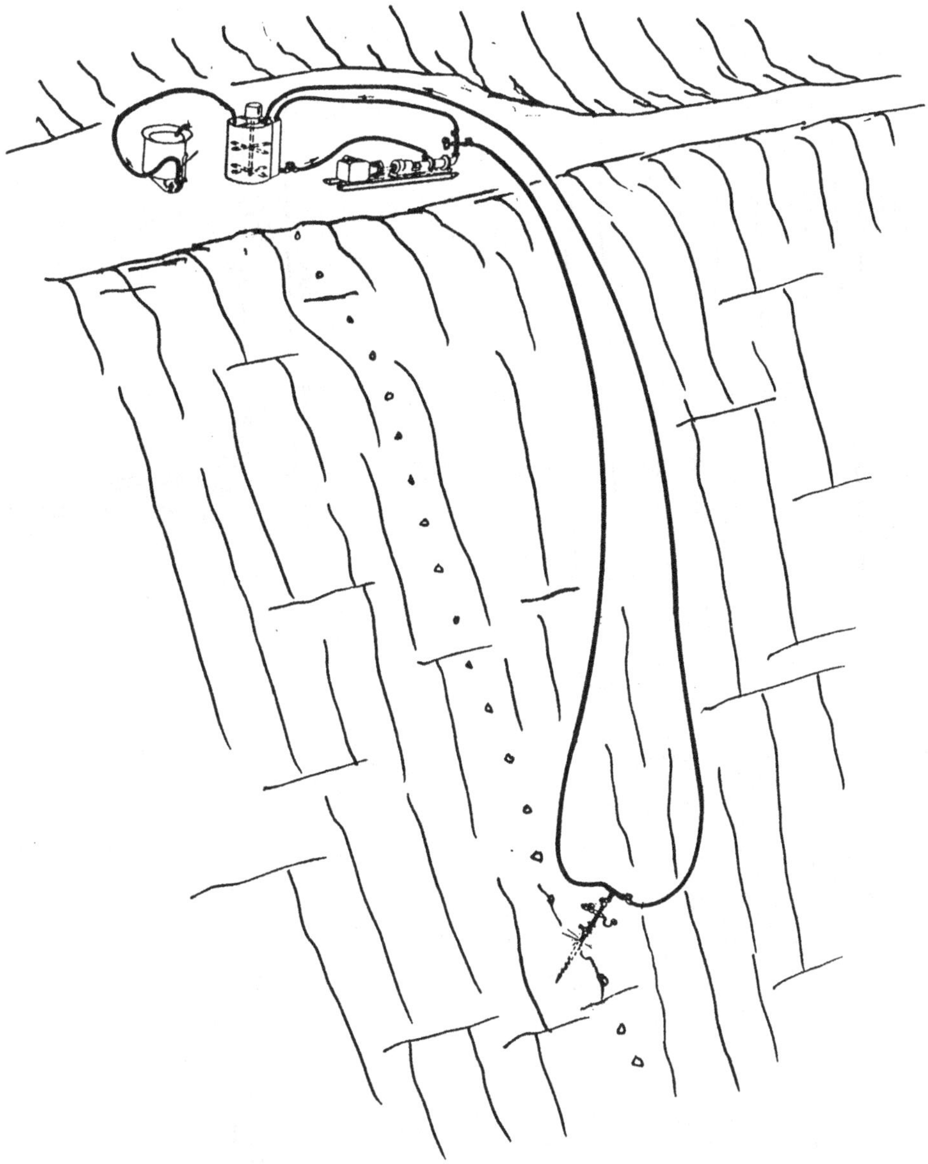

Figure 2.7-7A. *Grout delivery to holes in difficult situations.*

holes, a separate agitator, pump, and circulation line is needed for each hole, of course. The mixer is called on to produce a great variety of w : c ratios with this arrangement (refer to Section 4.1-8 for thickening mixes), but this is no problem. It means that the mixer, which is the most expensive plant item, is used to the maximum and is continuously in use. If the cement is in bags, keeping the mixer supplied is hard work—a very muscular character is usually appointed to the task.

Some proprietary packaged grout outfits come with one of these mixers to just one agitator and pump. Although this might be convenient from the maker's point of view, it is wasteful on most holes because the mixer is only working now and again. Further comments on these packaged outfits follow in Section 2.7-10.

Thinking now about mixers that are not the highly efficient ones just discussed, but are of lesser efficiency (Sections 2.2-3 and 2.2-4), it is not possible to use them on more than one hole at a time. They are too slow in their mixing cycles. On fast holes, some of them cannot even keep up with the demands of just one hole.

2.7-9 Do-It-Yourself Mobile Gear Setup

Convenient practices include setting up the mixer, agitator(s), and pump(s) on a skid platform or trailer for easy mobility around the site (see Figure 2.7-9A).

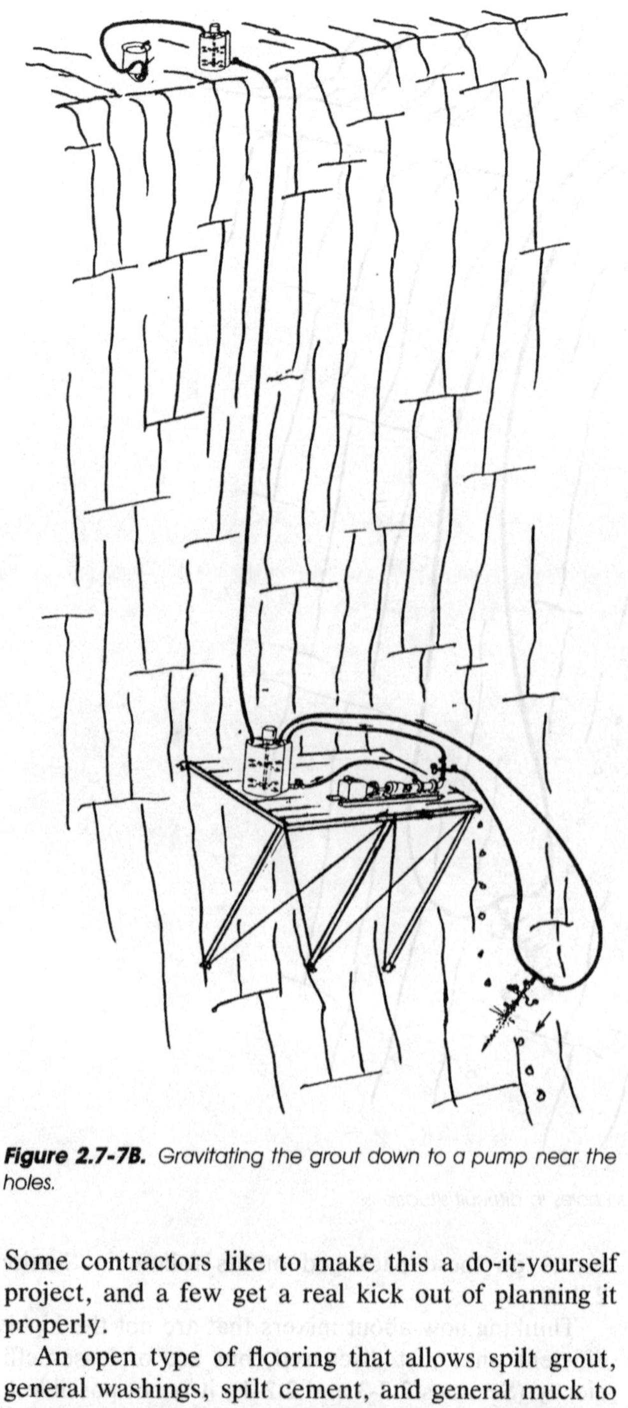

Figure 2.7-7B. *Gravitating the grout down to a pump near the holes.*

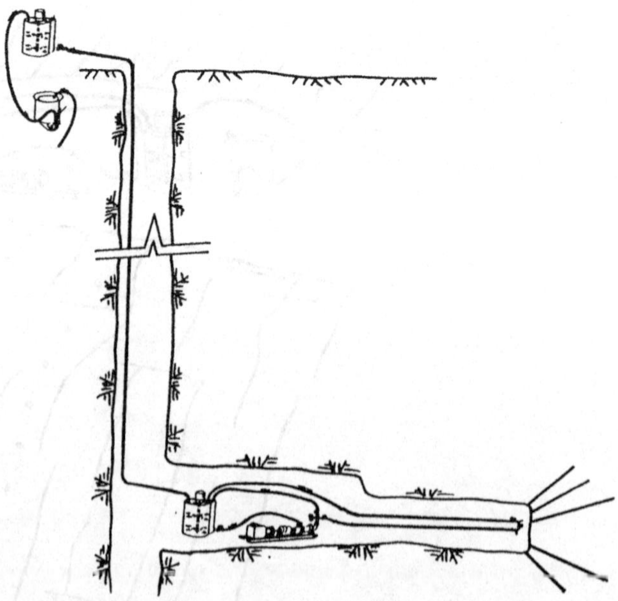

Figure 2.7-7C. *Gravitating grout down a shaft to an underground pump.*

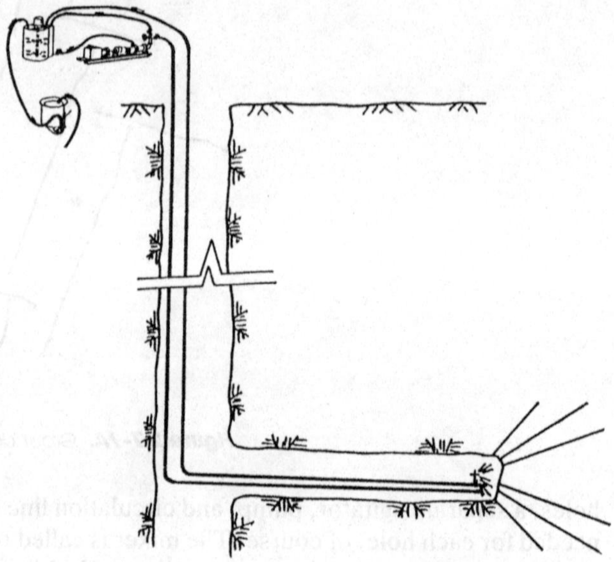

Figure 2.7-7D. *Grout reticulated underground from surface plant.*

Some contractors like to make this a do-it-yourself project, and a few get a real kick out of planning it properly.

An open type of flooring that allows spilt grout, general washings, spilt cement, and general muck to fall through it is important. This reduces slipperyness underfoot, a real hazard wherever wet cement in either the set or unset condition is repeatedly walked on.

If the gear is in one position for more than a few days, it is advisable to build a low sandbag dam around it to collect the washings and muck. When the cement goes hard in these, it can be easily broken up and shoveled away, leaving fresh capacity in the dam.

A roof over the gear is worthwhile, especially over stacks of cement and mixers. Some portable setups have a shed built over everything. This provides shelter not only to the gear but to the crew at times, and report sheets and documentation can be kept out of the weather. A word of caution though—if the roof is built of corrugated iron or sheet metal, do not make closed-in sides out of this material, otherwise the reverberation from the motors inside the confined space can drive everyone mad. In warm climates, sheds with

Figure 2.7-9A. Mobile gear shed, made up on site by contractor. Cemix mixer and its water meter at center, agitator to right of it, helical rotor pump with diesel motor at far right.

three sides open to the weather are better. Where it is too cold for this, some sort of sound-absorbent lining of the four walls will keep everybody happier.

2.7-10 Portable Packaged Outfits

Some of the equipment makers put the gear into compact packages and sell them as complete grouting plants (see Figure 2.7-10A). These can be convenient for a contractor starting out without a great deal of knowledge of grouting. For one knowing the ropes, however, they can be dearer than the do-it-yourself approach. Apart from the substantial initial expenditure, there is the waste of mixer potential (discussed in Section 2.7-8), the inability to plug in a separate standby item if one of the plant items fails (everything has to stop if that happens—a bad thing when grouting), and a rather congested mass of things to be washed out and kept clean. Even so, they are sufficiently popular to warrant manufacture.

Chemgrout make a range of ready-to-run outfits. One of them is pictured in Figure 2.2-2O. Current (1989) price of a typical set (CG-625D) is $22,500 from Chemgrout Inc., P.O. Box 1140 La Grange Park, Il 60525, U.S.A. (312-354-7112). Others makers include GKN Colcrete Ltd., Thorp Arch Trading Estate, Wetherby, West Yorkshire LS23 7BJ, England. (0937) 844066 and Häny & Cie. AG, Bergstrasse 103, CH-8706 Meilen, Switzerland. (01/9251131). See Figure 2.7-10B for an example of the Häny model.

Figure 2.7-10A. Ready-to-run outfit. Colcrete mixer at right, agitator at higher level on left with vertically mounted helical rotor pump directly below it. (Courtesy GKN Colcrete Ltd.)

Figure 2.7-10B. Another type of ready-to-run outfit. Mixer on right, agitator at center, ram pump at left. (Courtesy Häny & Cie. AG.)

2.7-11 Cement Supply

There are good reasons for using bagged rather than bulk cement on most grouting jobs:

1. Bagged cement is premeasured, usually accurately, and batching is, therefore, a simple matter of just counting the bags (never use part of a bag—see Section 4.1-6). Bulk cement requires reliable on-site measuring equipment and hence extra investment in plant.
2. Stacks of bagged cement on pallets can be handled easily by forklift from delivery trucks (or a store) to a platform alongside the mixers. This can be done over fairly rough and constricted access and requires no special equipment, whereas bulk cement requires erection of a silo at the mixers and good access for bulk trucks to it.
3. On a contract job, it is easier for inspectors to count (and then destroy) used cement bags than it sometimes is to keep a constant check on automatic bulk batching gear, which an unscrupulous contractor (let their numbers ever decrease!) might quietly tamper with. Since one of the payment items is based on the amount of cement used (refer to Section 9.2-1 for these items), the temptation exists.
4. Using bulk cement restricts plant portability. As is evident from the previous comments, the extras needed for handling it, and the access requirements, mean that grouting gear cannot be trundled all over the site, as it can often be when using bagged. The penalty for this is often greater cost, and the potential for more waste and troubles due to longer circulation lines than would otherwise be the case.

Having said all that, it might be remembered that bulk cement is cheaper than bagged in some countries and at some locations. Where the price differential is enough to warrant the installation of bulk gear and the access to it, then the matter is not as clear-cut. In addition, bulk cement saves the services of a hefty character to feed the mixer, but probably the same wage (or more perhaps) is paid for an operator for the bulk plant.

2.7-12 Central Mixing and Pumping Plants

As mentioned earlier, there are sites where carrying out all the mixing in one centralized location is obviously best. Examples include deep canyons without convenient access down the bottom, some underground work, sites where access around the job is too restricted to permit economical transport of bagged cement, and so on. In these, a central mixing plant sending the grout off in lines to agitators and pumps elsewhere can be a good solution. And, of course, setting up bulk cement facilities is usually worthwhile at such plants.

The next stage in the use of centralized equipment is the central pumping plant; this not only mixes the grout but also pumps it off to the holes through circulation lines. The concept comes from chemical grouting where some quite complex plants have been used, particularly where the grouting has been in a compact area and has required intensive placement of holes, an example being tunnel grouting. However, the concept does not translate quite as well to cement grouting if the work is spread out over a wide area, such as a dam site. This is because unlike chemical grouting, cement-grouted holes in a high-quality job should be individually grouted, with each hole receiving treatment appropriate to its particular grouting progress. To do this properly, each hole must have its own pump and circulation line all the way from the plant to the hole and back again (some comments about this follow shortly). This makes for a massive lineup of pumps and a plethora of lines over the site. Apart from their cost (they may be a matter of miles in total length), they can have other problems:

1. If earth-moving equipment is working on the site and passing over circulation lines, the lines must be protected by such methods as putting them inside larger steel pipes where they pass under haul roads or concentrated traffic areas. Even so, where lines wander across other areas, they can suffer a high casuality rate from the attentions of a certain mentality of earth-mover driver who seem to get great delight from accidentally (on purpose?) flattening them.
2. Each hole must have its own circulation line, and in the large numbers usually involved, it is very easy to get them confused, so that the wrong line gets hooked on at times.
3. The volume of all these lines has to be filled with grout several times a day after each washout, and the cement used in doing this is wasted. The contractor is not usually paid for it, of course (refer to Section 9.2-5 for contractural arrangements).
4. The grout reaching the hole is not as fresh as when mobile gear and short lines are used.
5. There can be difficulties in matching pressures in the lines with those required at the holes for the grouting.

And now for those comments about having to have individual circulation lines for each hole. Without doubt this is essential for really high quality work where every hole gets individual treatment. However, on some jobs it has been felt that the standard of work need not be as good as this, and so some degree of compromise has been reached by using merged circulation lines over some or all of the job. These carry just the one w : c ratio grout for all holes (irrespective of their real requirements), or perhaps there might be several common lines, each carrying a particular w : c ratio and connected to in turn with hoses from the holes as each hole's requirements change. Or perhaps manifolds may be used. These are little more than general group grouting, though, and as indicated in the section on group grouting (Section 2.7-5), produces second-rate grouting. However, considerable cost savings are possible if the central pumping plant need have only a few pumps and a few lines, instead of the many more needed for individual treatment of holes. On some jobs this cost saving may be the preferred option when traded off against lower quality grouting.

All these considerations have led to the suggestion given earlier in Section 2.7-2 that mobile gear is better for most sites. Nevertheless, some contractors prefer a central pumping plant or several semifixed versions of it at strategic positions around the site, particularly if computers are in use (discussed in Section 7.7). To descend to a human level (from the esoteric realms of grouting technology), maybe they get a kick from an impressive array of pumps, mixers, pipework, silos, and so fourth, perhaps with nice automated controls. Good luck to them if they are prepared to bear the possible extra cost and problems, but there is a lot to be said for keeping things simple and mobile in grouting, on the great majority of sites. Admittedly, a "Mighty Grout Plant" (MGP) is far more impressive to the visitor than portable gear spread here and there around the site. There is little enough to see anyway in grouting jobs, so maybe an MGP or two does the profession of grouting a service? Anyway, MGPs are very good for advertising and for impressing future customers of the contractor, so maybe they are warranted for these purposes?

2.8 WATER TESTING EQUIPMENT

2.8-1 Basic Gear

The basic equipment for carrying out a water test in a hole is shown in Figure 2.8-1A. Water for the testing is supplied from mains, or pump, or tanks located at least 30 ft ⟨10 m⟩ higher than the hole.

Control of the test pressure is by manipulation of the waste valve (to get rid of excess pressure) and the control valve.

The water meter indicates how much water has gone into the hole in the testing. It should have an easily read, large dial and sweep hand, or else have a large digital display. Dirt and poor light conditions combine to make meters hard to read sometimes and can result in errors unless reading is easy. Preferably meters should be of the type that can be set to zero at the start of the test.

The pressure gauge should be accurate and frequently checked for accuracy.

When the equipment is being used on a grout hole, the connection to the fittings on the hole should be of the snap coupling type and should be of the same size as those on the grout circulation line. This enables quick disconnection of the water test hose after completion of the water test, followed by immediate connection of the grout line if there is urgency for grouting to start.

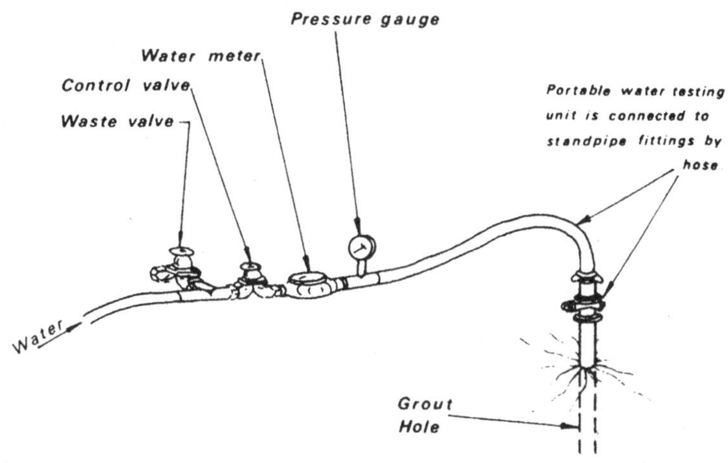

Figure 2.8-1A. *Water-testing gear.*

REFERENCES AND FURTHER READING

Gurley, A. W., and Carson, C. S. (1982). "Grouting plant and equipment." *Proceedings of Conference on Grouting in Geotechnical Engineering*. ASCE, New Orleans, 121–135.

Kravetz, G. A. (1959). Closure to paper "Cement and clay grouting of foundations: The use of clay in pressure grouting." *ASCE Journal of Soil Mechanics and Foundations Division,* Vol. 85(SM2), 109–114. Reprinted here with permission.

Mayer, A. (1959). Closure to paper "Cement and clay grouting of foundations: French grouting practice." *ASCE Journal of Soil Mechanics and Foundations Division,* Vol. 85(SM1), 41.

Parry-Davies, R. (1982). *Proceedings of Conference on Grouting in Geotechnical Engineering,* ASCE, New Orleans, Vol. 2 (Discussion volume, only available to participants at Conference), 33.

White, R. E. (1959). Discussion on paper "Cement and clay grouting of foundations: French grouting practice." *ASCE Journal of Soil Mechanics and Foundations Division,* Vol. 85(SM2), 90.

CHAPTER 3

Grout Materials

This chapter deals with cements for grout (Section 3.1), additives for saving cement (Section 3.2), and additives for producing particular effects (Section 3.3).

3.1 CEMENTS

3.1-1 Types of Cement

The majority of rock grouting jobs use Portland cement (Section 3.1-2). This is a hydraulic cement, which means that it reacts with water to set and form a hard product. Other cements used for grouting include blast furnace cement (Section 3.1-3) and microfine cements (Section 3.1-6).

3.1-2 Portland Cement

Portland cement is the cement used mostly when making concrete and when grouting. It is available in most countries in various types: normal, high early strength, low heat, sulfate resisting, and additional types in some locations.

CHEMICAL COMPOSITION In the dry powder form, the chemical composition of Portland cement is comprised of the following compounds:

Tricalcium silicate ($3CaO.SiO_3$) — C_3S
Dicalcium silicate ($2CaO.SiO_2$) — C_2S
Tricalcium aluminate ($3CaO.Al_2O_3$) — C_3A
Tetracalcium aluminoferrite ($4CaO.Al_2O_3.Fe_2O_3$) — C_4AF

Gypsum is added to slow the setting. Additionally, magnesium oxide, free lime, and silica are present in small quantities. The compounds react differently to each other: C_3S and C_3A hydrate rapidly, releasing heat and giving early strength; C_2S hydrates slowly and gives late strength. Gypsum reacts with some of the C_3A to form calcium sulfoaluminate. The proportions of the compounds are altered during manufacturing to give the different types of cements. Manufacturing involves grinding the ingredients and then burning in a long rotary kiln. A small variation in the nature or proportions of the raw materials can lead to significant variations in the compounds.

NORMAL PORTLAND CEMENT The American Society for Testing and Materials (ASTM C150) grouping for this cement is Type I. The British Standard (BS 12) calls it "Ordinary" Portland cement. "Normal" is used by the Japanese Industrial Standard (JIS) and "Type A" by the Australian Standard (AS 1315). The name varies in other countries, but the specification for it is usually similar to the ASTM spec. The American Petroleum Institute (API Spec 10) also ties in with ASTM but calls the cement Class A.

It is the cheapest and most commonly used cement for grouting, and on some sites it is likely to be the freshest type because it is probably also used for other work on site such as concreting.

Raw ingredients contain alumina and calcium; clay and chalk or limestone are common sources. In the kiln, the materials fuse to form clinker, which is later ground to a fine powder, and then gypsum is added.

It is not a fine cement. Specific surface area, which is a measure of the surface area of the grains, is required to be between 2,250 and 3,200 sq cm/g. The test is done by the Blaine air permeability method, which passes air through a specified amount of the cement after shaking it well.

Resistance to chemical attack is low compared with some of the other types but is usually adequate for most foundation conditions. Initial setting time is specified as 0.75 hour, and final set as 10 hours.

HIGH EARLY STRENGTH CEMENT Fine cements set faster. To get this early strength, these cements are ground finer. Specific surface areas are in the range of 3,250 to 4,200 sq cm/g. Chemical composition is similar to normal cement.

This cement is called Type III by ASTM, High Early Strength by BS and JIS, Class C (with variations) by API, and Type B by AS1315.

It is sometimes used where cracks are fine in order to take advantage of the finer grain size, but as explained in Section 6.1-5 this is not as crucial as the manner of mixing.

Although strength gain is faster, the eventual strength is only slightly more than normal cement. Heat is generated more rapidly during hydration (the chemical action during setting, explained in Section 4.4-3), but this has little effect on grouting unless the weather is very hot. If seepage is washing grout away, appropriate treatment does not include using this cement—the faster strength gain does not help enough to be any use (see Section 3.3-3 for runaway takes). Chemical resistance is about the same as normal cement. Initial set is 0.75 hour with final set at 10 hours.

LOW-HEAT CEMENT Different proportions of the compounds give lower heats of hydration, although more finely ground than normal cement. Early and final strengths are less and setting is slower.

It is called Type IV by ASTM, Low Heat by BS, and Type C by AS1315.

Specific surface area is specified as 2,800 to 3,600 sq cm/g. Chemical resistance is better than normal cement. Initial set is 1 hour: final set is 10 hours.

This type of cement is not often used for rock grouting, but sometimes it is useful in other types of cement grouting.

MODERATE SULFATE RESISTING This cement is less prone to attack by sulfates because of reduced C_3A content. Heat is generated more slowly than with normal cement.

ASTM calls it Type II, API lists it as Class B (together with High Sulfate Resisting), and AS1315 calls it Type D.

It can be useful for grouting in situations where this type of attack is possible.

HIGH SULFATE RESISTING Further variations to the compounds give this cement increased resistance against chemical attack.

It is designated Type V by ASTM and Class B (together with moderate resistance cement) by API.

It can be useful for grouting in severly aggressive environments.

OTHER PORTLAND CEMENTS Cements are available in some countries that are outside these classifications. For example, Super High Early Strength cement, Extra Rapid Hardening Portland cement, and Expansive cements (sometimes used in structural grouting to try to offset bleed shrinkage) are not applicable to rock grouting.

3.1-3 Blast Furnace Cement

Cement made from blast furnace slag can give more resistance to sulfate attack than sulfate-resisting Portland cement. The cement is made from granulated slag, gypsum, and not more than 5% of Portland cement clinker.

It is finer (4,000 to 5,000 sq cm/g) though less reactive than normal Portland cement. Final strengths are about the same. Twenty-five percent more water is required for hydration, and the set product is relatively light in weight. See Section 3.1-6 for its use in very fine form.

3.1-4 High-Alumina Cement

This cement is mentioned more for cautioning purposes than for suggested use. It has high early strength and good sulfate resistance, but is not durable in common site conditions. For example, it is attacked by alkalis.

Under moist conditions and heat, "reversion" can take place and strength is lost. Maximum service temperature recommended is 80°F (25°C). Surface rock conditions can be warmer than this on many sites.

It is widely believed that high-alumina cement should not be allowed to contact Portland cement in any form, particularly in concrete where it can cause deterioration. If the two types of cement are mixed, flash setting can result. This may be useful to plug leaks but may not be permanent. High-alumina cement can be used for grouting standpipes into position (Section 5.1-3) in rock or grout caps (Section 5.1-14) but not in structures. Any deterioration of grout caps resulting from this use is unlikely to be very serious.

The cement is a fused mixture of bauxite and limestone comprising essentially anhydrous monocalcium aluminate, which on setting becomes dicalcium aluminate hydrate with a release of aluminium oxide. Setting is accompanied by considerable heat. The cement is particularly sensitive to contamination. Water : cement (w : c) ratios thinner than 0.75 : 1 by volume are not advisable.

3.1-5 Cement Particle Size

Before going on to the very fine cements, a word about particle size would be in order.

Cements contain a wide range of particle sizes ranging from fine to exceedingly fine. Practically all particles pass No. 100 sieve ⟨150 microns⟩ in normal Portland cement, and high early strength cement is so fine that nearly all pass No. 200 sieve ⟨75 microns⟩.

These maximum sizes are relevant when considering the flow of grouts in fine cracks (Section 6.1-4).

In determining sizes of the very fine particles, sieves are no use of course. Electron microscope examination has found some as fine as 0.03 micron. High-speed, high-shear mixing of grout is necessary to ensure that all these very fine particles are individually separated and wetted. Section 2.2-10 discusses this.

The flow of grouts is influenced by the grading of the cement in them. Wide differences of grading are possible within the permissible limits of specifications. Some cements are produced by a continuous process that involves regrinding a coarse fraction by means of a feedback circuit. These have different gradings to those from a one-pass grind. Moreover, particular plant and material sources influence the final product. Mostly, these do not need to be taken into account in rock grouting, but a general knowledge of these factors can be useful if the grout develops peculiar properties. This sometimes happens.

It is not appropriate to go into gradings and cement properties in fuller detail in this book. They are readily available elsewhere.

3.1-6 Microfine and Other Very Fine Cements

Recently, there have become available cements that are appreciably finer than those mentioned. The impetus for their development came from a need to grout cracks finer than previously possible irrespective of the cost and also as a more durable and nontoxic method for grouting sands and gravels in lieu of chemical and clay grouts (Section 1.2-5).

These fine cements are available in a range of finenesses. They are relatively expensive though.

"FINE GRIND," "EXTRA FINE," ETC., CEMENTS Various names are given by manufacturers to this group of cements, which are Portland cements scalped or air separated to yield the finer fraction of their particles. Specific surface areas of about 5,000 sq cm/g can be reached in some plants. There can be problems in handling and bagging because of their talc-like nature.

These fine cements require more water than normal cements to make grout sufficiently fluid for pumping. My tests on an Australian product of this type found that a w : c ratio of 0.6 : 1 by volume was the thickest practicable. This compares with 0.5 : 1 for normal cement. More water can mean lowered durability, as Section 4.5-4 explains, and so can be undesirable.

In passing, it might be mentioned that the tests were directed toward development of a nonbleed grout for crucial posttensioning cables. The fine cements bleed less than normal ones do. However, the answer was found not in this type of cement but in API Class G (see Houlsby, 1988).

MICROFINE CEMENTS These are mainly mixtures of blast furnace and Portland cement (Section 3.1-3) ground exceedingly fine. Early work was in Japan by the Onoda Cement Co. Ltd. (Shimoda and Ohmori, 1982). This product is sold in the United States under the name MC-500 by the Geotechnical Corporation, Ridgewood, NJ. The same company has taken a lead in improving on it by research in the United States and Britain, leading to a recent U.S. patent (Clarke, 1988) for an even finer grout which consists of 100% blast furnace slag. Because of the absence of Portland cement, the microfine cement is alkali activated with 5% sodium hydroxide. This cement, known as MC-100, has a specific surface area "greater than 9,500, preferably greater than 11,000 and most preferably greater than 13,000 sq cm/g." Less than 3% exceeds 7.8 microns in size.

The blast furnace component is a by-product of steel production. It comes from the ball mills that grind the slag. A great deal of air is passed through these mills for cooling purposes and then has to be cleaned in large dust collectors. The material collected is used in microfine cement.

To put these cements in context, Figure 3.1-6A compares Microfine with Portland cements.

The Japanese cement, MC-500, is very fine, but U.S. sourced cement, MC-100, is even finer. Chemical

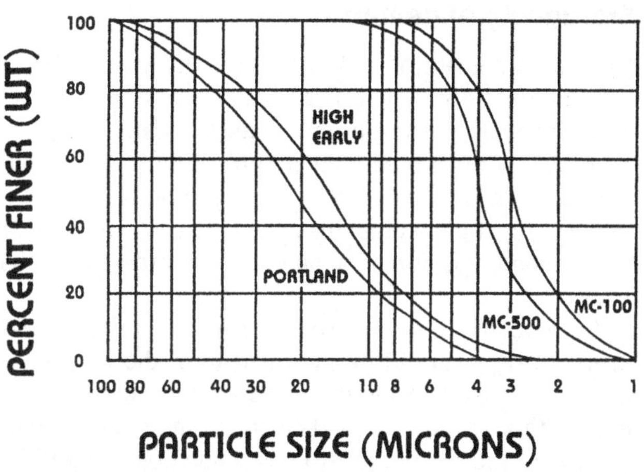

Figure 3.1-6A. Comparison of Microfine and Portland cements. (After Clarke, 1987)

	PORTLAND	MC-500	MC-100
SiO_2	22.	30.6	35.4
Al_2O_3	5.2	12.4	16.
Fe_2O_3	3.2	1.1	0.3
CaO	65.2	48.4	43.3
MgO	1.3	5.8	3.5
SO_3	1.9	0.8	0.3

Figure 3.1-6B. *Chemical composition of a Portland and Microfine cements. (After Clarke, 1987)*

composition of these cements is compared in Figure 3.1-6B.

MC-500 is a mixture of Portland and slag and therefore is intermediate chemically between them.

The advantage provided by the fineness of the Microfine cements is somewhat offset by their need for greater amounts of water to obtain fluid grouts. The amounts vary with the grading and type of Microfine cement, and the excessive water demand can present durability problems (Section 4.5-3) so a superplastercizer (dispersant) is used to make a flowable grout without the need for this extra water. The choice of superplasticizer can affect the working life of the grout: some lose their effectiveness an hour or so after mixing, and the grout then becomes a viscous sludge unable to penetrate. Any grout in agitator or circulation line has to be systematically dumped when this happens and fresh grout mixed and quickly used.

Other types and makes of very fine cements are available, and the growing market for them is likely to see additional ones probably accompanied by price reductions making them more attractive in fairly ordinary situations.

Long-term durability is uncertain as yet.

3.1-7 Freshness of Cement and Storage Arrangements

Cement for grouting must not contain lumps. Don't rely on a good mixer to break them up. Get fresh cement instead of using lumpy material. Good grouting practice is to use fresh cement only. In temperate climates this means use within about 3 months of manufacture. In tropical climates allowable age is much less and depends on humidity and how well the cement is packaged.

Store cement away from the weather. Weatherproof storage should be provided until the cement is about to be used. It should not be kept more than a day out of this storage, and then if the weather is wet, cement at the mixers must be in a shed or covered by tarpaulins.

Cement bags should preferably be plastic lined to make sure that moisture does not get into the cement before use. If cement is in bulk, transfer systems at silos should avoid contact with moist air.

3.2 CEMENT-SAVING ADDITIVES

3.2-1 Minimizing Cement Usage

In most countries, cement is the most expensive ingredient in grout. There is benefit, therefore, in using as little of it as practicable without sacrificing the aims of the grouting.

Ways for doing this vary from bulking the grout out with inert fillers to using reactive additives to take over some of the cementing functions.

3.2-2 Bulk Fillers

When voids are large and penetration is easy, fillers for bulking out the grout are usual. They weaken the grout, but strength is not very important in most of these applications. Examples of inert fillers include:

Sand
Clay
Lightweight aggregates
Heavyweight aggregates
Straw, sawdust, wood shavings, and so forth

SAND Sand is a cheap filler but requires care to avoid segregation in voids in the job and when handling. The easiest sand to pump is well rounded and well graded; the hardest is crushed, angular sand. Well-graded sand can fill out the mix to as much as three times the cement (by weight) if mixed in a high-shear mixer (Section 2.2-2) and still be pumped effectively, but angular sand will not usually bulk out as well as this.

Sand should conform to concrete quality requirements. The quantity of water for the mix varies depending on sand grading and type and should be found by trial mixes before starting. W : c ratios should be as thick as practicable in accordance with the principles discussed in Section 4.5 in the interests of durability.

The strength when set decreases with increasing sand and water content. It can be in the order of 3,000 to 4,000 psi ⟨21 to 28 MPa⟩.

There are tricks to successfully pumping a sand–cement grout around a circulation line (lines are discussed in Sections 2.7-3 and 2.7-4). It is very easy to get blockages doing this, and they can be more obstinate to clear than with unsanded grouts. When starting, don't immediately try to circulate the sanded grout. Instead, prime the line with an unsanded, watery grout, such as 2 : 1 by volume, and then thicken gradually, using a lightly sanded mix. Then slowly thicken it until the full sanded mix is reached. Don't turn the grout into the job while doing this thickening unless the job happens to require the same sort of progressive treatment. Then after finishing grouting, reverse the process and thin gradually right down to water alone. Make sure that circulation does not get interrupted when handling sanded mixes: awkward line blockages can result immediately.

A series of tests on sanded grouts were conducted by the U.S. Army Engineer Waterways Experiment Station and reported by Polatty (1961). The conclusions follow:

1. Sand deficient in material passing the No. 100 sieve can be successfully pumped in grout mixtures without the use of an admixture.
2. Concrete sands can be used in Portland cement grout provided material retained on a No. 16 sieve is removed.
3. An increase in fine material in the minus No. 100 size will permit an increase in the quantity of sand found pumpable.
4. Manufactured sands can be successfully pumped in Portland cement grouts.
5. The specific gravity of the sands tested had little or no effect on pumping characteristics.
6. In using sand deficient in material passing No. 100, the addition of a finely divided mineral admixture increases the sand-carrying capacity.
7. The addition of a finely divided mineral admixture to a normal cement grout will increase its sand-carrying capacity.

Admixtures used in the tests were diatomite, fly ash, pumicite, loess, and bentonite.

CLAY Clay can be used as a grout on its own (Sections 1.2-2 and 1.2-5), bentonite being the usual clay for this.

When added to cement the resulting low-strength grout can be suitable for some applications, such as filling large voids, but sand is often the preferred choice for this because of lower cost (usually) and easier use.

From the flow point of view, a cement grout containing enough clay as a filler to be worthwhile has the characteristics of a clay grout rather than those of cement grout and is outside the scope of this volume.

LIGHTWEIGHT AND HEAVYWEIGHT AGGREGATES The types used for making light or heavy concrete have the same purpose in grouts, but must be mainly in the sand size range in order to pass through the pumps and circulation lines. Trial mixes are necessary to establish proportions. Roller mixers may be necessary (Section 2.2-3).

STRAW, SAWDUST, AND WOOD SHAVINGS These can fill out low-strength grout for temporary purposes but because of possible organic decomposition are not suited to permanent work. They are mentioned in Section 3.3-3 in relation to sealing off fast-flowing seams. Trial mixes are necessary.

3.2-3 Reactive Additives

Additives that react on their own or with cement include the following:

- Pozzolans
  - Fly ash
  - Calcinated shale*
  - Pumicite*
  - Diatomite*
- Silica fume

FLY ASH Fly ash is a fine gray powder resembling cement. It is a residue from power stations burning black coal. The coal is pulverized and blown into the combustion chambers. After burning, the heavier ash particles settle downwards while the lighter (fly ash) particles remain suspended in the exhaust gases. Before going out the smokestacks, these gases pass through electrostatic precipitators or equivalents that remove the fly ash.

It is a pozzolan, that is, a siliceous and aluminous material, and is low in lime content. By itself it has little or no cementitious properties but can react in the presence of moisture with calcium hydroxide from lime or cement. The particles are mostly spherical with sizes from 100 microns down to 1 micron. Quality can vary and depends largely on the way the power station is operated.

Fineness is closely related to the operating condition of the coal crushers, the grindability of the coal itself, and to the efficiency of the ash collection system. Coarse ash is less reactive than fine.

Loss on ignition is perhaps the most critical characteristic. It is an indication of the unburned carbon present in the ash. Higher carbon content adversely affects the performance in grout.

* These pozzolans are mentioned for completeness but are too rarely used to warrant description here.

Chemical composition is dependent on the coal being fired. Coal from different sources is frequently blended at the power station and can affect ash chemistry even when no change in sources is made. The chemistry of an ash affects its performance.

The chemical reaction with Portland cement occurs when lime becomes free during cement hydration. The fly ash reacts with this free lime to form additional cementitious material. It has been claimed that this free lime makes very little contribution to strength and is susceptible to chemical attack by seepage, hence the fly ash reaction is beneficial in this respect. Fly ash can also improve resistance to sulfate attack.

It is important to obtain uniform fly ash. This is not always possible. Various specifications are available that try to control uniformity. An example is Australian Standard AS3582, which provides the following guidelines:

Fineness	75% finer than No. 200 (45 microns)
Moisture content	≦1%
Sulfuric anhydride content	≦2.5%
Magnesia content	≦4.0%
Available alkali content	≦1.5%

Care in selecting fly ash is required because effects on cement grout can vary widely—from providing improvement to providing the opposite: setting times and rheological properties can vary unpredictably unless the fly ash is from reliable and proven sources.

It has been used in ordinary rock grouting where available packaged with cement. However, unless bulk batching facilities are available, it is not worth separate handling into mixers on site. There can be exceptions to this where the fly ash is obtained at almost no cost (it is dumped at some stations) and labor is cheap, or where its chemical advantages are especially required.

The specific gravity of fly ash is between 1.9 and 2.4, compared with 3.15 for cement, so it settles more slowly in stationary grout.

Polatty (1982) mentions that the cost of fly ash in the United States has risen to the stage where it is losing competitiveness against sand as a bulk filler. He also points out that the water demand of a cement fly ash compared to fine sand would be more, so there is potentially more bleed with the fly ash.

SILICA FUME The application of silica fume to grouting is a recent development that appears to have considerable potential for further development. It can benefit the grout in its flow properties and its durability.

Silica fume, or "microsilica," is as fine as smoke. It is a residue from the manufacture of silicon or ferrosilicon alloys, made in electric arc furnace reduction of quartz by coal. Silicon oxide vapor condenses to form silica fume. It is up to 98% amorphous reactive silica in the form of spherical glasslike particles with an average diameter of 0.1 microns. Because it is many times finer than even the microfine cements (Section 3.1-6), its particles can move in between the cement ones, aided by the spherical shape of the silica fume.

It is a very reactive pozzolan, as well as being a filler to some extent. Lime liberated during cement hydration reacts with it to form compact secondary CSH. In use with concrete it reduces permeability and increases strength and resistance to chemical attack appreciably. In this regard, it is in use for preventing alkali aggregate reaction in concrete; in some Scandinavian countries it is a normal ingredient in cement for this reason. In Iceland, for instance, where this problem is widespread, all cement contains 7.5% silica fume, conveniently available locally from a refining plant. This cement has been used on a number of rock grouting jobs in Iceland with success.

Aitcin et al. (1984) found that over 10% of silica fume increases the viscosity of cement grout greatly. More water is needed to make the grout flowable. In 30 minutes the grout can develop marked thixotropic (Section 4.2-2) stiffening, but if stirred it returns to the liquid state. By way of digression, this property has been made use of to make "sticky shotcrete."

The use of a superplasticizer to reduce the viscosity was investigated by Aitcin et al. (1984). Up to 1% of the cement by weight brought the viscosity back to about the same as if silica fume had not been added.

3.3 ADDITIVES FOR PRODUCING PARTICULAR EFFECTS

3.3-1 Additive Purposes

From considerable experience doing without additives, my recommendation is that you should avoid using them at all except in unusual circumstances. Proper mixing is much better than resorting to additives to compensate for the deficiencies of poor mixing (Section 2.2-11).

The following discussion of additives is primarily for those unusual circumstances, but because not everyone shares my views on the subject, included is some comment on use for compensating for poor mixing.

3.3-2 For Rapid Setting

Special circumstances where additives may be required include the situation where seepage is washing grout away rapidly. Rapid setting is needed to get the grout to stiffen quickly to withstand this.

The simplest additives for this are sodium silicate and bentonite. When added to ordinary cement in small quantities, they quickly cause a gel. The time to reach gelling depends on the amounts used—the more of the additives the faster the set, but the weaker the set grout.

Calcium chloride is a well-known accelerator for cement but is corrosive and weakens the grout. There are situations though where these are of no concern.

High-alumina cement (Section 3.1-4) can give relatively rapid set but can be undesirable if Portland cement is later used in the same part of the foundation because of possible reaction.

3.3-3 For Stopping Substantial Seepage Flows

When grouting is called on to stop strong seepage flows through wide cracks or cavities, the technique is to plug off the flows temporarily using anything at all, and when once the flow has stopped, grout properly with normal grout. To achieve this initial blocking, resort has been made to any available material such as straw, sawdust, wood shavings, corn stalks, cane fibers, grass, hot asphalt (which cools and hardens underground), clay, sand, processed nut hulls, shredded rubber, fragmented plastic foil, finely ground fiber, shredded newspapers, flake materials, perlite, plaster of paris, chemicals, proprietary products, and others. The list is only limited by ingenuity. However, the material must be able to fit down the grout holes, although sometimes they have been enlarged to suit.

Adding bentonite and thick diesel oil to some of these materials can bring them into a gelling condition ("gunk"). This helps with the plugging process.

Foaming the grout by agitation, after modifying surface tension with an additive such as Teepol, can give a stiff grout suitable for blocking multiple runaways of relatively modest size. Certain chemical grouts can also be helpful, particularly those of the polyurethanes which react with water to form organic foams.

There have been graphic instances of runaway takes through such large spaces that hessian and plastic bags have been pumped down the grout holes. An apocryphal account alleges use of an old railway freight car for this purpose—it was supposedly shot down a hole (somewhat oversize grout hole!). However, the void was so large that the car didn't plug it but kept going and emerged downstream at high speed to the amazement of a fisherman, whose account was not believed at the local hostelry that night because of his reputation for insobriety and exaggeration. Believe it or not.

3.3-4 For Reducing the Water Content

The amount of water needed to obtain a thick flowable grout can be reduced by use of a superplasticizer. This is not a practical proposition generally, though, because of its high cost and the limited benefit of doing it.

The effect is to make the grout much more fluid, but the effect has a time limit. Some types only reduce water demand for an hour or so after mixing, others last longer, but most do not last as long as the average grout application.

A word of caution here: my testing of a wide range of these chemicals has found instances of false claims by makers. Thorough testing of proposed products is recommended before use. Some well-known ones when independently tested, have inhibited set or made gas voids, and other deleterious effects have occurred. Instances of test results where this happened are in Houlsby (1988).

3.3-5 For Improving Penetration

Here we have a range of additives that are not necessary if the grout is properly mixed (Sections 2.2-10 and 2.2-11).

There have been many proprietary products for improvement of penetration. Earlier ones of the lignosulfanite type, such as "Intrusion Aid" have largely been superseded by superplasticizers, plasticizers, and retardants. As suggested in Section 3.3-4, testing is recommended before accepting manufacturers' claims for any of these. Some of these claims are made in good faith but have been based on tests not really representative of site conditions or have been done by people without practical knowledge of grouting.

3.3-6 For Reducing Bleed

Bleed is discussed at length in Section 4.3, and the importance of high-speed, high-shear mixing is emphasized there and elsewhere (Section 6.1-5). However, lower quality mixing is prevalent in some countries and so is the use of additives to compensate. The terminology "stable" and "unstable" in relation to bleeding propensity has even been devised. My antipathy to this concept will have become obvious to the serious reader of this book.

The ostensible purpose of additives for reducing bleed is to control the settlement of the cement and thereby minimize the tendency for water to bleed upwards out of the grout. Grouts where bleed is slight are sometimes called "stable" grouts—in contrast to "unstable" grouts.

BENTONITE Bentonite is the commonest additive used for reducing bleed. When mixed with cement, sodium bentonite (a type of clay) changes to calcium bentonite, which then flocculates. The flocs are gelatinous and large and prevent sedimentation of the cement particles.

Drawbacks to the use of bentonite include the practical difficulty of ensuring thorough mixing with the cement, its weakening and thickening effect, and the possibility that the bentonite might separate out from the grout underground, leaving some cracks filled with bentonite only (no cement at all). I have seen this happen at an important dam, regrettably, and others have reported it. For instance, Deere (1982) in discussing grouting at dams in Brazil recounted the following:

> On two occasions we had to excavate the abutments a little deeper for changes in the design. We looked for the grout, and we looked for the bentonite, and we found some grout and the bentonite much farther in. It was such a loose structure that the bentonite went right out with the water and what we ended up with was a very irregular grouted surface of the precipitated cement and somewhat filtered out cement, and then back in another 5 to 10 m, all of the fissures were lined with bentonite. This happened on a second project. . . . this is why in Brazil a large number of engineers are specifying grout starting at either 1 : 1 or 0.6 : 1 [by weight] with anywhere from 0.5 to 2% bentonite. They want to end up with all of the big fissures filled solidly with a strong grout. They feel this is the only way to get a grout curtain that will resist leaching. (Vol. 2, p. 36).

Wyoming is the type of bentonite mainly used, but other types can give much the same effect. Bentonite is very messy to handle and mix. On contact with water it tends to form a sticky mess unless fed carefully into a high-quality mixer. It is unwise to mix bentonite in the grout mixer—the bentonite is best premixed 24 hours before needed and then added as a liquid to cement grout in the mixer—therefore an additional mixer is needed, and for thorough work this should be a high-speed, high-shear machine although it often isn't. Jefferis (1982) gives a useful guide to bentonite mixing.

Although bentonite has had extensive use in some countries for bleed control, there are grouting practitioners who like me, consider this as obsolete technology. The practice appears to have developed in countries to which the technology of high-shear, high-speed mixing (Section 2.2-2) had not spread. Grouting has, unfortunately, tended to be a rather parochial and inward looking pastime in some places. My close questioning of some leading bentonite users has shown an unfortunate absence of analytical thought about its effects.

This matter of stable grout has been the subject of misunderstanding in translation, as well as in import. For instance, a summing up of European thinking given by Lombardi (1985) discusses the rheology of cement grouts in terms of "stable" grouts, but when carefully examined discloses a different meaning for this adjective. Lombardi confirmed this in a private communication (February 1985) where he advised me that instead of "stable," he meant "thick." He further expressed agreement that it is better to avoid bentonite: it increases viscosity and shear stress (which he calls "cohesion"), it does not lubricate the grout as some earlier writers had suggested, and it is just a way to compensate for an excess of water.

Deere (1982) quotes Sabarly when defining the distinction between stable and unstable mixes as being 5% bleeding. On this definition and on his bleeding figures, any grout thicker than 1 : 1 is stable and does not benefit from bentonite addition. In the combined paper of Deere and Lombardi (1985) after outlining some usages, the conclusion reached is that "bentonite should only be used to increase the cohesion and limit the travel distance," (p. 162) that is, when less (not more) penetration is required. I have in fact used 6% bentonite (of the cement by weight) to stop grout penetrating through cracks in concrete and past faulty seals when doing contraction joint grouting.

Cautions against the use of bentonite emerged during the discussion at the 1982 ASCE New Orleans Grouting Conference. Greenwood (1982) in an oral presentation comparing rheological properties emphasized that the increased stiffness can invalidate any benefits of lesser bleed. As discussed in Section 4.3 the matter of bleeding of grouts in cracks after placement is a fact of life in grouting. It should be minimized as much as possible without diminishing penetration and reducing permanence.

In summary, then, I suggest that it is best to avoid use of bentonite. Instead, the grout should be properly mixed; bleeding will then be at a minimum.

REFERENCES AND FURTHER READING

Aitcin, P. C., Ballivy, G., and Parizeau, R. (1984). "The use of condensed silica fume in grouts." *Innovative Cement Grouting*. ACI Publication SP-83, Detroit.

Clarke, W. J (1987). "Performance characteristics of microfine cement." ASCE Presented at 23rd International Cement Seminar, Atlanta, Georgia.

Clarke, W. J. (1988). "Grouting composition comprising slag." U.S. Patent 4,761,183, Aug 2.

Deere, D. U. (1982). "Cement–bentonite grouting for dams." *Proceedings of Conference on Grouting in Geotechnical Engineering,* ASCE, New Orleans, 279–300. Also Vol. 2 (Discussion volume, only available to participants at Conference), 35, 36.

Deere, D. U., and Lombardi, G. (1985). "Grout slurries—thick or thin?" *Issues in Dam Grouting,* ASCE, Denver, 156–164.

Greenwood, D. A. (1982). Discussion on "Design of cement based grouts" by G. S. Littlejohn. *Proceedings of Conference on Grouting in Geotechnical Engineering,* Vol. 2 (only available to participants at Conference), ASCE, New Orleans, 11–13.

Houlsby, A. C. (1988). "Improvements in grouting of large ground anchors." *ASCE Journal of Geotechnical Engineering,* Vol. 114, No. 4, 448–468.

Jefferis, S. A. (1982), "Effects of mixing on bentonite slurries and grouts," *Proceedings of Conference on Grouting in Geotechnical Engineering,* ASCE, New Orleans, 62–76.

Lombardi, G. (1985). "The role of cohesion in cement grouting of rock." Fifteenth Congress on Large Dams, International Commission on Large Dams, Lausanne, Q58, R13, Vol. III, 235–261.

Polatty, J. M. (1961). "Investigation of sand–cement grouts." *ASCE Journal Of Soil Mechanics and Foundations Division,* Vol. 87 (SM2), 83–93.

Polatty, J. M. (1982). Discussion at Conference on Grouting in Geotechnical Engineering, Vol. 2 (only available to participants at Conference), ASCE, New Orleans, 3.

Shimoda, M., and Ohmori, H. (1982). "Ultra fine grouting material." *Proceedings of Conference on Grouting in Geotechnical Engineering,* ASCE, New Orleans, 77–91.

U.S. Army Corps of Engineers (1956). "Pressure grouting fine fissures," Technical Report No 6-437, October 1956. Waterways Experiment Station, Vicksburg.

CHAPTER 4

The Grout

In this chapter, the properties of the grout itself are dealt with. Section

4.1 deals with its thickness and how to describe this.

4.2 links this with flow properties, including thixotrophy.

4.3 explains bleeding and segregation.

4.4 discusses setting times.

4.5 goes into the important matter of durability.

4.6 describes ways for testing grout.

4.1 WATER : CEMENT RATIOS, BATCHING, AND MEASURING

4.1-1 Water : Cement Ratios

The most convenient way to specify the amount of cement in grout is to quote its "water : cement" ratio (w : c), that is, the ratio of the amount of water to the amount of cement. Common practice is to give the amount of cement as unity and to quote the water as the varying figure. For example,

Water : cement = 2 : 1, or
= 1 : 1, or
= 0.8 : 1, or
= 0.5 : 1.

Sometimes the 1 is omitted, similar to the practice in specifying concrete:

Water : cement = 2, or
= 1, or
= 0.8, or
= 0.5

Occasionally a reversed method is used, where the cement : water ratio is given: 1 : 0.8. Use of this is uncommon; Appendix D lists comparative values.

Specifying grout by its w : c ratio is not the only way. Drillers are used to describing drilling muds by a weight method, that is, "so-many-pounds grout" (pounds per U.S. gallon), and some extend this system to grout:

$$\text{Specific gravity of mud (or grout)} = \frac{\text{pounds per U.S. gallon}}{8.345}$$

Measurement is made on a mud balance (see Section 4.6-3).

4.1-2 W : C by Volume or by Weight?

This is another issue where practices vary. Some countries specify w : c ratios by volume, while others use weight. While in concrete work it is convenient to specify the w : c ratio by weight (or "mass") because weigh batching of all the ingredients is nearly always used, with grouts there can be advantages if volume batching is used. This particularly applies where the cement is used in bags—the most common situation on many sites. See Section 2.7-11 for comments on use of bagged cement. On these sites cement is measured into the mixer in whole bags, and the volume of the water is measured by a water meter. On the other hand there are large jobs using bulk cement where weigh batching is practicable, and specifying by weight is possible.

It is, of course, possible to batch by weight even when weighing equipment is not in use. This is done by using whole bags of cement whose weight is known. The weight of the water measured through the meter is also known.

The usual practice of specifying w : c ratios by volume is used in this book. The origin of the practice is fairly evidently attributable to the convenient happening that the loose volume of a 94-lb bag of cement is about 1 cu ft. In countries where cement is packed in sizes of bags other than this and particularly in those where metric usage preceded grouting development, specifying w : c ratios by weight has been long-standing practice.

Appendix D compares w : c ratios by the two methods.

A point to note about specifying by volume is that it is the loose volume of a bag that is referred to, not the absolute volume of the cement itself. The loose volume contains air, of course. The apparent crudity of batching like this is commensurate with the practicalities of grouting—it matches the order of accuracy (or is it inaccuracy?) of the activities.

There are various assumptions in use about the volume of a bag of cement. For example,

94-lb bag is assumed to be 1 cu ft.

40-kg bag is assumed to be 30 liters (or 27 liters more precisely).

50-kg bag is assumed to be 40 liters.

Further comments about these assumptions appear in Section 4.1-3. Comparisons are also given of loose volumes and absolute volumes.

General conversion factors to relate the volume and weight methods of showing w : c ratios are

To change w : c by *volume* into w : c by *weight*, multiply by $\frac{2}{3}$

1.5 : 1 by volume = 1 : 1 by weight

To change w : c by *weight* into w : c by *volume*, multiply by 1.5:

1 : 1 by weight = 1.5 : 1 by volume

These factors are not exact. The background to them is explained in Section 4.1-4, and Appendix D gives more exact relationships.

As mentioned earlier, water : cement ratios are given by volume in this book. Note that this means the loose volume of the cement. In certain cases the equivalent by weight is also given.

4.1-3 Assumptions about Cement Volume

As mentioned, when specifying w : c ratios, cement volumes are commonly taken as the loose volume of a bag of cement. Comments now follow about what that actually means in terms of usual sizes of bags.

The specific gravity (SG) of cement is typically 3.15. Dividing the weight (mass) of various sizes of bags of cement by the SG gives the absolute volume of the cement—as though the cement grains are fully in contact with each other to the extent that there are no voids at all. For instance, the absolute volume of

40-kg bag = 40/3.15 = 12.7 liters
50-kg bag = 50/3.15 = 15.9 liters
94-lb bag = 0.48 cu ft

The actual loose volume of a bag of cement varies with the degree of compaction. For instance, my measurements on a 40-kg bag have shown the following:

When contents are well compacted, loose volume = 25 liters.

When contents in very loose condition, loose volume = 30 liters.

When aerated by agitation (dry), loose volume = 35 liters.

Comparing these loose volumes with the absolute volume, for a well-compacted bag,

Loose volume : absolute volume = 25 : 12.7 = ~1.9 : 1

For a bag with contents in very loose condition,

This ratio = 30 : 12.7 = ~2.4 : 1

The loose volume, as far as batching is concerned, is intermediate between these extremes. A volume of 27 liters is a reasonable compromise (ratio = 27/12.7 = 2.1 : 1). Note though, the very approximate nature of this. It is tolerable, given the order of inexactitude of grouting. Some jobs have used 30 liters as the batching volume for this size of bag in order to simplify the mental arithmetic, but this is hardly warranted now with the advent of hand calculators.

In the case of 50-kg bags, the volume can be taken as 35 liters, which gives a ratio of 35/15.9 = 2.2 : 1, or it can be rounded off at 40 liters for simplicity. In this latter case, the ratio loose volume : absolute volume = 40/15.9 = 2.5 : 1, which is perhaps not very advisable as it assumes the cement to be very loose indeed.

The 94-lb bag with a loose volume as traditionally assumed at 1 cu ft gives a ratio of 1.0/0.48 = 2.1 : 1.

All this may appear confusing at first sight, but in practice it is not. Batching tables for 94-lb, 40-kg, and 50-kg bags and for bulk cement are given in Appendices B and C.

4.1-4 Converting between W : C Ratios by Volume and by Weight

The general conversion factors relating the volume and weight methods of showing w : c ratios were given earlier in Section 4.1-2. They are repeated here by way of reminder:

To change w : c by weight into w : c by volume, multiply by 1.5:

1 : 1 by weight = 1.5 : 1 by volume

To change w : c by volume into w : c by weight, multiply by $\frac{2}{3}$

1.5 : 1 by volume = 1 : 1 by weight

Appendix D lists comparative figures for commonly used ratios.

The conversion factors are not exact because they reflect rounding-off assumptions about the volume of bags of cement.

For instance, take w : c = 1 : 1 by weight. To find its equivalent by volume:

1. If 94-lb bags are in use,

   w : c = 1 : 1 = 94 lb : 94 lb

   Changing to volume = 94/62.5 : 1 bag

   = 1.5 : 1

2. If 40-kg bags are in use,

   w : c = 1 : 1 = 40 kg : 40 kg = 40 liters: 1 bag

   If volume of bag is taken at 27 liters,

   40 liters : 1 bag = 40 liters : 27 liters
   = 1.5 : 1

3. If 50-kg bags are in use,

   w : c = 1 : 1 = 50 kg : 50 kg = 50 liters : 1 bag

   If volume of bag is taken as 35 liters,

   50 liters : 1 bag = 50 liters : 35 liters

   = 1.4 : 1

The conversion factor is slightly sensitive to rounding-off assumptions. Appendix D uses 1.5 : 1.

The inverse of these factors is, or course, used when converting from volume to weight. Again see Appendix D.

4.1-5 Standard Mixes

There is no need to use any other mixes than

5 : 1

4 : 1

3 : 1

2 : 1

1 : 1

0.8 : 1

0.6 : 1

0.5 : 1

These mixes give adequate increments in viscosity. Mixes in between give negligible extra control over viscosity and are only a nuisance.

When thickening the mix during grouting, proceed down the list one mix at a time—never jump a mix. For instance, if thickening from 3 : 1, go to 2 : 1 for a while (see Sections 7.2-6 and 7.2-7) and then to 1 : 1 and perhaps on. Never go straight from 3 : 1 down to 1 : 1 or else the too sudden thickening is liable to prematurely block cracks in the job. Excessively fast thickening is one of the tricks used to choke off a hole.

4.1-6 Batching

Batching is the process of measuring out the ingredients for the grout for feeding into the mixer(s). Where bagged cement is in use, only complete bags should be used. These should have been accurately measured out by the manufacturer or bagger. The alternative is bulk cement, weigh batched (see Section 2.7-11).

Water should be measured through an accurate water meter (see Section 2.2-8).

Grout is normally mixed in batches, where the measured ingredients are put into the mixer, mixed, and then discharged as a batch. The measured quantities for each batch are for either:

1. Making starting grout (Section 4.1-7). This grout is put into empty agitators and lines at the start of grouting (perhaps at the start of a shift) or after washing out the gear or after a break in grouting.
2. Producing more of the same grout in the course of the grouting to replace grout taken by the job.
3. Producing the special w : c ratios needed when thickening the mix in the system to suit the requirements of the hole(s) (Section 4.1-8).

Batching tables for all cases are given in Appendices B and C.

4.1-7 Starting Grout

Starting grout is made at the starting w : c ratio for the hole(s) about to be grouted. It should be one of the standard mixes (see Section 4.1-5), and the proportions can be easily calculated or looked up in the thickening tables in Appendices B-1, C-1, and C-2 for the case of zero grout in lines and agitator.

4.1-8 Thickening Mixes

During grouting it is frequently necessary to thicken the mix circulating in the lines in order to provide the hole with different mixes after starting (see Section 7.2-6).

When thickening the mix, it is good practice to turn off the control valve at the hole while the new batch is being integrated into the previous grout. When the grout in the line has reached the new thickness, the hole is turned on again. Such care may not be warranted at a fast hole though.

When batching, the water is measured into the mixer first, followed by the cement and any other ingredients.

The batching quantities for this are not simple; tables have to be used. These are in Appendices B-1, C-1, and C-2. Before consulting them, the quantity of grout in the system (circulation line and agitator) needs to be known. The measuring methods described in Section 2.3-2 provide this quantity fairly easily. Care is necessary to ensure that makeup mixes are not too thick for the mixer to handle. Some mixers are better than others in this regard (see Section 2.2-2). To allow for different capabilities of mixers, the batching tables go down as far as 0.5 : 1 for the high-efficiency machines; less efficient machines will not be able to use those parts of the tables. Nevertheless, they can still cope by using batches with more bags per batch. The tables give a selection of mix sizes, from one-bag batches up to five-bag batches in some cases, and cover the range of mix thickening from 4 : 1 to 0.5 : 1. If the take is fast or if one of the thicker mixes is in use, it is often sensible to mix two- or three-bag batches, or even four- or five-bag batches sometimes, in order to get sufficient volume of makeup grout—as long as the mixer is large enough, of course.

When batching with bulk cement in pounds, the tables for 94-lb bags are close enough most times. For batching in kilograms, the 50-kg bag tables are exact. It is probable that someone has, or will shortly, make automated controls that are programmed to measure cement for thickening without requiring the operator to refer to tables.

The tables do not include additives. If these form a substantial proportion of the grout, it will be necessary to calculate appropriate tables based on the principles explained later. Note from the comments on additives in Section 3.3 that they are generally unnecessary, in my experience, and hence are not included in the tables given.

In deriving batching tables, the absolute volume of the cement is used. As noted in Section 4.1-3, this is calculated on the basis of a specific gravity of 3.15. Its use is necessary because if one bag of cement, assumed at 1 cu ft, say, is mixed with 1 cu ft of water, the result is not 2 cu ft of grout. The water fills the void spaces of the cement and produces a volume of about 1.48 cu ft.

As shown in Section 4.1-3, the following are absolute volumes of common bag sizes:

0.48 cu ft for contents of a 94-lb bag.
12.7 liters for contents of a 40-kg bag.
15.9 liters for contents of a 50-kg bag.

In the example just given, where the w : c ratio is 1 : 1 by volume, the cement in the 1 cu ft (94-lb) bag has an absolute volume of 0.48 cu ft. When this is added to the 1 cu ft of water, the total volume of 1.48 results.

This type of calculation is necessary when batching makeup grout for thickening because it is a matter of providing a certain volume of grout to alter the volume already in the system, while taking into account the amount of cement in that volume.

The basis of the calculations used for this when compiling the tables in the appendices is as follows:

If OldVol = volume of grout in system before thickening
MakeupVol = volume of grout in thickening batch
NewVol = volume of grout in system after thickening
Bags = number of bags of cement in thickening batch
AbsVolBag = absolute volume of cement in a bag (0.48 cu ft ⟨12.7 or 15.9 liters, etc.⟩)
MakeUpWater = volume of water in thickening batch
VolBag = loose volume of a bag (1 cu ft ⟨27 or 35 liters, etc.⟩)
OldCement = cement in OldVol
NewCement = cement in NewVol
OldWC = w : c ratio of grout before thickening
NewWC = w : c ratio of grout after thickening
MakeUpWC = w : c ratio of the thickening batch

Then the basic formula is

$$\text{OldVol} + (\text{Bags} \times \text{AbsVolBag} + \text{MakeUpWater}) = \text{NewVol}$$

The purpose is to calculate MakeUpWater. The other quantities then follow or are found on the way. Steps in the calculations are

OldCement := OldVol/(OldWC × VolBag + AbsVolBag)
NewCement := Oldcement + Bags
NewVol := NewCement × (NewWC × VolBag + AbsVolBag)
MakeUpVol := NewVol − OldVol
MakeUpWater := MakeUpVol − Bags × AbsVolBag
MakeUpWC := MakeUpWater/(Bags × VolBag)

The thickening tables list OldVol, MakeUpWater, MakeUpVol, MakeUpWC.

The calculations for this and for all other appendix tables were carried out in Turbo Pascal (Borland International Inc.).

4.1-9 Thinning the Mix

When grouting of a hole(s) has been finished and the next one(s) to be done require(s) a thinner mix in the line, water is added to the agitator to provide the necessary thinning. When this thinning water has been circulated and well integrated into the previous mix, the new, thinner mix is ready for use. Thinning should never be done while grouting of a hole is in progress.

Quantities of water for thinning are given in Appendices B-2 and C-3. Be careful when thinning that the new volume does not overflow the system. The tables extend up to capacities for quite large systems, so take care when applying them to small ones. If the system is fairly full when about to thin, it is best to waste some grout so as to avoid overflow when the thinning water is added.

The thinning tables are calculated on the same assumptions about the loose volume of a cement bag as the other tables, but there is no need for separate 40-kg and 50-kg tabulations because they both have the same proportion of cement in the water, almost. The same is true for bulk batching. Hence, there is one tabulation for traditional units and just one for metric in this instance.

The basis of the calculations is as follows:

If OldVol = volume of grout in system before thinning
ThinningVol = volume of water for thinning
NewVol = volume of grout in system after thinning
AbsVolBag = absolute volume of cement in a bag (0.48 cu ft ⟨12.7 or 15.9 liters, etc.⟩)
VolBag = loose volume of a bag (1 cu ft ⟨27 or 35 liters, etc.⟩)
VolCement = volume (absolute) of cement in the system
OldWC = w : c ratio of grout before thinning
NewWC = w : c ratio of grout after thinning

Then basic formula is

$$\text{OldVol} + \text{ThinningWater} = \text{NewVol}$$

The purpose is to calculate ThinningWater. Steps in the calculation are

VolCement := OldVol/(OldWC × VolBag + AbsVolBag)
NewVol := VolCement × (NewWC × VolBag + AbsVolBag)
ThinningWater := NewVol − OldVol

4.1-10 Measuring Grout

The main purpose in measuring grout is to find out how much has been taken by the hole(s). Other purposes include the measurements needed for thickening the mix (Section 4.1-8) or thinning it (Section 4.1-9).

A tally of the number of bags mixed provides quantities of the amount put into the grout. Not all of this is used down the hole, but the tally is a useful overall check on the more accurate measurements, which are taken by one or more of the following:

1. Dipstick measurements in the agitator.
2. Flow meter on the hole.
3. Flow meters on the outward and return parts of the circulation line.

These measurements may be taken at regular time intervals, such as every 15 minutes, or they may be taken continuously. Figure 4.1-10A shows how agitator measurements indicate the grout take.

Measurements at intervals indicate the amount of grout taken from the agitator, which is, of course, equal to the quantity taken by the hole(s). When a new batch of grout is to be added, dipstick readings are taken before and after it is put into the agitator. As well as providing a check on the batching volume (from the tables in the appendices), this assists with

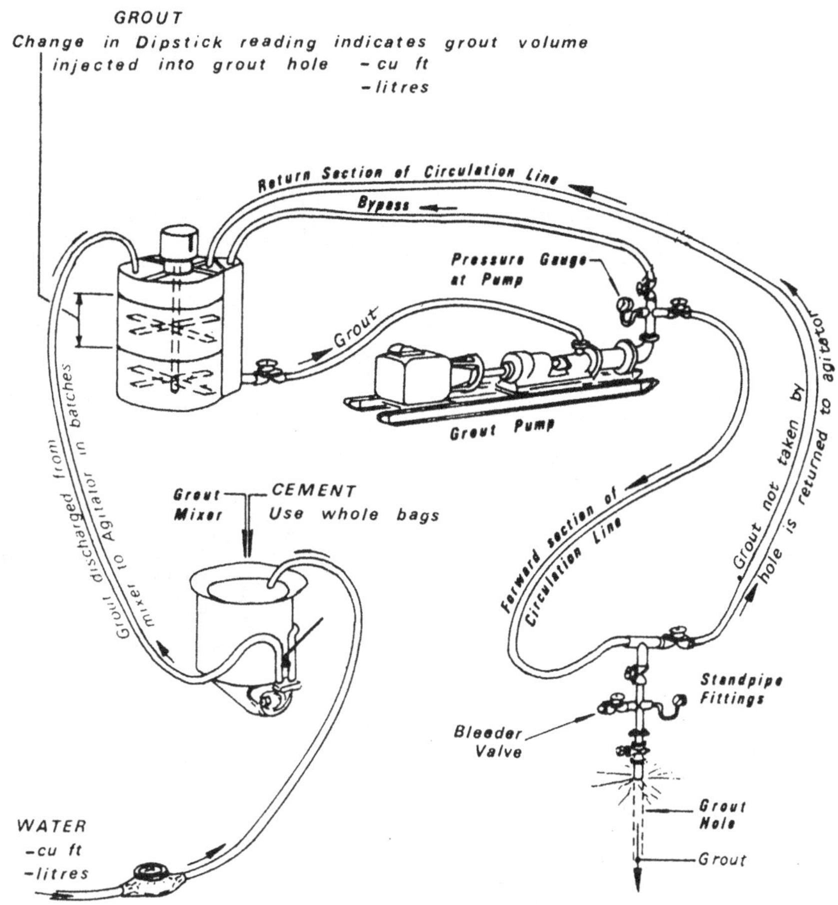

Figure 4.1-10A. *Dipstick measurements to find grout taken by the hole.*

keeping the tally of grout takes. This tally is simplified if the adding of a fresh grout batch can be deferred until the time for the regular reading.

Grout flow meters are discussed in Section 7.7-10 and a simple arrangement is shown in Figure 4.1-10B. They provide a continuous record of the rate of take. Unless mounted like this they can be troublesome on slow holes and when the hole approaches refusal: when this happens they should be disconnected and bypassed and readings taken by dipstick instead. They are of limited use, but can be beneficial when handling the grouting of rock prone to move (see Section 7.6) or when grouting near sensitive structures where the volumes of grout entering the hole must be strictly limited.

The arrangement of flow meters shown in Figure 4.1-10C is usually linked to computers (see Section 7.7). The meter on the outward line measures grout going to the hole, and the other meter measures returning grout. The difference between the two is the amount of grout going down the hole. A continuous measure is obtained, and this is beneficial, but the equipment is relatively expensive and can be troublesome in the rigors of a grouting site.

4.1-11 Calculation of Cement Take

After finding how much grout is taken by a hole, the next step is to work out how much cement was in that grout. This calculation cannot usually be avoided by merely quoting the cement mixed—there is waste to be considered, the amount to fill the lines, and grout left over from grouting of a previous hole or used in starting the subsequent hole.

The amount of cement contained in various volumes of grout is read off tables such as those in Appendices B-3, C-4, and C-5. These are calculated using the absolute volume of cement in a similar way to the batching tables and explained below.

After looking up the cement injected in each of the measuring periods, these amounts are totaled and given a rough check by comparing them with the quan-

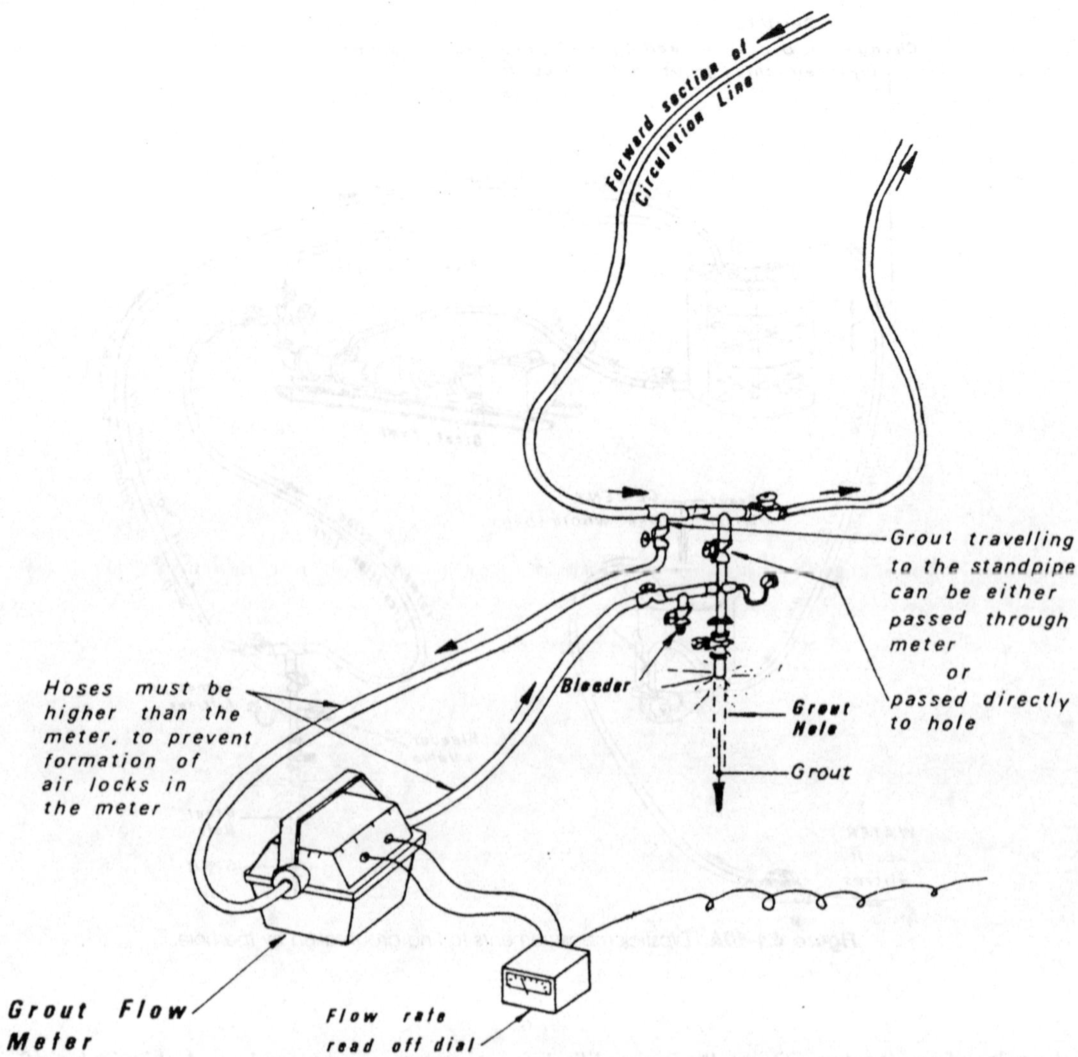

Figure 4.1-10B. *Flow meter fitted directly on the hole.*

tity of cement mixed after adjustments for wastage, amounts from a previous hole, or amount used for starting the next hole. Examples of this are given in Section 4.1-12.

General calculations for the tables in the appendices follow:

If VolGrout = volume of grout

VolCement = volume (absolute) of cement in the grout

AbsVolBag = absolute volume of cement in a bag (0.48 cu ft ⟨12.7 or 15.9 liters, etc.⟩)

VolBag = loose volume of a bag (1 cu ft ⟨27 or 35 liters, etc.⟩)

WtBag = weight (mass) of a bag of cement (94 lb ⟨40 or 50 kg, etc.⟩)

Then,

WC = w : c ratio

VolCement := VolGrout/(WC × VolBag + AbsVolBag)

WtCement := VolCement × WtBag

4.1-12 Reporting

There are various ways of reporting the results of measurements. Probably most are as good as each other provided they include the following:

- Grout measurements at the start and end of each period (Section 4.1-10) or incremental measurements if readings are continuous.
- Clock times for each reading booked.
- Water : cement ratio of the grout at each reading.
- Pressure on the hole at each reading.

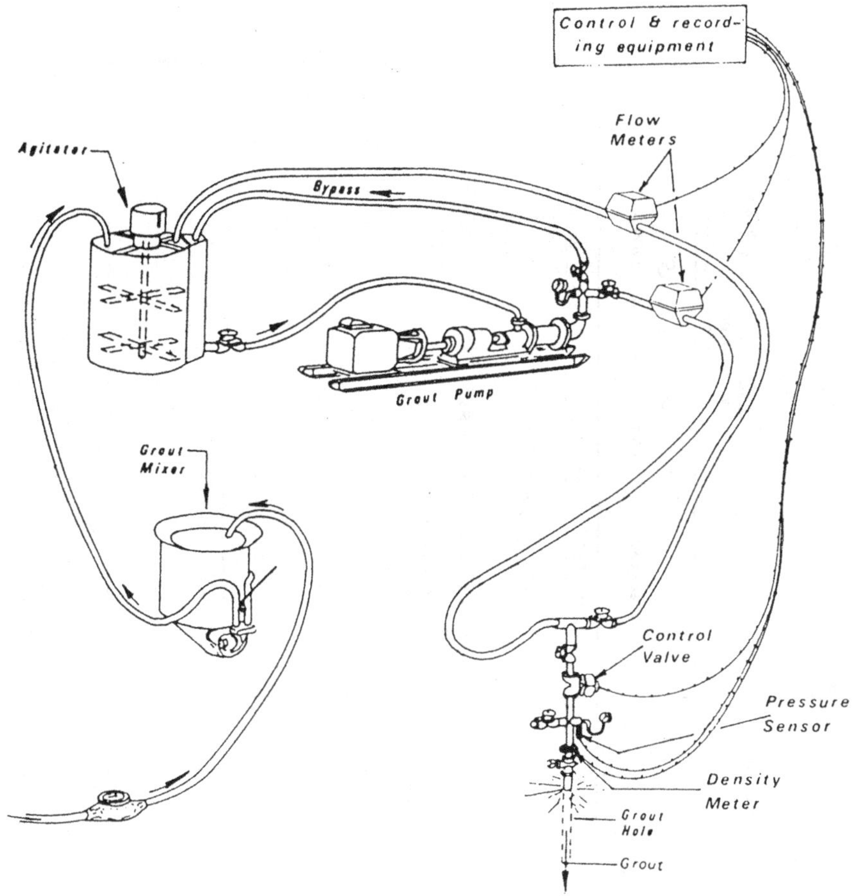

Figure 4.1-10C. *Flow meters on outward and return parts of circulation line.*

- Record of cement (in bags unless using bulk cement) batched into mixer.
- Details and times of all troubles such as leaks, connections, and rock movements.
- Results of water testing before grouting (Section 5.3-2).
- Comments or explanations of any unusual event.
- Cement and grout used to fill lines.
- Grout wasted by burst hoses or other misadventures on the surface.

Some reporting forms provide room for the subsequent calculations:

- Volume of grout injected in each period.
- Amount of cement in each of those volumes.
- Total cement injected down the hole (note that this includes losses through leaks or connections if the suggestions of Section 9.2-5 are adopted concerning methods of payment for grout.
- Amount of cement injected per unit length of hole.

Examples of suitable report forms are shown in Figures 4.1-12A and 4.1-12B.

Figure 4.1-12C is a blank report form of the same type, for use with figures entered into a desk PC computer by hand (see Section 7.7-3). The computer can be programmed to do all the calculations marked * automatically.

4.2 VISCOSITY AND THIXOTROPY

4.2-1 Viscosity

Consider the following explanation if you are not too sure what viscosity is: Visualize a thick mixture of cement and water—with not much more water than is needed to make it a fluid. Obviously, although it flows, it does so sluggishly if tipped out of a bucket. Water would tip out much faster. This grout has internal resistance to flow because of the cement particles in it. Actually, the effect described is a combination of viscosity and thixotropy effects, but for the moment we

TYPE OF GROUTING CURTAIN .... DAM

DAY LABOUR WORK

GROUTING REPORT

HOLE N° P 960-L

SEQUENCE PRIMARY

LOCATION CREEK BED

INCLINATION TO VERTICAL 30°

STAGE DEPTH 50 TO 100 FT

APPLICATION 1 OF 3

DATE	TIME FROM	TIME TO	TIME PERIOD (MIN.)	W/C BY VOL.	BAGS CEMENT ADDED TO MIXER	GROUT (CUFT) IN TANK START	GROUT (CUFT) IN TANK FINISH	GROUT (CUFT) INJECTED	CEMENT TAKE IN PERIOD (CUFT.)	PRESSURE AT STANDPIPE (P.S.I.)	REMARKS CONNECTIONS - UPLIFT - WASTE DETAILS - PLANT DIFFICULTIES, ETC.
24·5·73	2005	2020	15	W	–	0	7·1	7·1		15	WATER TEST UL = 8
24·5·73	2200	2205	5	3:1	3	10·50	8·25	2·25			FILLING HOSES
	2205	2220	15	"		8·25	2·79	5·46	1·56	15-70	
	2220	2235	15	2:1	2	4·35	1·17	3·18	1·26	70-15	SLIGHT SURFACE LEAK CH 954, 7 FT U/S AT 2225 hrs – CAULKING
	2235	2250	15	1:1	3	5·50	3·45	2·05	1·37	15	LEAK TAKING UP
	2250	2305	15	"		3·45	2·40	1·05	0·70	15	
	2305	2320	15	"		2·40	1·52	0·88	0·58	15-100	
	2320	2335	15	"	1	3·05	2·43	0·62	0·41	100-120	
	2335	2350	15	"		2·43	2·13	0·30	0·20	120	
25·5·73	2350	0005	15	"		2·13	2·03	0·10	0·07	120	
	0005	0020	15	"		2·03	2·03	NIL	–	120	
	0020	0050	30	"		2·03	2·03	NIL	–	120	HOLDING PRESSURE

TOTALS 9 BAGS 6·15 CUFT = 0·12 CUFT/FT

TYPE OF DRILLING WET PERCUSSION

SHEET 1 OF 1

Quantities are approximate only

EXAMPLE OF GROUTING REPORT (Day Labour Work)

E 14 HOLE N° P 960-L

Figure 4.1-12A. Example of report form completed by hand. Uses cubic feet and feet units. For "day-labor" (in-house) use.

TYPE OF GROUTING CURTAIN

..................DAM

CONTRACT Nº ....

GROUTING REPORT

HOLE Nº P107

SEQUENCE PRIMARY

CONTRACTOR'S FOREMAN PHIL M. UPJOHN

INSPECTOR I.C.B. LOW

STAGE DEPTH 0 m TO 10 m

STAGE LENGTH 10 m x 100 = 1,000 cm

APPLICATION 1 OF 1

LOCATION CREEK BED

INCLINATION TO VERTICAL 30°

DATE	TIME FROM	TIME TO	TIME PERIOD (MIN)	W/C BY VOL.	BAGS CEMENT ADDED TO MIXER	GROUT (LITRES) IN TANK START	IN TANK FINISH	INJECTED	WASTE PAID	WASTE UNPAID	CEMENT TAKE TOTAL IN PERIOD (BAGS)	NETT INC. PAY WASTE (BAGS)	PRESSURE AT STAND PIPE (BARS)	REMARKS CONNECTIONS - UPLIFT - WASTE DETAILS - PLANT DIFFICULTIES, - ETC.
30.2.77	1925	1930	5	W				100					1	20 UL } WATER TEST
	1930	1935	5	W				80					1	16 UL } (USE 15 UL)
	1935	1940	5	W				75					1	15 UL
30.2.77	1955	2000	5	3:1	3	310	270	40						FILLING HOSES
	2000	2015	15	3:1		270	130	140			1·36	1·36	1 - 2	
	2015	2030	15	3:1	1	235	115	120			1·17	1·17	2 - 3	
	2030	2045	15	2:1	1	160	65	95			1·31	1·31	3 - 2	SLIGHT SURFACE LEAK 2m u/s
	2045	2100	15	1:1	4	185	100	85			1·99	1·99	2	LEAK TAKEN UP
	2100	2115	15	1:1	2	190	130	60			1·41	1·41	3	
	2115	2130	15	1:1		130	90	40			0·94	0·94	3	
	2130	2145	15	1:1	1	135	110	25			0·59	0·59	3	
	2145	2200	15	1:1		110	100	10			0·23	0·23	3	
	2200	2215	15	1:1		100	100	NIL			0	0	3	
	2215	2230	15	1:1		100	100	NIL			0	0	3	
										140	3·20			WASTE (100 TANK + 40 HOSES = 140ℓ = 3·20 BAGS CEMENT)

TOTALS 12 BAGS 12·20 9·00 x 30 = 270 ℓitres of cement

TYPE OF DRILLING WET PERCUSSION SHEET 1 OF 1 CEMENT TAKE = 270/1,000 = 0·27 ℓitres/cm

EXAMPLE OF GROUTING REPORT

E 14 HOLE Nº P 107

Figure 4.1-12B. Another example of report form completed by hand. Uses metric. For contract work.

```
TYPE OF GROUTING:____*____JOB:__*____    HOLE No:____________
                                          STAGE:______ft to______ft.

                        WATER  TESTING

TESTING DATE:__-__-____                   TEST PRESSURE = 15 psi
  TIME         PERIOD     METER READING   WATER TAKE        LUGEONS
 (hr-min)      (min)         (cu ft)       (cu ft)
                              0.000
  ____
  ____          _*_         __.___           _*_              _*_
  ____          _*_         __.___           _*_              _*_
  ____          _*_         __.___           _*_              _*_
COMMENTS ON THE TEST:_____________________________________________
__________________________________________________________________
              SELECTED WATER TEST RESULT            = _*_ LUGEONS

                      GROUTING  DETAILS

GROUTING DATE:__-__-____
  TIME       PERIOD  W:C CEMENT  GROUT(cu ft)              CEMENT  PRESSURE
 (hr-min)    (min)       ADDED   In Agitator               TAKE
From  To                         From    To Injected       (cu ft)   (psi)
____ ____     _*_    _:1   _     __.__ __.__ __.__          _*_       _*_
____ ____     _*_    _:1   _     __.__ __.__ __.__          _*_       _*_
____ ____     _*_    _:1   _     __.__ __.__ __.__          _*_       _*_
____ ____     _*_    _:1   _     __.__ __.__ __.__          _*_       _*_
____ ____     _*_    _:1   _     __.__ __.__ __.__          _*_       _*_
____ ____     _*_    _:1   _     __.__ __.__ __.__          _*_       _*_

TOTALS   Cement Mixed=  _*_    Total cement take   =_*_
                               Average cement take=_*_ cu ft/ft
REMARKS:__________________________________________________________
__________________________________________________________________
```

Figure 4.1-12C. *Blank report form for use with desk computer.*

concentrate on the viscosity aspects—thixotropy comes later.

The definition of viscosity suggested by the ASCE Grouting Committee (1980) is "the internal fluid resistance of a substance which makes it resist a tendency to flow." (p. 814.)

Water has viscosity too, but a lot less than grout. The viscosity of grout varies, depending on the amount of cement in it. The cement particles act in abstruse physicochemical ways to produce this viscosity (and thixotropy also). Explaining them in detail is beyond the scope of this book, but a simplified description is that the particles have electrostatic attraction or repulsion as well as mechanical interlocking to various extents. The method of mixing makes a great difference to these effects, as discussed in Sections 2.2-10 and 4.5. Section 6.1-5 explains the effect on flow in fine cracks.

Viscosity dominates the behavior of chemical grouts (see Section 1.2-5) more than it does with cement grouts. It governs the penetration of chemical grouts in alluvium or sand, whereas it is only one of the factors controlling the penetration of cement grout in cracks (see Section 6.1).

Measuring viscosity is rather an uncertain business when it comes to cement grout because of the chemical reactions going on among the cement grains. If the tests are done immediately after mixing, this is minimized, but then the tests are not really representative of the grout going into the cracks, which is older. The methods of testing vary depending on personal preferences and availability of instruments. Fairly precise instruments of the rotating viscometer type (see Section 4.2-3) are not robust enough for use at the mixers; a site lab is really needed for them. This is not always practicable, and even if it is, time elapses between mixing and getting the sample to the lab. Therefore, field testing falls back on a second-rate method, the flow cone. This is a rugged, practical test, developed initially in the drilling industry and measures the time taken for a specific volume of grout to empty from a cone through an orifice. Unfortunately, there are several different sizes and styles of cones in use, giving different flow times. The different types of cones and the test method and results are discussed in Section 4.6-4.

The flow cone does not give a direct measure of viscosity. From laboratory correlations, this can be done indirectly to give an approximate result. However, inaccuracy is of little concern because it is relative values that are usually of interest.

But is it really necessary to measure viscosity on a rock grouting job? No it is not. Provided the mixers are working properly (Section 2.2-9) and batching of ingredients is accurate (Section 4.1-6), grout is sufficiently uniform on most jobs for all practical purposes. Testing its viscosity is superfluous. The test is more relevant on those jobs where uniformity of grout is so important that every batch from the mixer has to conform to preset values, for example, in grouting ground anchors or posttensioning cables (Houlsby, 1988). The subject is mentioned here for completeness and to lead up to thixotropy.

4.2-2 Thixotropy of Grout

Thixotropy is perhaps a little difficult to explain, so examples may help: The sketches in Figure 4.2-2A show cracks leading off a grout hole. Imagine that during injection of water or of grout the pressure is turned off for a minute or so. Flow will stop. Then the pressure is turned on again. Water and thin grout (i.e., 2 : 1 and thinner) will resume flowing immediately whereas thick (thixotropic) grout probably will not do this until the pressure builds up to a substantial amount. While this pressure is building up, the thixotropic grout is like a paste or sludge; this changes suddenly to fluid behavior when the pressure gives enough push to start the grout moving again. The point at which behavior changes from sludgy to fluid is called the Bingham yield point.

Take the case of ketchup ⟨tomato sauce⟩. It becomes thick after standing for a while, but give it a good shaking and it changes into a thinner, nicely behaved liquid. This is the subject of a ditty:

Take the sauce, and shake the bottle;
None will come, and then a lot'll.

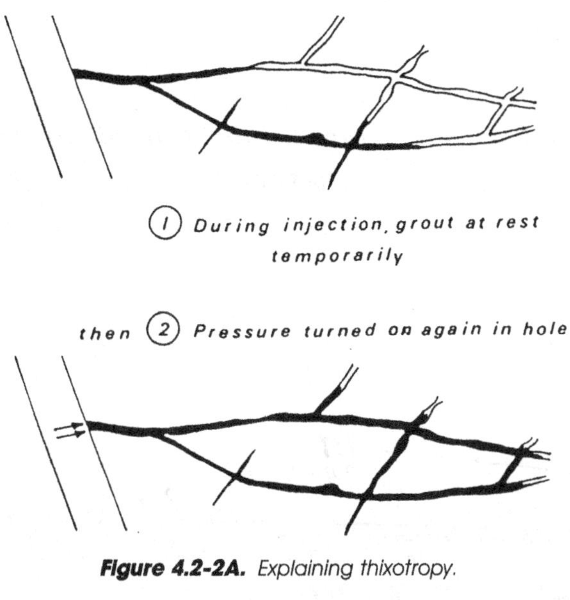

Figure 4.2-2A. *Explaining thixotropy.*

Mason (1986) gaily attributes this astute piece to folklore among those characters called colloid scientists or rheologists; he claims it's their standard definition of thixotropy!

To put things more formally, thixotropy is that property of wet grouts where they are gel-like when stationary and revert to liquid behavior when sufficiently agitated or pressured.

To relate the concept to actual conditions met on jobs, consider Figure 4.2-2B which shows velocity profiles for grout passing through wide and narrow sections of a crack in laminar flow.

The shear strain is a function of the velocity of the grout. If it is plotted against the shear stress (as in Figure 4.2-2C), which relates to the pressure causing the flow, water comes out on the plot as a straight line through the origin as shown by the dashed line. Water has negligible shear stress (i.e., negligible pressure buildup if flow is laminar) over the low range of shear strain shown. Contrast this with the continuous line, which shows a thixotropic fluid; flow does not commence until the shear stress (i.e., pressure) approaches the Bingham yield value. After this, the plot changes character and has a gradient related to the viscosity.

Relating this to grout, the next plot (Figure 4.2-2D) shows the curves for various grout mixes compared with water again. The w : c ratios in bold figures are by volume (as in this book generally), those in small print are by weight. Values of 1 : 1 and thicker show thixotropic properties.

The same plot is shown in Figure 4.2-2E with hatching added to indicate where most job usage relates. The intensively hatched zone indicates major usage. Thixotropic behavior becomes more pronounced as grout becomes thicker: 3 : 1 grout shows hardly any thixotropic stiffening when flow ceases, but 1 : 1 shows

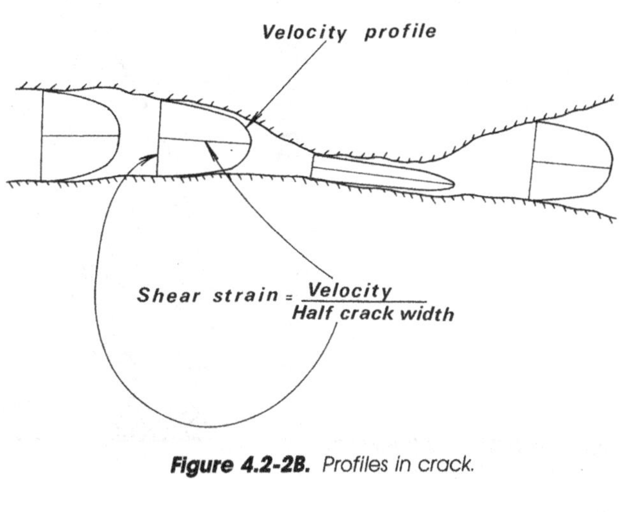

Figure 4.2-2B. *Profiles in crack.*

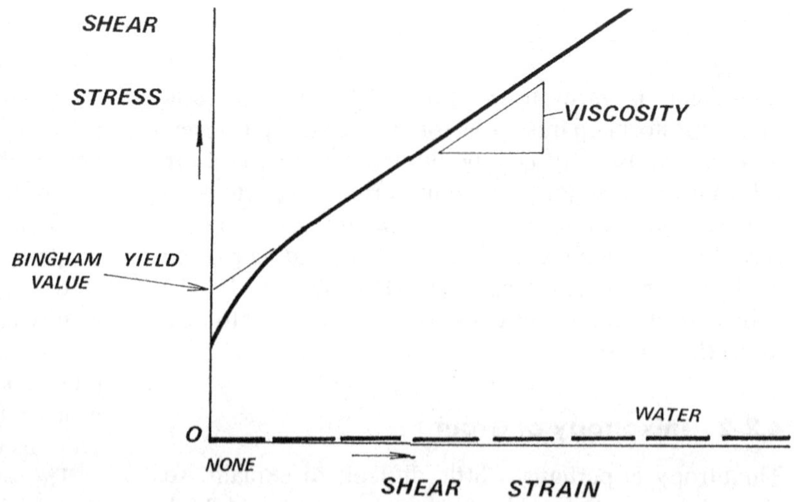

Figure 4.2-2C. *Plot illustrating thixotropy.*

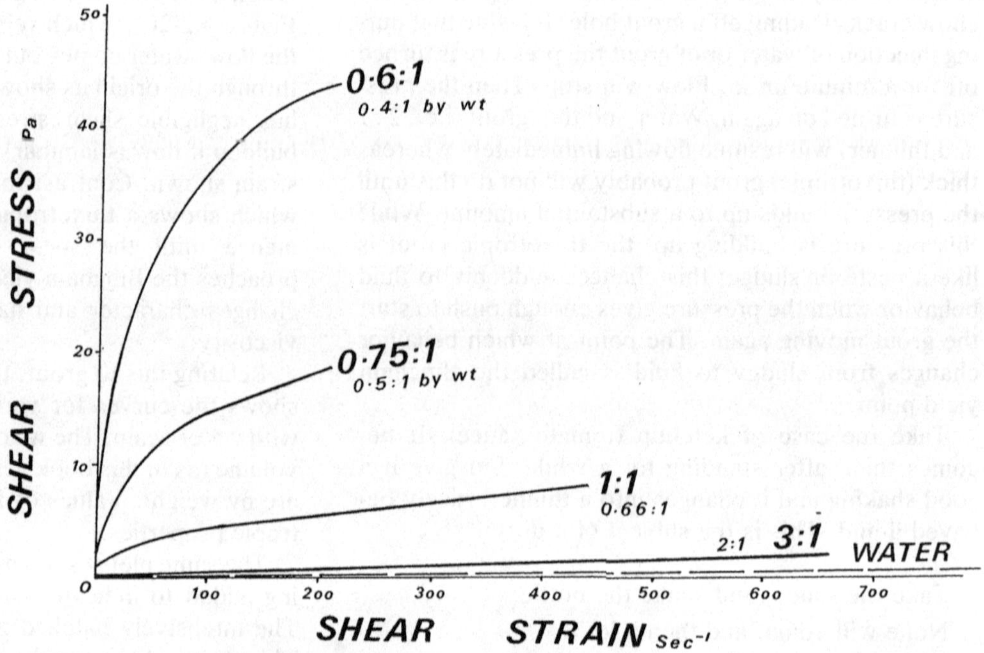

Figure 4.2-2D. *Thixotropic comparison of grouts.*

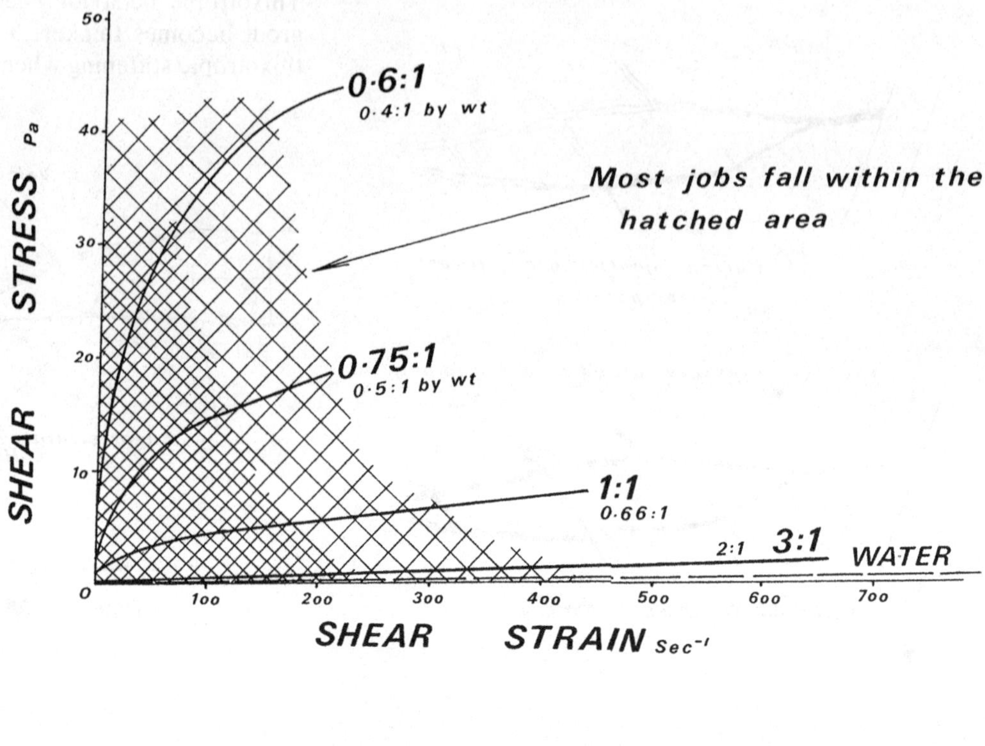

Figure 4.2-2E. *Thixotropy in relation to grout usage.*

it fairly readily, and in mixes thicker than this it dominates behavior. The source material for this plot is given later.

Thixotropy is a dominant factor in cement grouting. While it can be a nuisance if inadvertent interruptions to injection have occurred, it can be useful in resisting washing away of grout by groundwater flow, in handling of strong leaks (Section 7.3), and in handling of pressure-sensitive situations where its inability to transmit fluid pressures significantly enables safe use of (reasonable) injection pressures (Section 7.6). Apart from these controlled usages, it exerts uncontrolled influences when a mixture of crack sizes is being grouted simultaneously: flows in the various cracks change their pace, and perhaps even stop for a while in some places, while pressures and filling are equalizing. The ability of grout to get moving again after slowing or stopping is dominated by thixotropic effects.

4.2-3 Thixotropy Theory

Thixotropy theory is a complicated subject. Only enough of it is given here to explain its application to cement grout and to show how the graphs in the previous section were derived. Fuller treatments are in Ritchie (1965), Raffle and Greenwood (1961), and others. Rheology is the generic name for flow (or the lack of it in some cases).

SHEAR STRESS Various names are in use for this. Basically, shear stress relates to grout pressure, but because we are not dealing with water, there is an element of "overcoming a threshold" and then overcoming internal friction in the term. It is used as the ordinate in the graphs shown earlier. At the start of a grout application the grout is moving freely into cracks and the shear stress exceeds the Bingham yield value. This situation persists until grout slows down as refusal is approached or for other reasons; the shear stress then falls below the Bingham yield value and the grout changes to gel-like behavior. Pressure from the hole then ceases to be substantially transmitted through the grout.

Figure 4.2-3A shows what happens when grout is penetrating outwards from the hole.

Figure 4.2-3B shows how pressure gradually contracts back toward the hole once refusal is reached. These sketches are also shown in Section 6.1.

All this is very good if there is no need to start the grout moving again, but sometimes there is, such as when handling problems. In addition restarting may be needed in a foundation where a mixture of wide and narrow cracks are being grouted: at the start of an application grout will probably enter all the cracks (if

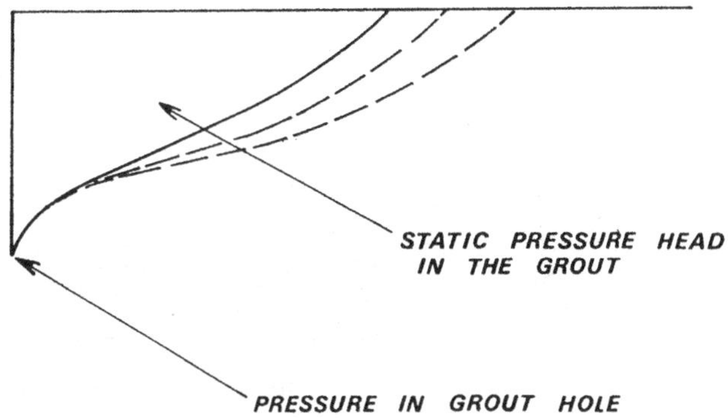

Figure 4.2-3A. *Pressure distribution from a grout hole while grout is active.*

the starting mix has been chosen sensibly—see Section 7.2-2), but then as the wider ones dominate by taking most of the grout (and most of the pressure too if they are very wide), flow in the finer cracks ceases after slight, but not adequate, penetration and the grout there becomes gel-like. Later, when the wide ones have had their fill and reached refusal, it should be the turn of the finer cracks to get their share of the grout, but maybe it is too late; the slight amount of grout near the hole may be too stiff to be made mobile again. It is for this reason, among others, that such a foundation should be given plenty of relatively thin grout, such as 3 : 1 or 2 : 1 before thickening (refer to Sections 7.2-5 and 7.2-6). Thin mixes have negligible thixotropic stiffening.

SHEAR STRAIN Shear strain relates to the amount of movement in the grout. It is also called shear rate. The concept is difficult to comprehend in relation to grout moving through a crack. One way of visualizing it is for the case of laminar flow (believed to be the usual type of flow for grout in the type of cracks considered here) in a crack of varying thickness. Consider the flow distributions show in Figure 4.2-3C. Velocity profiles are shown as parabolic; perhaps more precisely they should be shown with the noses of the parabolas flattened somewhat, but the message is the same. By dividing the peak velocity by half the crack width, a measure of the shear strain can be obtained. Although it is not a precise measure of this, it is sufficiently indicative for our purposes. Half the crack width has been used because it gives a closer analogy to deformation of a gel under nonsymmetrical pressure (which is the case) than use of the full crack width would: dividing by the full width would merely give the average velocity anyhow. The units cancel out to "per second," which is also hard to relate to grout movement.

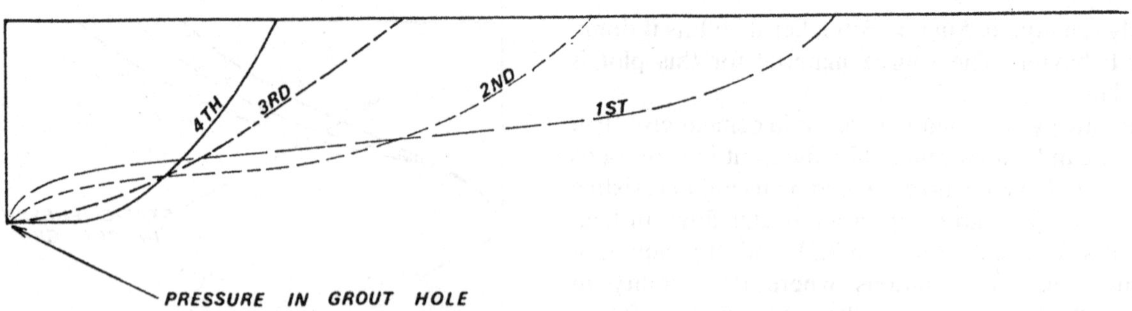

Figure 4.2-3B. *Pressure distribution contracts when grout stops.*

To get a sense of proportion, it may help to realize that the shear in high-speed, high-shear mixers such as those described in Section 2.2-2 is tens of thousands "per second," while for grout passing along cracks it is a hundred or so.

SHEAR STRESS VERSUS SHEAR STRAIN When shear stress and strain are plotted for water, almost no shear stress develops for the range of shear strain considered, so the plot is a straight line through the origin, as shown in Figure 4.2-3D.

Thixotropic fluids are shown by the full line; they do not pass through zero: their intercept on the ordinate is theoretically the "Bingham yield point," a measure of the degree of stress, or pressure, needed to start the fluids flowing. Their viscosity then governs further relationships.

BINGHAM YIELD VALUE Named after Bingham (1916), the nature of the Bingham yield point has been indicated earlier. Thixotropic fluids are sometimes called Bingham bodies in comparison to water, known as Newtonian.

One method for measuring the value uses a laboratory viscometer having two concentric coaxial cylinders. The outer cylinder has a variable speed of rotation, while the inner one is attached to a torsion wire. The grout to be tested is placed in the annulus between the two cylinders, and the torque transmitted through the grout to the inner one by rotating the outer at various speeds is measured. Calculation of the Bingham yield value is then carried out as shown by Mayer (1963). Further information on testing instruments for grout is given in Section 4.6.

APPLICATION TO GROUTS The following information is by no means precise and is only for giving a general impression for application to practical grouting. Figure 4.2-3E is the plot shown earlier in relation to this. The curves for the various mixes come from the work of Raffle (1960) and Hall, Edwards, and Skinner (1968). It should be noted that their work differs from the earlier, similar plot of Ritchie (1965) by a factor of 10 in shear strain. The later work has been accepted here for these illustrative purposes.

Hatching is used on the plot to show regions of applicability, with more intensive hatching for the principal regions and is based on the author's experience.

TIME EFFECTS Behavior in the vicinity of the Bingham yield point has a time component. If grout has become stationary for only a minute or so, it takes less pressure to get it moving again than if it had been stopped for appreciably longer. Plots of the type shown in Figure 4.2-3E can develop a hysterisis loop where the

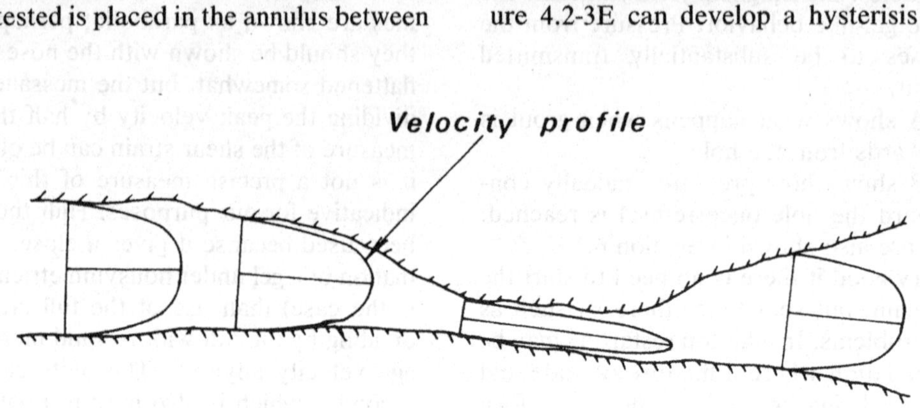

Figure 4.2-3C. *Velocity profiles in a crack.*

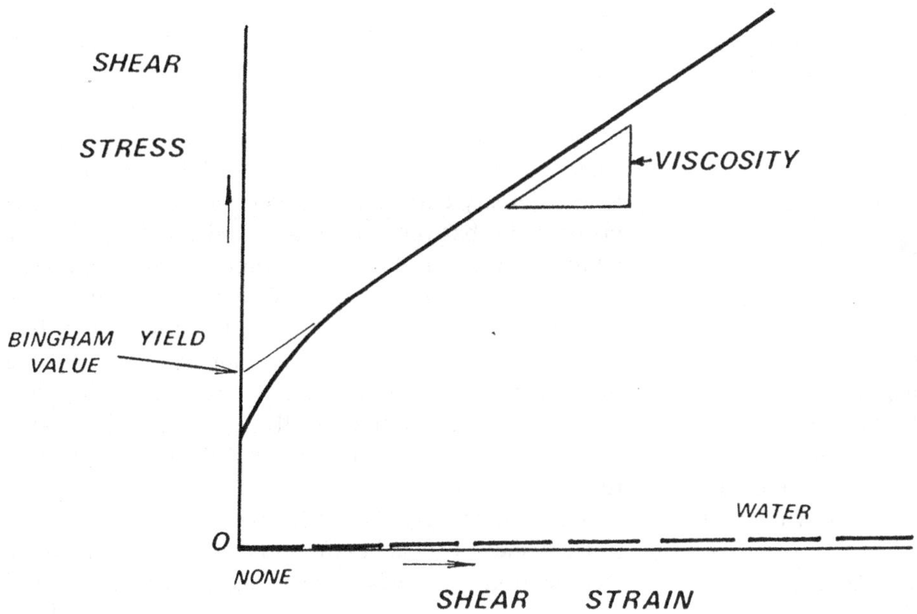

Figure 4.2-3D. *Thixotropy plot.*

curve for grout that is slowing down is not the same as for grout that is speeding up. Ritchie (1965) shows examples of this. The inference is that the sooner stationary grout gets moving again the better.

COHESION OF GROUT "Cohesion" is a term introduced by Lombardi (1985) and enlarged on in Deere and Lombardi (1985). He defines cohesion as the same thing as the Bingham yield point, without saying so. On a stress–strain plot (such as Figures 4.2-3D and 4.2-3E), he shows cohesion as the ordinate intercept to the yield point in an almost identical manner to the traditional way with which soil strengths are presented in geotechnical work. This can be confusing, as Lom-

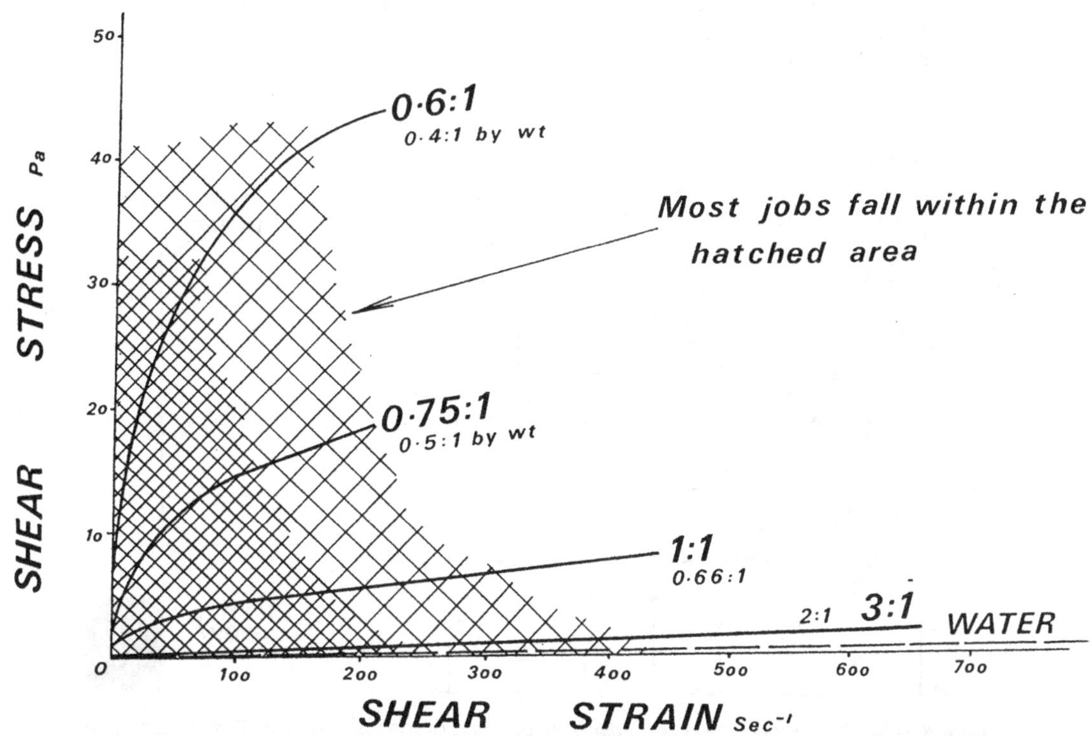

Figure 4.2-3E. *Influence of thixotropy on grouts.*

bardi has concurred during subsequent conversation with me. A different term was suggested as being preferable, but there seems to be no real reason for departing from the name used satisfactorily for many years: Bingham yield point.

4.3 BLEED AND SEGREGATION

4.3-1 Excess Water

Water is necessary to give cement grout mobility but becomes a nuisance after its transporting role is finished. The water needed for mobility exceeds the amount required for the chemical action of cement hydration. An axiom of grouting is that the minimum amount of water should be used consistent with getting adequate mobility.

Water-minimizing practices, tests, and case histories are given in Houlsby (1985). The basic issue is that of bleeding: once grout slows down or stops, cement particles gravitate downwards at the same time as water, and the thinner part of the grout floats upwards. In relatively large bodies of grout, such as in grout holes, density currents (see Section 6.1-3) result from this process and tend to structure it. If grout remains stationary long enough, cloudy or even clear water collects above a fairly distinct surface of settled cement.

The graph in Figure 4.3-1A compares the amounts of bleed water from various grouts at various times. The figures apply to large vertical bodies of grout where density currents have freedom to form, there is no bridging between the sides, and the grout has no additives or sand. A grout in common use for foundation grouting, such as 2 : 1, can lose 35% of its initial volume in bleeding after about 1.5 hours. For thin grouts, such as 5 : 1, bleeding can be 70%: and for excessively thin grouts, such as 12 : 1, it can be 85%; this is shown diagrammatically at the right-hand side of Figure 4.3-1A where the small amount of cement remaining is shown in black.

The thinner grouts settle more rapidly than the thicker ones. This is helpful in Displacement Grouting (see Section 6.2-8) where very high pressures are used to displace rock and where rapid bleed-water replacement is practiced. See also Section 6.1 where flow patterns in cracks are discussed. Note the approximate nature of the graph shown in Figure 4.3-1A. It should be regarded as typical. Bleeding varies with different finenesses and sources of cement.

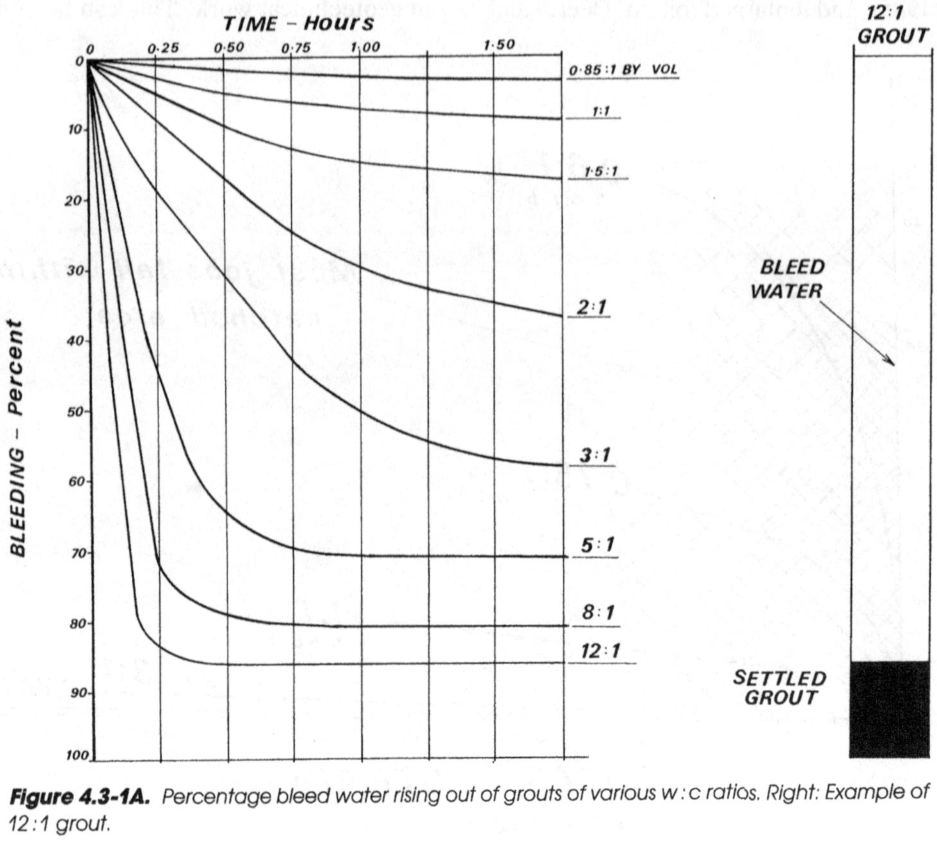

Figure 4.3-1A. *Percentage bleed water rising out of grouts of various w : c ratios. Right: Example of 12 : 1 grout.*

The method used for measuring bleeding also affects the results. See Section 4.6-7 for a discussion of the methods.

When cement settles in this manner, it develops a loose structure; for instance, the cement in the initial mixes listed in the following table settles to approximately the w : c ratios shown (not counting the bleed water:

Initial Mix	w : c of Settled Mix
1 : 1	0.9 : 1
2 : 1	1.1 : 1
3 : 1	1.3 : 1
5 : 1	1.4 : 1
9 : 1	1.7 : 1
12 : 1	1.9 : 1

These figures are derived from Hornby (1961). Note how thin initial mixes produce relatively thin settled grout. These have very loose structures that have low durability: the cement crystals are only lightly and loosely bonded to each other. This relatively open skeleton can permit the passage of water, and hence there is greater potential for attack and decomposition of the cement than in the compact, well-bonded structure produced by thick grout. Taking this to the ultimate leads to a preference for excessively thick grouts, but these can be impractical in many cases, and so a compromise is necessary.

It was once thought that grout pressure applied to thin grouts overcame this problem by densifying the skeleton of cement grains and expelling water, but the U.S. Army Corps of Engineers tests (1956) disproved this (these tests are described in Section 6.1-6).

4.3-2 Bleedwater Accumulation

During the course of a normal grout application, water and thin grout accumulate in the upper part of the grout hole (see Figure 4.3-2A).

Most grout holes contain some water before grouting. It usually comes from groundwater and from water testing and can often be pushed away through the jointing by incoming grout. However, this can be difficult in tight or near-tight conditions, and the water then remains in the hole. Even if it has been possible to expel all water at the start of grouting, there comes a time in the application when grout has filled the joints leading off the hole, and so bleed water from grout in it cannot now escape into the foundation. It must be removed from the hole by passing it out through the bleeder valve. This bleeding process is important but, unfortunately, tends to be skimped or overlooked unless operators are suitably enthused as to its importance.

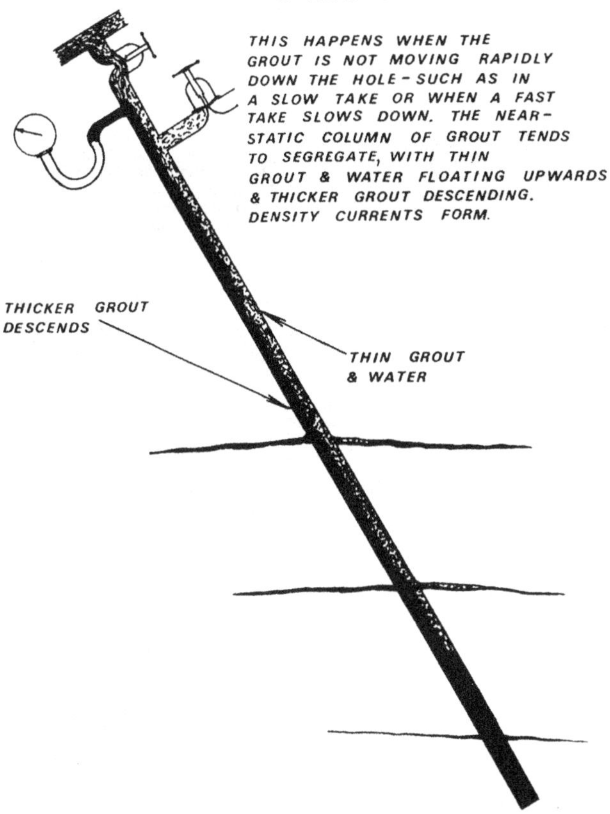

Figure 4.3-2A. Bleed water rising in a grout hole.

4.3-3 Bleeding a Grout Hole

Standpipe fittings should contain a bleeder valve, as discussed in Sections 2.6-2 and 2.6-5. This should be on the high side of the fittings on inclined holes to enable the water and thin grout to float out. The bleeder valve is very prone to blocking and should therefore always be of the diaphragm type (see Section 2.5-3) so that blockages can be freed by a few turns to and fro of the spindle. The steps in bleeding a hole are shown in Figure 4.3-3A. This bleeding is most effective with the thicker grouts. It is practical for grouts as thin as 3 : 1 but is probably of little benefit for grouts thinner than about 5 : 1 because such thin grouts are, in their fresh condition, not much denser than the discarded bleed products from the thick grouts. Furthermore it is difficult to visually distinguish between these grouts and their bleed water in the spray emerging from the bleeder valve. Such identification is needed when deciding when sufficient bleeding has been done.

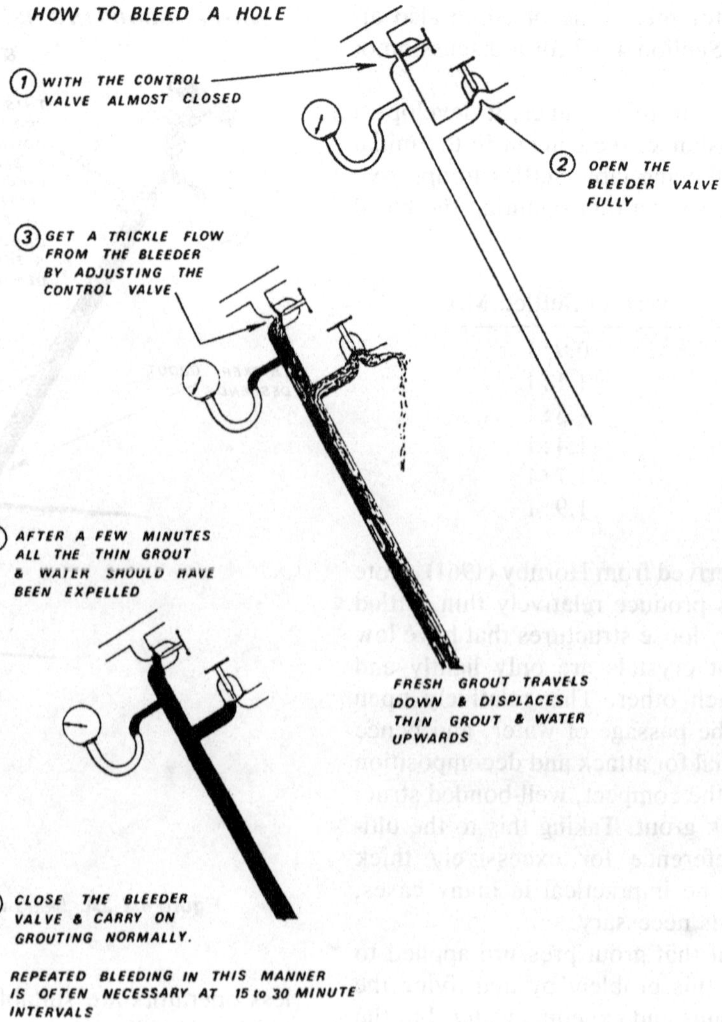

Figure 4.3-3A. *How to bleed a grout hole.*

Mixes of 3 : 1 and thicker can be individually identified visually (with practice), but mixes thinner than this can be difficult to distinguish: they have been scathingly and perhaps justifiably described as just "dirty water."

A simple way to check whether bleeding is being carried out properly on a job is to examine grout holes a day or so after they have been grouted. They should have been filled with grout right to the top; no topping up should be necessary, unless, of course, packers have been used down the holes.

4.3-4 Is It Really Necessary to Bleed Holes?

It can well be asked whether there is a real need to pay all this attention to the grout hole itself when the purpose of the grouting is, after all, to deal with the foundation, not just merely with the hole. However, the hole is part of the grout supply system and must be able to provide the designated w : c ratio to all the cracks it intercepts. If it is not properly bled, the situation could develop where upper cracks have only weak, watery grout available to them instead of the intended thicker mix.

Even with adequate bleeding, there is a tendency for cement to settle in the bottom of the hole and produce a density there that is greater than that of the mix supplied. If cracks in this region are too fine to accept this denser mix, grouting may be impaired. This is one of the factors to remember when deciding on stage lengths (refer to Section 5.4-2) and mixes (refer to Section 7.2-2 and 7.2-7).

4.3-5 Tests Investigating Bleed-Water Accumulation in Holes

On some sites where the need for bleeding of grout holes has not been appreciated, poor grouting has been experienced in the top few feet of curtains—this is, of course, the zone where the best grouting is needed and deficiencies here are particularly serious. Several instances of this led to an investigation in the early 1970s that included a series of tests simulating bleeding behavior in grout holes. The tests were described in a private communication to me from the late Charlie G. Flagg, lately of the U.S. Bureau of Reclamation (USBR) and formerly of the Corps of Engineers; fuller details are in Houlsby (1985).

The simulation of grout holes used 2.5-in ⟨63.5-mm⟩ diameter plastic pipe, some of it transparent, mounted vertically. Grout was injected through a tremie tube and therefore differed in this respect from actual holes, but once in place, the grout could be regarded as a reasonable simulation. Mixing was in a paddle mixer and therefore was likely to be of lesser quality than on some jobs (see Section 2.2-4) and more liable to show greater bleeding and segregation than with high-quality grout.

The first testing simulated a hole 32 ft ⟨9.8 m⟩ high with the lower 12 ft ⟨3.7 m⟩ being clear acrylic (see Figure 4.3-5A). Within 30 minutes of placement, some settlement and segregation could be seen in the trans-

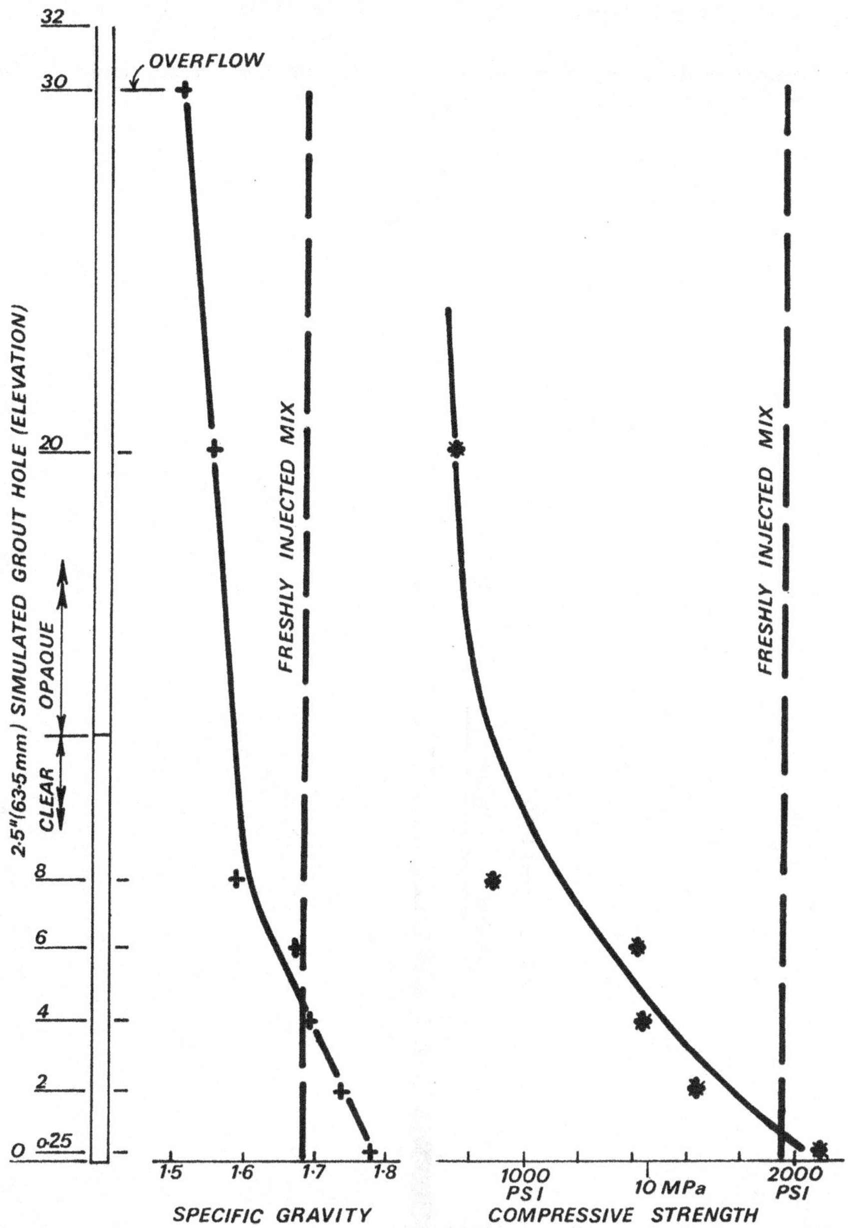

Figure 4.3-5A. *Bleeding tests in simulated grout hole. Specific gravity and compressive strength decrease in upper part of hole and increase in foot of hole, compared with freshly injected mix.*

parent section, and differences in appearance of the grout had become obvious. About 15 minutes later samples were drawn off and gave the specific gravity and compressive strength results shown in the graphs in Figure 4.3-5A. Notice how both these properties were lower than for the grout as originally injected in most of the upper part of the hole, as the result of the bleeding. The opposite happened at the bottom of the hole, where cement settled and produced a denser grout than first injected.

After several hours, voids in the grout could be seen in the transparent section; these resulted from a combination of arching across the pipe and settlement below the arches. Typical shapes are shown in Figure 4.3-5B.

In addition to these visible voids, there were similar voids in the opaque section. Voids in columns of grout have been experienced in other forms of grouting, particularly with posttensioning work; Houlsby (1981) gives examples. Means for minimizing voids in this context by using special cement or by adding superplasticizer are the subject of Houlsby (1988).

These tests prompted further ones in inclined pipes. These used 30-ft ⟨9.1-m⟩ pipes set up at 30° off the vertical. A 3 : 1 grout was used in one pipe and a 1 : 1 grout in the other. The grout was left undisturbed for 14 days and the tubes were then cut open longitudinally. They showed the following conditions:

- Voids were not as concentrated or as large as in the vertical tests.
- Numerous small voids or bubbles were spread

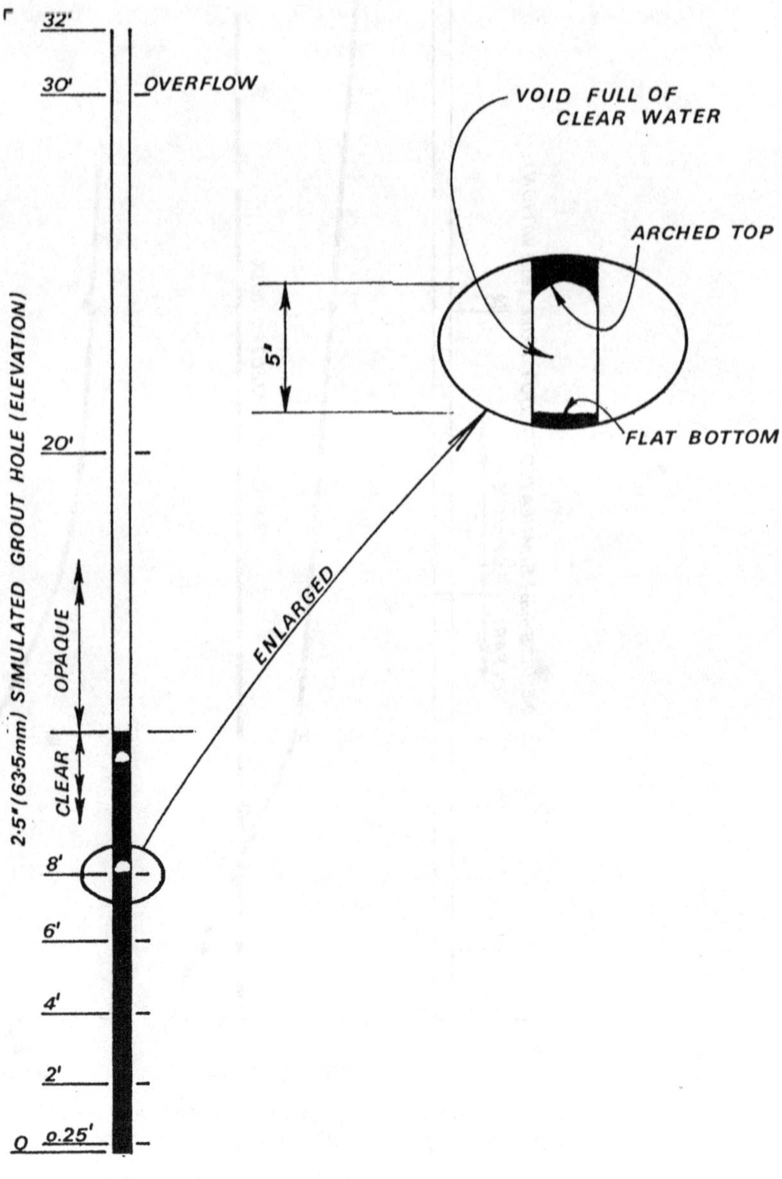

Figure 4.3-5B. *Voids formed in grout columns as the result of arching and settlement in simulated grout hole.*

along the upper side of the slope. These were more numerous in the 3 : 1 grout and were up to 0.75 in ⟨20 mm⟩ diameter.

- The quality of the grout varied in cross section. Channels on the upper side of the slope contained the lower density grout.
- Variations in the color of the grout indicated appreciable density current formation and segregation.

The findings of these various tests reflect the experiences of quite a number of grouting people and of myself.

The need for bleeding off from grout holes is not directly obvious and, consequently, has not been done on a number of jobs where it should have been. There has been confusion in understanding the effects of bleeding. This is illustrated by some comments by Cambefort (1977). He shows a photograph of grout from a hole with voids in it and with distinct streaking and density segregation. Comments accompanying the photograph suggest that the hole can gradually become blocked because of relatively quiet conditions in it, and that the voids are part of channels that remain active because the concentrated flow in them is forced to be turbulent. The comments then go on to suggest that these channels may become blocked before the cracks are properly grouted; remedial action suggested is to drill the grout out and start again. This line of reasoning fails to appreciate the rheology of grout in vertical holes (for extensive testing on this see Houlsby, 1988) and the benefits obtainable by regular bleeding off at the standpipe.

4.4 SETTING OF GROUT

4.4-1 Setting Times

When cement and water are mixed, chemical reactions start immediately and continue fairly actively for the next few hours and then with increasingly less activity for over a year. See Section 4.4-3.

The reactions should eventually produce setting and hardness. Setting can be thought of in terms of (a) initial set, where the grout firms up, but is not really hard, and (b) final set where it becomes hard.

The minimum w : c ratio able to provide the quantity of water needed for the cement to fully react ("hydrate") is around 0.2 : 1 by weight. This is 0.3 : 1 by volume and is too thick to be a practical grout. The thickest mix able to be handled by the best mixers is 0.5 : 1 (see Section 2.2-2), and this particular grout has limited application (see Sections 6.1 and 7.3). All the practical grouts have much more water than is needed merely for the chemical action. This excess water slows down the setting.

For instance, compare the ASTM Type I specified initial set time of 0.75 hours and final set of 10 hours (at paste consistency), with the graph in Figure 4.4-1A. This graph is very approximate indeed. It is based on Hornby (1961) and others and shows how very slowly the thin grouts set. Even 3 : 1, a grout recommended as the thinnest desirable for durability (Section 4.5-3) can take 16 hours for initial set. Firm set takes much longer. This is one of the reasons for using the thickest grout practicable.

In the case of the thinner grouts, setting time is critical if the grouting is being done in a foundation where groundwater is flowing. These grouts do not have the benefit of thixotropic stiffening (Section 4.2-2) and can be washed away if set is too slow. This has happened at at least one dam where 8 : 1 grout was unwisely used without realization of its slow set; all of the carefully placed grout was washed away in a matter of days. Sadly, this was not recognized until a great deal of grouting had been done.

However, it is not only those flowing water conditions that can be troublesome if grout is slow to set. A more common situation is disturbance from water originating from nearby drilling, water testing, and grouting operations; such water is liable to remove or damage grout that is still soft or perhaps even still fluid.

4.4-2 Stiffening

As part of the tests described in Section 4.3-5, measurements were made to get an idea of the rate at which the grout stiffened (see Figure 4.4-2A). This was done indirectly by measuring the drop off in hydrostatic pressure.

The first reading was taken an hour after placement and showed appreciable stiffening. This probably had a thixotropic component, but subsequent further stiffening over the next 3 hours had not.

This is just one example of the fairly self-evident axiom that the grout should be put into the cracks fast (and thick), as enlarged on in Section 6.1.

4.4-3 Chemistry of Setting—An Outline

The chemical reactions of cement hydration are complex and appear to be not fully understood by researchers. This is not the place for a dissertation on the subject. However, an outline might be of interest.

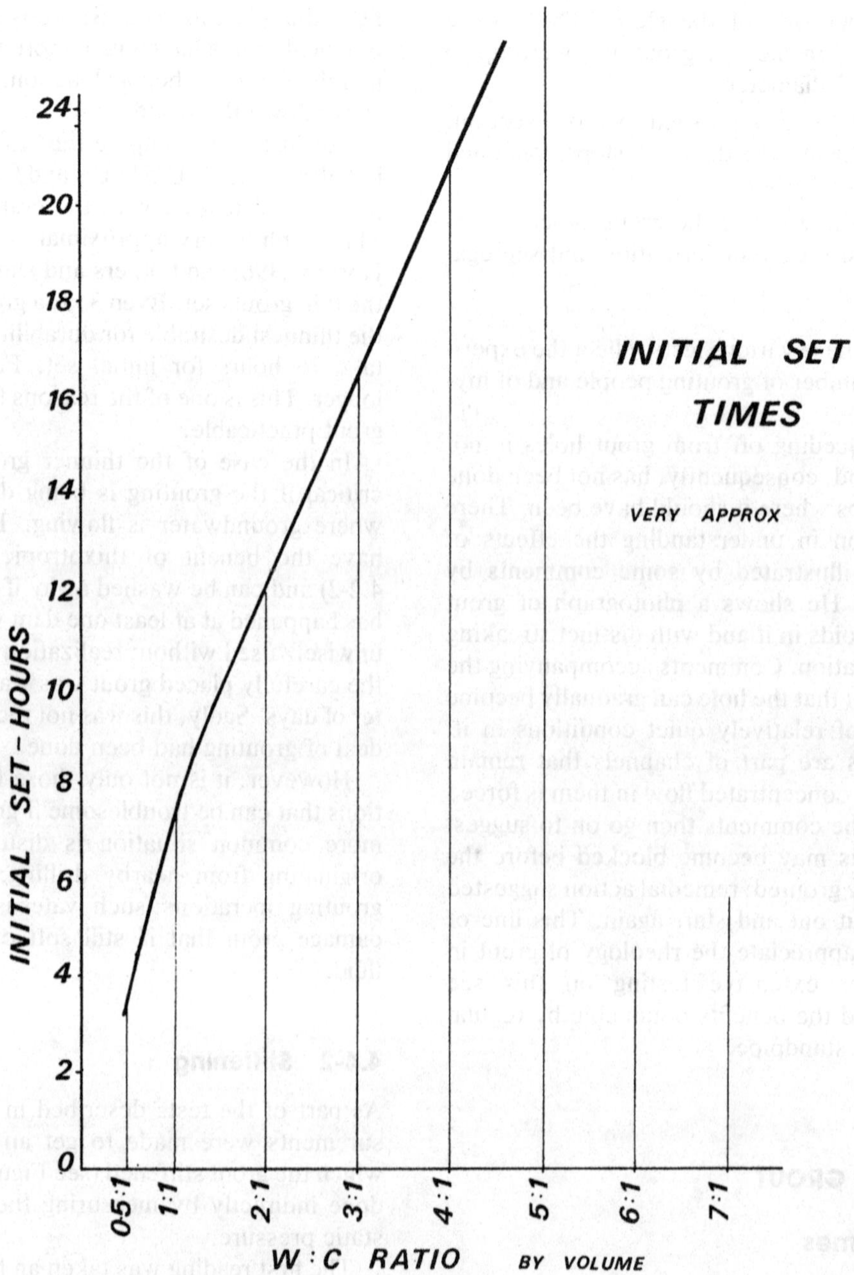

Figure 4.4-1A. *Effect of water content on initial set time.*

The subject is touched on in Sections 2.2-10 and 2.2-12 in relation to the effects of high-shear mixing and the manner in which it breaks off crystals.

Braun (1978) gives a description of cement reactions and crystal formation and summarizes the work of some other authors. He quotes Double and Hellawell:

> The individual grains of Portland cement become hydrated by a two-stage process. First a shell of watery, gel-like material is formed around the cement grain. Then fibre-like structures grow out from this. The spikes consist of hydrated calcium silicate. . . . As water gradually penetrates the anhydrous core of the cement grain, calcium ions leak out. Finally, when osmotic pressure ruptures the membrane, the fiber-like calcium silicate hydrate crystals commence to grow. (p. 30.)

These interlock to give strength to the hardened grout. These researchers considered "that tricalcium silicate crystals contribute to early strength while the compar-

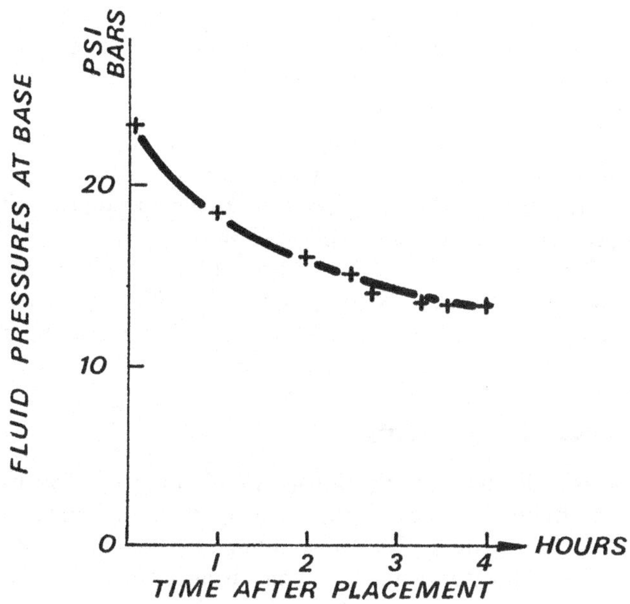

Figure 4.4-2A. *Test measurement of grout stiffening rate.*

atively slow reacting dicalcium silicate provides the gradual and continuous long term increase in strength." (p. 30, 31.)

Braun, in discussing grouts, speculates that the skeleton of interlocking crystals may grow better in a thixotropic grout than in the confinements of concrete. He suggests that cement reacts more actively in grout than in concrete.

When grout is agitated and kept moving around a circulation line (Section 2.7-3), the crystals tend to be broken off while they are trying to grow. Given that the potential for cement grains to continually replace the lost crystals is limited, the strength and durability of grout are gradually reduced the longer it is circulated. This is one of the reasons for insisting on fresh grout in quality work: grout should be discarded if it has been circulating for 2 hours (less time in hot weather); lines should be as short as possible (Section 2.7-12).

The aim is to have as much as possible of the crystal growth occurring when the grout has reached its final position in the cracks. Hence the aim is to get the grout in as fast as possible commensurate with practical constraints.

Crystal interlocking between the grout and the walls of cracks would be very desirable, but in reality rarely occurs to any great extent. This is because the passage of groundwater through crack systems has nearly always left a film of dirt or weathered material on the walls of cracks. Grout will not stick to this. Grout is not an adhesive—it will not stick the foundation together. It is mainly a filler.

4.4-4 Permitted Grout Temperatures

The chemical reactions of setting cement are affected by temperature. High temperatures accelerate setting, low ones can slow and even stop the reactions. Therefore, grout should not be allowed to get too hot or too cold. The exact limiting temperatures are not precisely laid down.

HIGH TEMPERATURES In hot climates, and particularly where circulation lines lie over bare rock in full sun, the grout can gain considerable heat traveling through the lines. Precautions can be taken, such as covering the lines (Section 2.7-3), but, besides being a nuisance, covering is not very practical over those parts of the lines that are moved from hole to hole, every hour or two.

If sufficient quantities of ice are available, batching water may be cooled with it, but the ice itself must not be put in mixers. Even so, the benefits of this are quickly lost if circulation lines are long. This is yet another reason for having the pumps close to the holes (as per the argument of Section 2.7-12).

As a guide to limiting temperatures, a desirable but not absolute limit is about 80°F ⟨27°C⟩. Grout above 100°F ⟨38°C⟩ should be discarded unless it can be injected within a few minutes of mixing. Although it may be reacting fiercely in the line, once it gets underground in cooler conditions the reactions should slow down. Any grout returned in the circulation line may have to be wasted.

LOW TEMPERATURES There are two aspects to be considered here: the slow chemical reaction, and frozen machinery. On sites regularly covered with ice or snow, large sheds have sometimes been built to house not only the machinery, but the grout holes and lines as well. This is expensive, but about the only way to ensure continuous production in such conditions.

It is best for the rock temperature and grout to be above 40°F ⟨5°C⟩. In temperate climates, where early morning temperatures can be about 40°F, the exhausts of air motors can freeze up rapidly if open to the atmosphere, and this can stall them. Exhaust heating and grout water heating are advisable then, perhaps using such crude methods as a brazier.

4.5 DURABILITY

4.5-1 Doesn't Grouting Last Forever?

The answer is *no* in many cases—regrettably. It depends on how well the grouting has been done. Poor grouting can be leached away by seepage or attacked chemically or biologically or weakened by erosion of soft materials around it. Investigation of case histories has indicated the principal deficiencies responsible:

Grout that is thinner than it should be (see Sections 4.5-4 and 4.5-5)
Grout mixed poorly (see Section 2.2-10)
Inappropriate injection techniques (see Section 7.2)

4.5-2 How Permanent Is Grouting?

Grout curtains at dams are probably the most likely form of grouting to deteriorate if the grouting has not been done properly. Although it can be difficult to monitor them convincingly, many have shown no indications of deterioration over many years, and it is reasonable to presume that this situation will continue indefinitely—say for a century or two. Silting of the storage floor can eventually make a grout curtain redundant anyway in the lower parts of many dams. This may possibly be not the case in high parts of storages having steep abutments where wave action may stop the accumulation of silt.

At the other extreme to this are cement grout curtains that have washed away within weeks of placement. And there are others where the grouting has disintegrated over a few years.

A common component of seepage in the lower part of dams is hydrogen sulfide (rotten egg gas). It is strongly aggressive to cement grout. The surest defense is to do a tight grout job so seepage can find virtually no paths through the grout and hence minimal surfaces to attack. Some jobs use sulfate-resisting cements (see Sections 3.1-2 and 3.1-3).

Coal and associated beds can be a threat to poor grouting. D'Appolonia, Shaw, and Erel (1982) report a case where the grout curtain at an old dam deteriorated and had to be replaced because seepage reacted with decomposing coal blossoms in the foundation and formed an acid that dissolved the grout curtain.

Refer to the tests discussed in Section 6.1-6 that indicate the manner in which grout flowing in cracks can leave behind bleed passages and sinuosities through which aggressive seepage and bacteria can travel and attack. Good grouting minimizes these.

4.5-3 The Key to Permanency

The aim is to fill all cracks completely. It may sound easy to get 100% filling, but it isn't necessarily so. Bleed passages, voids, and weak grout can occur all too easily if care is not taken.

This book is a guide for producing high-quality grouting but, as emphasized frequently, the suggestions given must be augmented by intelligent adaption to site conditions. Even then no guarantee of permanency automatically follows, but the chances of achieving it are better.

The deficiencies leading to lack of durability, listed in Section 4.5-1, are dealt with in appropriate sections of the book. Their effects are, however, discussed here; most arise from excess water.

On this score, 3 : 1 is a reasonable limit in the author's experience. See Section 4.5-5. Not everyone agrees, but it appears to be the cutoff line in a number of cases between durable and suspect grouting.

4.5-4 Excess Water

This is discussed in Section 4.3 in relation to the harmful effects of excess water, and Section 6.1 indicates how thin grouts give weak filling. Bleeding off thin grout and water from grout holes is necessary with the better grouts, and use of packers should be avoided (Section 2.5-16) because of the way they hinder this bleeding.

Figure 4.5-4A shows typical incomplete grout films as viewed through a borehole TV or periscope. They are actual cases. The grouting was done with 5 : 1 grout. Voids have been permanently left by the bleed water.

4.5-5 How Thin Is It Safe to Go?

If durable grouting is wanted, just how thin can we go before the grout becomes of doubtful permanency? Various people answer this differently. For instance:

- 3 : 1 is the thinnest mix recommended in this book, except in special circumstances (refer to Section 7.2-2).
- 3 : 1 or 4 : 1 (with the bulk of grouting done at 1 : 1) was U.S. Army Corps of Engineers practice as reported by Burwell (1958).
- 12 : 1 or even thinner ("dirty water!") used to be the practice of the USBR. After considerable investigation, current practice is 5 : 1 with this organization.
- There are many other opinions on the matter.

My extensive trials of all sorts of mixes and conditions over many years have shown that 3 : 1 is quite thin enough for almost all situations and is a reasonable compromise between penetration and durability.

Examples are given in Houlsby (1985) where jobs done with grout thinner than 3 : 1 have not proved durable or are suspect. Grouting suffers from the difficulty that the completed work can rarely be actually

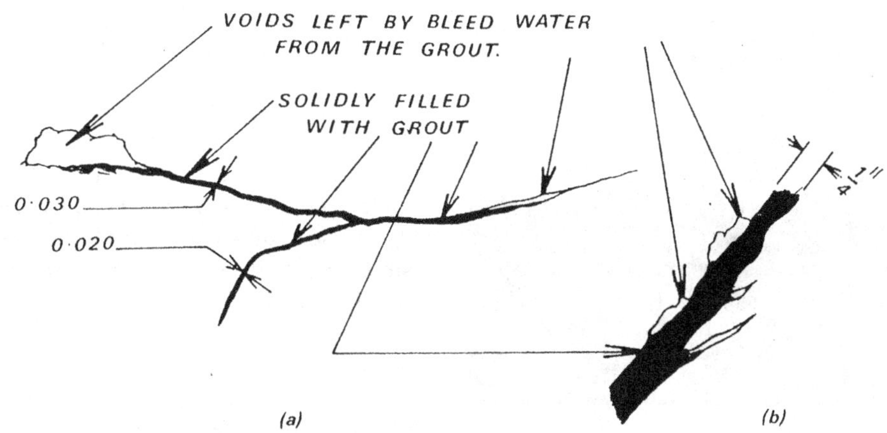

Figure 4.5-4A. *Bleed voids left in grouted cracks.*

seen. Evaluation of its success or otherwise is dependent on indirect and perhaps inconclusive performance monitoring. The examples given in Houlsby (1985) are therefore grouped as far as possible as "successful" or as "not as successful as was intended." There is inability to give definite success or failure verdicts usually, however, some have been obviously nondurable and are outlined in the next section. For additional information on these and for the "successful" examples see this paper.

4.5-6 Examples of Nondurable Grouting

BURRINJUCK DAM The Burrinjuck Dam is an old concrete gravity dam that leaked so much through the concrete that it was grouted over the period 1936 to 1948. The arrangement is shown in Figure 4.5-6A.

The grout curtain went down from the crest, through the concrete, and into the foundation. The concrete was of poor quality, and cracks in it took most of the grout. In accordance with the thinking at the time, grouts were very thin:

30 : 1 to 5.5 : 1. Average was 12 : 1.

Mixing was by paddle mixer, and injection methods were primitive.

The foundation tunnel shown in Figure 4.5-6A was excavated by nonexplosive methods after the grouting. It is only 3 ft ⟨1 m⟩ from the grouting and over the years has developed the sort of formations seen in limestone caves: stalagtites and shawls. The rock is pink granite. The source of the formations is obviously the grout curtain; seepage has leached the poor-quality grout by taking calcium into solution and then depositing it where it reaches air (see Figure 4.5-6B). A very intense pressure gradient amplifies this tendency. It is estimated that the volume of the formations is now about equal to the volume of cement put into the curtain, so there is no curtain left. Fortunately this does not matter to the safety of the dam because the rock was really too tight to need grouting in the first place;

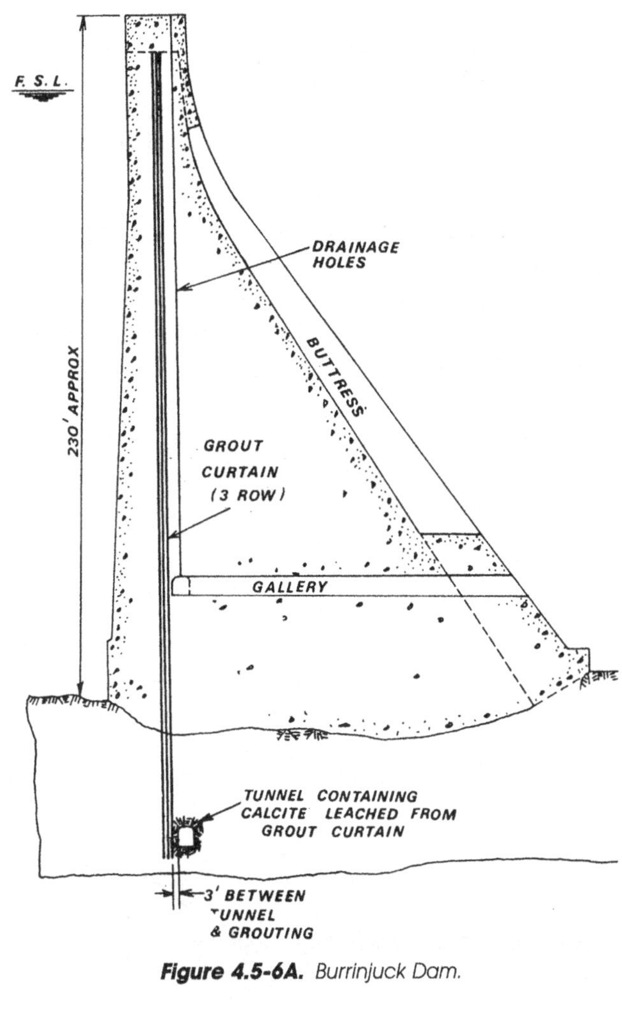

Figure 4.5-6A. *Burrinjuck Dam.*

Figure 4.5-6B. *Calcite formations in pink granite. Roof of tunnel, Burrinjuck Dam. Calcite is from washed-out grout curtain.*

extensive instrumentation confirms that there are no threats to the stability of the dam because of the loss of the grouting.

An additional attack on the grouting has come from biological action. Dark brown and black algae and apparent associated growths feeding on calcite can be seen in the tunnel and appear to penetrate back in the cracks. This type of growth is fairly common in wet galleries within gravity dams and in drain holes—which it can block at times. By feeding on striated and porous grouting, calcium is removed and grouting destroyed.

BLOWERING DAM The grout curtain at Blowering Dam, an earthfill dam, extends along a thin ridge forming one of the abutments. Much of this ridge curtain has slowly disintegrated. Consider the section shown in Figure 4.5-6C.

Seepage through the ridge is measured by a network of water table holes. The section in Figure 4.5-6C shows a typical situation where the water table has risen (in terms of a full storage) between 1973 and 1981. It has subsequently risen further at this section, and at some other places along the ridge it has completely gone. The mix was 5 : 1 mostly. Paddle mixers were used, and injection methods left a lot to be desired. Note, however, that some of the curtain under the dam itself has proved satisfactory, as discussed in Section 8.4-15 dealing with pressure profiles across curtains.

SOVIET DAMS Measurements of leaching from grout curtains at three dams in the Soviet Union have been reported by Petrovsky (1982).

From chemical analysis of seepage passing through the grout curtains, Petrovsky has deduced that loss of CaO from the grout amounted to 30 tons (30,297 kg) in the first 4 years at the worst case (dam 2). A total of about 25,000 tons (25,000 t) of cement had been used in the grouting, so the loss amounts to 0.1%, which is slight enough if the leaching is not concentrated in small areas. However, there can be doubt that it was not. The grouting was done with 5 : 1 to 1 : 1 grout.

4.6 INSTRUMENTS FOR TESTING GROUT

4.6-1 Is Testing Really Necessary?

Testing is not necessary in most rock grouting jobs. If cement grout is used without additives or bulking materials, and the mixers are the high-quality type detailed in Section 2.2-2 and maintained well, their prod-

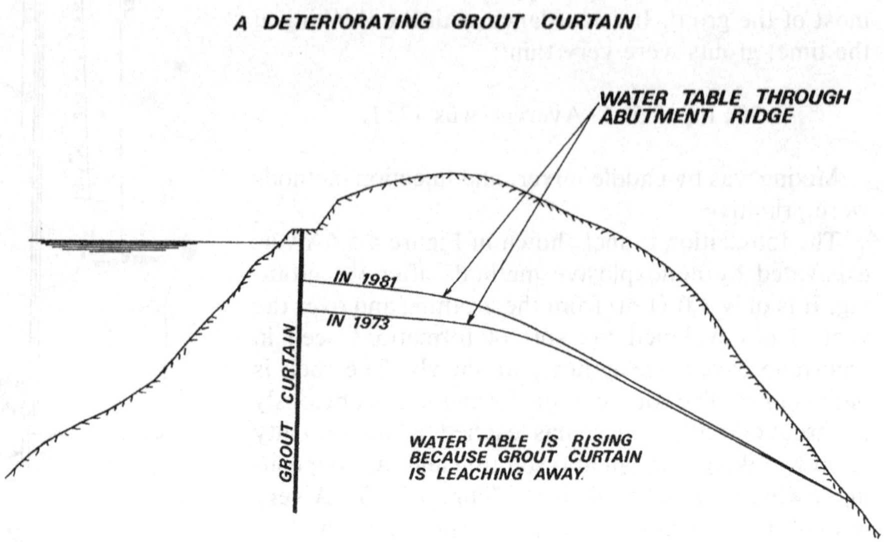

Figure 4.5-6C. *Blowering Dam abutment ridge.*

uct is sufficiently uniform to avoid the need for checks, bearing in mind the somewhat imprecise nature of this type of grouting. However, as mentioned in Section 4.2-1, there are indeed some other types of precision grouting that do need testing as a control on mixer output.

The information given here on instruments may be useful for contract administration purposes if a contractor insists on using poor equipment. As suggested in Section 9.2-8, specifications should not permit use of substandard gear, but sometimes it happens. Regular checks on the grout may help to keep the job on course, more or less.

Other uses for the instruments are laboratory research and on jobs where bentonite and other additives or bulking materials (Section 3.3) are used. Only the basic tests and instruments are described. Even then, some are almost outside the scope of this book.

4.6-2 Scope of Tests

The most useful site control test for cement grouts and additives is the mud balance. This is a simple, reliable test and is far better than any others. Next in usefulness, but a long way behind, is the flow cone.

To round off the description of site tests, the mixer test, mentioned in Section 2.2-9, is enlarged on in Section, 4.6-5. For research and laboratory work, bleeding tests can be useful (Section 4.6-7) and so can thixotropy and viscosity tests.

Further tests used for other forms of grouting but not described here because of almost no relevance to rock grouting are compressive strength, shrinkage, and water retention.

4.6-3 Mud Balance

The mud balance is an item taken from the driller's armory, where it tests drilling mud density. The Baroid mud balance (Figure 4.6-3A) is the type commonly used. It is basically just a simple calibrated weighing scale.

The grout is put into the cup and allowed to overflow. Then the lid, which is heavy and has a hole through it, is wriggled down onto its seating on the cup. Excess grout escapes out the hole, and the outside of the cup and top are wiped clean. This means that a known volume of grout is in the cup. The balance beam is then leveled by use of the sliding weight. A spirit level indicates when the level condition is reached.

Specific gravity (SG) is read off the calibrated beam against a mark on the weight. The beam has additional markings relating to mud, which are not applicable to grout except when it is specified by the driller's "weight" method (see Section 4.1-1).

Readings can be compared with the theoretical SG shown on the plot in Figure 4.6-3B or in the following table. The curve is calculated on the basis of the SG of cement, that is, 3.15. It is included here to indicate the sensitivity of SG to changes in w : c ratio but may be difficult to read off, so the same data is now shown in table form:

w : c (by volume)	SG	w : c (by weight)	SG
0.5 : 1	2.11	0.3 : 1	2.11
0.6 : 1	1.95	0.4 : 1	1.95
0.8 : 1	1.84	0.5 : 1	1.84
1 : 1	1.67	0.8 : 1	1.61
2 : 1	1.42	1 : 1	1.52
3 : 1	1.29	2 : 1	1.29
4 : 1	1.23	3 : 1	1.21
5 : 1	1.19		

Note that this table was calculated for w : c ratios by weight and thence for w : c ratios by volume, derived in accordance with the conversion factors listed in Appendix D. Accordingly, it involves assumptions about the loose volume of cement, as discussed in Section 4.1-4.

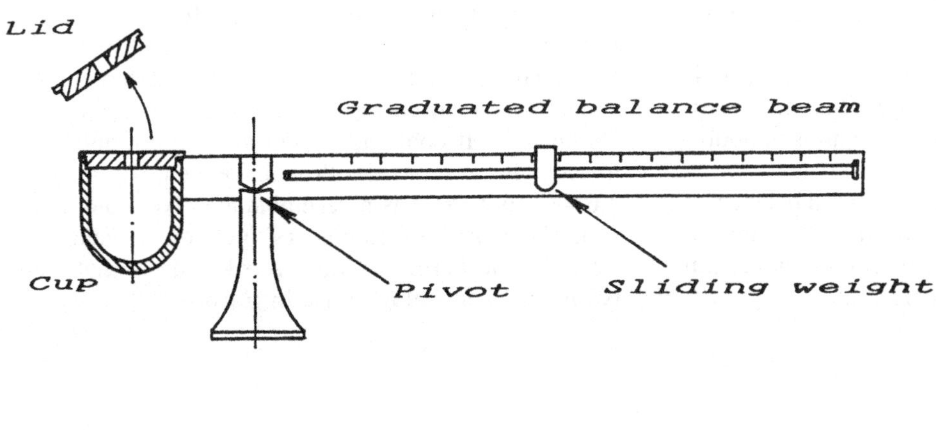

Figure 4.6-3A. Baroid mud balance.

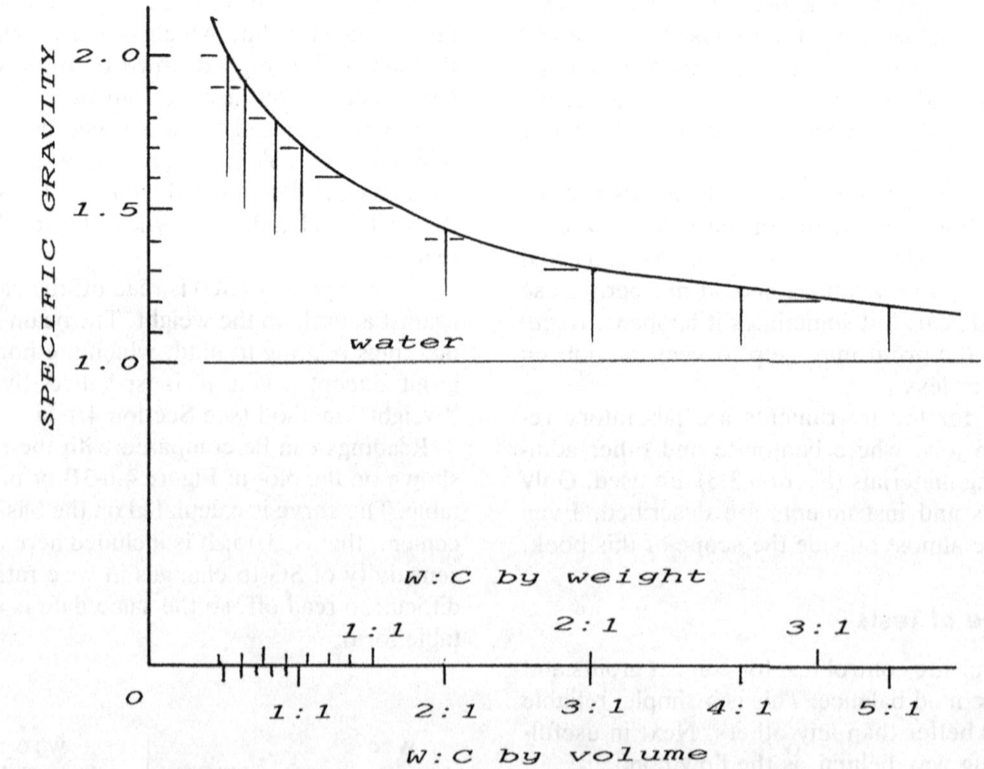

Figure 4.6-3B. *Theoretical specific gravity for cement grout (no additives).*

The mud balance test is particularly useful for checking mixer output for compliance with the theoretical or with some specified criterion.

Readings can be easily taken to an accuracy of 0.02 and are usually to 0.01. This accuracy means that the test can clearly distinguish between different w : c ratios; it can pick up incorrect measurement of ingredients and is very sensitive to the addition of any additives.

It can also detect bad mixing. For instance, in a series of tests examining mixing efficiencies in a Cemix mixer (these machines are described in Section 2.2-2), some were deliberately undermixed. The SG values for them all plot below the theoretical curve, some by as much as 0.03. Properly mixed grouts plot right on the curve or very slightly above it.

The test takes less than a minute, and the gear is tough and quite suited to on-site use. It is more positive and versatile than the flow cones, and is undoubtedly the best test for field use.

Baroid is a frequently-used brand of mud balance. It is better described as an implement than as an instrument in view of its massive construction—more akin to a hammer than a delicate balance.

4.6-4 Flow Cones

Flow cones are essentially a funnel. The time for a measured quantity of grout to flow through is measured. Unfortunately there is lack of standardization. Some common flow cones are shown in Figure 4.6-4A.

Of these, the U.S. Army Corps of Engineers' cone, subsequently used in some ASTM specifications, is probably the most popular for thick grouts in the United States. The Marsh cone remains the driller's favorite for muds; its use for grout follows from that. Lombardi (1985) names the Mecasol, Prepakt, and Marsh cones as those in use for grout in Europe.

Did the U.S. Army Corps of Engineers devise the cone known by their name? Well, it seems they didn't. They used it, among other things, for the extensive tests of grouting in fine cracks discussed in Section 6.1-6. However, in the report is the statement that "the flow cone used was one developed by the Prepakt Company." It contained 1,725 ml of grout and is not exactly the same as the Prepakt one shown in Figure 4.6-4A—but there is a resemblance. To avoid confusion the Corps of Engineers' version is called the ASTM cone here, and the Prepakt one is that Lombardi shows as being in use in Europe.

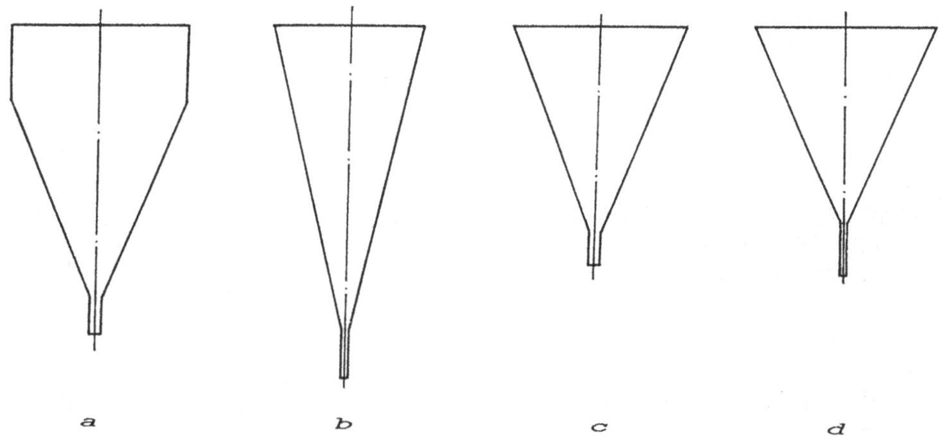

Figure 4.6-4A. *A collection of flow cones: (a) U.S. Army Corps of Engineers, ASTM type, (b) Marsh cone, (c) Prepakt cone, (d) Mecasol cone.*

ASTM FLOW CONE The size and condition of the discharge tube at the bottom are key factors in obtaining comparable readings in the ASTM flow cone (see Figure 4.6-4B). In specifications such as ASTM Designation C939 (''Flow of Grout for Preplaced Aggregate Concrete''), the cone is required to be made of stainless steel with an internal diameter of the dimensions as show in Figure 4.6-4B. The rest of the instrument may be made of the same material or of any essentially noncorroding material. The cone must be held steady in some supporting frame. Checking it is done by discharging 1,725 ml of water, which should take 8 seconds.

The test procedure is as follows (see Figure 4.6-4C):

1. Wet the cone by filling it with water and let this discharge 1 minute before testing the grout.
2. Block the discharge tube with a finger.
3. Put grout into the cone until it fills to the tip of the pointer, set at 1,725 ml.
4. Simultaneously start a stopwatch and remove the finger.
5. Stop the watch at the first break in continuous flow from the discharge tube.

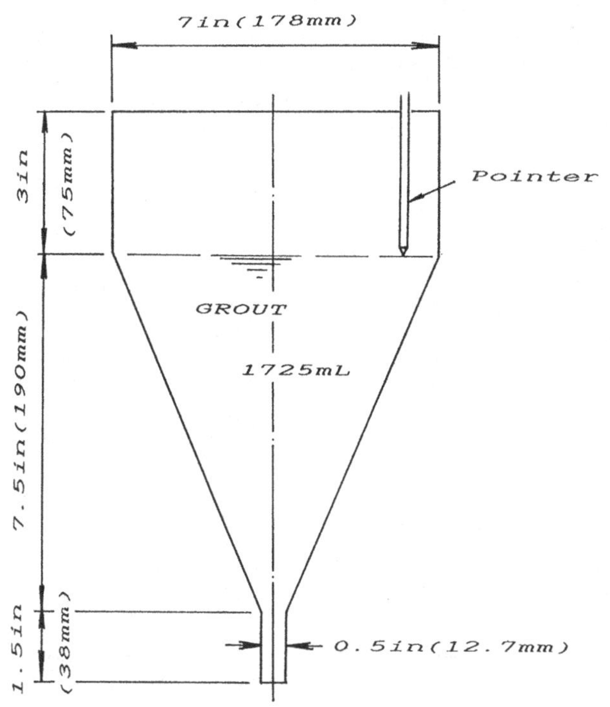

Figure 4.6-4B. *Dimensions of the ASTM flow cone.*

Figure 4.6-4C. *Conducting flow test in ASTM cone.*

For the w : c ratios used in rock grouting there should be no need for the further step required in the ASTM procedure of looking down into the cone to see whether light can then be seen through the tube. This is for very thick grouts with discharge times in the vicinity of 35 seconds or more, which might not clear through the discharge tube; the test is not suitable for such grouts. Any lumps in the grout can cause invalid results. The cone must be thoroughly washed out between each test.

One of the problems in standardizing flow cones is that this size of discharge tube loses its sensitivity if the grout is thinner than about 0.9 : 1 by volume. As shown later, flow times are then in the vicinity of 9 seconds—not much more than water. Thinner and perhaps longer, discharge tubes extend the range slightly. Some variations of this type are shown in Figure 4.6-4D.

The standard ASTM tube is shown on the left for comparison. The alternative Corps of Engineers' one on the right is from the testing on fine cracks.

The Brazil tube is reported in Deere (1982) as a development in that country, having a discharge tube intermediate in size between the two Corps of Engineers ones. It is primarily for testing grouts containing bentonite apparently (see Section 3.3-6 for my recommendations adverse to the use of bentonite).

MARSH FLOW CONE No large drilling rig is complete without one—or so it seems. Often made of plastic and with a handle on the side for holding while the test is in progress, their dimensions are shown in Figure 4.6-4E. The smaller tube opening and the slender cone obviously make it more suited to the thinner grouts than the ASTM instrument. Deere (1982) prefers it for grouts in general.

The procedure is similar to that for the ASTM cone except that only part of the contents is measured in the timed discharge. Discharge goes into a cup of a specified volume, and the time to fill this is measured. There

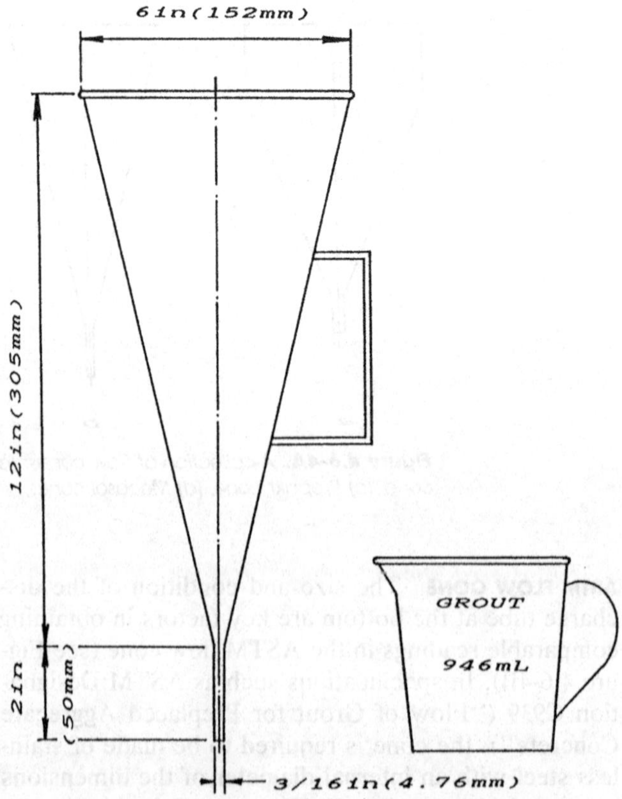

Figure 4.6-4E. *Dimensions of the Marsh flow cone.*

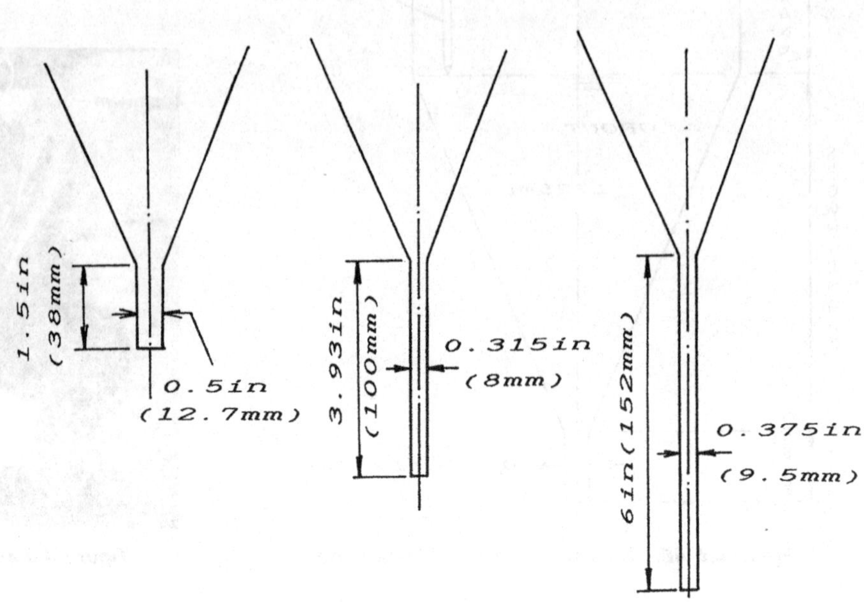

Figure 4.6-4D. *Different discharge tubes for the ASTM flow cone: Left: Usual size (as shown earlier). Center: Brazil version. Right: Alternative U.S. Army Corps of Engineers' size.*

are different practices about the measured quantity of grout. The standard U.S. practice is to use 1 quart ⟨946 ml⟩, but in some other countries this is rounded off to 1,000 ml, and sometimes a half sized test of 500 ml is used. The times of efflux of water for each of these volumes is:

1 quart ⟨946 ml⟩	26 seconds
1,000 ml	28 seconds
500 ml	14 seconds

These three times relate to the correct initial filling of 1,500 ml, which is 1 in ⟨25 mm⟩ from the top. Incorrect results will be obtained if this amount of starting mix has not been used. Deere (1982) draws attention to this and to the error of confusing results taken with 1,000 ml with those using 946 ml.

PREPAKT CONE This particular version is the one shown by Lombardi (1985), having the dimensions shown in Figure 4.6-4F. "Prepakt" is the name of one of the processes for making concrete by placing the coarse aggregate first and then intruding it with the cement and sand by grouting methods (see Section 1.2-1). This cone is for the sand–cement grouts used in the process. Accordingly, it has a relatively wide and short discharge tube having an internal diameter of 0.5 in ⟨12.7 mm⟩, as shown by Camberfort (1968) or 0.55 in ⟨14 mm⟩ as shown by Lombardi (1985).

MECASOL CONE The dimensions of the Mecasol cone are shown in Figure 4.6-4G. This is a European cone for grouts used in the same way as the others.

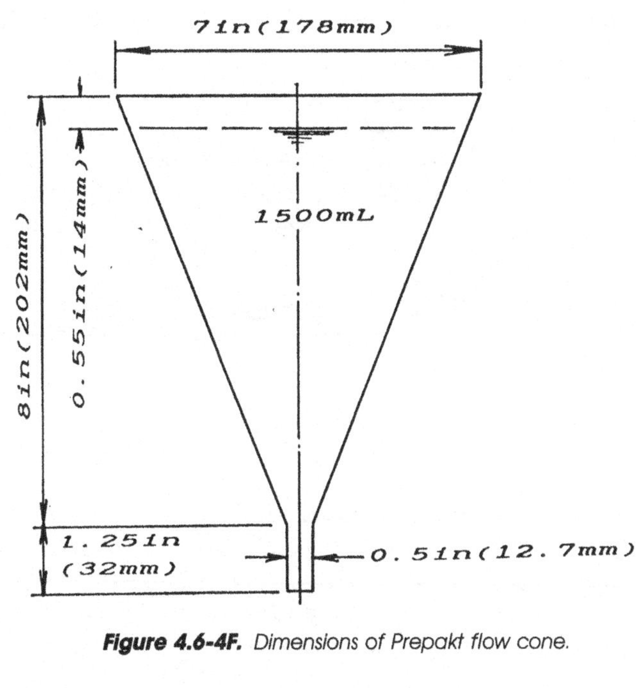

Figure 4.6-4F. Dimensions of Prepakt flow cone.

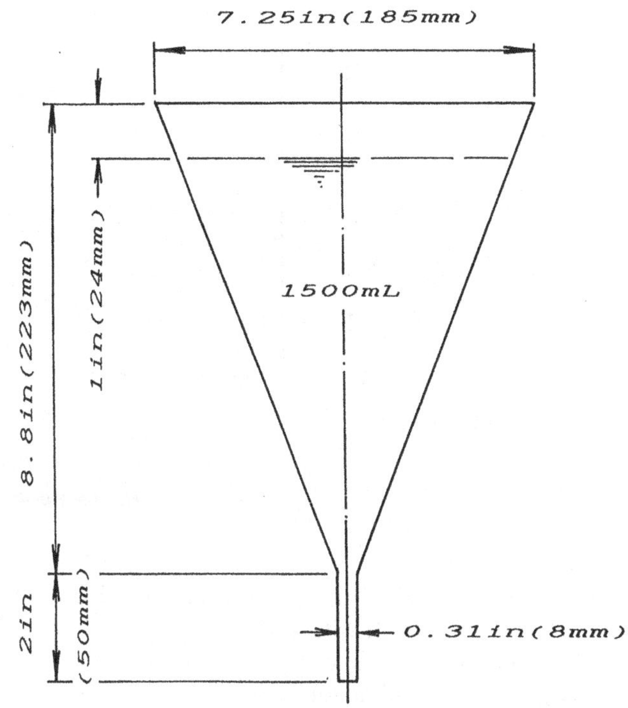

Figure 4.6-4G. Dimensions of Mecasol Cone.

RESULTS Examples of flow times obtained with the ASTM cone are shown in the plot in Figure 4.6-4H. The plot shows results by Klein and Polivka (1958) and by myself when testing a group of new and used Colcrete mixers (see Section 2.2-2 for description of these mixers). The envelope shown encompasses 18 tests. Results for the new mixers, which were straight from the factory, turned out to be intermingled among those from machines that had seen considerable field service.

The plot indicates the lack of sensitivity of this cone when grouts are thinner than about 0.9 : 1 by volume. It cannot distinguish between grout and water then. Use of one of the finer and longer discharge tubes, such as the Brazil one shown earlier, can extend this range a little.

A plot from the Marsh cone is given in Figure 4.6-4I.

Results shown come from three widely scattered parts of the world: New Zealand, United States, and Pakistan, and are perhaps indicative of different cement sources. The New Zealand data is from Clyde Dam (Foster, 1983). The U.S. results come from Deere (1982), and the Pakistan figures were reported from Tarbela Dam by Lowe and Sandford (1982).

Because the first and second set of results were carried out with 1,000 ml of grout instead of 946 ml, the very arbitrary method of subtracting 2 seconds from

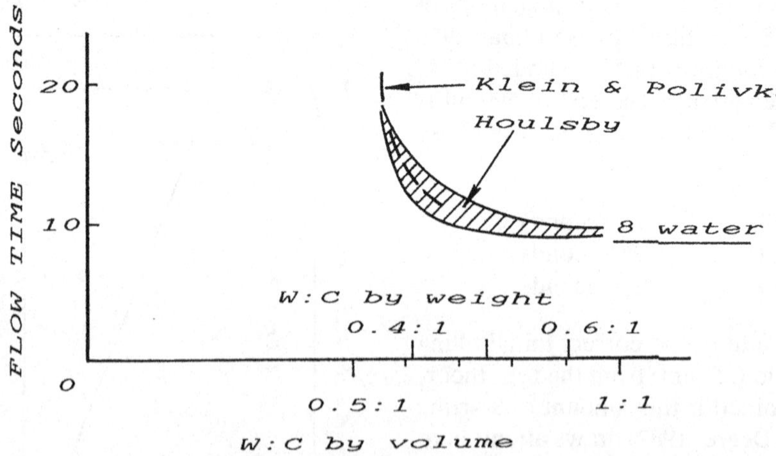

Figure 4.6-4H. *ASTM flow cone times.*

their curves has been used in the plot to give some degree of comparison with standard tests.

The Tarbela Dam curve has a different shape to the others, for some unknown reason.

The Marsh cone is useful in the range between 1:1 and about 1.5:1 by volume, but none of the cones is suitable for grouts thinner than that.

In summary then, suitability of cones could be split up as follows:

Grout w:c by Volume			
0.5:1 0.9:1	1:1	1.5:1	Thinner
ASTM (0.5 in tube)	ASTM (Brazil tube)	Marsh	Flow cones unsuitable

4.6-5 Simple Mixer Test

As indicated in Section 4.6-1, there is usually no need to go to the trouble of the mud balance and flow cone tests for checking the continued efficiency of grout mixers. A simple test for this, mentioned in Section 2.2-9 and given again here, consists of pouring some of the mixer output into a shallow tray to a depth of about an inch (25 mm). Let it go hard for a day or so, and then break it open across a vertical section to display the inside.

The section should be uniform without any streaking or banding which indicate poor mixing in which the various cement grain sizes have segregated out. Good mixing minimizes this sort of thing. Even the high-speed, high-shear mixers have been known to deteriorate to the extent that banding is found in this test;

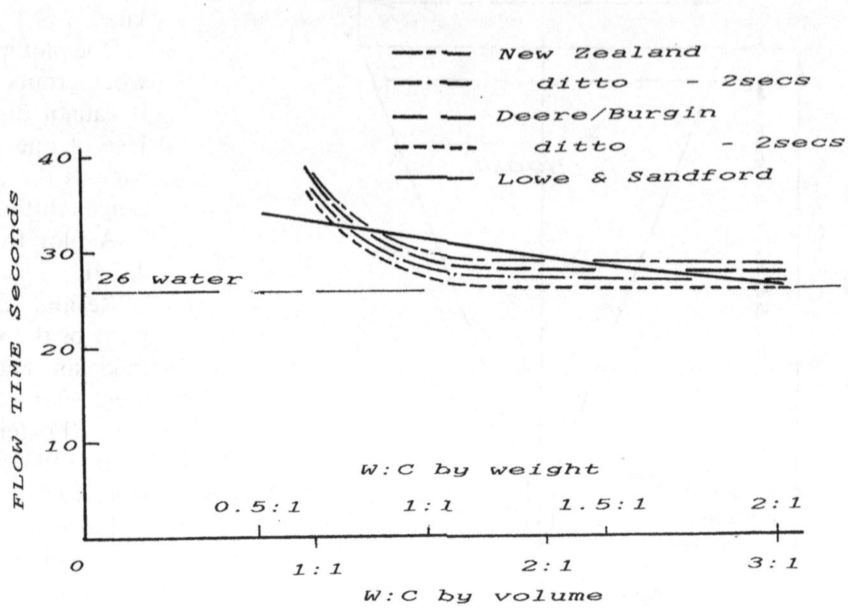

Figure 4.6-4I. *Marsh flow cone times*

prompt maintenance is necessary if this happens: perhaps a new rotor is needed or the machine has become too dirty.

Grout from efficient mixers is noticeably lighter in both color and density when set, compared with that from poorer mixers. Densities as low as 107 lb/cu ft ⟨1,710 kg/cu m⟩, and a very light gray color have been recorded from Colcrete mixers. These light densities can give good grout in place and are desirable.

4.6-6 The Gumboot Test

While thinking about simple tests, the gumboot test is one of mine, found very useful for judging the w : c ratio of grout at a hole when checking the mix in the line or checking grout coming out of the bleeder (Section 4.3-3).

It is most conveniently done when wearing smooth black rubber gumboots. Let the grout run onto the nose of a boot and then watch it as it runs off the sides of the nose. With practice, it is possible to identify w : c ratios by their appearance as they do this. Of course, experienced people can also identify w : c ratio by looking at grout in the agitator.

4.6-7 Laboratory Testing

Testing of grouts in a laboratory can be quite misleading if the mixing does not realistically simulate full-size site mixers. This simulation is not easy and in some cases has been solved by using an actual full-size mixer and discarding the remainder of the batch once the samples for testing have been taken. For details of laboratory mixers see Section 2.2-6.

BLEEDING TESTS Bleeding is discussed in Section 4.3-1 and examples given. Measuring bleeding is done by several methods, mostly improvised but nevertheless quite effective provided that:

- The container has its longest dimension vertically.
- The horizontal dimension of the container is at least several inches in order to eliminate any tendency for grout to bridge across from one side to the other while settling.
- There is no vibration of the container.

Measurement of the amount of bleed or of cement settlement is done by improvised means or by the very accurate method of ASTM Designation C827, which passes a horizontal ray of light over a sphere resting on the cement surface. The ray passes through a lens onto a vertical scale. The amount of cement settlement is magnified by the ray and so gives an accurate measure. The only uncertainty is whether the sphere sinks at the same rate as cement settlement or is bouyed up in laitance. Nevertheless, I prefer this test for bleeding, particularly with thick grouts.

VISCOSITY AND THIXOTROPY The flow cone, already described, is the simplest test of viscosity. However, thixotropy (Section 4.2-2) comes into it, and so does a gravity factor. Laboratory instruments that virtually remove the gravity component include various forms of consistency meter and viscosity instruments.

These mostly use the principle of torque measurement in a slowly rotating bowl of grout (described in Section 4.2-3). As indicated earlier, their use is more for special work than for foundation grouting, and so this is not the place to go into details.

REFERENCES AND FURTHER READING

ASCE Grouting Committee (1980). "Preliminary glossary of terms relating to grouting." *ASCE Journal of Geotechnical Division,* Vol. 106 (GT7), 803–815.

Bingham, E. C. (1916). "An investigation of the laws of plastic flow." Scientific Paper No. 278, U.S. Bureau of Standards.

Braun, W. I. (1978). "The solidification of slurries." *Ground Engineering,* Celebration Issue. Foundation Publications, United Kingdom, 30–33.

Burgin, C. R. (1979). "Investigation of the physical properties of cement–bentonite grouts for improvement of dam foundations." Thesis presented to the University of Florida.

Burwell, E. B. (1958). "Practice of the Corps of Engineers." *ASCE Journal of Soil Mechanics and Foundations Division,* Vol. 84 (SM1), Paper 1551.

Cambefort, H. (1968). *Bodeninjektionstechnik.* Bauverlag, Weisbaden, Berlin.

Cambefort, H. (1977). "The principles and applications of grouting." *Quarterly Journal of Engineering Geology,* Vol. 10, 57–95.

D'Appolonia, E., Shaw, D. E. and Erel, B. (1982). "Engineer's liability in dam inspections." *Structural Safety,* Vol. 1. Elsevier Scientific, Amsterdam, 27–51.

Deere, D. U. (1982). "Cement–bentonite grouting for dams." *Proceedings of Conference on Grouting in Geotechnical Engineering.* ASCE, New Orleans, 279–300, quoting Burgin (1979).

Deere, D. U., and Lombardi, G. (1985). "Grout slurries–thick or thin?" *Issues in Dam Grouting.* ASCE, Denver,156–164.

Foster, P. F. (1983). Ministry of Works and Development, New Zealand. Private communication.

Hall, E. W., Edwards, S., and Skinner, D. W. (1968). Uncirculated report. Cementation Ltd., London.

Hornby, L. (1961). Technical memorandum, Chem 12. Cementation Ltd., London.

Houlsby, A. C. (1981). "Grouting and sheathing of post tensioning tendons." *Bulletin of Australian National Committee on Large Dams,* No. 58, 33–35.

Houlsby, A. C. (1985). "Cement grouting: water minimising practices." *Issues in Dam Grouting*. ASCE, Denver, 34–75.

Houlsby, A. C. (1988). "Improvements in grouting of large ground anchors." *ASCE Journal of Geotechnical Engineering,* Vol. 114 (4), 448–468.

Klein, A., and Polivka, M. (1958). "The use of admixtures in cement grouts." *ASCE Journal of Soil Mechanics and Foundations Division,* Vol. SM1, 1547–1 to 24.

Lombardi, G. (1985). "The role of cohesion in cement grouting of rock." Fifteenth Congress on Large Dams. International Commission on Large Dams. Lausanne. Vol III, 235–261.

Lowe, J., and Sandford, T.C. (1982). "Special grouting at Tarbela Dam Project." *Proceedings of Conference on Grouting in Geotechnical Engineering*. ASCE, New Orleans, 152–171.

Mason, P. (1986). "The human thermostat." In *Ockham's Razor*. Australian Broadcasting Corporation, Sydney, Australia.

Mayer, A. (1963). "Modern grouting techniques." *Grouts and Drilling Muds in Engineering Practice*. Butterworths, London, 7–9.

Petrovsky, M. B. (1982). "Monitoring of grout leaching at three dam curtains in crystalline rock foundations." *Proceedings of Conference on Grouting in Geotechnical Engineering*. ASCE, New Orleans, 105–120.

Raffle, J. F. (1960). Research Report R.24/60. Cementation Ltd., London.

Raffle, J. F., and Greenwood, D. A. (1961). "The relation between the rheological characteristics of grouts and their capacity to permeate soil," Fifth International Soils Conference.

Ritchie, A. G. B. (1965). "The rheology of cement grout." *Cement and Lime Manufacture,* January 1965, 9–17.

U.S. Army Corps of Engineers (1956). "Pressure grouting fine fissures." Technical Report No 6-437, October 1956. Waterways Experiment Station, Vicksburg.

CHAPTER 5

The Grout Hole

In this chapter the emphasis is on the grout hole: methods are discussed of treatment at the surface (Section 5.1), drilling (Section 5.2), water testing (Section 5.3), stage grouting if deep enough (Section 5.4), and closure grouting (Section 5.5).

5.1 STANDPIPES, GROUT CAPS, AND SURFACE PREPARATION

5.1-1 Purpose of Standpipes

Each grout hole needs to have something on it for the connection of grouting fittings. This can be a pipe or it can be the tube of a packer (see Section 2.5-15). When pipes are used, the usual arrangement is as shown in Figure 5.1-1A. The various names for these pipes include:

Standpipe,
Nipple,
Connector,
Surface pipe,
Collar and others.

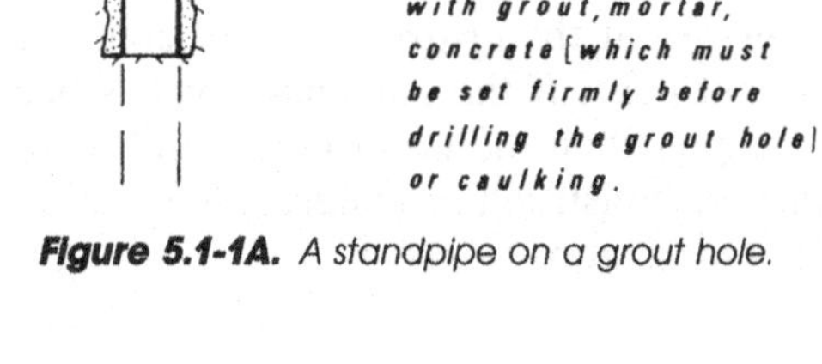

Figure 5.1-1A. *A standpipe on a grout hole.*

My preference for "standpipe" is because of its descriptive accuracy and perhaps also because of ingrained habit from long years of using it. Use whatever name you like, but in this book standpipe is the name for it.

5.1-2 When Are Standpipes Needed?

Standpipes are not always necessary, as shown in Figure 5.1-2A. A surface packer or a deep packer (Section 2.5-15) is used where pipes are not. A surface packer is seated in the mouth of the hole (where rock must be firm), and the fittings are attached to it. Refer to Section 2.6-2 for fittings. After grouting, if a packer is used it should be left in place until the grout gets firm but sometimes isn't, in the interests of speed and of not having to clean stiff grout out of it. This is a drawback of using a packer compared with a standpipe.

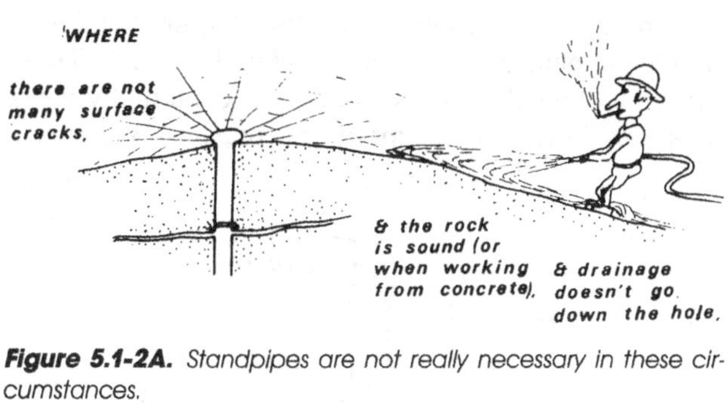

Figure 5.1-2A. *Standpipes are not really necessary in these circumstances.*

Figure 5.1-2B illustrates circumstances where standpipes are necessary. Standpipes provide:

- A firm collar for attaching the fittings.
- Control of leaks from the top of holes.

- Location for drills when setting up over the hole several times during stage grouting.
- Barriers against the entry of surface water and rubbish into grout holes.

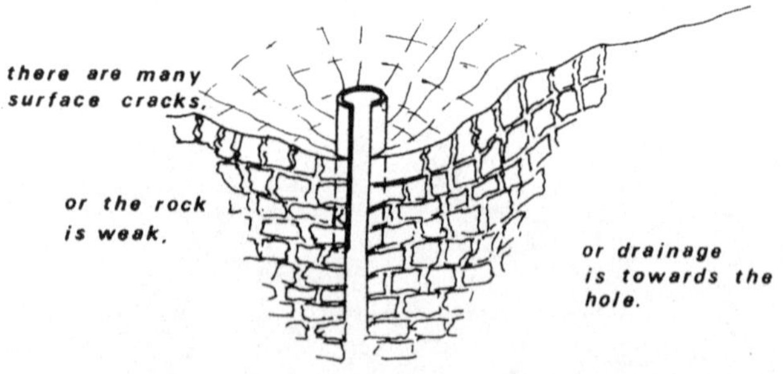

Figure 5.1-2B. In these circumstances standpipes are needed.

Standpipes must be embedded deeply enough to make them secure. They can get quite rough handling at times and must not come loose during grouting. The dimensions given in Figure 5.1-2C are usually adequate.

The amount shown for projection is a practical average. It must be long enough for the fittings to fit. On a sloping hole in a jagged foundation this amount can, at times, prove only just enough to swing the coupling followed separately by the isolating valve and then the rest of the fittings (these components are described in Section 2.6-2). In special cases, extra projection is sometimes directed and the contractor paid accordingly (contractural matters of standpipes are dealt with in Section 9.2-3).

The projection must also be sufficient to guard against entry of surface water and rubbish.

If a contractor carelessly or intentionally embeds standpipes to a greater depth than required, then he should be paid only for the stipulated amount. There can be technical reasons, such as exceptionally weak rock, or there can be monetary reasons, such as a

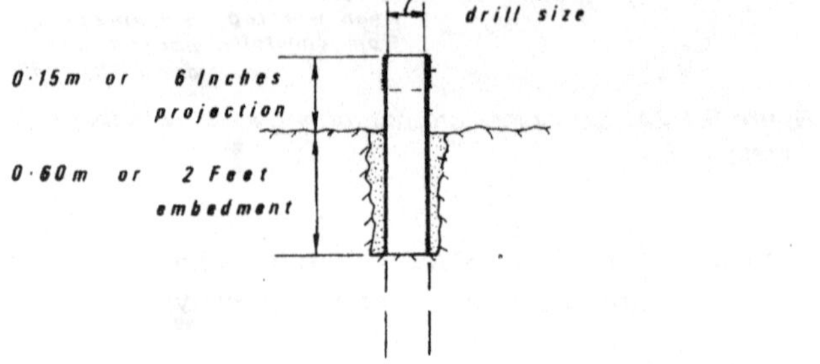

Figure 5.1-2C. Standpipe projection and embedment dimensions.

greater pay rate for standpipes than for hole drilling, for a contractor tending to go deeper (discussed in Section 9.2-3). By commencing the payment depth for drilling at this specified depth, dispute is minimized.

5.1-3 Placing Standpipes

Steps in placing standpipes in a rock surface are:

(a) Drill hole slightly oversize.

(b) Place pipe in hole and set it to the correct orientation and inclination for the grout hole. If hole is vertical plumb it with a spirit level. If the hole is inclined, use a template if there are enough identical holes to warrant making one; otherwise use an Abney level. Use of a theodolite may be necessary to give azimuth (direction) in complex or critical layouts.

(c) Grout the pipe into position or place mortar or concrete around it. If drilling through the pipe is required before the grout is about 24 hours old, add a quick-setting compound to it but avoid calcium chloride if structural concrete is to be placed over the hole later (undesirable reactions can occur on a long-term basis).

5.1-4 Attaching Fittings

The most convenient method for attaching grout fittings to standpipes is to thread the top of the standpipe to accept screwed fittings. Alternatives to this are to grip the standpipe in a clamping device attached to the fittings or even to put a surface packer inside the standpipe if threads have become damaged. Screwing is the preferred method, however, as it produces firm and reliable fittings if reasonable protection against damage during drilling is provided. The cost of threading is low when a large number of standpipes are machine threaded beforehand.

Protection of threads during drilling is easily provided by partly screwing a coupling onto the standpipe. Damaged threads can be difficult to fix in situ and cost money. Care pays.

Once a hole has been drilled through the standpipe, it must be protected against the entry of rubbish. In some high-risk locations, all grout holes are kept capped while not being worked on. This is not really necessary at well-drained situations. However, it is very necessary at the upstream toe of concrete-faced embankment dams if the embankment has been constructed higher than the grout holes, and there is the possibility of construction materials descending over the holes (see Figure 5.1-4A.)

Figure 5.1-4A. Trouble for the grouting crews! The embankment of this concrete-faced rockfill dam has been allowed to proceed higher than the grouting in the concrete toe. Rock descends onto the grout holes, drills, and crews.

5.1-5 Standpipe Size

Standpipes should be slightly larger than the hole to be drilled through them. Common sizes are:

3 in ⟨75 mm⟩ for 2.5 in ⟨60 mm⟩ trackdrill holes
1.5 in ⟨40 mm⟩ for small percussion drills
1.5 in ⟨40 mm⟩ for EX diamond drills

These are internal diameters (in accordance with usual conventions for specifying steel pipe sizes). Further comment on selection of the size in contract work is in Section 9.2-3.

5.1-6 Treatment at Finished Concrete Surfaces

Where grout holes commence from final concrete faces that may be decorative or high-velocity surfaces, standpipes should be recessed. This then allows the top of the hole to be filled over and the final surface made good (Figure 5.1-6A).

This is one of several methods. After grouting is finished, the socket is removed together with the tem-

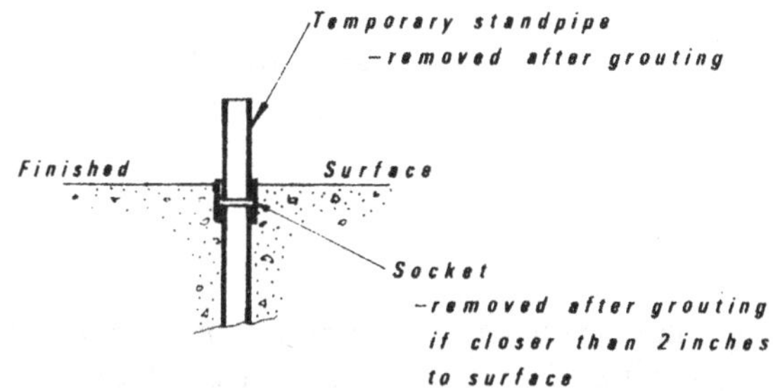

Figure 5.1-6A. Treatment of standpipe at a finished concrete surface.

porary extension, and the remaining recess is dry packed with mortar.

5.1-7 Standpipe Materials

The recommended standpipe material is steel. At least one inexperienced contractor has tried rigid polyvinylchloride (PVC) instead, resulting in much exasperation.

Galvanized steel is preferred to black because of the longer life available in screwed threads and the less chance of rust shards falling down the hole. Although each standpipe is used for only a short period, there are often long delays between site delivery and use and between installation and finished grouting through them. During this time they are in a very wet environment—from grouting activities if not from the weather—and rust can be rapid if they are not galvanized.

5.1-8 Payment for Standpipes

On contract work payment for standpipes and their installation, protection from damage, and so on, can either be by weight or by length. If paid by weight, contractors tend to use larger sizes than really necessary unless the size is stated in the specification, which then in turn limits the drill size, perhaps undesirably. Payment by length overcomes these difficulties and permits the contractor to use his own discretion. This reduces arguments. The payment depth and projection should be specified, as per Section 5.1-2, and the contractor can then choose the size of pipe to suit whatever drilling equipment is to be used. The size may vary over a site. See also Section 9.2-3.

5.1-9 Getting Rid of Standpipes

After grouting is finished, projecting standpipes may be a nuisance to other construction work. Under earth embankments and in some other locations, the projections interfere with proper compaction and rolling and usually have to be removed. Methods for doing this vary from just squashing the projections under the tracks of a heavy crawler vehicle or roller to cutting them off with oxy torches.

5.1-10 Grout Caps

Sometimes it is necessary to place concrete at the top of grout holes. This is known as a grout cap (Figure 5.1-10A).

Figure 5.1-10A. Example of a grout cap.

In this example, concrete is placed in a trench excavated along the line of a grout curtain.

A grout cap is not necessary where the surface rock is firm and competent enough to hold standpipes or surface packers (see Figure 5.1-10B).

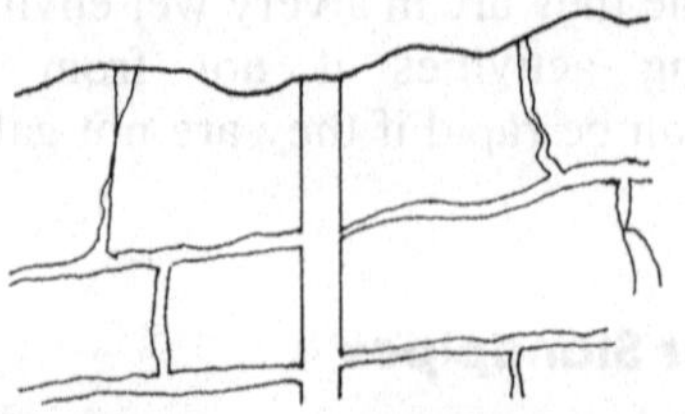

Figure 5.1-10B. Surface rock is firm and competent enough to hold standpipes or surface packers.

A cap may be necessary where the surface rock is closely jointed or rock is weak (see Figure 5.1-10C). A

Figure 5.1-10C. Surface rock is closely jointed or rock is weak.

cap is definitely needed where very weak rock or soft seams or weak patches are present in sound rock (see Figure 5.10D).

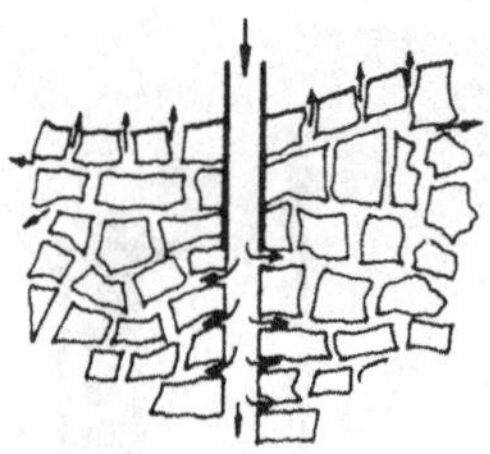

Figure 5.1-10D. Very weak rock or soft seams or weak patches in sound rock.

Without a cap, the standpipes would probably become loose and surface leaks alongside would be difficult to control. In a ''steps-and-stairs'' abutment (Figure 5.1-10E) the grout cap is recessed where possible. The slightly projecting parts are sloped for access and later compaction of fill.

Figure 5.1-10E. A grout cap in a "steps-and-stairs" type of horizontally bedded abutment.

5.1-11 Purposes of a Cap

Grout caps can facilitate the grouting in several ways. They hold standpipes firmly, they avoid surface leaks at the top of a curtain, and they provide a positive barrier against seepage. See Figure 5.1-11A.

It is a fact of life in grouting that the hardest cracks to fully grout are those in the surface region. This is because pressure tends to escape out the surface from these and also because pressures usually have to be kept low to avoid moving surface rock and opening up even more leaks. Ironically, this is the most vital part of a grout curtain; it lies directly under the dam or whatever structure is to be built over the grouting, in the very spot where seepage is most undesirable. A positive cutoff such as a grout cap therefore is a very

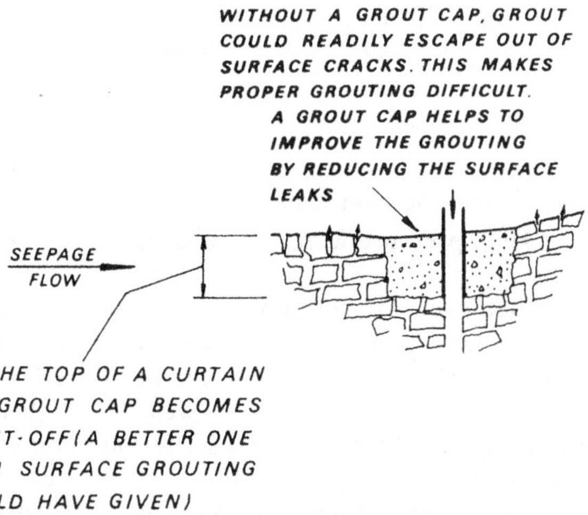

Figure 5.1-11A. *Beneficial effects of a grout cap in weak rock.*

useful expedient to handle this problem. In some cases of deeply weathered materials (perhaps hardly warranting the description "rock") where there are no cracks for grout to enter and grouting would be of doubtful efficacy, it has proved better to place an underground concrete wall to considerable depths rather than to primarily rely on grouting.

5.1-12 Shape of Grout Caps

The best shape for grout caps is shown in Figure 5.1-12A.

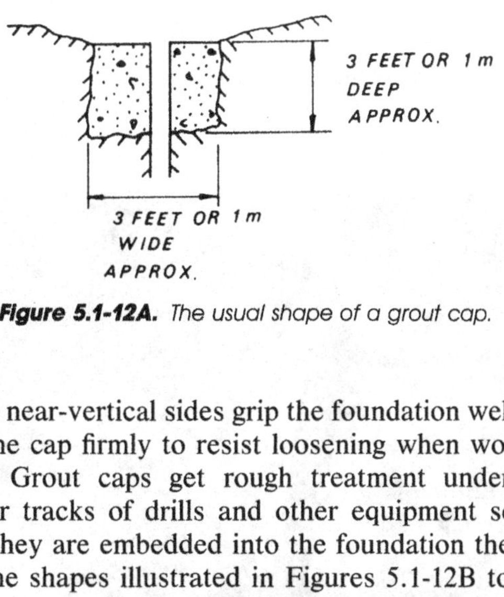

Figure 5.1-12A. *The usual shape of a grout cap.*

The near-vertical sides grip the foundation well and hold the cap firmly to resist loosening when working on it. Grout caps get rough treatment under the crawler tracks of drills and other equipment so the more they are embedded into the foundation the better. The shapes illustrated in Figures 5.1-12B to 5.1-12D are not embedded well into the foundation.

A concrete slab just sitting on the foundation (Figure 5.1-12D) can be a real nuisance when trying to grout through it. It has insufficient weight to withstand

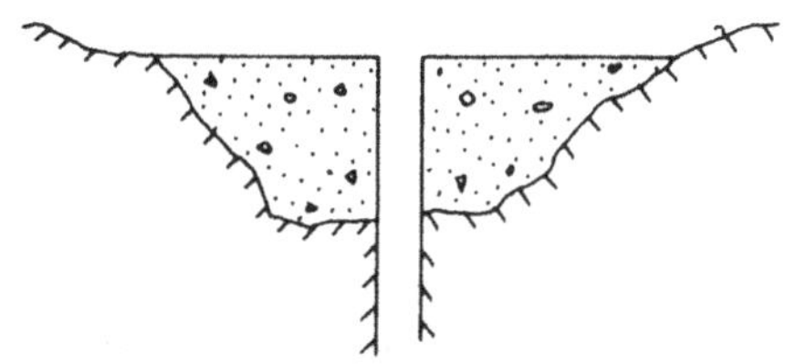

Figure 5.1-12B. *This cap is not well gripped by the foundation.*

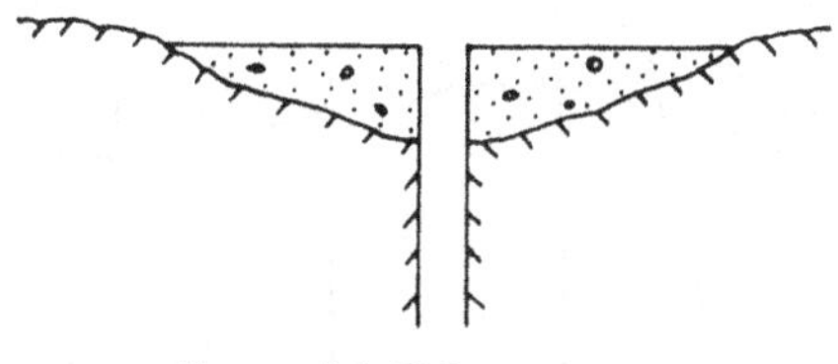

Figure 5.1-12C. *This is worse.*

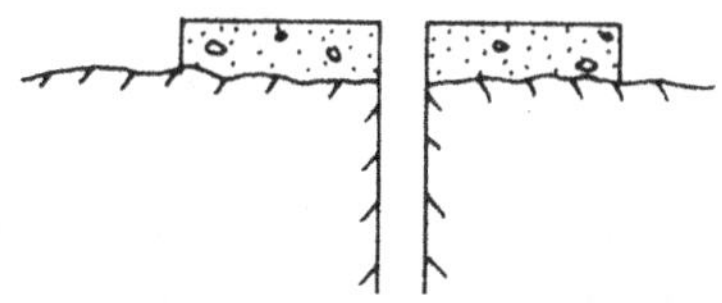

Figure 5.1-12D. *This is worse still.*

even low grout pressures and so is likely to lift as soon as grouting starts and to let grout leak out of both sides. Control of such leaks by caulking is messy, and the leaks are likely to break out again as soon as grout pressure is reapplied. The grout acts as a widespread hydraulic jack.

The only way to successfully use a slab such as the one shown in Figure 5.1-12D is to bolt it down. The bolts have to be of high capacity and must be well anchored at depth because uplift forces during grouting at ordinary pressures can be substantial. It can be successfully argued that such bolts are of less benefit than commonly supposed because they do not exert a restraining influence until elastic extension takes up the tension in them. This may permit the slab to lift slightly with the grout pressure; once this happens, the resulting crack between it and the foundation surface permits grout to travel along the slab and escape at the sides. Sometimes bolted slabs prove successful. An example of a bolted or doweled slab is at the upstream toe of concrete faced embankment dams as shown in Figure 5.1-12E.

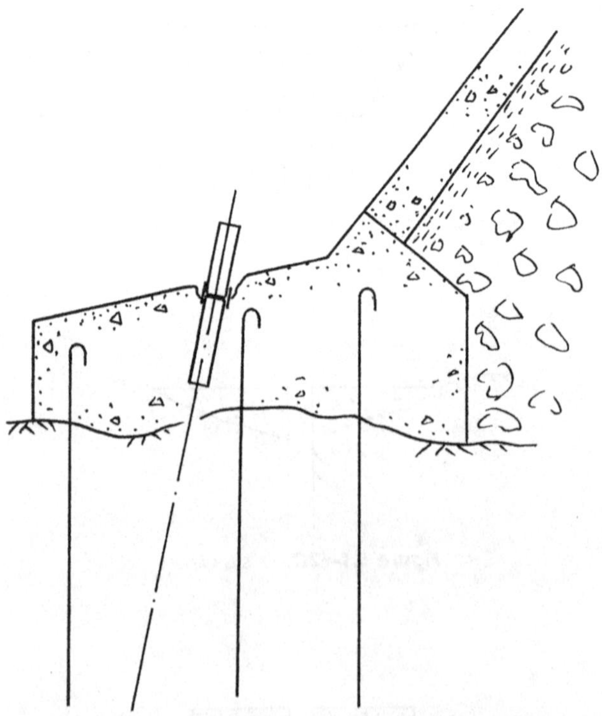

Figure 5.1-12E. Doweled toe slab acting as grout slab.

In this case, the doweling is not just for the benefit of the grouting but is part of the dam design. This is one of the reasons advanced by some designers for grouting through the slab instead of grouting from its foundation surface beforehand. The argument goes that if the doweling is required anyway, it might as well assist the grouting. On the other hand, better grouting is often possible from the final excavation surface at this type of dam before such slabs are placed. However, this delays the construction program at times and may not be welcomed by those with interests outside the grouting. For examples and discussion see Houlsby (1986) and also the photo in Section 5.1-4.

5.1-13 Excavation for Grout Caps

Explosives should not be used for excavating trenches for caps. If the rock is sufficiently hard that it requires explosives for excavation, then a grout cap is probably not really needed.

The disadvantage with explosives is that even light shots shake up the immediate area and open up cracks that were closed before (Figure 5.1-13A). This makes more surface leaks and weakens the foundation in the very area where it needs to be strongest.

Commonly used excavation methods are:

- Backhoe (Figure 5.1-13B), usually of the heavy crawler mounted type.
- Pneumatic hand-held tools, for example, jackhammers for plucking out the harder pieces with jackspades for the softer materials.
- Pick and shovel.
- Sawing with large circular saws is sometimes possible if the equipment is available.
- High-pressure water jet cutting.

When even weaker zones than the general foundation cut across the line of the cap or surface under it, treatment should be as illustrated in Figure 5.1-13C.

This gives better attention to weak seams that might not be groutable, and the concrete then provides the main treatment. Figure 5.1-13D is an example.

As discussed in Section 5.1-17, general foundation excavation at the line of a curtain should be carried down to groutable rock. The weak seams warranting this special treatment are then isolated features not needing further general excavation. On some sites, it

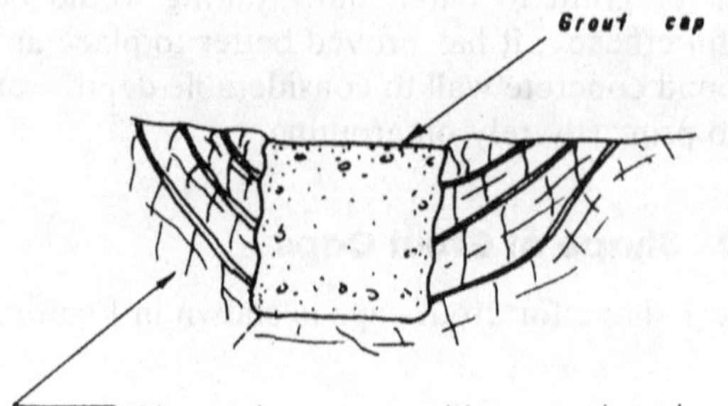

Figure 5.1-13A. Undesirable effects from blasting.

Figure 5.1-13B. Excavating a deep grout trench by backhoe.

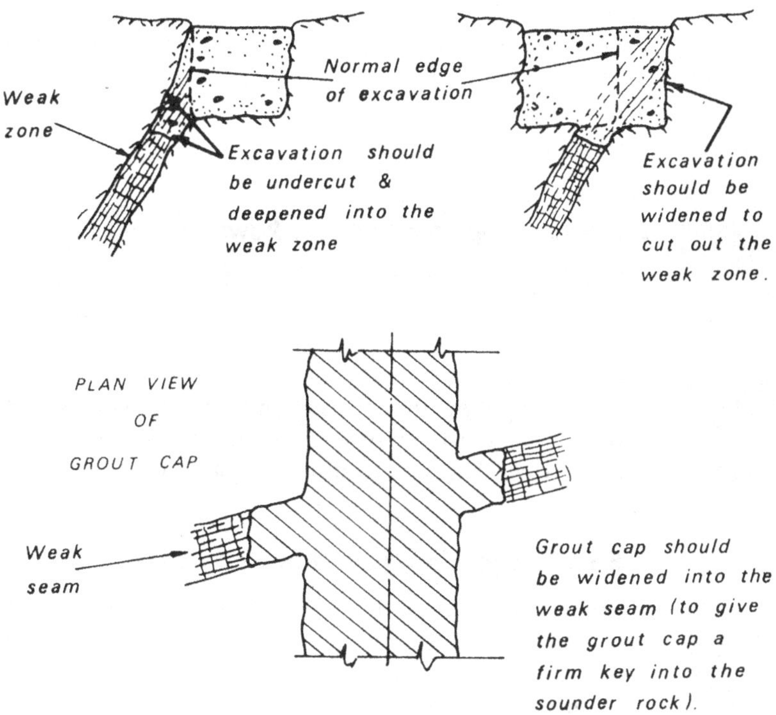

Figure 5.1-13C. *Treatment of weak seams.*

Figure 5.1-13D. *A grout cap widened into a weak seam.*

has not been economical or practical to reach groutable rock, and a solution has been to take the grout cap to much greater depth than usual until it bottoms out on groutable material. The cap then provides a cutoff through the ungroutable upper stuff (perhaps not justifiably dignified by the name "rock"), and the grouting commences lower down where it is in the right conditions. Powerful backhoes can usually dig the deep trenches required for this.

5.1-14 Standpipes in Grout Caps

Some options are discussed next.

(a) Place standpipes accurately in position before concreting the cap. Concrete getting inside the pipes does not matter if heavy percussion drills are used for the grout holes (Section 5.2 discusses drilling methods); these drills can easily drill it out. If diamond or light drilling is used, concrete can be kept out of the pipes by wrapping the bottom of the standpipes with old cement bags, hession, and so on. Standpipes need not go the full way down through the concrete; the remaining distance is drilled through.

(b) Place standpipes in the concrete of the cap after it has hardened using the same methods as for placing in rock (Figure 5.1-14A).

(c) Use no standpipe. Drill grout holes directly into the concrete and then use a surface packer set in it. This can be the cheapest method but has drawbacks, such as the need to have separate plugs put in holes to stop entry of rubbish when not being worked on and the difficulty of properly bleeding holes through packers (refer to Section 2.5-16).

The finished job is illustrated in Figure 5.1-14B. Standpipes are installed in a grout cap ready for grouting to start—not exactly in a straight line but near enough.

Figure 5.1-14A. Drilling holes for standpipes in a grout cap.

Figure 5.1-14B. Standpipes installed and ready for grouting to start.

5.1-15 Cap Surface Level

Usual practice is to leave grout caps in place as permanent features. It is therefore necessary to place them with their surface at a level that will suit construction of the structure above. Commonly this is a fill structure such as an embankment dam placed in layers with earth-moving equipment. In such cases, the top of the grout cap should be flush with the general foundation level to facilitate placement and compaction of earthfill. However, during grouting operations, the inevitable water and general mess associated with grouting can tend to deteriorate some types of surface material; any deteriorated materials have to be subsequently removed. The resulting surface must not be lower than the top of the cap or else the cap would stand proud of the foundation, and the hard knob this would produce could cause settlement cracking in the fill over it. Cracking here could be serious because of the steep hydraulic gradients at this point in the fill (refer to Section 8.4-9).

5.1-16 Caps for Multiple Rows

Where grout curtains have more than one row, it can be best to place a cap at the main row and do other rows without caps if the surface at these is tolerably adequate (or better). This allows the main curtain to be thoroughly done and backed up by the others. Where rock is worse than this, caps may be necessary on every row. This can lead to a lot of concrete.

5.1-17 Grouting from a Rock Surface

When grouting directly into a rock surface, excavation should be carried down to "groutable rock" if at all possible. Grouting is then done from this surface. A grout cap may be advisable see Section 5.1-10.

GROUTABLE ROCK What is groutable rock? This is difficult to define. Some guidance follows:

(a) It is predominantly firm, competent material.
(b) Cracks are wide enough to accept cement grout (refer to Section 6.1-5).
(c) Weak seams are few and do not contribute significantly to overall permeability.

Judging when excavation will reach or has reached groutable rock is another of those instances in grouting where experience is necessary, particularly as decisions about excavation depths often have to be taken when preparing designs based on exploratory information (see Section 8.4-14). Some amendment is usually possible when the site is opened up, but forecasts need to be reasonably close.

SURFACE PREPARATION UNDER DAMS When dams are being built on rock, removal of overburden and weak rock is taken down to competent foundation, which in most cases coincides with the groutable rock surface (see Figure 5.1-17A and 5.1-17B).

In other cases, only the core foundation is taken down to sound, groutable rock (Figure 5.1-17C).

IRREGULAR FOUNDATIONS Very irregular foundation profiles happen sometimes. It is advisable, but not essential, to knock any small knobs off along the line of a curtain to give a fairly uniform surface for grouting;

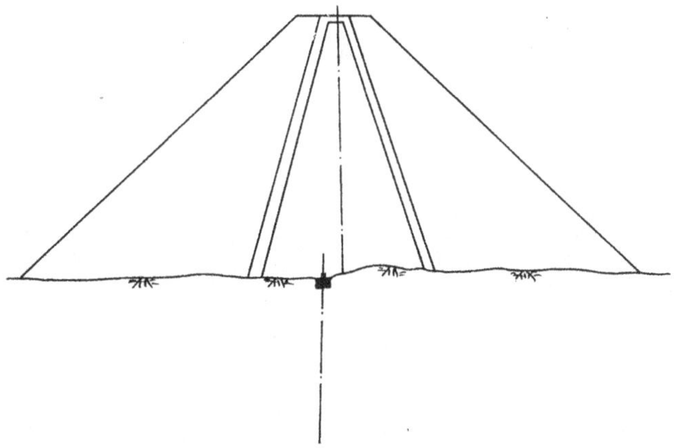

Figure 5.1-17A. *Typical section of an earthfill dam. Core and rockfill shells all excavated to groutable rock in this example.*

Figure 5.1-17B. *Dam foundation excavated to groutable rock uniformly. Grout curtain alignment is through the drill and the crane laying conduit pipes.*

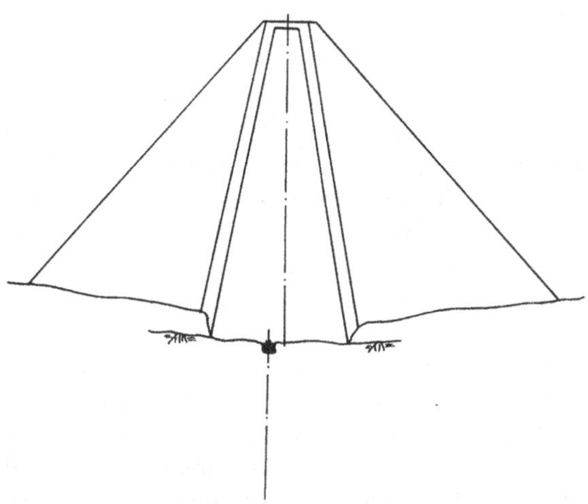

Figure 5.1-17C. *Section through dam where only the core foundation is taken down to groutable rock.*

this helps with the maneuvering of drills and gear. However, this is not always possible, and depressions and chases can occur that are too deep to eliminate by cutting off adjoining high spots. In addition, use of heavy explosives to remove hard knobs is often not desirable because of the shaking and cracking produced in the final surface. In these circumstances, dental concrete is used for filling the low spots after careful cleanup. This concrete must be well entrenched in the rock or bolted down to avoid dislodgement during grouting.

In Figure 5.1-17D, a deep hole and a shear zone have been filled with well-entrenched permanent mass concrete before starting grouting. Figure 5.1-17E is a photograph of a foundation prepared for concreting of deep pockets.

Note, however, that as a general rule, if dental concrete is required for other purposes, such as foundation shaping to facilitate fill placement, it is better to delay placement of it until after grouting is finished. Otherwise it may become dislodged during grouting (even if doweled). Extra grouting is then needed to rectify the problem after the resulting surface leaks are dealt with (see Section 7.3 for leak treatment), and replacement of the concrete may be necessary if it became damaged.

WHEN UNABLE TO REACH GROUTABLE ROCK In situations where excavation cannot be taken to groutable rock for some reason, the grouting can become more expensive because of the need to use measures such as:

(a) An unusually deep grout cap to reach suitable rock as described in Section 5.1-13.

(b) An underground concrete wall placed in a slurry trench down to groutable rock, with the grouting done before or after concrete placement. This is only possible where the material is soft enough for the slurry trench method.

(c) An underground concrete wall placed in large-diameter overlapping holes drilled in a row. Grouting is then done after concreting using grout holes drilled through the concrete. This method is suitable for all hardnesses of rock and for cavernous limestone but requires a large, expensive drill for the big holes.

(d) Cased grout holes through overlying material. In some foundations collapse might occur if holes are drilled and then casing run later, so methods such as Odex or Overburden (Section 5.2-8) provide for casing to be run down during the drilling. Once groutable rock is reached, the casing is stopped and normal drilling and grout-

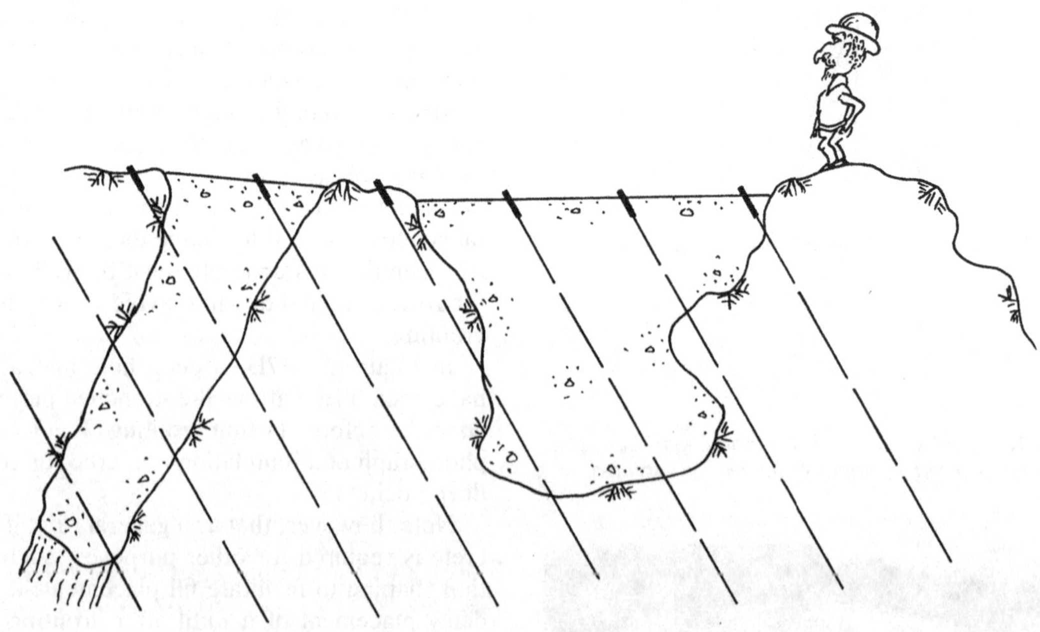

Figure 5.1-17D. *Deep holes filled with dental concrete to assist grouting.*

Figure 5.1-17E. *A rough foundation ready to receive concrete in deep pockets. Standpipes already placed.*

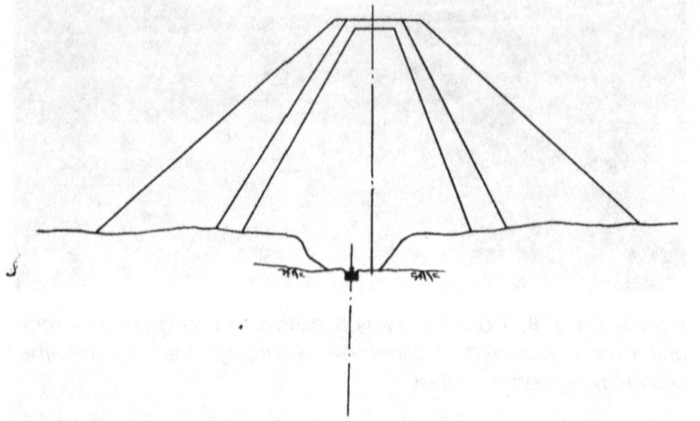

Figure 5.1-17F. *Typical cutoff trench for a grout curtain at an earthfill dam.*

> ing continued down through it. The cased section is not grouted, of course, and because this is not groutable material, its presence above a grout curtain can only be tolerated if sufficiently impermeable.

CUTOFF TRENCHES A simple way to reach groutable rock on most sites is to dig a cutoff trench such as the ones shown in Figures 5.1-17F and 5.1-17G.

A trench such as this permits localized deepening at sites where general deepening is not needed. When the trench is later filled, it provides a cutoff through the ungroutable material. Such trenches can give safety problems if they are too deep and have steep sides. The failure at Teton Dam had the trench profile shown in Figure 5.1-17H, which is believed to have contributed to piping and then collapse.

The cutoff trench at the failure location was unusually deep in comparison with the dam height. The Independent Panel reviewing the failure concluded that openings existed through inadequately sealed rock joints (i.e., inadequate grouting), but the prime cause of the failure was said to be the piping of the silt fill in the deep key trench into a multitude of very open foundation joints in the downstream face of this trench and its floor (Independent Panel, 1976).

Figure 5.1-17G. *Cutoff trench for grout curtain at an embankment dam site.*

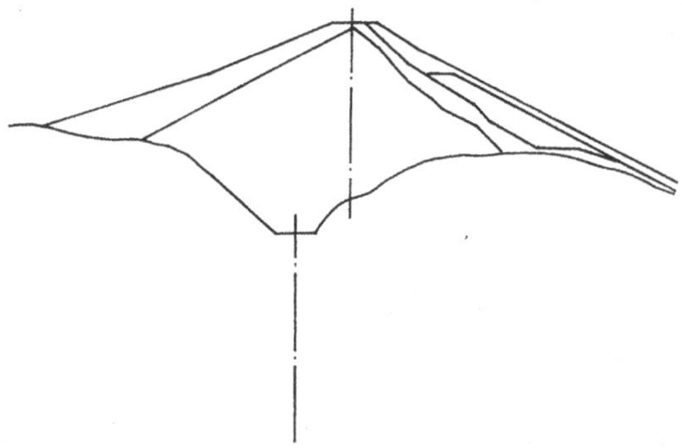

Figure 5.1-17H. *Cutoff trench under Teton Dam at location of failure.*

GROUTING FROM A CONCRETE SURFACE Examples of grouting from a concrete surface are:

(a) Grouting from inside a tunnel through a concrete lining.

(b) Grouting from galleries inside a concrete dam or in the toe slab of a concrete-faced dam.

Treatment of standpipes or holes through the concrete is shown in Section 5.1-6. Usually no other preparatory measures are needed as long as the concrete is old enough to have gained adequate strength to withstand grout pressures.

5.1-18 Surface Cleanup

If grouting is being done from a groutable surface, whether or not a grout cap is in use, the surface should be taken down to its final excavation level before grouting. It should also be cleaned by air and water jetting to get rid of any soft materials and to expose the final firm surface. The grouting then fills cracks up to this surface (presuming it is done properly, of course). This is preferable to doing further excavation after grouting for the purpose of getting down to the final surface; grout in the upper parts of cracks is liable to be damaged during any additional excavation, and there is not the assurance of proper grouting up to the final surface. Excavation to this surface should be done by methods which disturb the rock as little as possible. Preferably only light blasting, or none at all, should be used as excavation approaches final levels.

Grouting makes such a mess that a further air and water cleanup operation is needed afterward, but it should not be expected that this will remove further foundation material, except in the unusual case where rock has deteriorated after the initial stripping.

5.1-19 Removal of Weak Materials Underground by Washing

Foundations containing considerable seams or pockets of soft materials underground, such as clay or weathered rock, offer the possibility that the weak materials might be washed out and replaced with grout. While this seems easy, it isn't. The usual method tried is to pump water and sometimes air as well, down one hole to try to get it to come out a nearby hole after passing through the weak material. The idea is to wash out the material. Let me advise you not to waste money trying to do this. Although a connection may be established and plenty of dirt coming out looks dramatic, particularly if propelled by air, the results can be disappointing. The trouble is that once a connection is started, the washing just continues through it without necessarily enlarging into the rest of the soft material. Reversing the flow or applying it intermittently may help a little, but it is still not a panacea.

A cumbersome method possibly giving some degree of success is to alternatively wash and grout at daily intervals. The connection first made is filled with grout each day and allowed to set. The next cycle of washing and grouting finds and fills another connection, and so on. However, in most sites where weak pockets are a problem, this can take weeks to complete properly.

A pragmatic approach is to leave the soft material alone; just encapsulate it with extra good grouting in all surrounding open cracks. Then the soft material is protected against seepage. This method is not appropriate if the material is unduly weakening a foundation's capacity to carry a structural load.

An obvious improvement is to use jet grouting equipment (Section 1.2-6) down grout holes. This can rapidly remove all soft material within range.

5.2 DRILLING

5.2-1 Methods of Drilling

Drilling of grout holes is done by:

(a) Percussion drilling, where a toughened bit cuts its way through the rock by a hammering and chipping action.
(b) Rotary drilling using a diamond, tungsten, or roller bit. The rock is cut by diamond or tungsten inserts rotating in a relatively smooth manner. Roller bits cut by a combination of grinding and shattering.
(c) On rare occasions, downhole hammer drilling is used.

5.2-2 Selection of Type of Drilling

Controversy about the suitability of the various drilling methods has existed for many years, somewhat unnecessarily, I suggest. The fundamental issue is to select a drilling method suitable for the site based on:

The suitability of cuttings: Do they clog the cracks and prevent entry of grout?
Economics: Percussion drilling can be a quarter to a third of the cost of rotary methods.

Today percussion drilling predominates. This was not always the case. The advent of the modern heavy trackdrill has contributed greatly to this predominance. However, not all rock types are suitable for percussion drilling, and if uncertainty exists, site tests should be conducted with various drilling methods tried out side by side in representative geological conditions. Assessment of the results should be done as follows:

(a) Examine drill cuttings as they come out of the holes. If they are sharp and separate and free of slime and clumps, then the drilling method should be satisfactory. This is because such cuttings can be readily washed out of grout holes (see Sections 5.2-14 and 2.5-18 for washing of holes), leaving them clean and with the openings of cracks free from clogging. If, however, cuttings contain slimey, puggy, or adhesive material that might be difficult to dislodge from cracks with certainty, then a different drilling method would be better.
(b) Examine downhole. This is not essential but is helpful if the necessary borehole TV or periscope equipment is available (Sections 8.1-6 and 8.1-7). It enables in situ assessment of the cleanliness or otherwise of cracks after washing out.
(c) Compare drilling rates and costs, making allowance for when drilling is subsequently done on a larger scale.

Note that direct indications such as these are considered more reliable than indirect (and somewhat nebulous) methods such as comparing water test takes or comparing grout takes: rarely can enough holes be afforded in such tests to provide a valid statistical sampling (Section 8.3-5).

Although modern percussion drills have been found suitable for most rock conditions, there are exceptions. Conversely, diamond and other forms of rotary drilling have produced unacceptable cuttings in some sites, whereas percussion has not.

No attempt is made here to correlate rock types with drilling methods because of the variables found in even the one rock type. For instance, some types of fine-grained granites are quite unsuited to diamond drilling while other very similar types of the same rock are perfectly suited. Site experience or testing is the arbiter.

5.2-3 Percussion Drilling

The hammering action of percussion drills (see Figure 5.2-3A) is accompanied by a slight rotary motion. The hammer is at the surface mounted on a boom, and the blows are transmitted to the bit by means of sturdy drilling rods. Various types of bits are available (Figures 5.2-3B and 5.2-3C); selection should be based on rock type and drill characteristics.

A problem with cross and chisel bits in some rock types is that the wear is mostly at their outer edges. This can cause jamming in long holes and is troublesome if a worn bit has to be followed by a less worn one to complete a stage. They can give good drilling rates in rock which chips readily, however.

Button bits (see Figure 5.2-3C) tend to have less trouble with loss of diameter but are sometimes slower.

Cuttings are washed out of the hole by means of wash water traveling up the annulus outside the drill rods. It is essential to drill "wet" with grout holes. In other forms of drilling such as blasting work, air can be used for flushing instead of water, but the insistence on use of water for grout holes is to make sure that there is minimal chance for cuttings to be forced into cracks. Air can tend to do this, particularly in work below the water table, where it may produce a foamy slurry.

Another objection to air flushing is that any interruptions to the return of cuttings, such as a buildup of

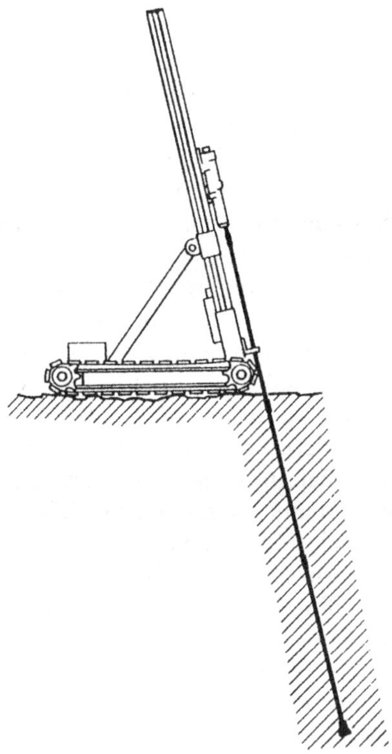

Figure 5.2-3A. *Typical percussion drill.*

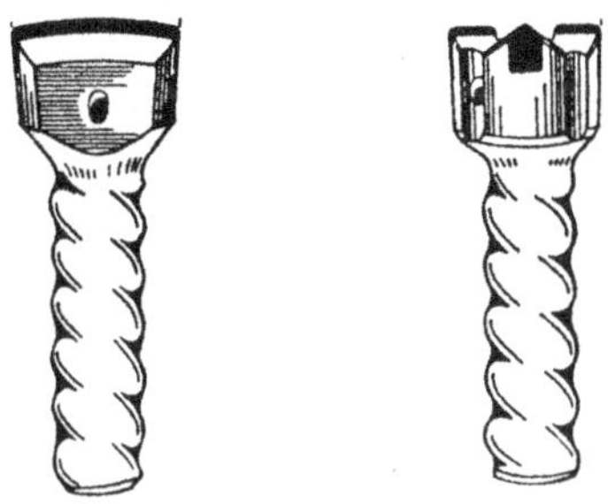

Figure 5.2-3B. *Percussion bits. Left: Chisel bit. Right: Cross bit.*

Figure 5.2-3C. *Button bit.*

temporary balling or plugging in the hole can result in compressed air entering foundation cracks and possibly penetrating them rapidly and extensively. This can cause rock movements and outbreak of surface leaks that would not otherwise occur. The surging action of pressurized air can be severe in a foundation. Nevertheless, there have been rare cases where drilling conditions have been so poor that air has been necessary to lift cuttings out of irregularly shaped, collapse-prone holes. The consequent risks to the foundation have then to be tolerated and perhaps later remedied by additional grouting. A further reason for avoidance of air flushing as a regular method is the clouds of dust erupting out of dry holes: hoods or other dust suppression methods may be needed to comply with industrial regulations in some countries.

Drill rods are screwed together and are hollow to provide passage for the drilling water. Fresh rods are added at the surface as the hole gets deeper. Their lengths are therefore governed by the travel available on the drill head mounting arrangements. The screw threads joining the rods have to be able to take the hammering action without distress as well as keeping the rods closely coupled under vibrating conditions. A coarse thread known as "rope thread" has been developed for this. However, this thread can be difficult to undo when breaking out the rods after drilling a deep hole, and improved threads are offered by some manufacturers. A thick grease called "rod dope" is wiped on the threads when adding a fresh rod to a string. This helps somewhat when breaking out the rods later but must be applied in moderation: too much of it left on the walls of grout holes can stop the grout getting into cracks. Some specifications have prohibited use of rod dope for this reason, but the prohibition is impractical; drillers need to use it and will continue to do so even if surreptitiously.

Hammers on percussion drills can be either pneumatic (air) or hydraulic (oil) operated. Pneumatic machines get air from compressors often coupled directly to the drill. Otherwise the air supply comes from job mains that must be of high capacity to be able to supply the large volumes of air demanded by these drills. Hydraulically operated drilling heads are powered through oil from a motor of some type: internal combustion, electric, or pneumatic. High efficiencies and lower running costs are reasons for their increasing popularity.

On most spread-out grouting sites, a dam foundation for instance, the most cost-effective drill is the large trackdrill. The crawler tracks of these machines enable maneuvering into difficult situations, and they travel relatively rapidly between holes with minimal manpower requirements. Their considerable weight gives them the ability to punch holes down rapidly with very few problems from the dancing- and jumping-around tendencies to which light machines, such as the old-fashioned wagon drills, were prone. Light machines have to be anchored to foundation dowels or some other restraint when drilling hard rock.

The doubly-tiltable type of boom shown in Figure 5.2-3D is able to be set up on just about any angle required. Fixed boom types are of limited use in steep and difficult terrain but can be satisfactory in areas of easy access. Both boom types are available in all sizes of trackdrill. Extendable booms are another useful feature for reaching holes in difficult places.

The maximum depth of holes drilled with large trackdrills (Figure 5.2-3E) is about 200 ft ⟨60 m⟩. Beyond this, the weight of the long string of rods tends to dampen the hammer blows and can tax the lifting power of the drill when breaking the string.

In tunnels, shafts, mines, or such locations where the large trackdrills cannot fit, small percussion drills are used for grout holes. These drills are mounted on mining bars or hand held if holes are short enough. Holes are usually smaller in diameter in these conditions.

The maximum depth for the smaller drills depends on the drill, how it is mounted, and how much space is available to feed new rods (Figures 5.2-3F and 5.2-3G).

A mining bar provided good support for the drill, but cramped conditions make use of relatively short rods necessary.

Air leg mounting (Figure 5.2-3H) has advantages if enough room is available. The drill is moved aside each time new rods are added so the rod next to the drill can be changed from short to long, then back to short again, and so on, as the hole advances.

If drilling underground in areas where high-pressure water might break into holes while drilling and flood a drive, arrangements should be provided to enable any such emergencies to be controlled. Precautions usually take the form of doing the drilling through an open valve or a stuffing box fitted to the standpipe and anchoring the drill with heavy chains. The stuffing box alternative lets drilling proceed even when passing through a high-pressure area. If instead, drilling is

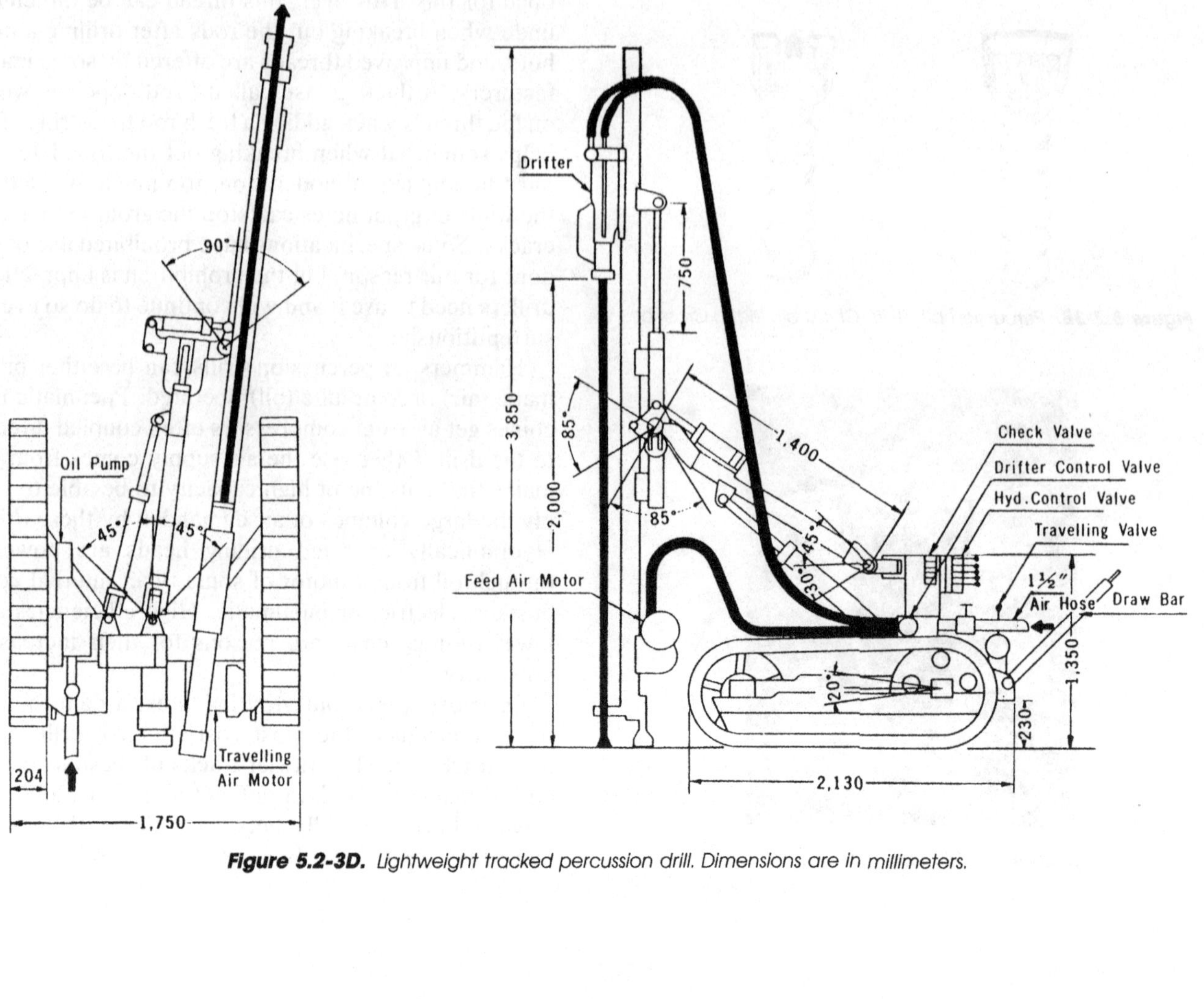

Figure 5.2-3D. Lightweight tracked percussion drill. Dimensions are in millimeters.

Figure 5.2-3E. *Large trackdrill drilling grout hole in rough foundation. Note the driller and offsider sitting down to the right of the photograph—the driller should be up at the controls.*

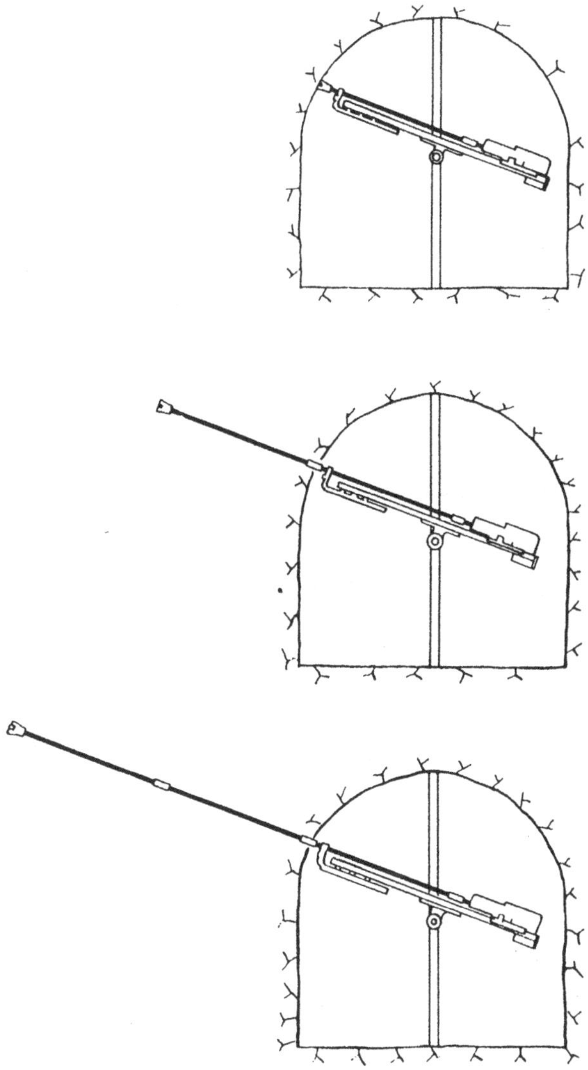

Figure 5.2-3G. *Percussion drill on mining bar in tunnel showing steps in addition of rods.*

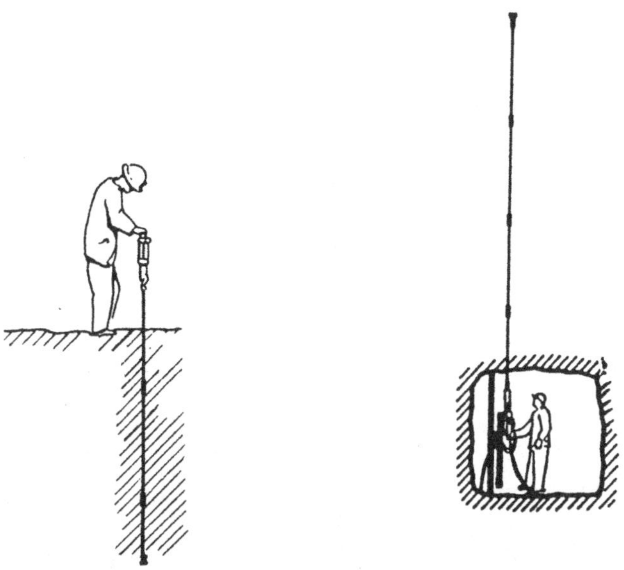

Figure 5.2-3F. *Small percussion drills. Left: Hand held. Right: On mining bar.*

through an open valve, which is closed off if an inrush occurs, time is lost pulling the rods. Drilling through a stuffing box can be troublesome though. It makes disposal of cuttings difficult, rods must have the same outside diameter as the bit, and wear on the gland can be severe. Anchoring the drill to the face is very necessary in some cases so that high pressures encountered suddenly do not propel the bit, rods, and drill forcibly back along the drive, possibly injuring someone.

5.2-4 Diamond Rotary Drills

Industrial diamonds provide the cutting in diamond rotary drilling. The diamonds are set in a matrix of softer, though tough, material. Rotary motion and pressure from the drill at the surface is transmitted through drilling rods to the diamond bit. The shape of the bit and the size and positions of the diamonds must be matched to the drilling conditions, and this requires considerable expertise.

Surface set diamond bits have the diamonds projecting from the matrix for about one-third of their diameter and have a pimply appearance. The size of diamonds and the hardness of the matrix must be varied to suit the rock and drill; otherwise expensive

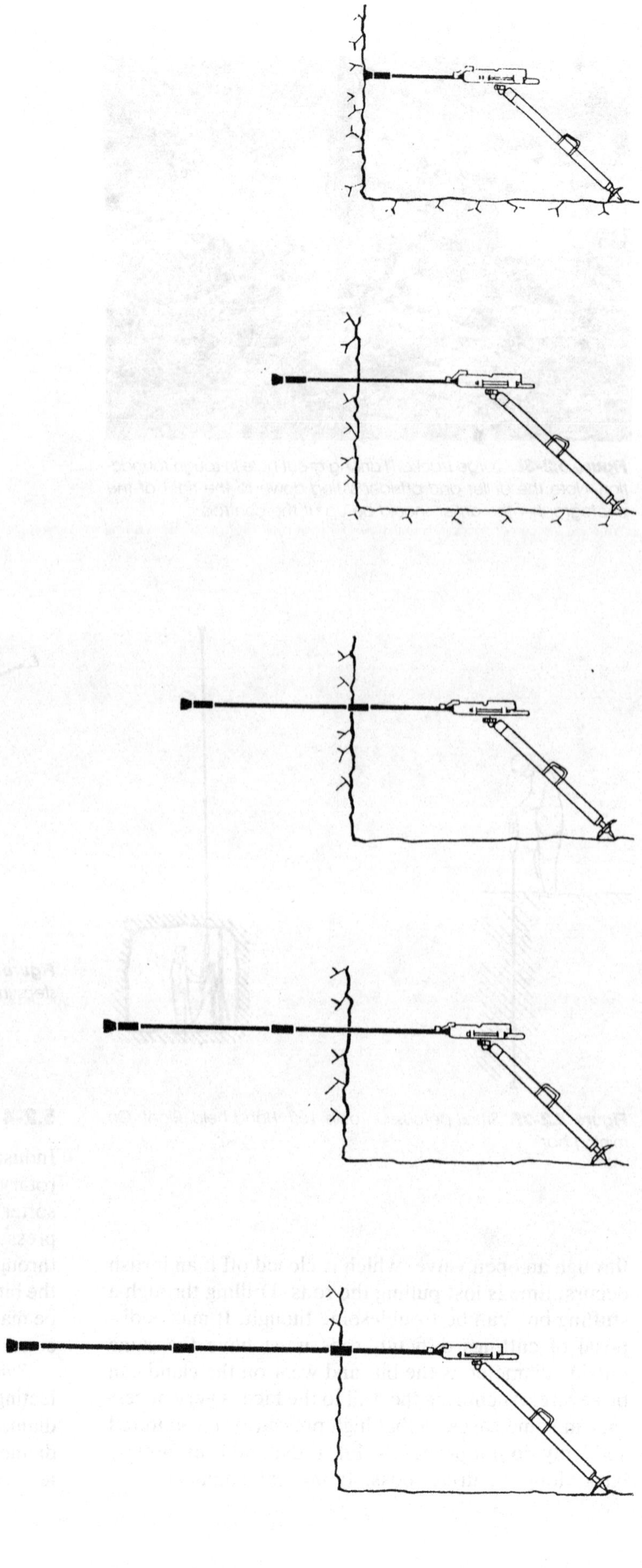

Figure 5.2-3H. *Percussion drill mounted on air leg showing steps in addition of rods.*

problems can occur, such as loss of diamonds by tearing out of the matrix; chipping of diamonds; and matrix polishing, which prevents exposure of fresh diamond faces to enable further cutting.

Various sizes of diamonds may be used in any one bit. After much use, when expert opinion indicates that the bit has reached the end of its economical life, it is returned to the manufacturer for recovery of what remains of the diamonds. Those salvaged are then used in new bits.

Impregnated diamond bits have small diamonds distributed in the matrix material somewhat like a grinding wheel. As diamonds are exposed and ultimately ground away, wear of the matrix exposes fresh diamonds.

NONCORING BITS For grout hole drilling, coring is rarely necessary so noncoring bits (also called "blasthole bits") are used. These provide faster and cheaper drilling than coring bits in most circumstances. The shapes of bits used for grout holes are illustrated in Figures 5.2-4A and 5.2-4B.

Specially shaped bits such as those shown in Figure 5.2-4B can be worth the extra fabrication cost when tailored to suit the particular rock requirements. Maximum footage per bit then results.

Figure 5.2-4A. *Various shapes of noncoring diamond bits. Left: For hard formations. Center: For medium formations. Right: For soft formations.*

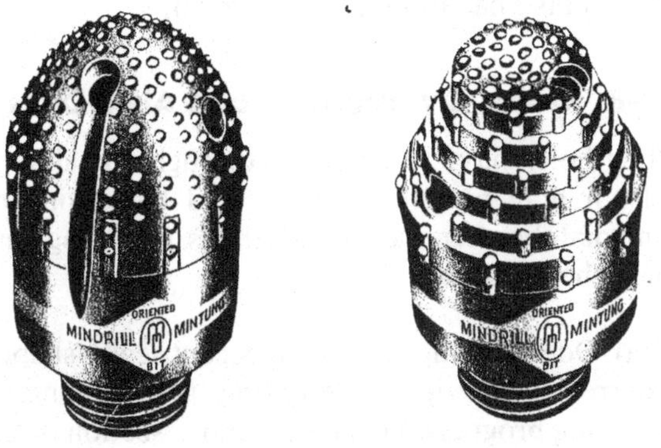

Figure 5.2-4B. *Special noncoring diamond bits. Left: Bullet nose. Right: Stepped.*

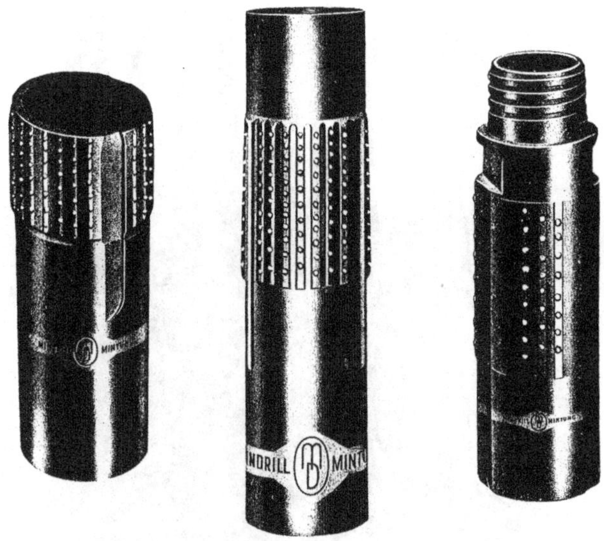

Figure 5.2-4C. *Typical diamond reamers.*

A reamer is usually mounted above the bit to maintain hole diameter and to assist with straightness. Reamers only cut the wall of the hole and variously have diamonds and/or tungsten inserts (Figure 5.2-4C).

All diamond drilling uses water to remove cuttings and cool the bit. Additives can improve drilling at times but must not clog cracks. Excellent publications are available on diamond bits and reamers, such as Heinz (1985).

Diamond drills are available in a range of capacities from small machines able to be carried by a couple of men to large truck-mounted units. The small units are sometimes clamped to the standpipe on the hole (Figure 5.2-4D), and this provides easy alignment when resetting over a previously drilled hole. The larger drills have to be maneuvered accurately into position, and this can take time, particularly compared with the rapidity with which trackdrills can be set up. The problem is even worse in rough terrain, where unless the diamond drills are mounted on crawler tracks, moving them from hole to hole can be an exasperating business.

5.2-5 Tungsten Rotary Drills

In tugsten rotary drilling the cutting is done by tungsten carbide or other very hard material in the form of tips or inserts. The drill provides rotation and pressure in a similar manner to diamond drilling. The type of cutting action depends on the rock type but is usually a combination of grinding and chipping. Bits are usually cheaper than the diamond bits and can vary from job-

Figure 5.2-4D. Small diamond drills mounted on standpipes drilling XRN grout holes.

improvised through a selection of factory-made products to expensive purpose-made units (Figure 5.2-5A).

Figure 5.2-5B shows drag bits with inserts. Similar inserts are sometimes also used in parts of diamond bits and reamers.

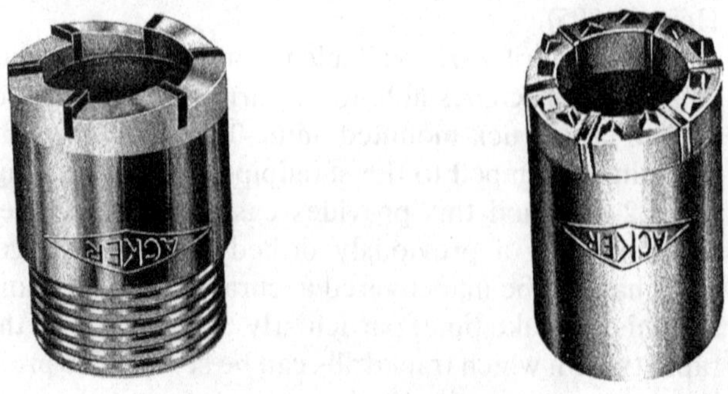

Figure 5.2-5A. Tungsten carbide bits.

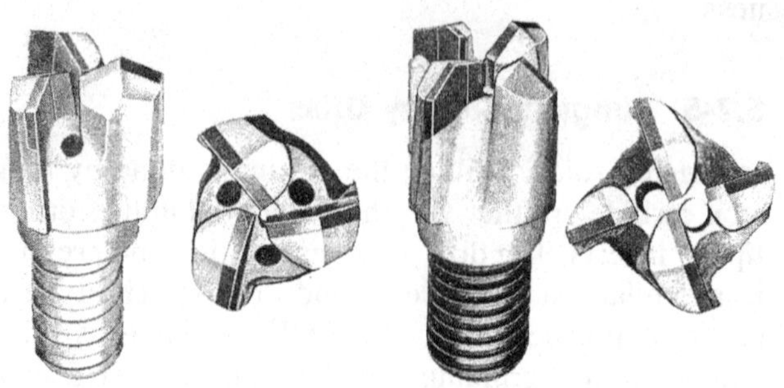

Figure 5.2-5B. Drag bits.

5.2-6 Roller Rotary Drilling

Rotation of the bit in a three-cone roller bit (Figure 5.2-6A) causes the cones to rotate; the cones are lightly meshed and give a grinding, shattering action. Two-cone bits are suitable for soft and broken foundations and provide greater clearance for removal of cuttings than the three-cone type. A variety of cone profiles and teeth sizes are available. For hard and abrasive foundations, three-cone bits are faced with tungsten carbide and have oversize ball and roller bearings.

Care is needed when roller bits become worn because if one of the cones becomes loose, it can fall out in the grout hole. Being made of toughened steel, it cannot be ground away by drilling, and fishing for such an awkward shape is no joy. The hole will probably have to be abandoned and a substitute one drilled.

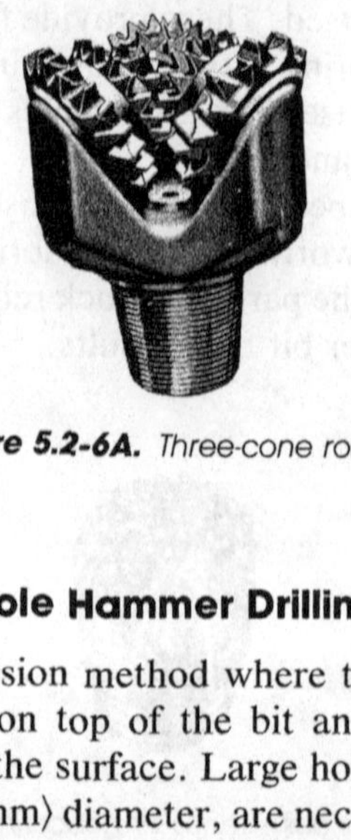

Figure 5.2-6A. Three-cone roller bit.

5.2-7 Downhole Hammer Drilling

This is a percussion method where the hammer goes down the hole on top of the bit and is rotated and pressured from the surface. Large holes, not less than about 4 in (100 mm) diameter, are necessary to accommodate the hammer, but holes as large as this are usually uneconomical for grouting. They are used on special occasions when holes are very deep or have to be very straight and the high cost can thus be justified and where noise has to be at a minimum.

5.2-8 Odex, Overburden, and Alvik Drills

When holes have to be cased through unstable materials, it may be necessary to feed the casing down immediately after the bit. Methods for doing this include the following:

(a) Odex drilling, where an eccentrically shaped bit underreams the tip of the casing, thus making space for casing progression. The eccentric section of the bit withdraws when not revolving and can be pulled back up through the casing. The casing is, of course, flush jointed (has no projections outside or inside at couplings).

(b) Overburden drilling (also known as Duplex) uses a special drill that can drive and rotate the casing independently of the main bit. The casing has a tip able to do a mild amount of cutting itself. The main bit excavates the hole to about the inside diameter of the casing, which is driven simultaneously or soon afterward, and cuts its way down around the opening.

(c) Alvik (Swedish) drilling uses either a retractable eccentric bit similar to the Odex method or a disposable bit that is large enough to open the hole to casing diameter and is left behind in the bottom of the hole (Lamberton, 1982).

(d) Other similar methods include the Klemm and Krupp double-head Duplex systems where the casing is rotated in the opposite direction to the drill bit. This assists with removal of cuttings and gives unusually straight holes. For a comprehensive review of overburden drilling methods see Bruce (1984).

5.2-9 Hole Sizes

The diameter of grout holes is not very important, within reason. Although some authors have attempted theoretical analyses purporting to show some significance for the larger length of crack exposed in large-diameter holes, experience does not confirm this thesis. Instead, practical issues dominate, such as the ready availability of certain sizes of equipment, the ability to flush holes effectively during drilling, the ability to thoroughly wash out afterward, whether packers have to be used in the holes, whether holes tend to collapse, and other site-specific matters.

Percussion Drilling. The hole diameter is closely related to the type of machine in use:

Large Trackdrills	2.5 in ⟨60 mm⟩
Small drills, mining drills, drifters, hand held, and so on	About 1.5 in ⟨40 mm⟩

Rotary: Diamond Tungsten. Holes are as small as possible because of the high cost of diamonds and/or tungsten:

Smallest used extensively	XRN or XRT	1.155 in ⟨29 mm⟩
Most common size	EX	1.5 in ⟨38 mm⟩ nominal
Largest	AX	2.0 in ⟨50 mm⟩ nominal

See the next section for origins of this nomenclature. European sizes are based on the hole diameter in millimeters.

Rotary: Roller The smallest hole size is dictated by the mechanical arrangements for fitting the cones sturdily in position:

Smallest usual size	2.5 in ⟨6 mm⟩ approximately
Usual size	3 in ⟨75 mm⟩ nominal

ORIGINS OF DIAMOND-DRILLING NOMENCLATURE Curiosity has been expressed about the lettering system used for diamond-drilling nomenclature. The following brief history of its derivation is based on Heinz (1985).

During the 1920s steps were taken to standardize drilling sizes in order to overcome many problems. The A and B sizes had already been in use for some time in the United States. When a third hole size, one smaller than A, was needed, rather arbitrarily E was chosen. When a fourth hole size was required, one somewhat larger than B, the letter N was selected primarily on the basis that it did not sound like E, A, or B. Thus the most common nominal sizes E, A, B, and N were introduced.

In 1929, after much preparation, a conference of manufacturers and users held in Chicago revised hole sizes and some drilling components and established tolerances. In order to differentiate the unchanged sizes from the previous ones, X was added to those changed. No subsequent changes have needed to be made to the standards established at that time, which is a tribute to the abilities of those who made them.

The R series, smaller than E, was introduced in 1930 in the United States by Boyles Bros', manufacturers. Various changes and improvements relating to coring barrels and rods have been made over the years but are not relevant here. The U.S. Diamond Core Drill Manufacturers Association (DCDMA) standards have been adopted in many countries, although some have their local variants. Some drillers have come to regard the nomenclature as their privileged jargon, unintelligible perhaps to those outside the fraternity.

To keep things nice and simple, therefore, the hole sizes listed in the preceding are given in their nominal sizes rather than in exact sizes, which, in any case, are not different enough to change the message.

5.2-10 Setting up Drills

A reasonable degree of accuracy is necessary when setting up the drills on grout holes. If vertical holes are required, a spirit level placed against rods or some suitable part of the mast or head is good enough when used in two planes at right angles to each other.

On inclined holes, an Abney level or bubble inclinometer sets angles sufficiently well enough. If holes are inclined in only one plane, the vertical component can be aligned by eying from a plumb bob string.

If holes are inclined in two planes, the design should specify the resultant angle of inclination and the azimuth so that the azimuth can be set with a prismatic compass and the resultant angle set with an Abney level. Good design practice is to consider drill setting up when working out the hole layouts (Section 8.4-3).

5.2-11 Hole Straightness

Most grout holes are not precisely straight. The bit can become deflected by harder or softer patches in the foundation or perhaps even if it partly passes through previously grouted areas. Long reamers and relatively large drill rods help to minimize bit drift but can restrict the return of cuttings in wash water, and this is undesirable in some rock types.

As previously mentioned, the limit of percussion drilling is about 200 ft ⟨60 m⟩, and holes drilled to this depth can be expected to drift as far as 10% off course and sometimes more. This does not matter in most grout curtains, where the precise position of holes is not important. However, when drilling down past a tunnel, for example, this drift may not be acceptable; downhole hammer or careful diamond drilling provides the answer. Although most diamond-drilled holes follow a slightly spiral course due to the crowd pressure from the drill, this can be kept slight by reducing this pressure to the minimum necessary for cutting and by using larger sized bits and rods than usual. For a hole that is 300 ft ⟨90 m⟩ deep, NX (3 in ⟨75 mm⟩) has been found a suitable size.

Downhole hammers can give straight holes if the expense is warranted. In critical cases, hole surveying has been used to map the course of holes but is rarely needed.

5.2-12 Logging the Drilling

When drilling grout holes, it is important to notice and report any events that are unexpected for the site, such as an exceptionally fast drill run, cuttings of a different color than usual, drill water loss, or appreciable vibration from the bit. A sudden fast run or even a complete drop of the drill for several feet can indicate that the drill has broken through into very soft material or into a cavity. This is information that needs to be taken into account when assessing the subsequent grouting (Sections 7.4 and 7.5); it also affects the driller as bits can spin off if the speed of the drill is not cut back promptly.

Similarly, loss of drilling water or a reduction in return can influence the positions chosen for stages. Complete loss warrants a special grouting stage (Section 5.4-2).

Any hole collapse around the rods also warrants investigation in terms of its effect on the grouting method selected (Section 5.4-3) unless it has been expected and appropriate provision made.

Events such as these should be reported in writing on a report form of some type (Section 4.1-12).

5.2-13 Drilling on Steep Faces

When grout holes are required in moderate and steep slopes, drilling methods vary according to the steepness.

SLOPES UP TO 45° Crawler-mounted drills can usually track up moderate slopes under their own power, but it is desirable to anchor them by cable to something substantial higher up the slope. This arrangement should hold the machine firmly in position to make sure it does not creep downhill during drilling, which could jamb the rods.

On slopes that are too steep for the machine to climb by itself, a heavy cable (two are better) operated by a power winch provides the pull. Descending should be under the restraint of the winch.

Some sites are too steep and irregular for this to be practicable. The answer then is to make a crude type of railway line up the hill (Figure 5.2-13A). The crawler tracks are taken off a drill and the rails placed to fit. Rails can be conveniently made of steel angle suitably supported or any available material. There is no need to use proper railway line. A winch at the top with two heavy cables over a snatch block provides the power for ascent and descent. When drilling, the machine is clamped down on to the tracks. Careful planning of each day's work (Section 5.5-6) is necessary to avoid interference between drill movements and washing and grouting operations.

SLOPES STEEPER THAN 45° Long holes should be avoided on very steep slopes (steeper than 45°), as recommended in Section 8.4-10. This confines the drilling to short holes that can be drilled with light, hand-held machines. On some jobs, operators have been roped down the cliff: on others they have been lowered in crane skips or scaffolding has been erected.

If long holes are unavoidable, heavy crawler drills can sometimes be mounted in large crane skips. For

Figure 5.2-13A. Trackdrill (minus the tracks) mounted on a crude railway system for drilling a steep face. Connected by steel cables to winch at top.

drilling, it is then necessary to clamp the drill to the cliff face (and make sure the crane does not creep).

5.2-14 Washing out after Drilling

When a drill has drilled to the required depth, the flow of wash water should be increased after the bit has been lifted slightly off the bottom. Then the bit should be run up and down 1 or 2 feet a few times to get cuttings into suspension and washed out as much as possible. Afterward, when the rods have been pulled from the hole and the drill has probably moved off to another hole, washout rods or pipes (Section 2.5-18) must be used with a strong flow of water to thoroughly wash the hole ready for grouting.

5.3 WATER TESTING

5.3-1 Types of Water Testing

There are two main types of water testing in relation to grout holes: *exploratory testing,* where permeability exploration of a site is carried out for design of the grouting (see Section 8.2), and *grout hole testing,* where each stage of every grout hole is water tested before grouting it. Grout hole testing is a simpler version of the exploratory test method and is discussed here.

The purposes of grout hole testing are:

1. To indicate the permeability of the stage. This information assists by:
   - Giving a guide to an appropriate starting mix (Section 7.2-2).
   - Giving a rough idea of what the cement take will be during the grouting; suitable programming and cement supply arrangements can then be made (Section 7.1-4).
   - Showing any effect of grout penetration from previous nearby holes. This needs to be known when assessing the effects of closure (Section 7.5-2).
   - Providing values for comparison with those assumed in the design (Sections 7.5-2 and 7.5-4). These are of greatest interest in primary holes before grouting and to a lesser extent in later holes. If significant discrepancies are found in comparisons, redesign may be called for.
2. To show the existence of surface leaks, connections between holes, and other possible happenings that will affect the grouting when it starts. Remedial treatment can then be planned such as plugging surface leaks (refer to Sections 7.3-1 to 7.3-5 for methods) or amending the program to deal with connecting holes (methods are in Sections 7.3-6 to 7.3-8).
3. To enable inexperienced crews to rehearse the grouting in a "dry run" (a rather wet one though!) using the grout circulation line for supply of water.

5.3-2 Water Testing in Grout Holes

The gear for doing water tests is explained in Section 2.8-1. A sketch of water test gear is repeated as Figure 5.3-2A.

The test for each stage of each grout hole should be carried out after washing out (Sections 2.5-18 and 5.2-14) for initial applications but is not necessary in second and successive applications (Section 7.2-10). The test length should be the same as for grouting; that is, if no packers are to be used to isolate grouting stages (refer to Section 5.4), none should be used for the water testing unless some special test is required. If packers are to be used for the grouting, they should be seated before the water testing. This has the additional purpose of checking whether they are properly seated. For sequences between testing and grouting in the various staging methods see Sections 5.4-4 to 5.4-6.

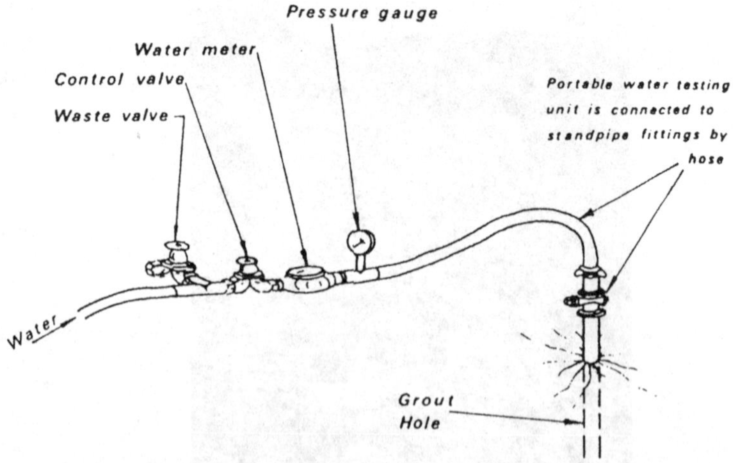

Figure 5.3-2A. *Water test gear.*

The water test method proceeds as follows:

(a) Connect the test gear to the hole (Section 2.8-1)
(b) With the control valve closed, turn on the water to the test gear.
(c) With the waste valve partly or fully open, cautiously open the control valve to build up to the test pressure (Section 5.3-3). Get rid of excess pressure through the waste valve if necessary.
(d) When the test pressure has been reached open onto the hole, start timing the test and simultaneously read the water meter (if it has not been set to zero).
(e) Read the water meter 5, 10, and 15 minutes later.
(f) After the 15-minute reading the test is usually complete. Turn off the water and remove the fittings.

5.3-3 What Pressure for the Test?

The recommended pressure is 15 psi. (1 bar) for the whole test unless the rock is too weak (a rare case).

On some jobs, water testing has been carried out at the same pressures as the subsequent grouting, which is usually greater than this. This is not recommended because water can penetrate more readily than grout can so any tendency for rock to move or dilate under pressure is made worse by unnecessarily high water test pressures.

My experience is that there is no need to use varying pressures for tests in grout holes, and the pressure need not be higher than suggested to give all the information needed. Note the difference, though, to water testing in exploratory holes (Section 8.2-7) where pressures are varied and can be quite high at times.

5.3-4 Calculations

The most convenient and direct way for reporting the water test results is in lugeon units (explained in Section 8.2-3).

Start by comparing the water taken in three runs of 5 minutes each (see Figure 5.3-4A).

If the water taken in all three runs is the same, as indicated in the top example by three bars of the same length, a representative test has been obtained and any of the three runs can be used to calculate the answer.

However, if takes decrease as in the center example, much of the water is filling empty, dead-end cracks and spaces and is not representative of the

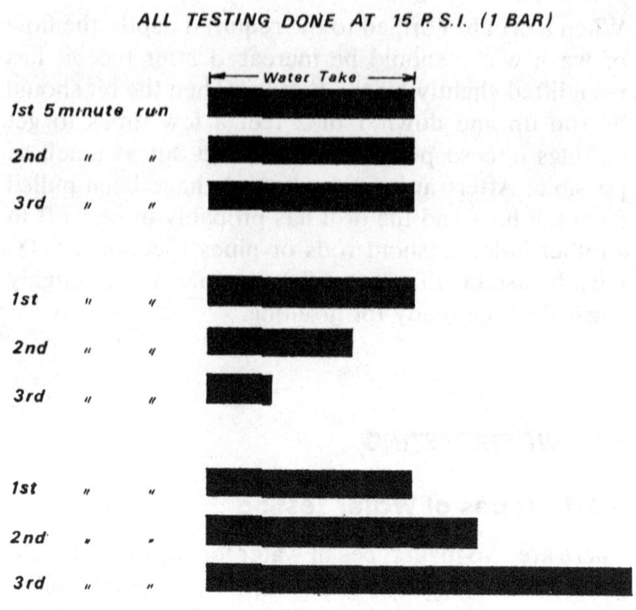

Figure 5.3-4A. *Bar diagrams illustrating interpretation of water test results.*

"through" permeability. It can then be wise to continue the test a little further until steady readings are reached.

In the bottom example, takes have increased during the test. This is undesirable. It indicates that washing out or movements are taking place. Pressure should be reduced in subsequent testing in the immediate area so as to avoid this.

Calculate the lugeon value for each 5-minute run by dividing the water taken by the length of the stage and then, if the single pressure of 15 psi ⟨1 bar⟩ has been used as recommended,

(a) when measurements have been taken in cubic feet and feet, read the lugeon value off the table in Appendix A, or
(b) if measurements are in liters and meters, multiply liters per meter per 5 minutes by 2 to get the answer in lugeons.

Example of (b): 50 liters taken in a 10-m stage in 5 minutes is 10 lugeons.

In most cases, the length of the stage used in these calculations is the same as that used for grouting; when the Downstage without Packer method (described in Section 5.4-4) is in use, the stage length for calculations is that of the new stage about to be grouted even though upper stages were exposed to the water test: with a little experience of the site, it is usually possible to get an idea of whether much of the water was taken in upper, previously grouted stages and, if it was, to do something about it.

Correction for hydrostatic head between the level of the pressure gauge and that of the test stage is superfluous, and so is any correction for water table level.

Lugeon values greater than 100 should be reported as ">100," as explained in Section 8.2-4. Lugeons should not be given with decimal values.

The formulas for the calculations are:

When test pressure is 15 psi and measurements are in cubic feet and feet,

$$\text{Lugeon value} = 175.928 \times \text{cubic feet per foot per 5 minutes}$$

Values for this are tabulated in Appendix A.

When test pressure is 1 bar and measurements are in liters and meters,

$$\text{Lugeon value} = 2 \times \text{liters per meter per 5 minutes}$$

If the testing has been done at several pressures or at a different pressure to that recommended,

$$\text{Lugeon value} = \frac{\text{cubic feet per foot per minute}}{0.0107620} \times \frac{142}{\text{pressure in psi}}$$

or

$$\text{Lugeon value} = (\text{liters per meter per minute}) \times \frac{10}{\text{pressure in bars}}$$

5.4 STAGE GROUTING AND METHODS

5.4-1 Why Grout in Stages?

When grout holes are longer than about 30 ft ⟨10 m⟩, it is usually best to divide them into shorter lengths, called stages, and grout each of these stages separately. This can produce better grouting because:

- Settlement of cement and rise of bleedwater is less. These affect the quality of the grouting, as explained in Section 4.3.
- Pressures can be controlled to a greater degree, and upper stages are exposed to less pressure than if stages are not used. Refer to Section 6.2-6.
- Leaks and connections can be handled with less trouble (Section 7.3).
- If drilling water is lost because the bit has broken into a wide crack or void, drilling can be stopped and a special stage made. This provides individual grouting of the feature. Then regular drilling and grouting can proceed with far less trouble.
- Similarly, if a zone is reached where soft or broken rock collapses into the hole appreciably, treatment can be provided by stopping drilling, making a special stage, and consolidating the bad area by grouting. However, if long lengths of hole collapse, special drilling and grouting methods may be needed. See Sections 5.4-11 and 5.4-12.

5.4-2 Stage Lengths

Common practice is to have relatively short stage lengths near the surface and to increase the length as holes get deeper. This practice largely reflects considerations about the effects of pressure. Long stages have greater pressures at their top than shorter ones do (explained in Section 6.2-6), and the surface regions are less able to stand substantial pressures than the deeper ones.

Lengths of planned stages should be multiples of drill rod lengths to make each rod change worthwhile. Rods usually come in 10-ft ⟨3-m⟩ lengths. In weak rock though, deep stages should not be much longer than surface ones.

5.4-3 Stage Grouting Methods

Methods for grouting a stage include:

- DOWNSTAGE WITHOUT PACKER
- DOWNSTAGE WITH PACKER
- UPSTAGE
- CIRCUIT GROUTING
- MPSP METHOD

Details of each of these methods follow here; later they are compared and discussed.

5.4-4 Downstage without Packer

Also called "Stage Down" and "Descending Stages." Figure 5.4-4A lists the steps in this method. As the name indicates, it is a method where the top stage is grouted, followed by the second, and so on down the hole, and a packer is not used to separate the stages.

5.4-5 Downstage with Packer

Also called "Stage Down," "Stage Grouting with Packer," and "Stage Grouting." This is the same as the previous method except that a packer is used to prevent grout entering stages above the one being grouted. Figure 5.4-5A shows the steps.

5.4-6 Upstage

Also called "Packer Grouting," "Stage Up," "Stop Grouting," "Stage Grouting," and "Ascending Stages." This grouts the hole from the bottom up. The hole is drilled full depth before starting grouting, as Figure 5.4-6A shows.

Practices vary about the time interval between grouting of stages. Some lift the packer as soon as pressure dissipates in the stage just finished and then immediately start the next stage, whereas others wait at least 6 hours between stages, as shown in Figure 5.4-6A. See Section 5.4-8 for comments on this method.

5.4-7 Circuit Grouting Downstage and MPSP Method

Circuit grouting downstage and the MPSP method are for perpetually collapsing holes, weak rock, and other difficult situations and so are left until later. See Sections 5.4-11 and 5.4-12.

5.4-8 Comparison of Methods

Which of the preceding methods is most suitable for a particular site? A comparison follows intended for fairly normal foundations, not difficult ones where circuit or MPSP or other special methods should be used. The following aspects are compared: Relative quality of work, costs, and time taken.

QUALITY OF WORK

Downstage without Packer This is the recommended method for highest quality work. The absence of a packer lets the grouting be fully controlled at the surface. Packer tube blockages are not a concern, and bleeding can be carried out properly (the need for this is explained in Section 4.3).

Apart from these aspects, a benefit of the method is the extent to which it gives self-checking of the grouting in upper stages when lower ones are being grouted. This is because these upper stages are directly subjected to the greater pressures used for the lower ones, pressures often considerably in excess of those at which they were grouted. If the grout in the upper stages has weaknesses, the weaknesses are likely to be found by these pressures and automatically regrouted. The need for this can be appreciated if the grouting is, say, part of a grout curtain under a dam where the storage head will be greater than the grout pressure used in the upper stages; if there have been imperfections in the grouting to the extent that they will not cope with the storage head, it is prudent to find them and fix them in the course of the grouting operations. The method does this fairly well without any additional work.

However, there can be annoying and repeated surface leaks in some relatively weak foundations. These are the very foundations where grouting is often difficult but most needed. Patience and persistence is recommended; the method is still the most suitable even though on some sites impatience has perhaps unwisely led to the use of packers to isolate persistently leaking surface stages.

Downstage with Packer The intention is to protect upper stages from high grouting pressures. This somewhat ambitious aim can be self-defeating if it leaves

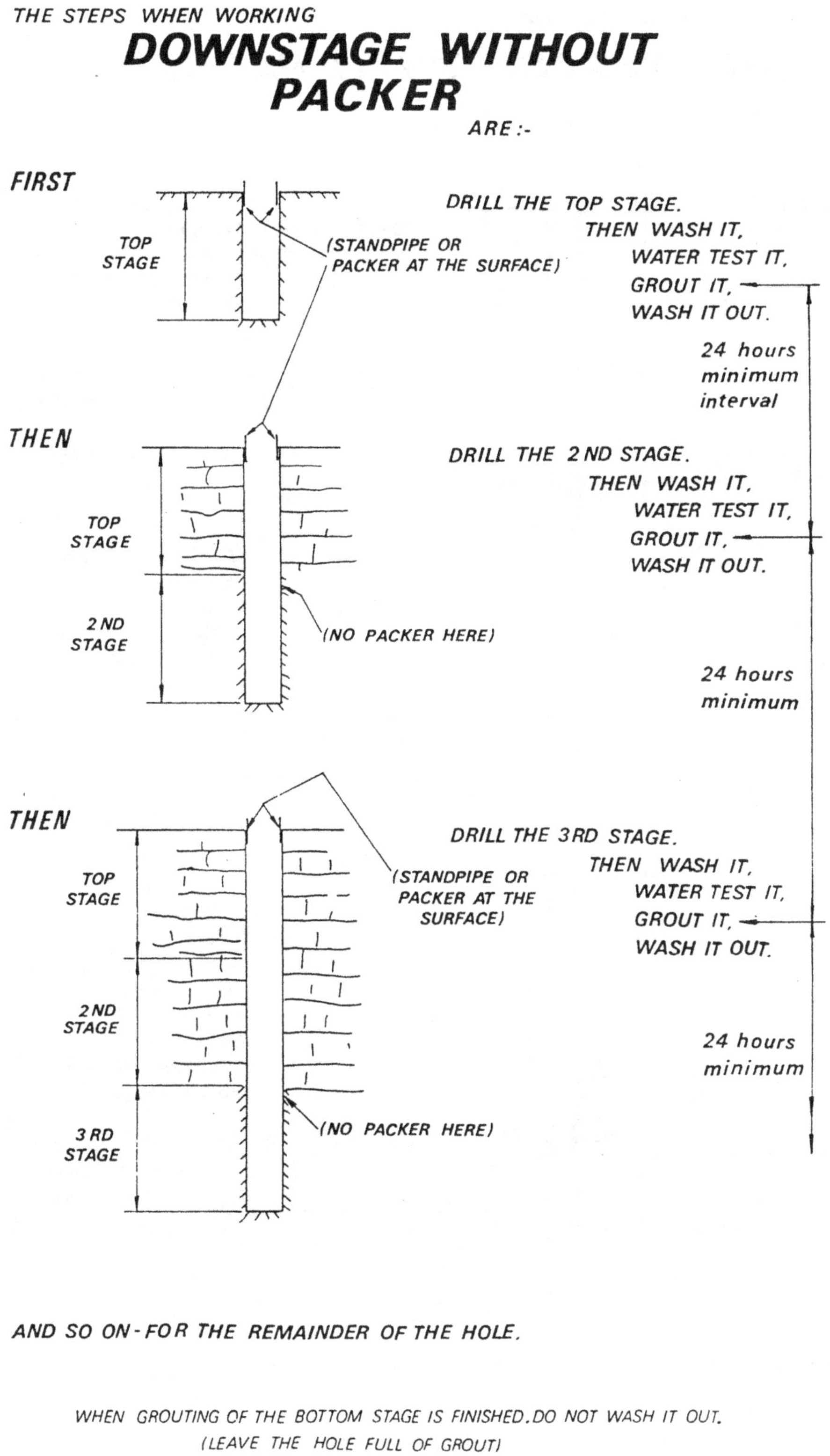

Figure 5.4-4A. Downstage Without Packer method.

THE STEPS WHEN WORKING

DOWNSTAGE WITH PACKER

ARE:-

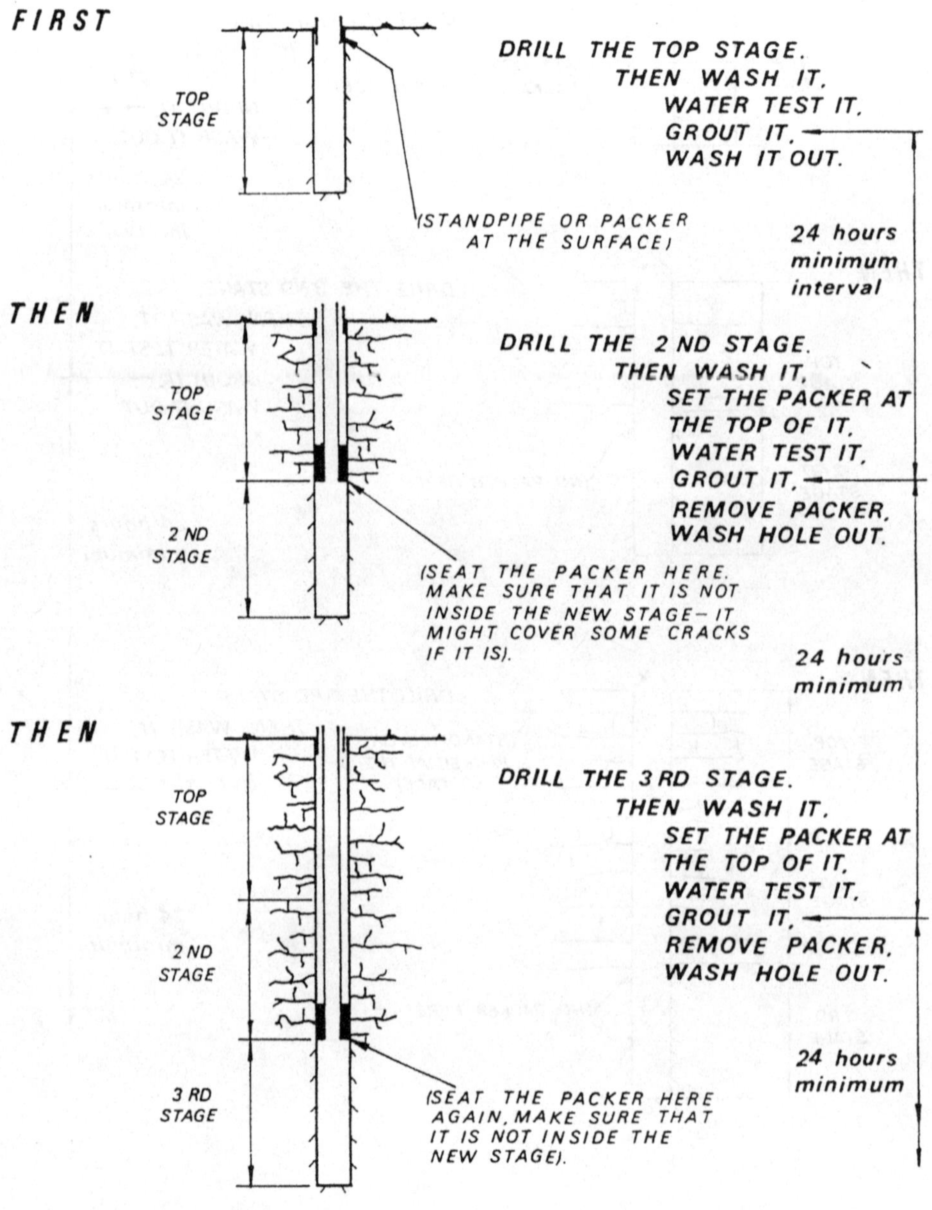

AND SO ON – FOR THE REMAINDER OF THE HOLE.

WHEN THE BOTTOM STAGE IS FINISHED, THE HOLE IS THEN FILLED BY ONE OF A NUMBER OF METHODS AS INSTRUCTED IN EACH CASE

Figure 5.4-5A. Downstage With Packer method.

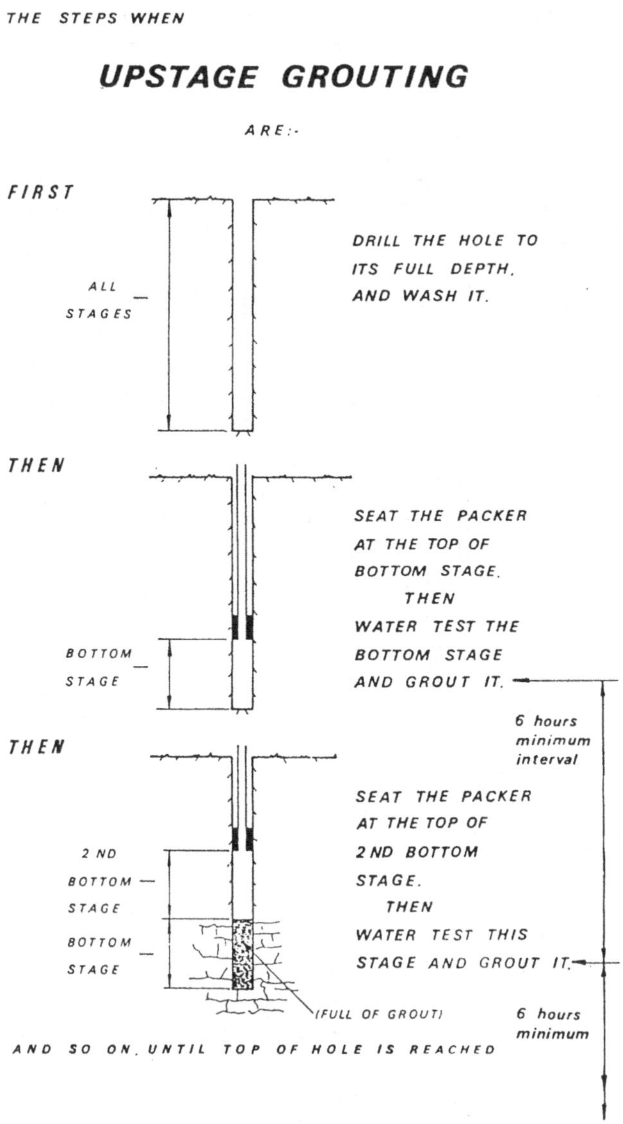

Figure 5.4-6A. *Upstage grouting method.*

poorly grouted upper stages undetected and umremedied.

The introduction of a packer opens the way for troubles such as blockages in the relatively small diameter tube passing grout down to the packer, difficulty in getting rid of bleed water (and thereby jeopardizing the quality of grouting in cracks at the upper part of the stages; refer to Section 4.3), and the possibility of inadvertently grouting packers into holes where they leak or where grout bypasses them (Section 2.5-16). Furthermore, it is not possible to effectively hold pressure through a packer after refusal (the need for this is in Section 7.2-9). These factors can contribute to poorer grouting. A further problem encountered in those sites having troubles with rock movements is that packers are often a nuisance when using the control methods (Section 7.6-4).

There is none of the checking of upper stages provided in the previous method. Weaknesses in upper stages have been known to pass undetected in some jobs until special check holes happened to find them.

This method cannot be used where holes collapse or are so noncircular or irregular that packers cannot be routinely seated properly. This situation is rather common.

Upstage This can only be used where holes do not collapse. It has the same limitations arising from the use of packers, with the additional one that because the rock above the packer has not yet been grouted, there is increased chance for grout to bypass packers and enter holes above them. If this happens, packers can become grouted in, and holes have to be replaced.

The pressures used in lower stages are sufficient to push grout to the surface if cracking systems lead there and to produce surface leaks with strong pressure behind them. The pressures are also great enough to move surface slabs and structures and do real damage there.

The quality of the grouting is usually inferior to the Downstage without Packer method because the use of packers means that pressure cannot be held after refusal. See Section 7.2-9 for comment on this method.

COSTS

Downstage without Packer Because drills have to be set up afresh for each stage, this method costs more than Upstage. However, modern tracked drills (Section 5.2-3) can be traveled and set up over holes so quickly that only a few minutes are occupied doing this, and for an efficient driller the interruption to continuous output is slight. In contract work there is therefore no justification for paying a bid item for drill setup (refer to Section 9.2-2 for contract arrangements), and so drilling costs for downstage methods should only slightly exceed those for Upstage if contractors are efficient. Other costs, such as grouting items, are much the same. Hence the overall cost of the Downstage without Packer method in keen competitive bidding should not be greatly more than for Upstage, and this price margin can often be offset if problems such as leaks, connections, and rock movements occur; these can be more effectively handled by the downstage method.

Downstage with Packer The extra cost of supplying, placing, and repairing the packers makes this the most expensive method of the three. In deep holes, packers can become so heavy that mechanical lifting methods are needed to handle them. This makes for slow, tedious work and can lead to trouble if the time taken to pull packers after grouting is so long that grout stiffens in their tubes before they can be recovered and washed out.

Upstage This is the cheapest method on sites where all goes well but not where they don't. Its apparent lower cost is often an attraction to specification writers who are trying to minimize costs and are keeping their fingers crossed that all will go well and holes won't collapse too often. A common strategy is to permit changing to the Downstage without Packer method if they don't go well. Better practice is for experienced advice to be obtained before contract preparation so that a suitable choice can be made at that juncture; this produces keener tendering because tenders do not have to be padded to cover uncertainties about which method will be directed. Refer to Section 9.2-8 in this regard for suggestions about contract preparation.

TIME TAKEN

Downstage without Packers On troublesome sites this is the quickest method because of its inbuilt ability to handle problems better than the others. Where there are no troubles, it can be slower than Upstage.

Downstage with Packers This is always the slowest method.

Upstage As noted before, this is the fastest method where there are no problems. The speed depends largely on whether there is a waiting time between grouting of stages. In some Upstage practice there is no waiting time; as soon as grouting of one stage is finished, the packer is lifted to the top of the stage above and grouting of that next stage gets under way. Arguments in favor of this include the idea that once grouting of a deeper stage is completed, grout in it will not be disturbed by immediate grouting of the stage above. Contrary to this are the difficulties encountered in sites where groundwater is at sufficient pressure to push grout back into the hole if the packer is lifted off a stage before the grout has had time to stiffen somewhat, sometimes a matter of several hours. However, packers cannot be left in place for this length of time; otherwise cleaning of their tube is well nigh impossible. This is an inherent drawback with Upstage grouting and, where severe, is better handled by changing to downstage work. Another aspect of waiting time between stages in Upstage work is that if intervals of about 6 hours are used, as indicated in Figure 5.4-6, stages can be water tested. This is beneficial, as explained in Section 5.3-1 (it finds problems in advance of grouting, and rational assessments of the grouting can be made), and is worth doing if the quality of the work matters.

With this method, a source of delay can be the need to finish grouting of holes completely before starting drilling of closure holes alongside. This is discussed in Section 5.5-4.

AN IMAGINARY EXAMPLE In the sketch shown in Figure 5.4-8A, jointing is simplified, cracks are shown enlarged, and the scale is indicated by the little man at the top. The question posed is to find which grouting method is the more suitable. If grouting is done by Upstage methods, what first starts to happen is illustrated in Figure 5.4-8B.

The hole has been drilled to its full depth and a packer placed at the top of the lowest stage. Grout passes out from the hole and moves downward as well as outward. Now turn to Figure 5.4-8C, where the next sequence of events is that lower cracks are now filled, and the grout moves upward. Some grout finds its way into cracks which lead into the hole above the packer, and this grout then enters the hole there. The operator at the surface is unaware of this and before long the packer has become solidly grouted into the hole, and later removal will be well nigh impossible. While this is going on, the high pressure used in a deep stage such as this, which is more than ample to push grout to the surface, does just this. Surface cracks there then eject grout at high pressure, and caulking is difficult against such a pressure. Then, possibly, some of the surface blocks of rock are not well attached, and the high grout pressure moves them off.

Now compare this trail of troubles with what happens if the grouting is done by the Downstage without Packer method (Figure 5.4-8D).

The top stage is shown being grouted here. Grout works its way downward and upward in much the same way and the same surface leaks and rock movements are encountered, but this time they are at low pressure and can be remedied somewhat more easily. There is no packer to get grouted in. Altogether the grouting is far easier and therefore done more thoroughly. Once this top stage is finished, the next stage is commenced and has the benefit of the earlier work in fixing the surface problems.

In summary, the Downstage without Packer method has a lot in its favor.

5.4-9 Combinations of Methods

Some jobs use a combination of methods so as to get maximum benefits. For example, all holes of the surface stage might be fully grouted right down to final

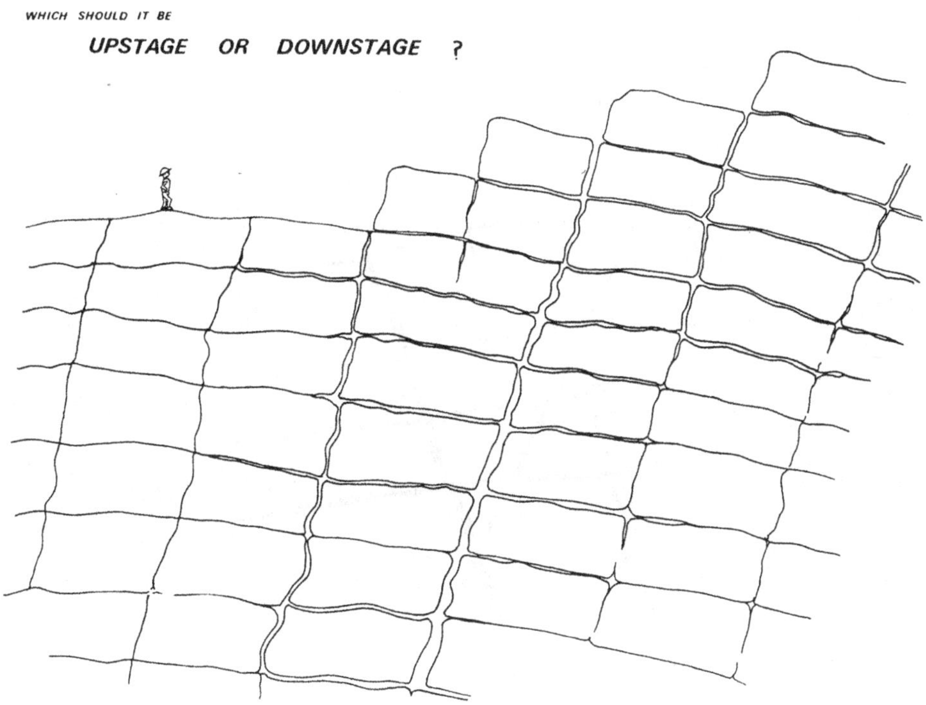

Figure 5.4-8A. *Should this site be grouted by Upstage or Downstage methods?*

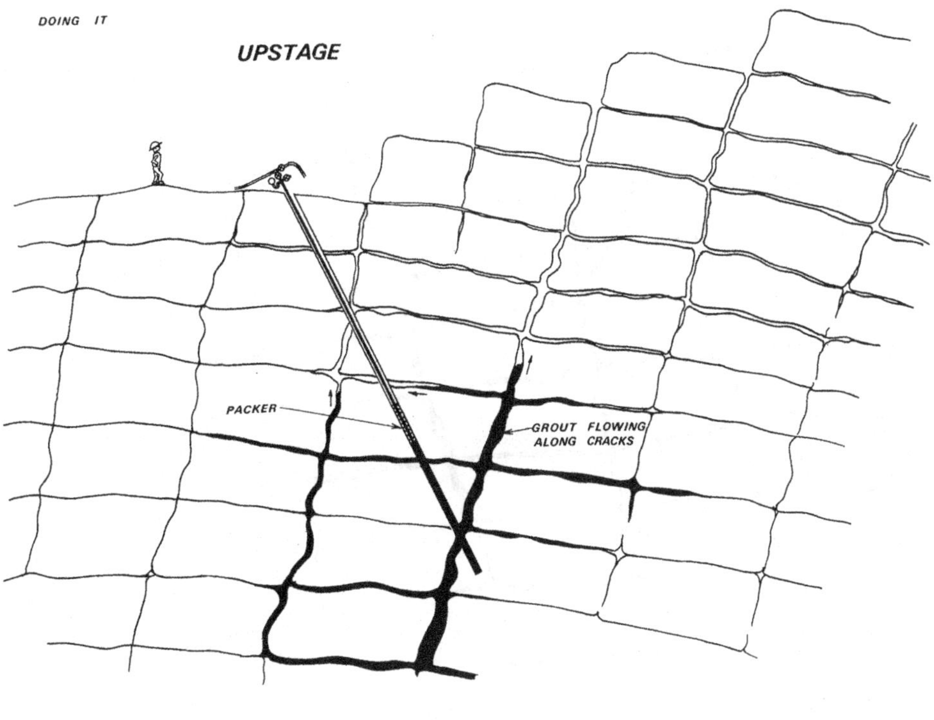

Figure 5.4-8B. *If Upstage methods are used.*

closure before starting the second stage anywhere; the second stage is also completed to full closure before starting the third stage. The third and subsequent stages can probably revert to the usual staggered pattern once the surface has been tightened. The name "Zone grouting" is sometimes given to this arrangement; the name also describes a method whereby every stage is finished to final closure before starting any work on deeper stages.

Another combination is to use downstage methods for upper stages followed by upstage in the lower stages of the same holes.

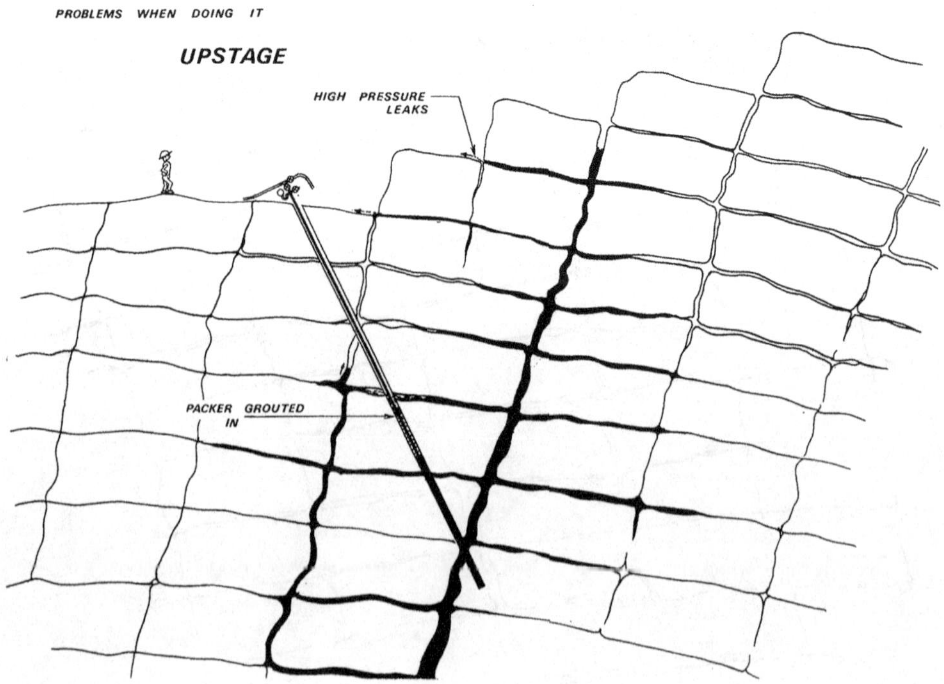

Figure 5.4-8C. Getting into trouble with Upstage method.

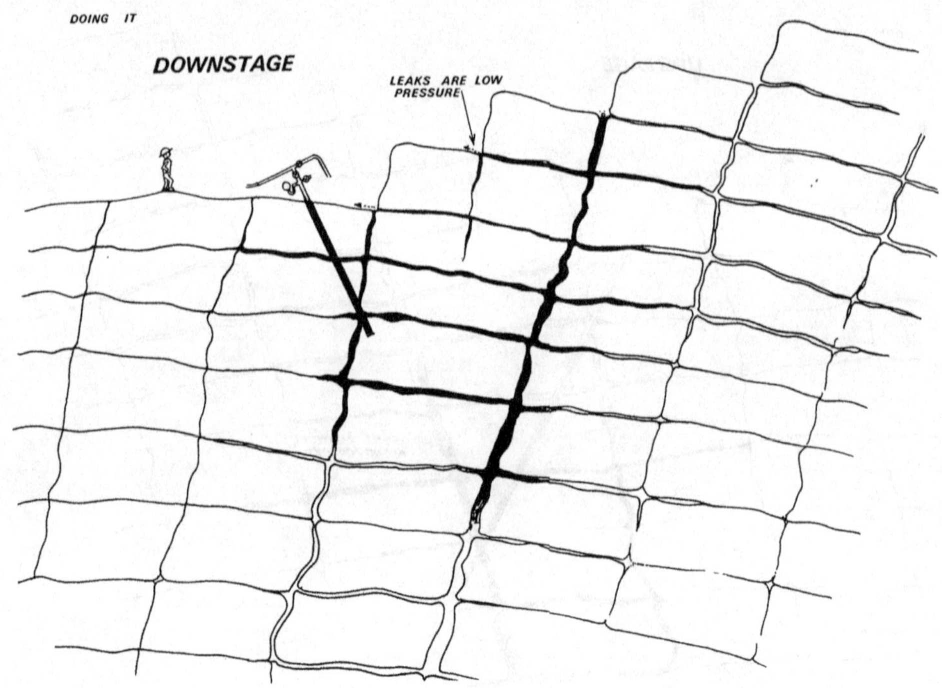

Figure 5.4-8D. Grouting the same site by Downstage methods.

5.4-10 What Can Go Wrong?

In case the impression is gained that the stage grouting methods are easy and straightforward, a few disconcerting reminders are in order. Remember that this book is a *guide*, and the comments and discussions are given in that vein; every grouting site has its own particular requirements, and adaptions are frequently necessary. Some of the particular features upsetting smooth, preplanned use of the methods described have been hinted at earlier. Others are:

1. A surface material may be so weak that despite such precautions as a deep grout cap (Section 5.1-17), low pressure (Section 6.2-1), slush grout over the surface (Section 7.3-9), and considerable caution, the Downstage without Packer method leaves ungrouted weaknesses when hole spacing has been closed down to about 5 ft ⟨1.5 mm⟩. The best way to handle this is to close further to 2.5 ft ⟨0.75 mm⟩ and possibly then to 1.25 ft ⟨0.38 m⟩. On some sites this still does not suffice; further closure should not be in the same row but slightly upstream and downstream. It is difficult to know when to stop in extreme cases. On some sites holes have rather foolishly been closed until they touch: this is taking things too far; maybe grouting was the wrong process there. See Section 8.4-7.

2. Irregular foundations where packers can only be seated here and there in holes and where these depths cannot be predicted. If use of packers is specified as mandatory (unwise, but it happens), this sort of site makes every hole a gamble: sometimes the packer works, sometimes it doesn't. Where it doesn't, fiddling around is necessary to find where it can and to locally modify the stage pattern as best as can be.

3. Grout connection going from a hole being grouted to one being drilled or one standing open after drilling, waiting to be grouted. Treatment of connections is detailed in Sections 7.3-6 to 7.3-8. If upstage methods are in use, some quick action including grouting may be necessary to save the hole to which the hole being drilled is connected. If it is only partly drilled, drilling may have to stop and a replacement hole drilled alongside after the incomplete hole has been (emergency) grouted. If downstage methods are being used, there need be no rush to save the hole connected to; the worst that can happen is that it may have to be washed out or that drilling is interrupted for a while. But probably it can be grouted in the routine way described in Sections 7.3-6 to 7.3-8. The moral of this is that holes should not be drilled too soon and left open in rock prone to connection (or in any rock for that matter).

4. An unexpected series of short stages found necessary to handle a concentration of major seams whose depths cannot be predicted in advance. Ideally the geological investigation (see Section 8.1) should be so good that their positions are known, so the design will have provided the short stages needed for them and there won't be any surprise. If this has not happened, the disruptions they can cause to the daily planning of work can be a headache.

There is plenty of other travail in the daily life of a grouting team.

5.4-11 Circuit Grouting

Now we come to really nasty sites.

Circuit grouting and MPSP grouting (Section 5.4-12) are methods for difficult foundations where hole collapse is prevalent. Circuit grouting (Figure 5.4-11A) makes use of a tube inserted down the full depth of the hole; this tube forms part of the circulation line.

Special fittings are needed. The method is difficult, prone to blockages, and relatively expensive. The blockages happen when material collapsing off the wall of the hole suddenly blocks the annulus outside the tube and this interrupts grout circulation. Then action has to be very quick to retrieve the situation and consists of quickly reversing the grout flow to try to

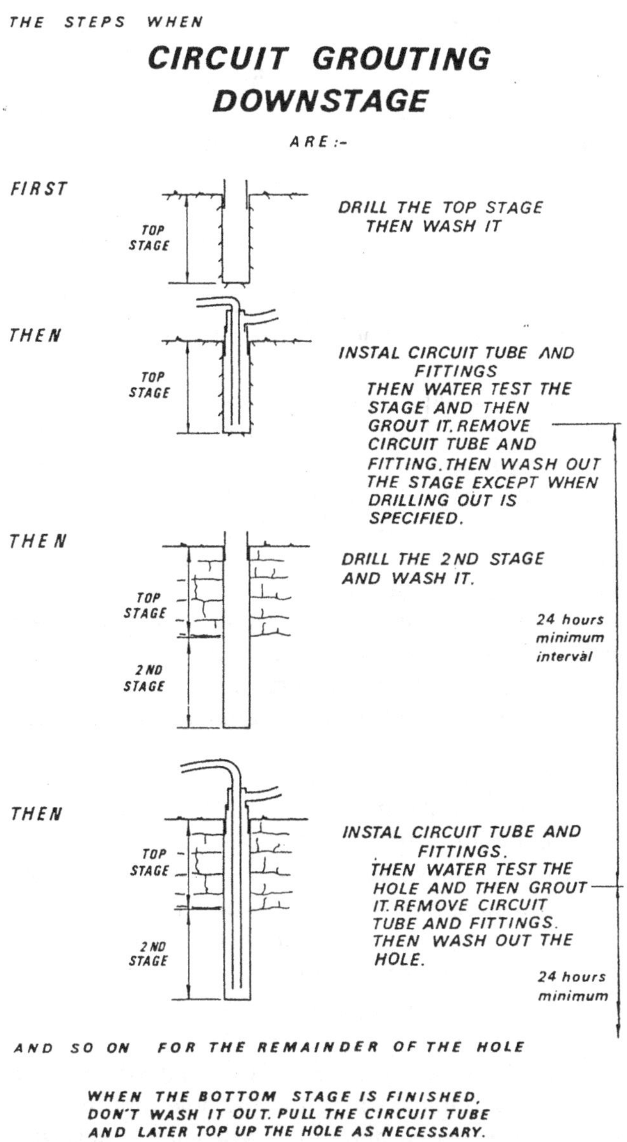

Figure 5.4-11A. Circuit grouting method.

dislodge the blockage; this is not always successful. The layout at the pump has to be organized suitably for these immediate reversals.

The quality of grouting is unlikely to be very good. When things go wrong, the crews are too busy handling the problems, for the actual grout injection. Changing the w:c ratio is slow, grouting cannot be taken to refusal as it should be (Section 7.2-9), and any groundwater picked up from the hole goes back in the returning grout and dilutes the mix in the agitator.

With the recent advent of the MPSP method, circuit grouting has become the least preferred method. It is unlikely to have much use in the future and is only included here for completeness and in view of its former extensive use by such organizations as the TVA.

5.4-12 MPSP

The multiple-packer, sleeved pipe (MPSP) method is the latest for collapsing holes, "sugary" rock, and foundations where simpler methods have run into trouble.

A tube with polypropylene fabric bags attached to it at each stage depth is placed down the full depth of the hole. These bags are inflated with cement grout to make permanent packers at each stage (the "multiple packers") and so define the stage lengths (Figure 5.4-12A).

Prone to collapse, the hole is cased. The multiple-packer tube can then be put down inside the casing and the casing pulled out of the hole to expose the walls prior to grouting the packers in place. Pieces of foundation material are very likely to fall out of the walls and settle around the tube. This does not matter very much provided grout can later find its way through the loose pieces (to get to the walls) and provided the pieces do not block expansion of the packer bags (which could be likely sometimes).

Expanding the packer bags is done this way (Figure 5.4-12B):

(a) The Multiple-packer tube is placed in position ready for grouting of packer.

(b) Grout supply tube with double packer is positioned opposite the port in the MPSP tube with small packers above and below to restrict the flow of grout to the vicinity of this port.

(c) Cement grout is pumped down the grout supply tube and lifts a rubber sleeve over the port in the MPSP tube. Grout then moves into the packer bag and inflates it against the hole wall. A hand pump is used due to the criticality of ensuring correct filling without rupture.

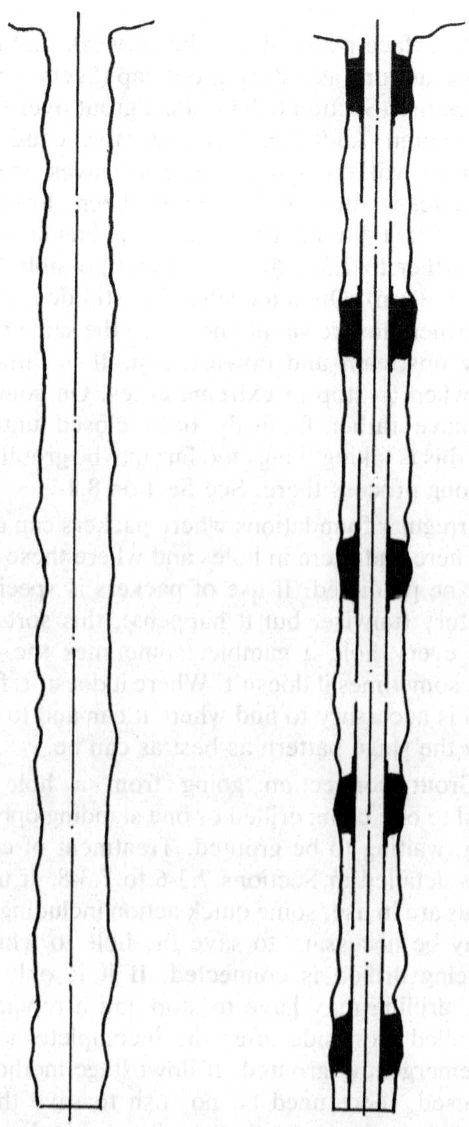

Figure 5.4-12A. *Left: Grout hole drilled full depth. Right: Multiple packer tube in hole with packer bags inflated with cement grout.*

(d) Afterward, the grout tube is lifted out, and the rubber sleeve closes back over the port. High early strength cement is used in the grout, which stiffens quickly when excess water is squeezed out of it through the fabric.

Once all the bags in the hole have been expanded and the grout has set, the water testing and then the rock grouting can take place (Figure 5.4-12C).

The same grout supply tube is used for this as was used to inflate the bags. It is positioned opposite the particular stage to be grouted and its packers seated onto the inside of the MPSP tube. The water test can then be carried out: water pumped down it lifts rubber sleeves located on the outside of two or three ports in

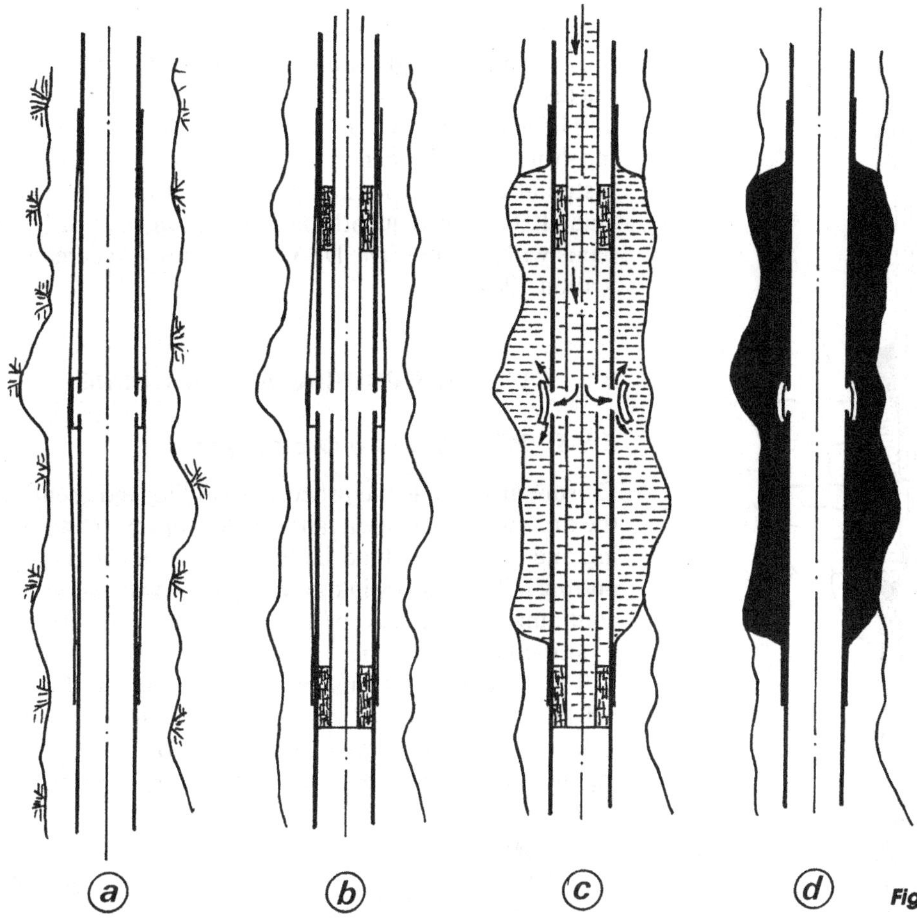

Figure 5.4-12B. *Grouting the packer bags.*

the MPSP tube (only one is shown in the figure). This allows the test water into the space outside this tube and so into the cracks in the rock. Grout then follows in the same way. Water and then grout are restrained by the packer bags from going outside the stage (in the hole at least). Once the grouting is finished and the pressure taken off, the rubber sleeves close over the ports and the grout supply tube can then be moved off to another stage.

After all stages have been grouted in a hole, the MPSP tube, which is left there permanently, is filled with thick grout to stop it becoming a drain hole.

Figure 5.4-12C shows the second top stage being grouted, but upstage is usual. Downstage can also be used. Once a stage has been grouted, giving it a second application can only be done with difficulty.

The process has been simplified in Figure 5.4-12C for clarity. Each stage is usually provided with three ports rather than the single one shown. The rubber sleeves over them are not shown.

The grout hole has to be larger than usual to accommodate all the hardware. Typically 4 in ⟨100 mm⟩ is used. The MPSP tube is about 2 or 3 in ⟨50 or 75 mm⟩ in diameter. The grout used to inflate the bags can have a w : c ratio of about 0.6 : 1 or 0.8 : 1 (by volume); the excess water from it is expelled through the pervious walls of the packer bags. The grouting pressure used for this is 30 psi ⟨2 bars⟩ above groundwater pressure. Higher pressures might rupture the bags.

Figure 5.4-12C shows a packer bag at the bottom of the hole. This might be dispensed with if groundwater pressure is not sufficiently great to push the grout back into the bottom of the MPSP tube after the grouting tube is withdrawn.

Planning of the work is perhaps somewhat easier than with other methods if everything works as it should. It is possible to drill and place the MPSP tubes into a number of adjoining holes before starting any foundation grouting and then to grout these in closure sequence (Section 5.5) to staggered depths (Section 8.4-6). This means that drills can be got out of the way for a while. A return might be necessary, of course, if further closure holes prove necessary.

The MPSP method was derived from the sleeve tube ⟨tube a manchette⟩ method for chemical grouting. This method fills the hole outside its equivalent of the

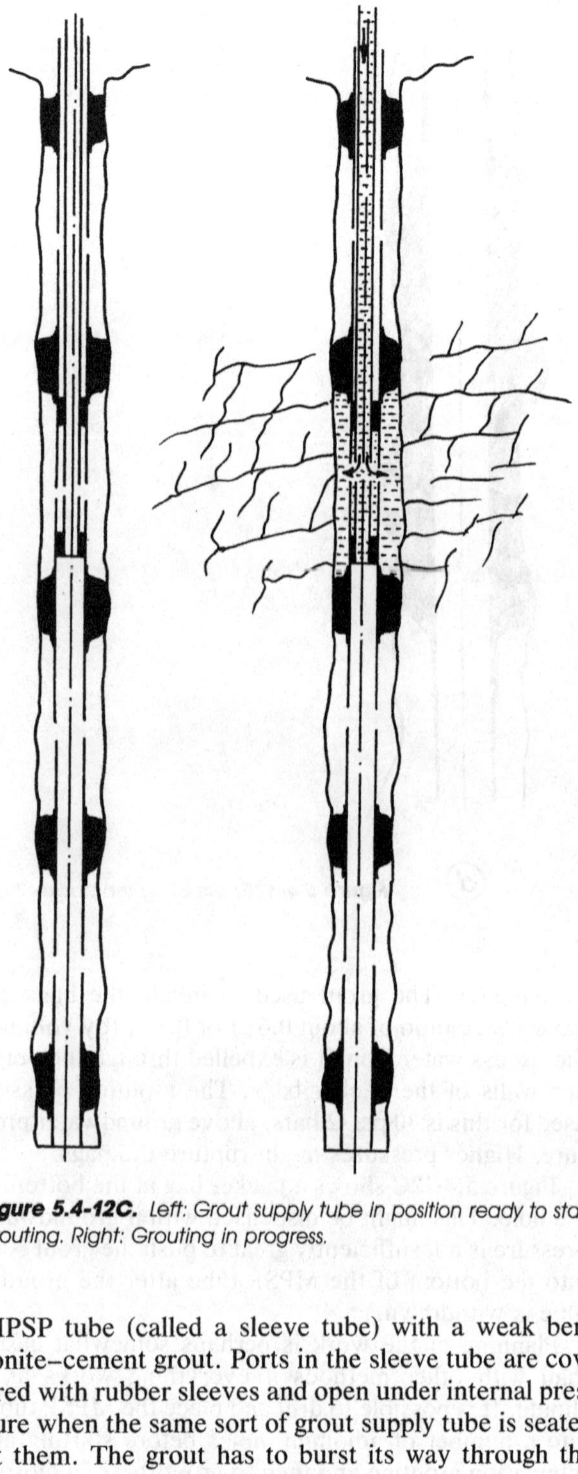

Figure 5.4-12C. *Left: Grout supply tube in position ready to start grouting. Right: Grouting in progress.*

MPSP tube (called a sleeve tube) with a weak bentonite–cement grout. Ports in the sleeve tube are covered with rubber sleeves and open under internal pressure when the same sort of grout supply tube is seated at them. The grout has to burst its way through the firm grout outside the tube. This works well enough with chemical grouting where no matter where the liquid grout reaches the hole wall, it can enter a void system (as in sand or gravel) and find its way through the whole void system. However, the method is unsuitable for cement grouting in rock because the liquid grout is restricted from getting into the cracks by the firm grout outside the sleeve tube (the chances of the burst cracks through this lining up with cracks is remote), and the high lateral restraint offered by rock makes opening the sleeves very difficult.

The MPSP method originated with the Italian grouting firm RODIO in the late 1970s and was initially tested at a dam in Turkey and at Tarbela Dam, Pakistan. It has subsequently been used on several large dams with success (Dr. D. A. Bruce, private communication 1989).

5.5 CLOSURE GROUTING AND PLANNING

5.5-1 What Closure Grouting Is

Closure grouting is the process of drilling and grouting the first holes in an area fairly widely apart, then drilling and grouting a hole midway between them, then putting holes midway in the remaining gaps, and so on, continuing this process of halving centers until enough grouting has been done.

The first holes are usually called *primary* (Figure 5.5-1A). The holes midway between them usually are given the name *secondary* (Figure 5.5-1B). Those halving the resulting spaces are called *tertiary* (Figure 5.5-

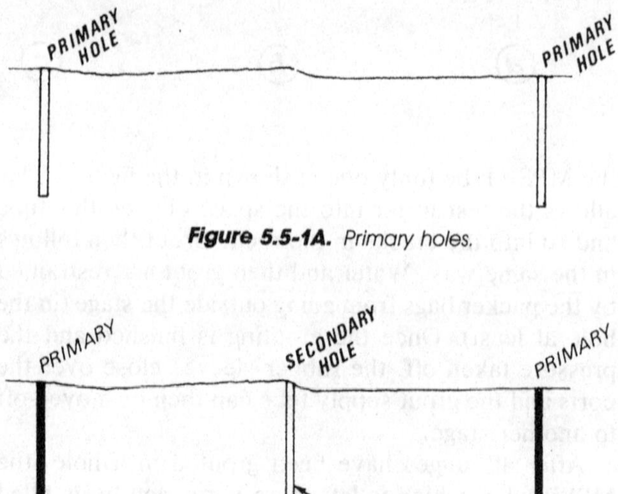

Figure 5.5-1A. *Primary holes.*

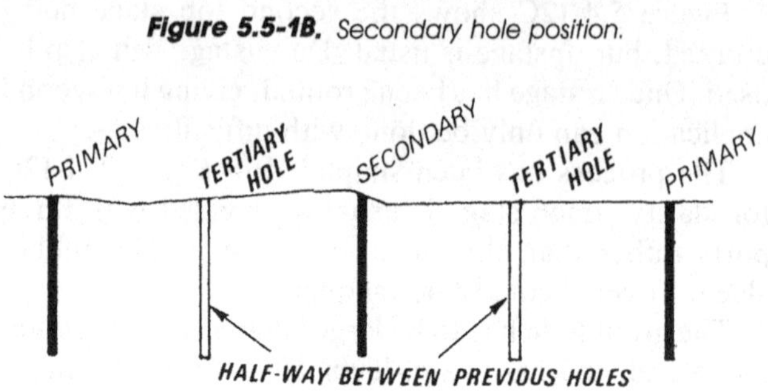

Figure 5.5-1B. *Secondary hole position.*

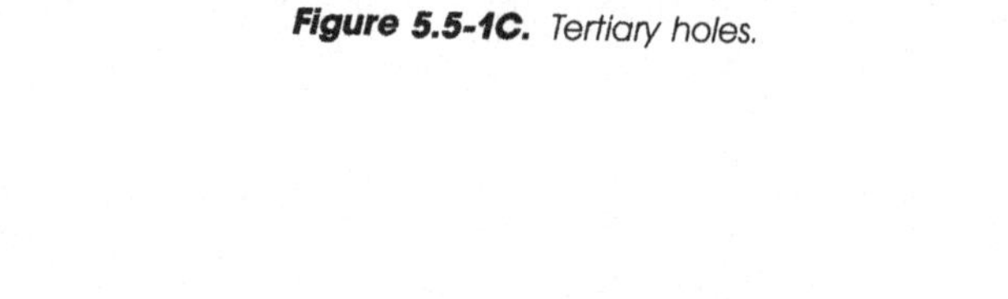

Figure 5.5-1C. *Tertiary holes.*

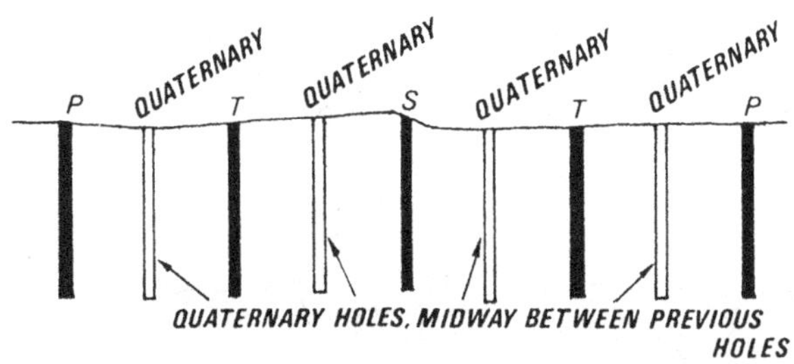

Figure 5.5-1D. Quaternary holes.

1C). The next sequence of holes is called *quaternary* (Figure 5.5-1D). If further closure is required, the names are *quinary, sextary, septenary, octary,* and *nonary,* and by this time the holes would be so close together that there wouldn't be room for any more, in that row at least.

There are other designations for closure holes, for example, A, B, C, D, etc. or 1, 2, 3, 4, . . . but these can get confused with curtain row designations if there are multiple rows (Section 8.4-8) similarly named or with hole numbers.

5.5-2 The Concept of Closure

Ideally, grout should penetrate outward from primary holes, as shown in Figure 5.5-2A. When a secondary hole is grouted, the added penetrations that should result are illustrated in Figure 5.5-2B. When the tertiary sequence of closure is reached, penetrations of grout may start to overlap (Figure 5.5-2C). Quaternary and later holes may be necessary to give complete overlap.

Note that the penetrations shown in these sketches are only indications, and actual penetrations may be very different and nonsymmetrical. The extent of grout travel from grout holes depends on many factors dealt with elsewhere, such as:

Factor	Section Number
Crack size	6.1-5
Grout mix	7.2-2 and 7.2-7
Mixing methods	2.2
Injection methods	7.2
Pressure	6.2
Crack configuration	6.1
Stage method	5.4
Rock movements	7.6
Connections between holes	7.3-6

and there are others.

5.5-3 Where Are Closure Methods Useful?

CURTAIN GROUTING Closure methods are particularly useful for grout curtains; the grouting is closed down to the required standard of tightness (refer to Section 8.3-3) or to some appropriate spacing (Section 8.4-7). By comparing water test results and cement takes in the holes as the closure proceeds, a measure is obtained of the grouting's level of success (or lack of it). This comparison may indicate that changes to tech-

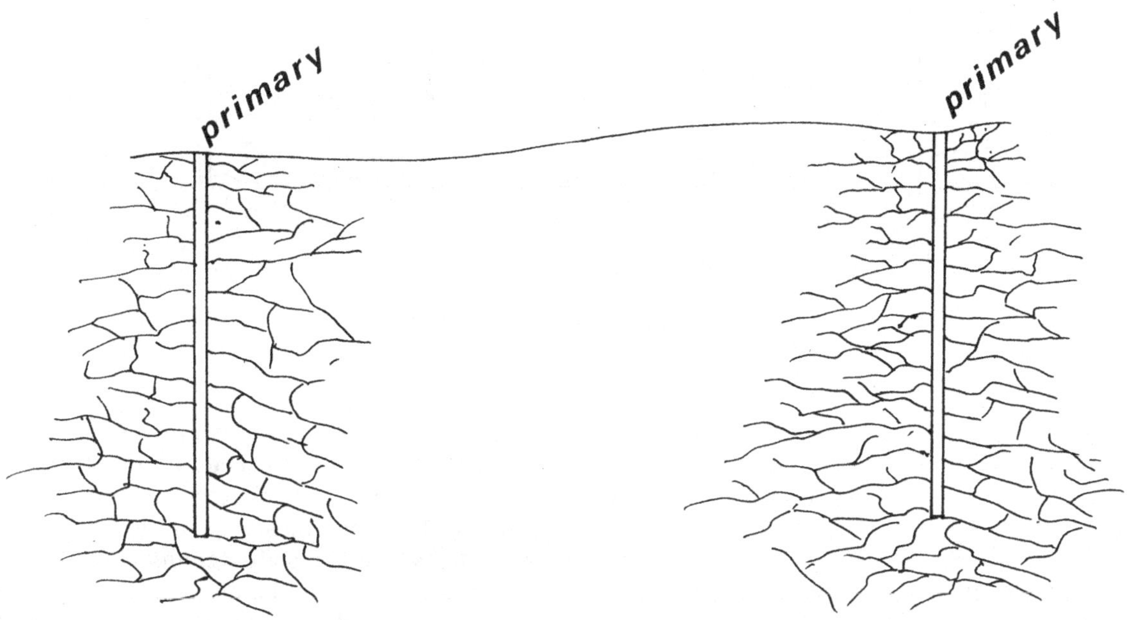

Figure 5.5-2A. Penetrations from primary holes.

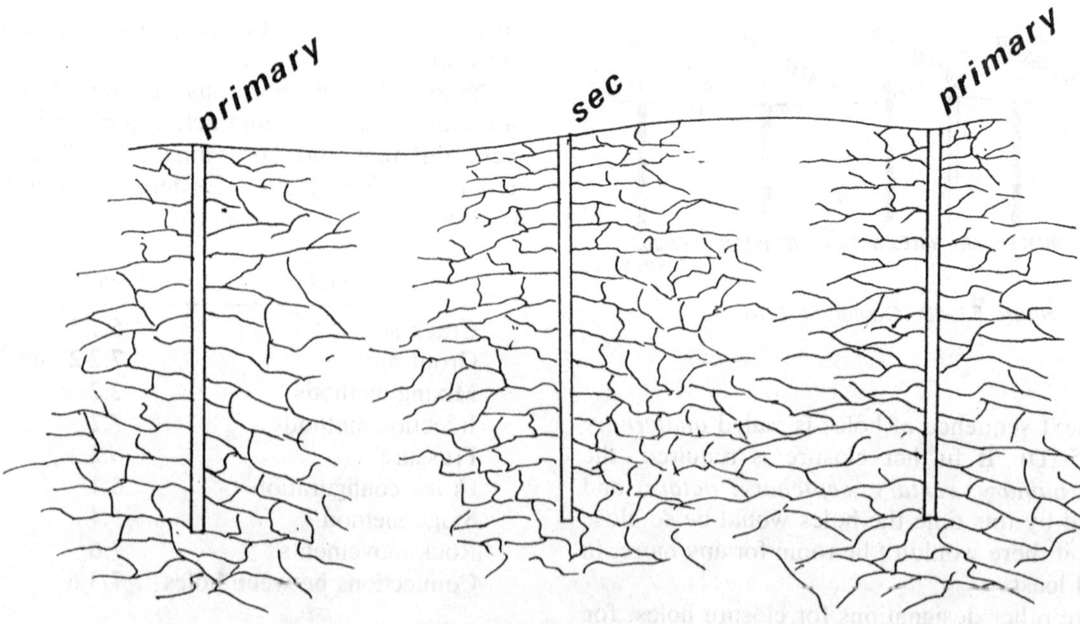

Figure 5.5-2B. *Added penetrations from a secondary hole.*

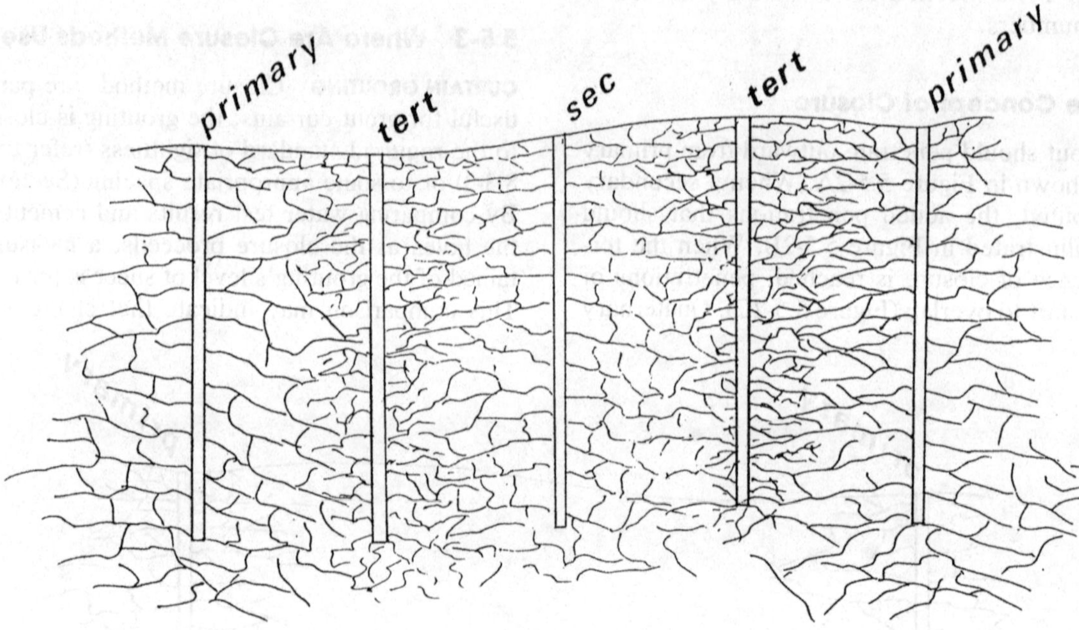

Figure 5.5-2C. *Penetrations from tertiaries.*

niques (see Section 7.5-2) are necessary; it also helps with decisions about when to stop grouting (Section 7.5 discusses how to decide when enough grouting has been done).

BLANKET GROUTING If consolidation or blanket grouting holes are planned on grid, hexagonal, or triangular patterns (Section 8.4-16), closure can often be organized into the layout and can provide useful information as the grouting proceeds. In this type of work it can be difficult to reliably evaluate the efficacy of the grouting by other means than closure assessments: water tests before and after grouting may help (Section 5.3). Cored holes can be quite misleading (limitations of coring as a tool for assessing completed grouting are explained in Section 8.1-5).

5.5-4 Applying Closure to Stage Grouting

As Section 8.4-6 explains, if a grout curtain is designed so that closure is applied at each stage depth, a flexible arrangement results having a pattern which can be curtailed or condensed in accordance with whatever the grouting discloses in the way of takes. When doing this, work on successive sequences has to be arranged to avoid confusion. For instance, when downstage methods (Sections 5.4-4, and 5.4-5) are in use, it is necessary to progress as in Figure 5.5-4A.

In this figure two stages of adjacent primary holes have been grouted, so drilling can start on the top stage of the secondary holes between them. Then when the primaries have been completed in the third stage and the secondaries in the second stage, the tertiaries can be started in the top stage (Fig. 5.5-4B).

The idea is to stay one completed stage behind holes of the previous sequence. In this context this means stage length as specified on the plans. It does not mean extra stages made because of drilling water loss, caving, making water, and so on (refer to Section 5.4-2 for comments on stage lengths).

When upstage methods (Section 5.4-6) are in use, Figure 5.5-4C shows how all stages of primaries have to be finished before starting drilling of a secondary, which then has to be completely grouted before drilling the tertiaries. This is so the complications and poorer grouting arising from connections are avoided as much as possible (Sections 7.3-6 to 7.3-8).

Closure grouting has the drawback that drills and grouting gear have to be moved back and forth over the area until such time as closure is completed. This is not a serious drawback usually, particularly if modern trackdrills (Section 5.2-3) are used, and is minor when compared with the advantages of the method.

5.5-5 Nonclosure Grouting

We briefly consider the alternative to closure—nonclosure grouting.

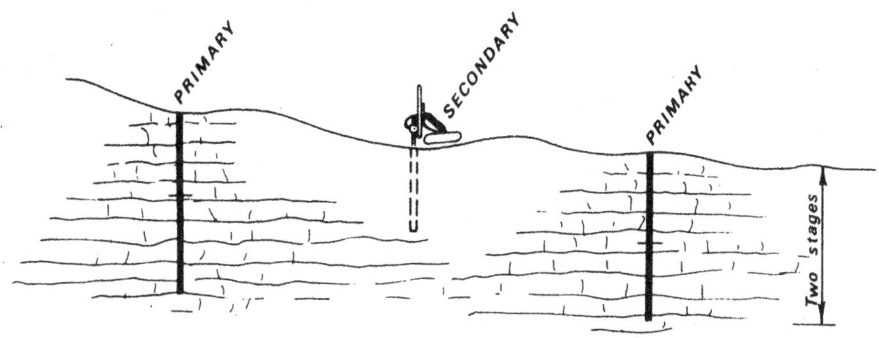

Figure 5.5-4A. *The minimum progress necessary in primary holes before the secondary can be started.*

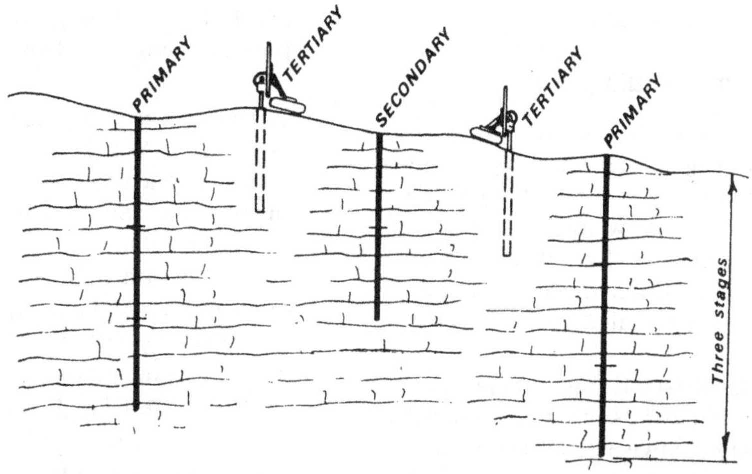

Figure 5.5-4B. *Progress necessary before tertiaries can be started.*

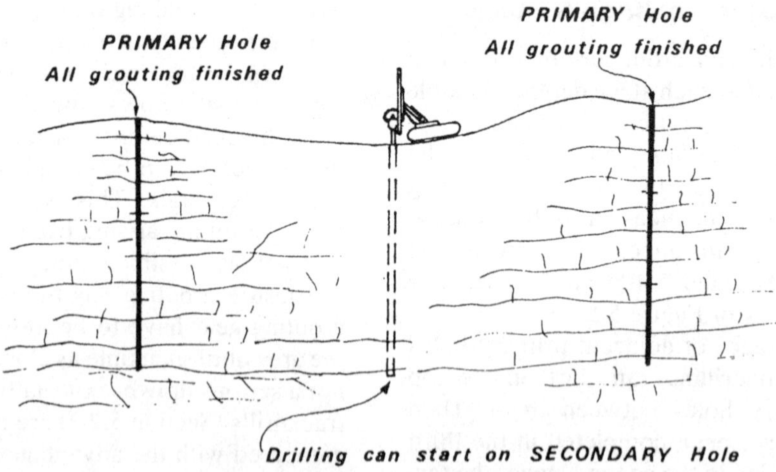

Figure 5.5-4C. *Necessary completion of holes in upstage work before starting next sequence.*

There are cases when lower quality work is permissible, and there are also situations where the fastest possible grouting is required. In these cases, nonclosure grouting can be tolerated. For this, hole spacing is decided in advance from experience, and the holes are drilled and grouted consecutively or on the skip-one-and-come-back system.

However, if work is pushed so rapidly that drilling is in progress on one hole and grouting on the next, there can be problems if grout connects to the drilling. Isolated instances of this are a nuisance, but if it happens repeatedly, progress is unavoidably slowed appreciably. A prudent precaution is to wait at least 4 hours after completion of grouting of a hole before starting drilling the one alongside it. However, this tends to inhibit the speed of the operation—hence the retrospective regrets of some of those who have hoped to save time this way.

5.5-6 Planning Closure Grouting

Considerable benefits moneywise and in morale can come from careful planning of the work as closure proceeds. Lack of planning can result in crews sitting around with nothing to do while waiting for the next hole to become available, wasteful plant movements and usage, and so forth. In addition to wasting money, morale drops when planning and direction are seen as confused.

Careful planning looks ahead. It forecasts progress at least a day in advance and makes sure that there will be minimal idle time and no unwise plant movements.

There are various ways for keeping track of progress (refer to Section 7.5-3). The method most commonly adopted for forward planning uses plots with holes shown in elevation. Progress is indicated in each hole by color schemes such as one to show that drilling is done, another to indicate completion of water testing, another for the first application of grout, and others for any subsequent applications. Problems such as leaks, connections, and so on, which require a further application, are indicated. Updating of plots should be carried out every few hours and forward predictions and plans modified as found sensible. Experience is necessary to make reliable predictions.

The forward planning process takes account of:

- The various times when holes will become available for action. These times are governed by stipulated time intervals (Sections 5.4-4 to 5.4-6) from previous action such as an earlier application (Section 7.2-10) of grout, from completion of the previous stage, or from treatment of a problem (Section 7.3).
- Progress in grouting closure sequences in the various stages. As indicated in the preceding, the idea is to stay one completed stage behind the holes of the previous sequence if downstage methods (Sections 5.4-4 and 5.4-6) are in use. If grouting upstage (Section 5.4-6), all stages of previous sequences have to be complete before proceeding to the next phase of closure.
- Any unplanned deepening of holes found necessary because of high water and/or grout takes in earlier holes (Section 8.4-6).
- Any unplanned additional holes found necessary because of high takes or because required stan-

dards of tightness have not yet been achieved (Section 8.4-6).

- Any unplanned deepening or additional holes found necessary because of connections between holes or leaks (refer to Section 7.3).
- Additional applications of grout to be given to stages where takes have exceeded stipulated amounts (refer to Section 7.2-10) and where treatment of very high takes is being given using the method of restricting the amount permitted to be taken each day (refer to Section 7.2-9).
- Special treatment where water testing has indicated that there will be connections to other holes when grouting starts.
- Any modifications to hole patterns and sequences in areas of difficult topography or adjacent to structures that require special care when grouting nearby or underneath. Examples are grouting alongside a tunnel from the surface and grouting under a concrete building or dam.
- Any unexpected problem causing delays in output, such as loss of drill rods down a hole, motor breakdown, burst hoses, pressure gauges found to be faulty, packer leakage or loss, and so on.
- Any shortages of plant occasioned by earlier breakdown or servicing requirements that will affect progress of one part of the work.
- Any shortage of grout ingredients warranting working on low take holes until adequate supplies are restored.
- The individual idiosyncrasies of main people in terms of their output and their effect on others. Drillers in particular vary widely in ability: some have high output but don't flush holes thoroughly (Section 5.2-14), which leaves more work for later washing out, whereas others produce good holes and minimize problems later.
- The nearness of the end of work for the day. If grouting of a hole is in progress, it is not permissible to stop until grouting is finished. Even with long experience it is difficult to predict just how long each hole will take, and some inevitably run longer than expected. On some jobs, arrangements are made for the crews to be paid overtime or to have time off in lieu. On others, care is taken to avoid starting holes that might run into overtime. Both methods produce wasted time, particularly, the latter. While some of the wasted time can be used cleaning up, it can eventually amount to a considerable cost. This is one of the reasons for working around the clock on some big jobs; incomplete grouting is merely handed over from one crew to another at the time of shift change.

There are many other possible factors to be taken into account, far too many to list here. Some are particular to a specific job or parts of a site.

Decisions often have to be made quickly. The planning should always be on top of the situation to ensure timely decisions.

Instructions to crews need to be clear and should preferably include a repeat of instructions to other crews nearby whose work may be influenced. A simple way to do this is to post a notice in some central location such as at mixers, on which the programs for all crews operating from those mixers are displayed. Invariably there are last-minute changes to instructions, and these should be clearly shown on such notices.

Traditionally, forward planning has been done by experienced people. An alternative is to do it by computer (Section 7.7). A loss of flexibility is inherent in doing so because it is not possible to write a program that caters to all of the many possibilities.

REFERENCES AND FURTHER READING

Bruce, D. A. (1984). "The drilling and treatment of overburden." *Geodrilling,* August and October.

Bruce, D. A., and Gallavresi, F. (1988). "The MPSP system: A new method of grouting difficult rock formations." ASCE Geotechnical Special Publication No 14. Geotechnical Aspects of Karst Terrains, 97–114.

Heinz, W. F. (1985). *Diamond Drilling Handbook.* South African Drilling Association, Johannesburg.

Houlsby, A. C. (1986). Discussion of "Progress in rockfill dams." *ASCE Journal of Geotechnical Engineering,* Vol. 112(2), 226.

Independent Panel to Review Cause of Teton Dam Failure (1976). "Report to U.S. Department of the Interior and State of Idaho on failure of Teton Dam." December. U.S. Bureau of Reclamation, Denver.

Lamberton, B. A. (1982). "Swedish valve tube grouting." *Proceedings of Conference on Grouting in Geotechnical Engineering.* ASCE, New Orleans, 907.

CHAPTER 6

What Happens during Injection

The flow of grout in various types of cracks (Section 6.1), and the pressures needed (Section 6.2) are amongst aspects of penetration dealt with in this chapter.

6.1 HOW GROUT FLOWS IN CRACKS

6.1-1 Difference between Grout and Water

Grout doesn't flow through cracks like water. The difference comes from the presence of cement particles, from density currents in the grout, from thixotropic stiffening of it, and from a few other things. Section 4.2 discusses viscosity and thixotropy aspects.

6.1-2 Ease of Grouting

In visualizing what happens when grout travels through cracks or fissures or bedding planes or shears or joints or whatever you like to call them, it is necessary to appreciate that some cracks are easier to grout than others. The easiest are near vertical, the hardest can be those near the horizontal. And note that the discussion here does not necessarily apply to voids, large spaces, or those very wide cracks where the grout can easily flow just about as freely as if it were in a hose. A very arbitrary dividing line could be drawn at about 0.1 in ⟨2.5 mm⟩; cracks and spaces wider than this and extending an inch or more ⟨more than 25 mm⟩ across the direction of grout flow can often pass grout more freely and in a rather different manner than now shown.

Figure 6.1-2A shows easily grouted cracks and ones not so easily grouted.

Penetration and the thoroughness of filling are helped if water and thin grout can be easily displaced by the incoming grout; this happens readily in most of those cracks that are near the vertical. They act as venting paths for these lighter fluids while denser parts of the grout can settle downwards at first. Then this denser, good grout builds up in the cracks while continuing to float off the lighter fraction, until eventually only dense grout remains.

Near-horizontal cracks do not provide this natural venting, but if upward cracks lead off from them at frequent intervals, some degree of venting can take place. However, if venting of this nature is not available, near-horizontal cracks can be the hardest to grout properly.

HORIZONTAL CRACKS The nature of grout travel in a horizontal crack is shown in Figure 6.1-2B. In this example, the worst case is considered: there are no upward-venting cracks. Grout travels radially outwards from the grout hole and has nowhere else to go but along the crack.

Initially, the grout moves freely under the pressure head in the hole. The pressure decreases as the distance from the hole increases, until at the tip of penetration it is relatively slight.

Eventually, the limit of grout penetration is reached when the pressure at the fringe of the grout has reduced so much that it cannot keep the grout moving any more (Figure 6.1-2C). The grout tends to push out in strings and meanders, among groundwater in the crack. The effect resembles the delta of some large rivers; photographs are shown in Houlsby (1985) and in Section 6.1-6. The strings of grout can penetrate appreciably out into groundwater regions, leaving very weak and uncertain grout depositions in the fringe area. Deposition patterns are believed to be affected by other factors besides pressure: density currents,

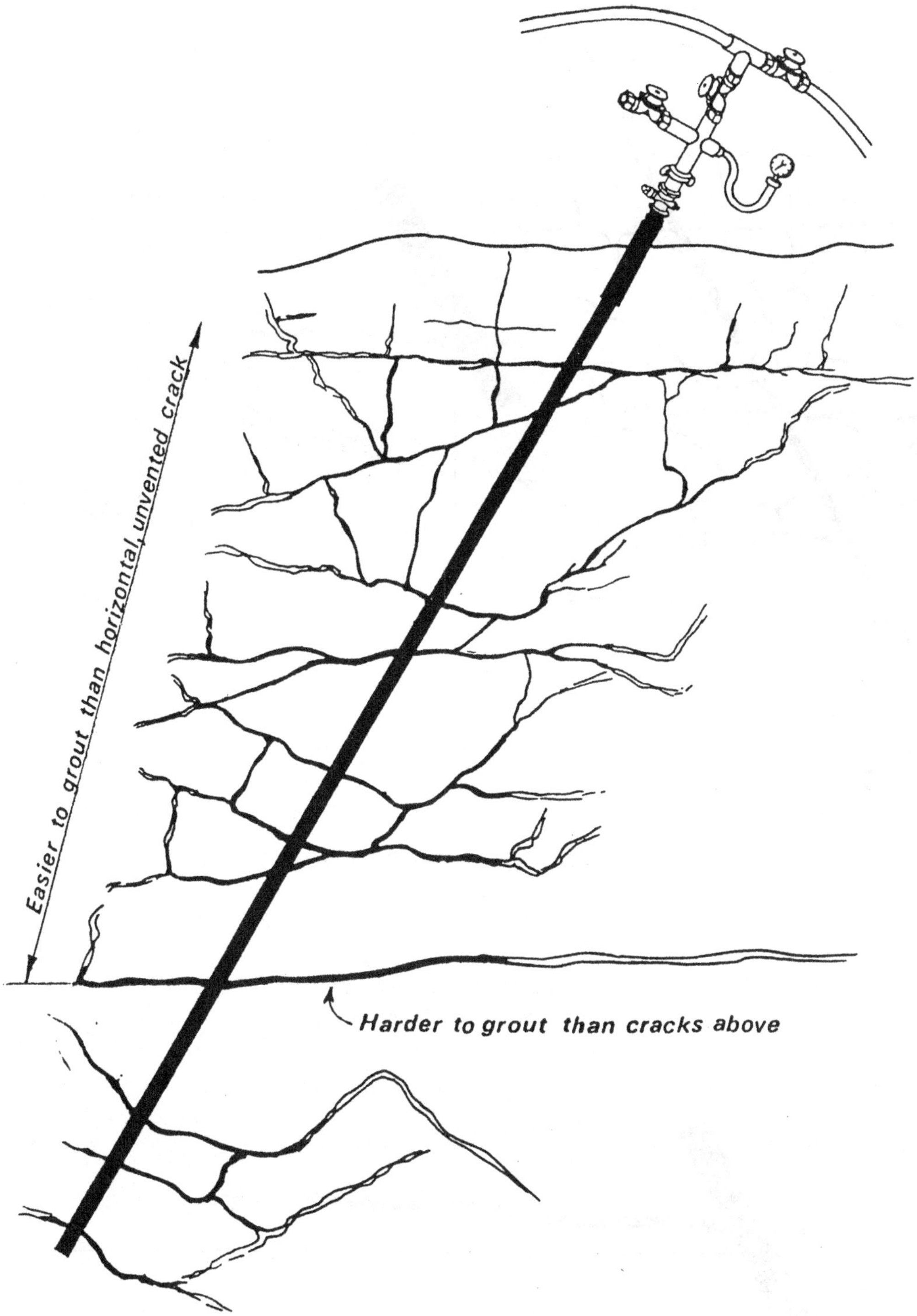

Figure 6.1-2A. *Some cracks are harder to grout than others.*

molecular attraction, and the nature of the crack's surfaces appear to contribute.

Up to the time when the limit of penetration is reached, grout further back in the crack has been in motion and additional grout has been entering the crack from the hole. There has been little opportunity for bleeding to start. Now the cessation of movement lets thixotropic stiffening occur (Section 4.2-2) and lets bleed water commence separating out from the grout (see Figure 6.1-2D). Because there is little room for this to occur in the upward direction, it takes place in a lateral fashion leaving ridges of grout separated by me-

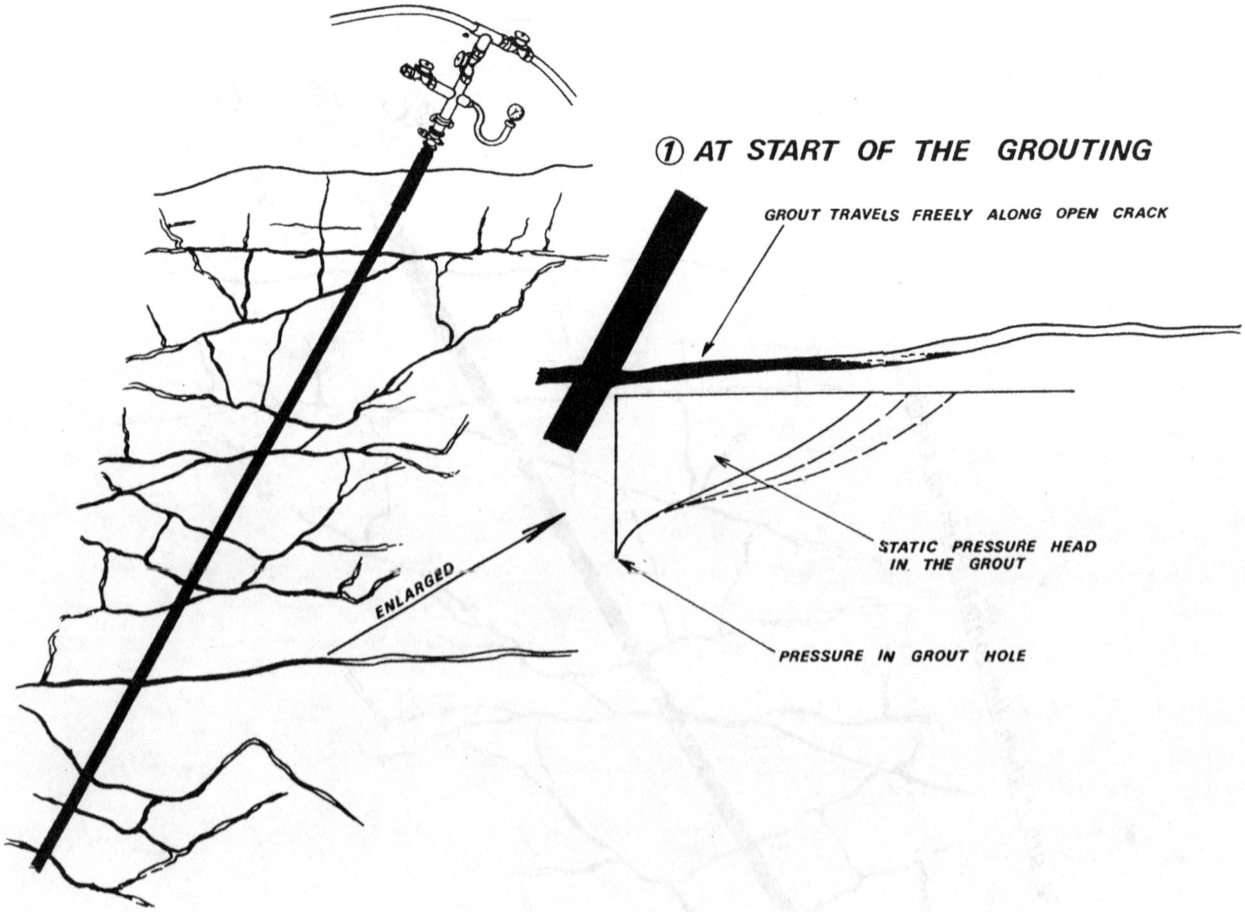

Figure 6.1-2B. *Initial penetration.*

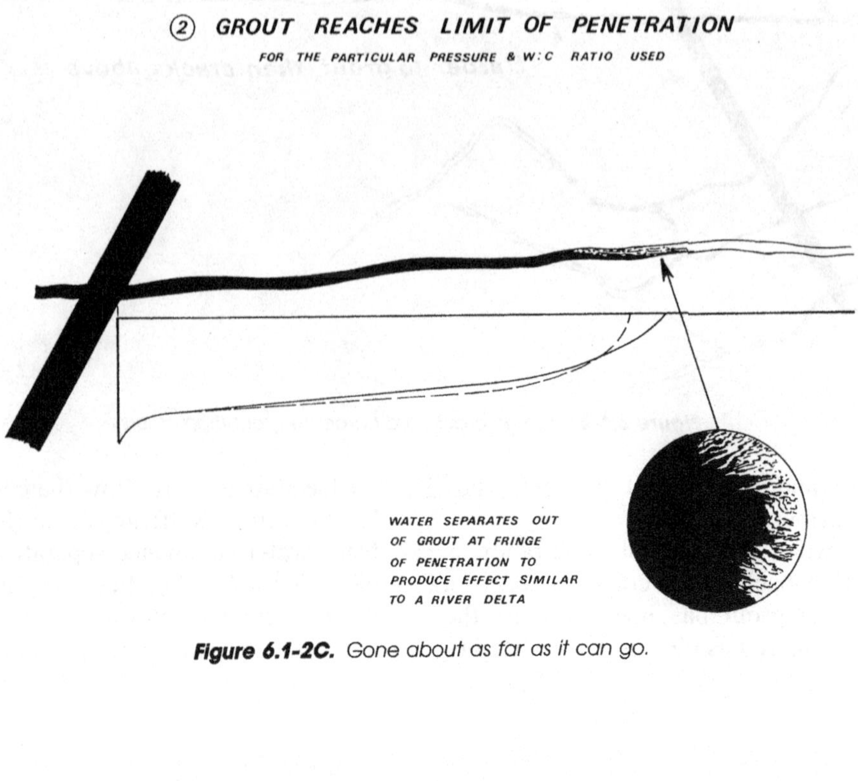

Figure 6.1-2C. *Gone about as far as it can go.*

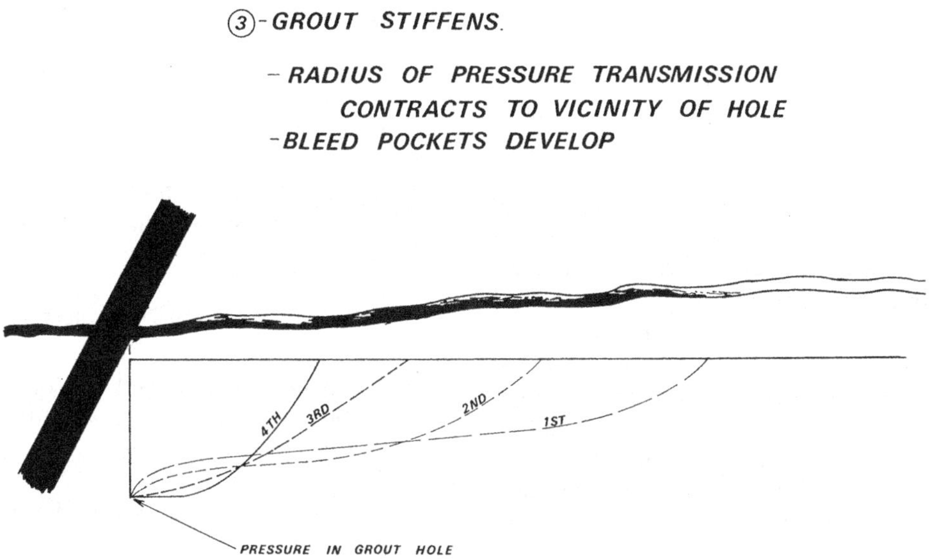

Figure 6.1-2D. *Movement ceases.*

anders of bleed water. This lack of a sound grout filling can lead to lack of durability, as discussed in Section 4.5. The nature of fillings is shown and commented on later. The pressure distribution in cracks of this type has been studied in extensive tests outlined in Section 6.1-6.

INCLINED CRACKS Now let's return to the more ordinary cases of inclined, intersecting jointing patterns: grout tends to have a generally outward radial movement from each grout hole, but follows the path of least resistance. This is usually through the widest cracks or through the wide part of cracks of variable width. Penetration into narrower regions then follows; sometimes this has to wait until penetration in the wider paths has reached its limit and then the resulting "bottling-up" raises pressures sufficiently to force grout into the narrower regions back nearer the hole. Other times if the wider paths are only slightly more free than the narrower ones, there is more or less simultaneous penetration. Of course, as the radial distance from the hole increases, good grouting takes place in the freer paths more than in the narrower ones because of the diminished pressures available at these distances. This is why closure grouting (Section 5.5) aims to provide overlapping penetrations from the final holes so as to make sure that narrower cracks receive adequate attention. Further comment on flow in crack networks is given in Section 4.2-3.

UNSYMMETRICAL FLOW In many situations the jointing is not symmetrical, and grout tends to go off in one particular direction from a hole rather than moving radially. The same thing can occur in more symmetrical foundations if groundwater is flowing through. Unsymmetrical penetration needs to be recognized and hole patterns adjusted suitably.

WATER TABLE AND SATURATION EFFECTS Below the water table, the push exerted on grout by pressure from a grout hole is reduced by back pressure from groundwater. If the groundwater is well below the top of the hole, back pressure is slight; at the other extreme it can be severe if the grouting is in an area of high water pressure.

When good-quality grout penetrates into water-filled cracks, it tends to push the water aside and remains as a cohesive entity itself, rather than becoming dispersed and thinned by the water. This is one of the reasons for preferring high-shear mixing (Section 2.2-2). However, if grouts far thinner than those recommended in this book are in use, some dilution even of high-shear mixed grout may be expected (i.e., in mixes of 5 : 1 or thinner).

Most cracks have water in them when being grouted. Even if above the water table, cracks become partly or completely water filled during grouting operations. This is not as pronounced in primary holes (Section 5.5-1) above the water table as it is in later holes of closure sequences when considerable amounts of water from drilling, washing out, and water testing have accumulated from operations on earlier holes.

ABSORBENT FOUNDATIONS Some sites in very arid regions contain porous rock able to absorb water from grout. Thinner grouts than usual are then necessary to provide workable w : c ratios after this partial water loss. However, saturation of the rock pores can eventually stop the water losses if joints are sufficiently numerous to give enough access for this to happen. Later holes in closure sequences can then be treated with normal w : c ratios.

A REMINDER–THIS IS ONLY A GUIDE Having outlined all these general concepts, it is now somewhat disconcertingly necessary to point out that departures from them are common in most sites. This is because of the highly variable nature of (ahem . . .) nature: geological origins produce an infinity of differences. And this is where it is necessary, in common with the descriptions of many other things in this book, to once again give a reminder that the concepts outlined (and they are no more than outlines—full descriptions of possibilities would be voluminous) are for guidance only. There is nothing like a good background of extensive experience to provide one with a healthy respect for the mercurial meanderings and capricious contortions of grout penetration.

6.1-3 Density Currents

If grout is moving slowly or is stationary, cement tends to settle and some of the water content rises. Thin mixes, such as w : c = 5 : 1, do this more strongly than thick mixes, such as 0.8 : 1: see Section 4.3-1 for details on bleeding. The cement as it settles makes a thicker grout than before, and this moves downwards whenever it can, rather like a sludge working its way down through dirty water in, say, the bed of a silt-laden river. This is the dense type of "density current." In rock grouting these currents are most relevant in the steeper cracks; they contribute to obtaining good grouting: the dense currents gravitate down, and the light ones float up, and this process continues until the dense, good grout finally displaces all of the light fluid.

Although not directly relevant here, understanding of the concept may be helped by visualizing the action of density currents in voids and large spaces where the sinking of the denser fraction down through the lighter is a dominant behavior pattern. For a study of their effects in, for instance, the large spaces around ground anchors and prestressing cables, see Houlsby (1988).

Keep in mind that these sludgelike properties can be transformed into much more liquidlike ones if the grout is energized enough. This is due to thixotropy (Section 4.2-2).

6.1-4 Grouting of Fine Cracks

The previous discussion is in terms of cracks in general. Thinking now in terms of fine cracks only, the question is often asked: what are the finest cracks that can be grouted with cement? The answer, from my extensive experience in actually doing it is:

Finest crack sizes able to be properly grouted are about

- 0.020 in ⟨0.5 mm, 500 microns⟩ without special care
- 0.015 in ⟨0.4 mm, 400 microns⟩ with extra care using high-quality grout, or a little finer under very experienced supervision.

In case these figures are a little hard to visualize, it is easier to remember that the thickness of an ordinary office pin (as used to pin pieces of paper together or in dressmaking) is about 0.025 in, that is, just a little fatter than the finest crack easily groutable.

It is certainly possible to get cement to move through finer cracks than these: even the coarsest cement grains are less than these widths. However, there is more than just this factor in the production of thorough grouting, as discussed shortly.

High-quality grout made with high-speed, high-shear mixers (Section 2.2-2) is able to produce better grouting in finer cracks than lower quality grout; the reasons for this are explained here, and the effect has been shown in tests described in Section 6.1-6.

The sizes listed are for grout made with ordinary cement such as ASTM Type I. As explained in Section 3.1, finer cements such as ASTM Type III (high early strength) or fine grinds of cement are able to penetrate better. Even better is Microfine cement, but it has some disadvantages (Section 3.1-6); it can penetrate cracks down to hairline size if grouting of them is really warranted. However, a realistic attitude about very fine cracks is that if the commoner types of cement are unable to effectively grout them, then grouting is probably not really needed; this reasoning is applicable to jobs such as most dams where some seepage can indeed be tolerated usually, but is certainly not appropriate if a high degree of tightness is needed such as in nuclear waste containment.

6.1-5 What Happens in Fine Cracks?

Consider the way the various sizes of cement particles move through a crack. Figure 6.1-5A shows the situation when they are moving freely. The finer grains move along through the constriction until some larger

Figure 6.1-5A. *Cement particles penetrating a crack.*

ones get there, and one or two may stick. Other large ones can then block up behind until the resulting "bridge" stops the flow (see Figure 6.1-5B).

Figure 6.1-5B. *Penetration stopped by bridging.*

The same thing happens if the grout has grains stuck together in clumps or flocs, instead of being separate grains (see Figure 6.1-5C). The clumps can become blocked in constrictions.

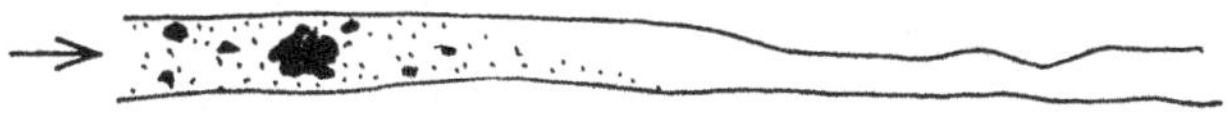

Figure 6.1-5C. *Clumps able to stop penetration.*

If individual grains bridge across a constriction, this results from a degree of interlocking at the (relatively) rough surfaces of the larger grains. If this mechanical interlocking can be prevented by coating each grain with a layer of adsorbed water, the grains are changed into smooth slippery objects that do not interlock and in which the actual cement particles do not touch each other. These are properties of high-quality grout produced by high-speed, high-shear mixing (Section 2.2-2), but lesser quality grouts may have them to some extent also. Additional benefit is obtained from the rounding of grains and production of further fines in the high-shear type of mixing. Figure 6.1-5D shows

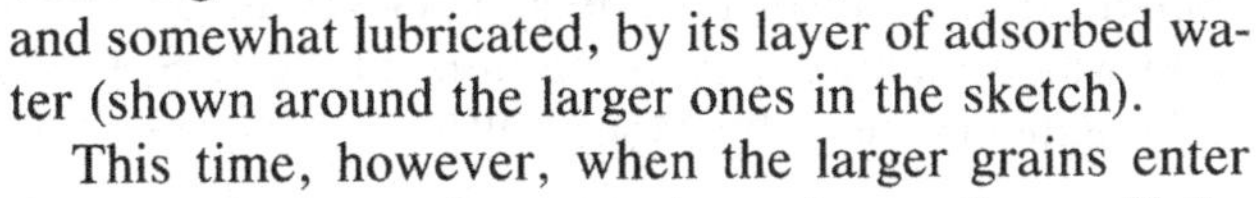

Figure 6.1-5D. *Individual lubricated particles from high-shear mixing.*

what happens in the case of converging cracks. The cement grains move along the crack, each surrounded and somewhat lubricated, by its layer of adsorbed water (shown around the larger ones in the sketch).

This time, however, when the larger grains enter the convergence, the water layer keeps them off the crack walls and reduces any tendency for catching there. Hence the individual large particles slip through without blocking (Figure 6.1-5E).

Figure 6.1-5E. *High-shear mixing causes less blockages.*

A feature of high-quality grout is the absence of clumps. The mixing action separates each grain (refer to Section 2.2-10). Hence blocking by clumps is avoided.

If blockages do happen to occur by bridging or by clumps catching, grout usually has the ability to bypass by traveling around the sides of the blockages. Anyway, most cracks are not precisely uniform in width so the grout moves fastest through the wide parts and then migrates sideways, and perhaps backwards, into the narrower parts later, in due course.

The previous sketches show grout as though it were water with inert independent cement grains scattered through it. Such presentation is not really correct and has only been given to illustrate the basics. Grout should rather be thought of as a complex where the grains relate to each other chemically and electrostatically, and there is some interaction between projecting crystals (Section 4.4). And in thick grouts the cement content is sufficiently high that distances between grains are not great, relatively. The effect of this produces what might freely be called a viscous soup; it is radically different in flow characteristics from just water with specks in it. The differences are elaborated in Section 4.2 in the general context of viscous flow and are the reasons why flow in fine cracks cannot be realistically thought of only in terms of specks of cement bridging across constrictions.

6.1-6 Tests on Penetration in Cracks

Speculation as to what happens when grout travels along cracks has long intrigued people associated with grouting, and this has fuelled a number of test programs. The most useful of these was probably the one conducted by the U.S. Army Corps of Engineers (1956), which commenced in 1949. Subsequent tests were carried out in 1962 by the author and associates in the Water Conservation and Irrigation Commission, N.S.W., Australia, and are reported in some detail in Houlsby (1985). There have been other, perhaps less extensive, tests in various parts of the world. Interesting extracts from these two large test programs follow.

U.S. CORPS OF ENGINEERS (1956) Grout was passed along an artificial crack between pairs of concrete slabs. The slabs were 52 in ⟨1,321 mm⟩ long, 7 in ⟨178 mm⟩ wide, and each was 3 in ⟨76 mm⟩ thick. The grout was injected near one end and traveled 48 in ⟨1,219 mm⟩ to an exit near the other end. The slabs were laid horizontally. Simulated crack widths tested were 0.010, 0.020, and 0.030 in ⟨0.254, 0.508, and 0.762 mm, or 254, 508, and 762 microns⟩. Shims provided the crack openings and were placed just inside the crack edges so as to give a seal against grout escaping except from the intended exit. Neat cement grouts of a range of w : c ratios were tested and so were a number of additives. Crack surfaces were smooth in some tests (the concrete was cast against glass) and rough in others. Pressures along the crack, flow rates, and other factors were measured. After some tests, the grout was allowed to go hard, and the slabs were later opened up for visual inspection of the quality of the filling.

Water: Cement Ratio versus Crack Size One of the features of interest in the testing was finding the thickest grout able to pass through the various crack sizes. The method used to find this was, unfortunately, somewhat irrelevant to real work: instead of trying various mixes to find which would enter the cracks and which would not, the testing started with thin mixes that were gradually thickened until refusal occurred. The results, therefore, have more in common with excessive thickening during an application (as deplored in Section 7.2-7) than with determination of starting mixes (Section 7.2-2). For what it is worth, the following tabulation lists the relevant results in terms of the thickest grout (without additives) able to enter:

- 0.010-in ⟨0.254-mm, 254-micron⟩ cracks:
  4 : 1 at 25 psi ⟨1.9 bars⟩
  2 : 1 at 50 or 100 psi ⟨3.4 or 6.9 bars⟩
- 0.020-in ⟨0.508-mm, 508-micron⟩ cracks:
  0.7 : 1 at all test pressures
- 0.030-in ⟨0.762-mm, 762-micron⟩ cracks:
  0.6 : 1 at all test pressures

The w : c ratios are quoted here by volume, as in the rest of this book (see Section 4.1-1); the actual test report (U.S. Army Corps of Engineers, 1956) gives them by weight.

Cement Grain Size versus Crack Size The tests used ASTM Type II cement, which had coarsest grains comprising 0.1% in the 50 to 100 mesh range, that is, between 0.010 and 0.006 in ⟨0.254 and 0.152 mm, 254 and 152 microns⟩. This is coarser than recent cements commonly used for grouting. Specific surface of some of the cement was 3,172 sq cm/g Blaine. Mixing was by a small paddle mixer; the grout would be classified as low quality in terms discussed in Section 2.2-4.

It was possible to pass this cement through the 0.010-in ⟨0.254-mm, 254-micron⟩ cracks, although having to resort to watery, excessively thin mixes unlikely to be durable (as discussed in Section 4.5). Note that some of the cement grains were about the same size as the cracks and yet bridging and blocking did not occur with these thin mixes; refusal, even in these rather narrow cracks, did not occur until excessive thickening was reached. The testing report rather surprisingly draws the conclusion (p. 22) that stoppage occurs when the crack width–grain size ratio is 1.7. Presumably this figure is based on the 0.006-in ⟨0.254-mm, 254-micron⟩ grain size; the reasoning is obscure. The report then goes on to state, in reference to this ratio, that "it is believed that a safe value should exceed 3." Note that the justification for this figure is not given by the writer of the report; it cannot be deduced from the testing. Some discussion is given of other work relating to grout penetration in sands (but not rock) carried out by Machis (1948). He found that the ratio of pore diameter to cement grain size necessary to permit penetration is at least 5. Perhaps this was kept in mind when the figure of 3 was (perhaps arbitrarily) put in the report.

Unfortunately, the ratio of 3 has been subsequently quoted by a number of authors, apparently without critically reviewing its source.

Pressure versus Crack Size Pressure influenced the thickness of mixes able to be passed through the finest crack but had little effect on the other two sizes, as tabulated earlier. One of the aims of using various pressures was to see whether water could be expelled from grout once flow ceased. For this purpose, pressure was maintained after refusal in some tests. The idea was that if excess water could be squeezed out of grout, then the remainder would be a better quality filling. This concept was popular at the time with those who liked to use thin grouts—they argued that pressure removed the unwanted water. The tests showed that this did not happen in the body of the grout at least up to 100 psi ⟨6.9 bars⟩. However, it may have contributed to filtering-type deposits found at the fringe of penetration in tests described later (Houlsby, 1985) and was brought out in discussions of the U.S. Corps of Engineers' tests by Moor (1959).

Pressure Distribution along Cracks A few of the tests had pressure-measuring arrangements at several points along the test crack. For the 0.010-in ⟨0.254-mm, 254-micron⟩ crack, the pressure distribution along the crack was not a straight line: it fell more rapidly at the beginning of the crack than along the rest of it. In the wider cracks pressure drop tended to approximate a straight line. It should be remembered that the grout was moving along a single uniform space and that under real conditions where the grout is flowing more or less radially out from a hole, this radial movement of itself affects the pressure distribution: one that would be uniform in the tests would change to something like the distributions in the sketches in Section 6.1-2.

The thixotropic behavior of thick grouts was demonstrated in some of the tests by closing the outlet and ceasing injection. Under these conditions water would take up a uniform pressure distribution along the crack; thick grout did not.

Quality of Grouting Only a few of the tests produced satisfactory filling when opened up for inspection a day after grouting. These had used 0.7 : 1 or 0.5 : 1 grout and were in the 0.030-in ⟨0.762-mm, 762-micron⟩ cracks. The quality of filling in the 0.020-in ⟨0.508-mm, 508 micron⟩ cracks was fair with 0.7 : 1 and 0.8 : 1, but poor, stringy filling followed from use of 2 : 1 in the 0.010-in ⟨0.254-mm, 254-micron⟩ cracks; thicker grouts could not be used in these.

The filling was regarded as "fair" when it could be scratched with the thumbnail or had numerous bleed-water channels remaining through it. It was "poor" when it was quite soft or when there was only partial filling taking the form of noncontinuous lenses or streaks. Almost no bonding to the upper slab occurred; bleed-water accumulation prevented this.

These results accord with the statement made earlier in Section 6.1-4 that the finest size able to be properly grouted is about 0.020 in ⟨0.5 mm, 500 microns⟩.

WATER CONSERVATION AND IRRIGATION COMMISSION TESTS As mentioned earlier, these tests are reported in some detail in Houlsby (1985). Their intention was to further investigate some aspects of the U.S. Corps of Engineers' tests and to test radial penetration from a grout hole. The effects of different mixing, pumping, and injection techniques were also examined.

Artificial horizontal cracks were grouted between pairs of reinforced concrete slabs each 3 ft × 3 ft × 2 in ⟨0.9 m × 0.9 m × 50 mm⟩ securely braced against deflection. Cracks were tapered with zero opening one side and either a $\frac{1}{16}$-in ⟨1.5-mm⟩ or $\frac{1}{8}$-in ⟨3-mm⟩ opening on the other side. At the center of the upper slab of each pair, grout was applied through standpipe fittings identical with those on regular grout holes, and then it traveled out from the hole. Mixing, pumping, and all other gear was full-size equipment as used on a large dam grouting job going on nearby; none of the gear was scaled-down laboratory-sized equipment (see Figure 6.1-6A). Use of full-size gear was regarded as important in achieving representative results.

Figure 6.1-6A. *Testing arrangement.*

Injection methods simulated actual procedures somewhat: water was passed through cracks first; grout pressure was kept at 30 psi ⟨2 bars⟩ for the first 15 minutes, and then increased to 50 psi ⟨3.5 bars⟩ until refusal; pressure was held on the wet grout for 30 minutes after refusal. Some tests were given several applications of grout at intervals of not less than 24 hours (Section 7.2-10).

After the grout had gone hard, upper slabs were taken off and the quality of the grout filling examined (see Figure 6.1-6B). Bleed voids and discontinuous fill-

Figure 6.1-6B. *Example of grout film revealed when upper slab lifted off. This was test 6, in a crack size widening from 0 to $\frac{1}{16}$ in ⟨1.5 mm⟩, using piston (reciprocating) pump, and paddle mixer. w : c = 1 : 1. Note the fringe around the limit of penetration at about 1.5 ft ⟨0.4 m⟩ and the many bleed passages left.*

Figure 6.1-6C. *Closeup (about quarter actual size) of test using 3:1 grout from a paddle mixer.*

ing were dominant apart from tests where thick grout had been used.

Assessment of the success of the grouting in terms of the degree of filling is given in Houlsby (1985) and a number of comments are made there:

: 1 gave optimum filling in the finer cracks such those in the narrow side of the 0 to $\frac{1}{16}$ in (1.5) tests; these averaged 0.015 in (0.381 mm, 381 ons) opening.

·out gave optimum filling in all cracks larger is.

ality grout mixed in a high-shear, high- xer (Colcrete) and pumped with a helical p (Mono) produced better filling than 'ity grouts mixed in a paddle mixer and ıth a piston pump for grouts thicker .. 5 : 1 (see Figures 6.1-6C and 6.1-6D).

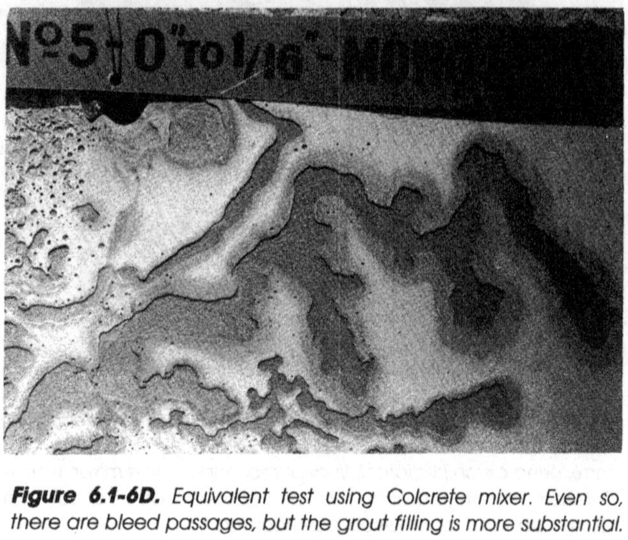

Figure 6.1-6D. *Equivalent test using Colcrete mixer. Even so, there are bleed passages, but the grout filling is more substantial.*

Figure 6.1-6E. *Almost zero filling resulting from use of 5:1 grout, paddle mixed. Just a coating of grout on floor of crack, with large areas of bleed passages and some bleed craters.*

- 8 : 1 grout left weak, powdery material that only occasionally reached the roof of the crack. Mostly, a paint-thin layer of grout was left on the floor of the crack (see Figure 6.1-6E). The rest of the crack volume was empty where it had been monopolized by bleed water at the time of injection.

My paper in which this testing is reported (Houlsby 1985) has the theme of minimizing water in grout in order to obtain durability commensurate with obtaining adequate penetration. The paper indicates the desirability of using 3 : 1 or thicker. The subject is dealt with further in Section 4.5.

LIMITATIONS OF TESTING Any tests conducted in artificial cracks, such as those just described, have limitations that make their relevance to real situations somewhat unsure. In particular, the use of horizontal cracks for the tests means that they are simulating the hardest conditions for grouting (Section 6.1-2). Their results may be on the pessimistic side of reality. As a follow-on from the Water Conservation and Irrigation Commission tests, an area of foundation nearby was grouted using the methods found best in the tests and was then dug open. Perfect grouting was found. These methods are incorporated in my recommendations in this book.

In Section 6.2-8, where grout pressures are discussed, the European preference for high pressures that deliberately jack cracks open is detailed. Cambefort (1977), one of the exponents of these high pressures, criticizes the U.S. Corps of Engineers' tests for restraining the test slabs from deflecting under grout pressure. He considers that the crack should have been permitted to widen with the pressure and that this would have then produced expulsion of the excess bleed water when the pressure was later removed. I am not impressed with this argument, as can be noticed from the discussion on the subject in Section 4.3-5.

6.1-7 Theoretical Ideas about Penetration in Fine Cracks

Some authors have suggested a permissible ratio between minimum crack width and cement size, by considering the case of bridging of individual particles. A ratio of 3 : 1 (crack width : a significant proportion of the coarse grains) has been advanced, apparently as a result of the U.S. Corps of Engineers' suggestion criticized in Section 6.1-6. As well as the objections given there, weaknesses of the case include lack of direct applicability to high-quality grouts or to grouts containing clumps. However, a main reason for its irrelevance is that other factors can dominate over cement properties as shown in the test work reported.

A slightly different approach to the relationship between crack and cement sizes has been taken by Littlejohn (1975) in dealing with the subject of preparation of holes for the grouting of rock anchors. Such preparation has the purpose of sealing the entrances to cracks that are wide enough to permit anchor grout to drain away through them; loss of grout like this can impair the strength of grouted anchors. In this context, his discussion starts with crack sizes of 0.010 in ⟨0.25 mm, 250 microns⟩ based on the proposition that this will pass the maximum particle size of 160 microns in ordinary cement (ASTM Type I) plus 50%. He then quotes experimental studies that suggested that the limit could be closer to 160 microns; this figure is then used in his subsequent papers (Littlejohn and Bruce, 1977). While that sort of grouting is not the subject of this present book, the concept has some relevance here. Following this line of thought a little further, it can be asked why think only in terms of maximum particle size? A significant proportion of the cement in grout is smaller than this and would drain off, leaving the larger grains stuck in isolation in the mouth of cracks, perhaps. Maybe if there are enough of these larger grains, they may eventually clog the cracks here and there in the hole and reduce and/or stop further flow, but this is by no means certain.

Theoretical speculations about the limit of groutability of fine cracks have appeared in quite a lot of studies over many years. Some have been completely abstract reasoning, others have in addition attempted correlation with some testing. Most appear to come from the academic sector. A study of many of these articles gives one the impression that most of the authors could benefit from a year or two's experience actually doing grouting. This all goes to underline opinions expressed at various places in this book that there is far more to grouting than any theoretical concepts can grasp, no matter how astute their authors. The practicalities predominate. The proof of the pudding is in the eating, not in how pretty the cook was.

6.2 PRESSURES FOR GROUTING

6.2-1 Maximum Allowable Pressures

Figure 6.2-1A is a graph of allowable grout pressures for normal conditions. The idea is to put as much pressure into the grout as the foundation can take, in order to get maximum penetration. But there are exceptions and variations; some foundations will stand greater pressures than others, and there are times when reduced pressures are sensible. The graph in Figure 6.2-1A shows suggested maximum allowable pressures for various foundation conditions and depths, provided nothing is going wrong. If it is, then much lower pressures are usually needed (see Sections 7.3 and 7.6).

The pressures shown on the graph are at the surface and are read off against the depth to the bottom of the stage or hole being grouted. For instance, in average rock conditions, maximum pressure for a stage that bottoms at 100 ft ⟨31 m⟩ is read off as 130 psi ⟨9 bars⟩.

Note that for inclined holes, holes that are near the horizontal, and upward holes, it is usually satisfactory to use the pressure corresponding to the length measured along the hole from its start in just the same way as for vertically downward holes; an exception though occurs if cover over the holes is light. Some jobs have gone to the trouble of making corrections for hole inclination and also for different densities of the various w : c ratios of grout, but this is somewhat pedantic and unnecessary apart from in very delicate situations.

The maximum recommended pressure is 150 psi ⟨10 bars⟩ for average and weak rock. Although opinions differ about this, the suggestion here is that there is no need to use higher pressures than this to get good grouting; this is based on considerable experience. A special exception is the treatment of clay in karstic voids, where pressures several times higher may be necessary to hydraulic fracture the clay, intrude and encapsulate it. See Watakeekul and Coles (1985) and Zuomei and Pinshou (1982).

The derivation of the graph of Figure 6.2-1A is explained later in Section 6.2-9.

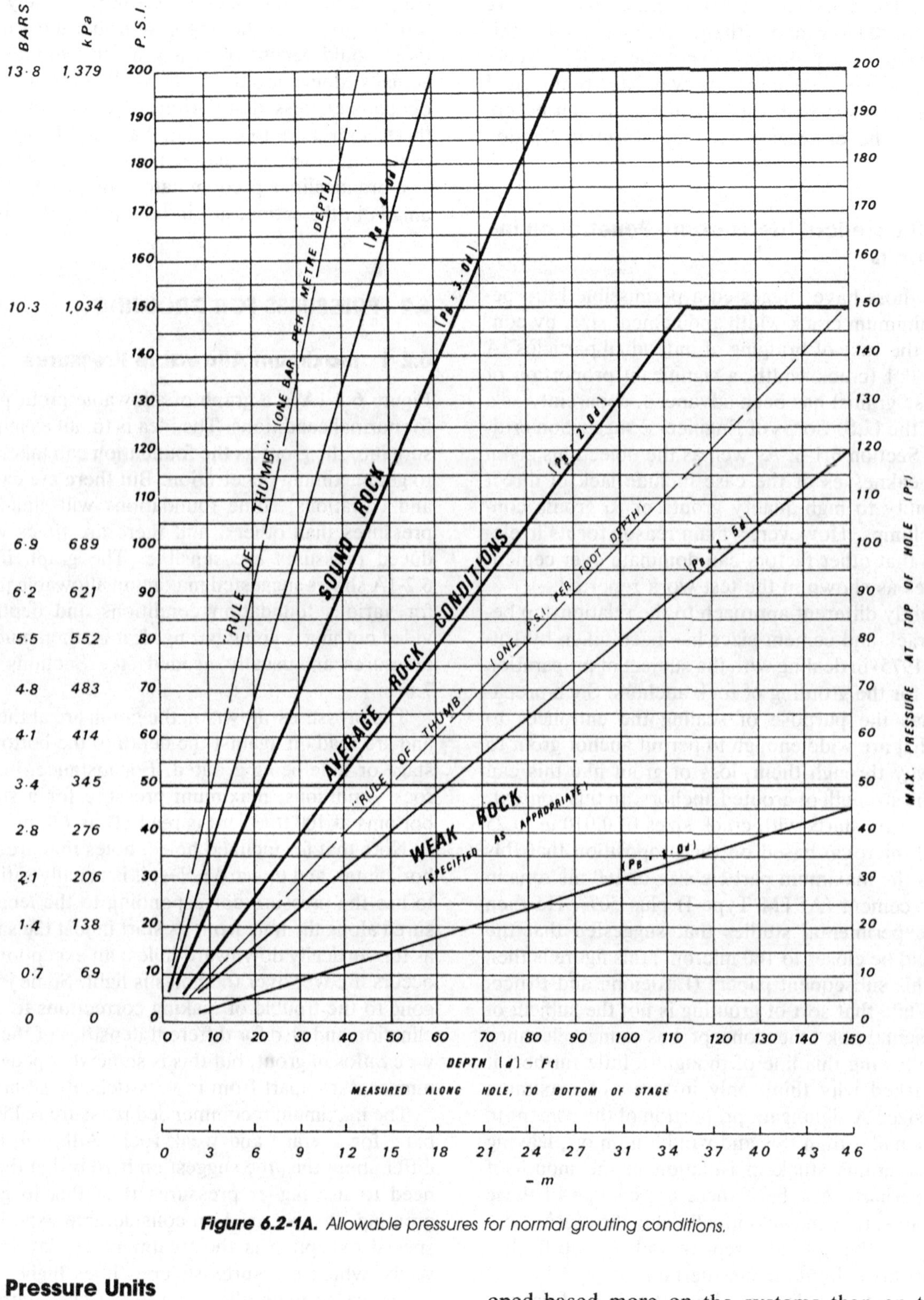

Figure 6.2-1A. *Allowable pressures for normal grouting conditions.*

6.2-2 Pressure Units

One of the nuisance complications in grouting comes from the two international systems of measurement. It is not just a matter of translating numbers from one to the other: in grouting "rules of thumb" have developed based more on the systems than on technical issues (see Section 6.2-8). In addition, when working in the metric system there are a variety of units for some things that have not been satisfactorily resolved by the particular choice adopted in official SI. Rumor has it that when those French academics at the time of the Revolution busied themselves inventing the metric system to stave off getting their heads amputated, they didn't even consider the implications for grouting! Shame on them. And then when the esteemed gentlemen on the SI committees sat down to update things, they didn't think about grouting either. As a result, grouting is a bit of a mishmash when doing it in metric. Curiously enough, if you say this to some of the people who do use metric, they don't see the problem at all—until you compare their "metric" with practice in some other "metric" countries. These differences are most evident when it comes to the various ways of specifying pressure, and to the concepts behind them.

In order to keep things (relatively) simple in this book, when pressures are mentioned, their units are given in only two of the possible alternatives: psi (pounds per square inch) and bars.

But these won't satisfy everybody. To also give the other metric equivalents would be annoyingly complicated.

For instance, on nonmetric jobs there is unofficial standardization on a single unit: psi. However, where metric is in use, alternative units for pressure used in various countries and at various times include the following:

bar
atmosphere
kPa
MPa
kg/sq cm
kp/sq cm (kilopund/sq cm)
kN/sq mm
N/sq m

and to some of these, "force" can be added if one wishes to be precise.

Approximate relationships between these are

15 psi = 1 bar
= 1 atmosphere
= 100 kPa
= 0.1 MPa
= 1 kg/sq cm
= 1 kp/sq cm
= 100 kN/sq mm
= 0.1 N/sq m

Although in SI metric kPa is the official unit, 1 kPa is a very tiny amount of pressure. Ordinary grouting pressures get up into the hundreds of kPa and this is not very practical. Bar has been the preferred unit for grouting since well before the introduction of SI and continues to find general acceptance in grouting. There are several issues involved; Houlsby (1977) discusses them. Even the unit bar has problems: pressures go into decimals, which can confuse poorly educated operators possibly leading to mistakes. This problem is not as bad as that with kPa, where slight changes of pressure are amplified into large numbers, giving oversensitive attempts at adjustments unless operators are told to ignore anything finer than 10 kPa.

Doing grouting in psi is miles easier!

6.2-3 Obtaining the Pressure

The pressure for grouting usually comes from a grout pump (Section 2.4). An alternative for minor jobs is to use gravity where the topography is suitable: the hydrostatic head of the column of grout in a grout hole then provides the pressure. If it isn't quite enough, a long standpipe can add a little more. Hydrostatic pressure of a grout column is approximately equal to 0.7 psi per foot of head ⟨approximately 0.16 bars per meter⟩, so deep holes can develop substantial pressures without pumping assistance; the grout can then be simply poured into them in these minor jobs, which avoids the need for much of the usual machinery.

Opinions differ as to whether grout pressures should be steady or pulsing. Refer to Section 2.4-12 for comment on this and on the recommendation preferring steady pressure.

6.2-4 Grout Pressure versus Weight of Rock

The specific weight of grout varies with the w : c ratio, but for typical mixes used in rock grouting, a convenient approximation is 100 lb/cu ft ⟨1,600 kg/cu m⟩. Rock is about 144 lb/cu ft ⟨2,300 kg/cu m⟩. In deriving permissible grout pressures, one concept used by some people is to think in terms of the weight of the rock being uplifted by the grout pressure.

Figure 6.2-4A does this. In such a comparison it is not really valid to consider rock as a "fluid" with hydrostatic pressure, nor can grout be regarded as a true fluid like water—grout develops a rather different pressure distribution after a while, as shown in Section 4.3-5, in response to bleeding, thixotropy (Section 4.2-2), and density currents. However, for illustrative purposes, the comparison has interest.

In Figure 6.2-1A a "rule of thumb" of 1 psi per foot depth is shown (slightly below the line for average

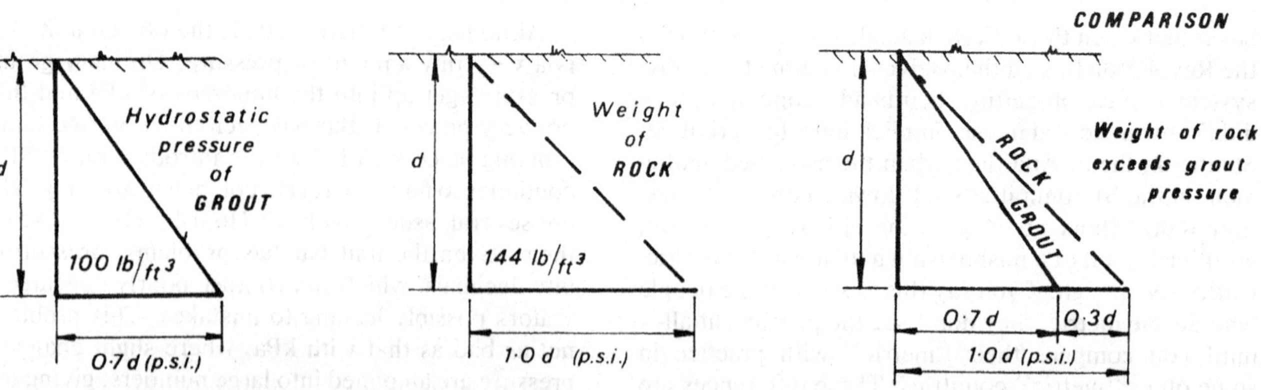

Figure 6.2-4A. *Left: Hydrostatic pressures of grout. Center: Weight of rock. Right: The two compared.*

rock conditions). This is a long-standing rule for deciding on grout pressures in countries that use psi. It has been the backbone of most pressure graphs in those countries and appears to have originated away back. It is compared with the metric version later, but for the moment consider its effects as shown in Figure 6.2-4B.

The grout pressure in the hole exceeds the weight of the rock. Thus the situation is not a balance between grout pressure and rock weight. The excess is carried by the structural strength of the rock as discussed later, and the situation is assisted by the fact that only sometimes does grout penetrate in a horizontal crack far enough to develop enough uplift pressure to actually exceed rock weight.

Noting that the rule of thumb is based on the practical experiences of many grouting practitioners, it is evident that there is much more to be taken into account than this form of theoretical analysis can provide (refer to Section 6.2-10 for more thoughts on this). Extending the results of experience, safe pressures for average rock conditions (compared in Figure 6.2-4C) have been found to be able to exceed this rule of thumb slightly.

The line shown in the graph in Figure 6.2-1A for average rock conditions has the following equation:

$$\text{Pressure at bottom (psi)} = 2.0 \times \text{depth (ft)}$$

Experience has further shown that in tough, sound rock, it is quite satisfactory to use higher pressures than these and conforming to the following equation:

$$\text{Pressure at bottom (psi)} = 3.0 \times \text{depth (ft)}$$

and in some cases this can be exceeded without moving rock. At the other end of the scale, weak rock has to have lesser pressure than the rule of thumb, for instance,

$$\text{Pressure at bottom (psi)} = 1.5 \text{ or } 1.0 \times \text{depth (ft)}$$

All these relationships are shown in Figure 6.2-1A.

6.2-5 Care with the Use of Pressure

The pressures shown in Figure 6.2-1A and discussed, are maximums. They should never be used when start-

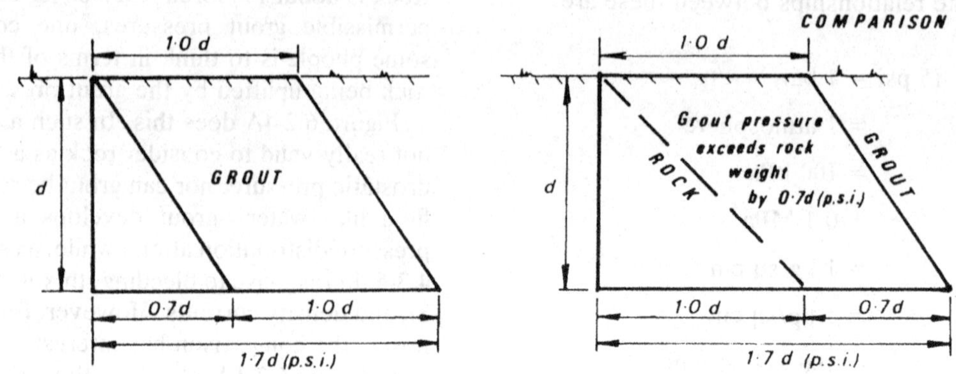

Figure 6.2-4B. *Hydrostatic pressures when pressure at surface (psi) is made equal to depth (ft).*

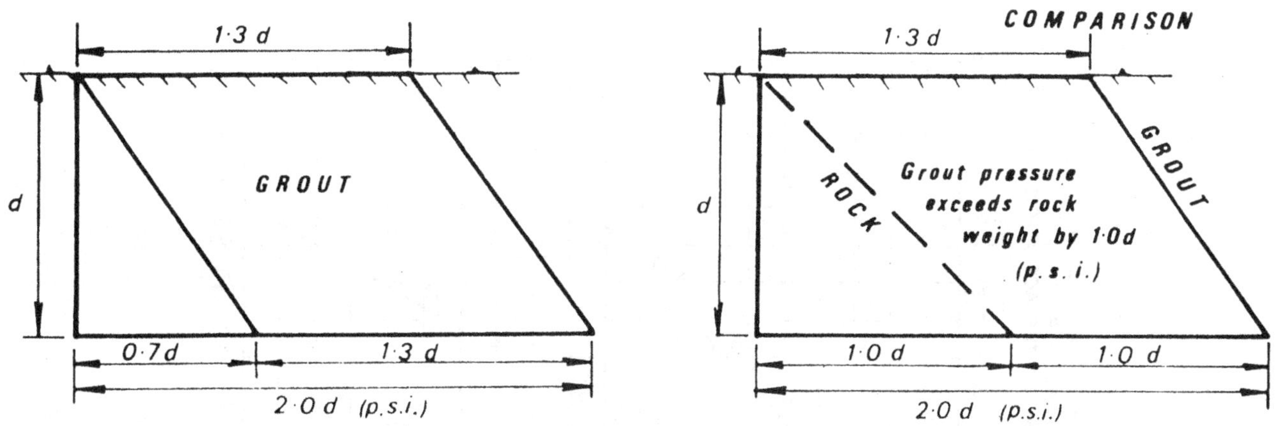

Figure 6.2-4C. *Hydrostatic pressures when pressure follows average rock conditions curve.*

ing a grouting application. As repeated elsewhere in several places (to emphasize an important message), for the first few minutes of a grout application keep the pressure down to a low value such as 15 psi ⟨1 bar⟩ while checks are made for surface leaks, connections to other holes, rock movements, and so on, and then only increase the pressure (gradually) if everything is all right.

6.2-6 Effect of Stage Length

The choice of stage length (see Section 5.4-2) can affect pressures considerably; see, for example, Figure 6.2-6A.

Where rock is weak, short stages are preferred. These produce lesser pressures than longer ones.

6.2-7 Grouting under Structures

When the grouting goes under or is down through a structure liable to be moved by the grouting, unless the movement is wanted (as in jacking up a building), reduced pressures are commonly used. The graph in Figure 6.2-7A gives an indication of this.

The graph is primarily intended for dams, particularly gravity structures, where excessive uplift from the grouting could lift part of the dam. This type of grouting is often done from galleries inside the dam (see Section 8.4-9), and grout may penetrate into the interface with the foundation or into horizontal cracks in the concrete. Figure 6.2-7A is read in terms of allowable interface pressures. These are added to the "plus" figure read off for the particular hole depth. If a packer (Sections 2.5-14–2.5-17) is placed down the

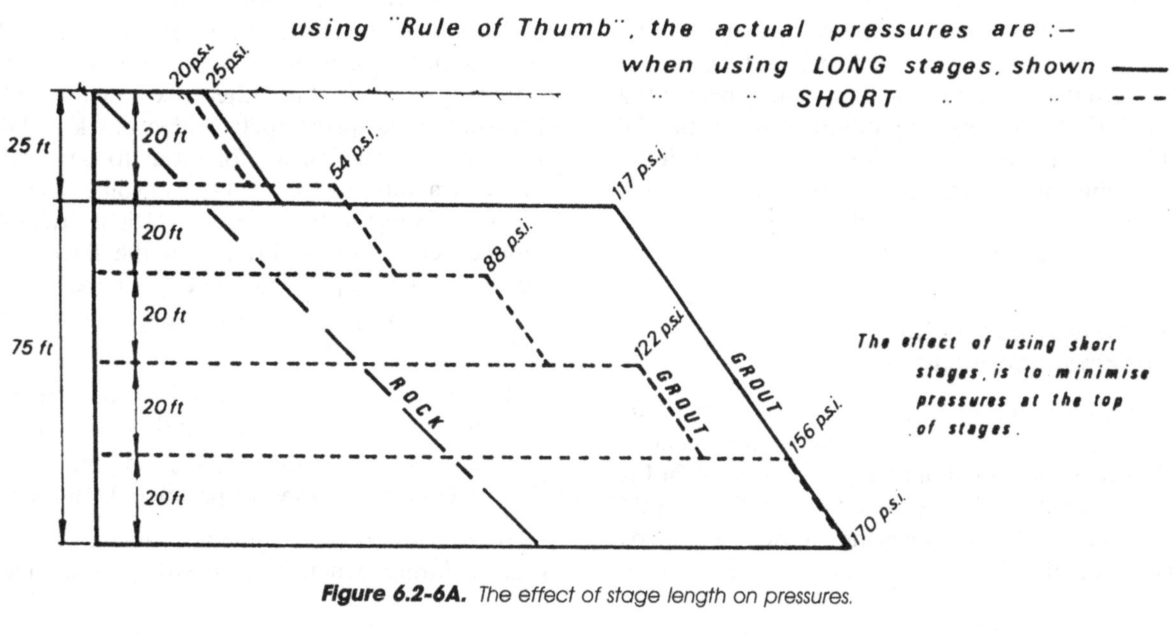

Figure 6.2-6A. *The effect of stage length on pressures.*

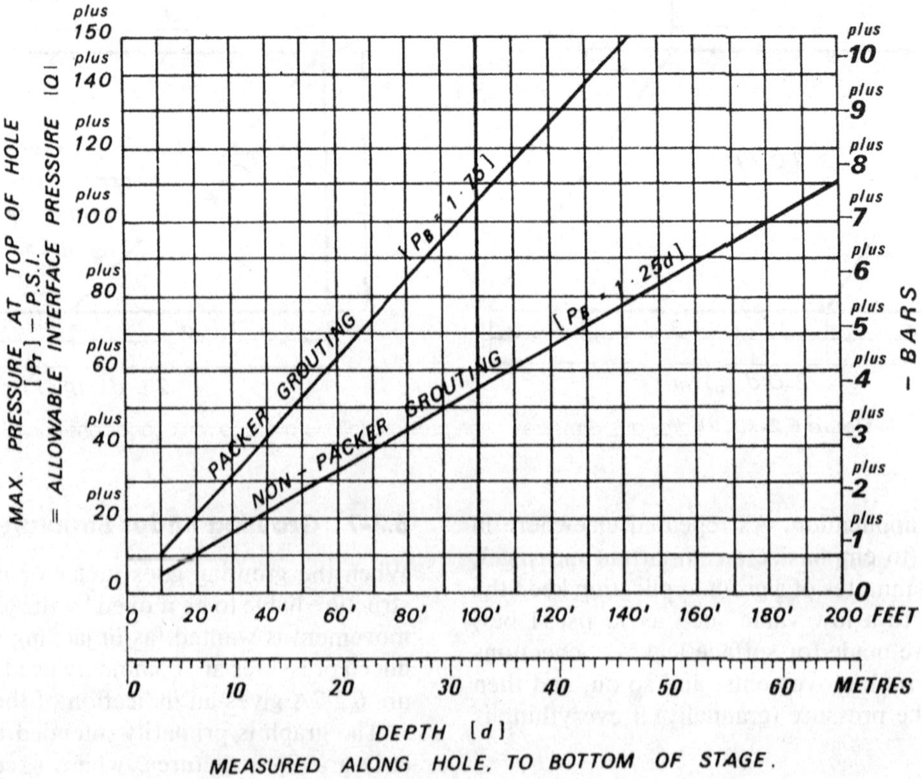

Figure 6.2-7A. Allowable pressures under a structure.

hole below the structure to reduce the chances of uplift, greater pressures can be used, as indicated by the upper line. This line is almost the same as the rule of thumb for working from an open surface (Figure 6.2-1A). The lower line, for nonpacker grouting, is more conservative and falls in the range of the "weak rock" discussed in Section 6.2-1.

Figure 6.2-7A is read in the following manner: if the bottom of the stage is, say, 100 ft below the gallery or point of grouting, the allowable maximum pressure at the standpipe for nonpacker grouting would be the "allowable interface pressure" plus 55 psi. The value of the allowable interface pressure varies with the weight of concrete above and whether other loads are present, such as stored water.

6.2-8 Penetration versus Displacement Grouting

Should pressures be so great that they move the foundation around? There are differences of opinion about this. Earlier comment about the pressure graph in Figure 6.2-1A dealt with the "one psi per foot depth" rule, but the graph also shows the metric rule: one bar per meter depth.

Notice how these two different and quite arbitrary rules produce vastly different results: for a stage ending, say, at 40 ft ⟨12 m⟩, the first rule specifies a pressure of 40 psi ⟨2.8 bars⟩, whereas the second rule gives 180 psi ⟨12.5 bars⟩ for the same depth. One rule gives 4.5 times greater pressure than the other, and yet both rules have widespread use. They reflect the differing philosophies about pressures: one school of thought prefers moderate pressures, which do not disrupt the foundation, while the other uses deliberately high pressures to displace rock, widen cracks, and hopefully make things easier for grout to penetrate. Hydraulic fracturing is induced and in some cases goes all the way from the usual initial vertical orientation of induced cracks right through to the final horizontal situation where uplift may be experienced.

The following names are used for these two approaches to grouting:

1. *Penetration,* or permeation grouting, where moderate pressures are used to avoid disruption.
2. *Displacement,* or claque grouting, where high pressures are used to produce disruption.

A fairly thorough treatment of how grout travels in cracks during penetration grouting (using moderate pressures in accordance with Figure 6.2-1A) is discussed in Section 6.1. For comparison, displacement grouting of a horizontal crack is claimed to do the following (note that Section 6.1-2 points out that horizontal cracks as shown in Figure 6.2-8A, without upward venting, are the hardest to grout properly).

At step 1, Figure 6.2-8A, the high pressure is expected to widen the cracks and thus permit better penetration. The grout has to be thinner than would otherwise be the case in order to transmit the pressure adequately; thixotropic grouts (refer to Section 4.2-2) will not do this. However, these thin grouts bleed appreciably (Section 4.3) and after the take has almost ceased (step 2), the application has to be continued long enough for much of the bleed water to be replaced by fresh grout (step 3). This can add hours to each application if done thoroughly. Once the grouting is finished and the pressure released, the cracks are expected to close and compress the grout in a sort of prestressing mode (step 4) and expel any bleed water still remaining. However, the practicality of this rather neat theorizing finds doubters, including myself.

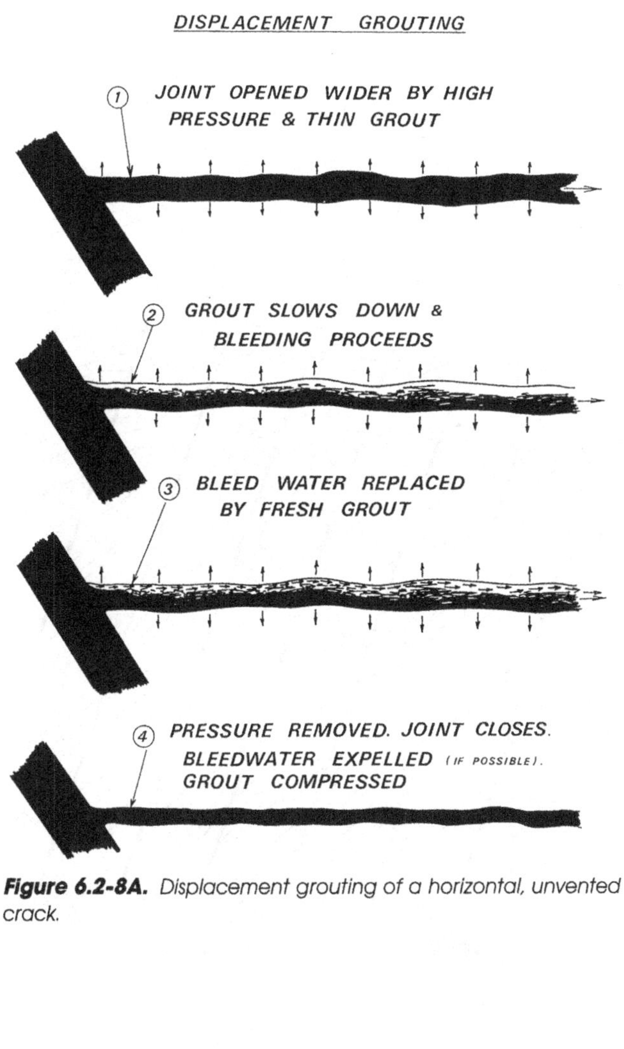

Figure 6.2-8A. *Displacement grouting of a horizontal, unvented crack.*

To compare the two approaches to grouting, consider a few situations. Using moderate pressures, three grout holes could be needed to grout this group of cracks (see Figure 6.2-8B), weathered seams, and a shear zone, but the natural materials filling some of these are not disturbed. Using displacement grouting, only one hole may be needed, but natural infilling is broken open; this may be tolerated if cement completely fills the ruptured remains, but if it does not, the grouting will leave the foundation in worse condition than it was originally (see Figure 6.2-8C).

At the surface, high pressures invite trouble in many sites. For example, see the situation in Figures 6.2-8D and 6.2-8E. This surface, containing slabs of rock able to be detached from the floor and side of the site, requires careful grouting using, say, five grout holes if handled with moderate pressures, in contrast to the possible disruptions and problems with surface leaks. Fewer holes will be needed if displacement grouting pressures are attempted.

In practice, however, high surface pressures would not be used at such a site. Their use would be delayed until the holes reached depths able to withstand the pressures. Hence in surface stages just as many holes are used in displacement grouting as in penetration grouting.

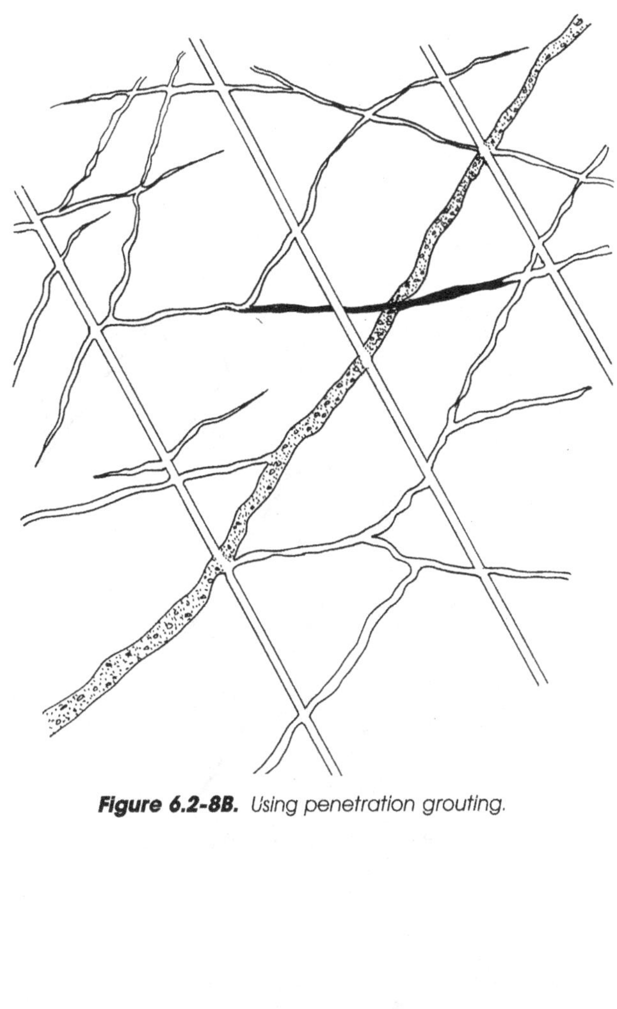

Figure 6.2-8B. *Using penetration grouting.*

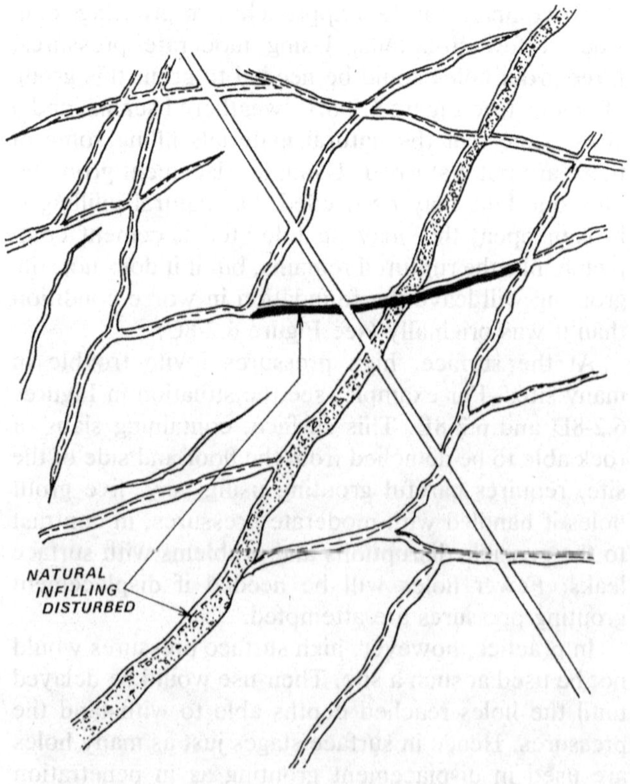

Figure 6.2-8C. *Using displacement grouting of same site.*

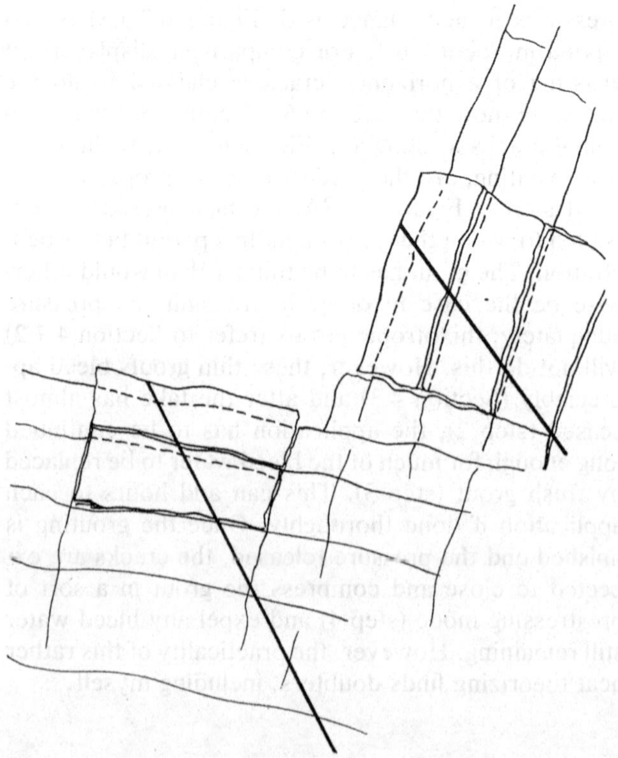

Figure 6.2-8E. *Displacement grouting.*

Figure 6.2-8D. *Penetration grouting.*

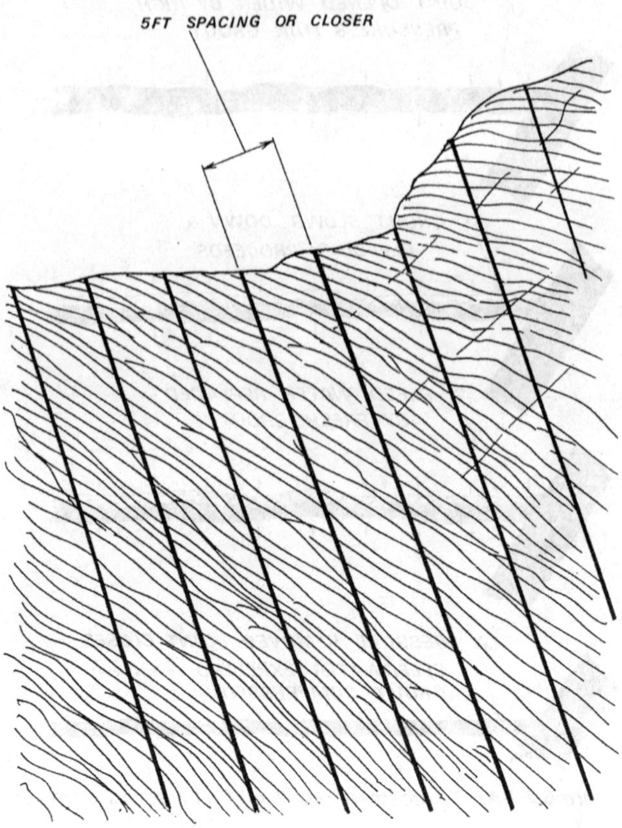

Figure 6.2-8F. *Penetration grouting.*

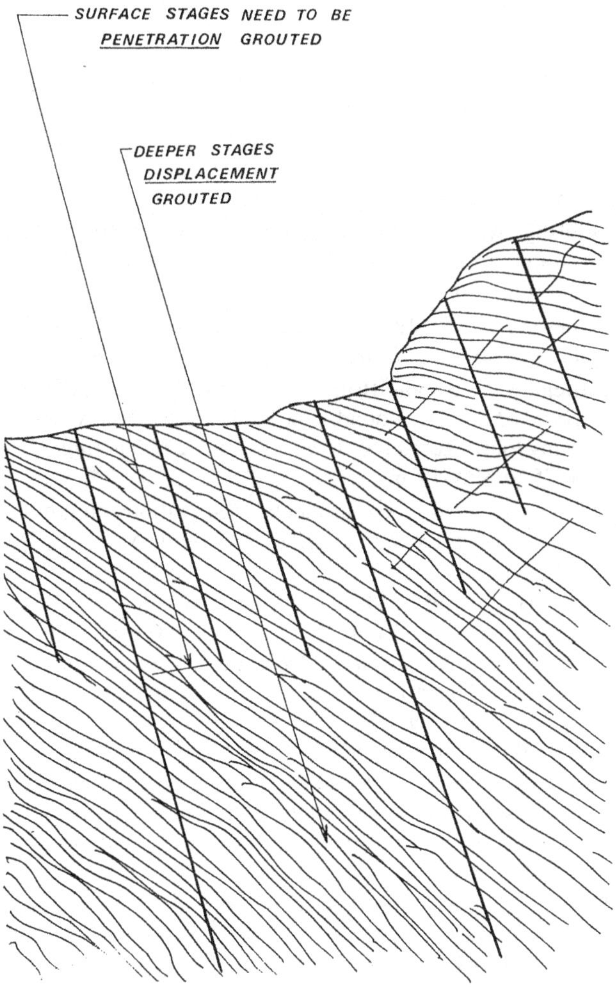

Figure 6.2-8G. Displacement grouting.

Where joints are almost too fine to grout, displacement grouting has been claimed to be able to open them sufficiently to achieve good grout travel (see Figures 6.2-8F and 6.2-8G).

If moderate pressures are used, close spacing of holes will be needed, but the cost of these will be offset to some extent by short grouting times. If displacement grouting of the same site is used, there are just as many holes in the surface stage, but less at depth. There arises uncertainty, however, as to whether these fewer holes have really given access to all the cracks there, and further holes to check this can result in as many holes as penetration grouting.

A critic of the displacement grouting concept (guess who?) has suggested that it "makes the patient worse in trying to fix him up."

Sabarly (1968) in comparing the two concepts has cannily suggested that the choice between high and low grout pressure depends not so much on technical issues as on which system of units is used: those who are accustomed to metric units go for the metric rule of thumb (one bar per meter) while the "Anglo-Saxon" (Sabarly's description as translated) prefer the other one (one psi per foot depth). The author is in the latter camp.

6.2-9 Derivation of Pressure Graphs

The graphs for maximum pressures shown in Figure 6.2-1A were derived from the following:

1. Graph shown in Creager, Justin, and Hinds (1945) having curved lines for massive rock and for sound stratified rock.
2. A similar graph suggested by the ASCE Task Committee on Cement Grouting (1962).
3. My study and collation of the pressures used for grouting dams throughout the world.
4. My study of the relationship between rock characteristics and successful pressures on many sites.

The curved lines shown on the early graphs (1 and 2 above) have been straightened. Analysis of the case histories showed no advantage in continuing the use of the curves; straight lines are equally valid practically and have the advantage of being able to be memorized (if working in feet and psi). These graphs were a development of mine in the mid-1960s and have been found satisfactory in extensive use by others and myself. The hardest part is in deciding under which rock type to classify a site. If in doubt, some tests early in the grouting are advisable, and, of course, there must be willingness to change to a different line if found prudent or to use several on the one site if the geology is sufficiently variable.

6.2-10 Structural Strength of Foundations

The capacity of a foundation to resist grout pressures depends on too many features to categorize. Theoretical analyses have often postulated horizontal cracks of infinite extent and have derived a number of pressure relationships for this situation. However, as earlier noted, such horizontal cracks are rare in nature. The majority of real cracks are inclined in various directions and intersect each other at all sorts of angles and positions. The theoretical relationships have little real relevance; practical experience is a surer guide.

The network of jointing found in many foundations provides an interlocking of the rock fabric in which dislodgement of part of it is resisted by surrounding elements. And when grout penetrates fully into all this jointing, pressures tend to become balanced; dislodgement of individual parts is thus inhibited. The limit to grout pressure is then a matter of the resistance to extrusion or rupture of any weathered materials present. If there are none and the rock is sound and tough, relatively high pressures are tolerable; where they are prevalent, pressures often have to be low.

Another factor is whether surface leaks or connections to other holes are liable to break out. If they are, initial pressures have to be low. With careful treatment (Section 7.3 gives details) it should be possible to eventually bring each application up to full pressure, but where frequent, it is prudent to aim at a lesser maximum pressure than would otherwise be the case.

In the relatively unusual case of horizontally bedded rock, where upward jacking is feasible as the effect of grout acting over a large area, there are often saving factors: thixotropic stiffening of thick grout reduces the active area of jacking compared to what would be the case if it was water doing the jacking, and cantilever strength of well-embedded slabs can give considerable resistance. On the other hand, if slabs are not very well embedded or are under tectonic stress (as discussed in Section 7.6), it may be necessary to bolt them down and then use low pressures and very thick grout.

Testing to get the feel of a site is of value in deciding safe pressures. On some sites testing has been carried on to the rupturing stage in order to find limiting grout pressures; an example was Balderhead Dam, reported by Morgenstern and Vaughn (1963). Testing was with both water and grout and deliberately induced hydraulic fracture. Every site is different, and results from one should not be transposed to another without due caution.

REFERENCES AND FURTHER READING

ASCE Task Committee on Cement Grouting (1962). Progress report. *ASCE Journal of Soil Mechanics and Foundation Division,* Vol. 88(SM2), 49–98.

Cambefort, H. (1977). "The principles and applications of grouting." *Quarterly Journal of Engineering Geology,* Vol. 10, 57–95.

Creager, W. P., Justin, J. D., and Hinds, J. (1945). *Engineering for Dams.* Wiley, New York.

Houlsby, A. C. (1977). "Towards appropriate metric units for grouting." *Ground Engineering,* Vol. 10(5), 39ff.

Houlsby, A. C. (1985). "Cement grouting: water minimising practices." *Issues in Dam Grouting.* ASCE, Denver, 34–75.

Houlsby, A. C. (1988). "Improvements in grouting of large ground anchors." *ASCE Journal of Geotechnical Engineering,* Vol. 114(4), 448–468.

Littlejohn, G. S. (1975). "Acceptable water flows for rock anchor grouting." *Ground Engineering,* Vol. 8(2), 46ff.

Littlejohn, G. S., and Bruce, D. A. (1977). *Rock Anchors—State of the Art.* Foundation Publications, Essex, United Kingdom. Reprinted from *Ground Engineering,* May 1975 to May 1976.

Machis, A. (1948). "Experimental observations on grouting sands and gravels." *ASCE Transactions,* Vol. 113, 181–205.

Moor, H. L. (1959). Discussion on "Pressure Grouting in Fine Fissures," by T. B. Kennedy. Original paper in *ASCE Journal of Soil Mechanics and Foundations,* Vol. 84(SM3), Paper 1731, August 1958. Discussion in *ASCE Journal of Soil Mechanics and Foundations,* Vol. 85(SM1), February 1959, 88–89.

Morgenstern, N. R. and Vaughn, P. R. (1963). "Some observations on allowable grouting pressures." *Grouts and Drilling Muds in Engineering Practice.* Butterworths, London, 36ff.

Sabarly, F. (1968). "Grouting and drainage of rock foundations in rocks of low permeability." *Geotechnique,* Vol. 18, 229–249.

U.S. Army Corps of Engineers (1956). "Pressure grouting fine fissures." Technical Report No. 6-437, October 1956. Waterways Experiment Station, Vicksburg.

Watakeekul, S. and Coles, J. (1985). "Cutoff treatment method in karstic limestone, Khao Laem Dam." *Fifteenth Congress on Large Dams,* International Commission on Large Dams, Vol. III, 17–38.

Zuomei, Z. and Pinshou, H. (1982). "Grouting of the karstic caves with clay fillings." *Proceedings of Conference on Grouting in Geotechnical Engineering,* ASCE, New Orleans, 92–104.

CHAPTER 7

Doing the Grouting

This chapter deals with aspects of grouting operations from the time when everything is set up ready to go to the time when some grouting has been done and assessment is necessary to decide whether more is needed. It also suggests methods for handling problems such as leaks, connections (Section 7.3), and rock movements (Section 7.6); the application of computers to grouting is also discussed (Section 7.7).

7.1 READY TO START?

7.1-1 Necessary Preparations

Before starting the grouting of each hole or stage check to make sure that everything has been adequately prepared.

Don't start grouting until everything is ready. After starting, don't stop until the grouting is finished, apart from deliberate interruptions to treat leaks as discussed in Section 7.3.

Before starting check that:

- The correct hole is connected.
- The circulation line is correctly arranged.
- An adequate supply of cement is ready.
- All machinery is working properly.
- Standpipe fittings are clean and working properly.
- Standby mixer, agitator, and pump are available.
- Caulking gear is at hand.
- No anticipated activity on the site will interrupt the grouting.

Each of these is dealt with in detail in the following sections.

7.1-2 Connected to the Correct Hole?

It is easy on most sites to accidently connect to the wrong hole. Among the multitudes of standpipes or holes, it is easy to connect to the wrong one occasionally (Figure 7.1-2A) unless identification is clear. This is particularly the case in poor lighting conditions, bad weather, messy jobs etc.

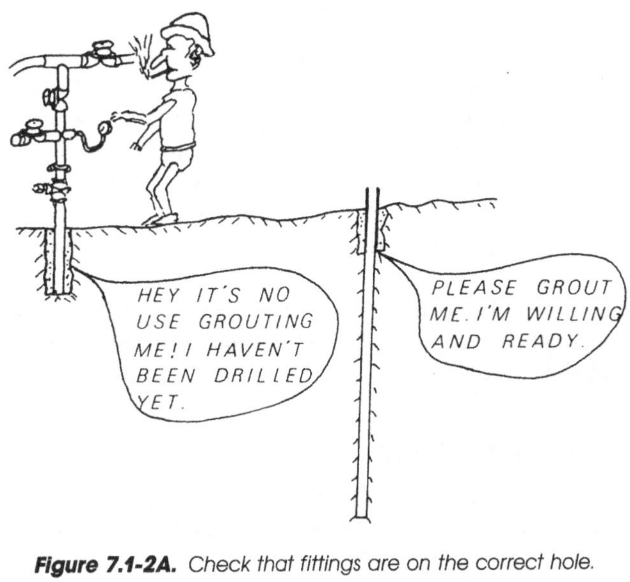

Figure 7.1-2A. *Check that fittings are on the correct hole.*

7.1-3 Is Circulation Line OK?

Most construction sites have many hoses spread around. In addition to the grouting hoses, there can be air and water lines. Check to make sure that the grout hoses forming the circulation line (refer to Section 2.7-3) do indeed form a circuit. Figure 7.1-3A shows

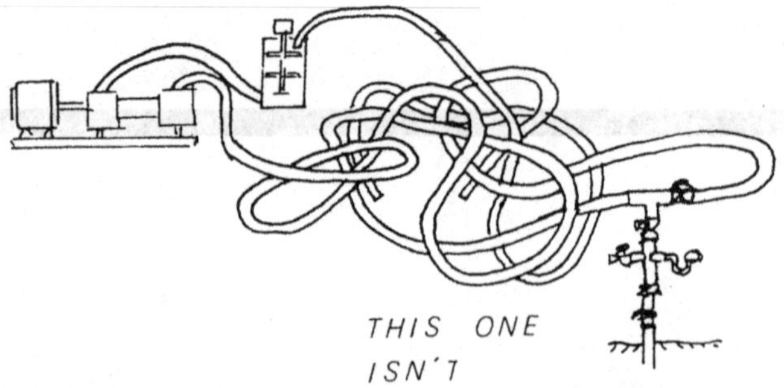

Figure 7.1-3A. Check that circulation line is connected correctly.

a confused mess somewhat typical of what actually happens on busy sites. Some experienced contractors minimize the problem by using hoses for grout lines which have a different appearance to those used for air and water, or are painted distinctively. Section 2.7-4 discusses ways to stop someone who is scrounging around the site for a hose for something else from taking one out of the circulation lines.

If several nearby holes are being grouted simultaneously, check carefully to make sure that each has its correct circulation line, unless of course some are being group grouted (see Section 2.7-5).

7.1-4 Cement Supply

One of the purposes of doing a water test in each hole or stage before grouting is to rehearse the grouting (refer to section 5.3-2) and to get an idea as to whether cement take will be large. If it will be, adequate supplies of cement should be made ready at the mixer, and follow-on supplies should be organized. Having to stop a grout application because cement has run out is not good.

Similarly, make sure ample supplies of water and any other ingredients (if any) of the grout are available at the mixer.

7.1-5 Machinery

Mixers, agitators, and pumps must be working properly. When they have internal combustion motors, make sure that there is enough fuel to last the duration of the grouting.

On particularly crucial or complex jobs, it is good practice to have standby units hooked up ready to swing into action to supply grout within a minute or so of any breakdown of the main equipment. The standby units are brought into action by the following means:

1. Connections from the main and standby pumps to a single circulation line so that switchover from the main to the standby is just a matter of opening the valve on the standby line and closing that on the main line, or
2. Two separate circulation lines, each with its own pump, ready for use. If the main pump or its line plays up, lines are switched at connections at the hole.

Each of these pumps has its own agitator. Mixers can be more flexible because they do not necessarily have fixed lines attached to them (see Section 2.7-8), and if there is no problem with the main mixer, it can continue to be used to supply a standby agitator and pump. Similar flexibility can permit a standby mixer to supply the main agitator if there are problems with the main mixer only.

7.1-6 Standpipe Fittings

The fittings on the standpipe (described in Section 2.6-2) must be clean and operating properly.

Pressure gauges must be reliable. As well as periodical checking for accuracy, as emphasized in Section 2.5-13, gauges on the standpipe fittings should be checked by making sure that they are undamaged and that they read zero before grout starts to circulate (unless groundwater is producing back pressure and giving a reading).

7.1-7 Standby Fittings

On all grouting jobs, even the simplest, there should be adequate standby valves, pressure gauges, and hoses available. To speed up any changeovers, complete sets of standpipe fittings (Section 2.6-2) should be ready near the holes.

Burst hoses can be common with certain types (usually inexpensive ones that are barely adequate to handle abrasive grout). Spares are essential. Couplings of the snap type allow rapid changeover of replacement hoses without too much loss of grout.

7.1-8 Caulking Gear

Surface leaks are common on some sites. Where caulking is the appropriate treatment, adequate materials and tools for this must be available. Refer to Section 7.3-2 for details.

7.1-9 Other Interruptions

Examples of interruptions are nearby blasting, plant movements over the grout hoses, adverse weather coming (perhaps, though, grouting crews are too tough to let a bit of rain or snow stop work?), and such things as the approach of knock-off time.

Blasting can cause surges in wet grout underground, particularly if the grout is under pressure at the time. The severity of this varies with the size of blast, delay patterns, toughness of the foundation, and other factors specific to the site. The effect on wet grout is usually deleterious. Relevant factors are too variable to specify here. They can be evaluated on specific sites by conducting tests in unimportant holes if the problem cannot be avoided by suitable construction planning; water tests before and after this testing help to indicate whether the blast shock in wet grout has upset the foundation.

The problems of protecting grout lines from damage by earth-moving plant is discussed in Section 2.7-12 (in relation to fixed plant). In addition to placing lines in large steel pipes, as suggested there, a simple alternative suitable for the passage of small vehicles is to lay large pieces of timber alongside the grout hoses so that wheels are held clear of them by the timbers.

The approach of knock-off time is a factor to keep in mind (the crews certainly do!). Its effect on a grout application depends on job policy about overtime and shift work.

7.2 HOW TO GROUT A HOLE

7.2-1 Summary

The following are basic steps for carrying out a grout injection:

1. Decide the starting mix (refer to Section 7.2-2).
2. Decide the maximum pressure (Section 7.2-3).
3. Commence grouting by opening the control valve to let grout into the hole (Section 7.2-4).
4. For the first few minutes use a low pressure and check for:

   Surface leaks and connections to other holes (Section 7.3).

   Rock movements (Section 7.6-1).

   Loose standpipe (Section 5.1-1).

   If none of these happen, gradually increase the pressure to the maximum allowable (Section 7.2-3). If any of these occur, keep the pressure low while dealing with the problem (methods are discussed in Sections 7.3 and 7.6) and omit step 5 unless the problem is too minor.
5. At about 15 minutes from start (or when grouting has been in progress sufficiently long to indicate trends), measure the grout injected so far (Section 7.2-6), then:

   Decide whether the mix should be thickened or not (Section 7.2-6).

   Continue to check for leaks, connections, or rock movements (see Sections 7.3 and 7.6).

   If none of these are occurring, continue grouting at maximum pressure with the thickest mix that the hole will take without it choking off (Section 7.2-7).
6. Periodically review the suitability of the mix by examining regular measurements of grout taken (Section 7.2-7). Thicken the mix if it is sensible to do so (Section 4.1-8), with the aim of getting as much cement in, as fast as possible.

   Most holes show a gradual reduction in take eventually. If this does not happen, review the mix (Section 7.2-7)—unless the hole is in a runaway condition (Section 7.2-9).

   Periodically bleed water and thin grout out through the bleeder valve (Section 7.2-8).
7. Carry on grouting until refusal (Section 7.2-9) is reached, except in holes having runaway behavior.
8. After refusal, hold the pressure on the hole for a further 15 minutes (Section 7.2-9).

These steps are explained in the following sections. The rationale behind them is given in Section 7.4. Methods for dealing with problems such as leaks, connections, and movements are discussed in Sections 7.3 and 7.6.

Once more the point is made and emphasized that the explanations and discussions are for *guidance:* they are *not instructions or rules* and should be regarded as basics—to be locally modified by site experience and trials. So many variations are required to suit the multitude of differences found in foundations that it is not possible here to go beyond basics. Even the basics are fairly elaborate, as will be seen.

7.2-2 Deciding the Starting Mix

GENERAL PRINCIPLES Start with grout that is the thickest possible consistent with getting adequate penetration in the finest groutable cracks. It is useless to start with a mix that is so thick that it immediately clogs cracks without penetrating them; on the other hand weak grouting (see Section 4.5) can result from a mix that is somewhat thinner than the cracks will accept.

Note that it is the finer groutable cracks that get consideration at this time, rather than the wider cracks (which get attention later).

The value of a thorough site investigation (see Sections 8.1 and 8.2) can become obvious at this juncture. It should have provided an assessment of crack sizes. Other relevant information comes from the water test in the grout hole prior to the grouting (Section 5.3-2) and from experience with previous holes in the area.

SOME GUIDELINES *Experience* is the best guide. If it is not available, try a starting mix of:

Water : cement ratio = 2 : 1 by volume
(loose volume—see Section 4.1-1)

unless there are obvious reasons for using something different, such as:

if cracks are relatively *FINE* such as 0.030 in ⟨0.75 mm⟩ or finer, try

3 : 1

if cracks are fairly *WIDE* such as 0.050 in ⟨1.25 mm⟩ or more, try

1 : 1

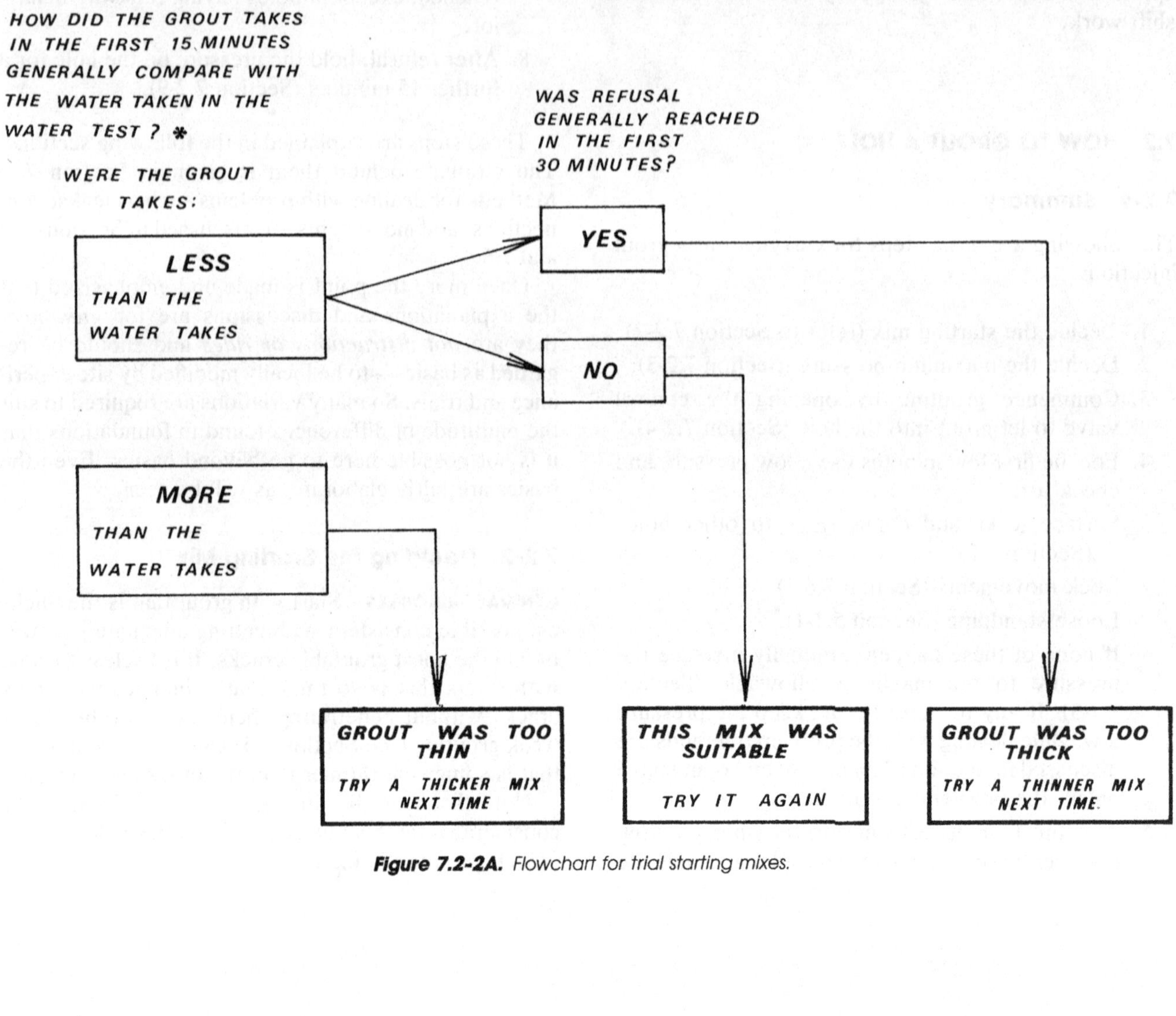

Figure 7.2-2A. Flowchart for trial starting mixes.

To check tries, compare them on the flowchart shown in Figure 7.2-2A after grouting a representative number of stages (at least six of them). The flowchart is intended mainly for the common situation where permeability is between about 5 and 30 lugeons (see Section 8.2-3), but it can be helpful outside this range if used thoughtfully. For the purpose of this exercise, do not thicken before 30 minutes, and measure the water take in the same units of volume as the grout take. The water test must be for 15 minutes at a pressure of 15 psi ⟨1 bar⟩.

SOME MORE CONSIDERATIONS *Thicker* mixes than usual may be necessary to handle surface leaks (Section 7.3-1), connections to other holes (Section 7.3-6), and rock movements (Section 7.6-1). Thicker mixes than usual can sometimes be used below the water table where the wet conditions may permit better penetration.

Thinner mixes than usual (i.e., they have more water) can be warranted in dry, absorbent foundations if the grout loses significant amounts of water into the foundation and hence loses some mobility. Usually this applies only to the initial holes, such as primaries and secondaries, because these tend to saturate the foundation in their vicinity so that later holes usually find a wetter environment.

Thinner mixes are also sometimes used when a stage or hole is given several applications of grout (at daily intervals; see Section 7.2-10): after the first application, the grout can be thinned by one step (refer to Section 4.1-5 for the steps) at the start of subsequent applications because cracks are probably a little finer as the result of being partly grouted already. The same logic can be applied to later holes in a closure pattern if penetrations from earlier holes partly reach to their locations.

RANGE OF STARTING MIXES Recommended limits for all these variations on starting mixes are

THINNEST MIX w : c = 5 : 1

Doubts about durability of this mix and of 4 : 1 make 3 : 1 preferable, though. See Section 4.5-5.

THICKEST MIX w : c = 0.5 : 1

Used only in extreme cases. 0.8 : 1 is a commoner limit.

ONCE EXPERIENCE OF A SITE IS GAINED When the "feel" of the site is gained, either by trials or from experience elsewhere, it is usually possible to decide on the usual starting mix, and then only depart from it where warranted by information from the water test or for other good reasons. It is important to regularly review the suitability of this usual mix. Methods for doing this are given in Section 7.4. Changes may be necessary when moving into different geological features.

7.2-3 Maximum Pressures

Refer to Section 6.2 for pressure graphs and suggestions about maximum pressures for different types of foundations and depths of hole. It is unwise to apply the maximum pressure straight away; as mentioned earlier in this chapter and in Section 6.2-5, it is better to start with a low pressure to make sure that obvious problems don't have to be dealt with before putting on the full pressure.

7.2-4 Commencing Grouting

Make grout having the required starting w : c ratio by:

- Mixing it in the mixer if there is nothing in the agitator (Section 4.1-7), or by
- Thickening or thinning any grout in the agitator, as appropriate (Section 4.1-8,9).

When the required mix is circulating past the grout hole, open the bleeder valve and then the control valve on the standpipe (these fittings are described in Section 2.6-2). Groundwater and/or water left over from the water test may come out the bleeder valve. Leave it open until grout of the required w : c ratio starts to come out.

7.2-5 The First Few Minutes

After grout comes out the bleeder, close it. Then watch the pressure gauge and adjust the control valve if necessary to keep the pressure down to a low pressure such as 15 psi ⟨1 bar⟩ (or less if the maximum allowable pressure is itself of this order). Keep the pressure at this low value for a few minutes while checks are made for

- surface leaks (Section 7.3-1).
- connections to other holes (Section 7.3-6).
- rock movements (Section 7.6-1).
- any other problem that could become worse if pressure is built up

If any of these problems occur, take appropriate action as described in the sections referenced. The procedure following here relates essentially to grouting without complications.

If there are no problems, increase the pressure gradually, all the time being alert to drop it back quickly if necessary. All being well, aim to build up to the maximum pressure over a few minutes. If the hole takes grout so fast that pressure will not build up, refer to Section 7.2-9 for steps to take.

Note whether the pressure is very sensitive to slight changes of the control valve: if the pressure shoots up with small changes, a "tight" hole is indicated, and grout take is likely to be small. If, on the other hand, the pressure stays low no matter how wide open the control valve is, a "fast" hole with high take is being grouted or, alternatively, problems such as leaks, connections, or movements are about to become evident or are happening unbeknown. Most holes are between these extremes.

7.2-6 About 15 Minutes from Start

At about 15 minutes from start of grouting, or when grouting has been in progress sufficiently long to indicate trends, some checks are necessary to make sure everything is going right. This review does not have to be done at precisely 15 minutes from the start; other periods ranging from 10 to 20 minutes are equally as good, but the use of 15 minutes is convenient for all the measurement "runs" (refer to Section 4.1-10).

Take a measurement of the amount of grout used so far. Look closely at the surface of the grout in the agitator. Decide which of the following situations is relevant and then follow the reasoning.

FAST TAKE

Indications

- The surface level in the agitator falls perceptibly when watched for about half a minute.
- There is lack of pressure sensitivity in the hole when the control valve setting is varied (as mentioned in Section 7.2-5).

Reasoning

- Did the water test before grouting have a relatively high take (experience indicates the relative magnitudes of takes)?
- If it did, the fast grout take is reasonable. Thickening of the mix may be necessary after a few more runs.
- If the water test did *not* have a relatively high take, the fast take probably indicates that a thicker mix could be used successfully without choking the cracks. Therefore, thicken the mix by one step now and possibly thicken further soon.

MODERATE TAKE

Indications

- The surface level in the agitator falls so slowly that it has to be stared at for about a minute to notice a drop.
- There is a definite, though mild and possibly sluggish, response on the pressure gauge when settings of the control valve are changed.

Reasoning

- This is the usual situation if the starting mix has been suitable. A thicker mix might give premature refusal.
- Keep on this mix for the next few runs at least.

SLOW TAKE

Indications

- Surface level in agitator falls too slowly to be seen.
- Pressure very sensitive to control valve settings.

Reasoning

- Did the water test before grouting indicate fairly tight conditions?
- If it did, slow takes can be expected on relatively thin mixes such as 3 : 1 or 2 : 1.
- Is 1 : 1 or thicker in use? If so, it is too thick for this hole.

7.2-7 Periodical Reviewing

Every 15 minutes, or at whatever other regular time intervals may have been chosen, review progress of the grouting for the purpose of seeing whether any changes are necessary. Do this by comparing the quantity of grout taken in the latest interval with takes in earlier intervals, making allowance for any relevant factors such as pressure changes or the outbreak of problems such as leaks, connections, or movements.

The quantity of grout taken in each interval should reach a peak and then decrease, along the lines of the graph in Figure 7.2-7A.

More about similar graphs and about variations from the ideal situation follows later in Section 7.4. As an introduction to the principles of reviewing, consider the following situations:

1. The Desirable Situation Figure 7.2-7B shows the situation where the grout taken from the agitator in each interval decreases until it dwindles to refusal. A simplified description follows (for a fuller description of the way grout moves in cracks see Section 6.1):

a. At the start of the injection, fresh grout should be moving fairly easily through the cracking system.

b. Then, when the easily reached cracks have become full with grout, the grout slows down and becomes sluggish, and the pressure drops off at locations away from the hole. The overall effect of this is to reduce the take gradually.

c. Eventually, the grout near the fringe becomes stiff and pressure becomes zero there.

d. After that, the remainder of the application is confined to injecting fresh grout into passages and voids that have formed where bleed water has risen up out of grout injected earlier in the application.

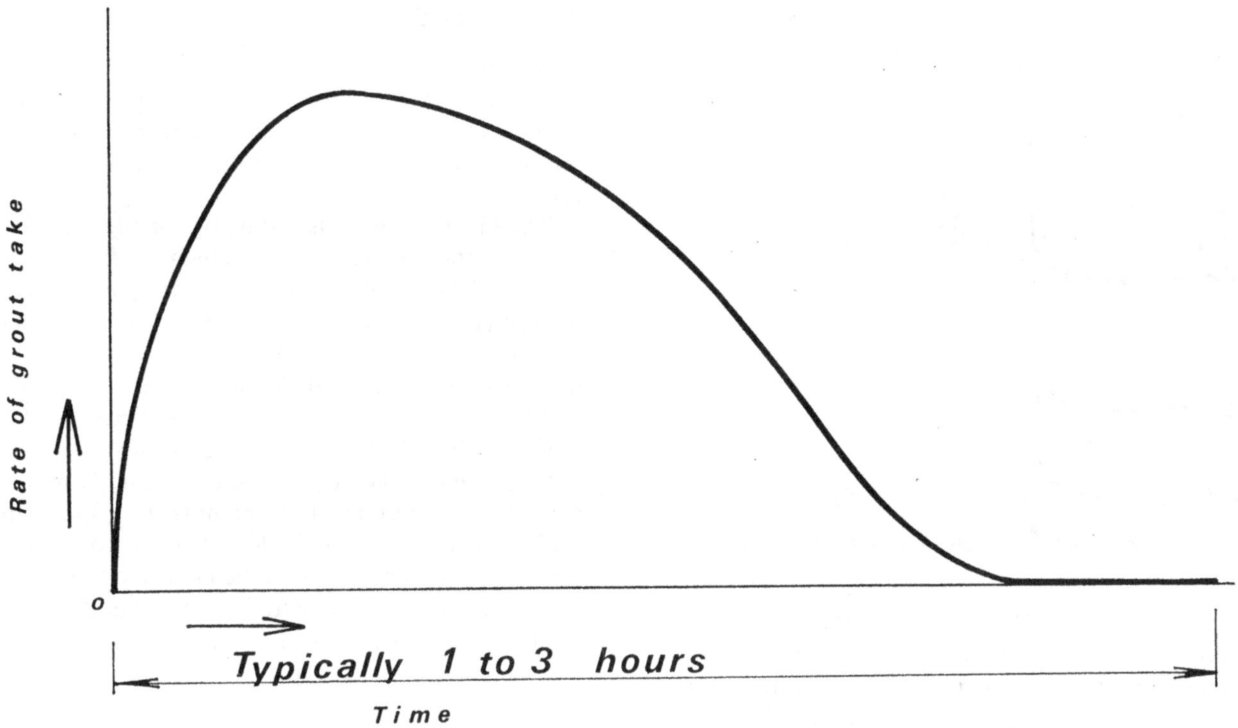

Figure 7.2-7A. *The rate of grout take should peak and then diminish.*

e. Eventually refusal is reached when all bleed passages accessible from the hole are filled with fresh grout.

2. *Sudden Takeup* The sudden takeup early in the application after the first 15 minute interval shown in Figure 7.2-7C indicates that the mix has been too thick. It has choked off the cracks.

3. *Sudden Increase of Take* The sudden increase in the second interval on Figure 7.2-7D probably indicates that trouble has occurred, such as rock has moved or leaks or connections have broken out. Vigilant observation should have noticed the trouble before it became apparent in the quantity measurements.

4. *Protracted Equal Takes* Unless refusal is near, a succession of almost equal takes, as shown in Figure 7.2-7E, indicates that the mix is probably too thin, because no limit to the penetration is being reached

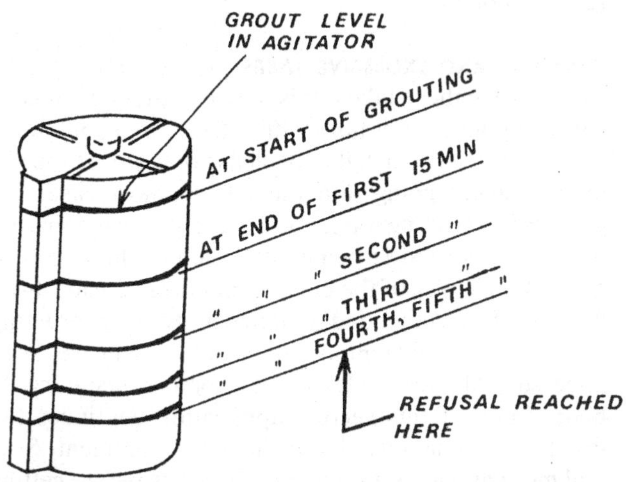

Figure 7.2-7B. *Schematic of slowing of grout drop in agitator.*

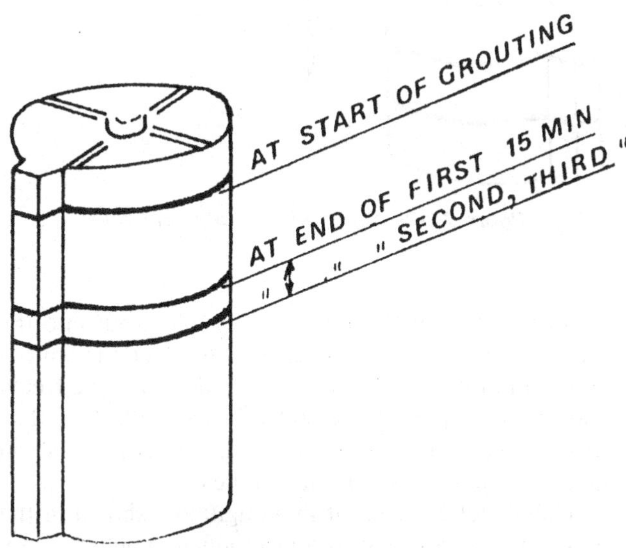

Figure 7.2-7C. *Appreciable sudden slowing of grout taken from agitator.*

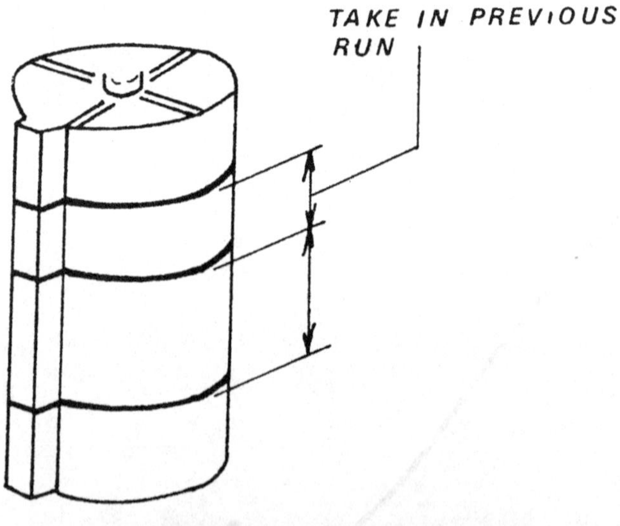

Figure 7.2-7D. *Increases in take from agitator.*

over an hour or so. This means that the grout is excessively mobile due to having too much water. If the mix is not thickened, a patchy job can be expected, and the application will take longer than would otherwise be the case.

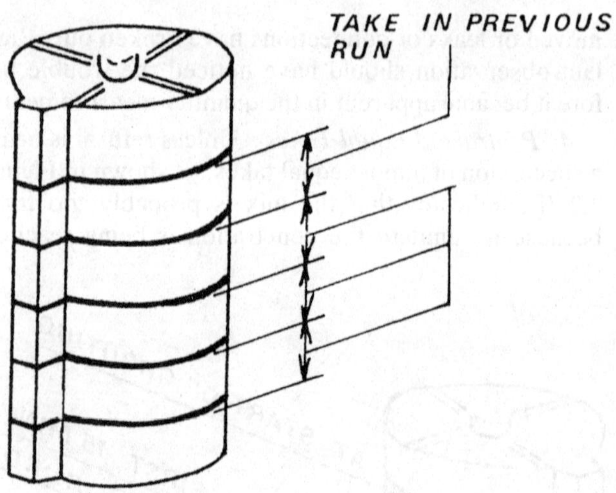

Figure 7.2-7E. *Agitator falling at uniform rate.*

In general, get the thickest possible grout into the cracks as quickly as possible so as to reach the limit of penetration while the grout is still able to move freely and transmit pressure effectively (Section 4.2). The time available for this ranges from 0.5 hours for small takes to 3 hours for fast, high takes.

The sketches of grout in an agitator, shown in the preceding figures, have been simplified to omit the topping up actually necessary usually.

7.2-8 Bleeding

While grouting is in progress, water and thin grout can collect as shown in Figure 7.2-8A. Every 15 minutes or sooner, the hole should be bled to remove these deleterious fluids. The process of bleeding is shown in Figure 7.2-8B.

The bleeder valve should be on the high side of the fittings if the hole is inclined. When the bleeder valve is opened, it might spray grout out: stand clear. If a packer is in use, bleeding is difficult to do properly (Section 2.5-15). When the pressure is released during the bleeding, check that the pressure gauge responds. For further comments on bleeding see Section 4.3.

Grout should never be thinned during grouting. If a hole appears choked off with thick grout, do not try to get it taking grout again by thinning the grout—this is useless usually, and if it does happen to work, the resulting grouting is likely to be poor due to the excessive water. The only time to do thinning is at the end of grouting a hole, before starting the next (Section 4.1-9) if a thinner mix is needed.

7.2-9 Approaching Refusal

Eventually, either refusal is approached or it is unlikely to be reached because runaway takes will obviously continue indefinitely if permitted to do so. However, note that use of special treatment to deal with problems may also result in not reaching refusal.

WHEN REFUSAL IS APPROACHING Keep on grouting until virtually no grout is taken in a 15-minute (or similar) measuring interval. Opinions differ about how close to absolute refusal the application should go. My practice is to go as close as the order of accuracy of the measuring methods will allow. This is not necessarily absolute refusal though.

RUNAWAY AND EXCESSIVE TAKES Large, long cracks can produce these takes. It is not good practice to keep supplying grout to a hole with a large take continuing indefinitely. Although it may please operators to see so much cement going into the job, in most cases the grout is penetrating too far and is partly wasted, particularly if a curtain is being constructed: the aim is to grout the vicinity of the curtain, not far and wide away from it. Recommended practice is to stop grouting when a fairly substantial amount of cement has been injected and then to let this set before resuming 24 hours later. If this second application continues the fast taking behavior, give it the same treatment. Several more applications may be necessary before getting

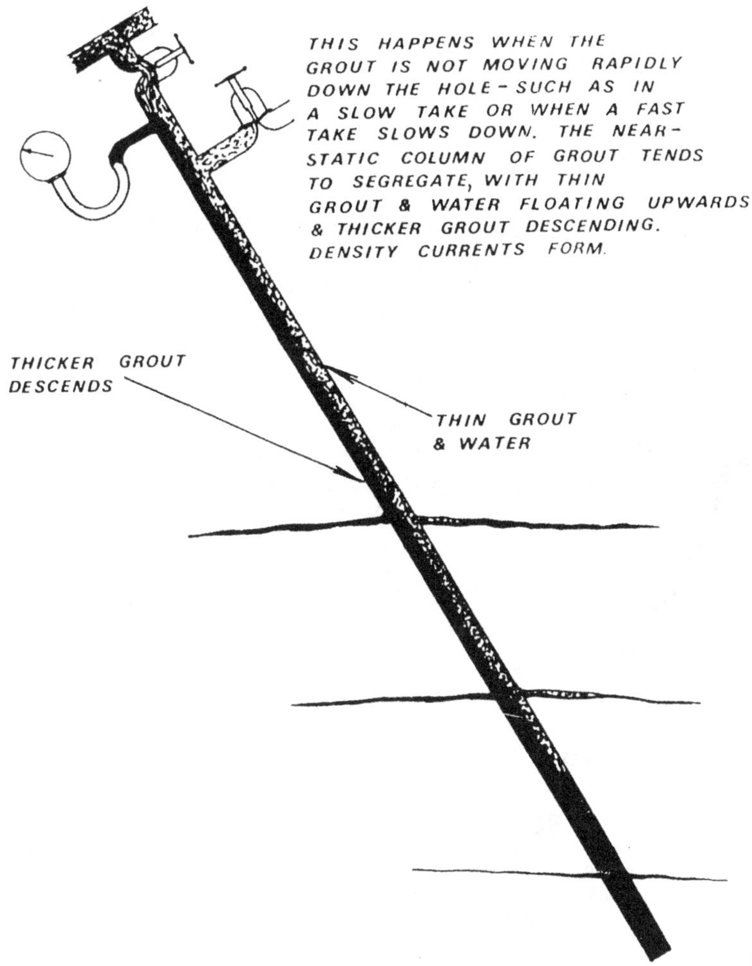

Figure 7.2-8A. *Bleed-water accumulation in a grout hole.*

refusal. In extreme cases of wide cracks, over a dozen applications to the one stage have been necessary.

The choice of numbers for the fairly substantial amount of cement at which the application is stopped varies with the site. Commonly used figures are 100, 150, or 200 bags of cement.

AFTER REFUSAL Once refusal has been obtained in terms of whatever definition is used, keep the pressure on the hole for a further 15 minutes or so. This holds the grout firmly in place in the cracks to ensure that thixotropic stiffening (Section 4.2-2), aided perhaps by some of the cement setting process (but not all the way to initial set—see Section 4.4-1), stiffens the grout sufficiently to resist removal by groundwater or by water from nearby grouting operations. Not everyone does this. I recommend it.

Then close the valve at the top of the standpipe (Section 2.6-2) and leave it closed until wash-out time. The rest of the standpipe fittings can then be taken away to do another hole (Section 2.6-4).

Then comes the completion of the bookwork (Section 4.1-12), the totaling of the cement used and the calculation of the take per unit length of hole. This is usually in bags of cement per foot of hole if the job is in traditional units. If metric is in use, by a quirk of the conversion factors, the same number will result if calculated in liters of cement per centimeter of hole. This is perhaps an unusual unit but is rather convenient. The cement is assumed to be in the same state of "looseness" as in 94-lb bags (Section 4.1-3). Takes can also be measured by weight. For comment on this method see Section 8.4-17.

7.2-10 More Applications of Grout

The process of grouting a hole or stage from start to refusal is termed an "application." Quite often one application is enough to produce the required standard of grouting, but at times it is not enough. One example is when rather a lot of water has been injected with the grout, and it can be anticipated that some of this water

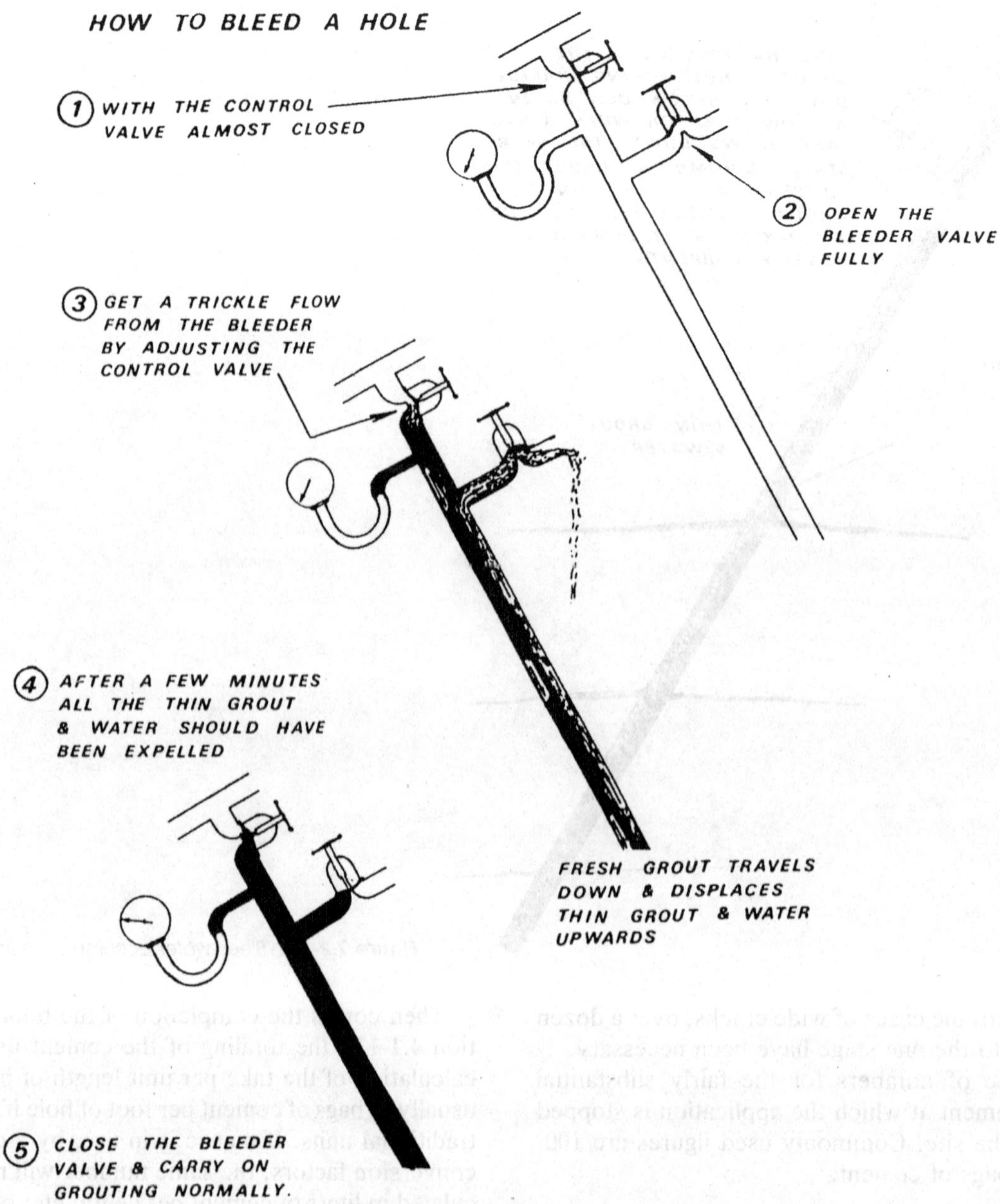

Figure 7.2-8B. Bleeding a grout hole.

has been trapped in the cracks and produced voids. Section 6.1 shows how this can happen in cracking systems.

The amount of water injected with grout is a function of the w : c ratio (thin grouts have a lot of water in them) and the quantity of grout taken. Tests have been made on this (see Section 6.1-6), and I have made trials on hundreds of grout holes at several dam sites to establish criteria for deciding when to give more applications and for deciding suitable values for the criteria. As a result, the following criteria have been developed and used extensively:

Give another application if the *cement* take has exceeded

- 0.25 cu ft of cement/ft ⟨0.25 liters of cement/cm⟩ when the final mix has been 2 : 1 or thinner or
- 0.50 cu ft of cement/ft ⟨0.50 liters of cement/cm⟩ when the final mix has been 1 : 1 or thicker.

These are convenient figures to use because the calculation of cement take per unit length of hole should have been made at the conclusion of the grouting, when all the figures are totaled (Section 7.2-9). Any need for another application can then be decided well before hole wash-out time. In some cases the hole would not need wash-out were it not going to be given another application.

Part of the research into multiple applications was to find out if they were really beneficial. The extensive use of the criteria has established that there is indeed real benefit. Substantial quantities of cement have been injected in second and subsequent applications. Analyses and cement quantities are given in Houlsby (1982b).

Another reason for giving more applications is to remedy deficiencies resulting from having to deal with leaks, connections, movements, and so forth. As suggested in Sections 7.3 and 7.6, this is an important aspect of treating problem areas.

At least 24 hours should elapse between applications. This lets the grout harden and water escape. Note that, as explained in Section 4.4-1, this is not really long enough for the thinner (2 : 1 and thinner) grouts to reach hard set but is a practical compromise. There is no need to do a water test (Section 5.3-2) before second and subsequent applications.

In handling runaway takes (Section 7.2-9), applications are given on successive days until refusal is reached. This has been known to go on for as long as two weeks, but in the end the area was tightly grouted.

In low-quality grouting only one application is given to each stage. It is an interesting reflection on differing attitudes to grouting that some people remain unconvinced about the need for multiple applications, even when they are trying to do good grouting. My experiences on the long-term performance of dams done with and without multiple applications is that they are essential (Houlsby, 1982b) for quality work.

7.3 WHAT TO DO ABOUT LEAKS AND CONNECTIONS

Leaks are dealt with in Sections 7.3-1 to 7.3-5 and connections in Sections 7.3-6 and 7.3-8.

7.3-1 Leaks

Surface leaks are encountered on many grouting jobs. Gear for dealing with them should always be available.

Leaks to underground openings such as tunnels, drives, and shafts are treated similarly to surface leaks. Note the distinction between these and leaks to other grout holes, called connections and handled differently (see Section 7.3-6 to 8).

Surface leaks emerge in situations such as the following:

- Distinct cracks coming to the surface.
- Indistinct cracks among weak rock and weathered materials.
- Among groups of countless fine cracks.
- Where weathered materials or soft rock have been dislodged by the grout.
- Where rock has been moved by grout pressure.
- Where concrete structures such as slabs or grout caps have been inappropriately placed.
- The most inconvenient places possible in accordance with Murphy's law.

Treatment to stop leaks varies depending on the nature of the leaks. Descriptions follow, starting with caulking and proceeding to situations where caulking is useless.

7.3-2 Using Caulking to Stop Leaks

If leaks are coming out cracks with firm edges, and are accessible and not numerous, caulking is the best treatment, Figure 7.3-2A illustrates such situations.

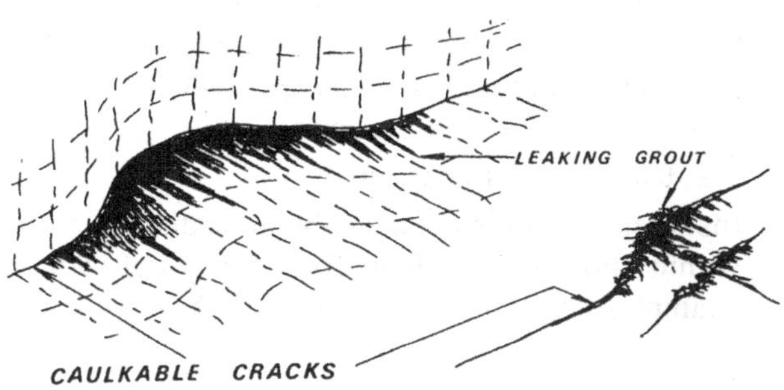

Figure 7.3-2A. Leaks that can be stopped by caulking.

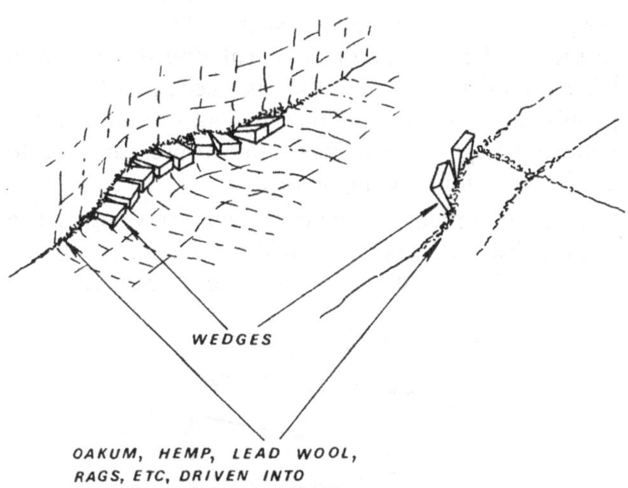

Figure 7.3-2B. Caulked leaks.

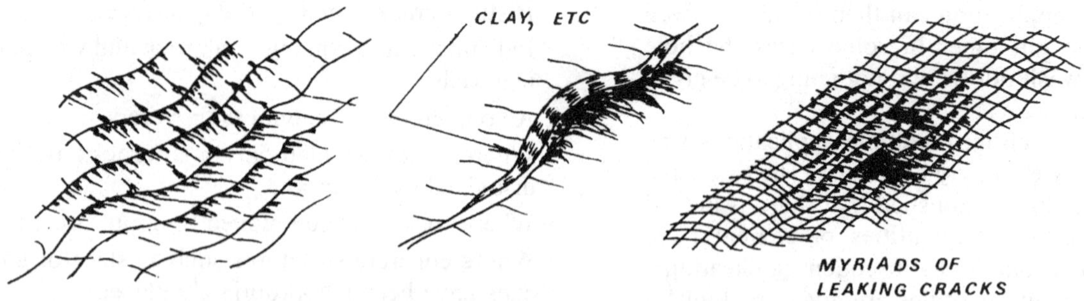

Figure 7.3-3A. Leaks where caulking is not practical.

Caulking is the process of driving soft material into the cracks, together with wedges in those parts of cracks that are wide enough for them. The cracks in Figure 7.3-2A will appear as in Figure 7.3-2B after they have been caulked.

7.3-3 Cases Where Caulking Is Unsuitable

Caulking is not practical in the following circumstances:

- Leaks are coming from cracks with soft or weak edges where wedges and caulking will not hold.
- Too many cracks are simultaneously leaking to be able to caulk them individually.
- The leaks are not accessible.
- Insufficient people are available to caulk the leaks within about half an hour from break-out. Some examples are shown in Figure 7.3-3A.

7.3-4 How to Caulk Cracks

1. Use the control valve on the standpipe to reduce the pressure to a low amount such as 15 psi ⟨1 bar⟩ and to reduce the volume of grout entering the hole to a trickle—not much more than is needed to keep the supply up to any cracks that are not leaking and are able to accept grout at this reduced pressure and to compensate for bleeding losses.

2. If the cracks are wide enough for wedges, use ones with a slow taper and slightly blunt end, as shown in Figure 7.3-4A. Wedges can be cut from hardwood, but softwood such as Douglas fir (Oregon) is better because of slight swelling when the wedges become wet in the cracks.

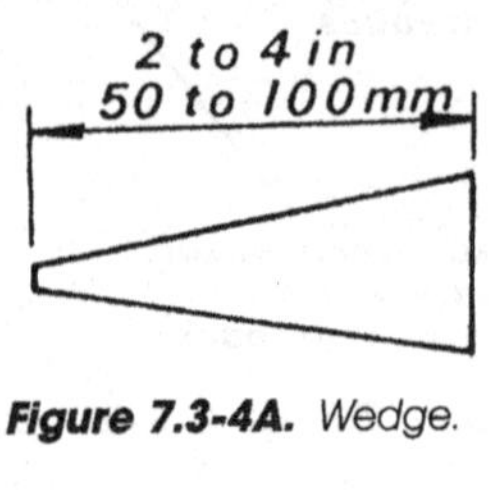

Figure 7.3-4A. Wedge.

Wrap the nose of wedges (see Figure 7.3-4B) with soft bulky material such as oakum, hemp, burlap, or rags.

Figure 7.3-4B. Soft material on wedge nose.

Commence caulking where the grout flow is the weakest. If cracks are too fine for wedges, drive the soft material in with caulking tools of the type boatbuilders use (blunt steel chisels of various sizes and shapes). Where wedges will fit into cracks, place them closely side by side. The gaps between them must be bridged by the soft material overlapping. The sketch in Figure 7.3-4C shows a crack partly caulked in this manner.

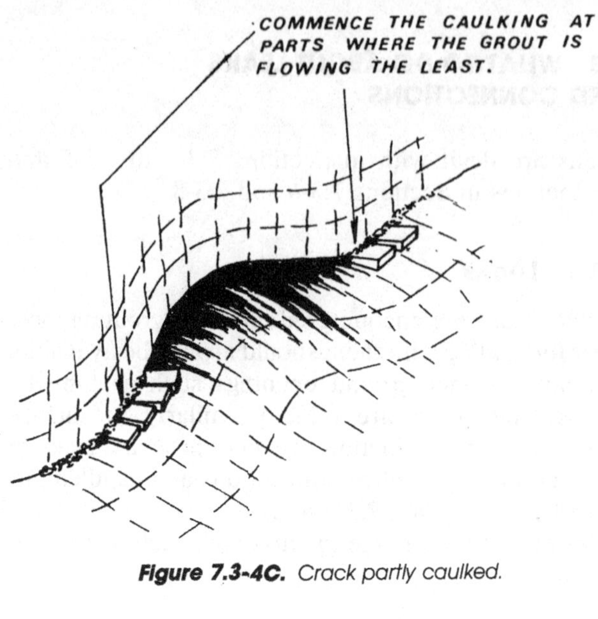

Figure 7.3-4C. Crack partly caulked.

3. When the caulking has been completed along all but the last foot or so, turn off injection completely and then complete the remaining caulking (see Figure 7.3-4D).

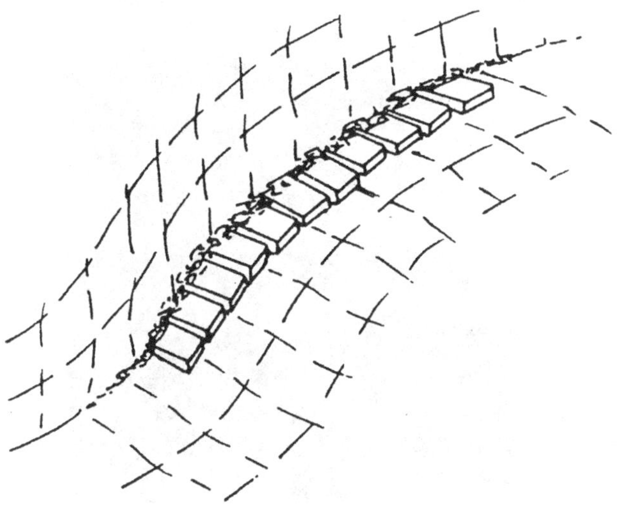

Figure 7.3-4D. Completed caulking.

4. If the crack is horizontal or slopes downhill to the caulking and is therefore difficult to vent by means of seeps through the caulking, caulk a short vent pipe into it with a valve on top (see Figure 7.3-4E). This valve is left open until the caulking is able to withstand some grout pressure and is then (later) closed when thin grout and water cease to come out.

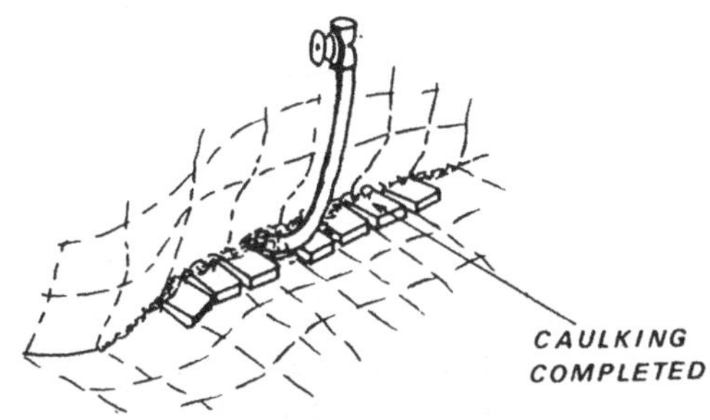

Figure 7.3-4E. Vertical pipe caulked.

5. The crack is now ready for grouting to resume. Thicken the mix to 1 : 1 and inject this for a few minutes to test the caulking and to feed thick grout up to it. Tighten any caulking leaking appreciably, but don't worry about mere seeps.

6. Thicken further to 0.8 : 1 and inject cautiously at low pressure to get this thick grout up to the caulking.

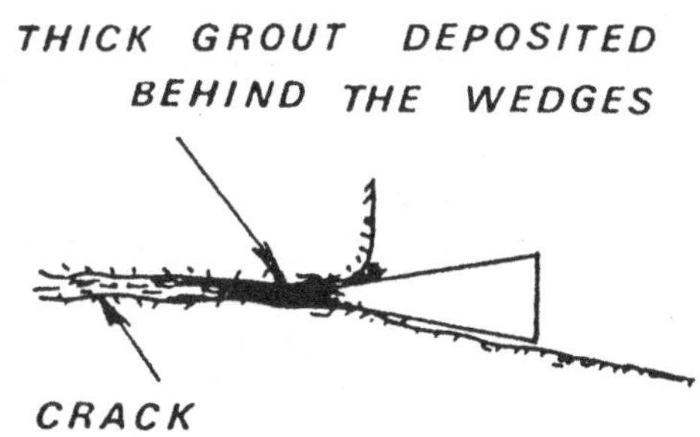

Figure 7.3-4F. How wedges work.

Figure 7.3-4G. Typical higgly-piggly run of wedges and oakum in a caulked crack.

Let excess water seep out of the caulking or bleed it off through a vent pipe. The aim is to achieve the situation shown in Figure 7.3-4F.

It can be helpful to stop grouting for a few minutes so that the stationary grout can develop thixotropic stiffening (Section 4.2-2) and so lose its fluid properties.

7. In difficult cases, further thickening of the grout may be necessary, with the grout injected intermittently.

8. Note that the leaks are stopped mainly by the stiff grout next to the caulking and not so much by the caulking itself.

9. Once the leaks are stopped, it should be possible to continue the grout application to refusal as usual, but with caution and without thinning the mix.

10. Because this treatment has probably plugged off the entrances to any fine cracks leading off the hole, it is usually desirable to give further applications (see 7.2-10) in the hole, or to drill and grout additional holes nearby.

CAUTIONS ABOUT CAULKING If a caulked crack opens up uniformly as shown in Figure 7.3-4H, it means that rock has moved after the caulking. Special treatment is needed to handle rock movements (see Section 7.6-4).

Always have sufficient caulking supplies on hand. They may be needed without warning.

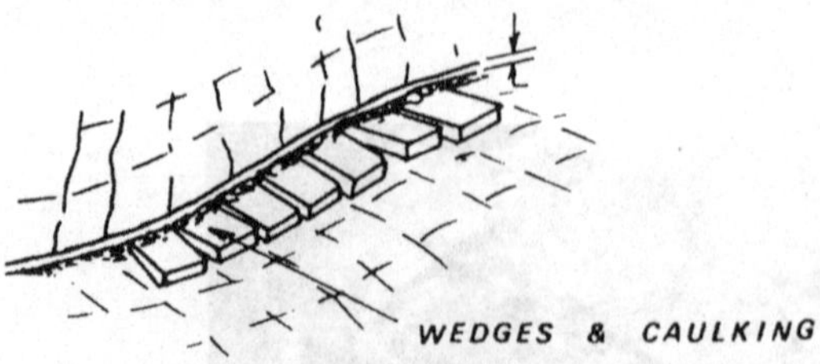

Figure 7.3-4H. *Beware if this happens.*

7.3-5 How to Stop Leaks Where Caulking Can't Be Used

BEST METHOD Place a coating on the foundation surface to cover all of the leaking area (Figure 7.3-5A). This binds surface chunks together and stops the surface unravelling during the grouting and is quite effective, to the surprise of people who have not previously seen it used. The flimsy-looking thin coating is able to hold against normal grout pressures as long as it remains integral.

The surface coating is made with slush grout, gunite, pneumatically applied mortar or similar, to produce a relatively flexible coat about an $\frac{1}{8}$ to $\frac{1}{4}$ inch thick ⟨3 to 6 mm⟩. Slush grouting is described in Section 7.3-9.

Figure 7.3-4I. *Caulking a firm crack with lead wool. This is in a foundation gallery.*

The very big disadvantage of this type of treatment is that it has to be placed several days before the grouting to give the coating time to set, and it is therefore unsuitable when immediate treatment is needed. This is where experience is needed to anticipate where leaks can be expected and thus treat the area in advance. On some sites where extensive areas have been expected to present this type of leakage problem, most or all of the site has been covered in this manner. Such treatment is also beneficial as surface preparation for the core foundation of some embankment dams.

NEXT BEST METHOD Note that this method and the next one are not as good as the previous one nor as caulking is for producing good grouting: they are very much "also ran" solutions, and the grouting can suffer in quality. However, at times they are unavoidable; afterwards further applications and extra holes should be grouted to deal with deficiencies incurred in the leak-stopping process.

Figure 7.3-5A. *The binding effect of surface coating.*

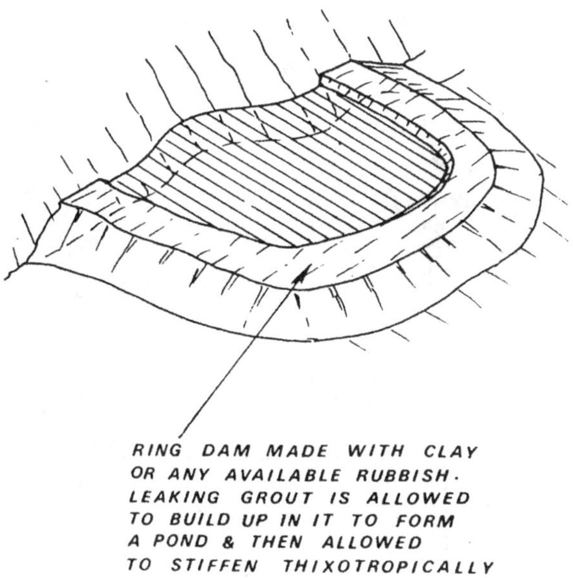

Figure 7.3-5B. Using a dam to stop a leak.

Ponding on the Surface Figure 7.3-5B illustrates this method. After grout has filled the little dam, injection is stopped or slowed to allow the grout in it to stiffen. Later, injection starts again, very cautiously, using thick grout.

LAST RESORT

Internal Blocking This method blocks the cracks underground by using a very thick mix which, while blocking off the leaking cracks not only neglects nonleaking finer cracks and may leave them ungrouted,

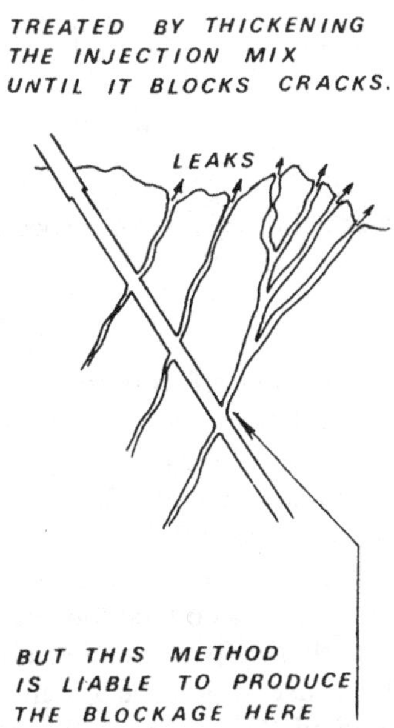

Figure 7.3-5C. Choking off a leak internally.

but the underground blockages in the leaking cracks may be only temporary and can be liable to leak once more when pressure is resumed. Grouting programs can be severely disrupted when one must resort to this method (Figure 7.3-5C).

7.3-6 Connections between Grout Holes

Grout connections between grout holes are usually a nuisance. Some idea that such connections are going to happen usually comes from the water test before grouting, and this gives time to think about what to do. Strong connections can substantially affect the grouting and require special treatment.

Treatment methods vary with the strength of the connection. The strength can be decided from the water test or from grout emerging from holes, as shown in Figure 7.3-6A.

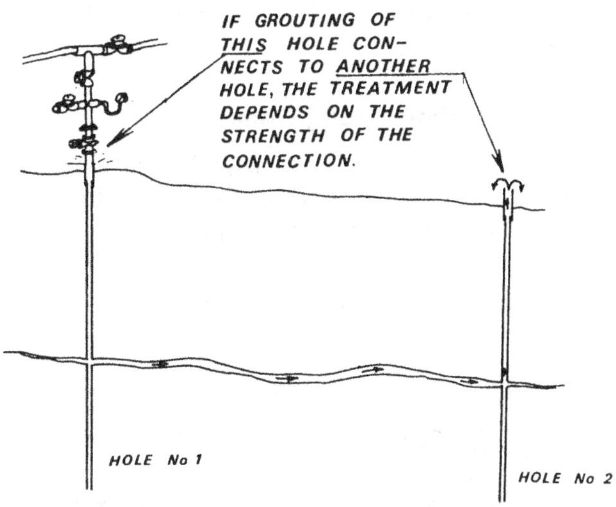

Figure 7.3-6A. A connection.

To decide what to do if grouting is in progress and an unexpected connection starts, examine what is coming out of the hole:

- If water or air is emerging but no grout, then the connection might take up of its own accord without any special measures.
- If grout is emerging but the flow is only a trickle and is thinner than the mix being injected, then this indicates that the connection is slight.
- If grout is strongly emerging, then the connection is strong.

7.3-7 Treating Weak Connections

The treatment method for weak connections is shown in Figure 7.3-7A.

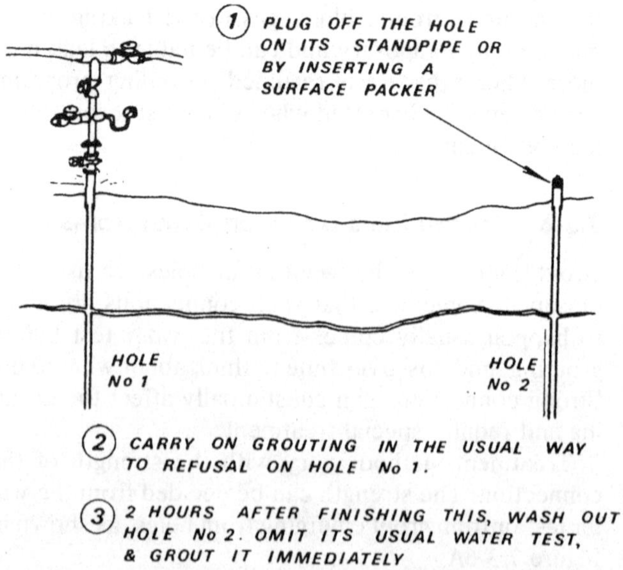

Figure 7.3-7A. *How to grout a weak connection.*

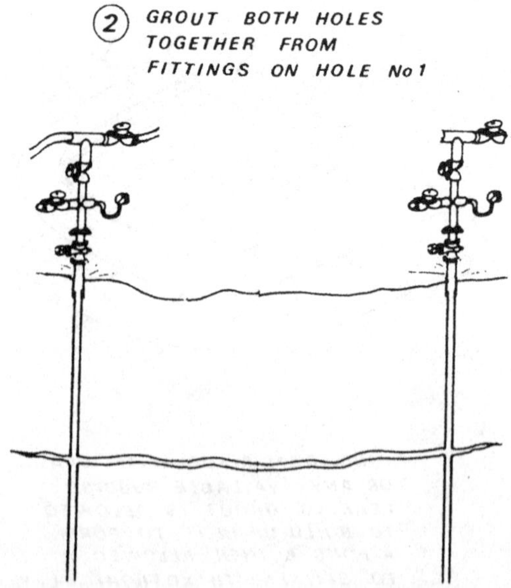

Figure 7.3-8B. *Grouting through the connection.*

7.3-8 Treating Strong Connections

Figure 7.3-8A shows the first step in preparing to grout both holes together because the connection is too strong for them to be grouted separately.

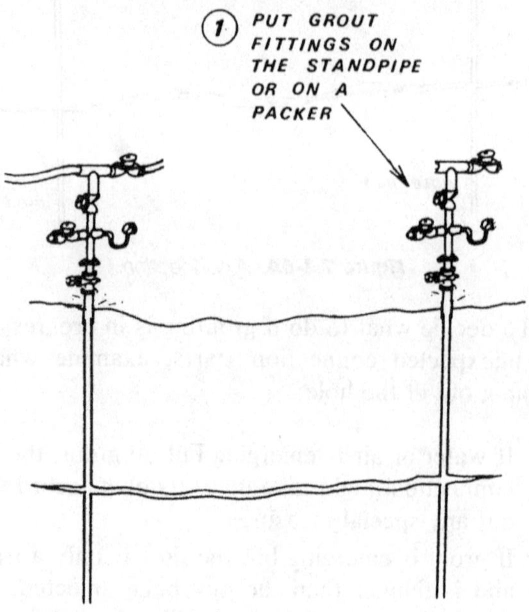

Figure 7.3-8A. *Fittings put on a connected hole.*

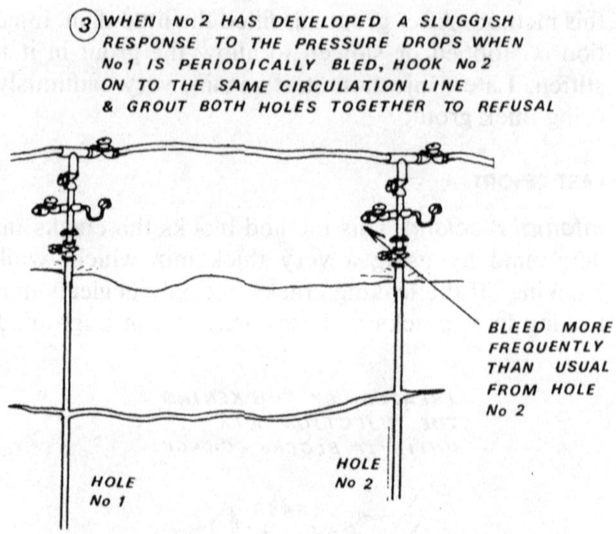

Figure 7.3-8C. *Grouting into the connected hole also.*

Grouting the left hand hole in Figure 7.3-8B then supplies grout for both holes. However, because the connection is not necessarily a clear path for passing good grout to the second hole, good practice is to hook on to the second hole eventually (see Figure 7.3-8C.).

Finer cracks leading off from the second hole have not necessarily had proper grouting treatment, so it is advisable to later put holes through the path of the connection to remedy this and check that the connection has indeed been sealed.

When grouting is being done by closure methods (refer to Section 5.5), if connections occur past holes grouted in an earlier step of closure, poor grouting is indicated. The cause should be sought and improvements implemented. This is one of the ways for finding faulty grouting.

7.3-9 Slush Grouting

A brief description of slush grouting follows. This form of grouting does not come within the scope of this present book and hence is not given in detail.

Basically, slush grouting is a coating of a cement–sand grout on the foundation. Preparation consists of cleaning the surface well and removing any soft material; air–water jetting is usual (Section 5.1-18). The surface must be moist at the time of grout application, without free water in the pockets.

Grout proportions should vary from

$$\text{Water} : \text{cement} = 0.5 : 1$$

for brooming into the fine cracks to

$$\text{Water} : \text{cement} : \text{sand} = 0.6 \text{ to } 0.8 : 1 : 2 \text{ by volume}$$

Some variations to these proportions are made if necessary to make suitable grout with locally available sand.

Mixing should be in a high-shear, high-speed mixer (refer to Section 2.2-2). On some sites it is possible to discharge directly from the mixer to the foundation through a single hose.

The grout should be applied to the surface by stiff brooming, spading, or spraying to get a fairly uniform coating.

Curing is desirable in warm and hot weather.

7.4 ASSESSMENT DURING GROUTING

7.4-1 Why Assess?

Assessment must be done because without it things can go very wrong very easily, and money and time can be wasted in large doses.

Assessment is needed (a) while the grouting is going on, and (b) afterwards.

This present section deals with making assessment while the grouting is in progress. See Section 7.5 for doing assessment after the grouting is complete. There is some intermingling between the two, but they are explained separately to make things as clear as possible.

Assessment during the grouting needs to be carried out on every grout application and done by someone familiar with grouting. Once more, experience pays off.

All this is because grouting is not a process where a recipe or prescribed procedure can be followed. Every hole has its own peculiar variations on the general theme, and in the case of large foundations, grouting characteristics can vary appreciably over the extent of the site. Successful grouting requires experience and initiative.

The basic concept is, of course, to inject maximum quantities of cement. The grout should be as thick as possible commensurate with reasonable penetration. Furthermore, the time when the peak rate of cement

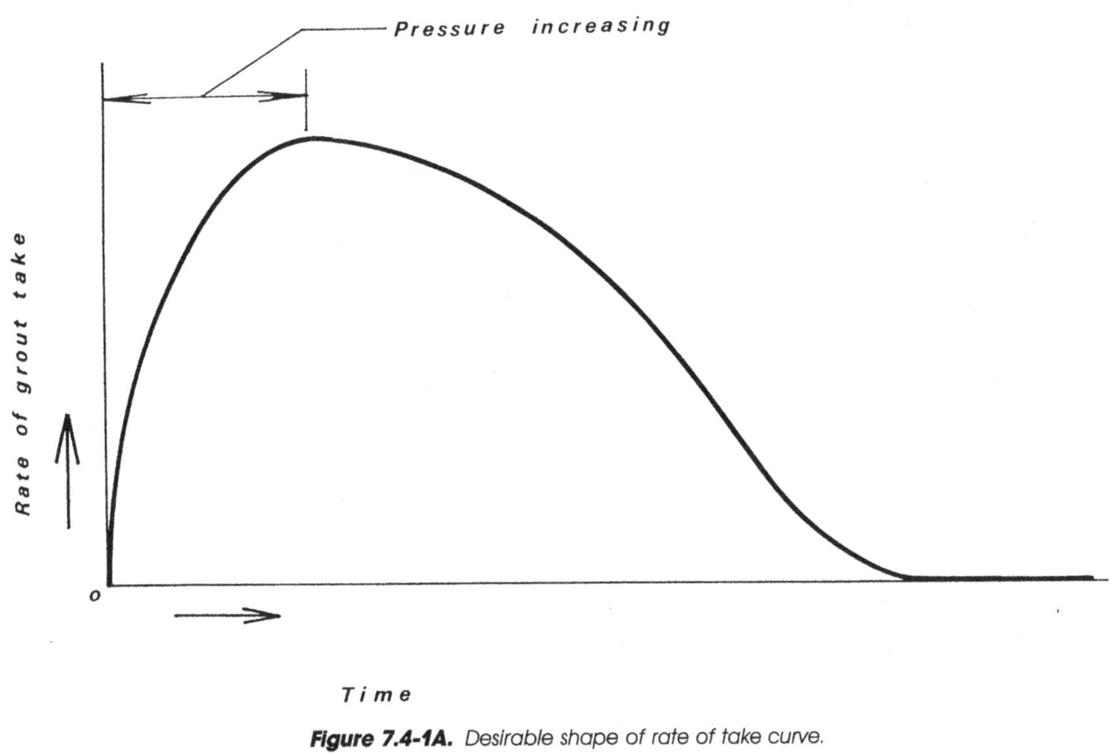

Figure 7.4-1A. *Desirable shape of rate of take curve.*

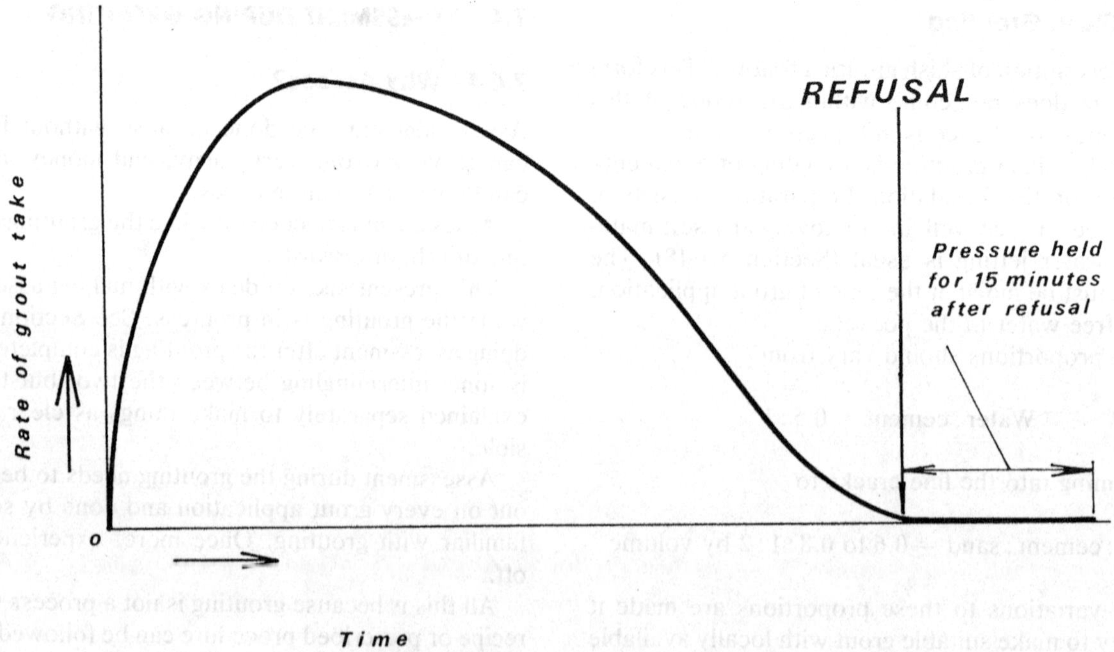

Figure 7.4-1B. *Refusal and afterwards.*

injection takes place should be as early as possible in the application, so as to get maximum mobility before the grout starts to stiffen. If a plot is made of rate of grout take (not just the cement take) against time from start of the application, the ideal shape of the resulting plot should be as shown in Figure 7.4-1A.

During the first quarter hour or so, the effect of slowly increasing the pressure can give a corresponding increase in grout take. Once the working pressure has been reached, the rate of take should have reached its peak, and the remainder of the application should show a lesser rate of take until finally refusal is reached. This ideal state of things does not often happen, though. Variations to the theme are dealt with shortly, but before doing so, some aspects of the ideal profile are worth a little more discussion.

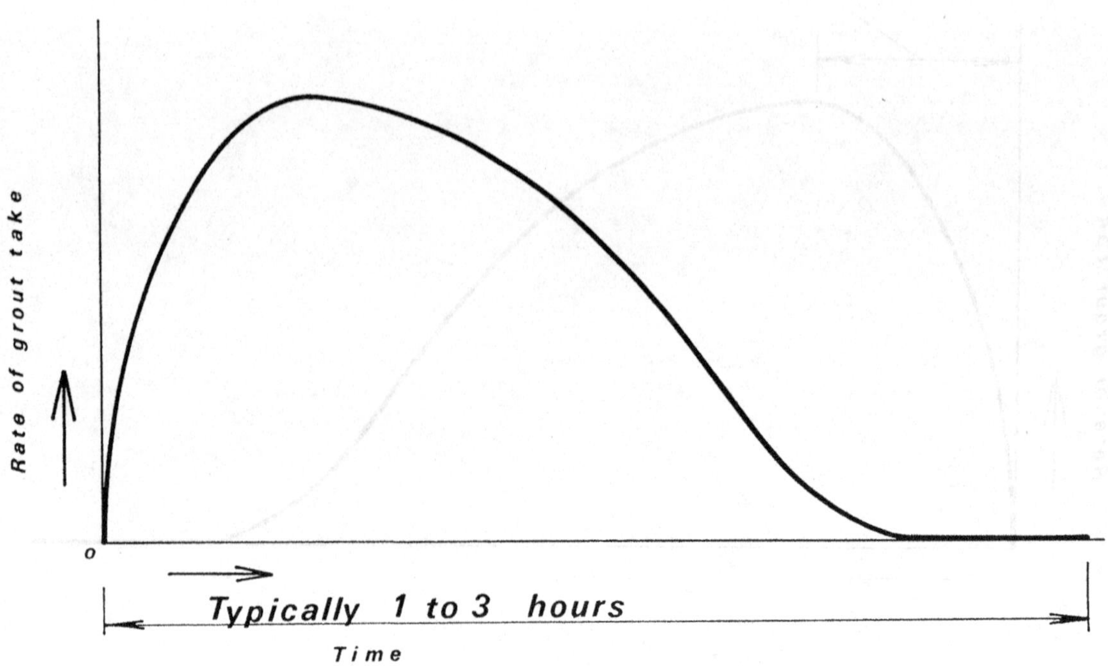

Figure 7.4-1C. *Typical time span.*

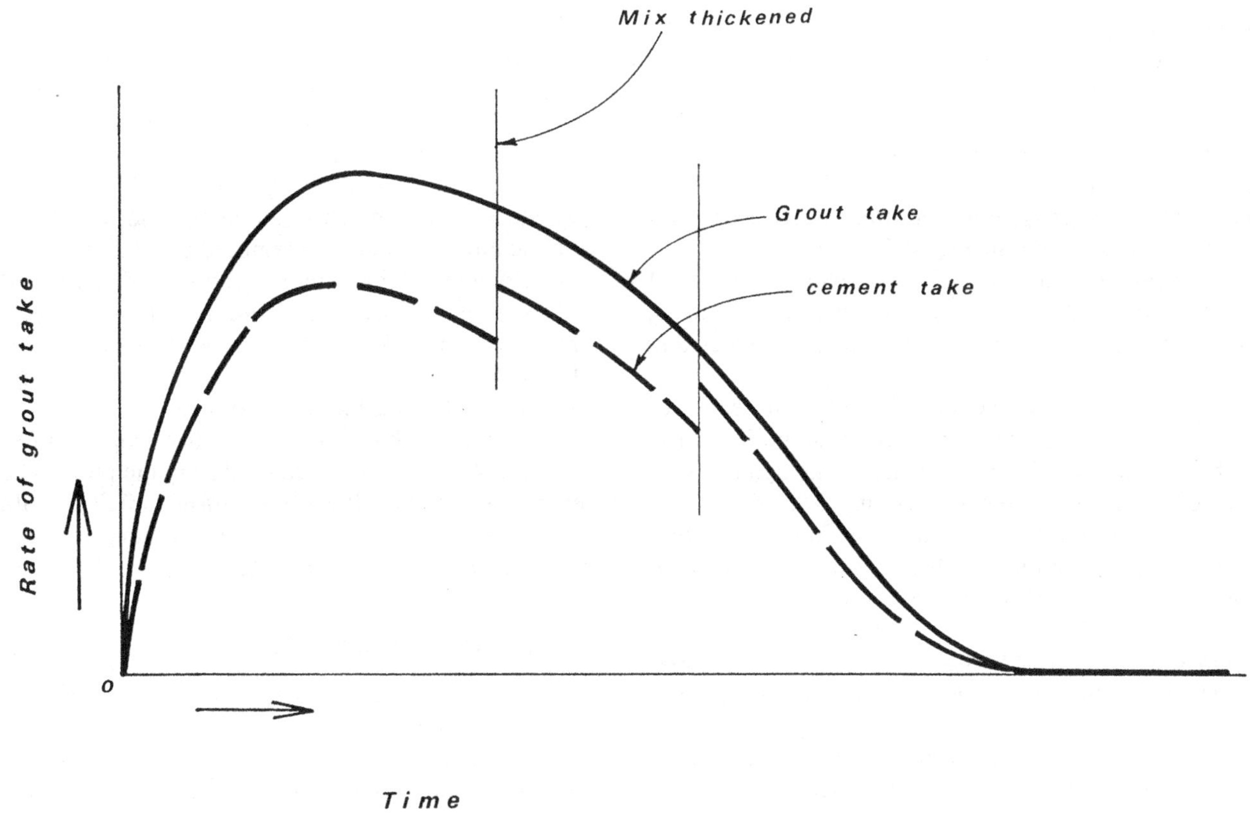

Figure 7.4-1D. *Effect of thickening on cement take.*

After refusal is reached, pressure should be held on the grout for a further 15 minutes, so the ideal profile has a tail on it as shown in Figure 7.4-1B.

Putting some figures on the time scale, the ideal graph is then as shown in Figure 7.4-1C.

Most grout applications last between 1 and 3 hours if carried out by the techniques given in this book. Unless there is an obvious reason, such as leaks or connections or rock movements causing a protracted application, finality ought to be reached more or less

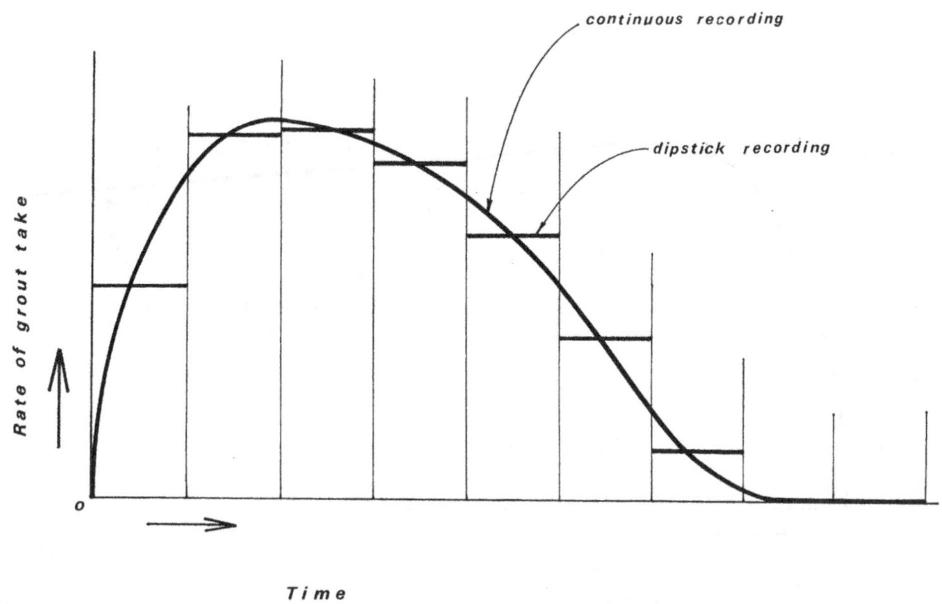

Figure 7.4-1E. *Effect of periodical reading.*

within about 3 hours. Perhaps this finality might not be refusal: it might be a deliberate "pulling off" because a high take has been reached or for some other reason (Section 7.2-9).

The profiles shown above consider the rate of *grout* take. If they are looked at a little differently to show the rate of *cement* take instead, the smooth curve has jumps in it, as shown in Figure 7.4-1D.

Each time the mix is thickened, even if the rate of grout take should happen to remain in a smooth curve, the amount of cement in it automatically changes and so gives a jump in the curve.

In practice, a smooth curve is modified to a series of jerks if readings of grout take are made periodically, such as by taking dipstick readings in an agitator (see Figure 7.4-1E) and refer to Section 2.3-2.

Continuous readings taken by a flow meter connected to a computer or other recording device can, however, give a smoother curve.

VARIATIONS TO BASIC PROFILE After having considered the ideal situation, some divergences from it are now discussed.

Returning to plotting of grout take, rather than of cement take, if there is negligible reduction in the take after several hours of grouting, the plot will resemble Figure 7.4-1F.

It can be seen that the hole is continuing to accept grout very freely. In most cases of this, but not all, the hole will accept a thicker grout almost as freely. On the principle that the thickest grout practicable should always be used, it is advisable to thicken rather than to continue with a grout that is too thin.

The situation shown in Figure 7.4-1G is undesirable. It indicates a runaway situation where the rate of take will continue to increase if the application is continued. Usually, the cause of this is that rock has moved or leaks are getting worse. Whatever the cause, the application should be given suitable treatment to arrest the situation. This can take the form of restricting the amount of grout permitted to go down the hole so as to achieve the situation shown in Figure 7.4-1H.

The effect of such restriction is to reduce any jacking effect on rock in the process of moving. A more drastic version of this treatment is to apply the grout intermittently (see Figure 7.4-1I).

This takes advantage of the thixotropic stiffening (Section 4.2-2) of thick grouts during the periods of 15 minutes or so when the grout is turned off. This stiffening tends to restrict the radius of pressure penetration and hence reduces the jacking effect.

7.4-2 Worked Examples

Assessment is usually a mental process carried out by the person supervising the grouting. If suitable recording equipment is available, the data for it can be partly graphed in the format of the graphs just discussed. For purposes of illustration, both tabulated data and the graphical form are used in the following examples. They start with straightforward cases and progress to more complex ones.

ASSESSMENT OF A MODERATE TAKE, TROUBLE-FREE APPLICATION Prior to starting the application, background data should have been obtained to give an idea of per-

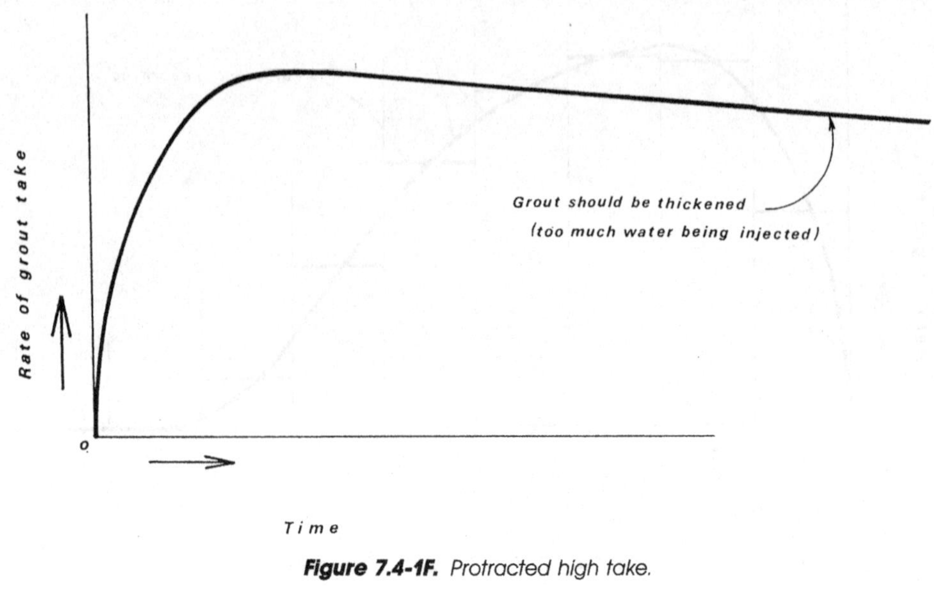

Figure 7.4-1F. *Protracted high take.*

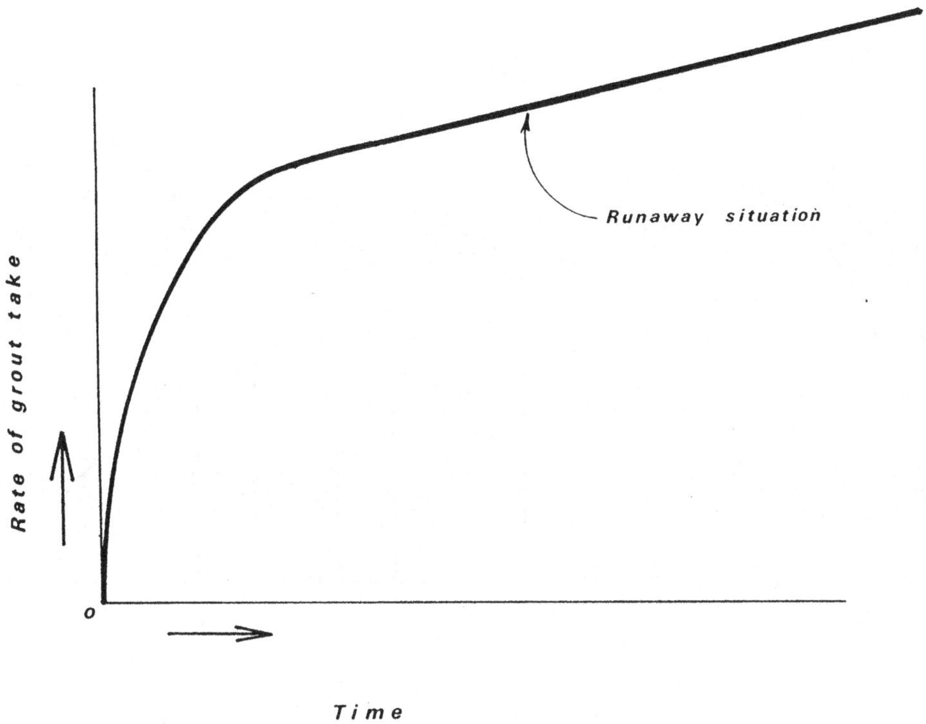

Figure 7.4-1G. Runaway take.

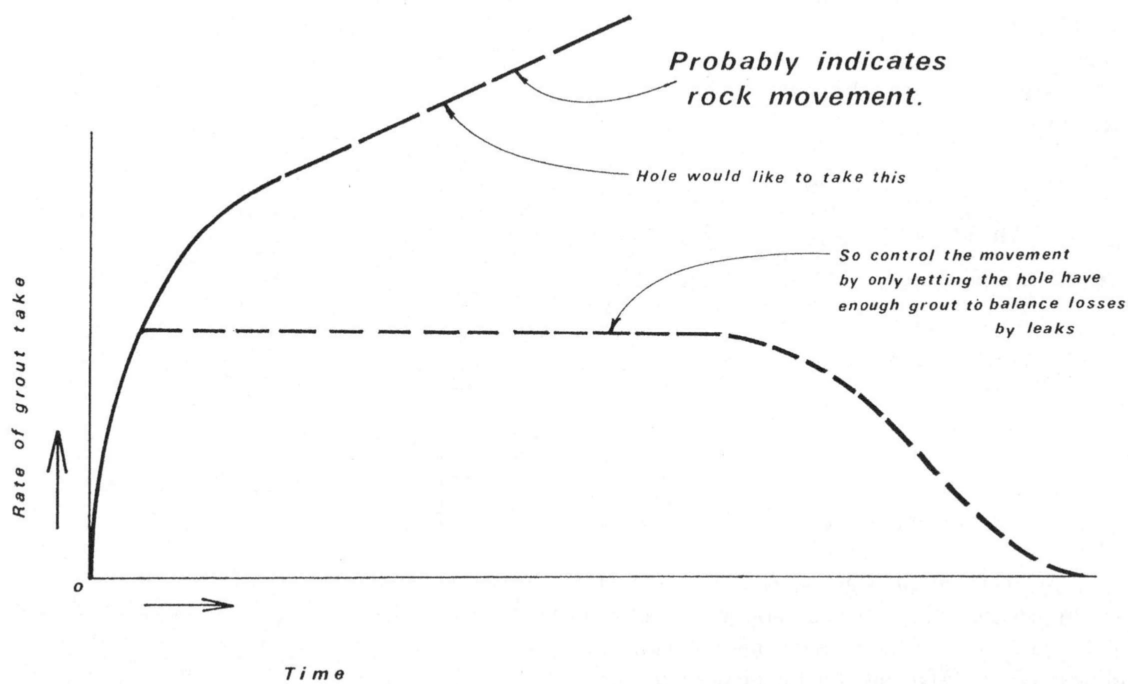

Figure 7.4-1H. Treating a runaway take.

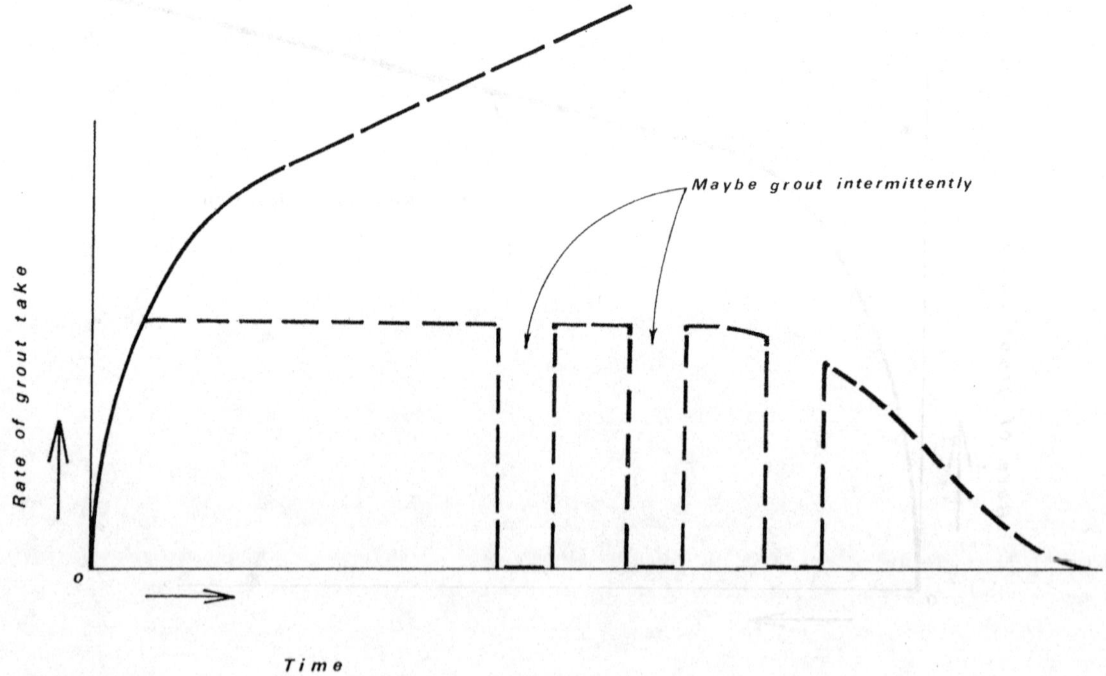

Figure 7.4-1I. *Intermittent grouting.*

meability (Sections 8.2 and 5.3-2), geological features liable to influence the grouting (Section 8.1), and any leaks or connections likely to cause grout loss (Section 7.3).

In this present example, the permeability is, say, 25 lugeons, the rock in the stage being grouted is assumed to be fairly uniform, and there are no leaks or connections to other open grout holes.

Information on the report form from dipstick readings at 15 and then 30 minutes from start of the application is as follows:

Time (min)	W:C	Grout Take (cu ft)	Grout Take (liters)	Cement Take (U.S. bags)	Pressure (psi)	Pressure (bars)
15	2:1	4.50*	127	1.80	15–45	1–3
30	2:1	6.50*	184	2.60	45–50	3–3.5

* Figures in this column show grout taken in 15-minute periods.

When graphed, these figures appear as shown in Figure 7.4-2A.

The commencement of a graph resembling those discussed in previous sections is emerging. It is in the form of steps because of the noncontinuous nature of the readings. When the results for the next 30 minutes become available:

Time (hr-min)	W:C	Grout Take (cu ft)	Grout Take (liters)	Cement Take (U.S. bags)	Pressure (psi)	Pressure (bars)
15	2:1	4.50	127	1.80	15–45	1–3
30	2:1	6.50	184	2.60	45–50	3–3.5
45	2:1	5.45	154	2.18	50	3.5
1-00	2:1	4.60	130	1.84	50	3.5

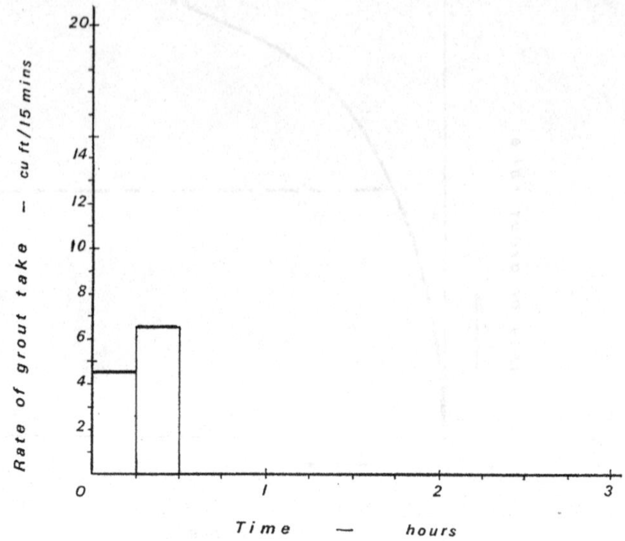

Figure 7.4-2A. *Take in first two periods.*

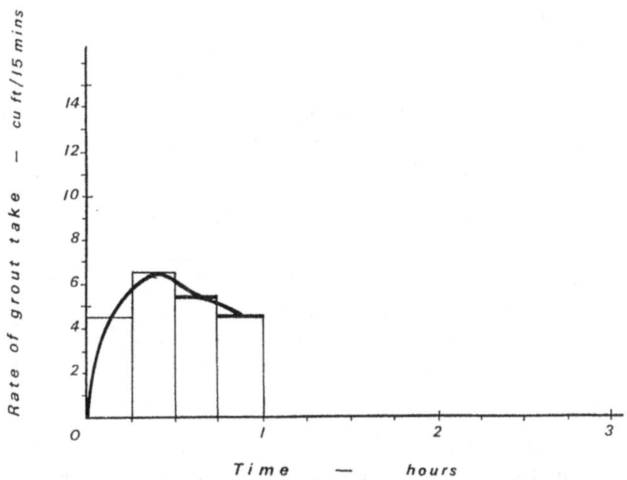

Figure 7.4-2B. Peak passed.

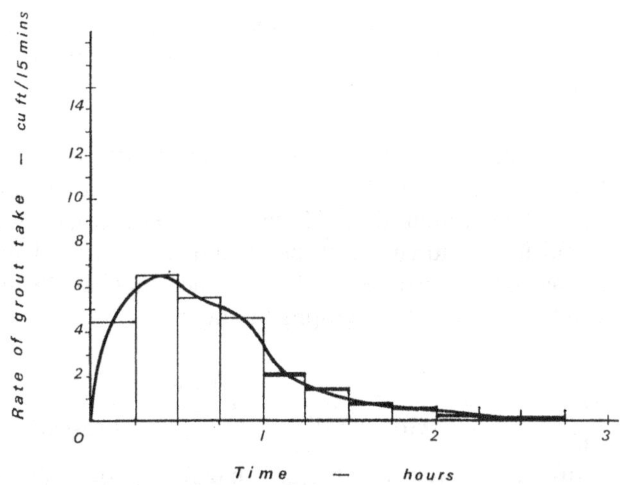

Figure 7.4-2C. The whole application.

The peak rate of take has been passed (see Figure 7.4-2B), most of the cracks have been filled in the vicinity of the grout hole, and the grout is having greater difficulty in penetrating further outwards. The pressure has remained constant. The quantity of cement being injected is reasonable when considered in relation to the water test result.

A mean curve has been put on this graph to lead the eye to the sort of graph discussed earlier. Proceeding with the application to refusal:

Time (hr-min)	W : C	Grout Take (cu ft)	Grout Take (liters)	Cement Take (U.S. bags)	Pressure (psi)	Pressure (bars)
15	2 : 1	4.50	127	1.80	15–45	1–3
30	2 : 1	6.50	184	2.60	45–50	3–3.5
45	2 : 1	5.45	154	2.18	50	3.5
1-00	2 : 1	4.60	130	1.84	50	3.5
1-15	2 : 1	2.10	60	0.84	50	3.5
1-30	2 : 1	1.40	40	0.56	50	3.5
1-45	1 : 1	0.75	20	0.50	50	3.5
2-00	1 : 1	0.50	15	0.33	50	3.5
2-15	1 : 1	0.25	10	0.17	50	3.5
2-30	1 : 1	NIL	NIL	NIL	50	3.5
2-45	1 : 1	NIL	NIL	NIL	50	3.5

Total cement take = 10.82 U.S. bags

The thickening at 1 hour 30 minutes from 2 : 1 to 1 : 1 produced a slowing of the grout take, but the quantity of cement injected did not suffer, hence the thickening can be judged as being judicious. There has been satisfactory reduction in take as refusal was approached (see Figure 7.4-2C). Pressure was held after refusal for a further 15 minutes. Total cement take was commensurate with the water test result insofar as relationships can be established (as indicated by experience).

A postmortem assessment of the application after completion could reach the following conclusions:

- The application was properly carried out.
- The cement injected was about the maximum possible for the hole.
- There were no leaks or connections, and hence additional remedial holes are not necessary alongside.

These conclusions are not completely self-evident from the data. They make use of long experience but are stated here for guidance.

Takes such as these are vaguely typical of those experienced in long, well-connected cracks open 0.040 to 0.100 in ⟨1 to 2.5 mm⟩ in a stage length of about 25 ft ⟨about 8 m⟩. These deductions come from noting that

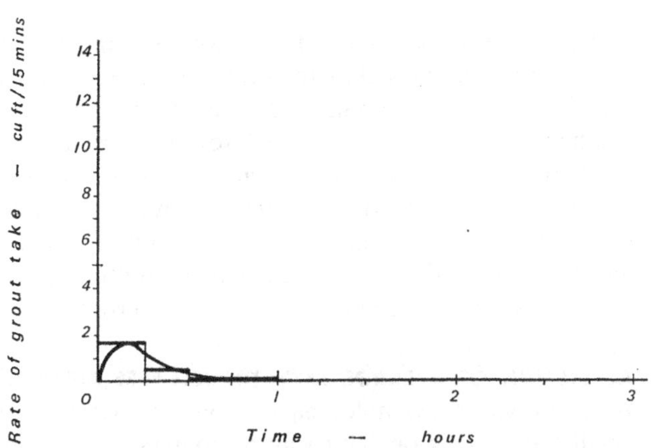

Figure 7.4-2D. A small take.

the take was moderate relatively to grout takes in general, and the time to do the grouting was on the long side.

ASSESSMENT OF A SMALL TAKE, TROUBLE-FREE APPLICATION In this example, the grout taken is not much more than required to fill the hole. The water test could have shown a low permeability, such as 4 lugeons, or alternatively a higher one, such as 20 lugeons. Figure 7.4-2D graphs the take.

Time (hr-min)	W:C	Grout Take (cu ft)	Grout Take (liters)	Cement Take (U.S. bags)	Pressure (psi)	Pressure (bars)
15	3:1	1.75	50	0.50	15–45	1–3
30	3:1	0.50	15	0.14	45–50	3–3.5
45	3:1	NIL	NIL	NIL	50	3.5
1-00	3:1	NIL	NIL	NIL	50	3.5

Total cement take = 0.64 U.S. bags

Because of the low water test, the application has started with the thinnest recommended mix of 3:1. The grout has quickly filled the hole and the few cracks. Note that the measured grout takes include the grout used to fill the hole. The quantity injected in the first 15 minutes was small, but the following take was even smaller. The hole was still taking at 30 minutes, but by the time the 45-minute reading was taken, refusal had been reached. Pressure was then maintained for a further 15 minutes.

A postmortem assessment carried out after the grouting could conclude that:

- There was no point in thickening the mix.
- There were no leaks or connections and hence remedial holes are not necessary.

The grouting behavior of this hole is typical of a foundation where cracks are either very fine, such as less than 0.020 in (0.5 mm) or are very few. The water test helps to decide which is the case: a lugeon value of 4 taken in conjunction with a grouting performance like this and with geological knowledge would indicate very few cracks, whereas a water test result of, say, 20 lugeons would show the presence of multitudes of cracks that are too fine to accept cement grout.

ASSESSMENT OF A LARGE TAKE, TROUBLE-FREE APPLICATION Previous examples have shown moderate and small takes. This one, by way of contrast, is a large take, and like the others, is trouble free.

The water test before grouting of this stage would probably have shown in excess of 100 lugeons. Geologically, the foundation would display obvious large open cracks. The following information extracted from the report form shows the start of the grouting:

Time (min)	W:C	Grout Take (cu ft)	Grout Take (liters)	Cement Take (U.S. bags)	Pressure (psi)	Pressure (bars)
15	0.8:1	24.00	680	18.46	0	0

This application has shown a great rate of take right from the start. The water test gave the clue that this would happen, and so the starting mix was appropriately 0.8:1. This is quite a thick mix. During the first 15 minutes the hole accepted this grout as fast as it could be supplied, without building up any pressure. It was obvious that an even thicker mix could be accepted by this hole without difficulty. Accordingly in the next 15-minute period, 0.6:1 grout (the next thicker practical increment) was used, as shown in the following figures:

Time (min)	W:C	Grout Take (cu ft)	Grout Take (liters)	Cement Take (U.S. bags)	Pressure (psi)	Pressure (bars)
15	0.8:1	24.00	680	18.46	0	0
30	0.6:1	19.75	550	17.96	0	0

At the end of 30 minutes, the hole was still taking as much grout as could be supplied, and the pressure had not started to build up. Further thickening was possible, and so the mix was taken to 0.5:1, which is the thickest practical mix for this type of grouting (see also Figure 7.4-2E):

Time (hr-min)	W:C	Grout Take (cu ft)	Grout Take (liters)	Cement Take (U.S. bags)	Pressure (psi)	Pressure (bars)
15	0.8:1	24.00	680	18.46	0	0
30	0.6:1	19.75	550	17.96	0	0
45	0.5:1	18.00	495	17.86	5	0.3
1-00	0.5:1	16.80	475	17.14	8	0.5

The thickening to 0.5:1 caused a slight drop in the rate of grout take but almost no change in the rate of

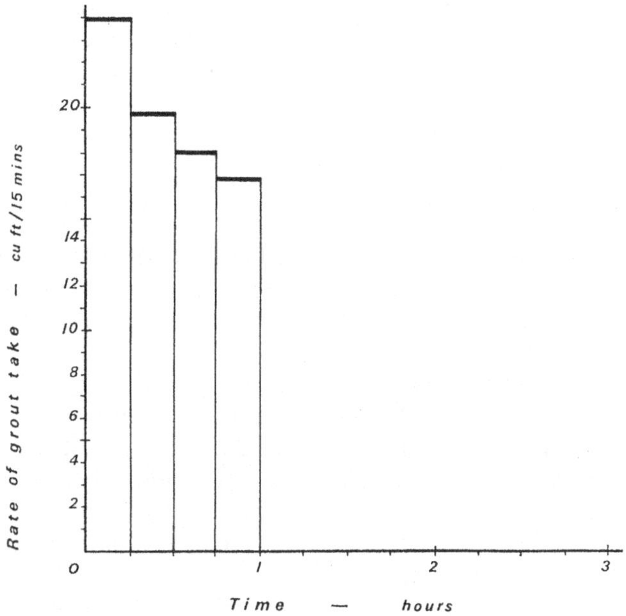

Figure 7.4-2E. *First four periods.*

cement take. The pressure started to build up, however, and this would indicate that the grout was no longer freely escaping in open cracks. The grout had filled all the wider ones, and the thixotropic stiffening, which can become dominant in these particular mixes, had started to influence the situation. The pressures listed are those recorded by the gauge in a situation where the control valve was wide open (and the return half of the circulation line was disconnected as discussed in Section 2.7-4). The pressure specified for the hole was 50 psi ⟨3.5 bars⟩, but this could not be applied until the hole started to show some restraint in accepting the grout:

Time (hr-min)	W:C	Grout Take ⟨cu ft⟩	Grout Take ⟨liters⟩	Cement Take (U.S. bags)	Pressure ⟨psi⟩	Pressure ⟨bars⟩
15	0.8:1	24.00	680	18.46	0	0
30	0.6:1	19.75	550	17.96	0	0
45	0.5:1	18.00	495	17.86	5	0.3
1-00	0.5:1	16.80	475	17.14	8	0.5
1-15	0.5:1	14.25	405	14.62	10	0.7
1-30	0.5:1	12.05	340	12.27	12	0.8

The rate of take has now started to show a more pronounced drop, and the pressure continued to build up. Evidently, there were still plenty of well open cracks yet able to pass this very thick mix:

Time (hr-min)	W:C	Grout Take ⟨cu ft⟩	Grout Take ⟨liters⟩	Cement Take (U.S. bags)	Pressure ⟨psi⟩	Pressure ⟨bars⟩
15	0.8:1	24.00	680	18.46	0	0
30	0.6:1	19.75	550	17.96	0	0
45	0.5:1	18.00	495	17.86	5	0.3
1-00	0.5:1	16.80	475	17.14	8	0.5
1-15	0.5:1	14.25	405	14.62	10	0.7
1-30	0.5:1	12.05	340	12.27	12	0.8
1-45	0.5:1	9.25	260	9.39	25	1.7
2-00	0.5:1	6.50	185	6.68	50	3.5

This situation is graphed in Figure 7.4-2F.

Now the reduction in take showed that refusal was in sight. The specified pressure was able to be applied at last, and this indicated that most cracks were full and the grout had stiffened appreciably:

Time (hr-min)	W:C	Grout Take ⟨cu ft⟩	Grout Take ⟨liters⟩	Cement Take (U.S. bags)	Pressure ⟨psi⟩	Pressure ⟨bars⟩
15	0.8:1	24.00	680	18.46	0	0
30	0.6:1	19.75	550	17.96	0	0
45	0.5:1	18.00	495	17.86	5	0.3
1-00	0.5:1	16.80	475	17.14	8	0.5
1-15	0.5:1	14.25	405	14.62	10	0.7
1-30	0.5:1	12.05	340	12.27	12	0.8
1-45	0.5:1	9.25	260	9.39	25	1.7
2-00	0.5:1	6.50	185	6.68	50	3.5
2-15	0.5:1	3.75	105	3.79	50	3.5
2-30	0.5:1	1.25	35	1.26	50	3.5
2-45	0.5:1	NIL	NIL	NIL	50	3.5
3-00	0.5:1	NIL	NIL	NIL	50	3.5

Total cement take = 119.43 U.S. bags

Refusal was eventually reached (see Figure 7.4-2G), and then pressure held for a further 15 minutes. The total cement take of 119.43 U.S. bags is relatively high for a rock grouting hole taken to refusal in one application. A more usual situation in this sort of foundation is for the hole to show no sign of approaching refusal at 100 bags, and so the application is terminated for the day once that take has been reached (or perhaps 150 or 200 bags might be chosen instead depending on the geology, see Section 7.2-9), and the hole is given another application the next day. Successive applications are given daily until refusal is reached.

Even though the application illustrated in this example reached refusal, a second application after 24

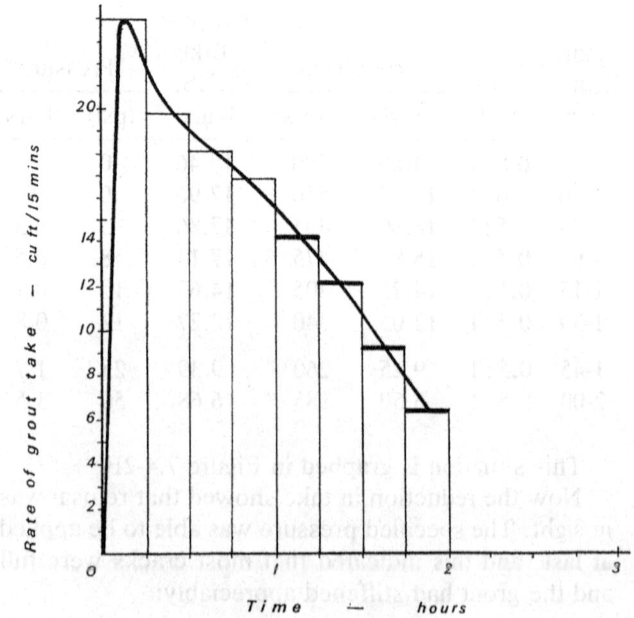

Figure 7.4-2F. *Take reducing well.*

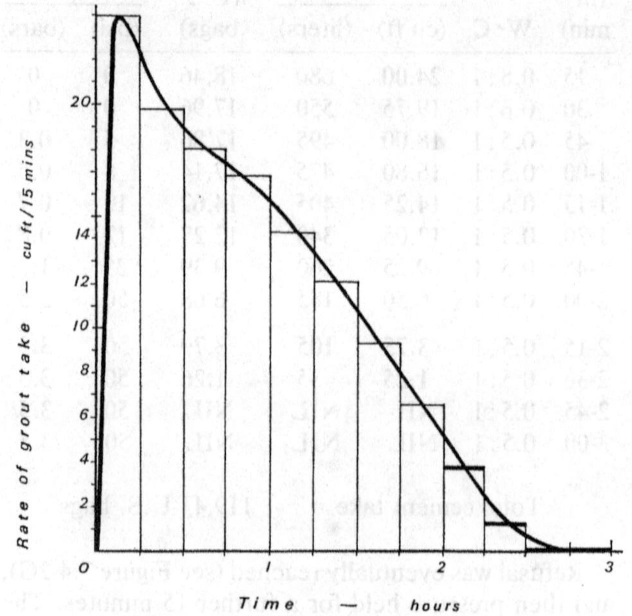

Figure 7.4-2G. *Refusal reached.*

hours (Section 7.2-10) would be warranted in order to compensate for the large amount of water injected.

The postmortem on this application could conclude that

- The mixes used were suitable.
- The hole would probably have taken grout even faster if the plant had been able to supply it rapidly enough.
- Because there were no leaks or connections, additional holes alongside would not have been necessary for remedial purposes. They would be necessary, though, as part of routine closure procedures where further holes are put down into areas where takes are as high as these.

ASSESSMENT OF TROUBLESOME APPLICATION The previous examples are of trouble-free cases. This present one is an actual case where surface leaks occurred and rock moved.

The water test before grouting showed a permeability greater than 100 lugeons, with many surface leaks. No pressure at all could be held during the test. Geologically, the foundation at this hole is heavily cracked, with many open cracks emerging at the surface. The application started with 2 : 1 in order to get some relatively thin grout to the finer cracks before changing to the much thicker mixes needed to deal with the leaks. The first 30 minutes of the application went as follows:

Time (min)	W : C	Grout Take (cu ft)	Grout Take (liters)	Cement Take (U.S. bags)	Pressure (psi)	Pressure (bars)
15	2 : 1	33.10	937	13.24	0	0
30	2 : 1	18.48	523	7.39	0	0

Take with the 2 : 1 grout was very fast, as expected, and no pressure built up. Water came out the various surface cracks first and was followed by grout in profusion. During the second 15 minutes the control valve was closed down a little to restrict the amount of surface loss. The grout was then thickened, but it could actually have been thickened a little earlier equally as effectively:

Time (hr-min)	W : C	Grout Take (cu ft)	Grout Take (liters)	Cement Take (U.S. bags)	Pressure (psi)	Pressure (bars)
15	2 : 1	33.10	937	13.24	0	0
30	2 : 1	18.48	523	7.39	0	0
45	1 : 1	19.49	552	12.99	0–15	0–1
1-00	1 : 1	10.88	308	7.26	15	1

Thickening to 1 : 1 was accompanied by some caulking of the worst cracks. These measures were beneficial: this was indicated by the start of a pressure buildup:

Time (hr-min)	W : C	Grout Take (cu ft)	Grout Take (liters)	Cement Take (U.S. bags)	Pressure (psi)	Pressure (bars)
15	2 : 1	33.10	937	13.24	0	0
30	2 : 1	18.48	523	7.39	0	0
45	1 : 1	19.49	552	12.99	0–15	0–1
1-00	1 : 1	10.88	308	7.26	15	1
1-15	0.8 : 1	8.86	251	6.82	15–20	1–1.2
1-30	0.8 : 1	8.51	241	6.55	20–28	1.2–1.5

The grout was then thickened to 0.8 : 1 and intermittent grouting commenced. During each period when the pressure was turned off, caulking was continued. A few minutes off and a few minutes on again were successful in producing a slight further increase in pressure. Unfortunately, there was a weak section of surface rock, and the increase in pressure caused this rock to move. This released pressure and leaks broke out widely again:

Time (hr-min)	W : C	Grout Take (cu ft)	Grout Take (liters)	Cement Take (U.S. bags)	Pressure (psi)	Pressure (bars)
15	2 : 1	33.10	937	13.24	0	0
30	2 : 1	18.48	523	7.39	0	0
45	1 : 1	19.49	552	12.99	0–15	0–1
1-00	1 : 1	10.88	308	7.26	15	1
1-15	0.8 : 1	8.86	251	6.82	15–20	1–1.2
1-30	0.8 : 1	8.51	241	6.55	20–28	1.2–1.5
1-45	0.8 : 1	4.87	138	3.74	28–7	1.5–0.5
2-00	0.8 : 1	4.52	128	3.48	7	0.5
2-15	0.8 : 1	3.88	110	2.99	7	0.5
2-30	0.8 : 1	2.12	60	1.63	7	0.5

The next hour was occupied in caulking and off–on grouting. This was beneficial in reducing the take and diminishing the leaks. However, the pressure was kept low in order to reduce the chance of moving the rock again:

Time (hr-min)	W : C	Grout Take (cu ft)	Grout Take (liters)	Cement Take (U.S. bags)	Pressure (psi)	Pressure (bars)
15	2 : 1	33.10	937	13.24	0	0
30	2 : 1	18.48	523	7.39	0	0
45	1 : 1	19.49	552	12.99	0–15	0–1
1-00	1 : 1	10.88	308	7.26	15	1
1-15	0.8 : 1	8.86	251	6.82	15–20	1–1.2
1-30	0.8 : 1	8.51	241	6.55	20–28	1.2–1.5
1-45	0.8 : 1	4.87	138	3.74	28–7	1.5–0.5
2-00	0.8 : 1	4.52	128	3.48	7	0.5
2-15	0.8 : 1	3.88	110	2.99	7	0.5
2-30	0.8 : 1	2.12	60	1.63	7	0.5
2-45	0.8 : 1	4.03	114	3.10	7	0.5
3-00	0.6 : 1	1.91	54	1.74	15	1

The reading at 2-45 hours showed an increase in take. While it is difficult to compare takes when intermittent grouting is in use, this increase warranted further caution, and so the mix was thickened. Leaks then slowed greatly and a slight increase in pressure resulted:

Time (hr-min)	W : C	Grout Take (cu ft)	Grout Take (liters)	Cement Take (U.S. bags)	Pressure (psi)	Pressure (bars)
15	2 : 1	33.10	937	13.24	0	0
30	2 : 1	18.48	523	7.39	0	0
45	1 : 1	19.49	552	12.99	0–15	0–1
1-00	1 : 1	10.88	308	7.26	15	1
1-15	0.8 : 1	8.86	251	6.82	15–20	1–1.2
1-30	0.8 : 1	8.51	241	6.55	20–28	1.2–1.5
1-45	0.8 : 1	4.87	138	3.74	28–7	1.5–0.5
2-00	0.8 : 1	4.52	128	3.48	7	0.5
2-15	0.8 : 1	3.88	110	2.99	7	0.5
2-30	0.8 : 1	2.12	60	1.63	7	0.5
2-45	0.8 : 1	4.03	114	3.10	7	0.5
3-00	0.6 : 1	1.91	54	1.74	15	1
3-15	0.6 : 1	2.26	64	2.06	30	1.5
3-30	0.6 : 1	NIL	NIL	NIL	40	2.5
3-45	0.6 : 1	NIL	NIL	NIL	40	2.5

Total cement take = 72.99 U.S. bags

The take then dropped off, and as refusal approached, further pressure buildup was possible. Absolute refusal was reached and pressure held for a further 15 minutes. All leaks had stopped, and the rock movement did not further propagate.

A postmortem of this application fairly readily concludes that the quality of grouting of all cracks in the vicinity of this stage is suspect because of the use of thick mixes and low pressures. Remedial holes were necessary on both sides of this stage to deal with the deficiencies of this grouting, even though a second application was given to this stage.

JUST HOW BIG IS A BIG TAKE? The previous examples give an idea of the relative magnitudes of takes, but even so, experience comes into this. As a very vague guide, my suggestion is as follows:

- A *big take* is in excess of 1 bag/ft ⟨1 liter of cement/cm⟩.
- A *moderate take* is between 0.3 and 1 bag/ft ⟨0.3 and 1 liter of cement/cm⟩
- A *small take* is less than 0.3 bag/ft ⟨0.3 liter of cement/cm⟩.

These correspond with the color scheme suggested in Section 7.5-3 for progress plots. The metric equivalents shown here are explained in Section 7.2-9.

However, as an instance of the wide diversity of opinion on the matter, Deere (1982) presents the following version:

	Cement Take		
	Bags/ft	lb/ft	kg/m
Very high	>2.9	>268	>400
High	1.4–2.9	134–268	201–400
Moderately high	0.7–1.4	67–134	101–200
Moderate	0.4–0.7	34–67	51–100
Moderately low	0.2–0.4	17–34	26–50
Low	0–0.2	0–17	0–25

7.4-3 Hole Assessment by Computer

The assessment processes illustrated in these examples can be tedious, and they require the services of an experienced person, who is usually fairly highly paid. Therefore, there is the obvious incentive to do as much as possible of the processing by computer. Only limited success can be expected by this. It is almost impossible to feed a computer all the possible nuances for analysis, and recourse has to be made to human computing for the remainder. Section 7.7 discusses this.

7.5 ASSESSMENT AFTER GROUTING AN AREA

7.5-1 Purpose

The need for assessment as well as how to do it during the grouting is explained in Section 7.4. This present section deals with assessment after the grouting of a group of adjacent holes to decide whether sufficient closure has been reached or whether further holes are warranted. It also deals with decisions about whether holes should be deepened.

After having just done the grouting of several neighboring holes of a closure sequence (closure is explained in Section 5.5), the question as to whether further closure holes are needed should be thought of in terms of the following:

- Lugeon values (see Sections 5.3-2 and 5.3-4) in each stage of each hole before the grouting.
- Cement takes in each of these.
- Were there leaks or connections?
- What the assessments during the grouting decided, i.e., (a) did the grouting achieve optimum cement injection? (b) were there deficiencies in the grouting? and (c) did any leaks, connections, or rock movements interfere with the quality of the grouting?

If first-class grouting is required, there is no substitute for this hole-by-hole examination. In second-rate work, averaging is sometimes used as an economy measure, but its submerging of untypical results can leave ungrouted or poorly grouted windows in the work. Careful examination can be tedious, but that is the way things are in grouting. In doing this reviewing, experience is necessary and can make a great deal of difference to the result. From experience one has to be able to decide, from the grouting data, how well the grouting is going and must know what action to take if it is not going well enough. There are far too many facets and nuances of this to be able to put everything into words, and accordingly the presentation here is a general treatment of the subject only, intended to guide those who might find it helpful.

7.5-2 Closure Assessment

Take a typical situation. Primary holes (Section 5.5) have been completed in an area, and then the secondary holes between them, followed by the tertiaries between those. It is necessary to decide whether quaternary holes are needed, or perhaps later holes also.

Suggestions about how to make decisions for this case and similar ones are considered in the following sections. Related matters are also mentioned, starting with the preparation of material for presenting the grouting data.

7.5-3 Plotting the Grouting

Various presentation methods are in use. They aim to show all the information needed to readily review the grouting in the holes. The usual way is to show each grout hole side by side on a plot, often not to scale. If the grouting data is given on these plots alongside each stage, a lot of paper is needed, using large overlapping sheets. Some people have felt they need to consult the plots frequently, and therefore have wallpapered the job office with continuous plots going around the walls, much to the mirth of visitors.

This plotting is simple enough to be done by hand—a few minutes several times a day is sufficient to keep things up to date—but can alternatively be done by computer if convenient, particularly if computerized measuring is in use (refer to Section 7.7). Figure 7.5-3A is an extract from one of these plots, done by hand in this case.

To enable thorough review, each stage of each hole should have all relevant information plotted such as water test results, cement takes, dates (to show sequences), pressures, the positions of leaks and of connections between holes, and any special comments. Notice the method used for showing takes in multiple applications (see Section 7.2-10) and the total cement take of those applications. By coloring the holes by some system, perhaps like that of traffic signals, the

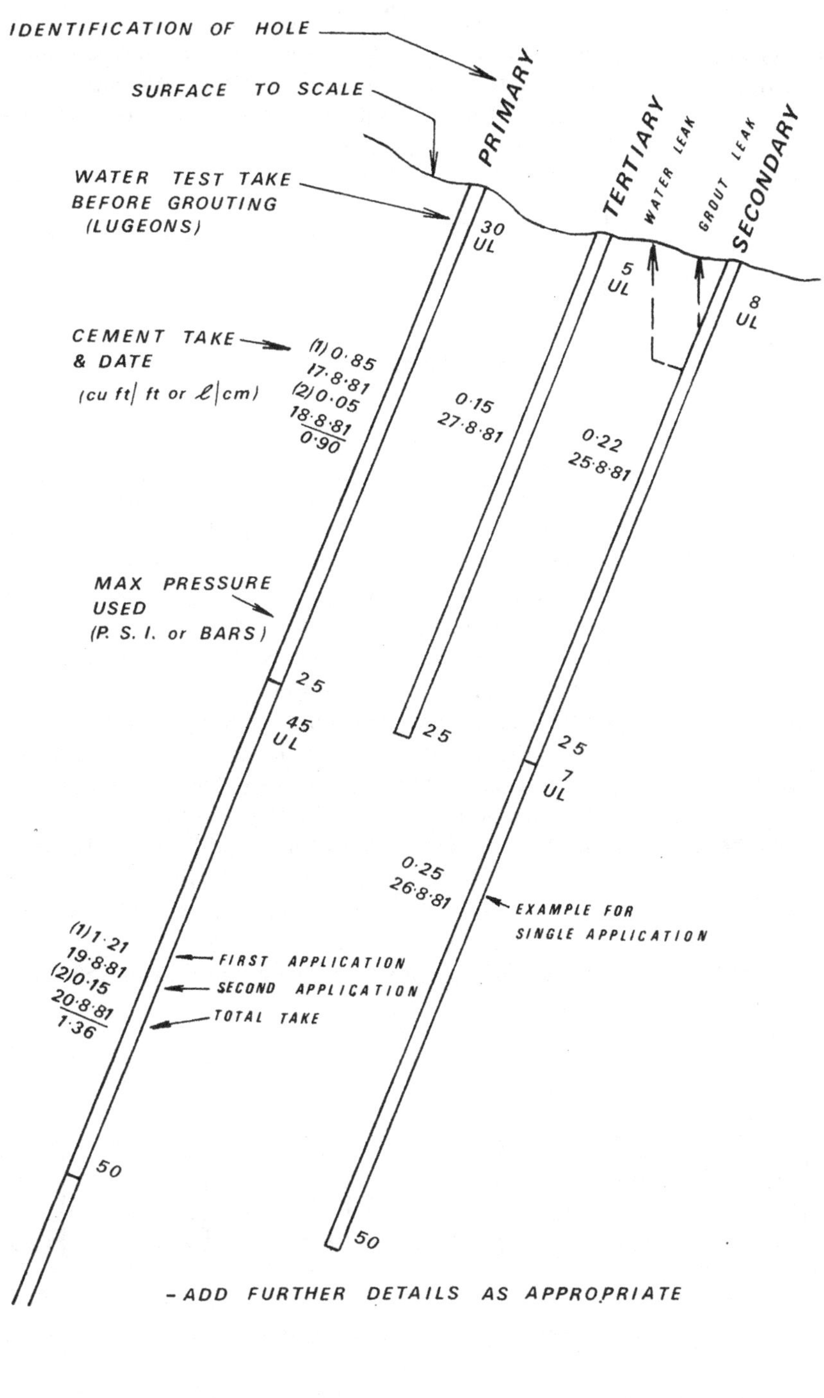

Figure 7.5-3A. Example of working plot showing grouting data.

general situation can be seen at a glance. A system found satisfactory is

Red	>1 bag of cement/ft ⟨1 liter/cm⟩
Yellow	>0.3 bag of cement/ft ⟨0.3 liters/cm⟩
Green	>Nil take
Blue	Tight hole (no take)

These takes correspond with the comments in Section 7.4-2 about describing the magnitude of takes. The metric conversion unit shown is explained in Section 7.2-9.

A large scale is necessary to show all this data clearly. Suitable scales are 1 in = 5 ft or 1 : 50.

The results of assessments during the grouting should be shown on the plot if unsatisfactory work was detected. The plot then provides a concise summary of all data needed to make an intelligent verdict about further closure. If a curtain is being grouted, the plot takes the form of an elevation of the holes; some examples follow later. Where curtains have more than one row, each row should have its own plot, but all the plots must be to the same scale and must be easily synchronized with each other to enable comparisons; sometimes the plots are put on transparent paper for overlaying. For blanket or consolidation grouting the plotting is done on a plan view with each hole shown as a circle or as a small "pie" where segments of the pie are filled in as different parts of the work progresses, and if the holes are deeper than one stage, each of the stages has its own plot.

7.5-4 Assessing

The analysis of the plots and assessing is carried out as follows, keeping in mind the standard (Section 8.3-3) set for the grouting.

Work through each stage in turn, preferably starting with the surface one. For each stage look at pairs of adjacent primary holes and all the holes between them. For convenience let us call this a "bay" of holes. Examine the details of the latest hole(s). These are not necessarily of the same sequence; as shown later it is possible to have closure to tertiary in one-half of the bay and at the same time have closure to quaternary in the other-half. In each of these latest holes, note the result of the water test carried out before grouting, and decide whether it is approaching the standard set for this part of the grouting. To do this look at the amount of reduction of permeability as closure proceeded from the primary holes to the hole under inspection. If the trend of reduction is such that after grouting the hole it is likely that the permeability will be down to the standard or better, then further closure is not necessary. Some examples of this estimation process follow.

In Figure 7.5-4A, permeabilities before grouting have reduced from 25 and 30 lugeons in the primaries to 15 in the secondary and then to 8 in one of the tertiaries and 9 in the other. At this rate of reduction, if the desired standard is 7 lugeons, it is reasonable to presume that the cement takes on the tertiaries are sufficient to reduce permeabilities a little more, and the target has been reached. There is no need for further closure on this score.

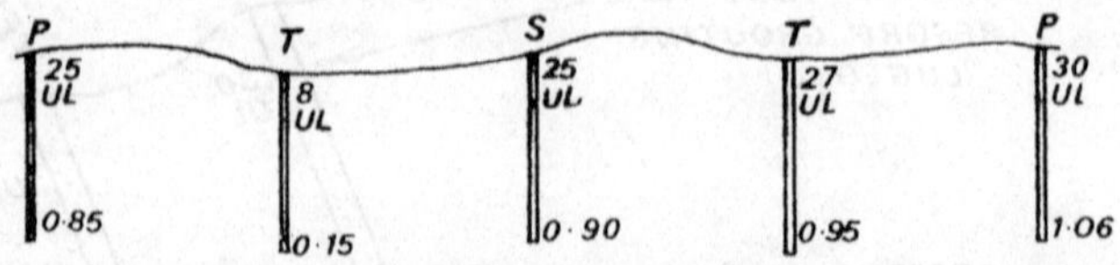

Figure 7.5-4B. Only partly effective closure.

Figure 7.5-4B shows an example where the secondary has takes that are much the same as the primaries. Evidently, the penetrations from the primaries did not reach as far as the site of the secondary. At the site of the left-hand tertiary, penetrations from the primary and secondary reduced permeability appreciably so that the tertiary only measured 8 lugeons before it was grouted. The cement take was sufficient to bring down the permeability after grouting to the target value of (say) 7 lugeons, fairly certainly, and hence further closure at this location is not needed. The situation is quite different at the other tertiary. Here the water test has shown no reduction from those of the primary and secondary on both sides of it. In addition to the high water take, the cement take is high. In these circumstances further closure is indeed necessary. Quaternaries would be placed on both sides of the tertiary, and even further closure may be necessary if the standard is not then reached.

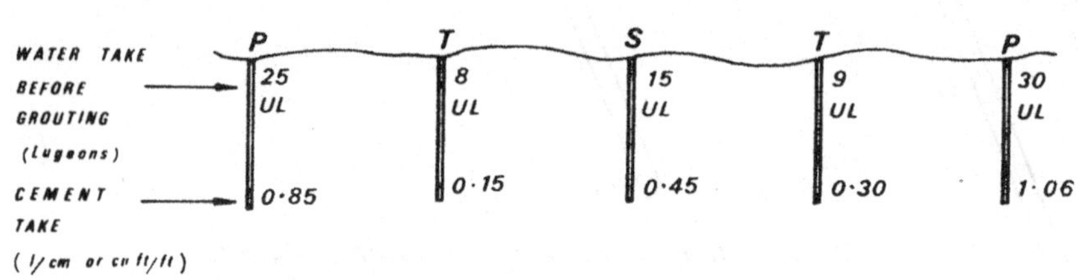

Figure 7.5-4A. Effective closure.

In Section 7.5-1 the comment is made that averaging of holes rather than individual assessment of each one is not good enough for quality work. This example illustrates the point: if the takes on the two tertiaries had been added together and averaged, an unrealistic impression of the situation would have resulted. This might even have led to the placement of quaternary holes at the left-hand tertiary; these would have been a waste of money. In other cases, though, the averaging may have gone the other way leading to omission of further closure where it was genuinely warranted and therefore leaving windows of poor grouting in the work.

Figure 7.5-4C shows no reduction at all by the time closure has reached tertiary sequence. Obviously, further closure is necessary. This situation usually indicates fine cracks, where the grout is not traveling very far, and so eventually closure to fairly close hole spacing is likely.

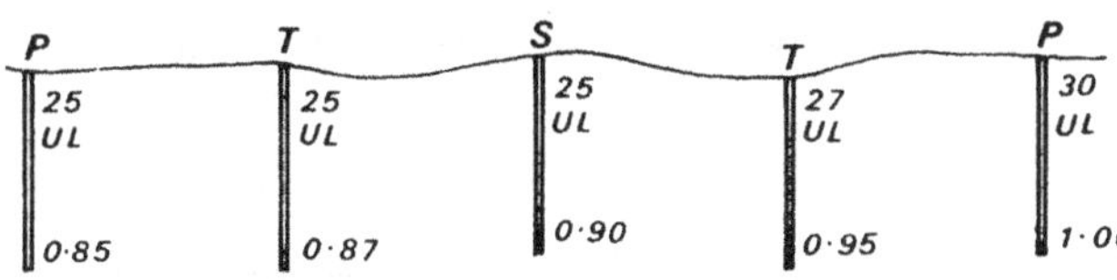

Figure 7.5-4C. *Further closure needed.*

The surface leak from the right-hand tertiary in Figure 7.5-4D raises doubts about the thoroughness of the grouting in cracks accessed from it because of the use of thick grout to stop the leak. The water test result and the cement take are the same as in the first example, and if there had not been a surface leak, there would have been no need for consideration of extra holes. Appropriate action is to put another hole on the side of the tertiary where the leak occurred. This hole may prove unnecessary, but the check is warranted. Sometimes a hole in this situation will need much more grouting. If holes connect, there is also the need to put more holes into the area where the connection occurred, and if it connected past previously grouted holes, their grouting records should be reexamined in order to find what deficiency permitted the connection.

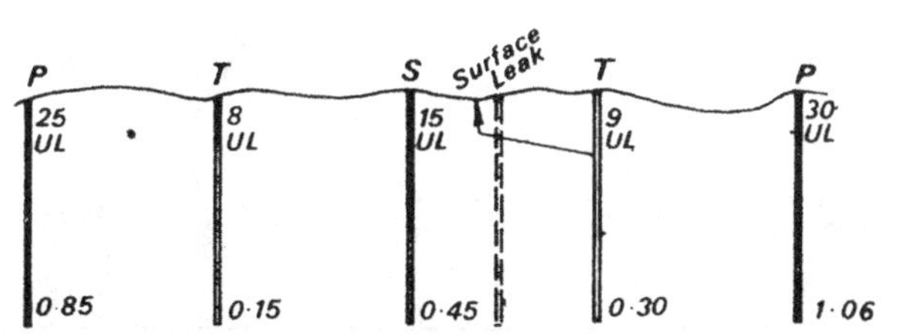

Figure 7.5-4D. *Extra hole to compensate for leak treatment.*

After tailoring the positions of extra closure holes along the lines of the examples given, the result is often a very irregular pattern of holes. Figure 7.5-4E is an example.

In extreme cases successive closure may mean that holes are becoming ridiculously close and might even be tending to touch each other. In such cases it is better to place some of the latest holes slightly to one side or the other of the main row. When this is done on a single-row curtain, it results in thickening of the curtain at such places, and this is a good solution because such weak areas usually warrant extra care. Further comment on this is found in Section 8.4-7.

Doubtfully grouted areas are sometimes checked out by drilling test holes into them. Direct and relevant testing for this is provided by pairs of water test holes straddling the grouting or spaced within it. Water is pumped into one hole of each pair and the amount of water, if any, passing through the grouting and emerging from the other is a measure of the efficacy of the grouting. Adding a dye such as fluorescence to the test water can assist identification of emerging water if there is uncertainty. An ultraviolet lamp may be needed to detect the fluorescene if dilution makes it too faint to pick up by eye.

Coring can be disappointing if used for this type of checking for reasons explained in Section 8.1-5, and there can be uncertainties if a foundation contains quartz seams which resemble grout. They can be distinguished by lightly painting doubtful parts of a core with phenolphthalein solution which reacts with any grout present to give a red tinge, quartz will not.

7.5-5 Relationships between Water Test and Cement Take

There have been many attempts by interested people to try to find a regular relationship between the permeability of a stage, as indicated by its water test, and the amount of cement subsequently taken in its grouting. Such attempts are futile, largely because they ignore the different rheological properties of water and thick grout (Section 4.2). Grout flows quite differently from water. Grout takes are not necessarily related to permeability as measured in terms of water flow. Having said that, it is necessary to point out that there can indeed be a vague relationship between them—a sort of family likeness. This can be noticed if a very large sample is taken; say a few hundred stages. Then the family likeness firms up a lot, but it isn't there if only a small sample is used. And in addition to all this, the experienced grouting practitioner uses a sort of mental calibration to relate lugeon values to cement take on each stage during assessment. However, this calibra-

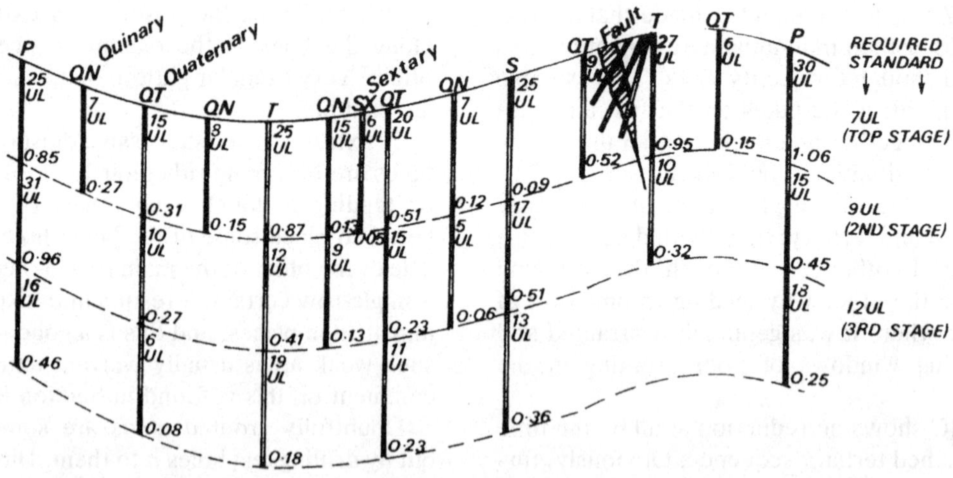

Figure 7.5-4E. *A final curtain.*

tion is tempered by the w : c ratios used and other factors and is mainly to set the alarm bells ringing if cement take is very different from that expected. Don't expect to be able to see that mental calibration expressed in the form of parameters or rules, though, it's just too complex and subjective.

Let's consider these relationships a little more fully now because of the prominence the matter continues to receive in the technical press. Probably the greatest concentration of papers on the subject of permeability testing and on trying to relate it to cement takes was at the Fifteenth International Congress on Large Dams (ICOLD) held at Lausanne, Switzerland, 24–28 June 1985. Question 58 of the Congress dealt with grouting and attracted 86 papers. In summing them up, the General Reporter for the Question, Mr. A Božović of Yugoslavia in Vol III, page 1544, agrees with the previous paragraph:

> There is a virtual concensus that the correlation between the water absorption and grout take is very weak. In Paper R13 (by G. Lombardi, Switzerland) it is shown by considering different rheological characteristics of the fluids used in water absorption tests and grouting, that such correlation cannot physically exist. So permeability and groutability are separate and distinct reactions of the fissured and jointed rock mass when exposed to the action of two different fluids under pressure.

Some comparisons of water and cement takes were included in the Proceedings of the 1982 ASCE Grouting Conference at New Orleans in Houlsby (1982b). They were given for three dams and showed the comparisons split up to accord with geological factors and depth of curtain; losses through surface leaks were also taken into account. For one of these dams, Copeton, the figures are given in Figure 7.5-5A for a further stage in amalgamation; they are all lumped in together for the whole dam and averaged out per sequence.

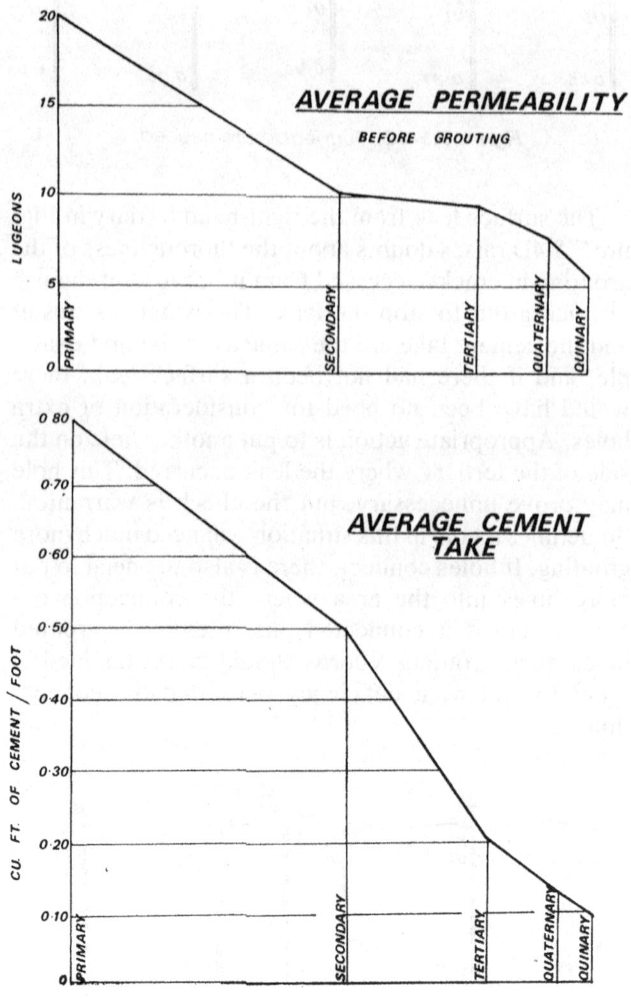

Figure 7.5-5A. *Water and cement takes at Copeton Dam.*

Note that with this large sample there is some sort of vague relationship between water and cement takes, but not even a firm one then. The effect of closure on the water and cement takes is obvious.

7.5-6 Overall Assessments

Once large areas of grouting have been completed on a job, overall assessment should be made. This should be on a larger scale than that described in Section 7.5-4. On some dams, an analysis is made at this juncture to derive statistical data about the grouting and to analyze various factors such as relationships between cement takes and geology, increasing stage depths, multiple applications (Section 7.2-10), pressures, and any special aspects. The analysis also totals the figures and comes up with averages of various groupings. Examples are shown in Figure 7.5-6A and 7.5-6B for four dams. Information about dams and details of their geology are in Houlsby (1982b).

The data is plotted on ordinates for each sequence of closure. Graphs at the bottom give overall average

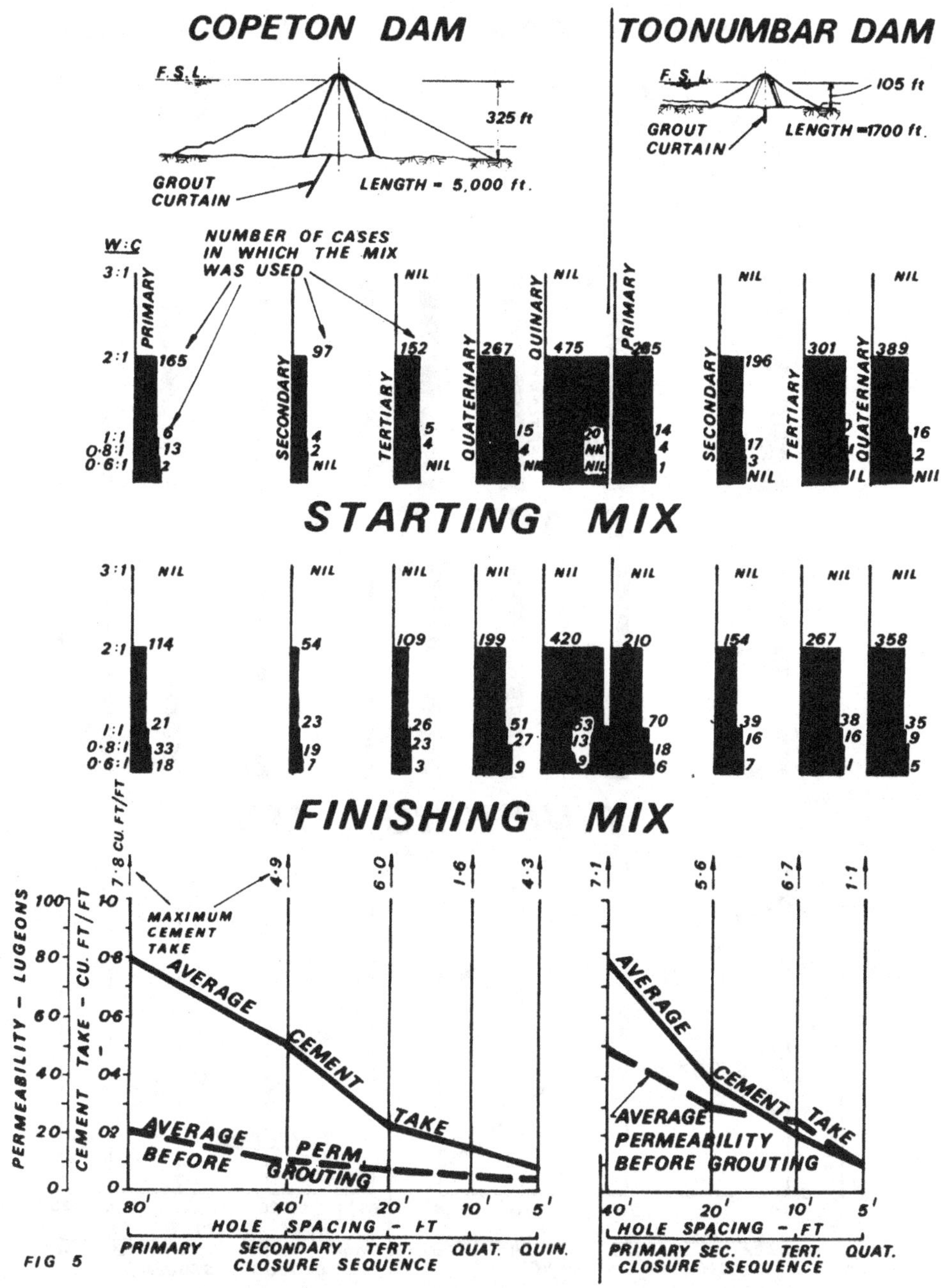

Figure 7.5-6A. *Grouting details—Copeton and Toonumbar dams.*

cement and water takes; the one shown for Copeton Dam has the same data as shown in Figure 7.5-5A, but is more condensed. Of particular interest are the figures for starting (Section 7.2-2) and finishing mixes (Section 7.2-7). Let's consider the values for Copeton Dam in some detail (Figure 7.5-6A). Most of the applications started with 2:1 grout, but 1:1, 0.8:1, and even 0.6:1 were used where the water testing showed that cracks were large enough to accept such thick mixes (and the geological investigation had shown, of course, that there were not many fine cracks that would be clogged by these thick mixes). In the primary holes, 2:1 starting mix was used in 165 stages, 1:1 was used 6 times, 0.8:1 was used 13 times, and 0.6:1 twice. In secondary holes, 97 stages were commended with 2:1, four with 1:1, and two with 0.8:1. Then as closure progressed, the thicker mixes were needed less until at the quinary sequence (5 ft ⟨1.5 m⟩ spacing between holes), 475 stages were started with 2:1, 20 with 1:1, and none with thicker mixes. The finishing mix on all these applications was either the same as the starting mix or was thicker. Some applications started

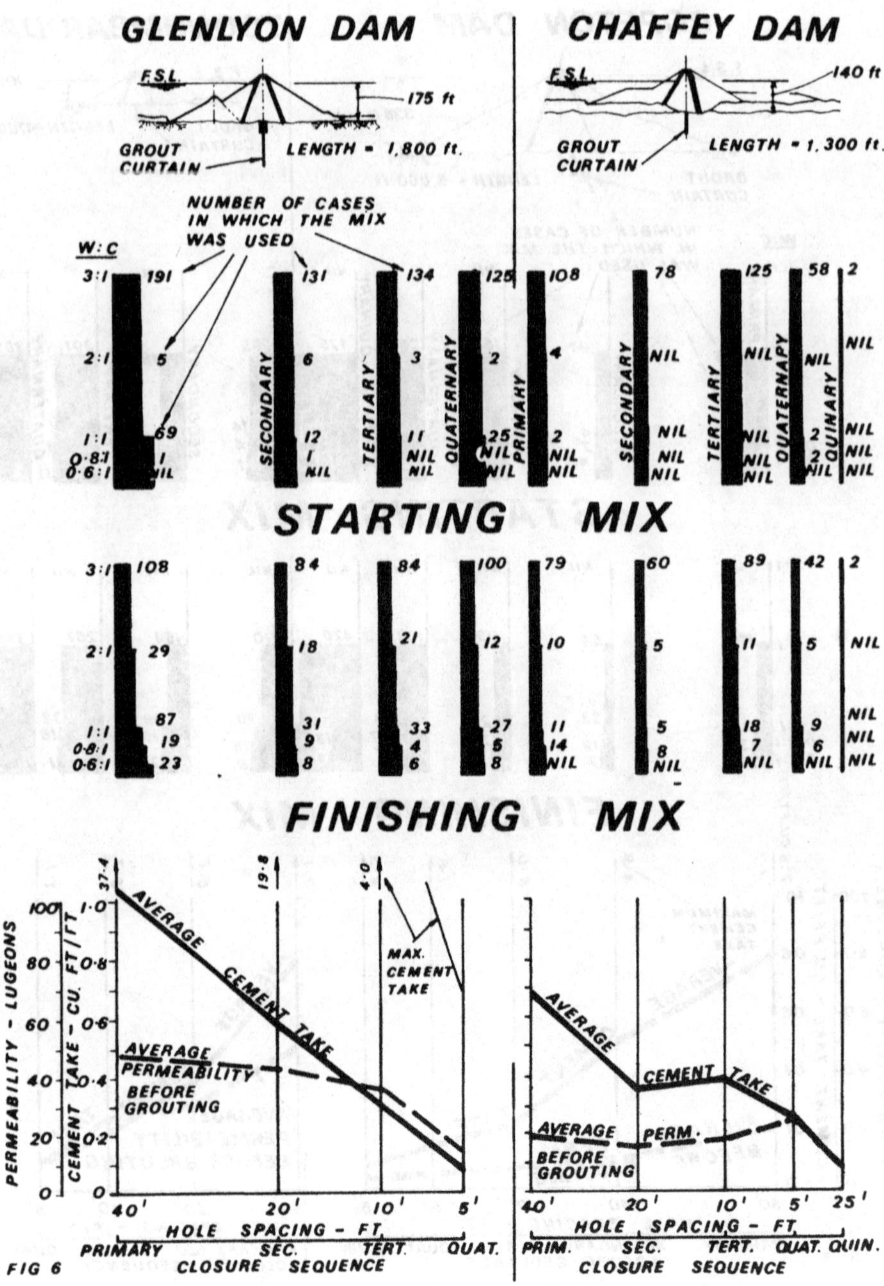

Figure 7.5-6B. Grouting details—Glenlyon and Chaffey dams.

with 2:1 and were found to require progressively thicker grout right through to 0.6:1, but the majority of thickening did not go this far. Additional comments on this and on the other dams shown in Figures 7.5-6A and 7.5-6B can be found in Houlsby (1982a).

7.6 GROUTING OF STRESSED ROCK

7.6-1 Rock Movements

If rock moves while it is being grouted, a change in procedures may be necessary.

Movements can be from several causes. Before going on to those resulting from rock stress, consider the examples in Figures 7.6-1A, 7.6-1B, and 7.6-1C unrelated to stress.

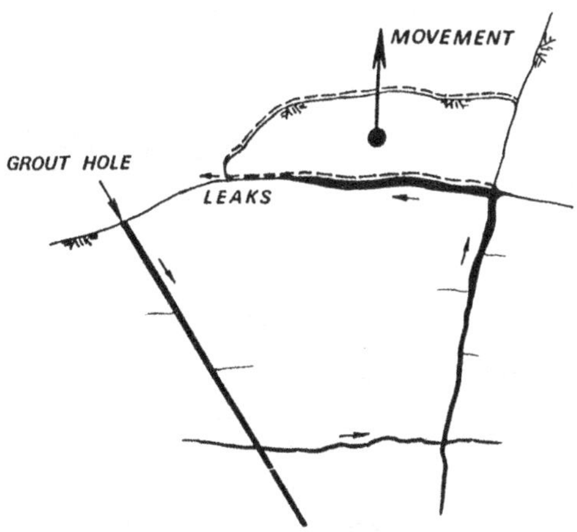

Figure 7.6-1A. *Surface slab lifted by grout pressure. Surface leaks result.*

Figure 7.6-1A shows a typical surface rock of the type that can be moved by grouting if not adequately gripped by the general rock mass. Relatively thin horizontal slabs are particularly prone to this; they might have insufficient cantilever or beam strength or perhaps are detached from the body of the rock by jointing.

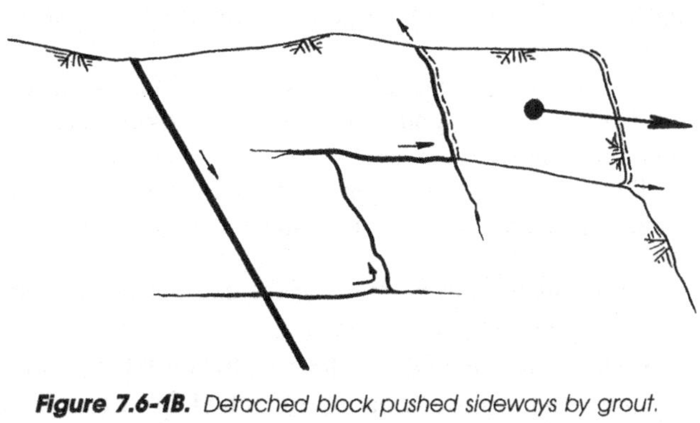

Figure 7.6-1B. *Detached block pushed sideways by grout.*

A slightly different situation is shown in Figure 7.6-1B where a detached block adjoins a face and sits on downward-sloping jointing. Surface leaks can be expected—if not before the movement, then certainly after it. Of course, blocks as bad as this should have been removed before grouting, but the situation is not always as clear-cut as the one shown.

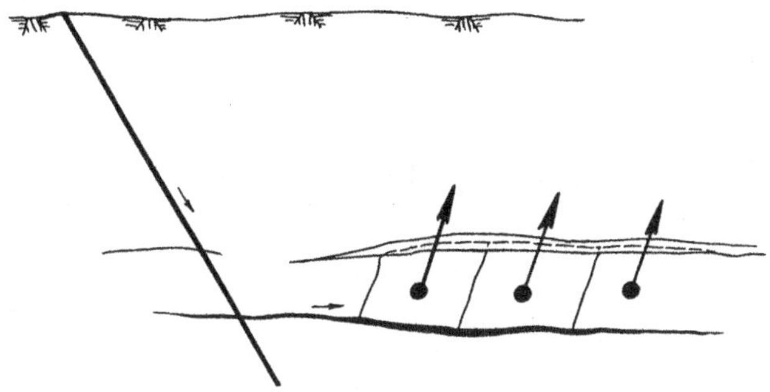

Figure 7.6-1C. *Underground movement.*

Underground movement of the type shown in Figure 7.6-1C has been known to happen occasionally. Wide joint openings between horizontal beds can permit rock to be moved into some if grout gets into only a few of the others. In this situation, instead of having more or less uniform pressure distributions in all joints, the uneven pressurization moves rock into the low-pressure regions. Local rupturing of jointing can be expected to follow, allowing grout to then pass through into the low-pressure regions and equalize pressures. It could be said that this sort of movement does not matter, and in fact helps the grout to get into otherwise inaccessible cracks, on somewhat the same philosophy as the use of very high pressures to deliberately rupture jointing, as discussed in Section 6.2-8. This type of movement is usually associated with widespread occurrence of well-open joints, which permit easy grout penetration far beyond the desired area of grouting and where it is necessary to curb grout takes (refer to Section 7.2-9).

7.6-2 Stressed Rock

Much of the earth's surface is under stress, usually related to tectonic plate activity. The subject of tectonics is outside the scope of this book; its implications as far as grouting is concerned are that valleys in particular may contain rock under stress that is liable to move under the stimulus of grouting. The resulting problems can tax the patience of grouting crews. To get a general idea of how stresses affect a valley, consider a

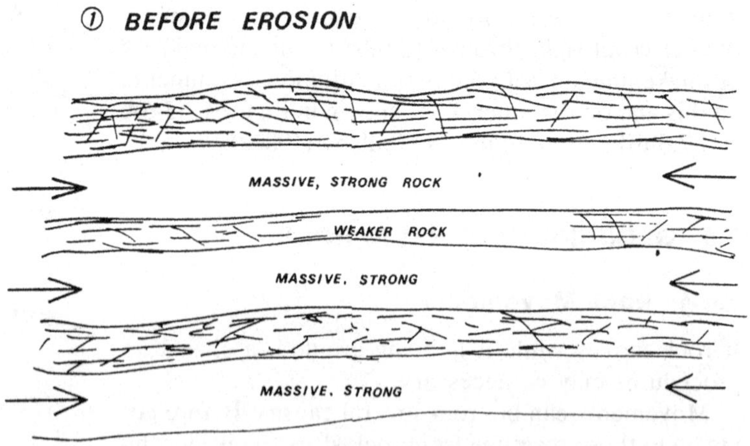

Figure 7.6-2A. *Before valley erodes.*

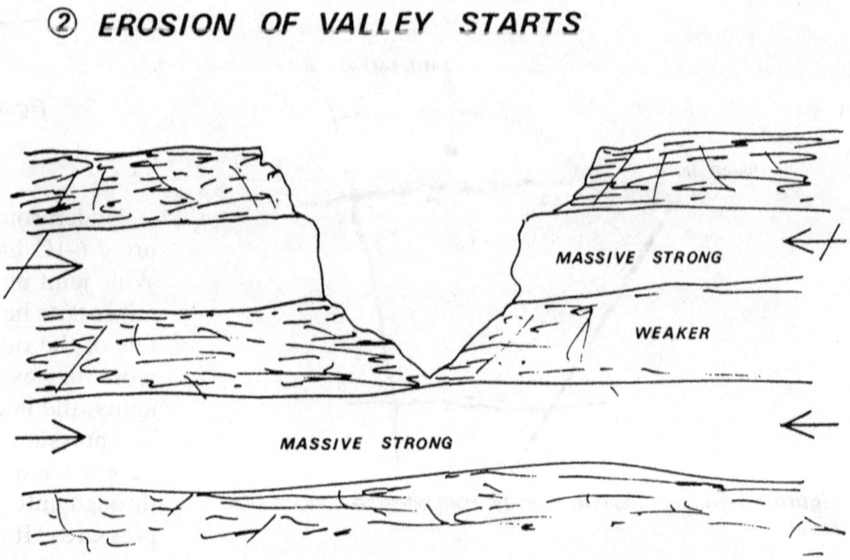

Figure 7.6-2B. *Valley cuts beds.*

following scenario in geological time. Figure 7.6-2A shows a foundation comprising alternating beds of massive, strong rock and weak, easily broken material. The massive beds attract any tectonic stresses while the weak beds don't—they shatter and stress relieve themselves to throw off any horizontal forces.

A valley now starts to erode through these beds (Figure 7.6-2B). After a while the top massive bed is cut and can no longer carry horizontal stresses across the site. Its stresses are then transferred to the next lower massive bed. This transferral takes place well back behind the valley walls, on a regional basis. However, since this lower massive bed already had stresses in it, it now has more.

When erosion eventually proceeds deeper and cuts this highly stressed massive bed (Figure 7.6-2C), its stresses are transferred in turn to the next lower massive bed, which already has its own stresses. The addition of the stresses shed from above may be too much for it and it may fail, as shown. This failure may take the form of valley bulging, a local anticline, a fault, or other feature.

Distortion of beds and the presence of highly weathered infill or fault gouge are the usual signs that this has happened.

There are easily recognized visual signs if stressed rock is present; they are indicated in Figure 7.6-2C:

- Valleys tend to be more of the canyon type rather than having gentle slopes.
- The massive beds, indicated by cliff formation, have large blocks bodily detached from the main

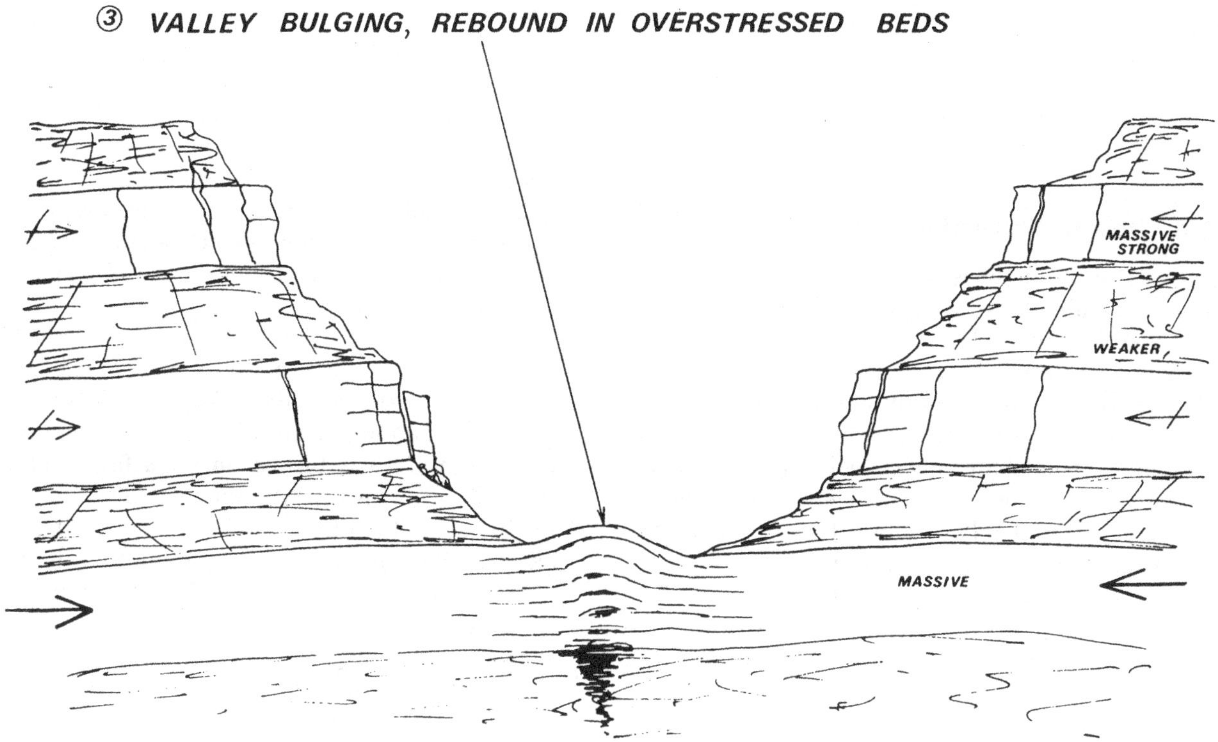

Figure 7.6-2C. *Overloaded beds bulge.*

mass behind. These blocks eventually tumble to expose fresh cliff faces.

- Major open jointing, more or less parallel to cliff faces, lies behind the faces. This eventually forms the fracture planes of future detached blocks.

Figure 7.6-2D. *Visual signs of stressed rock. High cliffs with vertical jointing open behind them (Blue Mountains, NSW, Australia).*

- The valley floor has weaknesses as earlier mentioned.

These visual signs are probably more useful than conducting stress testing and are certainly adequate for grouting purposes. Figure 7.6-2D shows visually observable stress in rock.

7.6-3 Movement Indicators

When grouting a hole, the following signs indicate that rock has probably moved:

- A sudden increase in grout take for no apparent reason.
- Sudden loss of pressure in the hole.

Grout operators should watch for these signs at all times, even on sites where movement is not particularly likely, and should always start each application at low pressure as recommended in Section 6.2-5. Normal pressure is then only applied if there are no problems.

If the signs of movement occur, the main control valve should be closed immediately while the cause is given some consideration. If the movement is in unstressed rock, there will probably be some telltale surface leaks indicating the offending rock. In severe cases it may need to be anchored by bolting; in less severe cases, surface leak treatment (refer to Section 7.3) may suffice to get refusal. When grouting resumes, it should be done generally as described in Section 7.6-4 for stressed rock.

7.6-4 How to Grout Stressed Rock

Some organizations insist that only very experienced operators handle the grouting of moving rock because of the high level of skill required.

When a site obviously has stressed rock, designs and specifications for its grouting should make provision for the special methods. Some of the beds may have almost reached the limit of their capacity to carry their stresses and only need the nudge provided by grouting to fail, probably by upwards heave in the case of the valley floor, as shown in Figure 7.6-4A.

The key to successful grouting in such cases is to limit the *energy* applied through the grout by limiting the volume and pressure. When multiplied together, volume and pressure can be loosely regarded as a measure of the energy applied. Control is not just a matter of keeping the pressure low (Wong and Farmer, 1973). The idea is to limit the energy supplied to an amount that does not precipitate stress relief or movement. This amount is found by experience on the basis that the rate of take should be restricted to the volume needed to fill the voids initially open, plus that subsequently required to compensate for leakage and bleeding losses. A grout flow meter (Section 7.7-10) gives continuous and accurate control of the volume and is essential in delicate situations.

Pressures should be kept as low as practicable, short of stopping the grout flow: 5 or 10 psi (0.3 or 0.6 bar) is the order of things for movements near the surface. Intermittent grouting may be helpful in bad cases: the grout is turned on for a few minutes and then off for the next few minutes, and so on. This gets the benefit of thixotropic stiffening during the off periods (see Section 4.2-2). Grout should be the thickest possible: w : c = 0.6 : 1 or 0.5 : 1 is advisable, and grouting should be taken to refusal unless there is no sign of approaching it after several hours' grouting.

Successive daily applications of grout may be needed in the initial holes in order to tighten an area

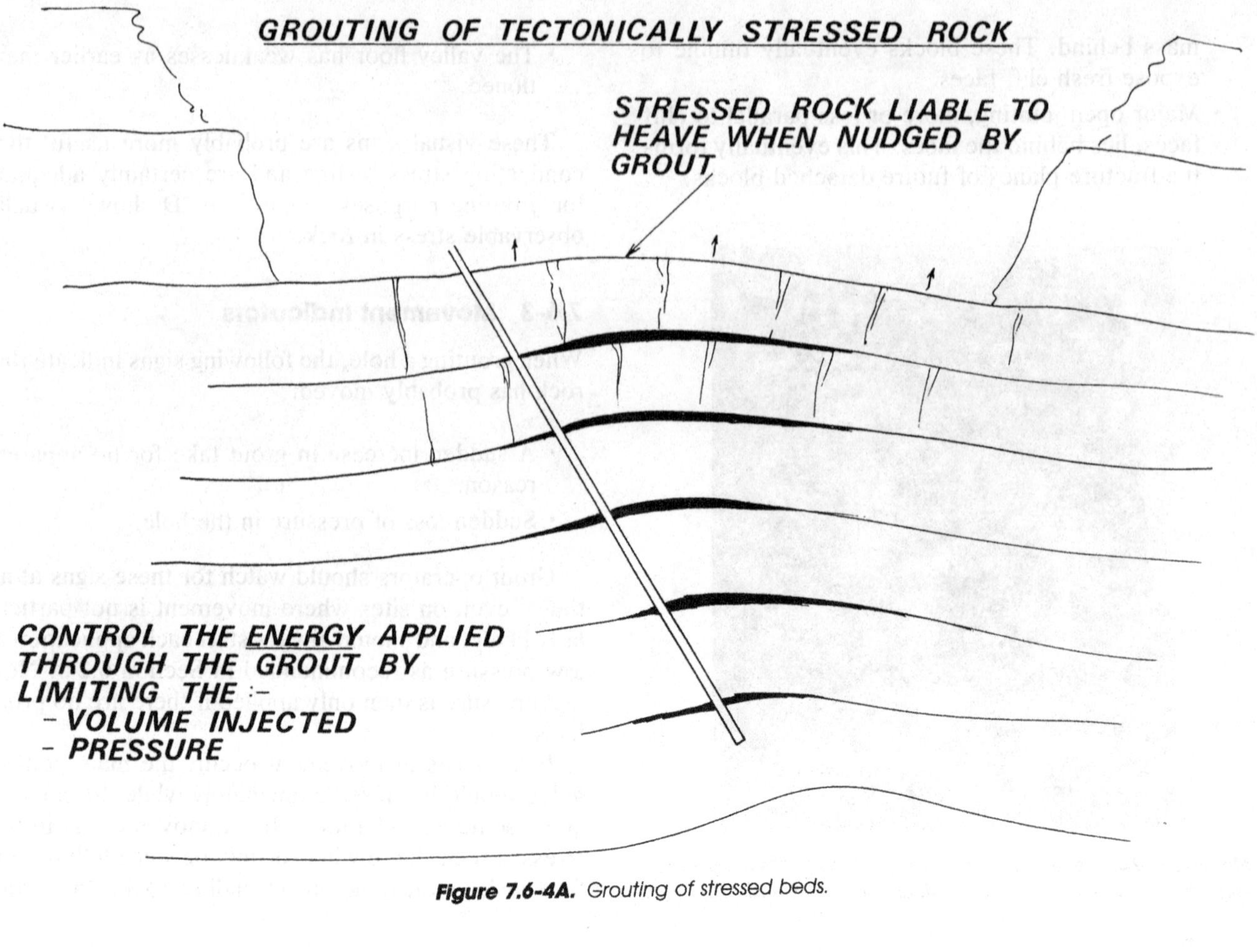

Figure 7.6-4A. Grouting of stressed beds.

sufficiently for ordinary grouting (with continued care, of course) to follow, for finishing the job properly. These successive applications compensate for volume lost by bleeding (refer to Section 4.3-3), for limitations on the volume injected, and for the effects of some stressed slabs that creep for hours after grouting and open up again later on the grouted cracks.

Grouting of stressed sites is necessarily a slow and careful job taking possibly twice as many holes as unstressed sites and costing more than this proportionally. Great patience is required!

7.6-5 Instrumentation to Detect Movements

Various ingenious arrangements have been devised to detect movement of surface rock. They range from the simple to the exotic. The simplest is a vertical rod, typically 30 ft ⟨10 m⟩ long, anchored at its base to deep beds; any uplift of the surface relative to its top can then be measured. Measuring has been done manually or with electronic warning devices of various sorts.

Sophisticated instrumentation such as acoustic emission detection represents the other end of the range. Considerable expertise is necessary with this method to enable separation of surface movement noises from others occasioned by grout travel in cracks.

Some jobs have tried survey methods, using a precise level (the ordinary dumpy level is not accurate enough) and stations embedded in the surface. However, reading is so slow that unless continuously taken in a very limited area, uplift is not detected until too late.

A great deal of money has been spent putting gauges in likely places on a number of sites, but rarely has the expenditure been justified. Often the uplift has been in an adjoining slab—not in one with a gauge. Many disappointments like this have led me to now omit instrumentation for the detection of uplift. Instead, a much surer and cheaper detection method is to pass all grout through a flow meter; this indicates sudden increases in flow and can be arranged, if desired, to sound an alarm. Where electronic monitoring of pressures is in use, alarms can similarly be arranged for sudden loss of pressure; Section 7.7-13 discusses computer control of this type.

7.7 COMPUTER-AIDED GROUTING

7.7-1 How Can Computers Help

Computers have been used or tried on cement grouting with varying degrees of success for the following, or if they have not yet been used there is obvious potential:

1. Recording grout takes, w : c ratios, pressures, details of troubles and relevant data (see Section 4.1-12).
2. Plotting graphs and diagrams of cement takes, water takes, and so forth (Section 7.4-1).
3. Analyzing the grouting (Sections 7.5-2 and 7.5-4).
4. Remotely operating the control valves (Section 2.6-2).
5. Batching the grout and operating the mixers (Section 4.1-6).
6. Controlling the pumps (Section 2.4).
7. Programming field work (Section 5.5-6).
8. Designing the shape of grout curtains in awkward configurations (Section 8.4-5).
9. Providing warnings of rock movements (Section 7.6-5).

Advances in computer technology and software can be expected to expand the possibilities in future years, and comments here will become dated in due course. However, they provide food for thought. No review of the current state of cement grouting would be complete without them.

7.7-2 Portable Site Recorders

The simplest type of computer activity is the use of hand-held recording devices at the grout holes and at the agitators. These record the grout takes, pressures, w : c ratio, and so on, which are manually punched in. An internal timing device records these entries against clock times. The units are battery powered. At the end of the grouting, or at the end of a shift, they are unloaded into a computer in the site office for processing. This method takes the place of written forms and the errors that may occur when the forms are subsequently transcribed.

These devices must be protected against entry of dirt and cement: the fingers and hands of those using them are often liberally coated with both and possibly with wet grout also. A watertight soft plastic sheathing is needed; the keys are operated through it, and it provides (hopefully anyway) protection to the electronics inside when the units have to be washed or scrubbed to remove cement deposits. The units must be able to take extremes of temperature without losing reliability. Large letters on the screen are a good idea so that they can be seen through smears on the sheathing. Units of this type are not as yet made specifically for grouting; some adaption is needed.

Ideally, these devices should be able to record textual messages also, such as details of any troubles as they develop and a record of treatment methods and of step-by-step progress in fixing the trouble. If they do not have this capability, their value is greatly diminished.

The portability of this method of recording and its lack of cables across the site are its main advantages compared with the methods described later. However, it could be said, with some justification, that a piece of paper and a pencil is just as good, and a whole lot less trouble. Be that as it may, information at hand at the time of writing (1989) is that trials of portable recorders have been on a limited scale, and organizations that have experimented with them, such as the U.S. Army Corps of Engineers (oral advice to me) have apparently not been sufficiently impressed to proceed with their use as a general practice.

7.7-3 Manually Loaded Desk Computer

Another option makes use of an ordinary personal computer (PC) located in an office, possibly remote from the job. During the grouting, details are entered by hand on forms such as those suggested in Section 4.1-12, and the information is later transcribed to the PC back in the office. The computer is programmed to analyze the data and make plots as required. Comments about these functions are in Section 7.7-11.

7.7-4 Direct Monitoring by Computer

Direct monitoring is the high-tech method. The computer is connected to equipment that transmits data to it for recording and processing. It may also control the grouting. The basic layout is (see also Section 4.1-10) as shown in Figure 7.7-4A.

Flow meters on the outward and return sections of the circulation line are connected to the computer; the difference between their readings is a measure of the amount of grout taken by the hole. At the grout hole, a pressure sensor and a density meter transmit pressure and w : c ratio.

An operational difficulty with this arrangement is the preservation of the connecting cables against damage from vehicular traffic, drill movements, tangles

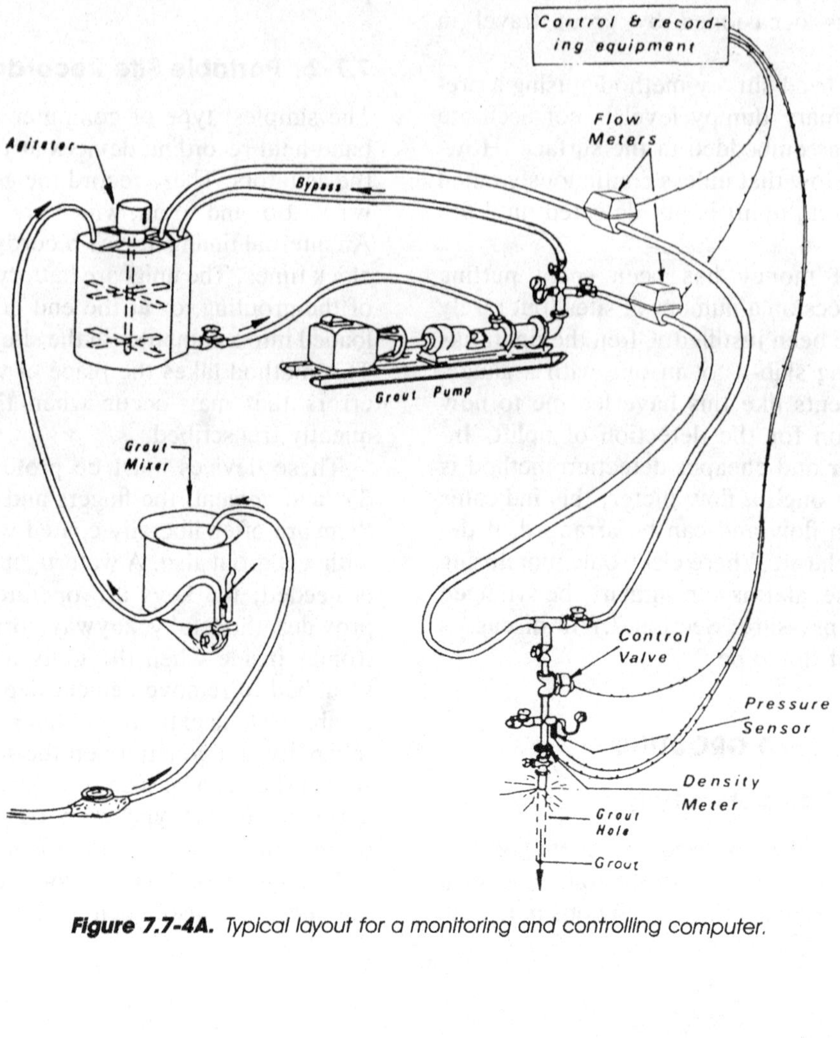

Figure 7.7-4A. *Typical layout for a monitoring and controlling computer.*

with hoses, and the general rigors of a construction site. To overcome this, radio transmission instead of cables is being tried by the United States Bureau of Reclamation (USBR). Personal reports at the time of writing indicate some success with this method after initial problems such as interference from other radio transmissions on site.

7.7-5 Remote Operation of Control Valve

In addition to showing monitoring arrangements, Figure 7.7-4A shows a remotely operated control valve. This valve can be electrically, pneumatically, or hydraulically operated; whichever method is used involves an additional hose or cable to the computer location and so amplifies the problems already mentioned. At Ridgway Dam (USBR) where hydraulic operation was chosen and installed, the remote facility was in fact not used: it was found preferable to manually operate the control valves.

However, the idea has had promotion in Japanese work using layouts such as shown in Figure 7.7-5A.

A single cable connects the computer to a "black box" containing all of the standpipe equipment shown in Figure 7.7-4A. An obvious technical objection to this layout is the short length of hose connecting the box to the grout hole. As explained in Section 2.6-5, this arrangement is asking for trouble in the later stages of an application, when flow through the hose almost ceases and can produce false refusal—not in the hole but in the hose. Another obvious source of probable trouble is the box itself: access to the various items is not as good as for "open-air" layouts, and this matters greatly in cement grouting work where quick removal and replacement of blocked fittings is essential. A lesser problem is the reduced portability of the

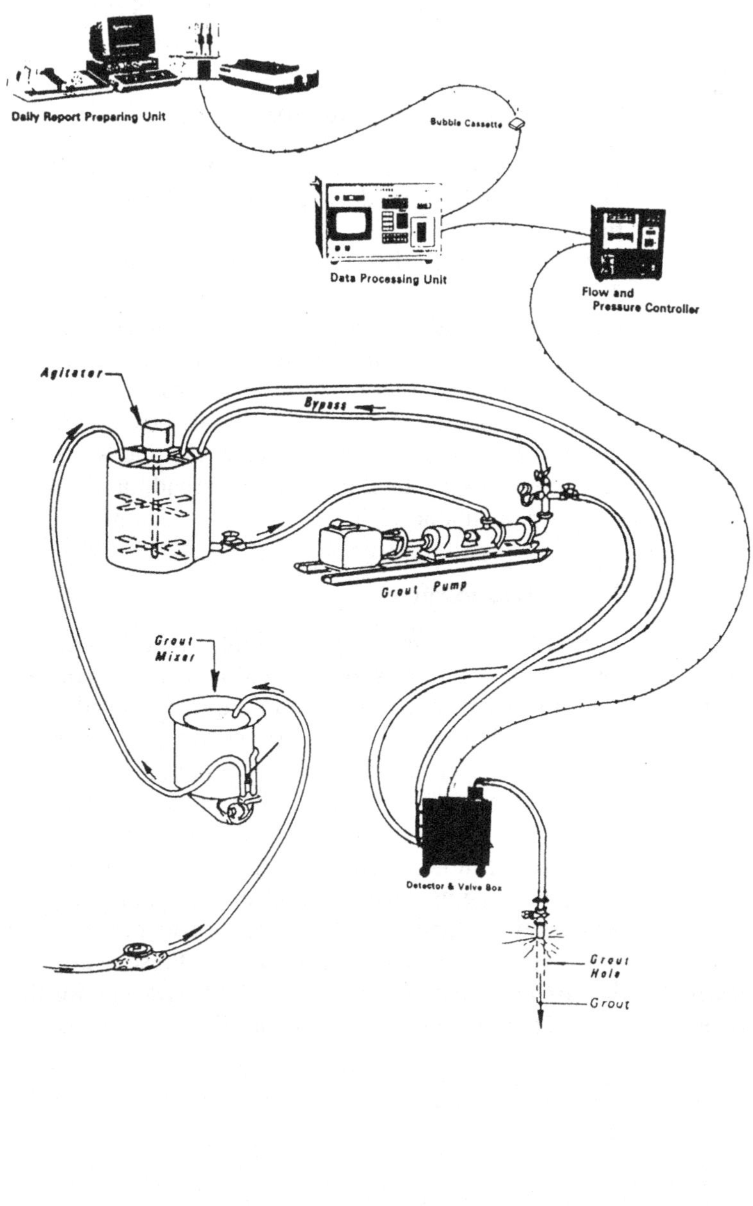

Figure 7.7-5A. *"Black box" control and monitoring.*

box around rough sites and on steep slopes. The method is more suited to chemical grouting, where requirements and problems are rather different to cement grouting.

7.7-6 Protection for the Computer

Grouting sites are notoriously messy, perhaps with a coating of cement dust windward of the mixers, splatters and splurts of grout in all directions resulting from hosing down the gear, and from bleeding and hose changing. These are not conducive to reliability on the part of delicate electronics. Therefore, computers must be housed in a protected environment to ensure reliability. Air conditioning, a humidifier, and electrostatic precipitation of dust are necessary (Davidson, 1983). Therefore, if a computer is used for monitoring, it has to be housed in some sort of trailer or shed, not always easy to arrange in the vicinity of pumps in rough terrain areas; and unprotected cabling can present a limitation to the portability of the grouting gear. The tendency is to use a central mixing plant in these circumstances. This detracts from the mobility and flexibility of grouting operations on spread-out and rough sites (as more fully explained in Section 2.7-12) and is not recommended. Reasons include inability to match circulation line pressures to hole requirements and the excessively long lines required.

Electric power for computers has to be free of line voltage variations or else malfunctions and loss of data can result. This can be a problem on sites where heavy electric motors are switched on or off without warning. A separate steady supply for computers may have to be provided by generators if line voltage is not sufficiently steady. Standby power should be available and automatically switched into circuit if the main supply is interrupted.

7.7-7 Reasons for Use of Computers

Much of the impetus for trying computers for the site work on cement grouting jobs has come from the current urge for high tech; a quite commendable urge in the right circumstances, of course. Chemical grouting had been handled successfully from remote control setups long before the days of computers, and their successful adaption to this form of grouting quite naturally led to the concept that they could equally as well be applied to cement grouting. However, in this context, these two types of grouting have fundamental differences.

An interesting summing up of expectations from their use with cement grouting is given by Davidson (1983) in a list of anticipated benefits, drawn up to justify trials:

> The ability to monitor additional holes more closely.
>
> A reduction in the time required to make a determination of meeting refusal criteria and, thus, a reduction in the time needed to complete the grouting program.
>
> Increased effectiveness of Inspectors.
>
> The potential for reducing the required inspection time.
>
> The capability for quick and accurate plotting of stage-by-stage results.
>
> The development of monitoring technology that will be available for future grouting work in the Bureau of Reclamation.

Note that subsequent experience somewhat dampened these hopes. Although computer monitoring has been used on subsequent USBR dams, its value is still questioned by some of the experienced people involved, and discussion with them suggests that persevering with it is by no means a certainty.

7.7-8 Benefits of Computer Records

There are very definite benefits from having a continuous record of events during grouting of each application: evaluation of the adequacy of the techniques used is enhanced; the effects of troubles and of remedial action can be evaluated more suitably; the time of refusal is indicated more exactly and hence grouting may be finished earlier (Grant, 1985); treatment of touchy sites (such as those prone to rock movement troubles or having structures that could be damaged by the grouting or have very weak rock, and so on) is improved technically if a continuous readout of salient figures is available.

7.7-9 Economics

At present, it is unlikely that the introduction of monitoring-type computers to a job will save money; indeed the opposite has been known to happen: considerable amounts of money have been lost in trying to transform cement grouting from its essentially low-tech nature to a high-tech one. The cost of setting up computers, their housing, and installing monitoring equipment can be a substantial part of the whole cost of a grouting job, unless these can be brought from a previous job, debugged. The benefits are not economic but technical; the question is whether these are worth the cost. Current contracts with the USBR call for the contractor to supply and maintain the sensing components for a monitoring system. The USBR supplies the computer and software.

On contract work, monitoring by computer may reduce the number of inspectors needed if one person sitting at a console can keep track of the progress of a number of applications. If, however, troubles arise with gear or in the form of leaks, rock movements, and so forth, an inspector has to be outside checking that the contractor is taking appropriate remedial action. Hence several inspectors are needed on site to cover such eventualities, and savings on this score by use of computers may be rather marginal.

Computers used in this manner are unlikely to lead to a reduction in the total number of people needed by a contractor; more may even be required because of the increased technicalities.

On the other hand, if the computer is a modest manually loaded desk PC, costs are negligible. Benefits can then include saving the time and patience of engineers. Well worth while. Portable recorders unloaded into a desk PC could also be useful (Section 7.7-2).

7.7-10 Recording Units

The simplest recording unit is the grout flow meter mentioned in Section 4.1-10. Their position in a regular type of circulation line is as shown in Figure 7.7-10A.

These meters are usually of the electromagnetic type that have no working parts and nothing to become grouted up. Their operation is based on electromagnetic induction: the voltage induced in any conductor as it moves across a magnetic field is proportional to

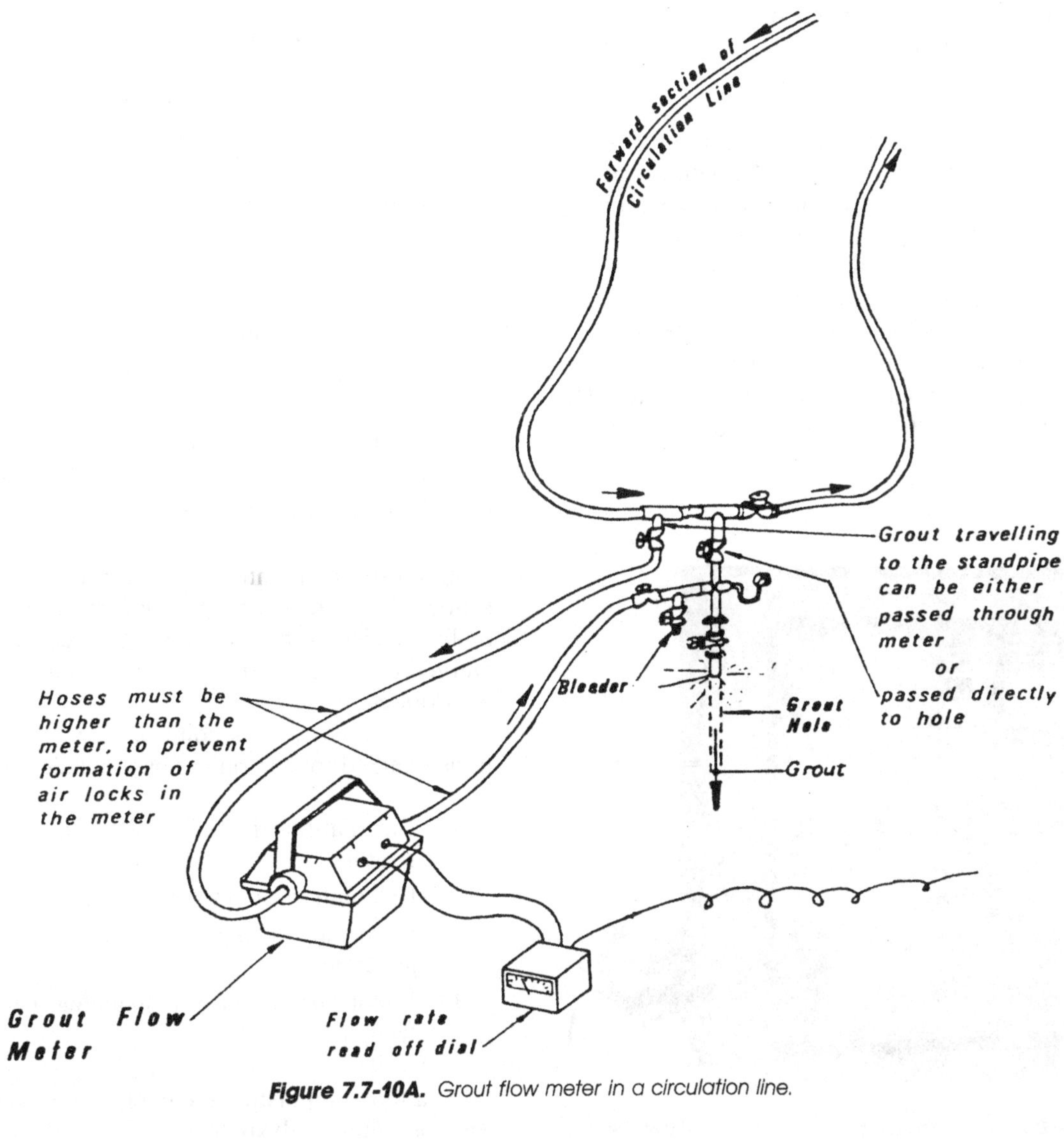

Figure 7.7-10A. *Grout flow meter in a circulation line.*

the velocity of the conductor. The grout is the conductor. It passes through a Teflon-lined tube about 1 in ⟨25 mm⟩ diameter of constant cross-sectional area; hence its velocity is a measure of the quantity flowing. The meter creates a magnetic field around the grout tube and small electrodes each side of the tube measure the induced voltage. Differing w : c ratios do not affect readings.

Brands of these meters available include Veriflux, made by Kent Instruments, England.

Turning now to more comprehensive equipment, several manufacturers offer complete recording units that can be either used on their own or in conjunction with computerized equipment. Among these are:

CRAELIUS GROUT RECORDERS These are portable units that hook into the circulation line and contain a Veriflux grout flow meter and a pressure sensor fitted in a stand on which recording equipment is mounted.

They are made by Diamant Boart Craelius AB, Bristagatan 13, Märsta, Sweden.

HÄNY RECORDING UNITS These are master-type recording units intended to monitor the grouting of single holes or of a number of grout holes being simultaneously grouted in an improved form of group grouting (Section 2.7-5). In the unit shown in Figure 7.7-10B each hole has its own flow meter, and the common pressure is measured by a sensor. The unit can also be used for the water testing (Section 5.3-2) of holes, and for this procedure the unit can be connected to a pressure transducer mounted down on the test packer. The manufacturer claims these units are protected against all the usual hazards on site—water and grout spray, burst hoses, bad weather, and so on, as well as against tropical conditions. The manufacturer is Häny & Cie. AG, CH-8706 Meilen, Switzerland.

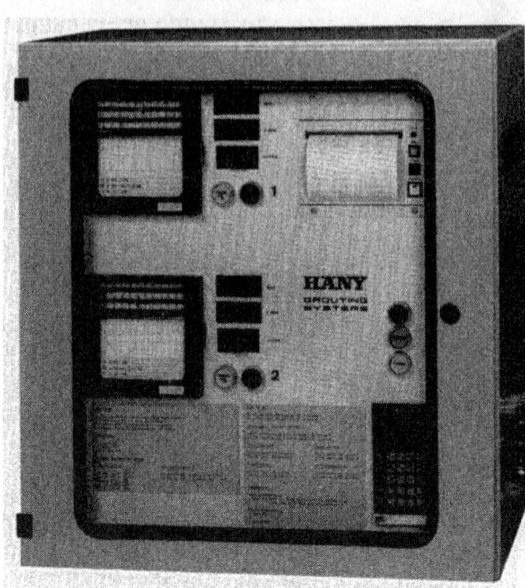

Figure 7.7-10B. *Häny recording unit. (Courtesy Häny & Cie. AG)*

7.7-11 Plotting and Analyzing

There is no doubt about the value of computers for plotting and analyzing data. Functions handled by even a simple desk PC, manually loaded, can include the following:

1. Calculation of lugeon values from water tests in grout holes before grouting them (refer to Section 5.3-4).
2. Assessment of the type of flow in these water tests and identification of problems that are likely to influence the grouting (refer to Section 5.3-4).
3. Calculation of the amount of cement in the grout injected on time basis.
4. Calculation of the total cement used in an application and checking that it matches the amount of cement mixed, making allowance for grout already in the system before starting and for grout remaining afterwards.
5. Calculation of average cement taken per unit length of stage.
6. Plotting of grout and cement take, against time.
7. Plotting of pressure against time.
8. Tallying of contract payment items such as lengths drilled, standpipe lengths, number of hook-ups, and bags injected (Section 9.2-1).
9. Printing of report sheets and summary sheets.

Much of this is interesting information, but not essential. It is unwise to produce more output than is really needed, otherwise operators can become submerged in a mass of data, and important things can be overlooked.

From a contractor's point of view, additional functions handled by computers can include the following:

10. Calculation of drilling rates and costs per unit length.
11. Calculation of standpipe placement costs.
12. Recording costs for water testing, hookup, and grouting.
13. Recording costs for handling each type of trouble.

At a more sophisticated level, supervising staff can use computer analysis for:

14. Recognition of the peak take of an application and grouping of that take in a magnitude field.
15. Classification of curve of take against time in a magnitude field.
16. Recognition of the appropriateness or otherwise of mix changes.
17. Recognition of the appropriateness or otherwise of the starting mix and of other mixes.
18. Examination of the approach of refusal and of the point of refusal.
19. Determination of the need for a further application.
20. Comparison of the cement take with the water take.
21. Statistical analysis of the grouting overall.

7.7-12 Designing Grouting by Computer

The need for assistance by computer is fairly rare in the design process, but there are circumstances where designing the underground shape of grout curtains can benefit from CAD application. An example is where the three-dimensional configuration has to be accurately contoured (see Section 8.4-5), possibly because holes are inclined in two planes, other curtains have to be intersected accurately, or perhaps the positions of tunnels or structures have to be taken into account. While it is perfectly possible to do all this by drafting methods, errors are easily made in the more complex situations, and CAD (if properly programmed, of course) can be of benefit.

7.7-13 Warning by Computer

Discussion in Section 7.6-5 mentions using computers to give warning if rock movements occur and suggests that this is best done by detecting sudden loss of pressure or sudden increase in take (the signs of movement), rather than by gauges to measure surface heave or movement. Such warning arrangements can be beneficial if pressure and take are being monitored anyway but are not worth establishing merely for rock movement control.

Warning devices for this can include whistles, sirens, and similar penetrating noise-producing arrangements, preferably at the grout hole although more likely at the pumps. The usual noise level on grouting sites makes the use of flashing lights and similar nonassertive arrangements somewhat unreliable. Where remote control is in use, the warning devices could include automatic cutoff of grout to the hole.

7.7-14 Automation of Cement Grouting

By way of pipe dreaming, I have for long speculated on how to do grouting automatically where the decisions about w : c ratios and pressures are made for you and where bleeding is carried out by the equipment whenever sensors find that it is necessary.

Ideally, for perfect grouting, there should be a means of control down inside the holes to tailor the mix and pressure to the optimum for each little group of small cracks and for each individual large crack. But this is quite impractical. As a compromise, the tailoring has to be suborned to the overall requirements of the various sorts of cracks in each stage. Note that this is not the same as the average crack size: the overall requirements take into account the range of sizes and types, and part of the art of grouting is to adequately cater properly for the whole range.

So the practical compromise is in terms of control at the top of the hole rather than having control devices down inside it (where they would get grouted in anyway).

Equipment that would automatically thicken the mix in response to the behavior of the hole has the potential to help inexperienced and uncertain grout operators produce better work. How would this equipment work? Well, here are some of my thoughts about it (and they are still thoughts—automation is not yet with us).

The options are to have some of the control at the mixers with the rest at the standpipe, or to have everything at the standpipe. Let's consider the first option.

AUTOMATION CONTROL AT BOTH MIXERS AND STANDPIPE

The layout of the gear would be much the same as shown in Section 2.7-3 except that an additional (and higher) agitator would be placed between the mixer and the main agitator. This additional agitator would have a remotely operated valve to meter its contents into the main agitator, as well as a flow meter (located in the bottom of a U pipe so as to keep it full of grout when discharging—these meters give false readings if air enters their measuring pipe) to measure flow from the agitator.

The mixer would only make one thickness of mix throughout a grout application. This would be a w : c ratio of about 0.8 : 1 except when thicker grout was needed to deal with leaks.

When thickening of the mix circulating to the hole is required, the computer-controlled valve under the agitator holding the thick grout opens and the required quantity of this grout, as measured by the meter next to this valve, discharges into the main agitator to thicken whatever the mix is there (not necessarily 0.8 : 1 though).

The remotely operated valve on the standpipe controls the amount of grout going down the hole as well as its pressure. On some sites, it could be worth while fitting the valve on the return line with remote operation controls to be able to vary this return pressure in a complimentary manner to the main control valve.

Programming the computer running all this would have to be done exceedingly well; the success of the whole setup would depend on it. The microprocessor would need a "smart" chip to provide an "expert system" using so-called fuzzy logic. This is a set of rules able to handle imprecise input by simulating expert reasoning. The programming would analyze the water test to get its "feel" for the application, and then, when grouting starts, would match the initial data and subsequent take and pressure relationships to enable it to search the memory for a match among the many scenarios stored there by an expert grouting practitioner. Once a match is found, the controls would be manipulated by the program to give a grouting pattern following the concepts of Section 7.4-1. Progress would be constantly monitored, and if the hole shows a tendency to behave differently from the pattern in use, the program would examine the situation and change to another pattern if it so decided. This examining would be done throughout the application, possibly making many changes. When refusal is reached, the program would hold the pressure for 15 minutes more before announcing completion. When leaks or other troubles occur, the program would be switched to manual override because preprogramming of all the possibilities for treatment of these is out of the question.

AUTOMATION CONTROL ENTIRELY AT THE STANDPIPE This would have two circulation lines to the grout hole, each with its own pump, and having circulation back to its own agitator. One line carries water only and the other has thick grout such as w : c = 0.8 : 1.

At the grout hole, each line branches off to its own valve (electrically or fluid controlled) and then to a small mixing unit so that varying proportions of water and thick grout can be drawn off the lines in response to the controls. The mixing unit is necessary because it would not suffice to omit it and just hope that there would be good enough mixing of the grout and the water by the time it got into the hole. High-quality grouts do not mix very easily with water anyway, so mechanical mixing would be necessary, and it must be carried out under line pressure.

Control of pressure for injection is done by the two valves above the mixer, so this function is in addition to controlling the w : c ratio. A small, very basic computer mounted alongside would have to be programmed very carefully to achieve this.

A flow meter mounted on the hole measures the rate of grout take, and this is the basis on which the computer functions.

The meter and mixer have to be mounted directly on top of the hole. A short hose connecting them from nearby is not permissible, as explained in Section 2.6-5.

The software would be the same as described previously for the alternative arrangement.

Should the computer be electric or fluidic? There is a lot to be said for the fluidic type (they are rare though) because of its use of water for its operation and the obvious appropriateness of this in the grouting environment, together with its ruggedness. The water for its operation would come from the water circulation line. However, electricity will be needed to operate the flow meter (unfortunately; electricity at grout holes is often a nuisance), and it might as well operate the computer.

The computer would record data on the injection on a small disk that is later removed and fed into a desk computer.

AUTOMATIC BLEEDING Automatic bleeding would be carried out in both systems, when a sensor located in the top of the hole near the hole wall indicates that too much water and thin grout has collected. This sensor could use alternating current passing between two electrodes about an inch apart; it would distinguish between water and the various thicknesses of grout. A bleed valve, projecting from the side of the fittings below the flow meter and operated remotely, would be opened to let water and thin grout escape. The computer would handle this opening control and would at the same time make appropriate adjustments to control valves.

Concluding Thoughts on Automation All the possibilities discussed could be made to work with enough development, but is it worth it? The cost to a contractor of each of these standpipe units would be far more than the usual job-made conventional fittings and are sure to require a lot of maintenance and fiddling. The advantage—that almost untrained people could be let loose on grouting—would be lost if breakdowns occurred while grouting is in progress leading to interruptions to the continuity of injection.

And because the software would require such a melding of expertise and art, could it be programmed sufficiently well to be reliable? If it can't, it would be better to forget automation.

So much for that old pipe dream, or is it a standpipe dream?

REFERENCES AND FURTHER READING

Belsterling, C. A. (1971), *Fluidic Systems Design,* Wiley-Interscience.

Davidson, L. (1983). "Micros step into dam grouting." *Engineering News-record,* December 15.

Davidson, L. (1984). "Electronic monitoring of cement pressure grouting at Ridgway Dam." Embankment Dams Branch, U.S. Bureau of Reclamation, Denver.

Deere, D. U. (1982). "Cement–bentonite grouting for dams." *Proceedings of Conference on Grouting in Geotechnical Engineering.* ASCE, New Orleans, 279–300, quoting Burgin, C. R. (1979). "Investigation of the physical properties of cement-bentonite grouts for improvement of dam foundations." Thesis presented to the University of Florida in partial fulfillment of the requirements for the degree of Master of Science.

Ferguson, H. F. (1967), "Valley stress relief in the Allegheny Plateau", *Bulletin of the Association of Engineering Geologists,* Vol. 4(1), 63–71.

Grant, L. F. (1985). "Computer-assisted grouting evaluation systems." In *Issues in Dam Grouting.* ASCE, Denver, 132–148.

Houlsby, A. C. (1982a). "Optimum water : cement ratios for rock grouting." *Proceedings of Conference on Grouting in Geotechnical Engineering.* ASCE, New Orleans, 317–331.

Houlsby, A. C. (1982b). "A digest of typical cement grouting takes." *Proceedings of Conference on Grouting in Geotechnical Engineering.* ASCE, New Orleans, 1000–1014.

Nichols, T. C. (1980), "Rebound, its nature and effect on engineering works." *Quarterly Journal of Engineering Geology,* Vol. 13, 133–152.

Wong, H. Y., and Farmer, I. W. (1973). "Hydrofracture mechanisms in rock during pressure grouting." *Rock Mechanics,* Vol. 5. Springer-Verlag, New York, 21–41.

CHAPTER 8

Site Investigation and Grouting Design

Before designing grouting for a new site it is wise to investigate enough of the foundation to know how it will relate to the grouting and to be able to decide how much grouting to do.

This chapter deals with the two main aspects of investigation—geology and permeability—and then goes on to the design of grouting:

Geological investigation	Section 8.1
Permeability investigation	Section 8.2
Necessity for grouting	Section 8.3
Design of grouting	Section 8.4

BENEFITS OF GOOD INVESTIGATION A good investigation preceding design should be able to provide the following:

- A reduction in that uncertainty present in all grouting work.
- Information for deciding whether grouting is really needed (Section 8.3).
- Information indicating which type of grouting is suitable (cement, chemical, clay, or combinations) (Section 1.2-5).
- Data for designing grouting (Section 8.4).
- Data for compiling an appropriate contract specification (Section 9.2).
- Information for preparing realistic bids.
- A basis for comparisons with what eventuates during the progress of the work (Section 7.5-4).

Investigations can be expensive but usually save money overall because they can allow the grouting to be more accurately tailored to conditions and therefore not wasteful. However, some designers prefer instead a policy of almost no investigation and a "find out as you grout" approach; if funds for investigation are short, this policy is convenient, but it can cost more in the long run because excessive grouting usually results. In addition there can be hefty claims for extras from contractors resulting from their need to make guesses when preparing bids.

8.1 GEOLOGY

8.1-1 Features Pertinent to Grouting

There is no need to have detailed geological knowledge to understand the effects of different types of foundations on grouting. A general idea of the subject is sufficient most times, perhaps helped with the details by a geologist in special cases.

The principal thing to understand is the cracking or void systems. After all they are the main concern of grouting. The next point of interest is the physical properties of the rock. However, this is not to say that rock identification and characteristics can be ignored—they shouldn't be—but they are not prime factors most times.

The features that can influence grouting design and construction are now discussed.

1. Spacing of Joints As far as cement grouting is concerned, it is the open, groutable joints that are of interest. If they are widely spaced, the grouting is usually easier than if closely spaced where troubles such as frequent surface leaks, collapsing holes and patchy penetrations can happen (see Figure 8.1-1A). These make for more expensive grouting, perhaps requiring special surface treatment (Section 7.3-9) or unusual methods (Sections 5.4-7, 5.4-11, and 5.4-12).

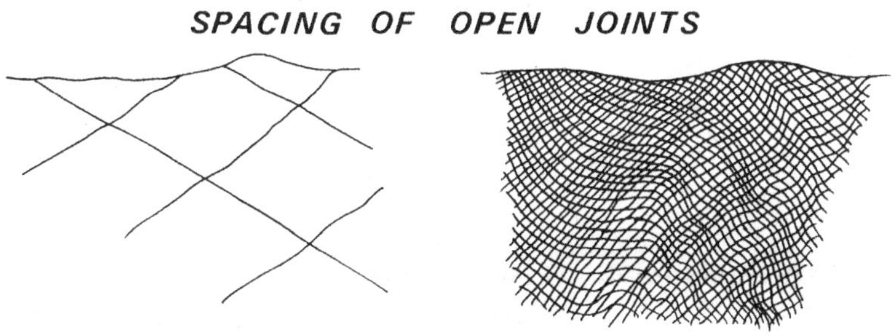

Figure 8.1-1A. *Left: Easier grouting. Right: Harder grouting.*

2. *Joint Widths and Continuity* The easiest joints to grout have widths in the range between about 0.250 in ⟨6 mm⟩ and 0.020 in ⟨0.5 mm⟩. Section 6.1-5 describes how grout penetrates in cracks at the finer end of this range. At the other extreme, joints wider than indicated can permit the grout to travel too easily, so precautions may be necessary to discourage grout from penetrating excessively from the required area; refer to Section 7.2-9 for control methods in these circumstances (Figure 8.1-1B).

Continuity of open jointing systems affects penetration: lack of continuity means that more grout holes will be needed than if grout can travel appreciable distances through the systems.

3. *Joint Inclination* Where dipping is mainly between 0° and about 60°, vertical grout holes may give optimum interception (see Figure 8.1-1C). These are the easiest to drill and are preferable. Steeper jointing usually requires use of inclined holes.

4. *Uniformity of the Site* Uniformity of jointing permits a regular layout of grout holes, whereas irregular jointing, dykes, disconformities, and so on may require placement of holes at various inclinations and

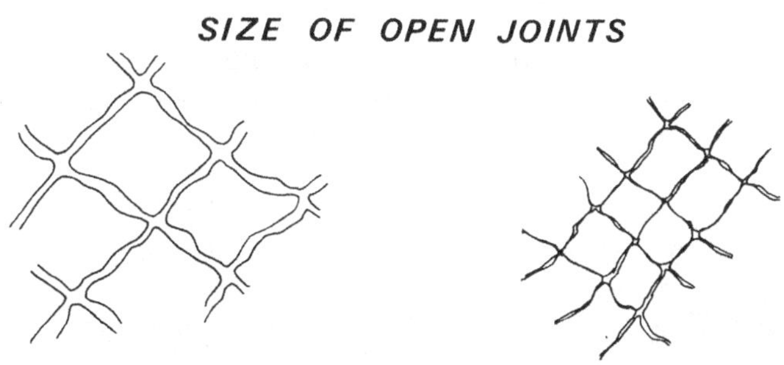

Figure 8.1-1B. *Left: Easier grouting. Right: Harder grouting.*

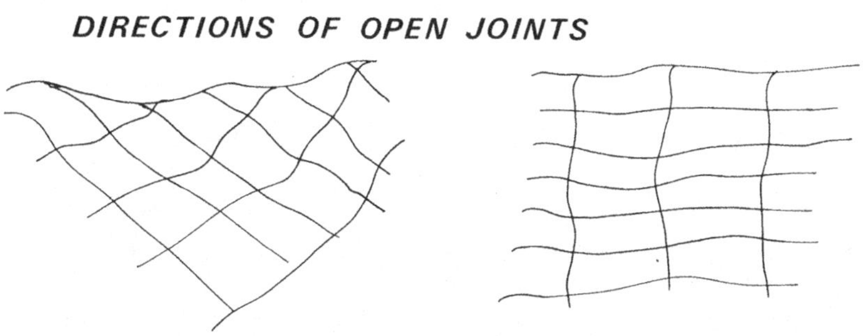

Figure 8.1-1C. *The jointing on the left requires different hole inclinations to the set on the right.*

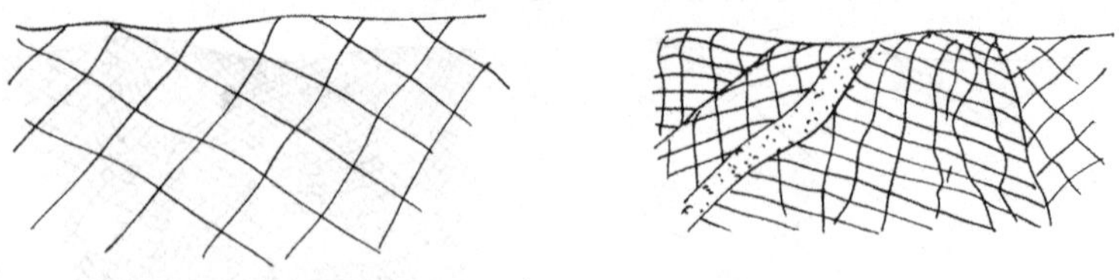

Figure 8.1-1D. *Regular holes suit the left jointing, but not the right.*

spacings (Figure 8.1-1D). Weaknesses may need to be treated especially intensively. Layout of holes is discussed in Section 8.4.

5. *Rock Soundness* Holes that don't collapse permit easier grouting than those that do (see Figure 8.1-1E). In the latter case, packers cannot be used (refer to Sections 2.5-15 and 2.5-16), and stage lengths may have to be shorter than usual (Section 5.4-1) if collapsed material blocks holes.

6. *Strength* Grouting of strong, massive, well-anchored rock (Figure 8.1-1F) is usually easier than when working in weak, broken, loose materials where holes repeatedly collapse, or where blocks move (Section 7.6-1).

7. *Stress in Rock* If a foundation contains tectonic stress (Section 7.6-2), the grouting will need to take account of this (see Figure 8.1-1G). Early recognition of the condition is necessary; the treatment methods are explained in Sections 7.6-2 and 7.6-4.

8. *Piping* Where material in joints can be removed by seepage, either by taking the material into solution or by eroding it, the grouting will need to be more intensive than otherwise in order to ensure that seepage through such joints is virtually eliminated (see Figure 8.1-1H). Section 8.3-3 indicates appropriate standards of grouting.

9. *Chemical Attack* The presence of coal or other carbonaceous material, or of other deposits that may provide chemically aggressive seepage can warrant provision of a higher standard of grouting than would otherwise be the case (Section 4.5-2).

10. *Karst and Other Voids* Large voids such as karst, and also old mines, shafts, and so on, require

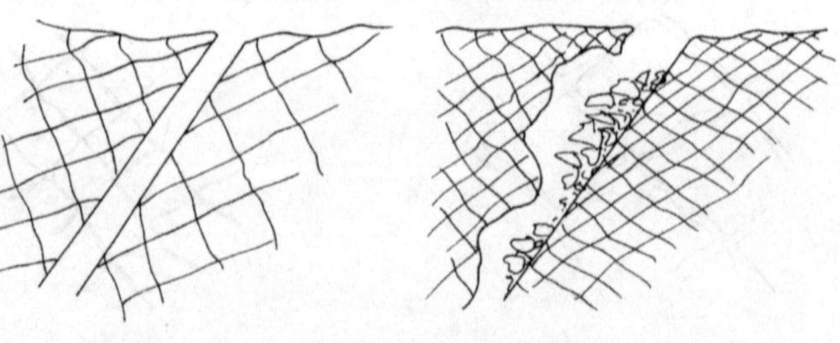

Figure 8.1-1E. *Collapsing holes can be a nuisance.*

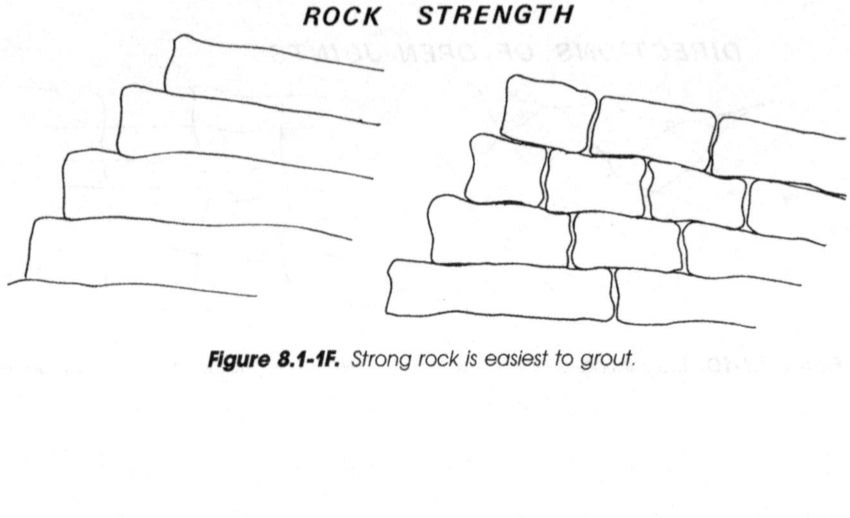

Figure 8.1-1F. *Strong rock is easiest to grout.*

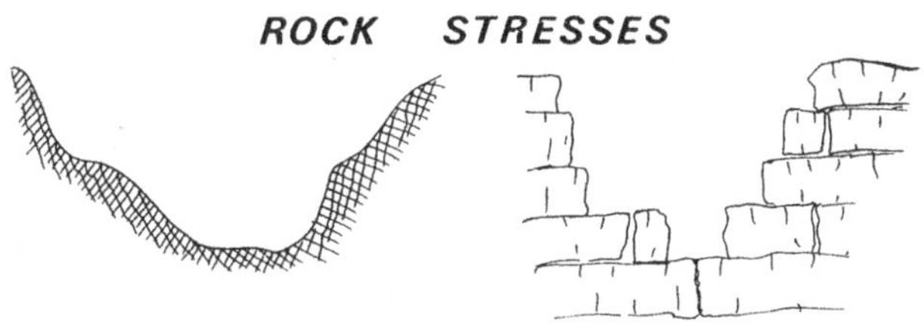

Figure 8.1-1G. *Left: Typical unstressed rock. Right: A canyon in stressed rock.*

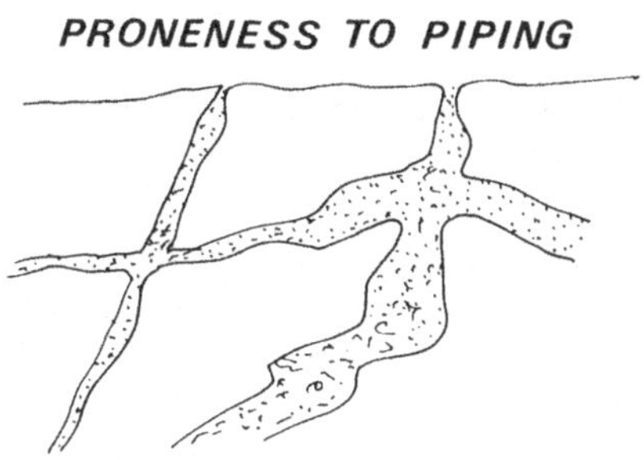

Figure 8.1-1H. *Joints containing soft material that can be piped (washed) away by seepage.*

Figure 8.1-1I. *Regular jointing with a slight dip. It is easy to design grout holes for a pattern such as this.*

special provisions when grouted, possibly using fillers in the grout (Section 3.2).

8.1-2 Geological Exploration

On most of the larger grouting sites geological exploration is necessary for such purposes as site suitability and structural assessment as well as for the grouting. It needs to be properly coordinated to grouting aspects including the extra permeability testing needed for grouting investigation.

Exploration should commence with regional geological mapping, leading through to detailed examination of the site in particular. Note that, as mentioned earlier, thorough identification of rock types is not imperative for grouting purposes. Strength and other physical features are more important. Joint details need to be carefully sorted out.

At dam sites, attention should be given not only to the dam site itself, but to abutments and the storage basin to ensure that any features which will threaten the waterholding capacity are found and assessed. There are some dams which have never been able to hold water because of lack of this attention. For instance, see Tschebotarioff (1951).

If karst or other voids are thought to be present, finding them might not always be easy. Sabarly (1974), in his amusing and perceptive essay on this and on other grouting topics, mentions the role of bad luck when drilling investigation holes. He says

> Everyone knows how boreholes hate to intercept karst

and he goes on to instance a site where an extensive karst system was not detected. Subsequently

> all the boreholes were re-examined only to find that the karst system wound its way cheerfully between them. One of the holes, which had gone through a large cave, was perverse enough to follow down a large stalactite which met up with a stalagmite!

8.1-3 Exploration Methods

Exploration methods in use include:

- Surface inspection.
- Core drilling.
- Borehole TV.
- Borehole periscope.
- Impression packer.
- Shafts and adits.

Each is outlined in the following sections.

8.1-4 Surface Inspection

Any naturally occurring outcrops provide visual information, but it is better to clean part or all of the site down to surface rock to enable thorough inspection. Sometimes this is done in costaines or dozer cuts with final cleanup by such means as air and water jetting.

Thorough air and water jet cleanup is particularly important to uncover jointing details; these cannot be properly seen, most times, if only hand tools and brooming are used.

8.1-5 Cored Holes

Core drilling has limited value in grouting investigation. It can help with finding rock properties and some of the jointing details but cannot reliably indicate joint widths or strikes in most cases unless cores are oriented. The process for doing this adds appreciably to the cost and with some methods is not always reliable. Fuller comment on these limitations follows later. Coring is, however, a popular method for general geological investigation and has been extensively refined. The holes are often convenient for the water testing as described in Section 8.2.

Equipment and methods for core drilling are well described elsewhere (such as Heinz, 1985). Brief relevant data follow here for coring. Sections 5.2-4 and 5.2-9 have equivalent data for diamond drilling of grout holes, which is usually noncoring.

Sizes In round figures, the following are standard sizes of holes:

U.S. DCDMA System

Designation	Hole Diameter (in)	(mm)
E	1.5	38
A	2	50
B	2.5	64
N	3	76
H	4	102

Actual sizes vary from these rounded figures. There are many offshoots from these basics and many variations in designation terminology. The system is based on the concept of neatly nesting each size of casing into or over the adjacent size. The diameter of cores varies with the type of bit and number of coring barrels. For a brief history of the origin of the designation nomenclature see Section 5.2-9.

Sizes used in some European countries are listed in the following table. This system is designed around core sizes. Hole sizes therefore vary with the number of tubes in the coring barrel.

Metric Sizes (Craelius)

Designation	Barrel Type	Core Diameter (mm)
B	Single (normal wall)	46,56,66,76,86,101
TB	Single (heavy wall)	46,56
TT	Double (thin wall)	46,56
T2	Double (normal wall)	46,56,66,76,86,101
T6	Double (heavy wall)	76,86,101
T6S	Double (diff. sampling)	76,86,101

WIRELINE DRILLING This drilling method recovers cores without having to pull the rods and bit out of the hole. The core barrel is recovered by unlatching it from an outer tube and hoisting it to the surface on a cable. Hole collapse is largely avoided, and drilling can be faster and cheaper than with conventional methods. Special packers are available for carrying out water testing downstage (refer to Section 8.2-7 for methods); after removing the core barrel and lifting the outer tube to expose the stage to be tested, a packer is lowered down beyond the bit and seated in the hole there for the test.

Size designation follows the U.S. system with the addition of "Q" usually, although some manufacturers have slight differences.

Number of Tubes in Barrels Coring barrels can be obtained with single, double, or triple tubes. In sound rock, single-tube barrels can give adequate core recovery, but double-tube units are in more use universally. When a foundation is generally weak, or where a sound foundation contains weak seams or beds, triple-tube barrels are preferable. These minimize damage to the core from flushing water, and if the innermost tube is split longitudinally, the core can be transferred to core boxes with minimal disturbance.

Deficiencies in Coring Coring is a deservedly popular method for general geological exploration, but it has deficiencies when applied to grouting exploration:

1. Even with the very best triple-tube barrels, 100% core recovery is rare when passing through soft seams. This is because of the grinding action of the bit as it enters such seams and the sluicing action of drilling fluid as the bit passes through them. Use of drilling foam or mud may reduce the problem in some materials, but not in all. Full knowledge about such seams is vital to planning effective grouting: if recovery is zero or very slight, there arises uncertainty as to whether the unrecorded section of hole is passing through a void or through materials. Furthermore, the nature of such poor material needs to be known to

check it for piping potential; if it washes out during drilling, it is very likely to be prone to piping.

2. If drilling mud is used, either necessarily or otherwise, entrances to cracks are likely to become constricted so that holes cannot be used for water testing. Additional holes then have to be drilled for this purpose, possibly by a different drilling method.

3. The weight of rods becomes so great, particularly in deep holes, that unless the driller is particularly skilled, weaker materials can become compressed and distorted. The need for good recovery in such materials was emphasized earlier. Furthermore, when this happens, it is not possible to distinguish it from drill drop resulting from entry into a void. The best that can be hoped for in such ambivalent situations is that voids will disclose their presence by loss of return water. Part of the driller's duties must always be to log any loss of drilling water and any drill drop.

4. Cores cannot usually convey data about joint widths, as noted earlier, nor can strikes be found. Dips can be, if the holes are vertical, but not with inclined holes unless they are oriented. Since joint information such as this is paramount in design good grouting, the lack of it necessitates recourse to other exploration methods as adjuncts to coring.

Note that coring is not a suitable method for use after grouting to find where grout has gone or for assessing whether enough grouting has been done. This is because of the grinding away of any grout that may be present when the bit enters it. Nearly always the grout is lost, unless it is very thick or happens to adhere unusually well to the roof of the crack.

8.1-6 Borehole TV

A number of manufacturers offer closed-circuit units with cameras mounted suitably for use in a hole (see Figure 8.1-6A). The camera is usually of the miniature type, viewing the hole axially. This produces a full circumferential view similar to looking inside a pipe and does not necessarily pick up cracks well. An optional sideways viewing mirror should be available in the field of view of the camera to direct a close image of the hole wall as Figure 8.1-6A shows.

The sideways viewing option permits looking into cracks, and their width and dip can be measured through the camera by comparison with graduated markings mounted at the side of the mirror. A light of controllable intensity mounted below the mirror provides a shadowing effect as the camera is slowly moved past each crack and this assists with determining its dimensions. Rotation of the mirror should be available on command from the surface to provide coverage of the full circumference of the hole.

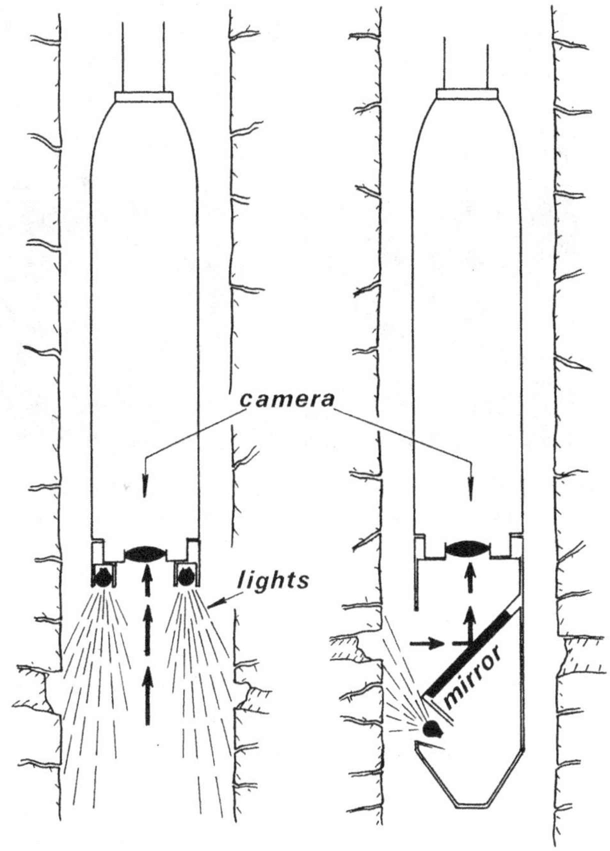

Figure 8.1-6A. *Borehole TV. Left: Viewing hole axially. Right: Viewing wall of hole in detail. Arrows indicate view to camera lens.*

The camera unit must be watertight. When in use in dirty water, visibility can be improved by surrounding the unit with clean water encased in a tough plastic bag or in a glass jar; care is required to leave enough waterway outside the bag or jar so that dirty water can get past when the camera is being moved, otherwise it becomes a "piston in a cylinder" situation making movement difficult.

A coaxial cable up the hole connects the camera with viewing and recording equipment. Some makers have portable units for this, powered by a car battery, that can be carried like suitcases (see Figure 8.1-6B). A hood around the viewing monitor may be necessary in bright sunlight.

Black and white cameras are adequate for most applications, although the extra information provided by color can be useful in doubtful interpretations. However, the cost of a color camera is appreciably more, and this is very pertinent if it gets lost down a hole. This threat is one of the main discouragements to use of TV in holes: it is easy for pieces of rock to fall out of

Figure 8.1-6B. Portable borehole TV unit powered by car battery. Camera being held, coaxial cable wound inside an old motor tire, readout and control equipment in small black box. (A. Dishington)

a hole above a camera and then jamb the camera against the hole wall. Recovery is not always certain. A prudent precaution is to fasten the camera to a strong plastic rope able to withstand considerable pull when jiggling to try to free the camera. Since the sort of hole mainly needing examination is the type most prone to disgorge pieces when flicked by the cable or camera, either very patient operators or a good insurance policy are necessary.

TV units are usually made to fit in 3-in ⟨75-mm⟩ holes.

8.1-7 Borehole Periscope

Borehole periscopes are simple instruments for use in dry parts of holes. They comprise a mirror set at 45° to the axis of the hole, and a telescope at the surface to view it (see Figure 8.1-7A).

The mirror reflects the wall of the hole. This is illuminated by a small globe above it, typically operated by a 12-V car battery. Lengths of tubing, usually plastic, hold the mirror and light and more are added as observations progress down the hole. A calibration device on the mirror enables crack widths, strikes, and dips to be measured. Recording is usually manually, although photography is possible through the telescope, which is preferably of the reflex type as shown in Figure 8.1-7A.

This is the best tool for examination of cracks above the water table or where the hole fills very slowly after pumping out. It can give magnified views of cracks, and if weak materials have been eroded back from the wall of the hole during drilling, it enables them to be viewed out to considerable distances from the hole. Periscopes are so simple that they can be made up by

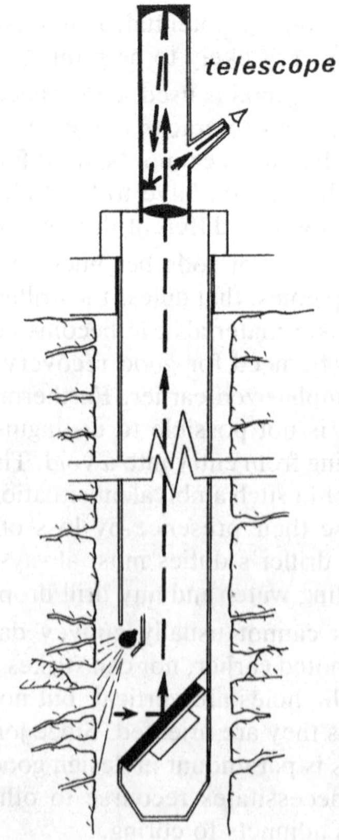

Figure 8.1-7A. Fundamentals of borehole periscope

Figure 8.1-7B. Borehole periscope in use. (U. Krause)

most local workshops without difficulty. Some ingenious refinements are possible.

Practical sizes of periscopes are suitable for 3-in ⟨75-mm⟩ holes or larger. There is little risk of jamming down the hole because the tubes almost line the hole over the full depth to the mirror. Mirrors can occasionally be damaged by dislodged stones, and it is desirable to make them replaceable by attaching them to metal backing held in place by a magnet in the nose of the tubes. A strong plastic rope down to the nose assists with handling and recovery. If by some chance the tubing becomes hopelessly stuck in a hole, the financial loss is negligible.

Fiber optics present a potential for improved downhole examination but appropriate equipment is not yet available, to my knowledge.

8.1-8 Impression Devices

Impression devices are long, soft rubber packers that are inflated very firmly against the wall of the hole for a short time. They are then deflated and quickly brought to the surface for examination. It is hoped that their surface then presents an impression of the wall of the hole including cracking. Considerable doubt exists about whether they effectively do this.

8.1-9 Other Borehole Devices

Prior to the advent of borehole TV, limited use was made of borehole cameras. These devices have a camera and flashlight made to fit in a hole. Some have two cameras, mounted one above the other, to take stereoscopic pairs of photographs; these can be useful for remotely measuring large voids such as karst or abandoned mine workings. A variation on this theme uses electronic transmission such as paired radar signals; when lowered down a hole into a void, these can survey it fairly well.

8.1-10 Shafts and Adits

In critically important sites where the cost can be justified, shafts and adits (i.e., tunnels) are sometimes driven. These enable direct visual inspection.

A method now largely superseded was the drilling of calyx holes of 3 or 4 ft ⟨0.9 or 1.2 m⟩ diameter; the geologist was lowered down on a rope to do the mapping. Not a job for the faint hearted!

8.1-11 Presentation of Geological Information

Results of exploration should be presented succinctly and clearly. They must be intelligible to designers who may not have had the benefit of intense geological training. The presentation should include the following elements.

- Drawings of sections with geological features shown simply, and with exploratory holes and their results (including lugeon values as discussed in Section 8.2) shown alongside them. Other helpful information includes the addition of anticipated stripping levels and the outlines of any structures. Although distorted scales are sometimes used for such sections to permit magnification of hole results, this is not advisable because of the need to replot them to a natural scale when designing grout curtains and hole layouts.
- Plots of joint concentrations and principal strikes and dips. A method for doing this is given in Section 8.1-12.
- Results of borehole information such as TV or periscope observations and drilling information.
- Photographs and logs of cores.
- A geological report.

8.1-12 Mapping of Joint Patterns

Recording and presentation of joint patterns for the various parts of a site can be done in a number of ways. For grouting purposes, those that give a quickly comprehended presentation are preferred, such as hemispherical equal-area projection; this method is now outlined.

It is well-summarized by John (1968) and in geological texts to whom reference should be made for further details if required. Sufficient information is, however, given here and in Section 8.4-3 (where its application to design of grout hole inclinations is explained), for routine usage.

Imagine a southern hemisphere (Figure 8.1-12A) with a joint p drawn through its center at the equator. From this point, a perpendicular, OP, to the plane touches the spherical surface at P. Therefore, OP is normal to the curved surface and is perpendicular to a tangential plane at P. This plane, p', will be parallel to the initial plane p and reproduces its strike and dip. These properties are unique to the position of point P on the hemisphere, and so a diagram showing P, such as in Figure 8.1-12B maps them clearly.

The diagram is imagined as being located in the horizontal plane at the equator. Dip is indicated by circular graduations and strike by the tangent to the radius at the point. This example has been set up with strike measured from the north toward both east and west, and with dip measured in the usual way below the

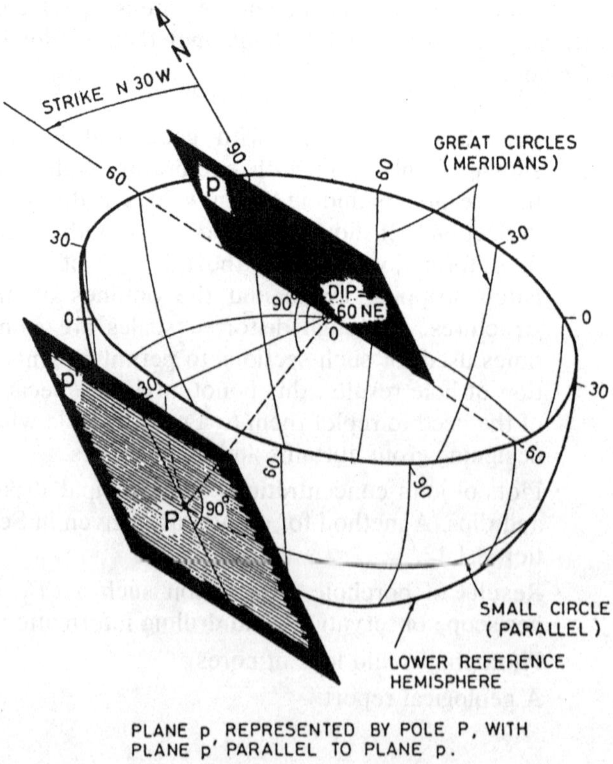

Figure 8.1-12A. Pictorial view to explain joint representation. (After John, 1968)

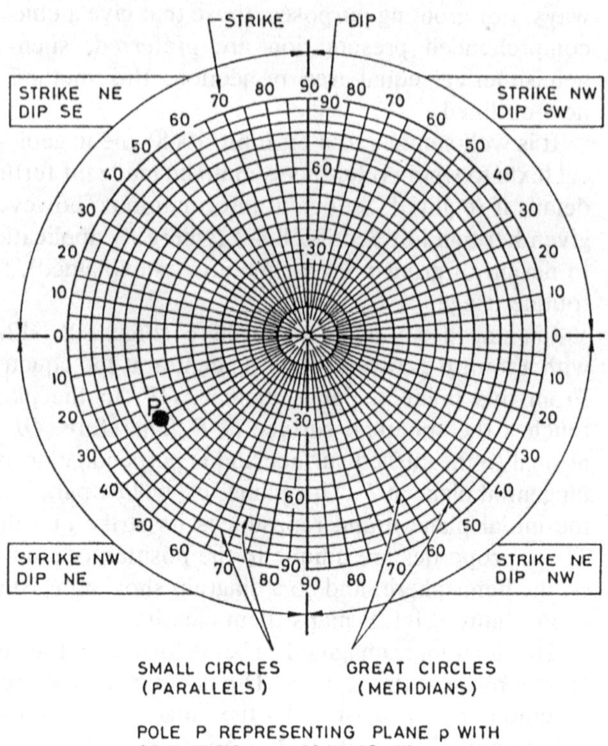

Figure 8.1-12B. Point represented on an equal-area diagram. (After John, 1968)

horizontal. Some users prefer to show strike as an azimuth, with north as zero, and the method works equally well for this.

By plotting measurements taken on each joint in an area, each of which appears as a point on the net, concentrations of points will appear at the dominant jointing positions. Contours of intensity can then be drawn to indicate the population densities as shown in Figure 8.1-12C.

In this example, more than 12% of the joints are in the small area of the plot indicated by solid black shading. The average strike of this area is 45° NW and dip is 30° NE. It is surrounded closely and more or less concentrically by others having lesser intensity, and before long a large area of white is reached that indicates that so few joints have strikes and dips in it that they can be ignored, probably. In a crescent on the right-hand side are the plots for most of the remaining joints. The most intense of these is an 8–12% group near the top of the diagram. These have about the same strike as the black-shaded group and have a dip of about 70°. A third grouping, this time of 4–8%, in the lower right, strikes 45° NE with a dip of 80° NW. These three groups of principal intensity represent, of course, the common situation where jointing patterns tend to follow three axes at about 90° to each other.

Points plotted near the rim of the net indicate nearly vertical joints. A scattering among these can produce plots at the diametrically opposite part of the net, as happens at the lower right and upper left.

Figure 8.1-12D shows another example, which also is from an actual site. A somewhat lopsided pattern exists on this plot. The main concentration is near the top of the net, to the right of north and indicates average strike of about 65° west of north and dip of 60° SW. The other group at middle and lower right strikes between N and about 45° NE with dips ranging from vertical to 30° NW.

Use of these plots for design of grouting is explained in Section 8.4-3.

8.2 PERMEABILITY

8.2-1 Why Investigate Permeability?

A good knowledge of a foundation's permeability is essential if one of the purposes of the grouting is seepage control.

Permeability data is also helpful in other forms of grouting, such as when strengthening a foundation; simple before and after water tests can be useful for assessing how well the grouting has been done (see Section 7.5-4).

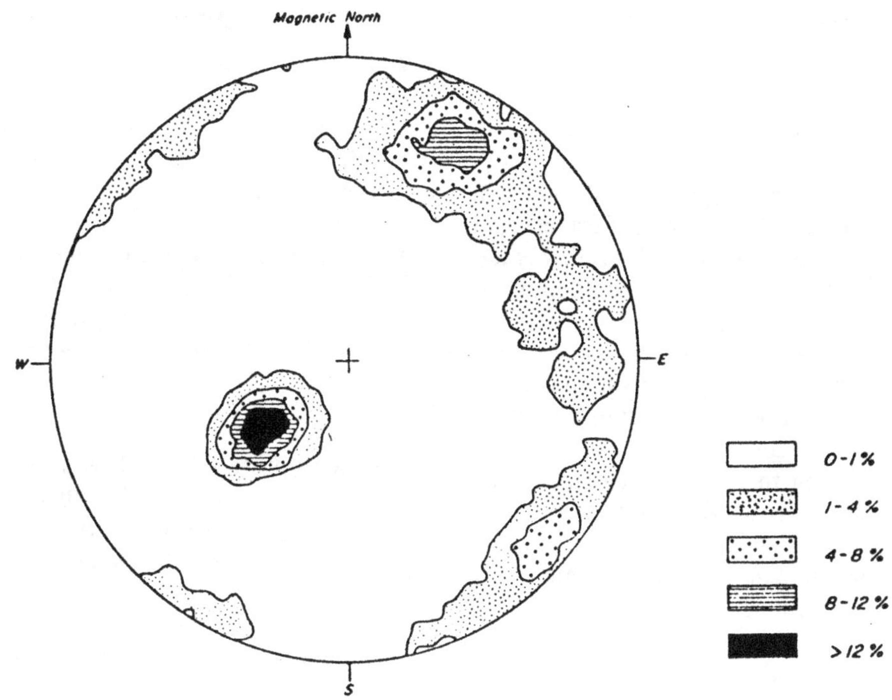

Figure 8.1-12C. *Equal-area net contoured to show intensities of joint occurrences. Key on the right indicates the percentage of population in each contour grouping.*

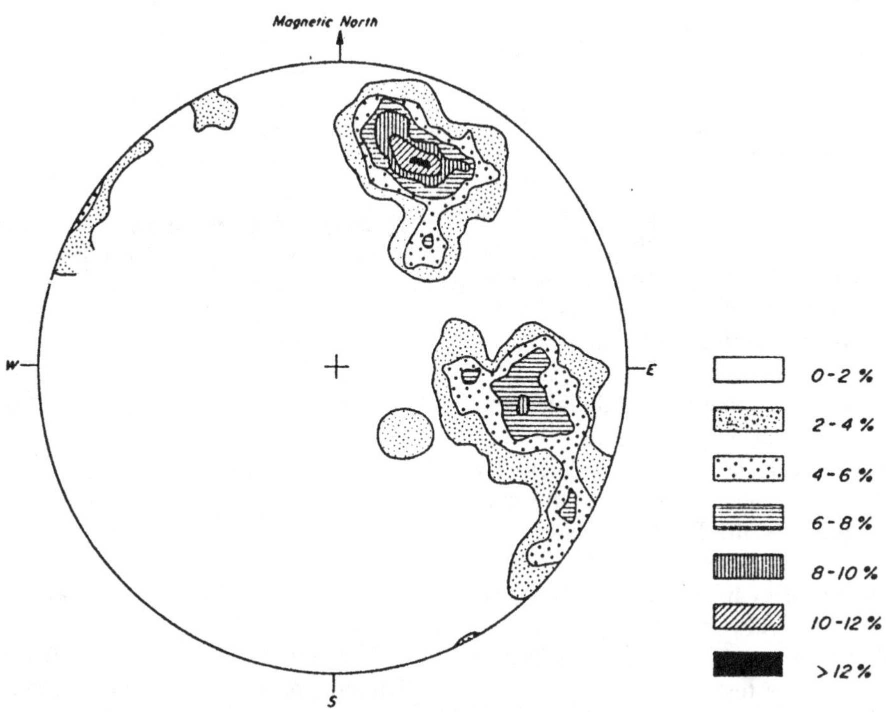

Figure 8.1-12D. *Another example of plotting.*

8.2-2 Different Ways of Describing Permeability

Ideally, the permeability of a site should be described in terms of how much seepage can pass through it under standard pressure head. However, some seepage will go downwards, some will go up, and some will move at a skew to the general direction of travel. And to make things even more complicated, all this is dependent on the pressure, which varies with depth, effect of groundwater, length and width of seepage paths and other things. Hence general geotechnical practice is to define permeability as a type of velocity unit measuring the rate of water flow under a standard pressure. Total seepage through a cross section is then calculated by multiplying this by the cross-sectional area. Sometimes variations of permeability across a site are taken into account, at other times an average is good enough. A considerable number of methods is available for such calculations. Some are discussed in Section 8.2-14.

This type of permeability unit is generally not as useful in grouting work as one that indicates the flow radially from the test hole. A test of this radial type used extensively in grouting is the lugeon test.

The lugeon test is a "pump-in" test where the volume of water taken in a section of test hole is measured during time intervals. It was derived by Lugeon in 1933, and the related unit of permeability has been named after him.

The lugeon test has found wide acceptance; most of the practices described in this book are geared to it, but not everyone is enamored with it. There are indeed alternative tests (see Section 8.2-12). At first sight the lugeon test gives the impression that it is not directly relevant to permeability issues, but on reflection and with familiarity, a different reaction usually emerges: its radial flow is more relevant to grout flow (which helps when doing the grouting and assessing it—Sections 7.4-2, 7.5-1, and 7.5-3–7.5-6), and the test is so simple that it can be done by relatively unskilled labor as part of the process of grouting each hole (Section 5.3).

The lugeon test is crude—there is no doubt about that—but it is commensurate with grouting, whereas the more sophisticated tests needed for the velocity type of unit can be tricky and more expensive and require a higher degree of skill. Their results are not much direct use to grouting, even then. Certainly, in the design phase it helps to be able to decide how much seepage would pass without grouting. However, once having decided to grout, it becomes sensible to switch to the lugeon type of test.

8.2-3 The Lugeon Unit

Definition of the lugeon unit:

1 lugeon unit = 1 liter/m/minute at 150 psi

or put more formally:

1 lugeon unit = 1 liter of water taken per meter of test length, per minute, at 10 bars pressure ⟨150 psi approx. For other pressure units see 6.2-2⟩

or

1 lugeon unit = 0.0107620 cu ft of water taken per foot of test length per minute at 142 psi ⟨exact conversion. 150 psi usually used though⟩

In terms of "velocity"-type permeability units, very approximate calculated relationships are

$$1 \text{ lugeon unit} = 10 \text{ ft/year}$$
$$= 1.3 \times 10^{-5} \text{ cm/second}$$

To give a sense of proportion for the unit:

- 1 lugeon unit is the type of permeability where grouting is hardly necessary.
- 10 lugeons warrants grouting for most seepage reduction jobs.
- 100 lugeons is the type of permeability met in heavily jointed sites with relatively open joints or in sparsely cracked foundations where joints are very wide open.

Tests at Other Pressures To calculate lugeon units from test results, the formula is

$$\text{Lugeon units} = \text{liters/meter/minute} \times \frac{10 \text{ (bars)}}{\text{actual pressure (bars)}}$$

or

$$\text{Lugeon units} = 0.01076 \text{ cu ft/ft/minute} \times \frac{142 \text{ (psi)}}{\text{actual pressure (psi)}}$$

The actual test pressure is often less than Lugeon's defined pressure of 10 bars because that is too high for many foundation conditions (Section 6.2-1). A straight-line relationship between the actual pressure and the definition pressure is assumed; this assumption is not strictly correct, but a more precise relationship is not worth the complication in view of the approximations made when using the results anyway (see Section 8.4-14).

8.2-4 Sensitivity

The lugeon test has its maximum sensitivity in low-permeability situations. This is appropriate because those are the ones where decisions about grouting can be marginal and therefore need greater accuracy than when permeability is greater and grouting is obviously needed.

The realism of lugeon units diminishes in a mode resembling logarithmic:

- In the 1 to 5 range, each variation of 1 unit is meaningful.
- Between 5 and 10, meaningful increments are about 2 units.
- Between 10 and 15, meaningful increments are about 5 units.
- Between 15 and 50, meaningful increments are about 10 units.
- Between 50 and 100, meaningful increments are about 30 units.

Although the scale has no upper limit, above 100 lugeons it is meaningless to distinguish further. All values have the same significance in terms of grouting decisions: grouting is very necessary. Recommended practice is to quote values in this range as >100 lugeons without further elaboration.

Lugeons should never be given to a decimal place. Always round calculated values off to the nearest whole number except for values less than 1; any value in this range that is above zero should be rounded up to 1.

WARNING Lugeon values must never be used on their own without some idea of the crack sizes and spacing to which they apply. This is because the same lugeon value can be obtained from a single large crack as from a lot of much finer ones.

The implication of this in terms of grouting decisions can be significant but nevertheless is quite able to be accommodated when using the unit if values are viewed in conjunction with crack properties.

Borehole periscope or TV inspection of cracks in situ is very desirable as an adjunct to water testing and can also help when deciding grout mixes (refer to Section 7.2-2).

8.2-5 Types of Permeability Testing

The two different types of permeability testing are:

EXPLORATORY TESTING, where permeability exploration is carried out before design of grouting; it is dealt with in subsequent Sections 8.2-6 onwards.

GROUT HOLE TESTING, where testing is carried out in grout holes before grouting; see Section 5.3.

8.2-6 Exploratory Testing

Exploratory testing is carried out at a new site, at one previously grouted, or where grouting is in progress and wider checking is required than can be provided merely in the grout holes. It is more thorough than grout hole testing (Section 5.3).

Exploratory tests are usually performed in core drilling holes (see Section 8.1-5), although sometimes special water testing holes are drilled in situations where the geology is sufficiently well known to not need coring or where the delays occasioned to drilling by stopping to do water tests are not desirable.

The length of test stage should be about 15 or 20 ft ⟨5 or 6 m⟩. If greater precision is required or if special geological features need to be investigated, this should be reduced to suit. While longer stages than those recommended provide useful information, relevancy suffers unless the cracking system is known to be very uniform. Lugeon specified a stage length of 5 m, but there is no need to stay strictly with this length.

Packers (Sections 2.5-15–2.5-17) are used to isolate the test length from the remainder of the hole. The best method is undoubtedly the downstage with packer technique (described in Section 5.4-5 in terms of grouting). This uses a single packer; drilling is stopped each time the end of a test length is reached and the packer is placed in position. The procedure is shown in Figure 8.2-6A.

Removal of the drill is not usually necessary, and drilling resumes an hour or so later, once the testing is done.

On some sites double packers have been used; these have two packers with suitable arrangements for passing out the test water between them. The hole is drilled full depth before starting testing when these packers are used. Although this may save drilling times, results can be highly dubious because there is no simple

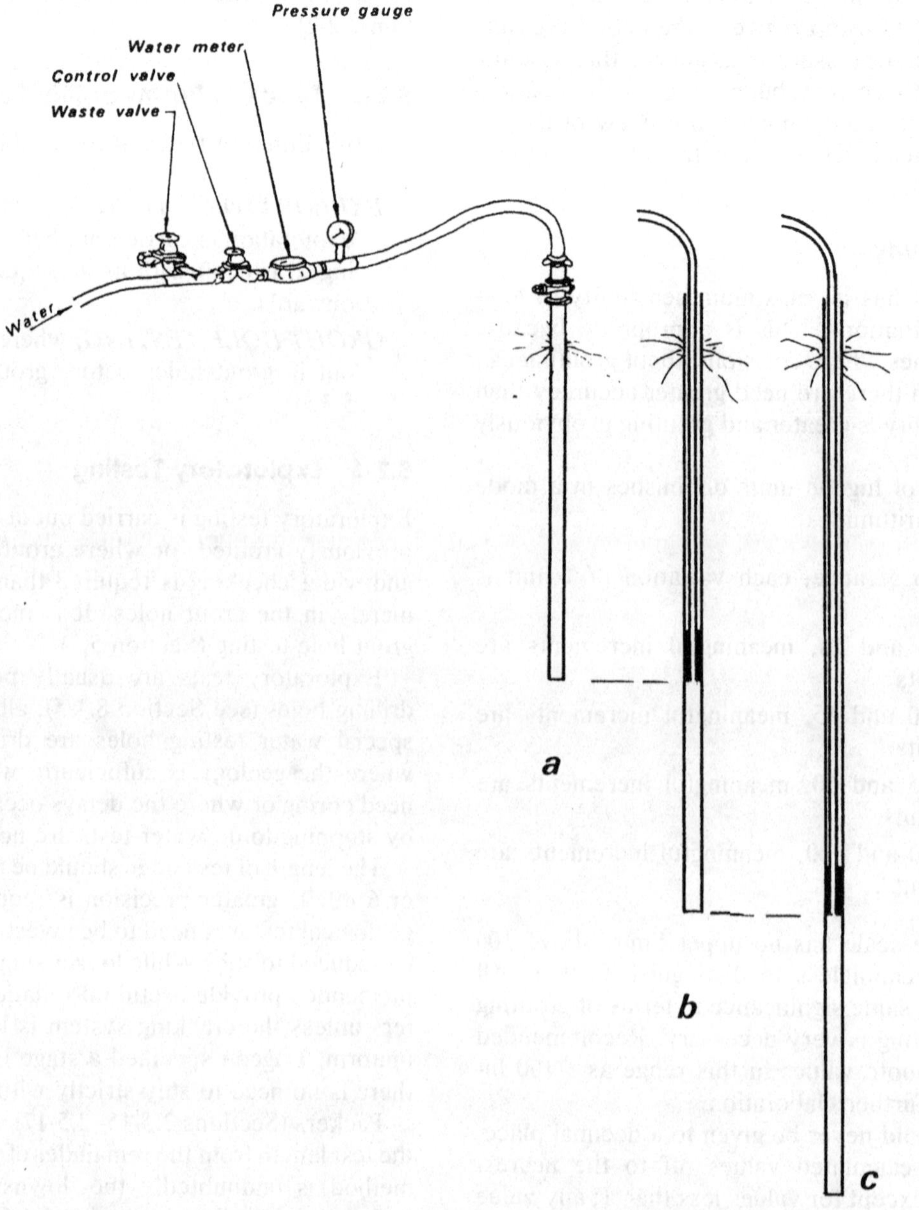

Figure 8.2-6A. *Steps in conducting exploratory water testing: (a) top stage drilled, washed out, then tested, (b) second stage drilled, washed out, packer placed at top, then tested, (c) third stage likewise, and subsequent stages.*

means of verifying that the lower packer is not leaking; if it is, water passing it becomes measured in the hole take, and this gives erroneous results (although on the high side). Upstage methods (see Section 5.4-6) have been resorted to in extremities, for permeability exploration, but the calculations are messy and the results somewhat uncertain.

Of course, when holes are no longer than 20 ft ⟨6 m⟩, and also in surface stages, there is no need to use a packer; the testing can be done from a standpipe if one is fitted.

When wireline drilling for cores is in use (Section 8.1-5), it is possible to lower a special water-testing packer through the rods after pulling the bit back clear of the test stage.

It is important to use the largest sized tubes possible down the hole and through the packer in order to minimize head loss in the test water. They should have an internal diameter of at least an inch ⟨25 mm⟩, and couplings should be flush internally. There should be no constrictions anywhere in the system, and the hose connecting the test gear to the hole should be at least this diameter and should be short, say 3 ft ⟨1 m⟩.

8.2-7 Details of Testing Method

The routines for conducting these tests vary among users. Some are unjustifiably complex; they give the impression of being more in the nature of demonstrations of mathematical ingenuity than of practical know-how. The method described in this section is my improvement in the late 1960s to earlier methods in order to overcome problems met on a number of dam sites. These earlier methods had used three pressure increments; this was found insufficient to give clear definition of events during the testing. The developed method uses five pressure increments and has been extensively job tested. It was published in Houlsby (1976) and appears to have found wide acceptance. Some users increase the sensitivity by using more than five increments.

The overall concept is that the test method must indicate:

- Whether the water flow through the cracks under test is essentially laminar or turbulent. The result reported for the test is strongly affected by this.
- If the test is dilating the cracks.
- If materials are being washed out or rock moved by the test water.
- If voids are being filled by the water.

The effects of these points on the test results are indicated shortly.

TEST METHOD

1. Apply water at a *LOW* pressure for 10 minutes. This is followed immediately by
2. A *MODERATE* pressure for the next 10 minutes, then
3. A *PEAK* pressure for the next 10 minutes, then
4. The *MODERATE* pressure again for the next 10 minutes, then
5. The *LOW* pressure for the final 10 minutes.

These five different pressure runs are done immediately after each other. They can be denoted as *A–B–C–B–A*.

The lugeon value is then calculated for each of the pressure runs using the formula given in Section 8.2-3. Then the five values are inspected to determine the flow regime or testing event. In the following examples, bars are used to pictorially convey relative magnitudes, but once the concept is understood, the process of deciding which lugeon value to report for the test can be done entirely mentally.

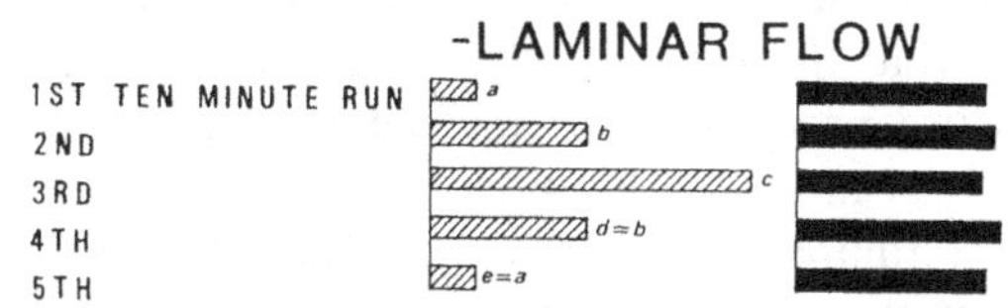

Figure 8.2-7A. *Similar lugeon values for each run indicate laminar flow.*

On the diagram in Figure 8.2-7A, the pressure increments are shown in hatched bars. The manner in which pressure rises to a peak in the third run and then reduces, in a symmetrical pattern, can be noticed. The bars shown solid on the right portray the lugeon values calculated for each of the pressure increments. In this example, they are more or less the same value for each of the pressures. The implication of this is that the flow was generally laminar (i.e., smooth flow along cracks, without turbulence, usually the result of quiet flow in finer cracks—see Section 8.2-11), but not exclusively, of course, considering the mix of crack sizes and conditions met with in most test stages.

When flow is of this type, any one of the five lugeon values can be accepted as the result of the test and reported as such, or their average used.

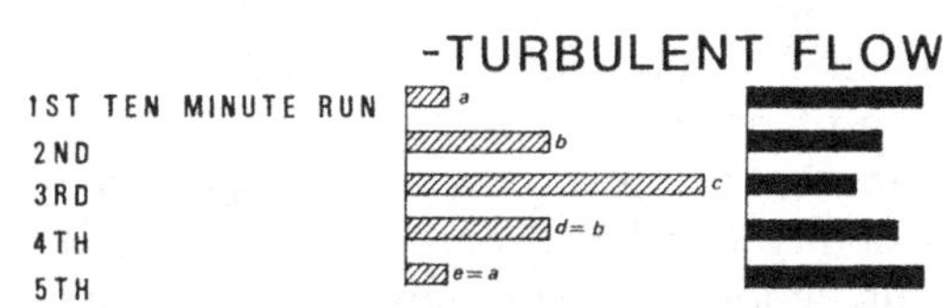

Figure 8.2-7B. *Symmetrically low lugeon values at higher pressures indicate turbulent flow.*

In the example shown in Figure 8.2-7B, the lugeon values for the various pressure increments are no longer the same. A symmetrical distribution with the lowest value occurring on the highest pressure is indicative of turbulent flow (i.e., rough, eddying, turbulent motion usually the result of fast flow in wider cracks) as explained later in Section 8.2-11. This lowest value should be reported as the result of the test. Note the considerable error that would result from averaging the values.

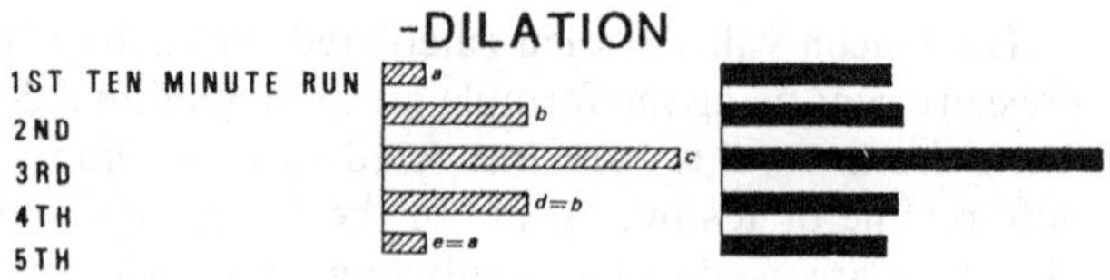

Figure 8.2-7C. *High lugeon value at peak pressure indicates dilation.*

The pattern of values shown in Figure 8.2-7C is somewhat the reverse of the previous one. The peak pressure produces a much higher lugeon value than before or after it. This pressure has been sufficient to locally dilate cracks by compression of softer materials or by closure of adjacent parallel cracks. The dilation has been temporary, and for this reason the reported permeability of the stage should be that obtained for the lowest pressures, or alternatively for the medium pressures if these are less than for the low pressures, indicating that turbulent flow was occurring prior to the dilation.

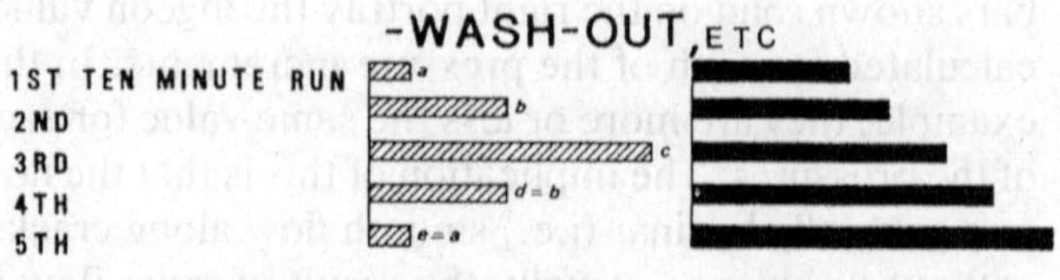

Figure 8.2-7D. *Lugeon values increasing even when pressure drops, indicates washout.*

Figure 8.2-7D shows permeabilities increasing as the test proceeds. They do not decrease when the pressure does. This indicates permanent washing out of joint filling materials, or movement of rock caused by the testing, followed by "propping" where rock fragments fall into joints and prevent closing after relaxation of pressure. Too much of this sort of thing indicates that test pressures are too high. The reported permeability can be taken as that measured for the final run (at low pressure); this presumes that the peak pressure bears some resemblance to eventual in-service pressures, which would produce somewhat similar results if not grouted.

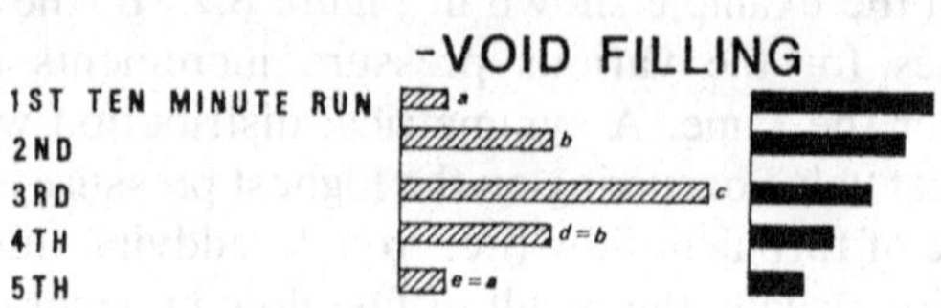

Figure 8.2-7E. *Decreasing lugeon values throughout the test indicate void filling.*

Figure 8.2-7E shows the reverse of the previous pattern. The lugeon values are progressively reducing, indicating that empty voids, cracks, and so on are being filled and that water cannot then progress further. Prudent action is to carry the test further (at low pressure) until the reducing trend ceases, indicating that all empty voids have become filled. Even better action is to saturate such a foundation before starting the test. The reported permeability should be the lowest measured; this is an indication of the flow-through permeability once the voids are filled.

8.2-8 Test Pressures

The "correct" pressure given by Lugeon for this type of water testing is 10 bars ⟨approximately 150 psi, 1 MPa⟩. One suspects that he found it to be a convenient round figure; there is certainly nothing magic in it. Rarely is there need to exceed this: only if groundwater back pressure is more than about half of it would a higher test pressure be warranted.

As mentioned earlier, the common situation is that it is too high for general use. Lugeon apparently had nice tough sites in the Swiss Alps to deal with; we lesser mortals have to contend with poorer sites mostly. Water at this pressure can do more damage in terms of foundation disruption than grout can, so although it is a reasonable maximum pressure for grout (refer to Section 6.2-1), it is not necessarily so for water. A satisfactory guide for this is the rule of thumb of one psi per foot depth (Section 6.2-4) when the testing is in sound rock, that is, the peak test pressure is a little less than would be used for grouting in the same rock. Likewise, in weaker foundations, test pressures should be less than those for grouting, as indicated in Section 6.2-1.

These pressures are the peak (maximum) pressures in a five-pressure test. The graph in Figure 8.2-8A gives them in relation to depth. It also shows the low and medium pressures as used in the first and fifth and in the second and fourth increments of the test.

Appendix E shows this same information in tables and also shows the pressures for weak rock. Jerks in the graphs in Figure 8.2-8A correspond with rounding off in the tables; dashed lines apply to measurements in feet and psi; continuous lines are for metric.

Some people make corrections for head loss in testing equipment and corrections for the depth to groundwater (e.g., Dick, 1975). Others measure the pressure down the hole with transducers or other instruments. While these refinements certainly improve the accuracy of the result, given that waterway through the packer is adequately large (see Section 8.2-6), advantage often cannot be taken of the improvements: when fed into the process of deciding whether grouting is necessary, they can become submerged in assess-

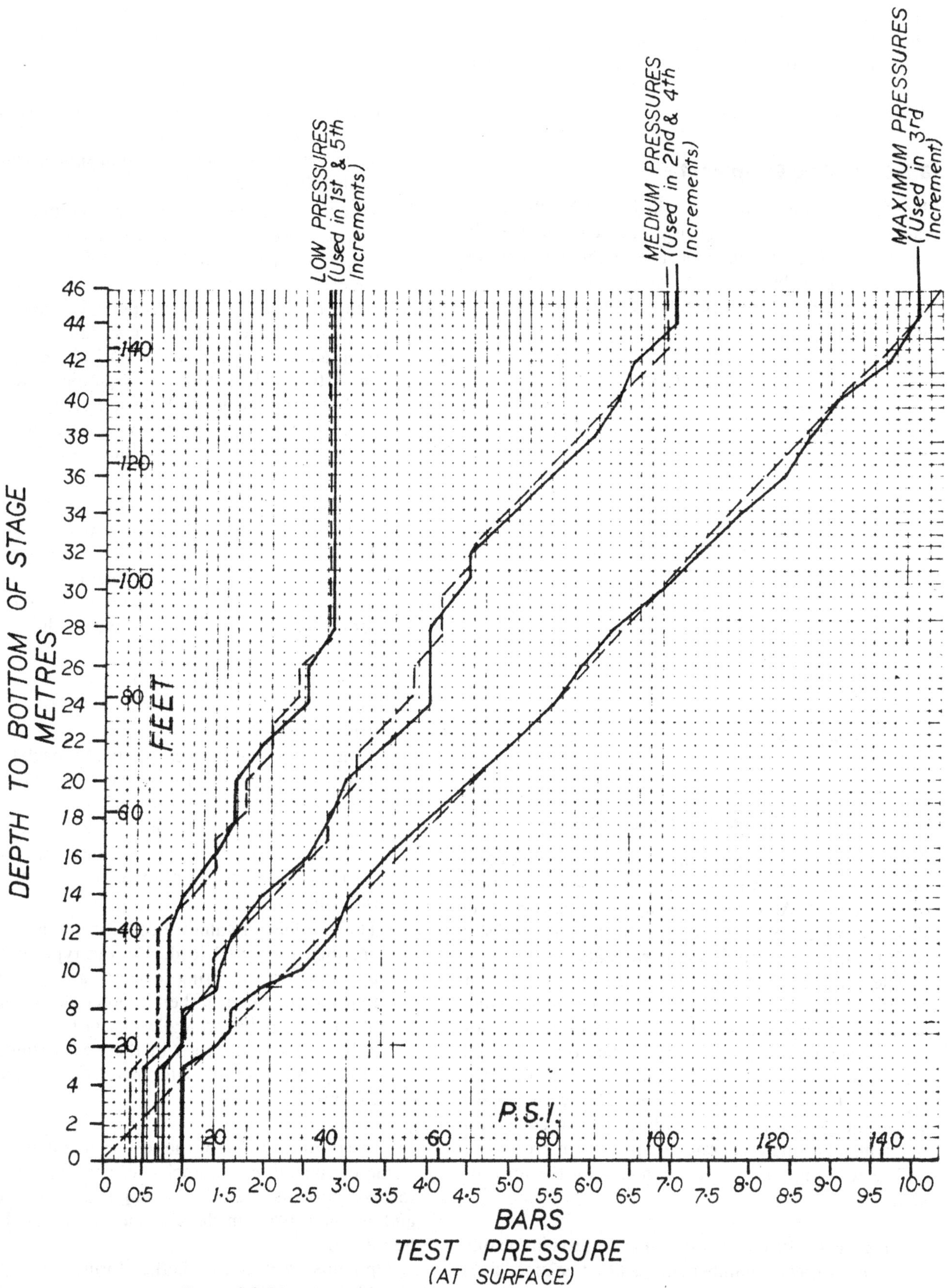

Figure 8.2-8A. Pressures for each run of a five-pressure test. Normal rock.

ments incorporating a mass of other data, as described in Section 8.4-14. For this reason, my practice is to omit pressure corrections, but to make mental adjustments during the assessment process to any results that are crucially marginal.

8.2-9 Testing Equipment

The testing equipment shown in Figure 8.2-9A is the same as for water testing in grout holes detailed in Section 2.8-1. More sophisticated equipment than this is used at times, but this is all that is really needed. For comments about using computers for analyzing results see Section 7.7-11.

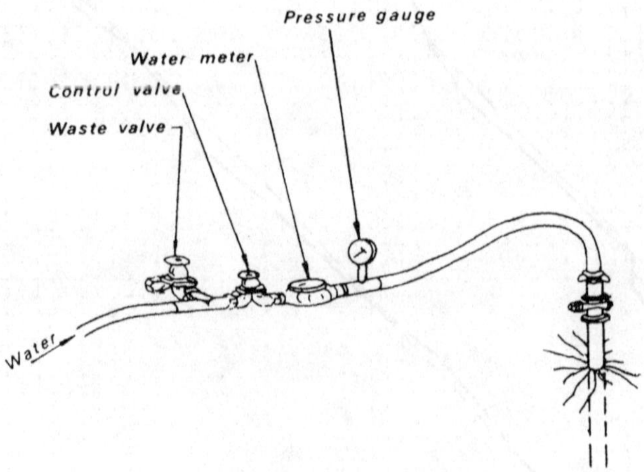

Figure 8.2-9A. *Gear for conducting water tests.*

8.2-10 Report Form and Calculations

Most large organizations doing this type of work have their own report forms. There is little standardization on the subject. An example of my particular choice is shown in Figure 8.2-10A. It was included in Houlsby (1976) and happens to be in metric units.

The form in Figure 8.2-10A contains all relevant entries and conclusions, right from the field testing results themselves down to the office inspections of lugeon calculations to decide the value to be reported for the tests. The example includes each of the main types of testing events and concludes with a case where the pump or water supply could not cope with the demand at the moderate and maximum pressures.

The following are some comments on the form itself:

- Figure 8.2-10A shows six separate tests in the one hole. A greater number can be fitted on most sizes of paper.
- The first column for dates provides space for an entry against each test in case they are done on different days.
- Stage depths are given in the second column. If irregular staging is used, such as overlapping (to cover some special feature) or repetition of a stage, an asterisk in this column should lead to an explanation below.
- Test times occupy the next four columns. In the first of these is the 10 minutes required time (as a reminder) printed on the form, followed by an entry of the actual time, which is almost always the same 10 minutes. An exception is shown in the last case where a shorter test run of 6 minutes resulted when it became absurd to carry the test on further. Clock times at beginning and end of each test run are entered; these enable checking to assure that the runs were done consecutively and enable coordination with other testing events in the vicinity.
- Pressure at the surface is shown, first as the required amount as read from Figure 8.2-8A or from Appendix E, and second as the actual pressure attained. In cases where the take is beyond pump capacity, or where some problem develops, the actual pressure might not reach the required amount.
- The water injected into the hole in the test period as read from the water meter is entered in the third from last column.
- A lugeon value is then calculated for each test run, usually in the office after the testing. The calculation uses the formula given earlier in Section 8.2-3 without correction for head loss, groundwater level, and so on. Values over 100 are written as >100 because greater refinement than this in this range is absurd.
- The five lugeon values (second from last column) are then inspected to decide the type of flow or event occurring during the testing. A representative lugeon value is then adopted after consideration, using the principles described in Section 8.2-7. This value goes in the final column. The examples shown indicate the relative simplicity of this process.

Note that no graphing of the test results is necessary. Some organizations routinely do use graphing nevertheless, perhaps without realizing it to be superfluous and perhaps even misleading, as discussed later in Section 8.2-13.

Opinions vary as to whether forms such as these should contain entries put there in the field or whether

DATE 19..........	STAGE METRES	TESTING TIMES: REQUIRED MINS.	TESTING TIMES: ACTUAL MINS.	CLOCK TIMES: FROM	CLOCK TIMES: TO	GAUGE PRESSURE BARS: REQUIRED	GAUGE PRESSURE BARS: ACTUAL	WATER INJECTED INTO HOLE IN TESTING PERIOD (TOTAL QUANTITY READ FROM WATER METER) LITRES	LUGEON VALUE: FOR EACH TEST	LUGEON VALUE: REPRESENT-ATIVE FOR TESTED STAGE	GROUP
	FROM	10	10	0820	0830	2·8	2·8	29	2		
	27·67	10	10	0830	0840	5·0	5·0	46	1		
	TO	10	10	0840	0850	7·8	7·8	81	2	2	A LAMINAR
	33·87	10	10	0850	0900	5·0	5·0	54	2		
		10	10	0900	0910	2·8	2·8	25	1		
	FROM	10	10	1230	1240	2·8	2·8	120	7		
	24·08	10	10	1240	1250	4·5	4·5	180	6		
	TO	10	10	1250	1300	6·8	6·8	198	5	5	B TURBULENT
	30·26	10	10	1300	1310	4·5	4·5	178	6		
		10	10	1310	1320	2·8	2·8	150	9		
	FROM	10	10	0855	0905	0·7	0·7	20	5		
	6·20	10	10	0905	0915	1·7	1·7	57	6		
	TO	10	10	0915	0925	2·7	2·7	122	8		C DILATION
	12·20	10	10	0925	0935	1·7	1·7	76	7		
		10	10	0935	0945	0·7	0·7	20	5	5	
	FROM	10	10	1315	1325	2·0	2·0	62	5		
	15·16	10	10	1325	1335	3·0	3·0	114	6		
	TO	10	10	1335	1345	4·8	4·8	198	7		D WASH-OUT
	21·29	10	10	1345	1355	3·0	3·0	153	8		
		10	10	1355	1405	2·0	2·0	119	10	10	
	FROM	10	10	1140	1150	2·8	2·8	118	5		
	22·57	10	10	1150	1200	4·5	4·5	151	4		
	TO	10	10	1200	1210	6·8	6·8	208	4		E VOID FILLING
	30·42	10	10	1210	1220	4·5	4·5	94	3	3 OR 1	
		10	10	1220	1230	2·8	2·8	28	1		
	FROM	10	10	1512	1522	0·8	0·8	269	56		
	3·60	10	10	1522	1532	1·4	1·2	432	60		
	TO	10	6	1532	1538	2·4	1·2	259	60	60	#
	9·60	10	10	1538	1548	1·4	1·2	432	60		
		10	10	1548	1558	0·8	0·8	279	58		

#PARTLY BEYOND PUMP CAPACITY

Figure 8.2-10A. *Example of test report form showing testing events.*

the field work should be copied onto them in the office. Possible errors in transcription make this latter option unattractive, even if forms become rather grubby in the field.

8.2-11 Some Comments on Turbulent Flow

Amplification of earlier (in Section 8.2-7) comments on turbulent flow and laminar flow follows to explain the basis of the methods used to deduce flow regimes.

Guerra, Weyermann, and Muta (1968), Arhippainen (1970), Lancaster-Jones (1975), and others have discussed the detection of turbulent flow. They have shown that for flow that is solely turbulent, test pressures have a square root relationship, compared with the direct relationship of laminar flows. They have also given instances of cases where the relationship was neither a square root one nor a direct one, but was intermediate.

Accepting that turbulent flow is indicated by the square root relationship, and that laminar flow follows a direct relationship, it follows that if the five lugeon values calculated for a test are not generally equal, but instead, show a lower value for the peak pressure run than for the medium pressure runs, while those for the lowest pressure runs are equal, then the flow is non-laminar. It is conveniently designated as "turbulent"; this is, however, not an accurate description, but suffices.

Because a test stage usually cuts many open cracks of various sizes, and because the finer ones are liable to exhibit laminar flow while wider ones have turbulent flow, the overall effect is likely to be a mixture of flows. Hence it is not surprising that the pure square root relationship is not solely experienced as the only alternative to laminar flow. Thus "turbulent" is a convenient all-embracing designation for all flows apart from solely laminar flow.

The reported permeability of a stage that exhibits such turbulent flow is taken as that for the peak pressure. It can be argued with some justification that if the flow is in the vicinity of genuine turbulence, the lugeon value should be recalculated on a square root basis. This would produce a slightly lesser value. However, the proportion of solely turbulent flow cannot be reliably assessed when a mixture of laminar and turbulent flow exists. There seems little realism, therefore, in recalculating on a turbulent basis unless the result is quoted as a range, with the solely turbulent value at one extreme and the solely laminar flow at the other. Rarely is this warranted.

8.2-12 Alternative Test Methods

As mentioned earlier, not everyone uses the lugeon test and the method described. Some alternatives are summarized here.

TRH, EIPD, AND ED TESTS These three complementary tests were developed by Andrade (1988a,b) in Brazil as a replacement for the lugeon test, which he sees as obsolete and unreliable. He considers that Lugeon's original concept has been unjustifiably extended for purposes never intended (true enough, but the method is, even so, quite valid and useful in my experience).

The TRH test (all initials relate to the Spanish wording) he calls "hydraulic register test." Its purpose is to find the positions of the main water flows into the foundation. The test is carried out somewhat like that shown in Section 8.2-6, preferably downstage, with a single packer but with the addition of a perforated tube running down the full length of the test section. This tube has a permeable sleeve over it, fairly close to the hole wall. When the water flow out through this sleeve is steady, dye is introduced. This then stains the sleeve wherever the water is flowing strongly through it. After the test, the sleeve is carefully extracted and the dye traces recorded by passing it through a photocopying machine (more than one use for the office equipment!). Examples given show separate, wide cracks. Just how well it would perform in multicrack and rough hole situations seems uncertain.

The EIPD test (water injection test under decreasing pressure) is a falling head permeability test where the pressures are sufficiently low to ensure that flow is laminar. The purpose is to avoid turbulent flow, which introduces uncertainty in the calculations.

The ED test (directional test) attempts to divide up the circumference of the hole into quarters by means of a 90° partitioning device that relies on the use of seals against the hole wall. The efficacy of this arrangement in any but perfectly drilled smooth circular holes seems doubtful. Such holes are rare.

The mathematics given for calculation of results from these implements appears more appropriate to sites with few cracks than for the more usual situations. The concepts appear to have had little job-proving as yet; perhaps further development may be useful.

AQUIFER TEST METHOD This test method is essentially the drill stem test as carried out by drillers to test transmissivity of an aquifer. It is particularly relevant for very deep exploratory holes such as those testing a site before sinking a shaft. Grouting methods for shafts are discussed in Section 8.4-12.

In deep holes below the water table, the lugeon test loses relevancy somewhat. It is more concerned with the vicinity of the hole than further afield, hence a transmissivity-type test can be of more value for indicating grouting requirements in terms of potential water flow into an excavation.

The method is described by Daw (1984) in relation to grouting. Basically, it consists of pumping out at a controlled rate followed by a recovery period. Behavior during this recovery period indicates the desired information.

The test is done with special equipment comprising a packer and a downhole pump, sensor, and valve.

OTHER TEST METHODS An interesting summary of the various opinions about water testing was given by Božović (1985) in his role as General Reporter for the Grouting Question at the ICOLD Congress held at Lausanne, Switzerland. Other methods of testing are given by some of the authors he summarizes.

8.2-13 Why Plot Graphs?

Some workers in this field prefer to read lugeon values off graphs rather than calculate them. In Section 8.2-10 I suggested that graphing is not necessary and can even be misleading. However, graphical concepts are outlined here for those who prefer them, but some critical comment is included.

Figure 8.2-13A shows the basic concept of this type of graphing. The water taken in the test is plotted as a rate against the pressure. In accordance with the presumption that tests at reduced pressures can be extrapolated to full pressure in a direct ratio, a line from the origin through the result of a test at such reduced pressure will meet the full-pressure ordinate at a point that would have been the result of the test if that full pressure had been used. The slope of this line is a measure of the lugeon value and is measured off plots by those who prefer to find it this way rather than by calculation. Typical plots of this type for four of the test events described in Section 8.2-7 are shown in Figure 8.2-13B.

Laminar flow plots as a straight line for the three values of pressure (one peak value and two at medium

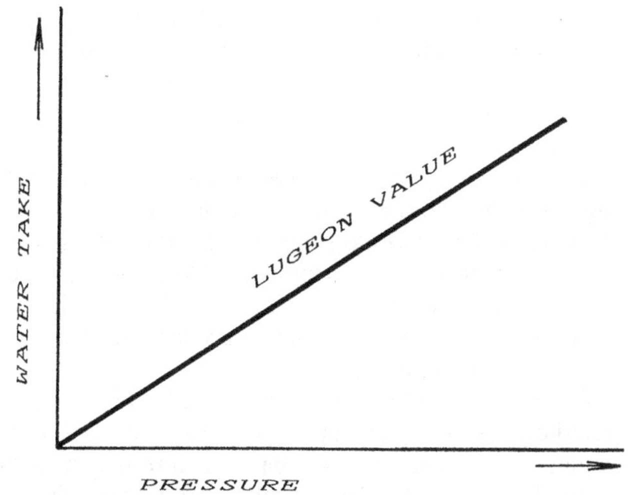

Figure 8.2-13A. Lugeon value is the slope of the line in this type of graph.

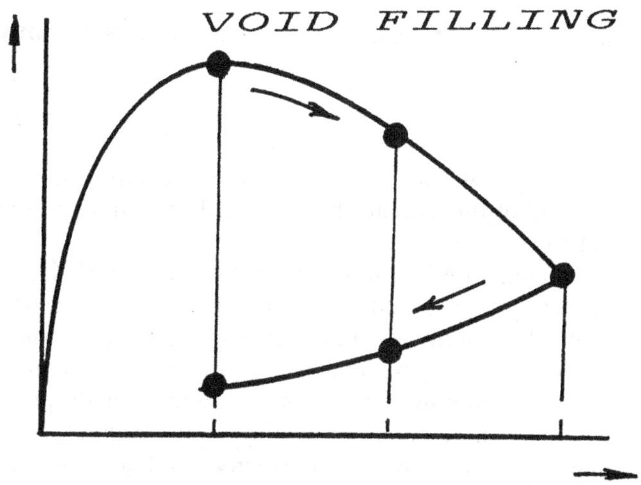

Figure 8.2-13C. Example of plot for void filling.

and low pressures in accordance with the five-pressure test of Section 8.2-7). Turbulent flow plots as a curve, flattening out at higher pressure. Dilation is a curve the other way and is shown repeating itself with falling pressures, but in reality often shows a hysterisis loop on the return. Wash-out plots upward even when pressures are reduced, but not necessarily in the curving manner shown.

All those plots are conveniently shown starting through the origin. What happens when we plot void filling? (See Figure 8.2-13C.)

The water takes plot ever downwards this time. But what form should the plot take prior to the first reading? It is shown as a freehand curve back to the origin, but there is no real basis for that.

Lugeon values are obtained by taking the slope of a tangent to a selected part of curves or the slope of a straight-line plot. Care is needed in the selection of a suitable point in the case of curves, and the answer can be sensitive to the choice. Not everyone makes the same choice or assumptions about zero corrections. Why risk the inconsistencies of graphing when it is quite superfluous?

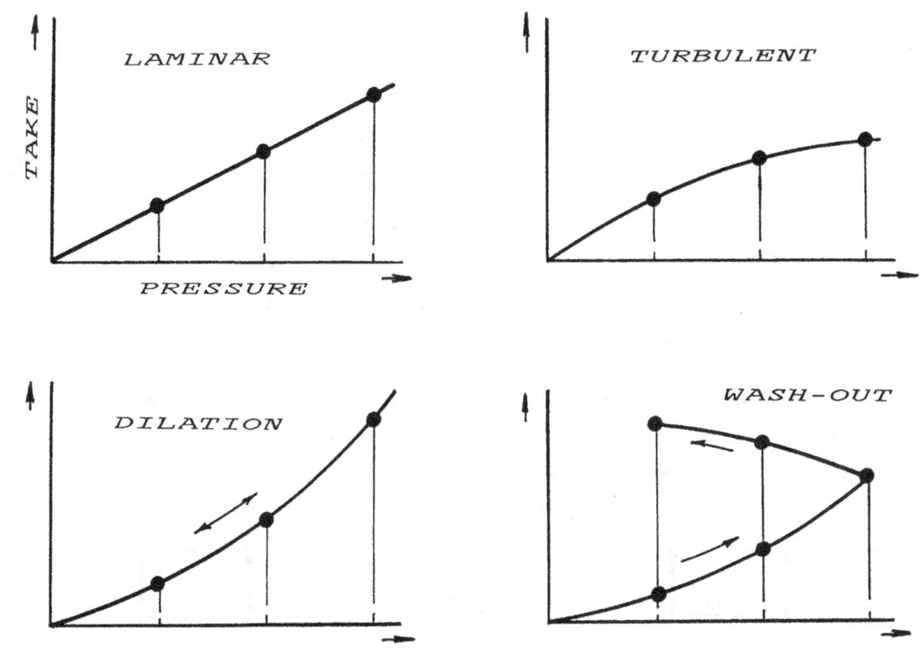

Figure 8.2-13B. Examples of plots for laminar and turbulent flows and for dilation and washout.

8.2-14 Theoretical Ideas about Water Testing

The theoretical aspects of water testing have been the subjects of a number of technical papers. It is not proposed to go deeply into the subject here—merely to mention some of the points. The listed references at the end of this chapter provide further reading on the subject.

Vaughn (1963), Snow (1968), Sabarly (1968), Hoek and Bray (1981), and Witherspoon (1986) present formulas for finding an equivalent (i.e., average) permeability of a test length in rock by relating the quantity of water taken in a water test to the test length, pressure, size of hole, and to the ratio of permeability normal to the hole over permeability parallel to it. The derived permeability is in conventional permeability terms. As indicated in Section 8.2-2, this type of unit is not very helpful in grouting work.

Sabarly (1968) discusses the applicability of lugeon tests to various grouting situations.

Louis (1969) also derived an equivalent permeability. One formula he gives is for laminar flow in smooth, clean cracks; another is for laminar flow in rough, clean fractures. Derived permeability is in terms of half width of a crack, spacing of cracks, surface and relative roughness, and some other minor factors.

Since turbulent flow is common when testing foundations most needing grouting, the applicability of these formulas appears to be very limited. Further limitations come from the need to assume the crack sizes and spacing.

During discussion at the 1963 symposium on grouting held in London, Vaughan (1963) offered some "speculative remarks on the permeability of fissured rock" in terms of a "rather simplified model." He presented a formula deriving an equivalent permeability relative to crack size and frequency, test length, pressure, and some minor factors.

Vaughn carefully stated that "he did not know whether this theoretical approach was of any practical value, but it might serve as a basis for interpreting field results" (p. 54). Perhaps he felt like me that it is not realistic to expect to be able to model the vagarities of nature.

The work of Baker (1955) is quoted and presented in the form of a nomogram by Littlejohn (1975). Baker's formula refers to saturated rock and relates the test pressure to the viscous drag of injected fluid in fine cracks. It provides for other fluids besides water, but not, of course, for cement grout, which is not a homogeneous fluid. The nomogram provided by Littlejohn simplifies use of the formula.

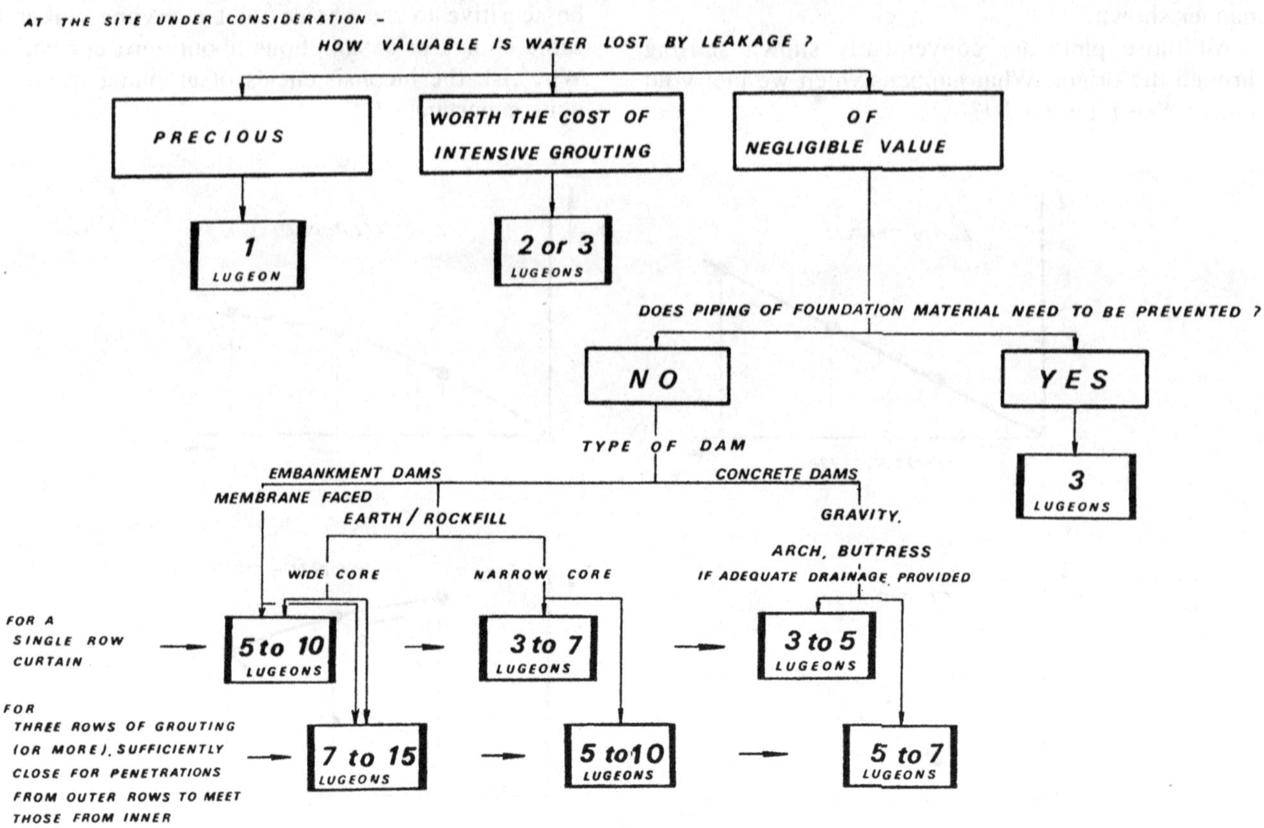

Figure 8.3-3A. Guide to the need for grouting at a dam.

8.3 WHEN IS GROUTING NECESSARY?

8.3-1 Obvious Situations

Many sites obviously need grouting so clearly that there is no question, so the subject here is more about marginal cases, where there may be some doubt, and also about standards of grouting. The application is mainly to seepage control.

8.3-2 Can Grouting Be Safely Omitted?

Yes, sometimes. But not often at dams if seepage has to be controlled. Hence in Section 8.3-3 some guidance is given about when grouting can be and when it can't be omitted.

8.3-3 General Guidelines

The chart in Figure 8.3-3A outlines the main issues. Don't forget that (like the rest of this book) it is only a guide; it is meant to direct the thinking and needs intelligent adaption to each actual application.

Now let's look at the details of this chart.

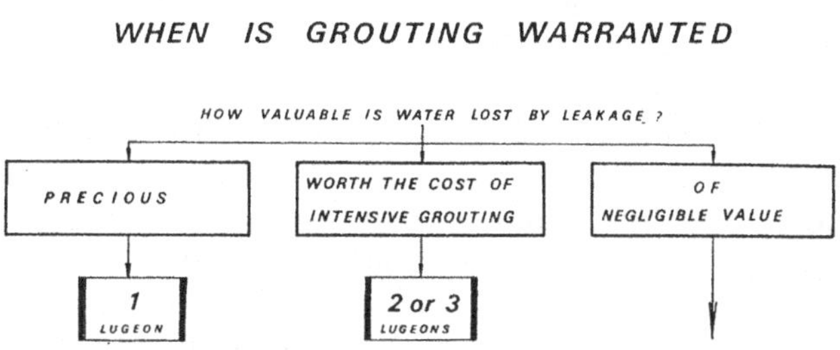

Figure 8.3-3B. *Need for grouting, related to water cost.*

The chart in Figure 8.3-3B gives guidance on the question "when is grouting warranted?" and gives it in terms of lugeons; these represent the standard of grouting to be considered for each of the situations. Any part of the foundation that has a natural permeability poorer than the standard may need to be grouted; often all of it has to be.

Starting with the issue of economics, how valuable is water lost by seepage? In some desert conditions or in other situations where water is scarce and therefore precious, almost no seepage at all can be tolerated. A tight curtain grouted to a standard of 1 lugeon will probably be necessary. This might also be the case where hazardous fluids or sanitary landfill have to be retained with virtually nil loss by seepage.

Where water will be pumped into the storage at considerable expense, such as at off-stream and pumped storages, water lost by seepage represents money lost. This cost, over the working life of the project, should be reckoned against the cost of grouting. If thorough grouting is opted for, it will probably need to be to a standard of about 2 or 3 lugeons to keep seepage down to trivial amounts (sometimes even tighter grouting may be needed to achieve this).

However, water lost by seepage usually has negligible value, in which case the chart leads on to the next question.

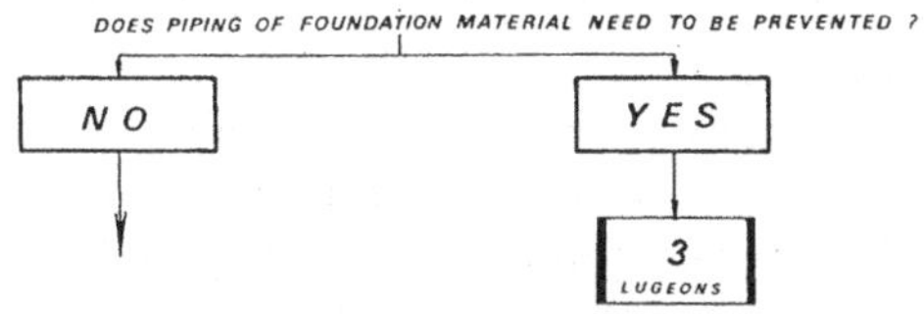

Figure 8.3-3C. *Need for grouting, related to piping.*

Figure 8.3-3C refers to possible piping of soft and weak material from seams as discussed in Section 8.1-1. If this is likely to happen because of the pressure in seepage, precautions usually are necessary to minimize it. A standard of grouting of 3 lugeons is suggested. Multiple rows in the curtain may also be advisable as explained in Section 8.4-8.

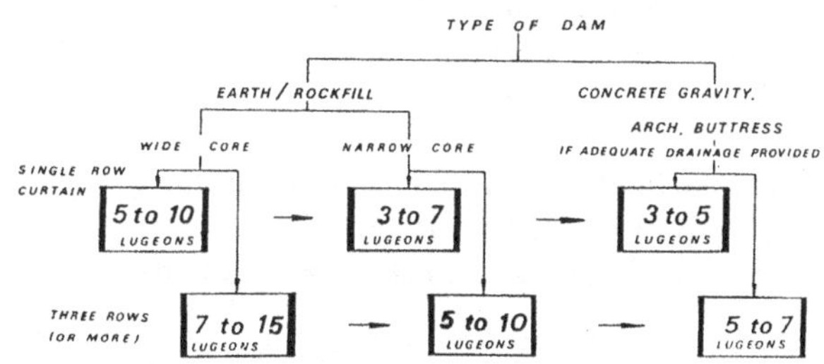

Figure 8.3-3D. *Need for grouting, related to dam features.*

If none of the previous issues applies consideration turns to the type of dam (Figure 8.3-3D). Embankment dams are on the left in the chart and concrete dams on the right. The embankment dams include (for these purposes) zoned earth/rockfill, homogeneous fills, and membrane-faced rockfill. The heading for wide core zoned fills includes these last two types of dams and suggests that if the curtain has a single row (the commonest arrangement—refer to Section 8.4-8), the standard could range between 5 and 10 lugeons. The actual figure selected within this range depends on local factors. The range may be relaxed to 7 to 15 lugeons if the curtain has three or more rows if these rows are carefully arranged so that penetrations overlap and form a thick region of grouting as shown in Section 8.4-8.

Under relatively narrow cores in zoned embankments, hydraulic gradients in seepage can be steep. Tighter standards of grouting are therefore called for, such as 3 to 7 lugeons for a single-row curtain and 5 to 10 for a good multiple-row curtain as described earlier. Gradients can be even steeper in the case of membrane-faced dams, but these do not present the threat from loss of core material into foundation cracks, and hence standards can be relaxed for them a little, as indicated earlier.

Concrete dams of all types are grouped together in the chart; this is a simplification that should be borne in mind. Adequate foundation drainage is necessary (refer to Section 8.4-9 for the relationship between grouting and drainage), otherwise tighter standards are advisable. Those shown for drained conditions are 3 to 5 lugeons for a single-row curtain and 5 to 7 for good multiple-row curtains.

The flowchart in Figure 8.3-3A shows all of these separate parts combined. It primarily applies to surface regions; at lower depths the grouting does not need to be as tight usually. It can be advisable, in important cases, to have widely spaced probe grout holes to check foundations even if the chart indicates no need for grouting.

Having set a standard for the grouting, decisions about when to cease closure (see Sections 5.5-3, 5.5-4, and 5.5-6) can be based on it.

For the background to the standards suggested in the chart, and for a discussion on them, see Section 8.3-6.

Note particularly that the figures suggested relate to curtain grouting done thoroughly, with extra grouting at deficiencies and weaknesses, as explained at Section 7.4.

8.3-4 Other Factors

When using the chart in Figure 8.3-3A, other factors must be kept in mind such as particular features and peculiarities of the site in relation, for instance, to cracking frequencies and sizes. A thorough geological (Section 8.1) and permeability (Section 8.2) investigation helps greatly when reviewing these things.

On some sites the investigations find that permeabilities are highly variable; possibly the geology may also be. Some generalizations are then necessary and decisions about the grouting must reflect this. Section 8.4-14 illustrates such situations.

The pragmatic approach to all this emphasizes that finesse is futile; common sense is paramount when deciding what to do.

8.3-5 Site Grouting Tests

To resolve doubts, full-size tests are done on some sites to find out whether grouting is beneficial. Such tests can also evaluate other features: finding the best drilling method (Section 5.2-2), checking proposed w:c ratios (Section 7.2-2), trying various injection techniques (Section 7.2-7), finding safe pressures (Section 6.2-1), and comparing equipment (Chapter 2). Sometimes tests have the prime purpose of such evaluations rather than of checking a site for grouting suitability.

The general idea is to investigate the properties of the test area before grouting it, and then later to find out the changes in these properties produced by the grouting. Changes are evaluated in various ways and to differing degrees of sophistication. The simplest, and possibly the most pertinent, is water testing: the five-pressure type of water test is necessary (Section 8.2-7) because the constant-pressure type (Section 5.3-2) is not good enough for this. To this can be added core drilling and inspection of the holes with borehole TV or a periscope (Section 8.1-6, 7). However, cores can be misleading because they frequently do not show grout in thin seams because it gets ground away and lost during drilling, even with triple-tube barrels (refer to Section 8.1-5).

In addition to giving technical information, site tests can also help with the obtaining of lower bid rates: this is because the test data can reduce the amount of guesswork by tenderers when preparing their bids. Less padding is therefore necessary to cover unforeseen contingencies.

Further financial benefit can come from arranging the testing suitably to become part of the final job, if this is possible, remembering of course to regrout any parts of the testing found to be unsatisfactory.

Site testing has to be extensive to be of any use. Just drilling and grouting a few holes is rarely good enough; sufficient holes have to be done to give a reliable sampling. This typically calls for at least a couple of dozen holes. And these have to be carefully located to be representative. When alternative methods are being compared, the alternatives have to be tested in sufficient quantities to overcome random effects.

A caution is necessary here: do not judge the testing solely on the basis of the amount of cement injected. This can be misleading because it is affected by side issues such as the suitability of the w:c ratios, the pressure, the mixing method, and the injection techniques. Since most of these will have been varied in the different phases of the testing, the cement takes are sure to have been affected—possibly unpredictably. It is better to judge the testing on the before and after evaluations mentioned earlier.

The layout of holes in the testing should simulate intended layouts for the actual job in general. Exceptions to this are where the grouting is for strengthening purposes such as consolidation or blanket grouting (refer to Section 8.4-16). A fairly closely grouped set of test holes may be called for to enable a small area to be treated and possibly subjected to a load test.

Opinions differ about the wisdom of conducting site tests. Some organizations do them as a matter of routine before designing the grouting for jobs of any substantial size, believing the money to be well spent for enabling construction of sound, economical grouting. Others rarely do testing, feeling that the same information can be obtained more reliably and with less hassle by making use of the services of a competent, widely experienced grouting specialist. This latter attitude goes well with organizations that have trouble getting funding in advance of the main grouting.

Even if extensive site testing is indeed done, a questioning attitude is advisable throughout the subsequent grouting. As emphasized in Section 7.5, continuing critical reviewing of every aspect of grouting is needed to produce good work. Site testing is only the start.

8.3-6 Background to the Guidelines

Derivation of the grouting standards suggested in Figure 8.3-3A commenced with the work of Lugeon (1933). He worked on the principle that grouting is necessary where the permeability exceeds 1 lugeon for dams over 100 ft ⟨30 m⟩ high and 3 lugeons for those of lesser height. He was a consultant dealing with concrete dams in the Alps and his criteria apparently reflect conditions there.

People following his advice found that these figures could be relaxed a little at other types of foundations with different types of dams and with better mixing equipment. For instance, Fergusson and Lancaster-Jones (1964) reported use of 2 lugeons for the Dokan Dam in 1955 to 1960. This is a concrete arch dam in Iraq; its height is 385 ft ⟨117 m⟩.

At the W.A.C. Bennett Dam, a 600-ft ⟨183-m⟩ earth/rockfill dam in Canada, 3 lugeons was used for the outer rows and 1 lugeon in the center row of the curtain (Pratt, McMordie, and Dundas, 1972).

The subsequent easing of criteria shown largely reflects my experience taken together with assessments of current published practice. An earlier, and slightly different version of the chart shown in Houlsby (1977) and elsewhere has been widely discussed, used, and reviewed; as a result the slight changes incorporated in the version shown here were proposed in Houlsby (1985). They comprise an easing of the upper limits for embankment dams.

Because of the fortunate circumstance that my work has actively involved the grouting of quite an extensive range of dams that subsequently came under my care for safety surveillance and monitoring for many years afterwards, it has been possible to determine which grouting standards proved adequate in the long term and which did not. Likewise, discrimination between satisfactory and unsatisfactory methods in grouting processes has been possible as described in other parts of this book.

The former practice of some organizations has been to specify grouting requirements in terms of cement take rather than permeability. This is not recommended because cement takes reflect side issues of grouting technology, and also for the fairly obvious lack of direct applicability to permeability considerations. Discussion of this is found in Nonveiller (1970), Arhippainen (1970), Acker and Jones (1972), and others.

8.4 DESIGN OF GROUTING

8.4-1 Introduction

The principal subject here is the layout of grout holes. Mention is also made of aspects of grouting not dealt with elsewhere. Reference should also be made to these related sections:

Deciding if grouting is necessary	Section 8.3
Selecting the standard for grouting	Section 8.3-3
Stage grouting and methods	Section 5.4
Durability and w : c ratios	Section 4.5
Closure grouting and planning	Section 5.5
Grout caps	Sections 5.1-10 to 5.1-16
Surface preparation	Sections 5.1-17 and 5.1-18
Effect of crack sizes	Section 6.1
Investigation—permeability	Section 8.2
Investigation—geology	Section 8.1
Equipment	Section 2.0
How to grout a hole	Section 7.2
Assessment	Sections 7.4, 7.5

This section deals with hole orientation first, in relation to all types of grouting; then turns to layouts for the main types of grouting and situations.

The treatment is an outline of the subject of design. Like most other things in grouting, design requires astute adaption to site conditions and so treatment of the many ramifications possible would extend this book unduly.

8.4-2 Should Grout Holes Be Inclined?

The aim is to cut as many cracks as possible. The hole shown in Figure 8.4-2A is slightly inclined in order to

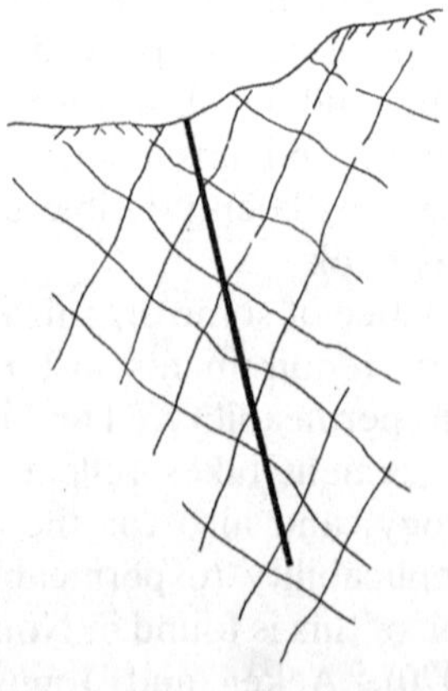

Figure 8.4-2A. Inclined grout hole intercepting jointing pattern.

give optimum crack interception. If, however, jointing is mainly vertical and horizontal (and is not intensive—see later), holes need to be inclined rather more than in Figure 8.4-2A (see Figure 8.4-2B).

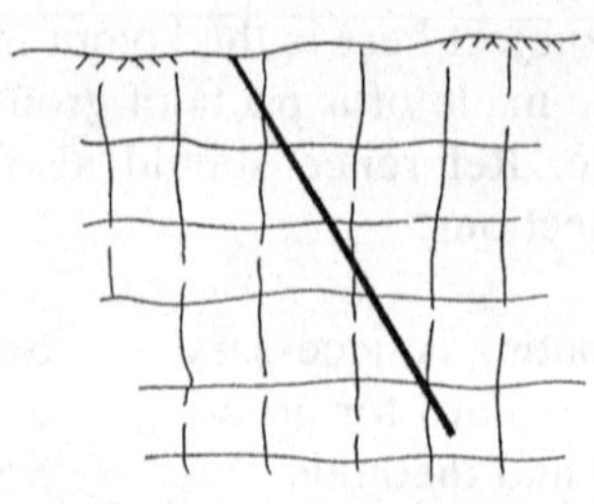

Figure 8.4-2B. Inclined hole for horizontal and vertical jointing.

If for some reason vertical holes are used in this type of jointing, they could miss many vertical cracks and so might not be as efficient as inclined ones as Figure 8.4-2C shows.

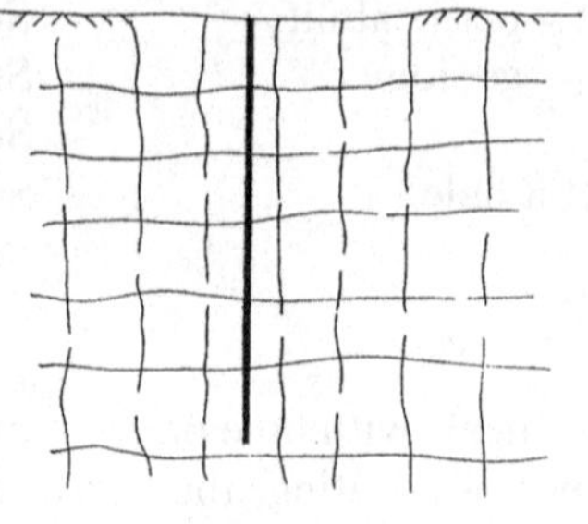

Figure 8.4-2C. Unsuitable use of vertical holes.

Jointing suited to vertical holes is shown in Figure 8.4-2D.

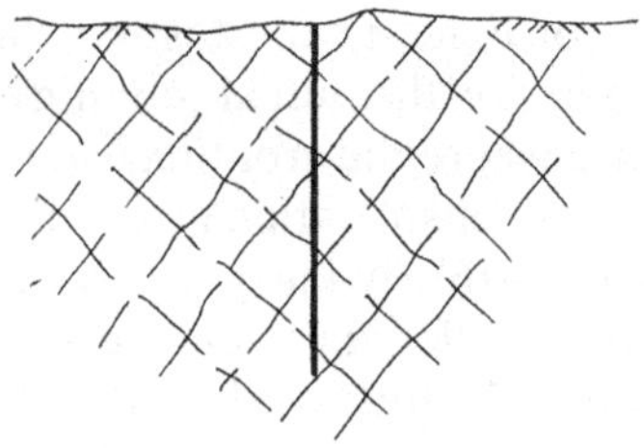

Figure 8.4-2D. Suitable use of vertical holes.

If some sets of joints are more important than others in, say, their potential for passing seepage, then hole inclinations should be biassed to give them priority (see Figure 8.4-2E).

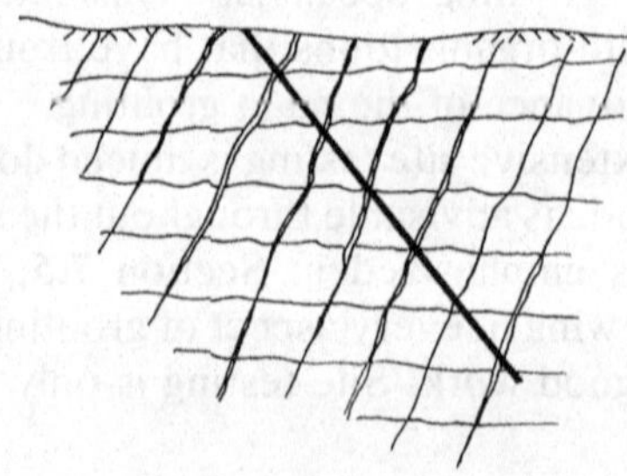

Figure 8.4-2E. Biassed inclination.

The near-vertical joints in this example are wider and more significant than the near-horizontal ones; holes are therefore inclined more than would otherwise be the case.

Vertical holes are the most convenient and should be used unless the jointing fairly evidently warrants resorting to inclined ones.

When cracks are closely spaced and grout can flow easily through the jointing system, vertical holes can often be used, even if the dip of some joints is also vertical or nearly so (see Figure 8.4-2F).

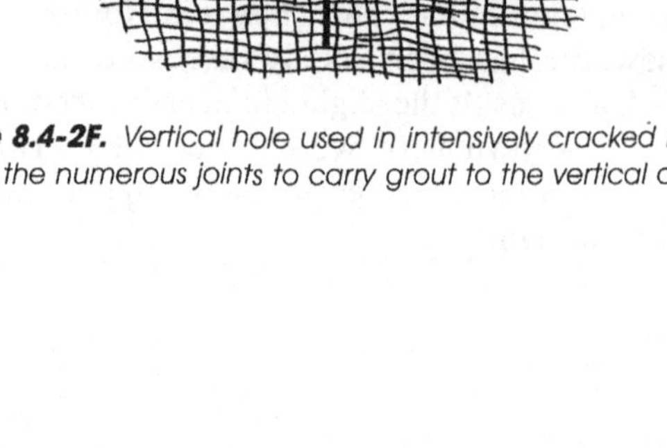

Figure 8.4-2F. Vertical hole used in intensively cracked rock; relies on the numerous joints to carry grout to the vertical ones.

The angle of inclination for sloping holes should be rounded off to the nearest 5° or 10°; it is pedantic to be more precise except in fans (Section 8.4-11). Angles are usually given from the vertical in grouting work. Note that this is different to geological practice for measuring dips (from the horizontal). Setting out of holes is easier when measured from the vertical because pendulum devices or plumb bobs are usually used for this (refer to Section 5.2-10). The angle most suited to vertical and horizontal jointing is 30° from the vertical (see Figure 8.4-2G).

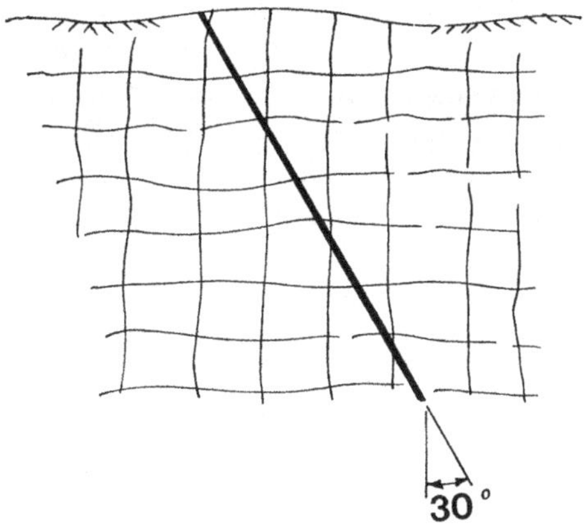

Figure 8.4-2G. *Most suitable inclination for vertical and horizontal jointing.*

Angles chosen must be practical; drills must be able to work on them. Thus 45° is about the maximum angle when working off a horizontal surface, but flatter-lying holes may be usable from steep surfaces.

In tunnels and at steep faces varying inclinations can be needed, as shown later, and at faults, unconformities, and so forth, special holes should be tailored to suit, also illustrated later.

8.4-3 Designing Hole Inclinations

The experienced grouting designer is usually able to decide the hole angles by looking the site over, if jointing is visible in outcrops or in stripped locations. This may be augmented by referring to geological information, particularly if significant areas are not represented in the visible parts, and also if the nature and directions of jointing changes below the surface.

If an experienced person is not available, the following guidelines should be helpful (some experienced people like to use them too).

The first step is to examine geological information (Section 8.1 suggests how to obtain this). It should indicate the nature and inclinations of jointing.

There are several ways of assessing the jointing information when deciding the best inclinations for grout holes. The one explained here is for the "hemispherical equal-area projection" method of representing jointing (see Section 8.1-12 for examples of jointing done by this method). It is relatively easy to work with.

Figure 8.4-3A shows a vertical section through a southern hemisphere along with a radius on it. The radius is, of course, normal to the tangent at its intersection with the curved surface. By definition (see Section 8.1-12) a tangent represents the dip of jointing. It is, therefore, convenient to represent a grout hole by a radius; because of the convention already mentioned whereby grout hole inclinations are best specified to the vertical, the surface graduation on the curve where the radius meets it is the inclination of the hole as well as being the jointing dip.

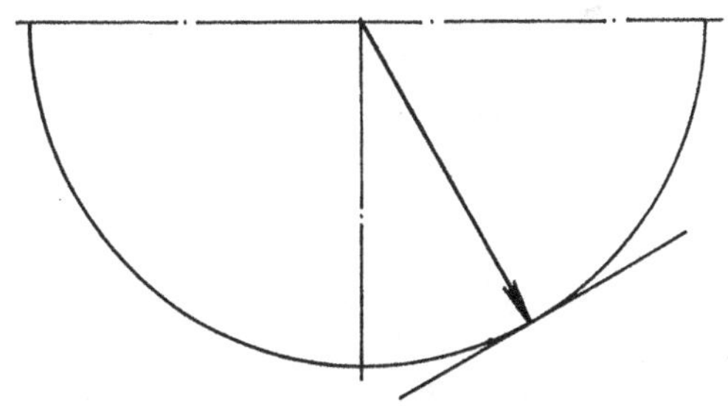

Figure 8.4-3A. *Radius normal to tangent representing dip.*

This representation can be used to visualize the range of dips amenable to satisfactory interception by a grout hole (see Figure 8.4-3B).

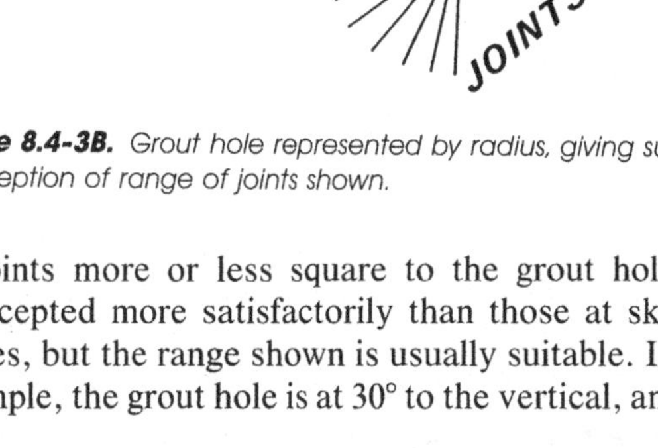

Figure 8.4-3B. *Grout hole represented by radius, giving suitable interception of range of joints shown.*

Joints more or less square to the grout hole are intercepted more satisfactorily than those at skewed angles, but the range shown is usually suitable. In this example, the grout hole is at 30° to the vertical, and the range of satisfactory joint inclinations is from vertical to horizontal in the right-hand sector and then around to a dip of about 30° in the left-hand sector, much as shown earlier. This amounts to about 60° each side of the hole inclination, and it is possible to use this figure of 60° without further refinement in rough design; it requires modification though, in careful design. Before proceeding to show why, the sketch in Figure 8.4-3C indicates this same range for the grout hole under discussion.

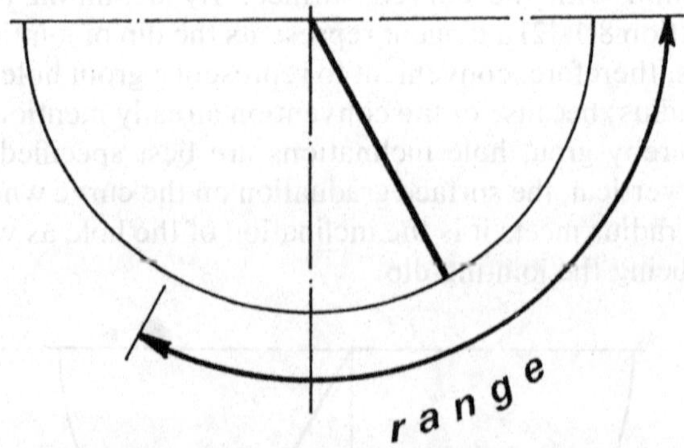

Figure 8.4-3C. *Range of joints suitably intercepted.*

It is fairly easy to visualize jointing directions and suitable hole inclinations in this manner. However, things can become complicated when strikes (the geological sort) are taken into account, as indeed they must be. It is necessary to consider the situation in plan view then and to superimpose the orientation of the dam, or structure, or whatever.

Figure 8.4-3D shows a simplified case using the same hole as in the last few sketches. This is a plan view of the equal-area net with jointing dip and strike population intensities indicated by contours as explained in Section 8.1-12. The blacked-in area has the greatest intensity, the horizontally hatched area has less, and the spotted area has even fewer. The white area has no joints in this example. For simplicity, joints are shown symmetrically disposed about the east–west axis, which in section looks like the previous few sketches, and the grout hole is the same, this time viewed from above.

The grout hole goes directly into the black area, the part of the net where the most joints are and is therefore ideally suited to intersecting these joints. It is not quite as ideally placed for joints in the hatched and spotted areas, but is probably good enough. The spotted area extends out to the same range limits as the sections shown before. Note what a simple pictorial representation is obtained with this method of presentation and design. It needs to be—now come the complications.

If, unlike the previous example, the joints do not lie square across the direction of the grouting, the complications start. In Figure 8.4-3E the alignment of the grout curtain does not run squarely through the main jointing. There are two possible alternative treatments for the holes:

1. Keep the holes in the same plane as the curtain, as shown by the heavy continuous line. Main jointing will be cut on the skew. (Some mental visualizing of the situation may be necessary to appreciate that this is so. Refer to Section 8.1-12 for basic principles if in doubt.) However, the range of suitable intersections is less than with square intersection, and cracks toward the northern end of the net will have poor interception or none at all.

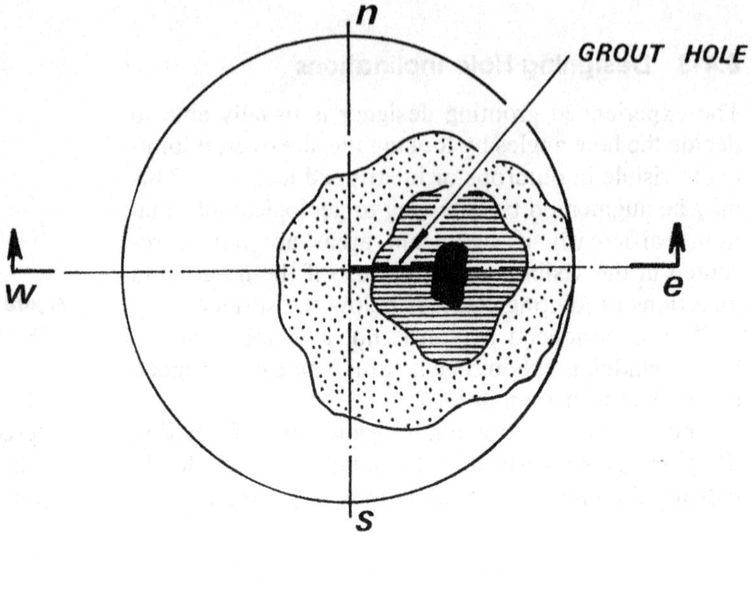

Figure 8.4-3D. *Intercepted joints shown in plan view.*

2. Use doubly inclined holes pointing outwards from the plane of the curtain, as shown by the heavy broken line. This goes squarely into the main jointing and will probably produce the best grouting. Note, however, that if the spotted area had extended further to the top left or the lower right, intersection there would be becoming poor or even bad.

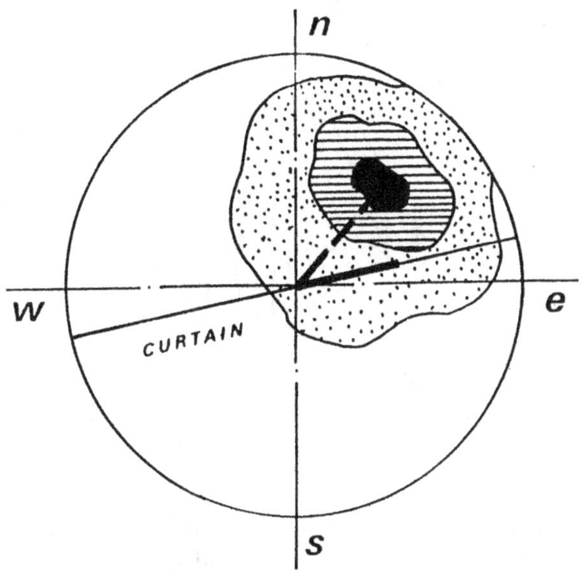

Figure 8.4-3E. *Joints striking differently to curtain.*

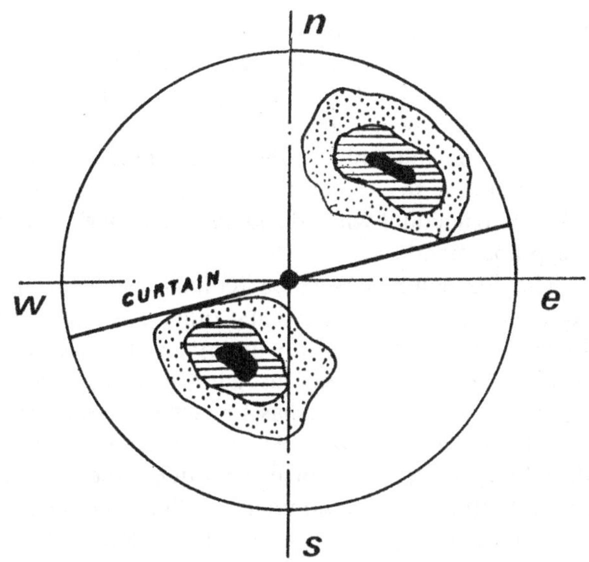

Figure 8.4-3F. *Complementary jointing patterns striking unsuitably for curtain.*

The choice of these alternatives would be influenced by the position of any other jointing. So far only one group of joints has been considered, but in real situations there are often two or three groups, at right angles to each other more or less. Consider first the example with two groups shown in Figure 8.4-3F.

The curtain does not happen to be oriented suitably for either group, and in any case use of holes inclined to suit one group only (as in the previous example, for instance) would be nearly parallel to the other group and so would be unsatisfactory. In this particular case a reasonable compromise would be to use vertical holes. These could give acceptable interception of both groups and are shown as a dot at the center of the net. As mentioned earlier, vertical holes are usually more convenient than sloping ones anyway, so their use in this situation would be very convenient. However, the situation may be different if there is a third set of joints, which is the usual case (see Figure 8.4-3G).

In this example, this third set is nearly vertical and has the common situation whereby its strike is about normal to the strikes of the other sets. Notice how plots at the edge of the net, such as this set, overlap to

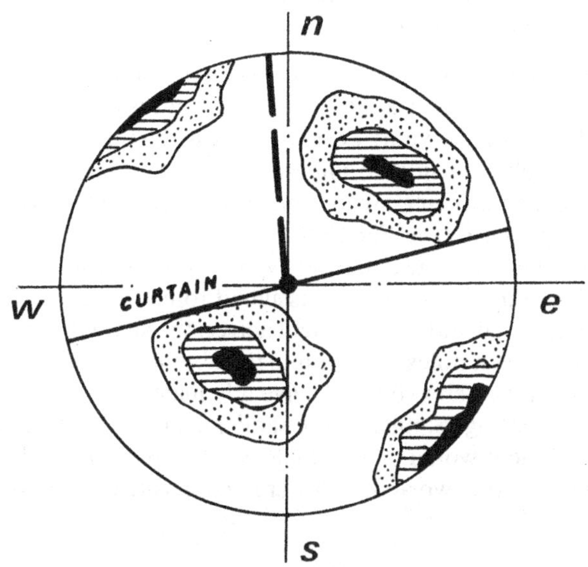

Figure 8.4-3G. *Three sets of joints.*

the diametrically opposite side somewhat. This is because the edge of the net denotes vertical joints, and in a population of near-vertical joints there are usually dips on both sides of vertical.

The vertical holes found satisfactory in the previous example are no longer ideal because they are unlikely to cut this third set at all well. A compromise would be to incline the holes a little toward the third set, as shown by the thick dashed line, but this means holes are out of the plane of the curtain and are doubly inclined; disadvantages of this are discussed shortly. Can vertical holes still be used even though they neglect the third set? Probably they can in this case: the third set of joints is nearly parallel to the curtain so that their potential to carry seepage through the curtain is not great and a lower quality of grouting may therefore be tolerable in them. However, the width and nature of these joints and their extent should be taken into account when deciding, particularly in high, steep abutments.

The foregoing may seem tedious, particularly to someone studying it for the first time; there have been attempts to do it all by computer instead. However, because hole layout is not just a mechanical process of analyzing joint sets, but has to include other factors specific to each site, as well as practical issues such as drillability and ease of correct setout of holes, it is not possible to program a computer with all the nuances. However, computers can be useful for some other aspects of grouting, as the discussion in Section 7.7 indicates.

8.4-4 Using Hole Inclinations in a Curtain Layout

Having decided on suitable hole inclinations to suit the various parts of a site, it is then necessary to merge them into an overall design. In doing this some of these carefully considered angles may have to be varied in order to produce a workable design. Let's start with a simple case (see Figure 8.4-4A).

This is a typical dam site valley. For the purposes of the example, jointing is assumed to be approximately horizontal and vertical right across the site. On the left side, grout holes sloping into the abutment at about the inclination shown would give adequate interception of the jointing. Similarly, the same inclination in the right abutment would be satisfactory. However, the inclinations in the two abutments are in opposite directions to each other; so what should be done where they meet in the floor of the valley? There are two options:

1. Transition smoothly from one direction to the other as shown in Figure 8.4-4B. Notice how poorly the vertical joints are intercepted. And these are quite liable to carry seepage through the line of the curtain if not properly grouted.

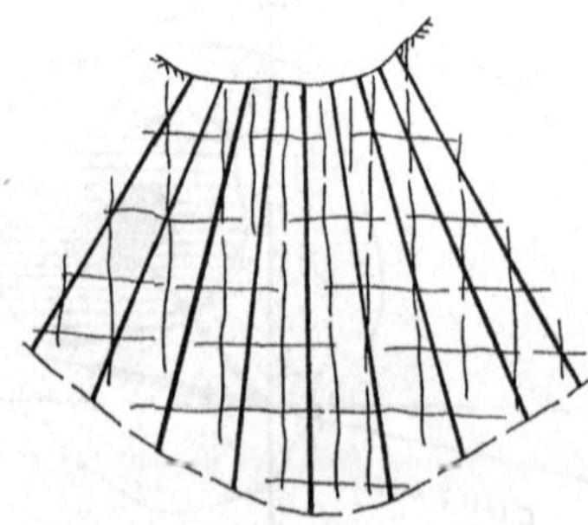

Figure 8.4-4B. Smooth transition from one inclination to the other.

2. Crossover from one direction to the other while maintaining the best inclinations throughout (see Figure 8.4-4C). Vertical cracks are adequately cut in this method, and although it appears to have greater length of holes, it is the recommended method. In order to check out both alternatives thoroughly, I grouted several comparable dams using each method. Even though the transition method started with less holes than the crossover method, so many additional holes were needed to bring the standard of grouting in the vertical cracks down to an acceptable figure that the final number of holes exceeded those in the crossover method in all cases, and far more time was lost in fiddling around in confined working areas.

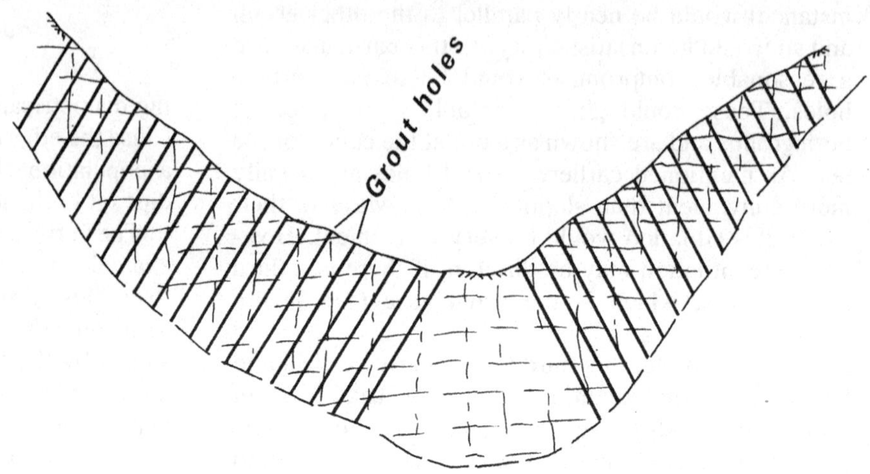

Figure 8.4-4A. Commencement of a design for typical valley.

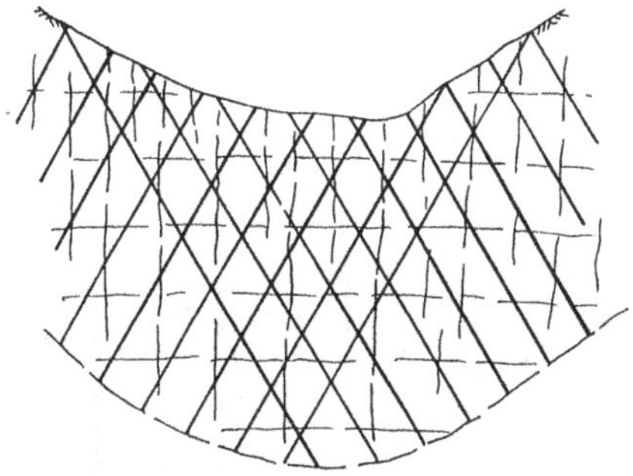

Figure 8.4-4C. A crossover from one inclination to the other.

Another important factor favoring use of the crossover method is that it gives doubled-up treatment of geological weaknesses in the valley floor.

Nearly every valley has some weakness in its floor—that is why the valley is there—water has followed the weakness to erode the valley. This may take the form of a fault, shear zone, anticline, graben, or some other feature and presents a specific feature warranting special attention. This is best provided by the crossover layout.

Of course, if unlike the previous example, the jointing is amenable to vertical holes, things are simplified in the valley floor. The vertical holes simply continue across—unless weaknesses there specifically require locally inclined holes.

Curtains do not have to be symmetrical. If the jointing in one of the abutments is quite different to the other, hole inclinations should also differ. See, for example, Figure 8.4-4D.

The left abutment requires fairly well angled holes, whereas vertical holes will suffice in the right. A fault separates the two and is crossed by holes from both directions in the crossover. The extreme right-hand hole of the crossover is in an awkward location for drilling; this sometimes happens on real sites but might be avoided by relocating the hole and changing its angle to still provide adequate intersection in the area it deals with, which is down near the base of the curtain. Dips on the left abutment gradually vary, but it is unwise to gradually vary hole inclinations to match (complications and inadvertent errors in hole layout can be introduced if they are). If the curtain happened to be longer than the one shown, a change to another angle for part of it would be warranted.

In this example the curtain has not been deepened in the vicinity of the fault. If water testing had found permeabilities there in excess of the adopted standard for the curtain, then deepening would indeed be needed.

The cases so far have been in valleys where sloping abutments have dictated the direction in those having sloping holes. However, alternative directions are possible if the surface is fairly level.

The horizontal and vertical joints shown in Figure 8.4-4E can be intercepted equally well by the inclined

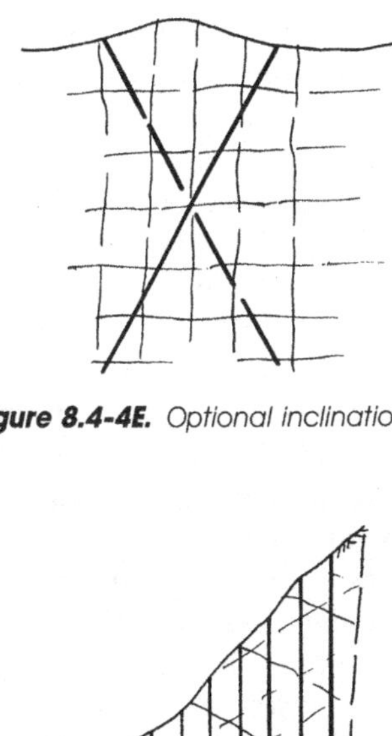

Figure 8.4-4E. Optional inclinations.

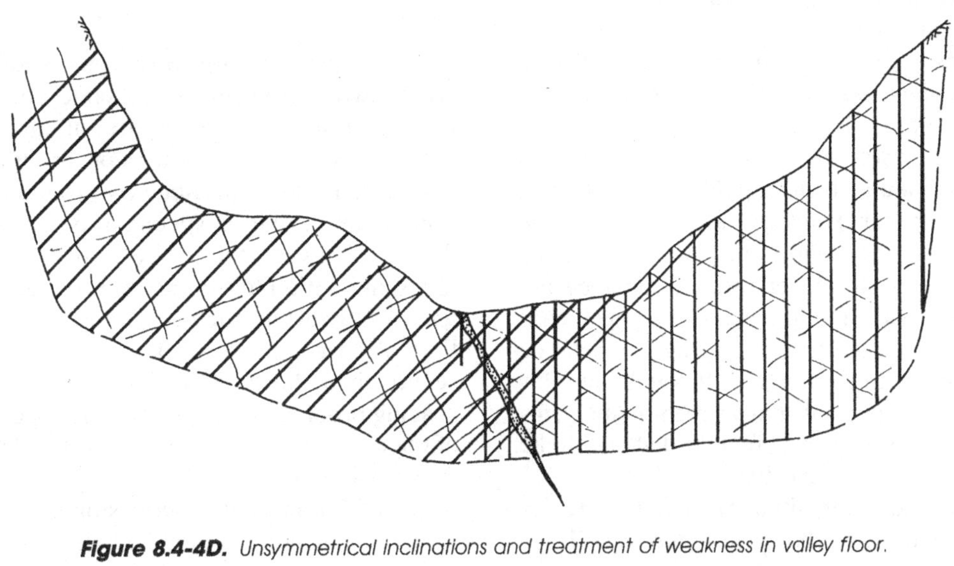

Figure 8.4-4D. Unsymmetrical inclinations and treatment of weakness in valley floor.

hole shown in full line or by the hole at the same angle but in opposite direction shown dashed. This concept is used in some multiple-row curtains where one row is at one of the angles and the next is at the other angle. If there are more than two rows, the alternation arrangement continues in each.

8.4-5 Curved Configurations and Kinky Curtains

Simplicity in layout is desirable: the chances of inadvertently leaving gaps in curtains and errors in hole setout are then minimized. There are also less headaches when doing the assessing (as per Sections 7.4 and 7.5).

Nevertheless, there are times when simplicity is not possible. Designing then requires extra care. When heavily curved configurations are unavoidable or where the curtain has to have kinks or warps or has to meet up with curtains from somewhere else (a tunnel maybe) or where structures have to be avoided or carefully grouted under, complications can abound. Some people enjoy overcoming such challenges.

In such cases the designer has to have a well-developed sense of three-dimensional geometry. A useful method for checking intersections of curtains, plotting warps and relationships to structures, or designing complex hole layouts to relate them to topography and geology is the use of "contours on the curtain." Contours of levels at regular intervals of height are drawn on elevations, plan views, and sections of the curtain(s). They then provide a clear picture of curtain planes and positions and can be changed and adjusted as necessary to achieve the desired design aims. Some computer-aided design (CAD) systems may assist with this (see Section 7.7-12).

Even with relatively simple curtains, there are catches for unwary designers. One of the most frequently met ones is the treatment of curtains where a kink or a curve coincide with a sharp change of surface level, such as in Figure 8.4-5A.

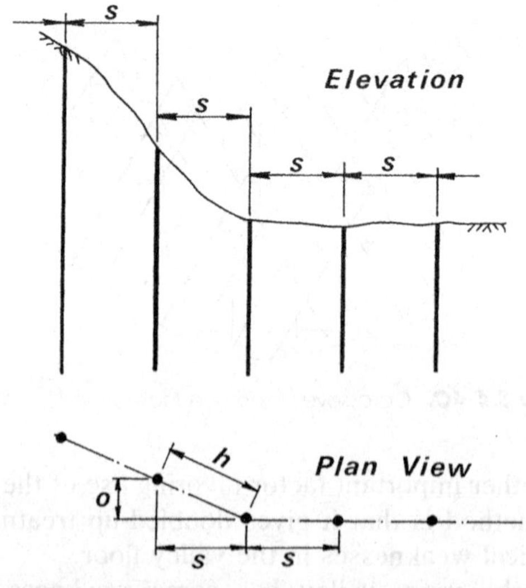

Figure 8.4-5A. *A trap for unwary designers.*

The elevation shows vertical holes that have been unthinkingly specified as being equally spaced s apart in the plane of the paper. However, a kink in plan view moves some holes into a different plane so that the desired spacing, s, is no longer the real distance between holes. Instead, the offset, o, is significant enough in the example to make h, the actual distance between them, more than intended. In some jobs this might not be very important, but in high-quality curtains the hole spacing is a design feature that should be adhered to. It is discussed shortly.

Consider now the same situation if holes are inclined in one direction, as shown in Figure 8.4-5B.

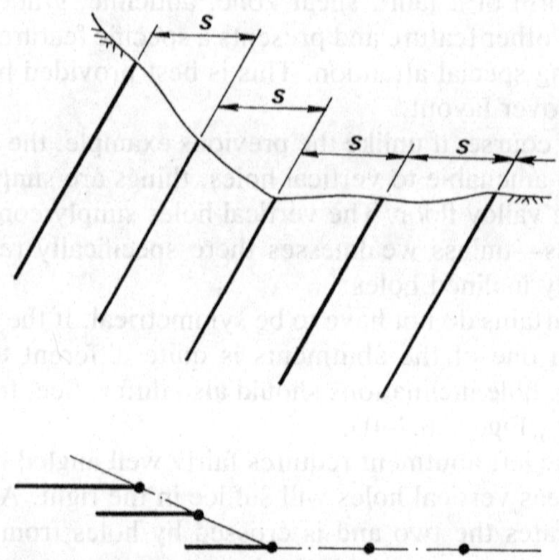

Figure 8.4-5B. *Inclined holes in the same sort of trap.*

The same elevation, plan view, and hole spacing s are shown and produce the same effect. It can also happen at heavily curved parts of curtains and where the line goes over high knobs or through deep pits in a foundation. This is particularly pronounced if the position of the curtain is specified as a proportion of the distance between the axis of a dam and its upstream toe or in relation to a zone (refer to Section 8.4-9). The line of the curtain then responds to changes in elevation; if these are sudden, kinks or curves can be more severe than those shown here. The remedy for this spacing problem is to specify hole spacing in the plane of the curtain, no matter where it wanders, rather than in the plane of the paper.

An additional catch with sloping holes can come from the manner in which the hole spacing is specified.

In Figure 8.4-5B, spacing s is measured horizontally so that on slopes there is a sort of stepped effect. This is the recommended method. The alternative is to mark out the spaces at equal intervals on the foundation surface without regard to slope, but this produces uneven distances between holes when going up and down slopes. In complex layouts, if the designer does not have the final levels of surfaces, it can be best to show hole spacing normal to the holes rather than horizontally.

Layout of doubly inclined holes can get into complications if not thought out carefully. For a curtain with regular hole spacing, it is usually best to specify this spacing horizontally, and then to give the resultant of the two angles of inclination (rather than each individually) with a bearing for azimuth. This azimuth should be kept at the same figure even when the curtain makes small changes in direction.

It is rarely that a designer has actual final surface levels to work with at the design stage. Some estimation is necessary from exploratory data. In order to cover uncertainties about this, the dimensioning should be sufficiently flexible that a reasonable range of levels can be accommodated. If holes are vertical, this is easy; but if they are inclined, the final positions can vary depending on foundation level. This may be unimportant in some sites, but in others, where curtains have to meet or where structures have to be precisely avoided, dimensioning has to be suitably specific. Sometimes holes are specified where they pass through a particular elevation, even if this lies on a projection of the hole above the foundation.

In regularly spaced curtains, common (and quite satisfactory) practice is to specify the horizontal distance between primaries (measured along any bends in the curtain) and then to divide up this space evenly for the subsequent holes (see Sections 5.5 and 8.4-6 in reference to primary, secondary, and other types of holes).

8.4-6 Staggering Sequences

For curtains with several stages, the recommended layout is shown in Figure 8.4-6A. This type of layout usually combines economy with quality work. It avoids doing needless grouting, while at the same time it enables extra grouting to be given to bad spots found while the work is in progress. The foundation is, in effect, probed and checked more thoroughly than scattered investigation holes can provide (Section 8.1-5).

By showing a layout such as this on working designs, and providing suitable clauses in the specification permitting considerable variations to the design, any weak spots or unexpected features can be readily given additional attention as found warranted. The concept is as follows:

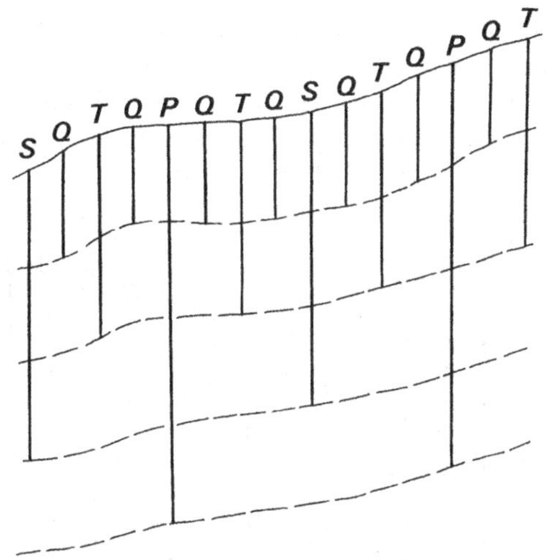

Figure 8.4-6A. *Staggered stage lengths for the various sequences.*

1. Primary holes (P) are taken down to the base of the curtain.

2. Secondary holes (S) go down to the second bottom stage (or not as far if the rock tends to be tight) but are extended to the bottom stage on both sides of any primary that shows relatively high water and/or cement take (refer to Section 7.4-2 for magnitudes of such takes in relation to standard of grouting). For example, see Figure 8.4-6B.

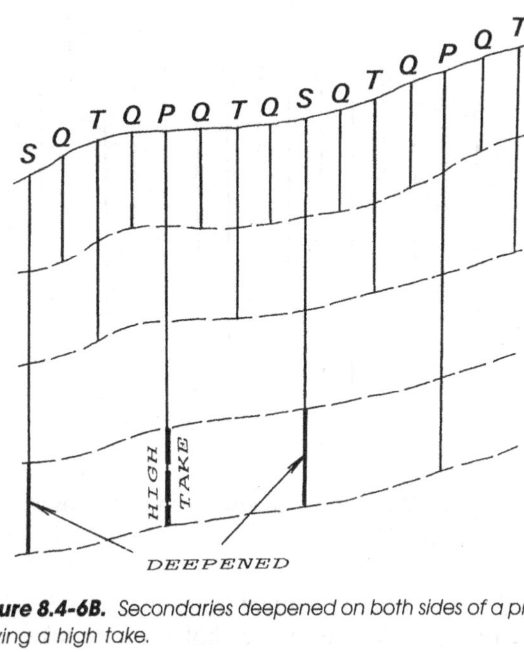

Figure 8.4-6B. *Secondaries deepened on both sides of a primary having a high take.*

3. Tertiary holes (T) go down to the third bottom stage but are extended below this on both sides of any secondary that has shown relatively high takes or is too far above the target standard (as described in Section 8.3-3) as shown in Figure 8.4-6C.

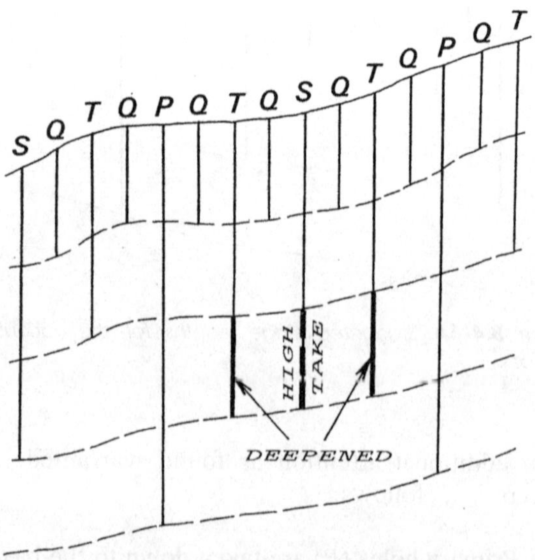

Figure 8.4-6C. *Tertiaries deepened on both sides of a secondary having a high take.*

4. Quaternary holes (Q) go down for only one stage or so but are extended down on both sides of any tertiary that has shown relatively high takes or is too far above the target standard (see Figure 8.4-6D).

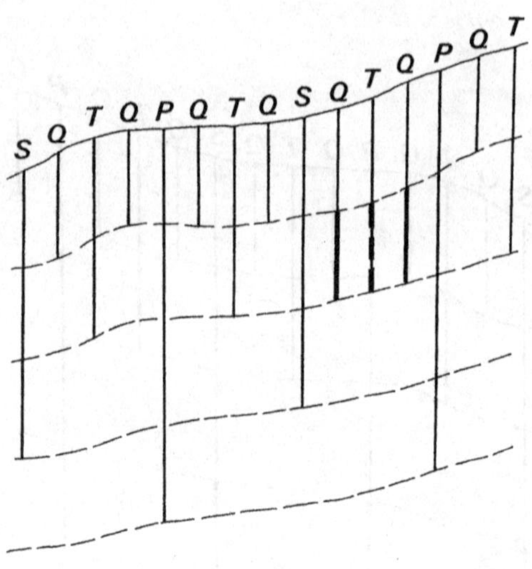

Figure 8.4-6D. *Quaternaries deepened on both sides of a tertiary having a high take.*

5. Quinary and subsequent holes may also be warranted on both sides of earlier holes where takes are too far above the target standard (see Figure 8.4-6E).

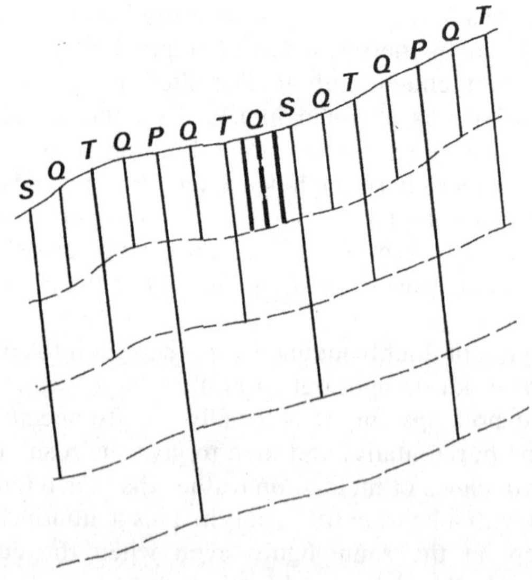

Figure 8.4-6E. *Quinaries placed alongside a quaternary with a high take.*

6. Quinary (and maybe subsequent holes in extremely difficult sites) may need to be shown on the design drawings where primaries are widely spaced or the rock is very weak.

The layout discussed applies particularly to foundations where the permeabilities decrease with depth and the bottom stage is in rock having a permeability not greatly in excess of the required standard. There are many variations on the theme: one extreme is where permeabilities remain high for the full depth of the curtain coupled with short penetration distances—in such a case it would probably be fairly evident at the start that tertiary and possibly even subsequent holes will need to go to the full depth.

8.4-7 Spacing of Primaries

The distance between primaries in a curtain should be as regular as possible in order to simplify setting out and thus minimize errors. Primary holes should not be so close that they connect with each other; connections at this juncture can jeopardize the quality of grouting and interfere with proper evaluation of takes. On the other hand, primary holes should not be so wide apart that it is necessary to routinely close down to such remote sequences as sextary or septenary in order to get a reasonable spacing of final holes.

The following are some typical spacing of primaries:

40 ft ⟨12 m⟩	For most sites
20 ft ⟨6 m⟩	Where permeabilities are low enough to ensure very few connections

Note that primary spacing should be divisible by four to simplify spacing of the subsequent holes (as far as quinary anyway).

With primary spacing of 40 ft ⟨12 m⟩, quaternaries will bring hole spacing down to 5 ft ⟨1.5 m⟩, and with primaries at 20 ft ⟨6 m⟩, tertiaries will give this spacing. It is not usually wise to specify a final hole spacing on designs because this inhibits proper intensive treatment of any surprise feature found in the work. On the other hand, it can become absurd to continue closure until the holes nearly touch each other: recommended treatment when continued closure does not achieve target standards is to cease closure in the original curtain row when about 2.5 ft ⟨0.75 m⟩ is reached and then to place any further holes necessary slightly to one side and the other of the row. This gives a widening of treatment at the bad area, which probably suffers from lack of penetration and will accordingly benefit from such widening.

8.4-8 Multiple-Row Curtains

Opinions vary about whether a single row of holes is adequate or whether several rows are needed. Probably the issue is best resolved in terms of the quality of the grouting. If the grouting is done to a good quality with thorough assessment (see Sections 7.4 and 7.5) and all weak spots given appropriate attention, single-row curtains have proved adequate generally. If, however, it is suspected that these will not be adequately attended to, several rows may be advisable.

An additional reason for using multiple-row curtains is to gain a thicker curtain in weak foundations where it is thought that the steep hydraulic gradient (refer to Section 8.4-15) through a single-row curtain may be too severe for the weak material to withstand (see Figure 8.4-8A).

The steepness of the gradient through a single-row curtain depends largely on how far the grout penetrates upstream and downstream from the curtain. This penetration can be irregular. Its extent can be gauged from the sideways penetrations in the plane of the curtain as judged from the routine assessments during closure (Section 7.4). Commonly, the thickness of grouting (upstream to downstream) exceeds 5 ft ⟨1.5 m⟩, but there are exceptions to this, particularly where crack openings are almost too fine to grout (Section 6.1-5): several rows may be necessary to get this final thickness of grouting.

It is important to make the spacing between rows in multiple curtains a distance that is less than about twice the general penetration distance. The aim is to make sure that penetrations meet (see Figure 8.4-8B).

If penetrations do not meet, leaving ungrouted corridors between the rows, the effect of the multiple rows is largely lost because full hydrostatic head can build up in the corridors under the right conditions. This can induce steep hydraulic gradients across the downstream row if it is tighter than the other two and completely negates the purpose of using multiple rows—gradients are steepened rather than eased in this case. Even if the central row is made the tightest, as is usually done, the other rows provide little assistance if ungrouted corridors remain between them: seepage passing through weaknesses in one of the curtains (and there are sure to be some weaknesses in the sort of foundation requiring multiple curtains, in spite of every care) merely has to run along the corridor to weaknesses in the next row, and so on to escape fairly readily through the whole curtain.

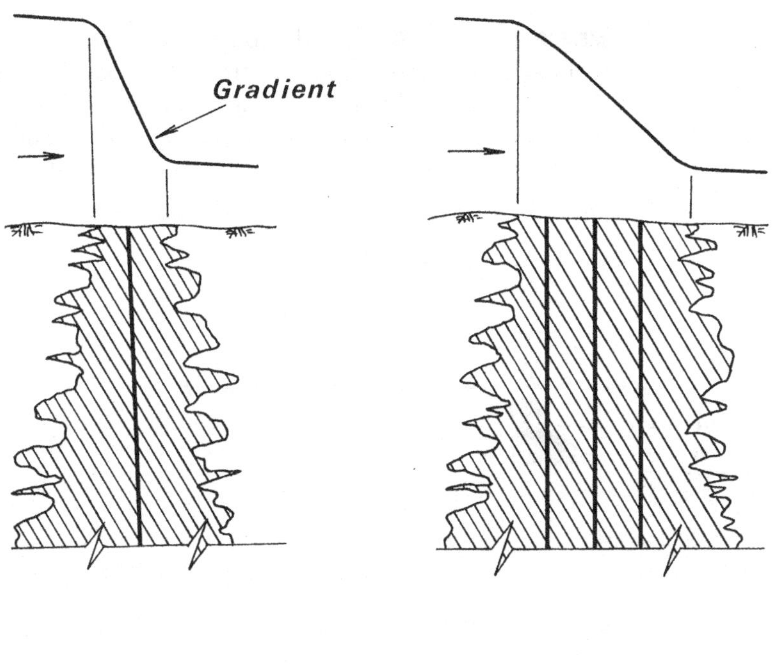

Figure 8.4-8A. *Comparison of hydraulic gradients through (left) single-row curtain, and (right) three-row curtain. Hatching shows grout penetration from curtains.*

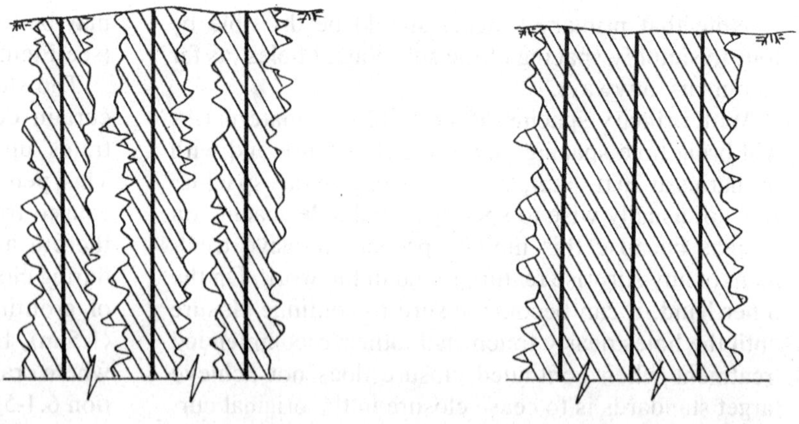

Figure 8.4-8B. *Left: Curtain rows too far apart—penetrations do not meet. Right: The required situation—penetrations from the three rows meet and provide a thick grouted area.*

Spacing of rows is often about 5 ft (1.5 m) or 3 ft (1 m), but sometimes has to be closer in which case equipment working on one row can get in the way of that working on the adjoining one; careful job planning is necessary to minimize delays.

When row spacing is 3 ft (1 m) or closer, hole congestion can occur if closure in any of the rows comes down to such close intervals that further holes need to be placed slightly upstream and downstream as suggested in Section 8.4-7. This has happened on some jobs; again, careful planning is necessary to handle the situation.

Closure (see Section 5.5) in each row should be planned to match progress in other rows. Outer rows should be completed in each area before starting interior row(s).

If inclined holes are used, it is sometimes beneficial to have the inclinations in opposite directions in adjoining rows. This helps give better crack interception sometimes but can make for delays in commencing the next row because of the longer length of work to be completed first.

8.4-9 Positions of Curtains at Dams

For storage dams, grout curtain positions vary, depending on the type of dam, the geology, the designers's preferences, and so on.

EARTH CORE EMBANKMENT DAMS Curtains should meet the surface not further downstream than the center line of the core and not further upstream than about a third of the way between the center line and the upstream toe of the core (see Figure 8.4-9A). The reasoning behind this is explained in Section 8.4-15. The most common curtain position is in the vertical plane at the center line, but vertical curtains are located in many instances slightly upstream of this and some have an upstream slope in order to suitably intercept vertical jointing located more or less parallel to the axis of the dam (discussed in Section 8.4-3). When dams have a curved alignment, it can be convenient to lay out the curtain on a series of chords to simplify field setout; the space between the curtain and the center line then varies along the curve slightly.

Some very weak foundations are blanket grouted to various extents upstream and downstream from the curtain. Where this is done there is sometimes a gradual deepening of blanket holes as they approach the curtain so that adjoining it they form minicurtains of their own to amplify the main one.

MEMBRANE-FACED DAMS Membrane-faced dams have a sloping membrane on the upstream face (see Figure 8.4-9B). The membrane is the water barrier and is usually either concrete, asphalt, or plastic. The embankment is permeable, such as compacted rockfill, hence

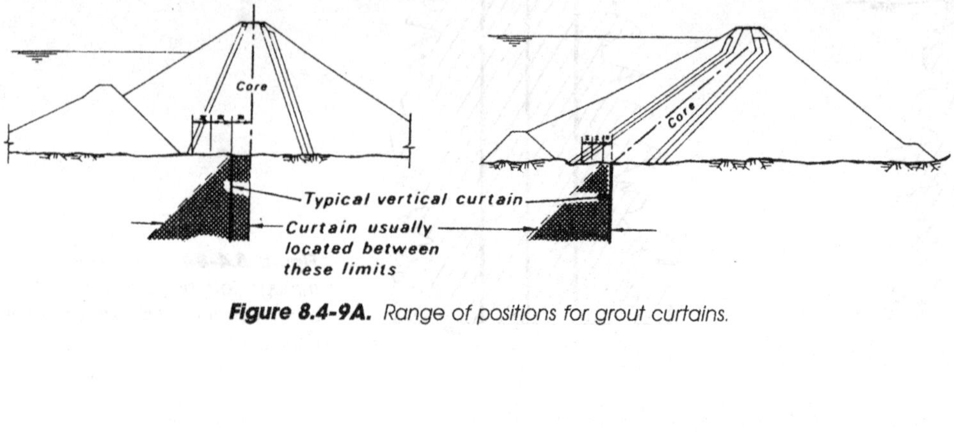

Figure 8.4-9A. *Range of positions for grout curtains.*

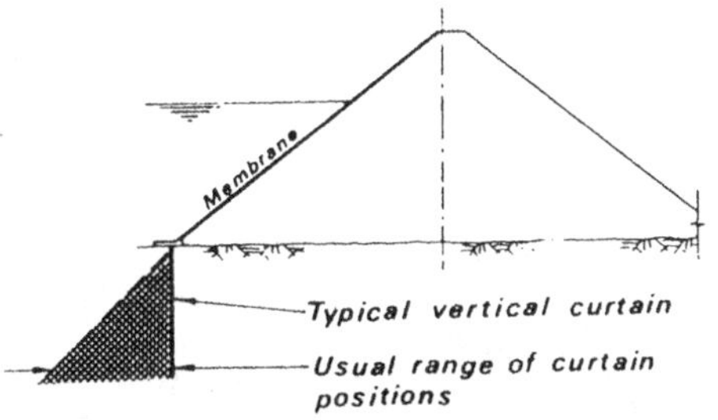

Figure 8.4-9B. *Range of curtains at toe of membrane.*

the grout curtain has to be at the junction between the membrane and the foundation. A concrete toe slab or plinth here provides a suitable top for the curtain. Details of curtain hole arrangements in the concrete are given in Section 5.1-12.

Seepage gradients are steep at the top of the curtain, and the distance for seepage to exit is usually less than most other types of dams. Accordingly, as well as having inverse filters placed over exit positions, some designers have used multiple-row curtains to ease gradients, as discussed in Section 8.4-8. Others have used a single-row curtain and have strengthened it where necessary by additional holes placed in weak features.

A gallery has been incorporated in the toe slab of some of the higher dams of this type to provide access for further grouting if later found to be necessary and for monitoring purposes. This is a reassuring, though expensive, precaution if doubts exist about the adequacy of the grouting. Drilling from such a gallery after storage commences should be done through an open valve or a stuffing box, as for some underground work (see Section 5.2-3) to give control over blowins, which if severe, might otherwise flood the gallery and give safety and remedial problems.

HOMOGENEOUS FILL DAMS These dams are made primarily from relatively impervious materials. The principles of grout curtain location are the same as for the core of a zoned dam. Attention has to be paid, however, to the position of any internal drains or pervious parts to ensure that adequate distance separates them from the top of the curtain.

CONCRETE GRAVITY DAMS Figure 8.4-9C shows the range of positions for curtains drilled and grouted from a gallery. This is the most used arrangement, however, there are variations such as (a) doing the curtain from the rock surface before placing the dam (if the construction schedule permits) or (b) doing a preliminary curtain (and possibly blanket grouting also) from this surface followed later by one from the gallery or (c) doing the grouting from the crest (small dams).

The purpose of grout curtains under concrete dams

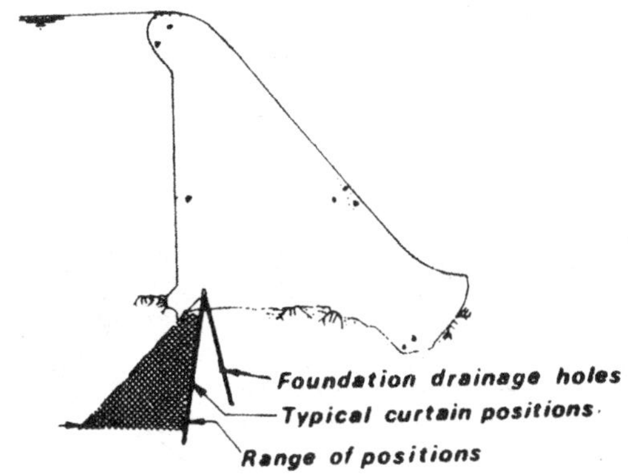

Figure 8.4-9C. *Curtain from gallery in concrete gravity dam.*

is to limit the amount of seepage reaching foundation drainage holes so as to ensure that these holes are not overloaded and can therefore perform their intended role of minimizing uplift pressures. The curtain is not the prime safety control—the foundation drainage holes are. It is important to keep them clean and discharging considerably less than their capacity. The standard of tightness needed in grout curtains for adequate performance of their role is discussed in Section 8.3-3. In low permeability foundations, grouting may not be necessary at all—provided drainage is adequate. In foundations of greater permeability, if a curtain is constructed to the suggested standard, uplift pressures are under the control of the drainage system, no matter how great the natural permeability. For discussion on this see Sabarly (1968).

ARCH AND BUTTRESS DAMS Both arch and buttress dams have a relatively short length of contact between the water retaining concrete and the foundation (Figure 8.4-9D), and both flex somewhat under changing service load conditions. Hence the grout curtain must surface in a region that remains in compression during all phases of loading cycles.

Some highly contorted grout curtains are necessary when arch dams are of the thin, doubly curved type. As mentioned earlier, the prime mechanism for controlling uplift pressure is the curtain of drainage holes, so these are given the prime position in relation to base pressures. The grout curtain is then put in upstream of the drainage, where physically able to be fitted. In steep abutments and where special consideration has to be given to geological weaknesses, the grout holes can unavoidably finish up in some very awkward positions. Some large arch dams have several grout curtains and blanket grouting as well. Tunnels have been constructed into abutments to enable grouting and draining to be carried deeply into them where a curtain from the surface would not suffice.

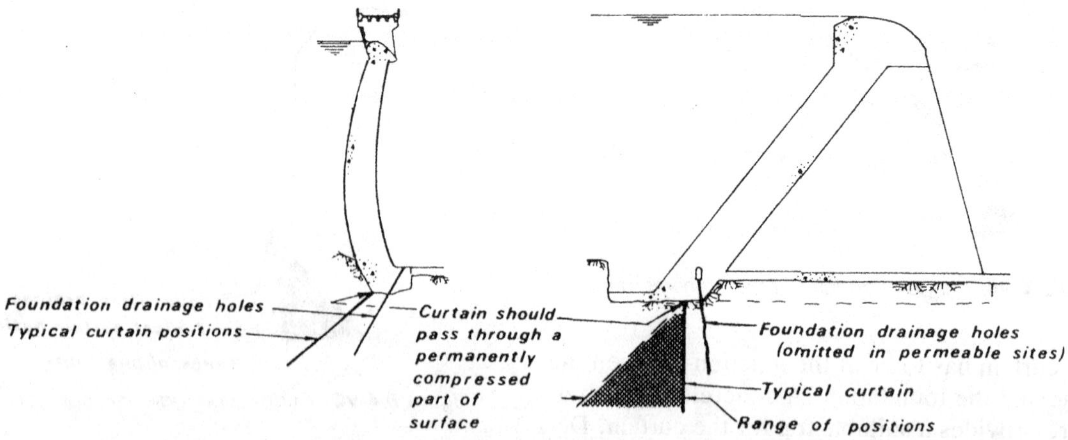

Figure 8.4-9D. *Left: Arch dam curtain and drainage arrangements. Right: Buttress dam arrangements.*

Buttress dams vary greatly in the thickness of their upstream face; some are thick enough to have a gallery to which foundation drains discharge, others merely rely on high permeability through the rock between the buttresses. Grout curtains are located near the upstream limit of the face, and like those of other types of dams, vary in inclination from vertical to sloping.

ABUTMENT AND STORAGE BASIN TIGHTNESS Grout curtains sometimes have to be extended into abutments; additionally, isolated sections of curtain may be needed in saddles or very permeable parts of the storage basin. Criteria for deciding the need for these vary widely. Generally they are slightly relaxed from those of Section 8.3-3, but there are cases where no relaxation is advisable. For instance, a slender ridge abutment retaining storage should be grouted to the same standard as the dam, particularly if emerging seepage can threaten ridge stability. On the other hand, a thick, competent abutment can be permitted to have considerable seepage flow along long seepage paths, particularly if the lost water has no monetary value.

Tunnels are sometimes driven into deep or especially pervious abutments to provide access for grouting, and possible later regrouting if found necessary.

8.4-10 Layouts at Steep Faces

When steep abutments or faces are to be grouted directly into the rock and access is difficult, an arrangement similar to that shown in Figure 8.4-10A often facilitates the work. The arrangement would be part of a grout curtain. Work from the face itself is minimized by having short, single-stage holes there, possibly drilled by lightweight drills (drills are discussed in Section 5.2). Long, multiple-stage holes are drilled down in behind the face from convenient access above using heavy drills. A fan of holes at the foot of the slope provides a transition from the slope treatment to the normal curtain to the right of it.

An alternative is to lay rail tracks up the slope (Section 5.2-13) and equip the drills with doubly flanged wheels and clamps; stair or ladder access for crews may also be necessary. With this setup the regular pattern of curtain holes can be used on the slope, but work is slower than on, say, gentle slopes.

Surface leaks can be a problem if they come out on steep slopes away from access. Refer to Section 7.3 for treatment. If jointing behind faces parallels the

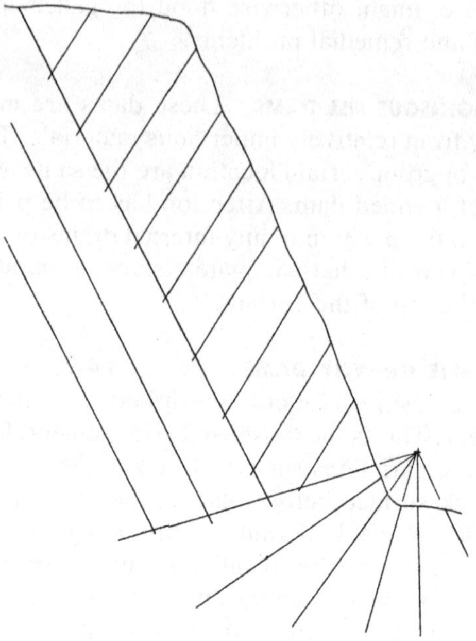

Figure 8.4-10A. *Hole layout at steep, high face.*

faces and is extensive, surface rock may move outwards under the influence of grout pressures. Bolting of faces may be necessary. Section 7.6 deals with rock movements during grouting.

8.4-11 Fans

Hole layouts where all, or most, of the holes radiate from one point are a convenient solution where access is limited. A typical fan is shown in Figure 8.4-11A. This is a vertical section through an awkward place where any other layout would be difficult. The holes all radiate from one point, commonly about 5 ft ⟨1.5 m⟩ above the floor and the same distance out from the face, to provide room for a drill to be mounted at the pivot point.

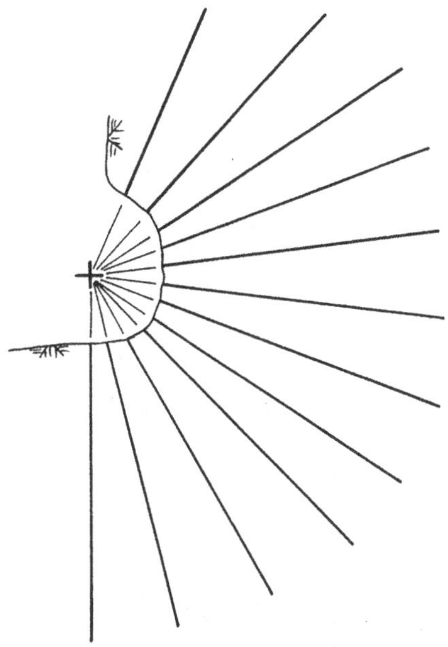

Figure 8.4-11A. *Example of a fan of grout holes.*

Figure 8.4-11B. *Fan layout for blanket grouting from a gallery in a confined area. Starter pipes placed for each grout hole before pouring concrete.*

Figure 8.4-11C. *Detail of grout hole starter pipes at gallery formwork.*

If a fan is part of a curtain being closure grouted, it will be important to also grout the fan in a closure manner (Section 5.5) because holes are unusually close at the surface and are more prone to connect there than usual. Alternatively, fans in blanket grouting work are sometimes done by nonclosure methods to speed progress in congested areas.

Where access is only available from an adit or end of a tunnel, fans have been used in plan view as well as in elevation: each of the plan view radiations is itself a fan (Figures 8.4-11B and 8.4-11C). Extensive blanket grouting is possible in this manner, from just one location. However, cramped conditions make the work slow and unpleasant, and costs are much greater than in open-air work.

A serious drawback with fans is that at least some of the holes are liable to give poor interception of jointing patterns (see Section 8.4-2). All that can be done is to increase the intensity of holes where this happens. An example is shown in Figure 8.4-11D.

This fan becomes parallel to jointing in two places. At each, the spacing of holes has been reduced, giving more holes than would otherwise be the case. This is the design solution. However when the fan is constructed, it will be important to give special attention to the inadequacies when doing routine evaluation (Section 7.4) and maybe put in even more holes if warranted.

Apart from these special considerations, the regular spacing of holes in a fan should be commensurate with

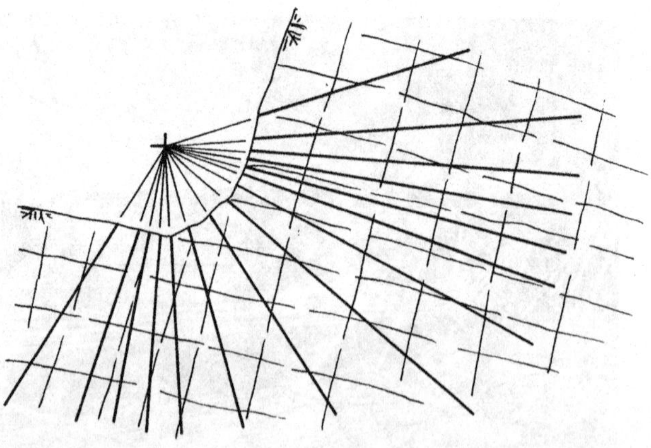

Figure 8.4-11D. *Increased intensity of holes where they parallel the jointing.*

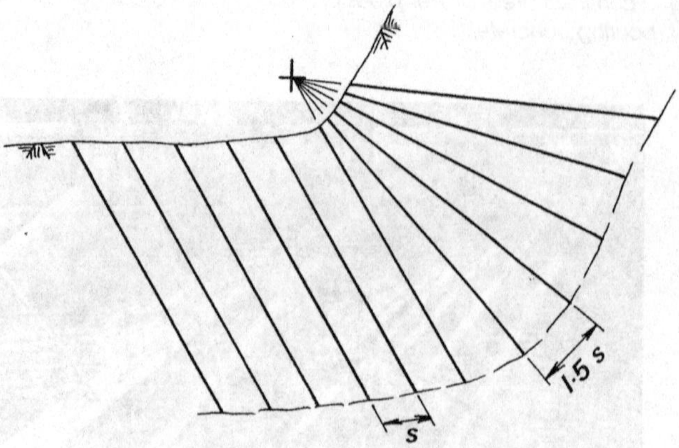

Figure 8.4-11E. *Spacing of fan holes.*

that in nearby nonfanned sections (see Figure 8.4-11E).

If spacing of a nearby section is *s*, laying out fan holes so that the maximum space is 1.5*s* gives a commensurate spacing over much of the fan hole length. It produces a spacing of *s* at about 65% of hole depth, which is reasonable in terms of relative importance at various depths. Another example of the use of fans is shown in Figure 8.4-11F.

In tunnel grouting, fans can be used for holes placed ahead of a drive to strengthen or reduce the permeability of rock to assist with excavation (see Figure 8.4-11G).

For blanket grouting of the walls of a large tunnel, groups of fans can be useful. In Figure 8.4-11H a two-deck jumbo (moving frame for drilling from) is in use probably for other drilling such as the blast holes. It doubles as a platform for the grouting. Three pivot points are shown to speed up the drilling—provided, of course, that the overlapping holes are not drilled simultaneously—drill bits or rods colliding in holes gives no joy to anyone!

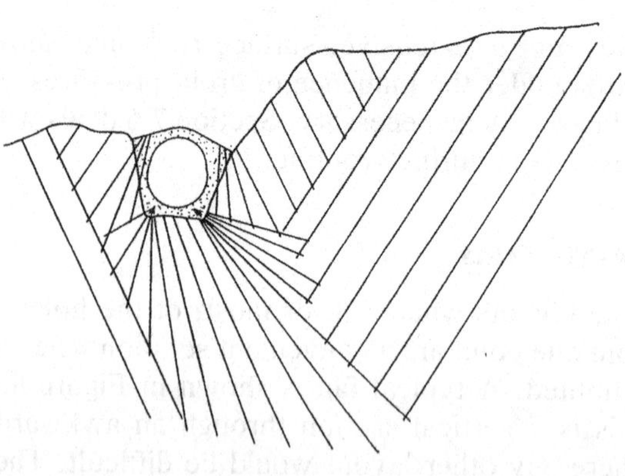

Figure 8.4-11F. *Fans are used here to carry a curtain across a deep excavation for a conduit. After placement of the concrete conduit, further fans are run down through it.*

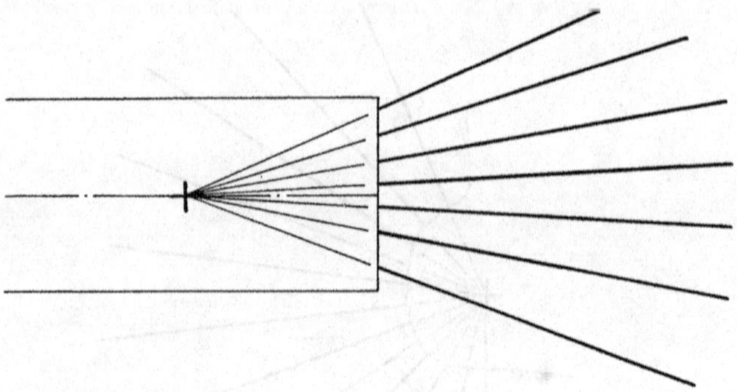

Figure 8.4-11G. *A cone of fanned holes to improve rock in advance of a tunnel drive.*

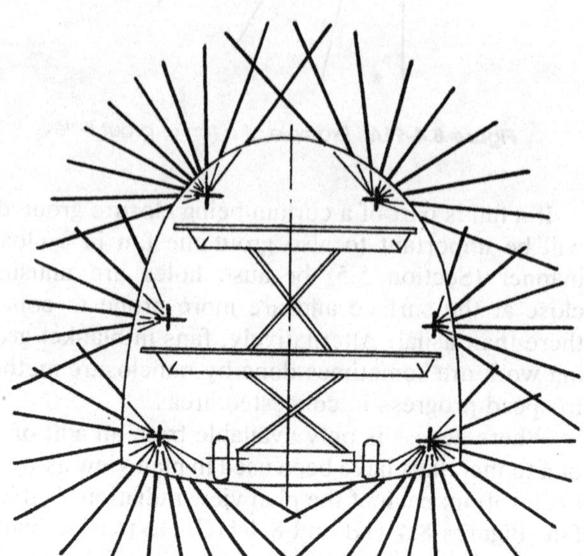

Figure 8.4-11H. *Blanket grouting a large tunnel using fans from several decks of a mobile platform.*

In drilling fans in tunnels extra care is necessary in setting the drills on line. Even so, holes are unlikely to be really straight (Section 5.2-11). Remember drillers are working in cramped conditions with deafening noise, dirt, dust, and maybe water and compressed air spraying around. They aren't in the right frame of mind for finesse. So don't design needlessly complicated fans or have holes in impractical places. Keep fans simple.

Design fans with the type of drill in mind. Lay out the holes to suit the drill. For example, Figure 8.4-11I shows a small drive not much higher than the driller. The drill is clamped on a mining bar and can be swiveled around from the center of the clamp to get to the angle for each hole. The design of this layout is shown in Figure 8.4-11J.

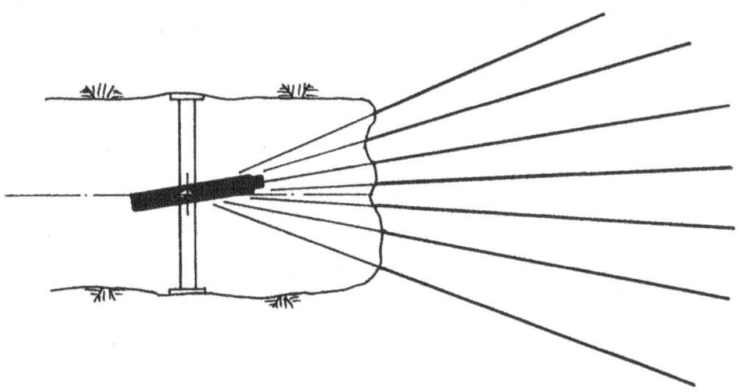

Figure 8.4-11I. *Drill mounted on mining bar to drill a fan.*

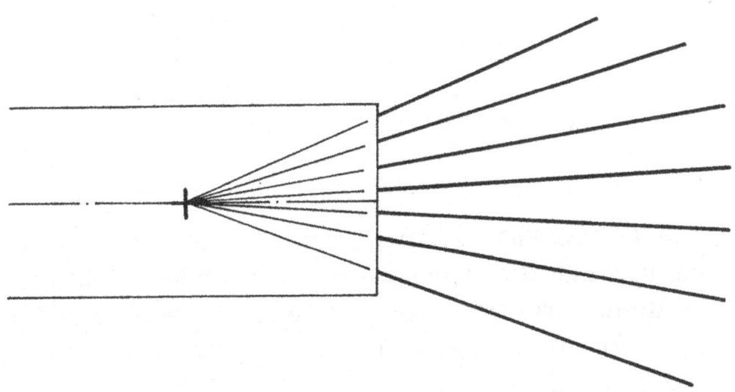

Figure 8.4-11J. *Fan design showing pivoting point.*

The designer has to specify a sensible distance back from the face to the pivot point. It must suit the drills and access and need not be at mid-height but must not be too close to the roof or floor for the drill to fit or for the driller to work.

8.4-12 Grouting to Assist Tunnel and Shaft Excavation

If tunnels or shafts are being excavated in rock that is passing inconveniently large water flows or has weak

Figure 8.4-11K. *Fan of blanket grouting holes into a specific geological feature. Close spacing because of poor interception with dominant horizontal jointing.*

areas, grouting can be a useful tool in treating these problems.

TUNNEL GROUTING This type of grouting has been touched on in Section 8.4-11 in relation to fans. Grouting for tunnel excavation proceeds ahead of a drive (see Figure 8.4-12A). Repeated use of this arrangement should produce some overlap, as shown in Figure 8.4-12B.

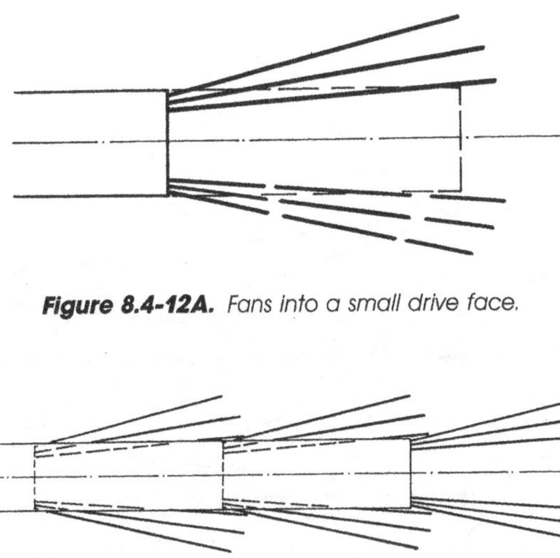

Figure 8.4-12A. *Fans into a small drive face.*

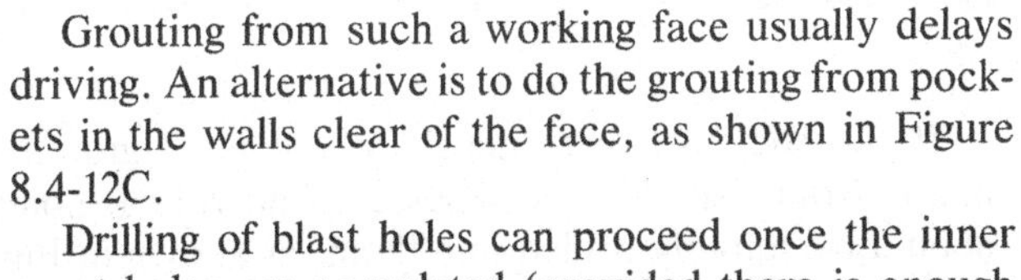

Figure 8.4-12B. *Repetition pattern.*

Grouting from such a working face usually delays driving. An alternative is to do the grouting from pockets in the walls clear of the face, as shown in Figure 8.4-12C.

Drilling of blast holes can proceed once the inner grout holes are completed (provided there is enough room for the gear). At this juncture the chances of

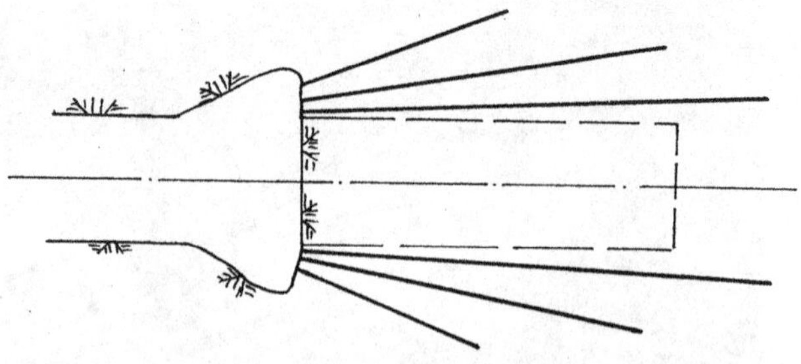

Figure 8.4-12C. *Grouting from pockets to keep face clear.*

grout connecting to the blast holes is minimized. Quick-setting grouts (Section 3.3-2) may be warranted.

An alternative layout (Figure 8.4-12D) where a long length of bad foundation has to be grouted before excavation ventures into it, surrounds the area to be excavated by a ring of grouting carried out from long holes.

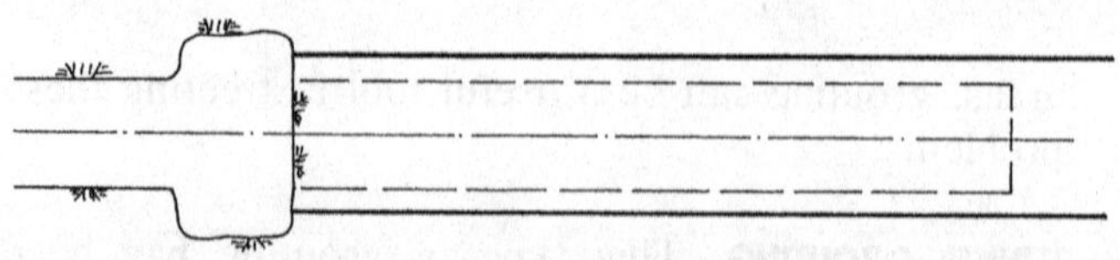

Figure 8.4-12D. *Grouting well ahead from pockets.*

For grouting a large face ahead of excavation, the holes may need mobile access, as shown in Figure 8.4-12E.

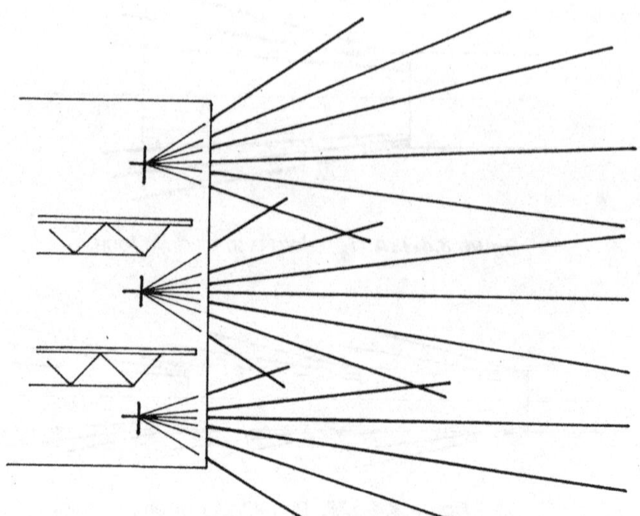

Figure 8.4-12E. *Fans into a large drive face from a mobile platform.*

If the rock is good enough to not need grouting apart from isolated weaknesses, individual holes should be suitably drilled into them, as shown in Figure 8.4-12F.

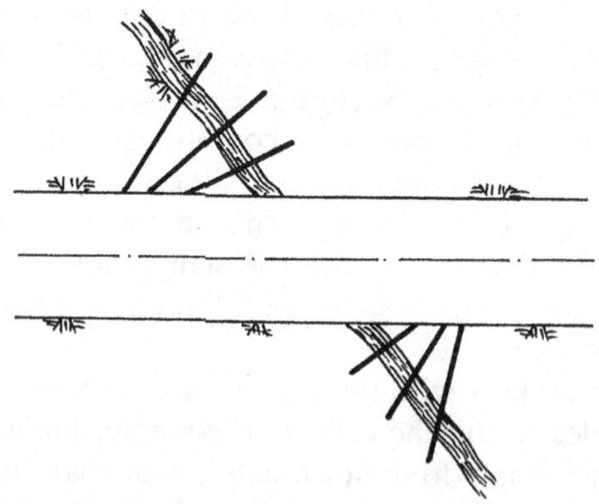

Figure 8.4-12F. *Grout holes arranged to intercept weakness.*

After excavation, grouting may be necessary to repair shaking from blasting or for general strengthening or water reduction purposes. This is done by blanket grouting in rings of holes, as shown in Figure 8.4-12G.

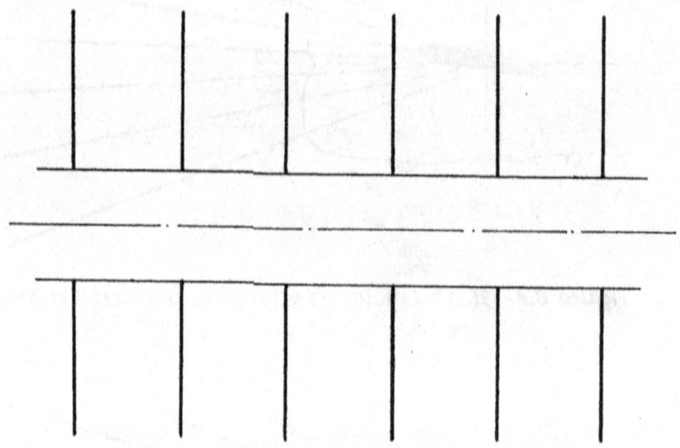

Figure 8.4-12G. *Blanket grouting in a tunnel.*

SHAFT GROUTING Shaft sinking can make use of grouting in much the same way and for the same purposes as tunnel grouting. For instance, a typical layout of holes for grouting in a shaft sunk in stages is shown in Figure 8.4-12H.

It is a relatively slow process to stop and grout each stage before proceeding further, and so with large deep shafts it is usually better to grout the full depth of the shaft before starting it (see Figure 8.4-12I).

A ring of deep grout holes spaced around the site of the shaft is drilled and grouted ready for the excavation. These grout holes are rather bigger and deeper than most and have to be carefully drilled to ensure that they do not drift away from the area to be treated. Sometimes downhole surveying and directional drilling is used to effect this. Selective stage grouting can be used, with only those stages needing it being grouted.

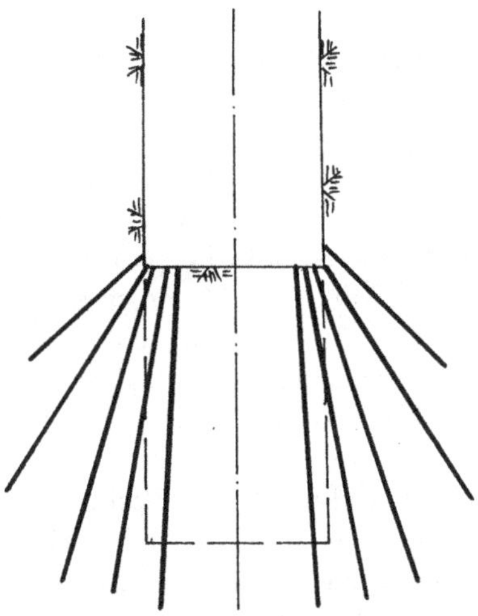

Figure 8.4-12H. *Fans of grout holes in shaft sinking.*

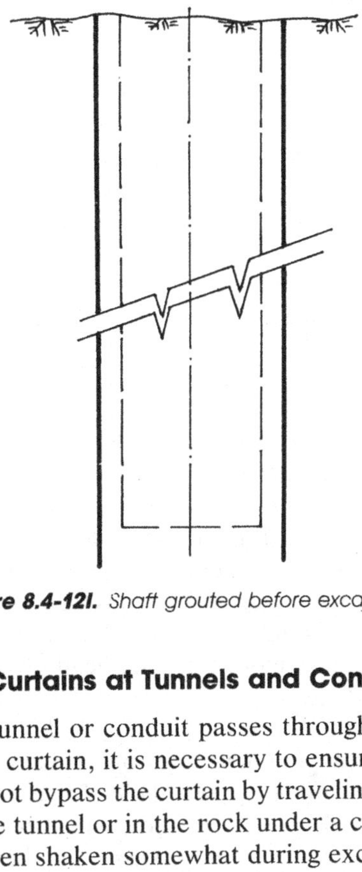

Figure 8.4-12I. *Shaft grouted before excavation.*

8.4-13 Curtains at Tunnels and Conduits

Where a tunnel or conduit passes through or under a dam grout curtain, it is necessary to ensure that seepage does not bypass the curtain by traveling in the rock around the tunnel or in the rock under a conduit. This rock is often shaken somewhat during excavation and so provides an easy potential seepage path. Except where the rock is very sound and has remained tighter than the standard selected for the curtain (refer to Section 8.3-3), a grout curtain is then needed at the tunnel or conduit, as shown in Figure 8.4-13A.

Both curtains are vertical and in the same plane in this example. This need not necessarily be the case. Inclined curtains and warped curtains are sometimes

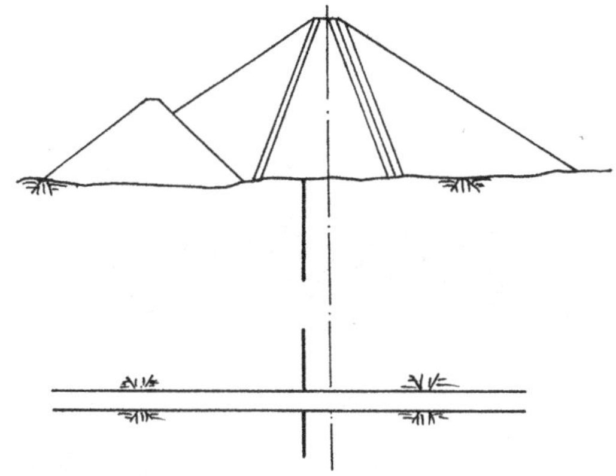

Figure 8.4-13A. *Grout curtain from a tunnel under a curtain from the surface.*

necessary to suit jointing patterns and skewed tunnel alignments.

If the foundation between the tunnel and dam is sufficiently permeable to warrant joining the curtains, or if the tunnel passes through the surface curtain, the two curtains should be merged into each other, as shown in Figure 8.4-13B.

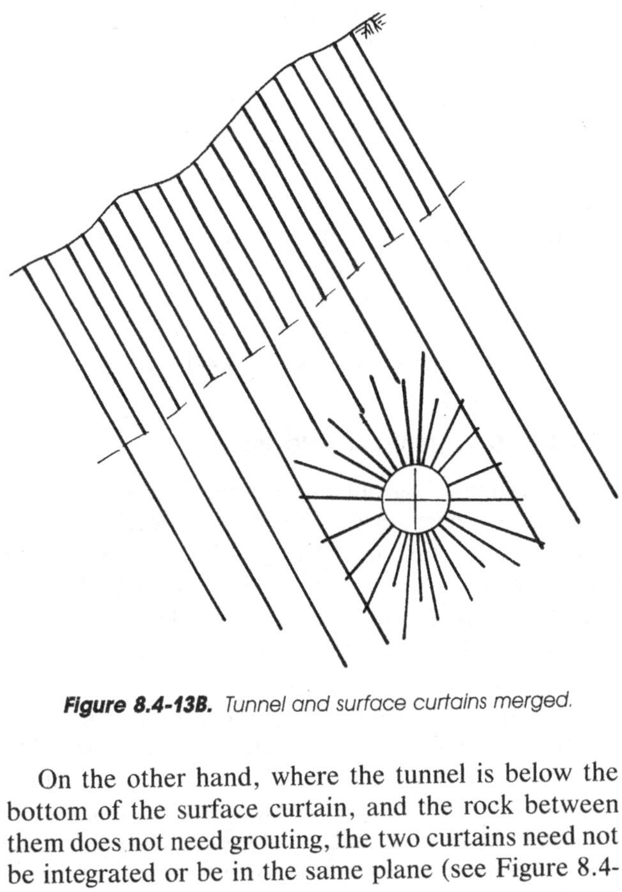

Figure 8.4-13B. *Tunnel and surface curtains merged.*

On the other hand, where the tunnel is below the bottom of the surface curtain, and the rock between them does not need grouting, the two curtains need not be integrated or be in the same plane (see Figure 8.4-13C).

There is no need for the holes in the tunnel curtain to be evenly spaced. If the jointing warrants it, non-

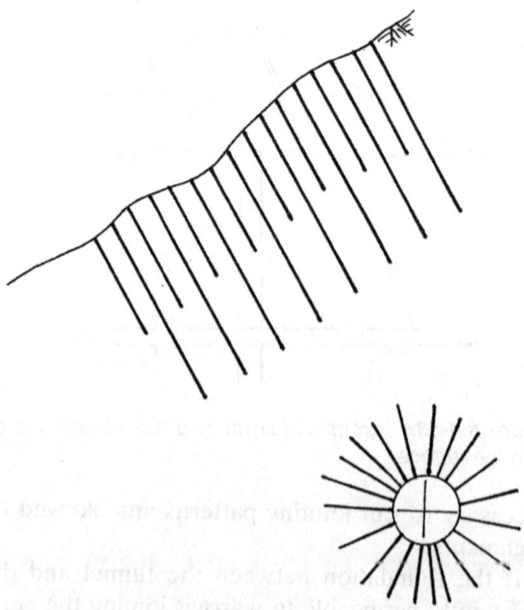

Figure 8.4-13C. Separate tunnel and surface curtains.

symmetrical hole layouts like those in Figures 8.4-13B and 8.4-13C should be used. However, unless tunnels are sufficiently large to permit drilling of nonradial holes, some are inevitably near parallel to one or more of the jointing directions. The same situation occurs with fans (refer to Section 8.4-11) and the same solution applies: place holes closer than usual in the areas of poor interception.

Grout curtains at a conduit should usually be done from holes outside rather than inside because of easier working conditions in the open. Holes can fan under the conduit from both sides unless it is very large such as the one discussed in Section 8.4-11.

8.4-14 Designing a Curtain

When designing a grout curtain from exploratory information, it is necessary to decide how deep to make it and to what standard. The standard can vary with depth: deeper stages can often be done to a lesser standard than surface requirements. The concepts of Section 8.3-3 help to decide just what ones to adopt for the various depths.

Having settled the standards, the next issue is that of depth. On some sites the matter is clear-cut: low natural permeabilities mean that a shallow curtain of only one or two stages will be adequate; or at the other extreme, high permeabilities at great depths mean that the curtain will never bottom on anything tight and so should be taken down a depth equal to the head of water above it (a very arbitrary criterion based on satisfactory use). However, there are many cases where an intermediate situation exists, and careful consideration of each part of the site is then necessary.

Planning is most conveniently done on an elevation drawn through the proposed curtain alignment, showing relevant geological and permeability information. If possible, the anticipated stripped surface (Sections 5.1-17 and 5.1-18) should be plotted, as well as any structures that will affect the positions of grout holes. The plot must be to a natural scale, even though distorted scales are commonly used to show geological information. The natural scale permits plotting grout holes on the section.

This plot should not be overloaded with geological data. Just the relevant essentials. The permeability data (Sections 8.2-6 and 8.2-7) is of the most importance usually and should be shown clearly alongside each hole, plotted accurately. The recommended method is to show the permeability in lugeons (Section 8.2-7) as a bar having a length to scale representing the number of lugeon units and with the value written as a number at the end of the bar. There can be an advantage in shading the bar as far as the selected standard (i.e., 5 lugeons or whatever—see Section 8.3-3), so that a general scanning gives an overall impression as to how many tests exceed the standard. Where permeability exceeds 100 lugeons, the bar can have a jagged end to indicate indefinite continuation; it is not sensible to show the precise value of lugeon values above 100 (see Section 8.2-10); they should just be marked as >100.

With this data in front of you, the first problem is usually lack of uniformity or of pattern in the positions of the greater permeabilities: they can have quite a random positioning. It then becomes necessary to decide which of the deeper ones really matter. This decision needs to reflect the purpose of the grouting and of the dam. If possible, preferred treatment is to take the curtain down to cover all locations where the standard is exceeded, provided that this does not go down an amount which is more than the head of water above the final surface.

The bottom of the curtain should be more or less smooth. Avoid an uneven bottom. It isn't worthwhile just to save a little footage of hole. The complexities of variable hole depths can introduce possible errors during construction. Certainly, stages can be curtailed a little or some holes omitted in definite tight areas at depth, but keep the general smooth bottom and regular pattern (assuming it is possible to have one, that is).

Notice how this generalized use of the exploratory data glosses over much of the detailed information collected. It is not possible to cater for all nuances. The resulting curtain would often be a shambles. Here then we have the philosophy behind some of the comments relating to water testing where it is suggested that precision is out of context (Section 8.2). It is pointless, most times, to go to great trouble to get very accurate results—they just can't be used in all their glory. However, not everyone is convinced of this!

Having profiled the bottom of the curtain, the next step is to lay out the pattern of grout holes on the plot, taking into account any inclination necessary to give optimum crack interception (Section 8.4-3). Set out the primaries and work out a suitable dimensioning system (Sections 8.4-5 to 8.4-7). Then decide on stage depths (Section 5.4-2), making them as uniform as possible (Section 8.4-6), preferably following the surface profile, but making variations where necessary at uneven surface locations and when approaching the bottom.

Secondary and subsequent holes are then outlined on the drawing and their depths decided, remembering that they can be deepened if necessary when the work is in progress (Section 8.4-6).

That description of the designing process is rather sketchy, as indeed it must be in view of the considerable variations existing from site to site, but it gives an idea of how to start. The rest is up to you.

8.4-15 Effects of Grout Curtain at Earth Core Dams

The following discussion compares the effect of a grout curtain on (a) hydrostatic pressures in a foundation and (b) pore pressures in an earth core dam.

The discussion simplifies the issues and is not intended as a precise presentation; for instance, flow nets are not used.

In the pictorial view shown in Figure 8.4-15A, a vertical section is shown through a dam with zoning indicated. The stored water is on the left of the dam and imposes full hydrostatic head in cracks in the foundation until the upstream toe of the impervious core is reached. Then the head at the interface between foundation and core starts to reduce because the storage head no longer has direct access there. This slow reduction continues until the grout curtain is reached. Further head loss occurs through the curtain—the amount of loss depends on the tightness (or standard) of the grouting. Downstream of the curtain, pressure reduction continues until tailwater level is reached.

Now comparing this with what happens in the dam itself just above the foundation. Pore pressure in the core (see Figure 8.4-15B) drops from storage head to tailwater. The drop is shown here as a straight line; it actually is curved, as shown later. Superimpose this on the foundation pressures as shown in Figure 8.4-15C.

Notice that in this example pressures in the core exceed those in the foundation. This is an ideal situation: there is no tendency for seepage to jump the top of the curtain or for seepage to move from the foundation upwards into the core.

Looking at actual dams, an instance of this ideal situation is shown in Figure 8.4-15D.

Long-term pressure measurements taken by piezometers in the base of the core and in the foundation are plotted here. This dam has a single-row curtain flanked by blanket grouting to varying depths extending over the whole base of the core. Permeabilities (in lugeons) after grouting are shown for the various

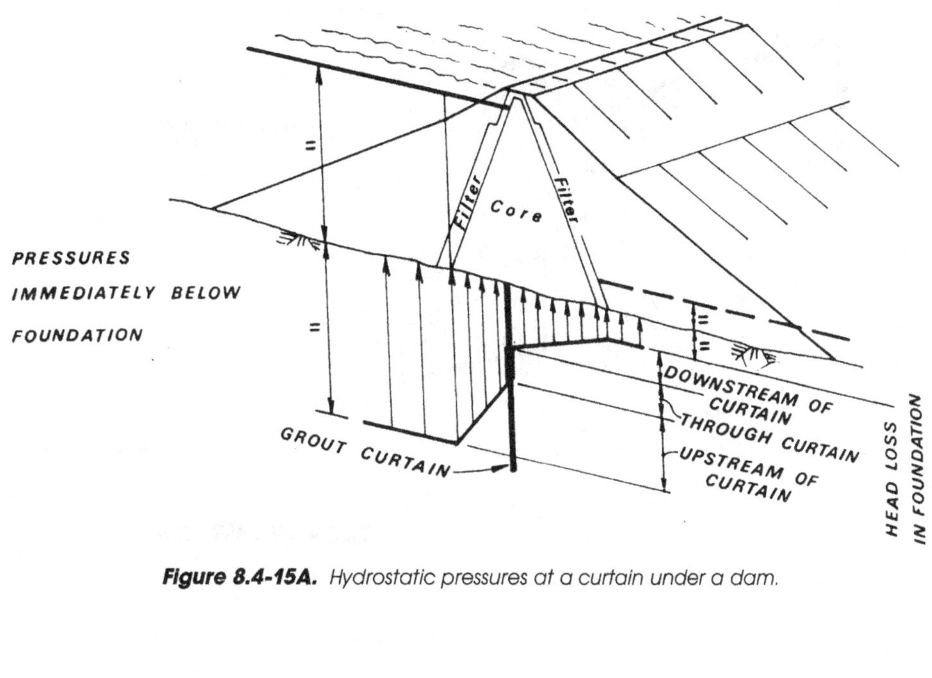

Figure 8.4-15A. *Hydrostatic pressures at a curtain under a dam.*

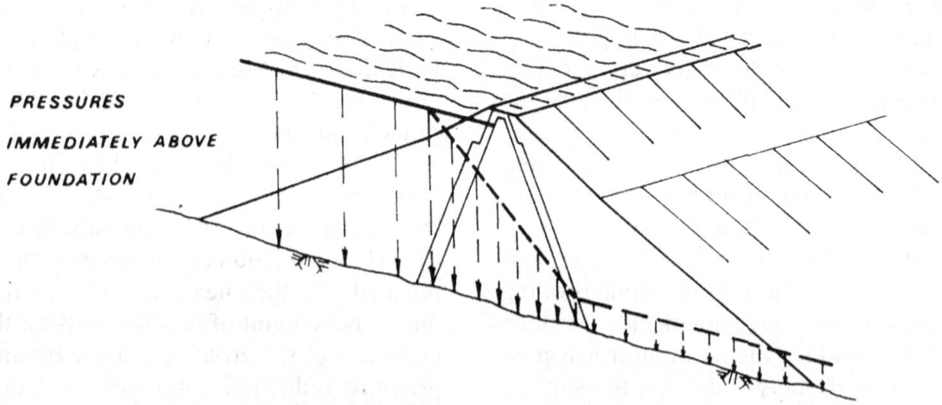

Figure 8.4-15B. *Pore pressures in earth core dam.*

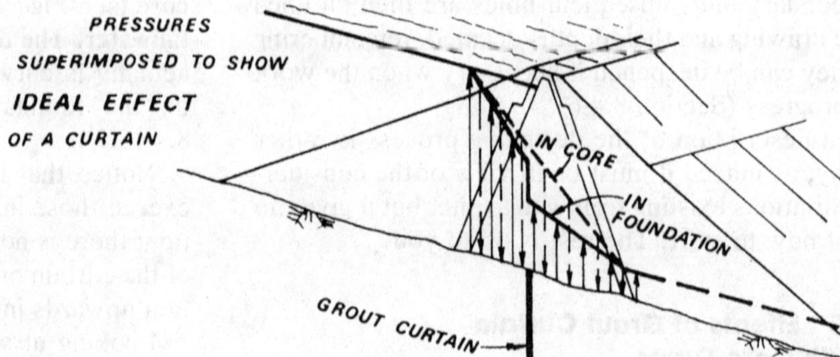

Figure 8.4-15C. *Pressures in foundation and core compared.*

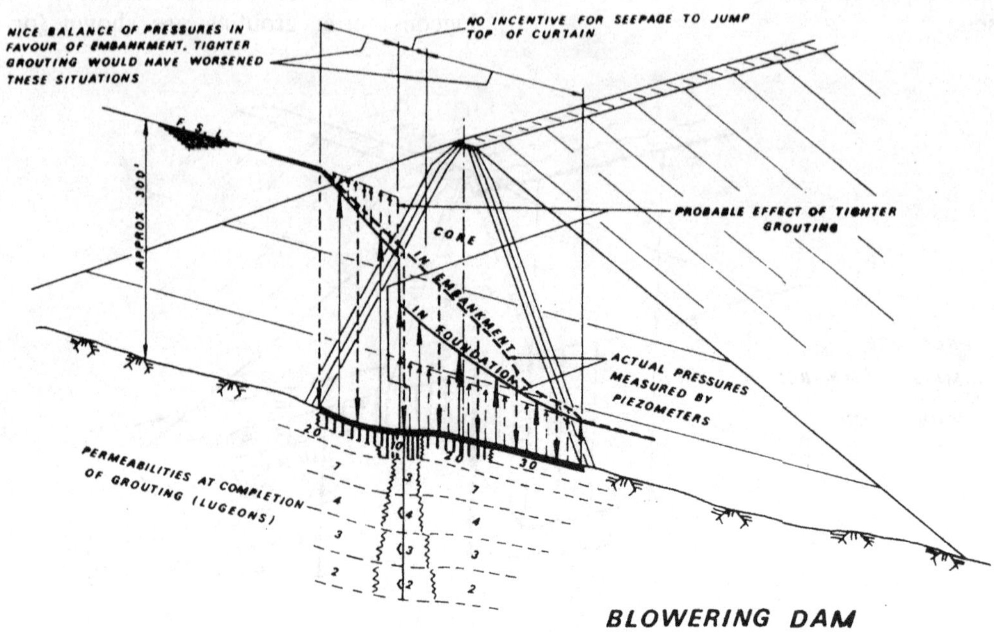

Figure 8.4-15D. *Comparison in an actual dam.*

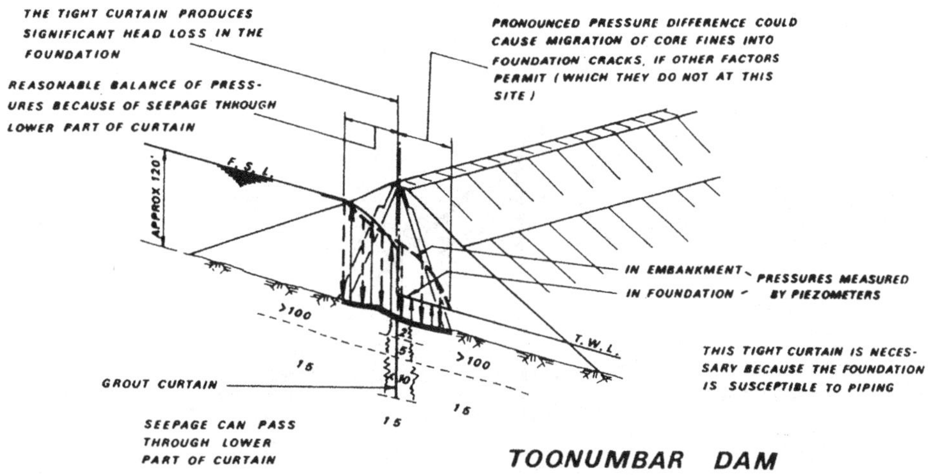

Figure 8.4-15E. *Another actual case.*

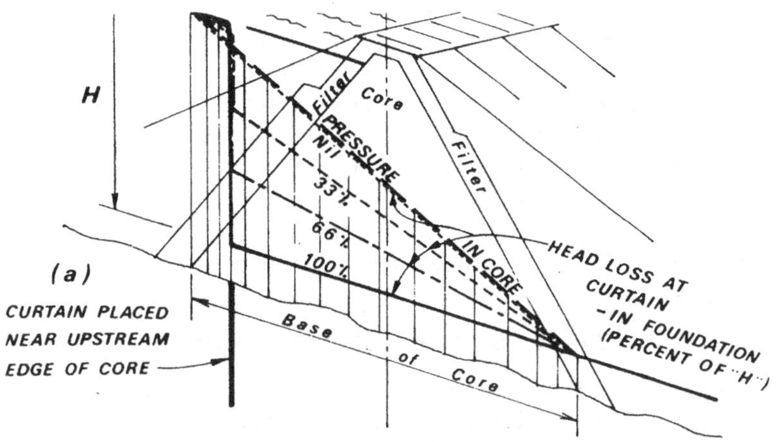

Figure 8.4-15F. Pressure distributions from a curtain as far upstream as possible.

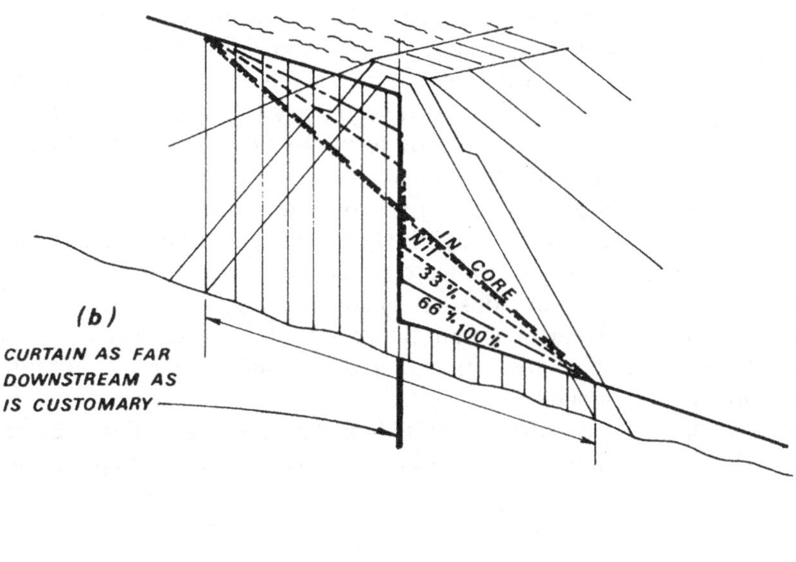

Figure 8.4-15G. Pressure distributions from a curtain placed at downstream position.

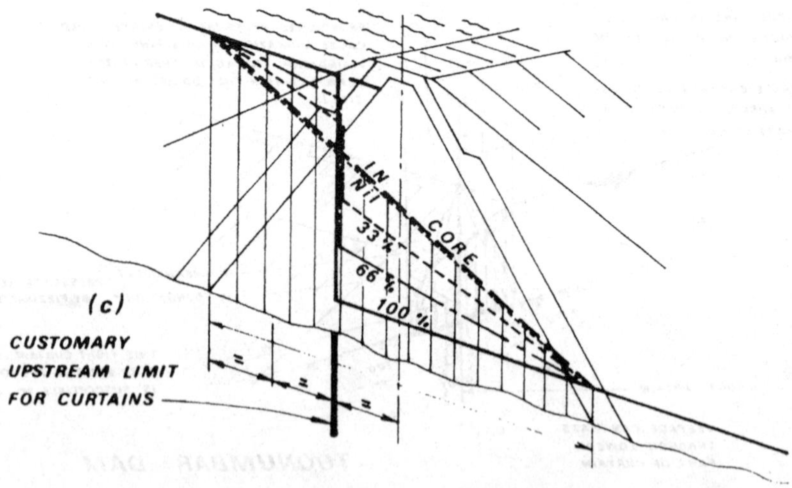

Figure 8.4-15H. *Pressures at compromise position.*

stages of the curtain; pregrouting permeabilities are shown upstream and downstream.

In the dam shown in Figure 8.4-15E the curtain has a standard of 2 lugeons in its surface stage but is not as tight in lower stages, so any seepage passing it has to first descend to the lower stages and loses a lot of head in doing so. Head loss in cracks just below the surface is considerable at the curtain and produces a marked contrast to pressures in the core. Upper parts of the foundation had a permeability in excess of 100 lugeons before grouting and contained material susceptible to piping; this is why the curtain was taken to the unusually tight standard.

Where is the best place for a curtain under the core of a dam? If it is placed near the upstream edge of the core where shown in Figure 8.4-15F, this figure compares pressures in the core with those in the foundation for various standards of curtain. If the curtain is absolutely perfect so that the head loss through it is 100%, an unfavorable relationship exists with the core: foundation pressures upstream of the curtain exceed core pressures, giving the potential for seepage to jump the crest of the curtain if surface arrangements there are not adequate. This tendency may be aggravated by the pronounced drop in foundation pressures downstream. Lesser standards of tightness at the curtain reduce this tendency. In the ultimate, if the curtain is completely useless, the pressure balance is good; this is, of course, an absurd situation.

For a curtain placed at the center line of the core (which is as far downstream as is customary), the effects of the various standards are not as severe, but the incentive for jumping the curtain crest is much greater for the better standards of grouting (see Figure 8.4-15G).

A reasonable compromise is when the curtain is in the position shown in Figure 8.4-15H. Figure 8.4-15I examines it for various standards of grouting and shows core pressures somewhat more realistically.

For 100% head loss, there is minimal incentive for jumping the curtain. With lesser standards, this incentive disappears. The difference between core and foundation pressures downstream of the curtain is also less with the lesser standards.

The standards of grouting suggested in Section 8.3-3 for various situations reflect these considerations.

8.4-16 Blanket Grouting Layouts

Blanket (or consolidation) grouting strengthens a foundation. It does this by filling open cracks and voids. As well as reducing any tendency to settle under structural load, this can eliminate dilatency in a fractured foundation.

This form of grouting uses holes spaced over a surface on a grid pattern. Hole arrangements vary. A typical rectangular pattern is shown in Figure 8.4-16A.

With this layout primary holes (P) are drilled and grouted first, then secondary (S) diagonally between them, followed by tertiaries (T) to complete the grid pattern. Additional holes are placed, if needed, in any locations that are below the required standard. One complete "bay" is shown; the pattern can be extended in all directions to cover the area to be grouted.

Some jobs have used a hexagonal pattern or other departures from the rectangular, but complexities can then arise in organizing closure and in interpreting results.

At dam sites, if blanket grouting is used in conjunction with a curtain, the blanket holes might be ar-

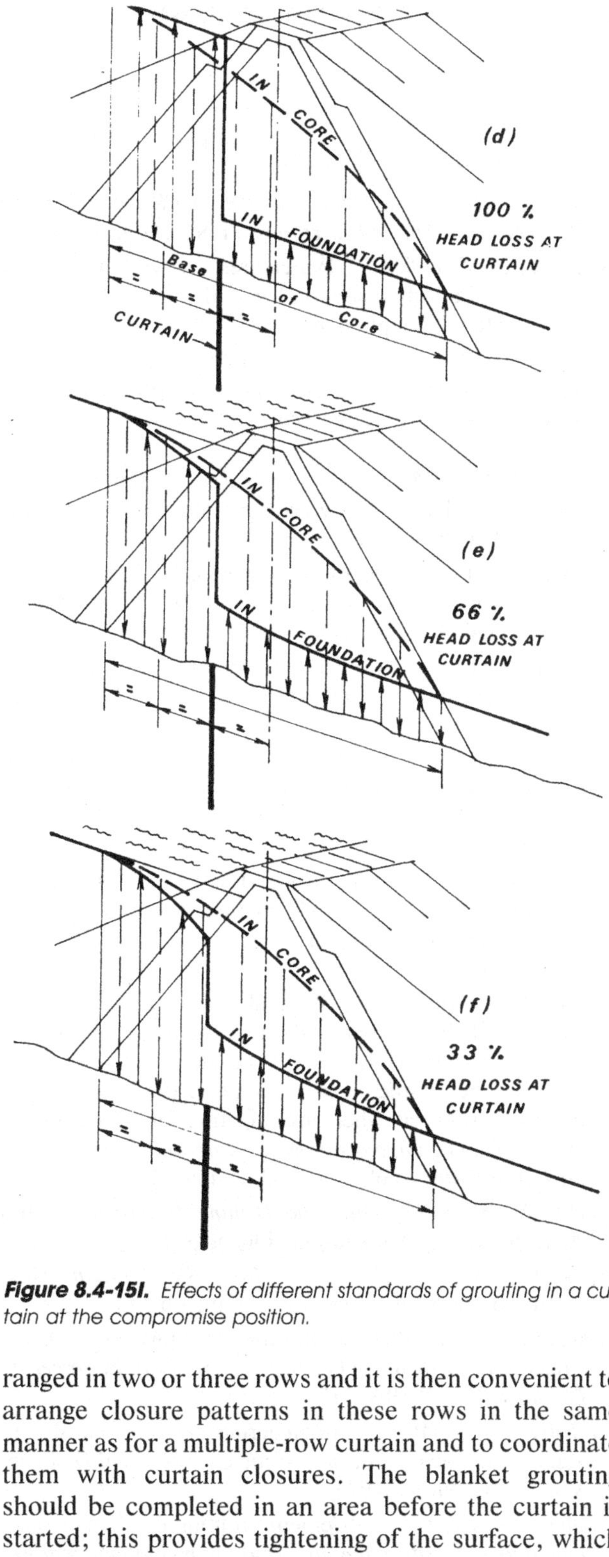

Figure 8.4-15I. *Effects of different standards of grouting in a curtain at the compromise position.*

ranged in two or three rows and it is then convenient to arrange closure patterns in these rows in the same manner as for a multiple-row curtain and to coordinate them with curtain closures. The blanket grouting should be completed in an area before the curtain is started; this provides tightening of the surface, which assists subsequent curtain grouting.

Holes may need to be inclined to suit jointing (Section 8.4-3). They should all be inclined the same direction to avoid confusion when assessing results (Sections 7.4 and 7.5).

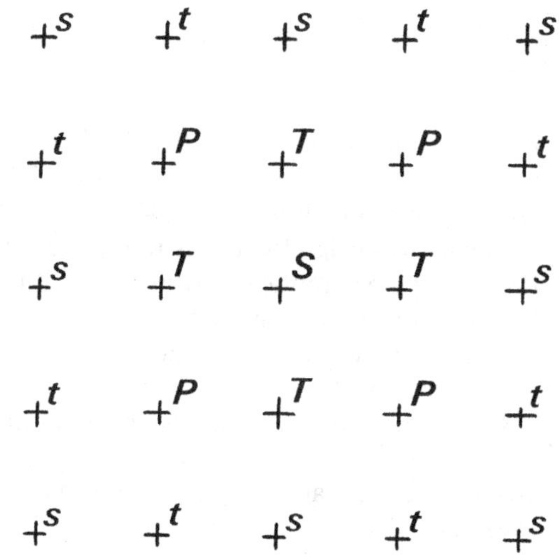

Figure 8.4-16A. *Blanket grouting pattern. Closure sequences for one "bay" shown in capital letters; adjoining ones in lowercase.*

Primary holes in the rectangular pattern shown would be about 10 ft ⟨3 m⟩ apart on most sites. However, if connections are too numerous for this spacing, the primaries should be widened out to twice this distance and the intervening holes given additional closure sequences until final spacing is about 5 ft ⟨1.5 m⟩ or less.

Stage grouting (Section 5.4) is sometimes used with blanket grouting perhaps amounting to two or three stages. Downstage without packer is the recommended method for this but upstage has been persevered with on some sites in spite of the consequent numerous surface leaks.

Where the quality of the grouting is considered to be subservient to speed and cost, group grouting is used (Section 2.7-5). Instead of grouting holes individually, several holes are grouted simultaneously. As many as eight have been grouted in one group. They are usually adjoining holes, so if the foundation is prone to surface leaks or rock movements (Section 7.6-1), the quality of the grouting can suffer while dealing with these problems.

8.4-17 Estimation of Grouting Quantities

Part of the design process is the estimation of quantities and costs for the grouting. Here we enter on matters that are almost entirely in the realm of experience.

People tackle the estimates in different ways. Some base them on anticipated cement takes per unit length of hole, as I do. Others prefer working on the vertical area of a completed curtain, a method that is not applicable to blanket grouting. Some work on the weight of cement used per unit length of curtain, a method with obvious drawbacks.

The following steps explain my method as far as is possible:

1. After preparing the design, the number of holes of each sequence is counted, and the total footage of holes shown in the design is calculated.
2. Because the design footage is rarely the same as the actual final footage, estimates of this final footage are made from experience, taking into account the particular geological and permeability features of the site. Note that it is not good policy to prepare the design to this final footage because of the tendency it would induce to overgrout—the estimates may prove too high in the event anyway.
3. An estimate is made of the average cement take over the whole job. This is in bags of cement per foot length or (loose) liters of cement per centimeter. These two units give the same number, by a freak of metric conversion.
4. This average cement take, multiplied by the estimated final footage, gives the total cement for the job for the "grouting" item (Section 9.2-5).
5. The design footage plus some, but not all, of the estimated extra is used for the "drilling" item (Section 9.2-2), so this item becomes a compromise between the minimum (the design footage) and the estimated final footage.
6. The item "hookup" comes from a count of the number of design stages plus an estimate of expected additional ones (Section 8.4-6), and also plus estimated second and subsequent applications (Section 7.2-10).
7. The "standpipe" item comes from a count of the number of holes plus an estimate of the number of extra surface stages arising from closure.

You will no doubt perceive from all this just how subjective the process is. Regrettably it cannot be passed on. It takes many grouting jobs to become "calibrated" on all these factors.

When it comes to costing a job, a background of costs for previous jobs is needed. Any adjustments for particular ease or difficulty on the site need to be made, together with all the usual adjustments for labor matters and contractural arrangements. There is no point in giving Australian grouting costs here. They are not relevant elsewhere. Note, however, the comments on relative pricing structures in Section 9.2-6 and those on the bid items in Sections 9.2-2 to 9.2-5.

REFERENCES AND FURTHER READING

Acker, R. C., and Jones, J. C. (1972). "Foundation and abutment treatment for rockfill dams." *ASCE Journal of the Soil Mechanics and Foundations Division*, Vol. 98(SM10), Paper 9303, 995–1015.

Andrade, R. M. (1988a). *Mechanics of the Flow in Fractured Rock Masses Applied to Dams.*" ENGEVIX, Brazil.

Andrade, R. M. (1988b). "Water pressure test, an obsolete tool." ISRM Symposium on Rock Mechanics and Power Plants, Madrid, September 1988.

Arhippainen, E. (1970). "Some notes on the design of grouted curtains on the basis of water pressure tests." Tenth Congress on Large Dams, Montreal, Vol. II, 137–145.

Baker, W. J. (1955). "Flow in fissured formations." Proceedings of the World Petroleum Congress, Rome, Section II, 379–393.

Božović, A. (1985). General Reporter for Question 58, *15th International Congress on Large Dams*, Lausanne, Switzerland, Vol. III, 1467–1582.

Daw, G. P. (1984). "Application of aquifer testing to deep shaft investigations." *Quarterly Journal of Engineering Geology*, Vol. 17, 367–379.

Dick, R. C. (1975). "In situ measurement of rock permeability: influence of calibration error on test result." *Bulletin of the Association of Engineering Geologists*, Vol. XII(3), 193ff.

Fergusson, F. F., and Lancaster-Jones, P. F. F. (1964). "Testing the efficiency of grouting operations at dam sites." Eighth Congress on Large Dams, Edinburgh, Vol. 1, 121–139.

Guerra, J. R., Weyermann, W., and Mota, O. S. (1968). "Le caractère de la percolation d'une roche d'après les observations préalables faites pour le project de l'écran d'étanchéité" The character of percolation through the rock mass as observed when investigating for design of the cut-off. *Proceedings of the Ninth International Congress on Large Dams*, Vol. 1. Istambul, 109–122.

Heinz, W. F. (1985). *Diamond Drilling Handbook.* South African Drilling Association, Johannasburg.

Hoek, E., and Bray, R. (1981). *Rock Slope Engineering.* Institution of Mining and Metallurgy. London.

Houlsby, A. C. (1976). "Routine interpretation of the Lugeon water-test." *Quarterly Journal Of Engineering Geology*, Vol. 9, 303–313.

Houlsby, A. C. (1977). "Engineering of grout curtains to standards." *ASCE Journal of the Geotechnical Division*, Vol. 103(GT9), 953–970.

Houlsby, A. C. (1982). "Cement grouting for dams." *Proceedings of Conference on Grouting in Geotechnical Engineering*. ASCE, New Orleans, 1–34.

Houlsby, A. C. (1985). "Design and construction of cement grouted curtains." *Proceedings of the Fifteenth Congress on Large Dams.* Lausanne, 999–1015.

Iwai, K. (1976). "Fundamental studies of fluid flow through a single fracture." Ph.D. Thesis, University of California, Berkeley.

John, K. W. (1968). "Graphical stability of analysis of slopes in jointed rock." *ASCE Journal of Soil Mechanics and Foundations Division*, Vol. 94(SM2), Paper 5865.

Lancaster-Jones, P. F. F. (1975). "The interpretation of the Lugeon water-test." *Quarterly Journal of Engineering Geology*, Vol. 8, 151–154.

Littlejohn, G. S. (1975). "Acceptable water flows for rock anchor grouting." *Ground Engineering*, Vol. 8(2), 46–48.

Louis, C. (1969). "A study of groundwater flow in jointed rock and its influence on the stability of rock masses." Imperial College Rock Mechanics Report No. 10. September 1969.

Lugeon, M. (1933). *Barrages et Geologie*. Dunod, Paris.

Nieto, A. S. (1977). "Significant engineering-geology features at damsites in flat-lying sedimentary rocks." *8th Ohio River Valley Soils Seminar on Earth Dams and Embankments Design, Construction, Performance* Louisville, KE.1-1 to 1-16.

Nonveiller, E. (1970). "A rational approach to design of grout curtains." Second International Congress on Rock Mechanics, Beograd, Vol. 3, 217–222.

Pratt, H. K., McMordie, R. C., and Dundas R. M. (1972). "Foundations and abutments—Bennett and Mica dams." *ASCE Journal of the Soil Mechanics and Foundations Division*, Vol. 98(SM10), Paper 9290, 1053–1072.

Sabarly, F. (1968). "Grouting and drainage of rock foundations in rocks of low permeability." *Geotechnique*, Vol. 18, 229–249.

Sabarly, F. (1974). "Engineering Geology and Dam Foundations." *2nd International Congress of the International Association of Engineering Geology*," Brazil, Vol. 2, VI–GR1.

Sato, K., Watanake, K., and Kotajima, N. (1984). "Fundamental study on flow resistance in rock fissures." *Soils and Foundations*, Vol. 24(1), 1–8.

Snow, D. T. (1968). "Rock fracture spacings, openings and porosities." *Proceedings of the ASCE*, Vol. 94(SM1), 73–91.

Tschebotarioff, G. P. (1951). *Soil Mechanics, Foundations, and Earth Structures*. McGraw-Hill, New York, 334.

Vaughan, P. R., (1963). *Grouts and Drilling Muds in Engineering Practice*. Butterworths, London, 54.

Witherspoon, P. A. (1986). "Flow of groundwater in fractured rocks." IAEG Bulletin No. 34, 103–115.

CHAPTER 9

Grouting by Contract

Contracts for cement grouting are arranged in various ways. They depend on the practices and preferences of whomever draws up the documents and often reflect national practices—construction industries in different countries do things differently. Briefly, the basic alternatives are:

A. The contractor provides gear, personnel, and materials only. All technical direction and responsibility lie with the "principal" (a convenient name for the owner, client, consultant, "Engineer," or for whomever the job is being done).

B. In addition to the provisions in alternative A, the contractor provides experienced supervisors and organizes the grouting in detail. The role of the principal is to provide inspectors who watch what is going on and do the bookwork but don't interfere unless foolish things are about to be done or provisions of the contract are breached.

C. The contractor does everything. The principal might not even have a representative on site. The contractor is required to produce prescribed results from the finished grouting, and methods for achieving this are not specified in the contract.

Note that some grouting is not done by contract at all. The principal does all the work without engaging a contractor. Known variously as "in-house," "direct labor," "day labor," and so forth, it can be an advantageous method for tricky jobs or where funds are uncertain or sporadic. The method avoids arguments and gives maximum flexibility for making radical changes if found necessary as the work progresses. In the case of government authorities, the choice of doing grouting this way may be politically motivated: the current government might prefer its construction work to be done by contractors or by its own forces. The latter option requires a continuity of large jobs in order to retain the requisite skills and possibly the maintenance of a large plant pool, although much of this might be rented. It has been said that without the profit motive, this method of working tends to produce slower work; on the other hand, my considerable experience with it has produced the opinion that the work is generally of better quality, particularly where there is enlightened and skilled engineering supervision directing the work in close detail. It is the method of choice for hazardous or exceptionally difficult grouting; the use of a negotiated contract (alternative C) is the next best resort. There are sites where the principal has done the tricky grouting and contractors have done the rest.

9.1 CONTRACTURAL ARRANGEMENTS

9.1-1 Which Type of Contractural Working Is Best?

The answer to which type of contract is best depends, of course, on circumstances that are too variable and intangible to exhaustively explore here. However, some generalizations may be of interest.

Common practice is to merge alternatives A and B to some extent. This is usual on work for governmental authorities: inspectors are provided who take more than a passive interest in operations but not to the extent of alternative A. In such an arrangement it is difficult to predetermine just how much involvement the inspectors are to take, and in the course of the work this degree of involvement has been known to change in response to the degree of competence (or lack of it) evidenced by the contractor, the degree of personal harmony between the parties, the emergence of unforeseen technical problems, and so forth. However, it is rarely possible to realistically define the degree of inspector involvement in contract documents, and recourse is usually made to somewhat ambiguous statements to the effect that the contractor will do what he is told. This inhibits the contractor from using highly paid skilled supervision, and the work can suffer because no matter how eagle-eyed the inspectors are, they cannot see everything. Ideally, the contractor's crews should comprise skilled and dedicated personnel who don't need to be inspected at all. Only a few contractors have such crews, but they fairly regularly suffer from having to work with inspectors who know far less than they do and who sometimes peevishly direct foolish action and consequently provoke disharmony and consequent claims for extras.

When contracts are organized in accordance with alternative C, disputes with supervising staff are avoided provided the contractor is sufficiently skilled to really know how to do the work at the time of tendering. However, if he only has vague ideas on how to achieve the prescribed results and intends to learn as he goes or if investigation was incomplete and missed some important factor that will greatly affect the work, then this type of contract can lead to major litigation, poor work, and possibly bankruptcy. Such a contract is the favorite type in some European countries where large specialist companies operate with top-level technical expertise available on call within the company. Such companies also operate in other parts of the world. Some show more interest in doing a good job than in making excessive profits, and these are of course preferred where exceptionally tricky grouting jobs are awarded by negotiation rather than by tendering.

9.1-2 Specialist Grouting Contractors

In addition to previous comments about companies that specialize in grouting work, it should be noted that these companies fall into several categories:

1. Small firms with possibly just one team comprising, typically, an experienced engineer or supervisor, two or three experienced grouting operators and a driller, and perhaps one or two general laborers. Equipment is of the highest quality and is well maintained. Probably there is no office; the home of the owner suffices. The gear may be kept in the backyard. Overhead is low, and tendered rates can be very competitive particularly if the owner is the engineer or supervisor and actively participates in the field work. Small firms of this type have been known to produce better quality work than some of the larger contractors.

2. Slightly larger companies with several teams. Grouting is possibly one of a number of associated specialties.

3. Large companies, some with branches overseas from their home base. Research laboratories and a core of experienced technical people back up their work. Principal activities lie more in the chemical grouting sphere than in cement grouting, and some companies have patents on their proprietary grouts. Publicity sections actively advertise jobs they have done. Some jobs are indeed worth lauding, but there are others where close investigation or personal involvement has indicated a disappointing standard of work; caution is therefore advised in accepting claims of "expertise." Some of the problems in this group of companies arise from the understandable human tendency for supervisors in far-flung locations to battle through problems on their own, reluctant to call on the skill available at the head office, so as to demonstrate their own ability. This has led to some atrocious grouting. However, as indicated earlier, these companies have also done some fine work.

9.1-3 Nonspecialist Contractors

General civil engineering contractors sometimes take on grouting work when it forms part of their main contract or when they are looking for a fill-in job to keep their forces occupied between contracts. Post-tensioning contractors also take on cement grouting at times, using the equipment they have for their grouting work on tensioning cables or bars.

In such occasional grouting operations, supervisors usually come from other types of work and are unlikely to have much cement grouting experience. Therefore, if the principal decides to permit an engagement on grouting, the role of the inspectors becomes more important, and their experience must be able to compensate for the contractor's lack of it.

Bid rates from nonspecialist contractors such as these are commonly lower than those from specialist companies. This is because (a) the nonspecialist does not have to maintain dedicated grouting crews and/or (b) he either is looking for fill-in work between larger contracts or is trying to break into the grouting scene, and hence he is not looking for a high profit on the grouting. This should be borne in mind when comparing nonspecialist tenders with those from specialist companies; if the quality of work is a strong consideration, the lowest bids should not necessarily be accepted.

One method used for obtaining better competence from a nonspecialist contractor is to specify that experienced supervisors will be brought on site who have adequate experience and skills, as determined by the principal. Grouting will then only be permitted to continue while those approved supervisors are personally directing it. This arrangement presupposes that the principal has in turn sufficient expertise available to be able to decide on the suitability of candidates. The method has been used on a number of dams with varying success; problems arise when an approved supervisor leaves suddenly without provision of a replacement and when untrained, unmotivated people form the grouting crews.

9.1-4 Preselection of Tenderers

Some principals only invite bids from a list of competent contractors who have been preselected on the basis of past performance for that principal or for others. This weeds out the incompetents and saves the principal a lot of headaches, and possibly, inspecting staff does not have to be as numerous or as skilled. There should also be less claims for extras and less disputes given that contractors want to stay on the list with subsequent work in mind. However, this restraint does not always do the trick.

Legal considerations prohibit some organizations from accepting any but the lowest tender. This can cause difficulties if the grouting contractor thus chosen is not competent.

9.1-5 Subcontractors

Some contracts permit a general contractor to employ a subcontractor for grouting. The principal then has little direct contact with the subcontractor except at the inspector level. The principal should reserve the right to approve subcontractors before commencement and to discipline them (through the main contractor) if necessary.

Ideally, subcontracting should be confined to specialist grouting firms, but even if contracts specify this, difficulties have been experienced in effectively implementing it. Persistent prevarication by main contractors in the interests of getting their favorite "el cheapo" subcontractor for the grouting have resulted in reluctant submission by busy principals or their consultants.

9.1-6 Role of the Principal

The principal has to decide the standard required for the grouting or some other definition of the prescribed results. This may amount simply to a statement of permeability required after completion of grouting, as for alternative C of organizing a contract, as discussed earlier, or this may form part of requirements within a detailed design, such as for alternative A and B. In this latter case, the design provided by the principal (or by his consultant) varies in scope from jobs where only the general arrangement of the grouting is given to the other extreme where each hole is detailed (a desirable practice in the majority of contracts) subject to revisions found desirable as the work proceeds. (For details of design refer to Section 8.4; for the provision of extra holes as closure proceeds, see Section 8.4-6; and for methods of assessing the work as it proceeds, refer to Section 7.4.)

The principal has to also draw up the contract documents. These are discussed in Sections 9.2-1 to 9.2-8.

Provision and possibly training of inspectors, if used, are also responsibilities of the principal. With large organizations inspectors are on permanent staff and have other duties if there are lapses in the continuity of the grouting work. Smaller principals might hire them for specific jobs; some consultants have inspectors with various talents on their payroll. Further comments on inspectors are in Section 9.1-1.

9.2 CONTRACT DETAILS

9.2-1 Bid Items

Practices vary widely on bidding for grouting. A lump-sum price is usual for alternative C contracts, although some have been done on a cost-plus basis, as have some tricky cases of alternatives A and B. A schedule-of-rates contract is, however, the most workable arrangement for alternatives A and B. Organizations and consultants tend to differ greatly from each other as to the items and what they cover; some people rather uncritically continue to repeat old (and perhaps outdated) practices. Rather than presenting a comprehensive summary of all practices, the following list contains the results of much refining of those various practices and contains my "job-proven" recommendations for bid items:

1. Drilling (refer to Sections 5.2 and 9.2-2).
2. Standpipes (refer to Sections 5.1 and 9.2-3).
3. Hook-up (refer to Sections 7.1 and 9.2-4).
4. Grouting (refer to Sections 7.2, 7.3, and 9.2-5).
   Special extras:
   Slush grouting (refer to Section 7.3-9),
   Division of the "grouting" item into

Curtain grouting, and
Blanket grouting:
Casing grout holes (refer to Section 5.2-8).
Tunnel grouting items.

Items 1 to 4 are discussed in the sections that follow.

Surface preparation for grouting, such as grout caps, is more conveniently included among other excavation and concrete items in general contracts if they have such activities.

9.2-2 Bid Item: Drilling

The drilling item should be divided into separate items for the main stage depths if stage grouting (refer to Section 5.4) is in use. This permits lower rates for the shallow stages, which is fair to the contractor (drilling is easiest in the shallow stages) and helps reduce costs to the principal, these stages being more frequent than deeper ones in most curtains. A high rate can then be bid for the deep stages; this is fair to both parties, remembering the greater cost of deep drilling and the relatively low frequency of these deep holes.

The alternative to this method is to bid the drilling item as a figure to cover all drilling irrespective of depth. Quite apart from the issues just mentioned, the problem then arises that tenderers must judge how much of each drilling depth, and consequent different costs to themselves, will be encountered in the job and then they must average the costs out over the whole contract to strike their tendered rate. This is not fair to either party: if the contractor's estimate is on the low side and he subsequently loses money on the rate, the quality of the drilling can suffer; if the winning contractor's rate is higher than it need be, then the principal loses out a little. These reasons are why it is better for both parties if the rates are in terms of depths; tenderers have less guessing to do, and the principal is freer to order more, or perhaps less, stages in holes.

The drilling rate should be in terms of unit length of the grout hole. In the case of inclined holes this is measured along the slope. The length should exclude a standpipe (refer to Section 5.1-8) and any casing, which are paid for separately. Measurements for payment are usually made by a tally of the drill rods with adjustment for partial drilling at the last rod. Measurements are not taken until the hole has been drilled to the specified depth and washed out to the inspector's satisfaction. This washing is done through the drill rods while they are still in the hole (refer to Section 5.2-14) and must not be confused with the hole washout done later with a washout hose or bit (Section 2.5-18). If holes collapse due to poor drilling practices or for some other cause attributable to deficiencies on the part of the contractor, the unusable parts should not be paid for. Likewise, parts of holes unusable because the contractor has lost drill rods, packers, and so on, in them should not be paid for.

The drilling rates should cover the cost of furnishing all labor, materials, tools, equipment, and operations required for setting up drills; drilling the holes; washing out the holes after drilling; maintaining them free from obstructions until grouted; and all associated work.

Note that there is no separate payment for setting up drills as this is an insignificant part of the cost of drilling if modern equipment is used (Section 5.2-3). The drills can be quickly moved to a hole and set up in a short time; the costs for this should be sensibly absorbed in the drilling rates.

Additional discussion of contractural aspects of drilling is in Section 9.2-7.

9.2-3 Bid Item: Standpipes

Some principals pay for standpipes by length, others by weight. Paying by length encourages contractors to use the smallest sized pipes practicable and is the preferred option on technical grounds, whereas paying by weight can encourage use of unnecessarily large pipe unless the size is stipulated in the contract or directed by the inspectors. The drawback in specifying or directing sizes is that this in turn limits the size of the drilling equipment that has to pass through the standpipe. There are no technical reasons for limiting the maximum diameter of grout holes (Section 5.2-9), and the contractor may have equipment that exceeds the size of pipe selected. Hence there is needless restriction of the contractor's options, which in turn probably leads to greater costs to the principal.

In addition, payment by weight may lead to use of standpipes that have an internal diameter appreciably greater than the grout holes, which can be poor technically. This requires a greater reduction when bushing down to the control fittings (refer to Section 2.6) than would otherwise be the case and presents obstacles to the free entry of bleed water up into the control fittings when routine bleeding is being carried out (see Section 4.3-3).

The contract should specify how far the standpipes are to project from the surface and how far they are to be embedded. (Section 5.1-2 gives recommended amounts.) Measurements for payments by length should be to these limits, which means that in the majority of cases it is the same length for each hole. If the contractor installs some too deeply, payment below the specified depth is made at the drilling rate, not the standpipe rate. This usually costs the contractor money and so encourages correct placement next time. It is not good practice to have standpipes any deeper than absolutely necessary for strength because the entrances to cracks can be obstructed. This does not apply within grout caps, of course. Payments by length are made at the bid rate per foot ⟨meter⟩. Incorrectly aligned pipes are not paid for, nor are loose ones or any damaged beyond use when drilling through them or in other ways.

The standpipe rate should cover the cost of furnishing all labor, pipes, and other materials; tools; equipment; the setting up and drilling of holes for the standpipes; fixing the pipes in position; and all incidental work related to the installation.

The contract should permit the placement of standpipes (and grout holes also in some cases) over geological features away from the general grouting locations. Examples are springs, crevices, faults, and so on, where seepage control may be necessary for construction purposes.

As discussed in Section 5.1-2, it is not always necessary to place standpipes on grout holes. Where the principal permits their omission, no payment is made for the standpipe item. Drilling payments then commence from the surface, and costs associated with surface packers (Section 2.5-15) are included in the hook-up item.

9.2-4 Bid Item: Hook-up

"Hook-up" is a conveniently expressive name for the process of getting everything ready for grouting to start. Section 7.1 details the necessary preparations.

Payment practices for this vary widely among grouting organizations, as do definitions of its ramifications. My recommendations are:

Hook-up should include payment for:

- The water testing in the grout hole prior to grouting (Section 5.3-2). Note that this is not the same as exploratory water testing, described in Section 8.2-6.
- Supply and setting up of the grouting gear and providing any necessary access to the hole.
- Connecting the circulation line, standpipe fittings, and everything else necessary to enable grouting to start (refer to Section 2.7 for details).
- Cleaning sufficient area around the hole to enable prompt detection and appropriate treatment of any surface leaks (refer to Section 7.3).
- Supply and installation of packers whether surface or deep type (refer to Section 2.5-15).
- Cleanup after grouting and removal of equipment.

Hook-up payments should be made on the amount bid per hook-up rather than on some sort of unit length basis.

One hook-up payment should be made for each application of grout to each stage, where holes are individually grouted. This means that if a stage is given several applications (Section 7.2-10), each application will be separately paid for. The example shown in Figure 9.2-4A applies this to a two-stage hole.

At first sight, this may seem an excessive number of payments. However, the method works well. It is fair to both parties because there is no need for guesswork by tenderers when preparing bids, and hence they can bid lower and more realistically than when they have to make guesses and cover their uncertainties by submitting padded bids. Guesswork comes in if hook-up is paid differently, as some principals do, perhaps paying only once for each hole no matter its depth or number of stages or perhaps paying only once for each stage.

Where holes are grouted in groups (refer to Section 2.7-5), hook-up should be paid for each application to the whole group.

The inclusion of water testing in this item avoids needless additional items. The cost of setting up the

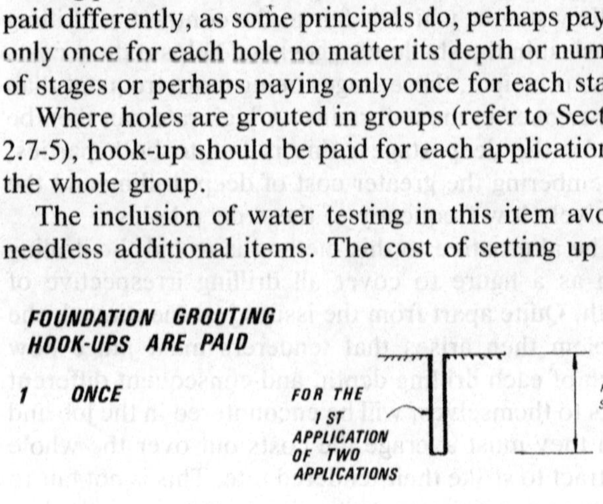

Figure 9.2-4A. Hook-up payments.

gear for a test and doing it (as described in Section 5.3) is almost a constant, and if every stage is water tested before grouting (as recommended), water testing becomes a regular part of each hook-up. An exception is at second and subsequent applications, where a water test before grouting is not usually needed (Section 7.2-10). However, the number of multiple applications is far less than first applications at most jobs, and so the pricing structure should not be greatly influenced by the omission of the water test.

If a water test shows that the stage or hole is tight, or so nearly tight, that grouting would be a waste of time (negligible quantity would be injected), grouting is not proceeded to, and payment for hook-up should be 25% of the bid rate.

Note that there is no separate item for mobilization and demobilization. These are included in the hook-up. This is because of the easy mobility of modern grouting equipment. An experienced contractor should be able to set up mobile mixers, agitators, pumps, and the rest of the equipment in a few hours. The cost of doing this is not worth payment as a separate item and can easily be merged into hook-up payments. An exception could be very large jobs where a central grouting station might be preferred by the contractor (discussed in Section 2.7-12) and there might also be computer recording by means of radio transmission from the grout holes, as commented on in Section 7.7. It can be argued that at some sites central mixing and computer involvement are not necessary (see Sections 2.7-2 and 2.7-12) and that therefore the contractor should not be reimbursed for needless complexity (or even ostentatiousness in some cases, regrettably).

Another cost nominally included in hook-up is that of obtaining the gear. In the case of specialist grouting contractors, it is usually brought from a previous job, with little cost for new items. Nonspecialist contractors usually have to rent the main items (mixers, agitators, pumps) and buy the rest (hoses, gauges, valves, pipe fittings, etc.). However, some tenderers prefer to put these costs in the "grouting" item, as discussed in Section 9.2-6.

9.2-5 Bid Item: Grouting

This item covers the actual grouting. It should include the supply of cement (purchase and handling into the mixers), supply of water and any additives directed, mixing, injecting the grout, washing out after grouting where required, treatment of surface leaks except by slush grouting (see Section 7.3-9), and supply of all necessary labor and equipment for these operations except where included in the hook-up.

The most convenient method of payment is on the number of bags of cement injected or on the weight of cement injected in the case of bulk cement. Measuring methods are detailed in Section 4.1-10. The fairest method to both parties is to pay for cement that goes down the hole even if it then escapes from surface leaks (provided, of course, that the contractor promptly takes suitable action to throttle the leaks—refer to Section 7.3) but not to pay for cement needed to fill circulation lines or lost from burst hoses or leaking fittings. This encourages short circulation lines, which in turn means fresher grout at the hole. The length of these lines is a matter of satisfying a number of factors, including the nature of the terrain, how often the contractor is disposed to move the plant around the site, ease of getting cement supplies to mixers, and so on (see Section 2.7-1).

In this arrangement, cement wasted in the regular bleeding process is paid for provided that excessive quantities are not bled off. Thus there is no disincentive for carrying out bleeding (refer to Section 4.3 for a discussion of the need for bleeding). In addition, this cement is included in the measurements taken (Section 4.1-11), and its exclusion would be difficult.

Payment should be withheld for improperly mixed grout. This includes grout having a different water : cement ratio to that specified (see Section 4.1 for w : c ratios), grout not mixed long enough (Section 2.2), grout that is not uniform, grout containing the wrong type of cement, lumpy or stale cement, and so forth. Nor should payment be made for grout that is too hot or too cold (Section 4.4-4 gives permitted temperature limits) or too old. Cement must be fresh and stored and handled appropriately (refer to Section 3.1-7), and water and any other ingredients must comply with specified requirements.

In contracts where different types of grouting are required, such as curtain, blanket, tunnel, and so on, different bid rates should be provided for each of the types.

9.2-6 Relationship between Hook-up and Grouting Items

These two items tend to overlap in the ultimate analysis. This is no detriment—tenderers can compatibly bias their bids for them. For instance, if a tenderer considers that the job will have a low cement take per unit length of hole (a fairly "tight" foundation), the bids could be biased toward the hook-up rate and the main costs covered by this, knowing that not much money will be made on the grouting item. At the other extreme, in a foundation that will gobble up the cement, the grouting rate will be the profitable one, and the hook-up rate can then be bid on the low side. Differing attitudes to the relationship between these two items have been regularly noticed when comparing tenders from different contractors.

Tender documents should make it very clear that any quantities shown in schedules of rates are only put there to enable tenders to be compared, and they bear little relation to actual quantities. This is necessary to guard against claims for extras later if the estimated quantities prove to be wide of the mark. This sort of claim is prevalent with nonspecialist contractors or subcontractors, who may not have enough knowledge to form their own opinion of what actual quantities will turn out to be and so base their costs on the quantities in the schedule. In other forms of engineering, schedule quantities may bear some relationship to actuality, but rarely do they do so in grouting. This is not to say that they never have; some grouting specialists, including the author, have derived some satisfaction from accurately forecasting the final figures at times (but not always!). See Section 8.4-17 for suggestions about preparing such estimates.

For similar reasons, it is important for contract documents to emphasise that the pattern of grout holes shown on the drawings, or the depths of holes, or stage details, or closure details, or other aspects of the layout only give tenderers a general indication of the work required and can be changed drastically, if found prudent, as the work progresses. In grouting, it can be frustrating to deal with an inexperienced contractor who argues about every slight departure from the initial design. As emphasized elsewhere (Chapter 1), the nature of grouting is different from much other construction work: everyone must be flexible and prepared to make prompt changes in accordance with whatever is found sensible as the work proceeds.

In an attempt to be fairer to both parties, the rates for the hook-up and grouting items have sometimes been adjusted up or down if actual quantities turn out to be vastly different from estimates. For example, if the number of holes or the cement take is appreciably less than expected, contracts have made provision for increased payments to compensate the contractor somewhat. On the other hand, to cover the case where the contractor may be raking in unexpected profits from a surfeit of holes and high cement takes, contracts have provided for payment at lower rates once a certain limit has been passed. This is fairer to the principal.

9.2-7 Further Comments on the Drilling Item

The choice of drilling method should be made by the principal. Selection between percussion and rotary is discussed in Section 5.2-2. If the principal is not sure which method is the more suitable when the contract is drawn up, advisable action is to get bids for both methods and retain the prerogative to direct the contractor as to which to use. Sometimes both are used on the same job, particularly if access constraints make one type of drill more practicable than another. This is discussed in Section 5.2.

The minimum permissible hole size can be specified; refer to Section 5.2-9. There is little point in specifying maximum sizes. Actual sizes adopted by a contractor usually depend on available equipment, and it is not competitive for a contractor to increase the size beyond commonly accepted ones. He also loses money on the standpipe item if holes are too large when standpipes are paid by length (refer to Section 9.2-3).

During drilling, an adequate flow of wash water is required to carry cuttings out of holes. Unexpected large quantities of clay or soil brought up in the wash water and any loss of water return must be drawn to an inspector's notice for logging. Likewise, if the drill runs into unexpected soft material or drops without resistance, it is necessary to immediately advise the inspector (Section 5.2-12).

Excessive use of "rod-dope," grease, or other lubricant on drill rods should not be permitted, but a small amount is often necessary to be able to "break" couplings when pulling rods (Section 5.2-3).

When a drill has reached the required depth, the hole must be flushed, as described in Section 5.2-14. Afterward, when the rods have been pulled from the hole and the drill has probably moved off to some other hole, washout rods or tubes must be used with a strong flow of water to thoroughly wash the hole and make it ready for grouting. Section 2.5-18 describes washout gear.

It is sometimes advisable to require tenderers to describe proposed drills and drilling equipment to ensure suitability for grouting work. Unsuitable gear is then rejected.

To avoid dispute and claims, the specification should advise tenderers if closure grouting (Section 5.5) is to be carried out; the contractor should not be given grounds for complaining about the number of drill movements. The documents should also point out that the number and depths of holes shown on the drawings are subject to revision and change during the work, and no extra payments will be made as a consequence.

Additional stages should be made if drilling water is lost or if a hole makes water substantially or collapses (Section 5.4-2). Some methods for treating leaks may also require special stages.

If drilling equipment is lost in a hole and cannot be recovered, the contractor should be required to drill a replacement hole and bear the expense.

After drilling, holes must be capped to protect them against entry of anything that can clog them or obstruct crack entry.

If the contractor chooses to drill hard grout out of holes instead of washing it out from those lengths where washing is required after grouting (Section 5.4-4 and Section 5.4-5), he should be required to do this drilling at his own expense.

9.2-8 Specification

When compiling contract specifications, there should be mention of matters listed below in addition to matters previously dealt with:

- Grouting method
- Closure grouting
- Stages
- Grout proportions
- Washout
- Leaks and connections
- Grout pressure
- Tight holes
- Equipment

Suggested wording and comment for these now follows.

GROUTING METHOD Where the contract is for curtain grouting, the grouting method(s) (refer to Section 5.4-3) should be stated. This enables tenderers to bid more accurately and hence lower.

Downstage without Packer See Section 5.4-4.

The work shall be carried out by drilling, then washing, then water testing, then grouting, then washing, each stage in succession, completing these five operations before commencing the next stage. Minimum times between completion of this grouting and commencement of work in the next stage shall be:

Drilling of next stage	*6 hours*
Grouting of next stage	*24 hours*

Work shall commence in the stage nearest the surface and shall proceed down the hole in stages until the full depth is reached. Washing out will not be required after grouting of the final stage. A packer shall not be used to isolate any stages except where directed for the purpose of dealing with leaks or specific features. However, a packer may be used at the surface instead of a standpipe, if directed.

Downstage with Packer See Section 5.4-5.

The work shall be carried out by drilling, then washing, then seating a packer near the bottom of the previous stage, then water testing, then grouting, then removing the packer, then washing out the hole from the surface, completing these seven operations in a stage before commencing the next stage. Minimum times between completion of this grouting and commencement of work in the next stage shall be:

Drilling of next stage	*6 hours*
Grouting of next stage	*24 hours*

Work shall commence in the stage nearest the surface and shall proceed down the hole in stages until the full depth is reached. After completion of the final stage, the hole shall be filled by pouring thick grout into it.

Upstage Grouting See Section 5.4-6.

The work shall be carried out by drilling the hole to its final depth, then washing, then seating a packer at the beginning of the lowest stage, then water testing, then grouting the stage, then removing the packer. After a minimum period of 6 hours, these same operations shall be carried out in the stage above.
Note: As discussed in Section 5.4-8, an alternative is to omit the water testing and go straight on to grouting of the stage above, but this is not recommended.
Subsequent stages shall be likewise treated until the surface is reached.

Full-Depth Grouting This is used in single-stage curtain holes and in much blanket grouting.

The work shall be carried out by drilling the hole to its full required depth, then washing, then water testing, then grouting.

Closure Grouting Most curtain and blanket grouting is carried out by closure (Section 5.5). To avoid misunderstanding, the specification should have wording such as the following:

Unless otherwise directed, the grouting shall be carried out in closure sequences whereby the initial holes, designated as "primary" holes, shall be widely spaced, as directed, and subsequent holes, designated "secondary," shall be located midway between primary holes. Further sequences of holes, designated "tertiary," "quaternary," "quinary," and so on, such that each halves the spacing between previous holes, shall be continued until the particular final sequence is reached, which brings the grouting to the required standard of tightness, or is directed.

The drawings show a typical pattern of closure sequence. The principal may direct extension of holes, placement of additional holes, omission of holes, changed inclination, and any other changes considered necessary to achieve the required final standard.

(a) When grouting is by the methods designated "full depth" or "upstage," drilling shall not commence on a hole until grouting is completed in both adjoining holes of the previous sequence.

(b) When grouting is by a "downstage" method, drilling on a stage shall not commence until grouting has been completed in one stage ahead of both holes of the previous sequence.

(c) If there are areas that are not to be closure grouted, drilling may commence in each hole 4 hours after completion of grouting in the prior adjacent hole, unless directed otherwise.

Stages Suggested wording where curtain holes are stage grouted (Section 5.4) is:

Curtain grouting holes shall be individually grouted. Holes deeper than . . . shall be stage grouted with the stages at . . . , . . . , . . . , . . . depths, or at such other depths as directed. Shorter stages may be directed if drilling water is lost, or a large flow of water emerges from the hole, or significant collapse of the hole occurs, or for any other reason as determined by the principal.

GROUT PROPORTIONS Because of differences in opinions about suitable w : c ratios (Section 4.5-5), tenderers should be given an idea of those expected to be used on the job. Suggested wording to match the recommendations of this book for a typical site is:

Grout proportions shall be as directed and may vary in the same application or from time to time as the work progresses. Proportions shall be within the following limits (water : cement by loose volume):

Thinnest grout	*3 :1*
Thickest grout	*0.5 :1*

Note: This thickest mix will require the use of mixers able to produce high-quality grout (explained in Section 2.2). If lower quality grouting can be tolerated for the job, this mix can be relaxed to about 0.8 : 1, which is within the capacity of the lower quality machines.

The proportions of the grout shall be in terms of volume where one 94-pound bag of cement is assumed to have a loose volume of one cubic foot. ⟨Metric equivalents vary in different countries as explained in Section 4.1-3.⟩

(a) Grouting of each application shall be commenced with the injection of 2 :1 or 3 :1 (water : cement ratio by loose volume) grout, except where commencement with a thicker mix is directed because of the known presence of large cracks, voids, leaks, latent or active rock movements, or other considerations warranting such action.

(b) If the quantity of grout taken with the commencing mix exceeds an appreciable volume, as determined by the principal, or the hole is taking grout rapidly, or the mix has been in use for one and a half hours, or for other reasons, thickening of the mix may be directed. Further thickening operations may be directed depending on the progress of the grouting.

(c) Grouting shall continue until (1) refusal is reached, in which case pressure shall be held on the grout, by means of grout in the circulation line, for a further fifteen minutes, or (2) the principal directs that grouting is to cease. This might be consequent upon rock movements, leaks, and/or connections; reaching a take of 100 bags or some such other quantity, as the principal determines.

These requirements accord with recommendations in Sections 7.2-9, 7.3, and 7.5.

(d) Throughout the grouting, holes shall be frequently bled to remove all thin grout, air, and water.

See Section 4.3 for comments on bleeding.

Further applications of grout shall be given in any of the following circumstances:

1. *When the cement take on an application exceeds 0.25 cu ft/ft ⟨0.25 liters of cement/cm⟩ in an application that has concluded with w : c = 2 :1 or thinner.*
2. *When the cement take on an application exceeds 0.50 cu ft/ft ⟨0.50 liters of cement/cm⟩ in an application that has concluded with w : c = 1 :1 or thicker.*
3. *When use of significantly reduced pressure has been necessary for treatment of leaks and rock movements or for other reasons.*

These criteria are discussed in Section 7.2-10.

The minimum time between successive applications to any one stage shall be 24 hours.

On completion of an application, a valve shall be closed on the hole and shall be kept closed until washout or for six hours, whichever comes first.

This is the on-or-off valve immediately above the standpipe (if one is used), shown in Section 2.6-2. With this valve closed, the rest of the grout fittings are taken to another hole.

If the hole had been making water before grouting and if the grouting has been carried out through a packer, the packer shall be turned off and left in place for at least 3 hours after grouting.

This is rather extreme action which will probably result in loss of the packer, but it is necessary in holes that make water, as otherwise the grout may be pushed out by the back pressure in the water and the grouting spoilt. This emphasizes the point that packers are unsuitable in such situations (refer to Section 2.5-16).

WASHOUT With some grouting methods it is necessary to wash grout out of the hole once it has firmed sufficiently in the cracks. This leaves the hole open for another application of grout or for drilling the next stage. Methods are dealt with in Section 5.4-3.

There is often uncertainty about how long after grouting washout should start. If too soon, useful grout may be flushed out of cracks adjoining the hole, whereas if left too long, grout in the hole can become too firm for washout gear to penetrate. Temperature has effects too. In hot weather, about 2 hours is usually suitable, extending to 5 hours in very cold conditions. Some flexibility is desirable in setting the time.

In deep holes, however, it sometimes takes so long to properly flush the upper part of the hole that grout has become too stiff in the lower parts by the time the washing reaches there. There is then no option but to let the grout go properly hard and then drill it out. It may take 12 to 24 hours for the grout to firm sufficiently for this; bogging of the bit can result if grout is drilled before it is properly hard. Contractors faced with having to drill out the deeper stages may elect to drill the upper stages too; in this case there is no additional payment for such drilling (refer to Section 9.2-2).

Unless directed otherwise, the hole shall be washed and/or drilled clean of grout between 2 and 5 hours after completion of grouting of each application, or at such earlier time as may be permitted, except on the final application of the final stage of a hole.

An optional additional clause could read something like this:

However, in holes deeper than 100 ft ⟨30 m⟩
(This depth should be subject to site evaluation and may differ greatly from this figure)
washing out will not be required below this depth. Instead, the contractor shall drill the holes out after the grout has become sufficiently hard, and payment for this will be made at half the bid drilling rate for the depths concerned.

LEAKS AND CONNECTIONS

If during the grouting of any hole grout is found to connect to ungrouted grout holes, action shall be taken as directed. The contractor may be required to immediately wash out all holes to which grout has connected.

If grout flows from any point in the surface, such leaks shall be immediately and effectively plugged or caulked by the contractor. While caulking, reduction of pressure or thickening of the grout may be required.

Action shall be taken as directed to minimize rock or concrete movements and may include pressure reduction, an increase in frequency of measurements of grout take, restriction of the amount of grout applied to the hole, cessation of grouting, and/or interruptions to the continuity of the grouting and may include the actual operation by the principal or his representative of the valves on standpipe fittings.

Any occurrence of sudden loss of pressure or a sudden increase in grout take shall be immediately reported.

These treatment methods are explained in Section 7.3.

GROUT PRESSURE Caution in applying pressure too quickly at the start of each application is advisable (Section 6.2-5). The following suggested wording deals with this and provides a statement of maximum pressures; the format is only shown here; refer to Section 6.2-1 for pressures to insert for the particular site conditions.

Grout pressure shall be measured at the surface at the hole. For the first five minutes of each application, the pressure shall be 15 psi ⟨1 bar⟩ or such lesser pressure as may be directed. If there are no leaks, connections, movements, or other happenings, pressure may be gradually increased to the following maximum pressures as permitted for the stage:

Stage	*Maximum Pressure*
A *to* B *ft* ⟨*m*⟩	M *psi* ⟨*bars*⟩
B *to* C *ft* ⟨*m*⟩	N *psi* ⟨*bars*⟩
C *to* D *ft* ⟨*m*⟩	P *psi* ⟨*bars*⟩
D *to* E *ft* ⟨*m*⟩	Q *psi* ⟨*bars*⟩

TIGHT HOLES In some foundations, the water testing may show certain stages to be too tight to warrant grouting. The permeability of these will probably be about the design standard (Section 8.3-3), or better, without grouting. In the following wording, the permeability is left blank; the figure inserted should be the design standard or 1 or 2 lugeons more.

If the water test in any stage or hole shows a permeability less than ____ lugeons:

(a) The stage, if in a curtain grouting hole, need not be separately grouted but may, if there is to be one, be grouted as an extension of the subsequent stage.

(b) The hole, if for blanket grouting, need not be grouted unless it can be conveniently included in group grouting of other holes. If not grouted, it shall be filled by pouring thick grout into it.

EQUIPMENT Good grouting depends a lot on good equipment, as emphasised in Chapter 2. It has therefore been found prudent to evaluate the contractor's proposed gear; this needs to be done well in advance of proposed use, in good time for any unsuitable items to be replaced.

All equipment will be subject to approval by the principal in relation to its suitability and ability to function efficiently. The contractor shall submit details of proposed equipment sufficiently prior to commencement of its possible use to enable consideration of the proposals and to enable the contractor to obtain acceptable replacements of any items regarded as unsatisfactory.

All grout shall be mixed in effective high-speed, high-shear mixers operating at a mixing speed of more than 1,500 rpm of the "Colcrete" type or equivalent. Mixing time shall be adequate to obtain grout that is uniform and effectively mixed. A sufficient number of mixers shall be available, ready for immediate use, to produce grout at the rates required by the hole or holes being grouted and without interruption due to mixer breakdown.

After mixing, grout shall be discharged from the mixer to an agitator, and then a pump shall draw grout from the agitator and shall pump the grout around a circulation line past the grout hole(s) and then back to the agitator. Grout fittings on the hole(s) shall be directly connected to the circulation line without intervening hoses. All grout used shall be drawn from the circulation line.

Agitators shall be fitted with baffles to reduce vortex formation and shall be provided with accurately marked dip sticks to enable measurements of their contents.

Grout pumps shall be efficient and capable of running continuously for at least 12 hours under normal load. They shall be capable of delivering all necessary quantities and pressures up to the maximum pressure specified and be capable of continuously pumping grout having a water : cement ratio of 0.5 : 1 by loose volume.

As noted earlier, this limit, which is a quite thick grout, can be relaxed to about 0.8 : 1 for lower quality work. Discussion of grout pumps is in Section 2.4, and locating grout pumps is discussed in Section 2.7.

Grout pumps shall be of the helical-rotor, progressive-cavity type.

They shall be located suitably to obtain pressures in the circulation lines at the holes that are not excessively greater than those required for grouting.

Comments on circulation lines are in Sections 2.7.

The maximum internal diameter of circulation lines shall be one inch ⟨25 mm⟩.

Control of grout injection shall be exercised by means of valves mounted in fittings directly on the standpipes of holes, or on the grouting tubes of packers, without the use of intervening hoses. A pressure gauge shall be mounted between the main control valve and the hole together with a valve, which can be discharged to waste, through which thin grout, water, and/or air are to be bled off periodically. A valve may be mounted on the circulation line immediately beyond each offtake to a hole.

The previous wording requires expansion on jobs where computer monitoring of takes, w : c ratios, and pressures is required. See Section 7.7.

Pressure gauges shall be of an approved type and shall be checked against a master gauge at maximum intervals of 7 days. Each gauge shall be fitted with an approved device to prevent entry of grout to the gauge mechanism. Each gauge shall be identified by a number.

Pressure gauges are dealt with in Section 2.5-9.

Valves in contact with grout shall be of approved diaphragm and/or plug cock types.

See Section 2.5-1 for comments on valves.

Packers, if used, shall make use of expansible tubes or rings of rubber or other suitable material. When expanded in a hole, they shall be capable of withstanding, without leakage, water pressure equal to the maximum grout pressure used. The tube conveying grout through them shall have a minimum diameter of one inch ⟨25 mm⟩.

Packers are described in Section 2.5-15 and their problems (somewhat discouragingly) in Section 2.5-16.

Adequate standby equipment shall be available in order to ensure that no interruptions to continuous grouting occur.

All equipment shall be maintained in a clean and functional condition.

Effective means of communication shall be maintained between personnel at the hole and at the pumps.

OTHER MATTERS The suggestions given are by no means exhaustive. Most jobs require quite a number of additional matters to suit local considerations. A good specification writer anticipates them.

CHAPTER 10

History of Grouting

The history that follows does not pretend to be exhaustive; it concentrates on matters relevant to cement grouting and shows how modern grouting has improved in some ways but not others. One of the purposes of presenting it is to reduce the amount of "reinventing the wheel" to which the grouting profession appears very prone. I have been dismayed by this on a number of occasions: people without much grouting background wading once more through all the elementary methods, with no cognizance of the improvements that have been made by others. Admittedly the paucity of published information on the subject has contributed to such situations. Hopefully this book will redress the matter somewhat.

The history deals with development of various types of cement grouting; it is not confined to rock grouting.

Glossop (1960) presents an excellent study of the subject, from which much of this chapter is drawn. The reference has not been repeated where material is taken from his work but is given for other sources.

10.1 CEMENT GROUTING HISTORY

10.1-1 Origin of the Name "Grout"

Several different origins of the word "grout" have been claimed. They include:

1. The Middle English word "grút" meant coarsely ground meal. The word was later used for porridge and, by analogy, came to be used for a liquid mortar of similar consistency. Glossop (1960) follows this explanation of the derivation, noting that Smeaton used it in this context in his book of the Eddystone Lighthouse, and then goes on to quote from a 1776 book, *A Treatise on Building in Water,* by George Semple:

> My Father, (who was a Workman about the Year 1675) often told me, and my own repeated Observations convince me, that the Method Masons practised in former Times, in building Churches, Abbeys, Castles, or other sumptuous Edifices in this Country was to this effect. After they laid the Out-side Courses with large Stones laid on the flat in swimming Beds of Mortar, they hearted their Walls with their Spawls and smallest Stones, and as they laid them in, they poured in plenty of boiling Grout or hot Lime-liquid among them, so as to incorporate them together, as if it were melted Lead whereby the heat of it exhausted the Moisture of the outside Mortar, and united most firmly both it and the Stones and filled every Pore (which as the Masons term it) set, that is, grew hard immediately, and this Method was taught to our ancient Masons, by the Romish Clergy that came to plant Christianity in these Countries. (p. 98)

Glossop states that the word "grouting" was used only by masons and bricklayers until adopted by tunnel miners for the method of injecting a lime slurry behind cast-iron liners, invented by Barlow in 1864 and improved by Greathead in 1886. The term then spread to cover a range of injection processes.

2. A slightly different origin, but with some resemblance, is attributed to the old Dutch word "grute," a plant used for the flavoring of beer (Golder, 1958). He states that "it is cognate with grits and groats and used to mean sediment, dregs or lees" (p. 1396) that is, a suspension of small solid particles.

3. The French for grout is *coulis,* a gravy or soup that has been strained through a sieve or cloth. Another term is *lait de ciment.*

4. The Portuguese and Spanish names for grout also refer to milk.

Other points of note are:

The word "cementation" has been used interchangeably with "grout" in England until quite recently. A large grouting company there uses the name in its title.

The verb "to grout" dates from the early nineteenth century.

10.1-2 Some Interesting Dates

ANCIENT ROMAN TIMES THROUGH THE MIDDLE AGES Grouting was used in the strengthening of walls, bridges, aqueducts, and so on.

1802 Grouting was used to repair a sluice at Dieppe, France, by Charles Berigny. This structure had been out of commission since 1797 due to piping of the alluvial foundation at a depth approaching 20 ft ⟨5 to 6 m⟩ below the low-tide level. Measures proposed for repairing it were first tried on a nearby timber weir and consisted of drilling holes at 1-m centers into the foundation. A crude type of piston pump was used to force clay grout into these holes. The pump was a tube of about 3 in ⟨80 mm⟩ internal diameter and converged to approximately a 1-in ⟨30-mm⟩ nozzle at the discharge end. A removable wooden piston was driven by hitting it with a heavy mallet. After each stroke of the piston, it was lifted out, and the tube was partly filled with clay grout. The piston was then put back in again and driven down to expel the grout into the foundation. Initial trials at the weir proved successful, although the grouting had to be tightened in subsequent years.

The method was then tried on the sluice. This was more difficult because its masonry base was more than 5 ft ⟨1.6 m⟩ thick and voids under it had become filled with silt. This silt was removed by drilling a large-diameter hole through the base and working a piston in it consisting of a wad of tow on the end of a rod. This agitated the silt, and the silty water then pushed past the tow wad on downward strokes and was lifted out by upward strokes. Mortar grout was used rather than clay grout so as to improve the bearing capacity of the foundation. Italian pozzolana was used because hydraulic lime was not available in France (although it had been developed in England during the previous century) and Portland cement had not yet been discovered. After several trials of the tube pump used earlier, it was discarded because the holes drilled through the masonry base were found adequate to act as the tube of the pump. The holes were filled with the liquid grout, and a tow ball was placed on top. A wooden piston then drove the grout into the foundation. After completion of grouting, some excavations through the grouted area showed the grout to have set perfectly.

FROM 1802 TO 1809 Grouting was used for the repair of locks at several ports in France and for construction purposes in the building of a lock at Dieppe. This lock required dewatering to a depth of 40 ft ⟨12 m⟩ below high tide. In order to minimize blow-ins, Berigny injected pozzolana mortar grout into the foundation as the work proceeded.

1818 Attempts were made to stop leaks into a 130-year-old naval dry dock at Rochefort that had been almost completely rebuilt in 1775. Serious leaks made constant pumping necessary. Grouting was carried out using pozzolana mortar grout injected with a tube pump having a 6.5-in- ⟨160-mm-⟩ diameter barrel and a leather-lined piston driven by a pile-driving type of monkey. The work was not completed because of a decision to rebuild the invert but was found to be successful where done.

1820 Mary grouted a lock on the canal of Saint Quentin. He (not she) used hydraulic mortar grout.

1831 In "Concreting by Injection," Charrie complained of the lack of information about grouting. A case history concerned a lock on the Rhone–Rhine canal built in extremely permeable ground on the bank of the Rhine. A concrete foundation raft was placed underwater with underfloor drains that were grouted on completion of the concreting with a mortar of three parts cement and one part hydraulic lime. A wooden percussion pump was used similar to those used in earlier jobs but of square cross section. The grouting was a success, although slight leaks remained.

In most of these grouting jobs it was found necessary to have relief holes available from which grout, water, mud, laitance, and so on, could freely escape while grout was injected into each grout hole.

1837 Raynal wrote a paper on the use of grouting for repairing masonry. His pump was made of elm wood 27.5 in ⟨700 mm⟩ long bored out to 2.5 in ⟨60 mm⟩ over most of its length to accommodate the piston. The remaining length was bored to only 1.3 in ⟨30 mm⟩ and formed the lower end, leading into a nozzle of smaller size that fitted into the grout hole. He used hydraulic quick lime for sealing leaks through masonry in lock walls. The lime was stored in an airtight bin and was ground and screened immediately prior to mixing 0.4

parts of it with 0.6 parts of pozzolana and sufficient water to make it semiliquid.

1838 Collin invented a pump for grouting cracks of capillary size in the masonry of a dam. He found that previous pumps were not satisfactory for this. His pump was made of cast iron with a piston operated by a lever on a rack and pawl. He regarded the steady pressure produced as better than the pulses of the earlier pumps. Comment on steady pump pressures is in Section 2.4-12.

1839 Beaudemoulin gave an example of the use of grouting to strengthen a bridge foundation. A bridge at Tours had had various collapses and settlements during construction. Large cavities were found under timber caisson bases beneath piers. The cavities were grouted through holes drilled from the roadway, with relief holes nearby. The grout used hydraulic lime, without sand, at the consistency of nearly liquid porridge. A static head was sufficient for placement of most of the grout and was followed by pumping using a metal piston with a clack valve. The piston was on a heavy drill rod, on which was threaded a heavy monkey operated by seven men. Beaudemoulin, like Collin, concluded that better grouting is produced by steady pressure than by the pulsing of a piston pump. He also expressed a preference for omitting sand from grout.

1840 A much longer pump was used in remedial grouting at the Cherbourg lock, where 345 holes drilled through the masonry were used to inject a grout of moderately hydraulic lime and cement. The pump had a barrel of elm 15.5 ft ⟨4.7 m⟩ long and a 4-in ⟨100-mm⟩ bore strengthened with iron bands. The piston was timber with leather packings.

1845 In the United States William E. Worthen grouted the foundations of a flume. He "put holes along wherever it was unsound and made a syringe to inject the cement under the foundation" (p. 100). In 1854 he grouted a masonry pier on the New Haven road at Westford.

FROM 1856 TO 1858 In England W. R. Kinipple carried out experiments to prove the practicality of grouting loose stones together under water to form concrete in situ and the suitability of grouting for sealing cracks and for the repair of marine damage to structures. Kinipple regarded himself as the inventor of cement grouting. He had trouble convincing others of the soundness of his methods and subsequently wrote: "In order to make sure and to give confidence, if possible, to those who, up to 1882, had doubted my system of grouting—and I am sorry to say almost every engineer I know still doubts it—some experiments were carried out under my direction . . . at Aberdeen in July 1883." He further wrote:

> As the inventor of the grouting system, I may be inclined to attribute more importance to this method of obtaining sound and substantial subaqueous work than is in reality due to it, but knowing from experience the excellent work which I have done by this means, without accident or loss of cement, sometimes under the severest conditions, and always at low cost, I do not think I can speak too highly in its favour, or that I can make it too widely known, for I feel sure that it only requires to be known and thoroughly tried to be adopted in the majority of cases (pp. 273, 274)

He used two methods: the first was to apply neat cement grout under gravity head, and the second was use of puddle clay forced in by a "stock rammer" in those cases where large channels permitted too much loss of cement before it could set. The stock rammer was a hollow wooden piston with a clack valve over the foot of the central hole. On upward strokes, the clack valve opened and admitted air and so broke any tendency for a vacuum to form. Grout holes were lined with a wrought iron pipe in which the rammer operated. Puddle clay was put into the pipe in lumps.

Kinipple made frequent use of grouting, including a coffer dam at Greenock in 1871.

1864 At one of the Rhein-Preussen mining shafts near Homburg in Germany, a break in the brickwork lining was sealed "by milk of cement being run in from the surface with the aid of a small hand pump."

1869 When building the first cast-iron-lined tunnel under the Thames, the Thames Subway, Greathead filled the space outside the liners with a lime grout injected with a hand syringe. He wrote that "the result was not satisfactory owing to the lack of pressure, and to the too great fluidity of the mixture" (p. 275). In 1940 this tunnel was damaged by a near miss from a bomb. During repair it was found that the old grouting was satisfactory up to the axis and above it, but the grout had often failed to fill the space above the crown of the liner.

In his patent of 1886, Greathead improved on this by the use of a compressed air grout pan (Section 2.4-8). This was a pressure vessel containing grout. Compressed air applied to the grout expelled it into the job through a hose. The method found widespread use in tunnelling work. About 1888 Greathead improved the pan, increasing its capacity and including hand-operated stirring blades. This improved pan has continued in use until recently.

1876 The first known usage of the type of Cement grouting forming the subject of this book was by Thomas Hawksley to seal fissures in the rock beneath an earth dam. This was at the Tunstall reservoir of the Wearsdale Company in England (Durham water supply). In 1877 and 1878 he used the same process on the Cowm reservoir of the Rochdale Corporation (Manchester). He put a number of holes down to rock and injected them with cement. This measure was successful until 1886, when leaks reappeared. Further treatment by grouting cured them. The cement used was a Portland cement manufactured since 1836 by the firm of G&T. Earle of Hull and known as "Earle's cement." A tradition avers that Hawksley said "he wished he had never seen Rochdale." Glossop comments: "This procedure of rock grouting, which is now standard practice in dam construction, was an invention of the greatest importance to engineering practice, but its adoption by civil engineers was slow" (p. 259). However, "A few years after Hawksley's work, rock grouting was introduced to the coalfields of Northern France, and from that time the development and use of injection processes proceeded continuously and with increasing speed, at first in mining and later in civil engineering practice" (p. 259).

1882 Reumaux, when sinking a colliery shaft at Lens, France, sealed off a water-bearing fissure by injecting cement into it through a borehole under gravity head. Grouting during shaft sinking then developed into an extensively used process. Powered pumps were developed.

1882 TO 1890 Colson, while constructing a dock at Malta, checked inflows through fissured rock by cement grouting, using "a double-barrel force pump . . . specially constructed with ball valves that could be easily cleaned" (p. 259).

1882 Leaks in a graving dock at Greenock that had been built by James Watt in 1785 were completely stopped by Kinipple using cement grouting.

1884 APPROX. Kinipple, as consulting engineer, used cement grouting extensively at St. Helier, Jersey, in the lengthening of a breakwater. To construct the foundation, which was at a depth of 60 ft ⟨18 m⟩ below high tide, he removed overburden to expose bedrock and then backfilled the resulting trench with shingle. After leveling by a diver, the shingle was grouted in place to form concrete. The grouting was done through a standpipe slung from a platform hung from a crane. This platform accommodated the mixing vats, 40 bags of cement, and eight men. The work was successful.

1891 R. L. Harris, while building new piers on old foundations of a bridge carrying the New York and Northern Railway across Croton Lake, made a coffer dam by cement grouting loose rock of the old foundations to form grouted concrete. This was successful. He used a grout of equal parts Portland cement and sand injected through a 1.5-in ⟨38-mm⟩ pipe by means of a No. 2 Douglass hand-operated pump. He wrote:

> A small quantity only of grout was slowly forced through, and the nozzle was then withdrawn, but the hole maintained . . . seldom returning to any hole on the same day; my belief being that in quiet water the cement would accrete on the surface of irregular stones at and below the level of injection, and that by consecutive slight accretions at proper intervals of time the voids between them would be filled. (p. 274)

Harris said that the idea of "successive accretions" occurred to him when he was camped beside a spring "where twigs became encased [with a lime incrustation] in a single night. My belief is that my mode is different from ordinary grouting, that it is new, and also is an application of the principle of accretion, or adding, little by little, similarly to the process of nature" (p. 274).

FROM 1892 TO 1893 Water pressure tests in diamond-drilled holes were used at a dam site on the Genesse River, New York State. At first the packers used to isolate test stages were made of cylindrical bags filled with seeds that swelled when wetted, but subsequently rubbers were used. Test pressures were 100 to 200 psi ⟨7 to 14 bars⟩. Following testing, bran was added to the water. If this reduced the take, it was assumed that cracks were fine. This appears to have been the first use of water testing as a site investigation method. It was developed by Rafter.

1893 At New Croton Dam, New York State, the grouting process seems to have been applied for the first time systematically and on quite a large scale to deal with fissures in the limestone foundation and to strengthen it. Compressed air drills were used to put test holes down. These were usually 12 to 14 ft ⟨3.7 to 4.3 m⟩ deep with a few holes to 100 ft ⟨30 m⟩. When fissures were met, an attempt was made to wash them out with water. They were then grouted using a No. 2 Douglass pump operated by four or six men. In some cases gravity head was sufficient. Another account of grouting at this dam reports that fissures were "stock rammed" with clay at the suggestion of the contractors. A small pile driver forced the clay down a 3-in ⟨75-mm⟩ pipe.

1896 Leaks at Egyptian dams at Damietta and Rosetta and locks at Shubrah were stopped by "stock ramming" with clay, hydraulic lime, and cement on Kinipple's advice.

The Delta Barrage of the Nile had developed underscour and cracks. Remedial grouting was carried out with Kinipple as consultant. Voids below the main barrage were successfully grouted through holes from the crest using Portland cement applied under gravity head. A subsidiary weir was built downstream to reduce the head across the barrage and was constructed entirely as grouted concrete under water. Several thousand tons of cement were apparently used in this.

1901 A grouted cutoff was constructed for Hinckston Dam, United States, an 85-ft ⟨25-m⟩ high cinder fill. Grout was poured into the grout holes and then pushed down with compressed air.

1908 Severe leaks from two Walshaw Dean dams of the Halifax Corporation, England, were treated by H. P. Hill, who put down grout holes into the foundation below the cutoff trench. The commencing grout mix had a w : c ratio of 20 : 1, and this was thickened slightly to 10 : 1 as injection proceeded. The amount of cement used was 1386 tons ⟨1400 t⟩.

1910 Grouting of Estacada Dam commenced. The dam is located on the Clackamas River 25 miles southeast of Portland, Oregon, on a volcanic breccia that is highly permeable. In 1915 the grouting was reported in the first technical paper devoted to the grouting of a dam foundation. This was presented to the ASCE by Rands (1915). It caused much interest and produced a discussion (also in Rands, 1915) much longer than the original paper. The discussion indicates that grouting on a small scale had been carried out on a number of dams in the United States as far back as 1899, and the great interest taken in grouting was due to a large number of dam failures occurring in the early years of the twentieth century.

The paper dealt with an early case in which a grouted cutoff was used. The idea of using grouting for this purpose was proposed by the consulting engineers, Sellers and Rippey: "Thorough enquiry failed to disclose any precedent for the use of grout in the general treatment of foundations, although it was found that cavities in limestone rock under the New Croton Dam had been filled with grout much as a dentist would fill a cavity in a tooth" (p. 277 of Glossop, 1961). In the discussion (Rands, 1915), Rippey referred to overseas reports of successful grouting of dam sites. The general idea at Estacada

> "provided for drilling a double line of holes of an average depth of 50 ft ⟨15 m⟩ under the heel of the dam across the entire valley to and under the shore abutments and the subsequent forcing into each of these holes of grout of such consistency as to percolate through the entire substructure. . . . After drilling the double line of holes, the program contemplated the test of each hole with water pressure, a record being kept of the quantity escaping and the pressure applied to each hole. Upon completion of the tests, the cement grout was to be pumped into the holes under pressure and after allowing time for hardening, a third line of holes was to be drilled midway between the first two or outer lines and tested with water pressure. The idea was that if water pressure was applied to the center holes at slightly above the hydrostatic pressure to which the rock would be subjected by the water in the reservoir after completion of the dam and no appreciable leakage occurred, it would be reasonably certain that the cement grout had proven effective in making the entire foundation impermeable. . . . The plan of grouting was that of two lines of holes, 6 ft ⟨1.8 m⟩ apart, with 6 ft ⟨1.8 m⟩ between holes in the same line. (pp. 263, 264 of Glossop, 1961)

Using diamond and calyx drills, 34,000 ft ⟨10,000 m⟩ of holes were drilled. The grout was mixed and injected with a Canniff "grouting machine," a device used at a number of U.S. jobs at the time. This machine had two identical drums. Mixing commenced by putting cement and water in one and then bubbling compressed air up through it to provide a crude form of mixing. This batch was then sent down the grout hole by applying compressed air to its surface in much the same way as the Greathead pan. While this batch was being injected, the next batch was mixed in the other drum, and when ready, valves were switched to send this batch down. The process continued in the same way, alternating between drums. Injection was not continuous. In the discussion (Rands, 1915), Hulse suggested that it might be operated with water instead of air. This was later done on the Catskill Aqueduct. The air operation had tended to produce balls of grout.

Drills were steam-driven Sullivan diamond rigs and shot drills electrically operated. The starting grout pressure on each hole was 25 psi ⟨1.8 bars⟩, which was increased to 200 psi ⟨14 bars⟩ finishing pressure. W : c ratios varied from 1.5 : 1 to 15 : 1, but best results were reported with ratios of 3 : 1, 4 : 1, and 5 : 1. Circuit grouting was tried but discarded. A grout cap was found necessary.

Differences of opinion developed between the consultants and client about the dam. It was another case of the well-known story of a client complaining that too much money was being spent on the grouting while the consultant considered that not enough was being spent, and anyway blaming the contractor for not doing the work according to instructions.

1911 At Hatfeld colliery near Doncaster, England, the process known as "silicatization" was successfully used in shaft sinking. The process was developed by Albert François, a Belgian, and used a solution of aluminum sulfate and sodium silicate followed by cement grout. François regarded the chemicals as lubricants enabling the injection of a greater quantity of cement. He missed the point that they were a chemical grout in their own right. However, the method spread and was widely used in shaft sinking, particularly in South Africa.

1919 Asphalt grouting was used by G. W. Christian at Hales Bar Dam, Tennessee, in cavernous limestone where in spite of the use of several thousand tons of cement in grouting, the dam leaked seriously on first filling. Attempts were made to stop the flow using "heavy rags, old carpets and burlaps, covered with cinder concrete reinforced with strips of wire mesh . . . but . . . as soon as one inlet was sealed, another would open nearby" (p. 278). In an attempt to remedy this, asphalt was injected hot through a heat pipe. When it left the pipe, it cooled and stiffened quickly. 11,000 barrels of asphalt were injected, and this was partly effective. The method was subsequently used to a limited extent on other sites.

1925 Joosten patented the chemical grouting process now known by his name. It enabled grouting to be carried out in alluvium, down to sand sizes. The process consists of injecting two chemicals successively. These react in situ to form a gel. He used sodium silicate and a strong electrolytic saline solution. Guttman later invented an improved process. The subject of chemical grouting is well covered elsewhere (Karol, 1983).

1932 Hoover Dam (formerly Boulder Dam) marked the development of a systematic approach to dam grouting. During design, engineers of the U.S. Bureau of Reclamation studied current practices to draw up the specification. This job marked the acceptance of cement grouting as a normal feature at dams for both foundation grouting and for grouting the contraction joints between the concrete blocks forming the dam. Additional grouting was necessary in 1938 to 1947 to overcome deficiencies in the original work. A. Warren Simonds, a Bureau engineer engaged on this work, presented several papers on the work and on subsequent dams: Simonds (1947, 1958) Simonds and Boggess (1950), and Simonds et al. (1950). After retirement, Simonds became a grouting consultant; among other activities, he visited Australia twice. I recall with pleasure the helpful and interesting discussions on grouting held with him on those occasions.

1933 The Tennessee Valley Authority commenced an extensive dam-building program. Many were on difficult foundations, such as limestone, and development of appropriate grouting methods was necessary. Clay–cement grouts were found useful in dealing with cavernous foundations (Leonard and Grant, 1958).

1933 Maurice Lugeon published his permeability criteria for deciding whether grouting is necessary in rock foundations for dams. The permeability unit he introduced has been named after him (Section 8.2-3).

By 1933 grouting had achieved general acceptance, and so many jobs have used it since, it is not appropriate to try to list them here. Instead, a few happenings and developments are mentioned that gave direction to improvements.

1949 TO 1956 The U.S. Army Corps of Engineers conducted a series of tests to study the behavior of cement grout penetrating fine cracks. The tests used artificial cracks between slabs of concrete (Section 6.1-6) and confirmed the advisability of using the thickest mixes possible. (USCE, 1956; as summarised in Kennedy, 1961).

1950S The old process of mud-jacking for lifting structures and road pavements was revolutionized by the work of James Warner, Douglas R. Brown, and others with the introduction of compaction grouting. This process uses a zero-slump grout made of cement, sand, and water that compresses (i.e., compacts) any weak materials in a foundation and can permanently raise roads and buildings and compensate for settlement (Section 1.2-1). It is at the other end of the grouting spectrum from penetration grouts and can be a very useful method for these purposes (Brown and Warner, 1973; Warner, 1978; Warner and Fera, 1977).

1958 A series of papers outlining current practices of large organizations was published by the American Society of Civil Engineers under the heading "Symposium on Cement and Clay Grouting of Foundations" (1958).

1958 Vacuum grouting, a development of mine, was introduced to enable inaccessible uphill spaces to be grouted from below through a single hole in locations where space limitations prohibit insertion of a vent tube. The development was found necessary when carrying out contraction joint grouting at Warragamba Dam, Australia (largest concrete gravity dam in the Southern Hemisphere). The process consists of using an ejector to remove air and water from the void followed by admission of grout into the resulting vacuum. Vacuum grouting is used for grouting small voids under steel penstocks in tunnels (called "interface" grouting) and other specialized applications (Section 1.2-1).

1963 A symposium was held in London on "Grouts and Drilling Muds in Engineering Practice" at which some advanced research and case histories were presented.

1979 An annual Short Course on Grouting was commenced by the University of Missouri-Rolla at St. Louis for practicing engineers and geologists. This course disseminated know-how to hundreds of attendees from various parts of the world and has additionally been given in other countries.

1982 A significant grouting conference sponsored by the American Society of Civil Engineers was held at New Orleans, at which leading grouting practitioners gave a thorough review of the state of the art as well as many case histories. This was the largest grouting specialty meeting to that time (and may hold the record for some time to come). Its proceedings, edited by Wallace Hayward Baker, a leading grouting engineer, are now a text on the subject.

This brings us to the present and provokes the thought as to what will be the next significant factor in the history of grouting. To stimulate thoughts on this, try Section 7.7-14.

10.2 HISTORY OF GEAR

10.2-1 Introduction

An essential part of all good grouting is the right gear. As the previous section indicates, grouting development has depended on gear development. This section enlarges on the history of selected items of equipment. It does not pretend to be thorough.

10.2-2 History of Grout Mixers

The elementary mixers were stirring devices of various kinds, often improvised from a drum and some sort of paddle, hand worked at first and powered later. Paddle mixers of this type are still in use (see Section 2.2-4), although now regarded as inferior to better engineered machines (Section 2.2-2).

HISTORY OF JET MIXERS In 1920 a young man named Erle P. Halliburton, pioneering the cementing of oil wells, introduced the jet mixer (Section 2.2-5) as part of his cementing service. He recalled the method used while he was working as a seaman for steam jetting of cinders from the boiler rooms on steam ships and developed this into a simple means of mixing cement and water (Millikan, 1971). He applied for a patent in 1922 and used jet mixers subsequently in his business, which has now become extensive.

HISTORY OF COLCRETE MIXERS One of the earliest improvements was the development of the Colcrete mixer (described in Section 2.2-2). This development was not so much for grouting purposes as for making higher strength concrete and commenced in the late 1920s in England. J. S. Morgan was looking for ways to do this by improving on the traditional mixing methods for concrete. His aim was to improve wetting of the cement particles and get better distribution of the cement in the water than is possible in conventional tumbling concrete mixers. He made trials with a colloidal mill; these obtain the wetting of powder by means of a rubbing and shearing action. Each particle is caused to roll through the liquid in such a way that any films of air or other gas are disturbed so that the full surface of the particle is brought into contact with the water. Clumps of particles are broken up.

These tests were promising and gave improved concrete when the cement was mixed in a mill and then added to the sand and coarse aggregate. Morgan was granted a patent for the process in 1930. J. C. Gammon then became associated with Morgan and initiated the idea of adding sand in the mill and of forming concrete by grouting coarse aggregate with this grout. This makes grouted concrete (concrete where the coarse aggregate is placed in its final position in the formwork and then a sand–cement grout is pumped into the voids; shown in Section 1.2-1). The process is known as Colcrete in the United Kingdom (Prepakt is the American equivalent). A provisional patent for making concrete in this manner was applied for in 1932, and the final patent was granted in 1934. Gammon continued the development when Morgan left in 1935 to do other work. However, in 1939 demands for the mixers from the War Office led to Morgan's return. Up to this time, one mixing unit made the cement grout and another was used for the addition of sand. Both units were colloidal mills and each required an 8-hp motor.

The system was cumbersome, and only very fine sand could be used.

The breakthrough came in 1941 when a vortex drum was added. This was the most important advance and made the machines able to handle coarse sand and provide very efficient mixing.

Early usage was largely confined to grouted concrete, but when the superior quality of the cement grout produced became appreciated, the machines came to be used for rock grouting and for special types of grout such as lightweight and foam grouts. Today their use has spread even further into virtually all forms of cement grouting.

British contractors used them overseas after World War II, and usage spread to other countries, including Australia, in the early 1960s. They, or copies of them, are now in use in most parts of the world.

HISTORY OF CHEMGROUT MIXERS D. C. Dahl and Ben Schatz started the company Chemgrout in 1963 after gaining on-site grouting experience with various jobs. The company initially had a contractor and consultant role but later became a large manufacturer of grouting equipment.

Early mixers were of the paddle and propeller types. Their first "colloidal" type (pictured in Section 2.2-2) was of the modified centrifugal pump variety recently replaced with a new type (also mentioned in Section 2.2-2).

HISTORY OF HÄNY MIXERS Häny & Cie AG was founded in Switzerland in 1875. Primarily a maker of centrifugal pumps, the first grout pump was sold in 1930. Based on the firm's knowledge of centrifugal pumps, the first "colloidal mixers" were made in 1973 for the mixing of bentonite and cement for slurry walls. Success led to the design of a range of standard mixers for cement, bentonite, fly ash, coarse sand, and other additives.

10.2-3 History of Pumps

The history of pumps is a story of ingenuity and skill.

As described earlier, in 1802 Berigny used a crude type of piston pump where the pressure for each stroke was provided by a hefty blow from a mallet, and the grout was put into the cylinder between strokes.

The rather limited output of this one-manpower pump was improved on by recourse to use of a monkey, the pile-driving type.

Various derivatives of these improvisations (which is what they were) found use in the first half of the nineteenth century. A variation came along in 1838 when Collin had the job of grouting cracks of capillary size in masonry. He found it necessary to develop a new type of pump. This was described earlier and featured a piston operated by a lever on a rack and pawl. Output was slow.

About 1845, in the United States, William E. Worthen (1891) used what he called a "syringe" to inject grout. This was a simple form of pump.

In England in 1856 to 1858, Kinipple's "stock rammer" made early use of a valve, although its purpose was to avoid formation of a vacuum in the job on the upward stroke rather than for the subsequently developed arrangement whereby valves permit a fresh charge of grout to be drawn into cylinders at each stroke of the piston pumps.

A departure from the piston concept came when Greathead, in England, invented the compressed air grout pan (also mentioned earlier in this chapter and in Section 2.4-8). This was a pressure vessel in which grout was exposed directly to compressed air. The air pressure pushed the grout through a hose and into the job. Although simple, the method has drawbacks, for example, compressed air may enter the grout, so that the job may become grouted with air as well as grout. This is usually not satisfactory, particularly if high-pressure air bursts away from the grout and damages the job and possibly its surroundings.

Toward the end of the nineteenth century, power-operated piston pumps came into use on the larger jobs then undertaken. There were several different manufacturers making pumps mostly intended for other duties but adapted for grouting service. Brands included Burton, Francois, Weisse and Monsky, Guyan, Cameron, Douglass.

However, the concept of using compressed air to inject grout retained some popularity, and the Canniff "Grouting Machine" was an example used at several U.S. jobs at the time. It was described earlier (Section 10.1-2).

Around this time, some pumps had pressure gauges that automatically recorded the pressure. Others limited the pressure to preset amounts.

One of the main problems with reciprocating pumps is their valves. As discussed in Section 2.4-9, these can be various types of balls operating in seats or they can be diaphragms. Additionally there are special types. Maintaining valves in an operable condition throughout a long grout application can be difficult with poorly designed valves, and much research has gone into the matter of improving them.

As a complete alternative, the helical rotor types of pump, which has no valves, came into use for grouting work in the late 1950s.

The advantages these have over earlier pumps are detailed in Section 2.4-2. They have had wide acceptance.

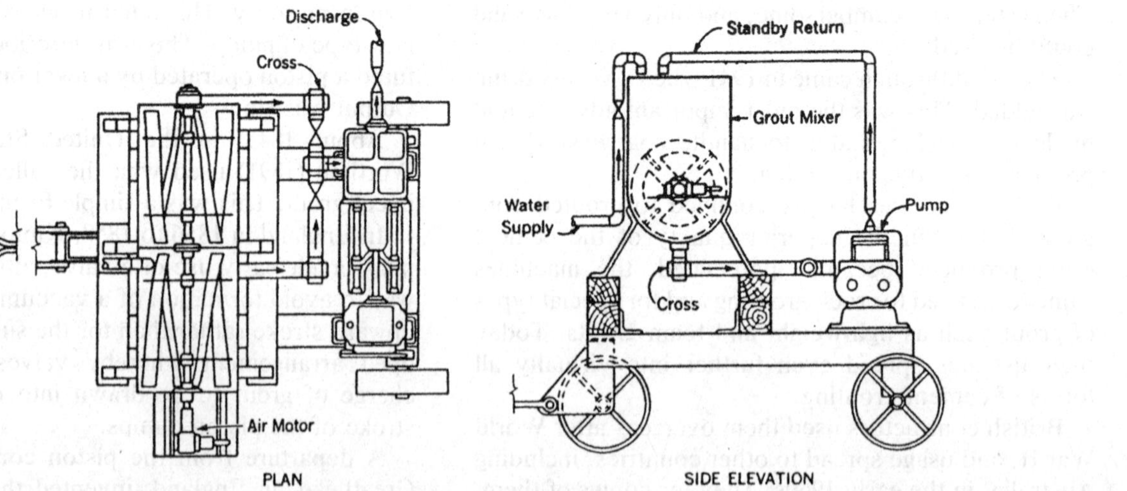

Figure 10.2-4A. *Norris Dam plant.*

10.2-4 Examples of Old Plant

A combined mixing and pumping unit used for the grouting of Norris Dam is shown in Figure 10.2-4A.

The paddle mixer had a horizontal shaft turned by an air motor. There was no agitator; the pump drew directly from the mixer. A larger and more versatile arrangement used at Boulder (Hoover) Dam had two pumps and a large agitator (Figure 10.2-4B).

This looks primitive compared with modern plants, but it did the job and formed a link in the evolution of the gear we have today. These two illustrations are from TVA (1949).

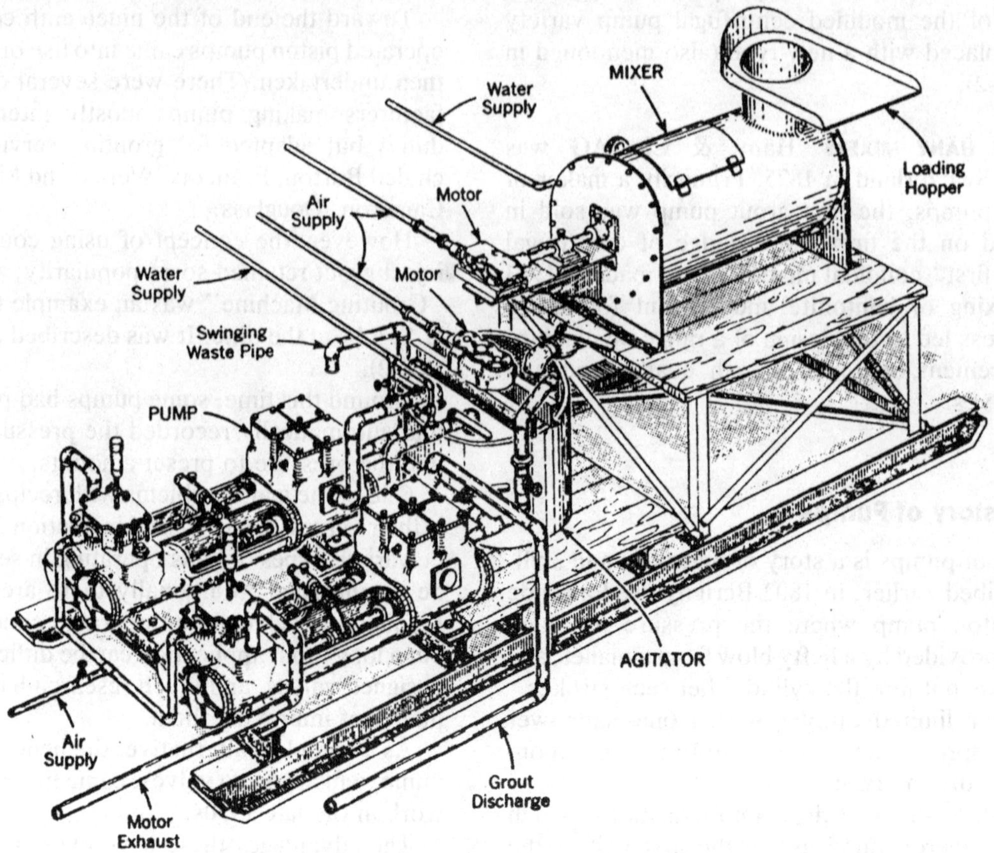

Figure 10.2-4B. *Boulder Dam grouting plant.*

REFERENCES AND FURTHER READING

Brown, D. R., and Warner, J. (1973). "Compaction grouting." *ASCE Journal of Soil Mechanics and Foundations Division,* Vol. 99(SM8), August, 589–601.

Glossop, R. (1960/1961). "The invention and development of injection processes. *Geotechnique.,* September 1960, 91–100; December 1961, 255–279.

Golder, H. Q. (1958). "Grouting and injection techniques." *Civil Engineering and Public Works (U.K.),* Vol. 53(630). p. 1396.

Grouts and Drilling Muds in Engineering Practice (1963). Butterworths, London.

Grouting in Geotechnical Engineering (1982). Conference Proceedings, February 1982. ASCE, New Orleans.

Karol, R. H. (1983). *Chemical Grouting.* Marcel Dekker, New York.

Kennedy, T. B. (1961). "Research in foundation grouting with cement." *ASCE Journal of Soil Mechanics and Foundations Division,* Vol. 87(SM2), 55–81.

Leonard, G. K., and Grant, L. F. (1958). "Experience of TVA with clay-cement and related grouts." *ASCE Journal of Soil Mechanics and Foundations Division,* Vol. 84(SM1), 1958, Paper 1552.

Millikan, C. V. (1971). "Cementing," in *History of oilwell drilling* J. E. Brantly, p. 1316.

Rands (1915). "Grouted cut-off for the Estacada Dam." *ASCE Transactions,* Vol. 78, 447–482. A discussion of the paper follows on pp. 483–546.

Simonds, A. W. (1947). "Contraction joint grouting of large dams." *Proceedings of the ACI,* Vol. 43, 637–652.

Simonds, A. W. (1958). "Present status of pressure grouting foundations" *ASCE Journal of Soil Mechanics and Foundations Division,* Vol. 84(SM1), Paper 1544.

Simonds, A. W., and Boggess, O. E. (1950). "Additional grouting at Hoover Dam, 1938–1947, inclusive." Technical Memorandum 639. USBR, Denver, CO, March 1950.

Simonds, A. W., Lippold, F. H., and Keim, R. E. (1950). "Treatment of foundations for large dams by grouting methods." *Transactions of the ASCE,* Vol. 76. Paper 2439.

(1958). "Symposium on cement and clay grouting of foundations." *ASCE Journal of Soil Mechanics and Foundations Division,* Vol. 84(SM1).

TVA (1949). "Geology and foundation treatment." Technical Report No 22. U.S. Tennessee Valley Authority.

U.S. Corps of Engineers (USCE) (1956). "Pressure grouting fine fissures." Technical Report No 6-437, October 1956. Waterways Experiment Station, Vicksburg.

Warner, J., and Fera, T. (1977). "Rehabilitation of concrete pavement." *ASCE Journal of Construction Division,* Vol. 103 (CO3), 395–403.

Warner, J. (1978). "Compaction grouting—a significant case history." *ASCE Journal of Geotechnical Engineering Division,* Vol. 104 (GT7), 837–847.

Worthen, W. E. (1891). Discussion on paper by R. L. Harris, "A coffer dam or caisson without timber or iron in the construction." *ASCE,* Vol. XXIV. Reference quoted by Glossop (1960) p. 100.

APPENDIX A

Table of Lugeon Values for Grout Hole Water Testing: Traditional Units

See Section 5.3-4 for explanations and methods of use.

WATER TAKE (cu ft/ft/5 minutes)			LUGEONS
0.0028	to	0.0085	1
0.0085	to	0.0142	2
0.0142	to	0.0199	3
0.0199	to	0.0256	4
0.0256	to	0.0313	5
0.0313	to	0.0370	6
0.0369	to	0.0426	7
0.0426	to	0.0483	8
0.0483	to	0.0540	9
0.0540	to	0.0597	10
0.0597	to	0.0654	11
0.0654	to	0.0711	12
0.0710	to	0.0767	13
0.0767	to	0.0824	14
0.0824	to	0.0881	15
0.0881	to	0.0938	16
0.0938	to	0.0995	17
0.0995	to	0.1052	18
0.1051	to	0.1108	19
0.1108	to	0.1165	20
0.1165	to	0.1222	21
0.1222	to	0.1279	22
0.1279	to	0.1336	23
0.1336	to	0.1393	24
0.1393	to	0.1450	25
0.1449	to	0.1506	26
0.1506	to	0.1563	27
0.1563	to	0.1620	28
0.1620	to	0.1677	29

WATER TAKE (cu ft/ft/5 minutes)			LUGEONS
0.1677	to	0.1734	30
0.1734	to	0.1791	31
0.1790	to	0.1847	32
0.1847	to	0.1904	33
0.1904	to	0.1961	34
0.1961	to	0.2018	35
0.2018	to	0.2075	36
0.2075	to	0.2132	37
0.2131	to	0.2188	38
0.2188	to	0.2245	39
0.2245	to	0.2302	40
0.2302	to	0.2359	41
0.2359	to	0.2416	42
0.2416	to	0.2473	43
0.2473	to	0.2530	44
0.2529	to	0.2586	45
0.2586	to	0.2643	46
0.2643	to	0.2700	47
0.2700	to	0.2757	48
0.2757	to	0.2814	49
0.2814	to	0.2871	50
0.2870	to	0.2927	51
0.2927	to	0.2984	52
0.2984	to	0.3041	53
0.3041	to	0.3098	54
0.3098	to	0.3155	55
0.3155	to	0.3212	56
0.3211	to	0.3268	57
0.3268	to	0.3325	58
0.3325	to	0.3382	59

APPENDIX A
Continued

LUGEON VALUES FOR GROUT HOLE WATER TESTING

WATER TAKE (cu ft/ft/5 minutes)			LUGEONS
0.3382	to	0.3439	60
0.3439	to	0.3496	61
0.3496	to	0.3553	62
0.3553	to	0.3610	63
0.3609	to	0.3666	64
0.3666	to	0.3723	65
0.3723	to	0.3780	66
0.3780	to	0.3837	67
0.3837	to	0.3894	68
0.3894	to	0.3951	69
0.3950	to	0.4007	70
0.4007	to	0.4064	71
0.4064	to	0.4121	72
0.4121	to	0.4178	73
0.4178	to	0.4235	74
0.4235	to	0.4292	75
0.4291	to	0.4348	76
0.4348	to	0.4405	77
0.4405	to	0.4462	78
0.4462	to	0.4519	79

WATER TAKE (cu ft/ft/5 minutes)			LUGEONS
0.4519	to	0.4576	80
0.4576	to	0.4633	81
0.4632	to	0.4689	82
0.4689	to	0.4746	83
0.4746	to	0.4803	84
0.4803	to	0.4860	85
0.4860	to	0.4917	86
0.4917	to	0.4974	87
0.4974	to	0.5031	88
0.5030	to	0.5087	89
0.5087	to	0.5144	90
0.5144	to	0.5201	91
0.5201	to	0.5258	92
0.5258	to	0.5315	93
0.5315	to	0.5372	94
0.5371	to	0.5428	95
0.5428	to	0.5485	96
0.5485	to	0.5542	97
0.5542	to	0.5599	98
0.5599	to	0.5656	99
0.5656	to	0.5713	100

Any values exceeding 0.5713 should be reported as

">100" (see Section 5.3-4)

APPENDIX B

Batching Tables For Grout: Traditional Units

See Appendix C for metric units

The tables in this Appendix are calculated on the basis of a 94 lb bag = 1 cu ft (loose).

APPENDIX B - 1 **94 lb BAG**

THICKENING TABLES

From 5:1 to 4:1
(by loose volume)

See Section 4.1–8 for explanations and methods of use.

1 BAG BATCH

Grout in Line and Agitator (cu ft)	MAKEUP MIX Water (cu ft)	MAKEUP MIX Mix Vol (cu ft)	MAKEUP MIX Mix w:c	Grout in Line and Agitator (cu ft)	MAKEUP MIX Water (cu ft)	MAKEUP MIX Mix Vol (cu ft)	MAKEUP MIX Mix w:c
0.0	4.00	4.5	4.0:1	2.5	3.54	4.0	3.5:1
0.1	3.98	4.5	4.0:1	2.6	3.53	4.0	3.5:1
0.2	3.96	4.4	4.0:1	2.7	3.51	4.0	3.5:1
0.3	3.95	4.4	3.9:1	2.8	3.49	4.0	3.5:1
0.4	3.93	4.4	3.9:1	2.9	3.47	4.0	3.5:1
0.5	3.91	4.4	3.9:1	3.0	3.45	3.9	3.5:1
0.6	3.89	4.4	3.9:1				
0.7	3.87	4.4	3.9:1	3.1	3.43	3.9	3.4:1
0.8	3.85	4.3	3.9:1	3.2	3.42	3.9	3.4:1
0.9	3.84	4.3	3.8:1	3.3	3.40	3.9	3.4:1
1.0	3.82	4.3	3.8:1	3.4	3.38	3.9	3.4:1
				3.5	3.36	3.8	3.4:1
1.1	3.80	4.3	3.8:1	3.6	3.34	3.8	3.3:1
1.2	3.78	4.3	3.8:1	3.7	3.32	3.8	3.3:1
1.3	3.76	4.2	3.8:1	3.8	3.31	3.8	3.3:1
1.4	3.74	4.2	3.7:1	3.9	3.29	3.8	3.3:1
1.5	3.73	4.2	3.7:1	4.0	3.27	3.8	3.3:1
1.6	3.71	4.2	3.7:1				
1.7	3.69	4.2	3.7:1	4.1	3.25	3.7	3.3:1
1.8	3.67	4.2	3.7:1	4.2	3.23	3.7	3.2:1
1.9	3.65	4.1	3.7:1	4.3	3.22	3.7	3.2:1
2.0	3.64	4.1	3.6:1	4.4	3.20	3.7	3.2:1
				4.5	3.18	3.7	3.2:1
2.1	3.62	4.1	3.6:1	4.6	3.16	3.6	3.2:1
2.2	3.60	4.1	3.6:1	4.7	3.14	3.6	3.1:1
2.3	3.58	4.1	3.6:1	4.8	3.12	3.6	3.1:1
2.4	3.56	4.0	3.6:1	4.9	3.11	3.6	3.1:1
				5.0	3.09	3.6	3.1:1

APPENDIX B - 1 — 94 lb BAG

Continued

THICKENING TABLES

From 5:1 to 4:1

1 BAG BATCH

Grout in Line and Agitator (cu ft)	MAKEUP MIX Water (cu ft)	MAKEUP MIX Mix Vol (cu ft)	MAKEUP MIX Mix w:c
5.1	3.07	3.5	3.1:1
5.2	3.05	3.5	3.1:1
5.3	3.03	3.5	3.0:1
5.4	3.01	3.5	3.0:1
5.5	3.00	3.5	3.0:1
5.6	2.98	3.5	3.0:1
5.7	2.96	3.4	3.0:1
5.8	2.94	3.4	2.9:1
5.9	2.92	3.4	2.9:1
6.0	2.91	3.4	2.9:1
6.1	2.89	3.4	2.9:1
6.2	2.87	3.3	2.9:1
6.3	2.85	3.3	2.9:1
6.4	2.83	3.3	2.8:1
6.5	2.81	3.3	2.8:1
6.6	2.80	3.3	2.8:1
6.7	2.78	3.3	2.8:1
6.8	2.76	3.2	2.8:1
6.9	2.74	3.2	2.7:1
7.0	2.72	3.2	2.7:1
7.1	2.70	3.2	2.7:1
7.2	2.69	3.2	2.7:1
7.3	2.67	3.1	2.7:1
7.4	2.65	3.1	2.6:1
7.5	2.63	3.1	2.6:1
7.6	2.61	3.1	2.6:1
7.7	2.59	3.1	2.6:1
7.8	2.58	3.1	2.6:1
7.9	2.56	3.0	2.6:1
8.0	2.54	3.0	2.5:1
8.1	2.52	3.0	2.5:1
8.2	2.50	3.0	2.5:1
8.3	2.49	3.0	2.5:1
8.4	2.47	2.9	2.5:1
8.5	2.45	2.9	2.4:1
8.6	2.43	2.9	2.4:1
8.7	2.41	2.9	2.4:1
8.8	2.39	2.9	2.4:1
8.9	2.38	2.9	2.4:1
9.0	2.36	2.8	2.4:1
9.1	2.34	2.8	2.3:1
9.2	2.32	2.8	2.3:1
9.3	2.30	2.8	2.3:1
9.4	2.28	2.8	2.3:1
9.5	2.27	2.7	2.3:1
9.6	2.25	2.7	2.2:1
9.7	2.23	2.7	2.2:1
9.8	2.21	2.7	2.2:1
9.9	2.19	2.7	2.2:1
10.0	2.18	2.7	2.2:1
10.1	2.16	2.6	2.2:1
10.2	2.14	2.6	2.1:1
10.3	2.12	2.6	2.1:1
10.4	2.10	2.6	2.1:1
10.5	2.08	2.6	2.1:1
10.6	2.07	2.5	2.1:1
10.7	2.05	2.5	2.0:1
10.8	2.03	2.5	2.0:1
10.9	2.01	2.5	2.0:1
11.0	1.99	2.5	2.0:1
11.1	1.97	2.5	2.0:1
11.2	1.96	2.4	2.0:1
11.3	1.94	2.4	1.9:1
11.4	1.92	2.4	1.9:1
11.5	1.90	2.4	1.9:1
11.6	1.88	2.4	1.9:1
11.7	1.86	2.3	1.9:1
11.8	1.85	2.3	1.8:1
11.9	1.83	2.3	1.8:1
12.0	1.81	2.3	1.8:1

APPENDIX B - 1
Continued

94 lb BAG

THICKENING TABLES
From 5:1 to 4:1

1 BAG BATCH

Grout in Line and Agitator (cu ft)	MAKEUP MIX Water (cu ft)	MAKEUP MIX Mix Vol (cu ft)	MAKEUP MIX Mix w:c
12.1	1.79	2.3	1.8:1
12.2	1.77	2.3	1.8:1
12.3	1.76	2.2	1.8:1
12.4	1.74	2.2	1.7:1
12.5	1.72	2.2	1.7:1
12.6	1.70	2.2	1.7:1
12.7	1.68	2.2	1.7:1
12.8	1.66	2.1	1.7:1
12.9	1.65	2.1	1.6:1
13.0	1.63	2.1	1.6:1
13.1	1.61	2.1	1.6:1
13.2	1.59	2.1	1.6:1
13.3	1.57	2.1	1.6:1
13.4	1.55	2.0	1.6:1
13.5	1.54	2.0	1.5:1
13.6	1.52	2.0	1.5:1
13.7	1.50	2.0	1.5:1
13.8	1.48	2.0	1.5:1
13.9	1.46	1.9	1.5:1
14.0	1.45	1.9	1.4:1
14.1	1.43	1.9	1.4:1
14.2	1.41	1.9	1.4:1
14.3	1.39	1.9	1.4:1
14.4	1.37	1.9	1.4:1
14.5	1.35	1.8	1.4:1
14.6	1.34	1.8	1.3:1
14.7	1.32	1.8	1.3:1
14.8	1.30	1.8	1.3:1
14.9	1.28	1.8	1.3:1
15.0	1.26	1.7	1.3:1
15.1	1.24	1.7	1.2:1
15.2	1.23	1.7	1.2:1
15.3	1.21	1.7	1.2:1
15.4	1.19	1.7	1.2:1

Grout in Line and Agitator (cu ft)	MAKEUP MIX Water (cu ft)	MAKEUP MIX Mix Vol (cu ft)	MAKEUP MIX Mix w:c
15.5	1.17	1.7	1.2:1
15.6	1.15	1.6	1.2:1
15.7	1.14	1.6	1.1:1
15.8	1.12	1.6	1.1:1
15.9	1.10	1.6	1.1:1
16.0	1.08	1.6	1.1:1
16.1	1.06	1.5	1.1:1
16.2	1.04	1.5	1.0:1
16.3	1.03	1.5	1.0:1
16.4	1.01	1.5	1.0:1
16.5	0.99	1.5	1.0:1
16.6	0.97	1.5	1.0:1
16.7	0.95	1.4	1.0:1
16.8	0.93	1.4	0.9:1
16.9	0.92	1.4	0.9:1
17.0	0.90	1.4	0.9:1
17.1	0.88	1.4	0.9:1
17.2	0.86	1.3	0.9:1
17.3	0.84	1.3	0.8:1
17.4	0.82	1.3	0.8:1
17.5	0.81	1.3	0.8:1
17.6	0.79	1.3	0.8:1
17.7	0.77	1.3	0.8:1
17.8	0.75	1.2	0.8:1
17.9	0.73	1.2	0.7:1
18.0	0.72	1.2	0.7:1
18.1	0.70	1.2	0.7:1
18.2	0.68	1.2	0.7:1
18.3	0.66	1.1	0.7:1
18.4	0.64	1.1	0.6:1
18.5	0.62	1.1	0.6:1
18.6	0.61	1.1	0.6:1
18.7	0.59	1.1	0.6:1
18.8	0.57	1.0	0.6:1
18.9	0.55	1.0	0.6:1
19.0	0.53	1.0	0.5:1

APPENDIX B - 1

Continued

94 lb BAG

THICKENING TABLES

From 5:1 to 4:1

Grout in Line and Agitator (cu ft)	MAKEUP MIX		
	Water (cu ft)	Mix Vol (cu ft)	Mix w:c
2 BAG BATCH			
0.0	8.00	9.0	4.0:1
0.1	7.98	8.9	4.0:1
0.2	7.96	8.9	4.0:1
0.3	7.95	8.9	4.0:1
0.4	7.93	8.9	4.0:1
0.5	7.91	8.9	4.0:1
0.6	7.89	8.9	3.9:1
0.7	7.87	8.8	3.9:1
0.8	7.85	8.8	3.9:1
0.9	7.84	8.8	3.9:1
1.0	7.82	8.8	3.9:1
1.1	7.80	8.8	3.9:1
1.2	7.78	8.7	3.9:1
1.3	7.76	8.7	3.9:1
1.4	7.74	8.7	3.9:1
1.5	7.73	8.7	3.9:1
1.6	7.71	8.7	3.9:1
1.7	7.69	8.6	3.8:1
1.8	7.67	8.6	3.8:1
1.9	7.65	8.6	3.8:1
2.0	7.64	8.6	3.8:1
2.1	7.62	8.6	3.8:1
2.2	7.60	8.6	3.8:1
2.3	7.58	8.5	3.8:1
2.4	7.56	8.5	3.8:1
2.5	7.54	8.5	3.8:1
2.6	7.53	8.5	3.8:1
2.7	7.51	8.5	3.8:1
2.8	7.49	8.4	3.7:1
2.9	7.47	8.4	3.7:1
3.0	7.45	8.4	3.7:1
3.1	7.43	8.4	3.7:1
3.2	7.42	8.4	3.7:1
3.3	7.40	8.4	3.7:1
3.4	7.38	8.3	3.7:1
3.5	7.36	8.3	3.7:1
3.6	7.34	8.3	3.7:1
3.7	7.32	8.3	3.7:1
3.8	7.31	8.3	3.7:1
3.9	7.29	8.2	3.6:1
4.0	7.27	8.2	3.6:1
4.1	7.25	8.2	3.6:1
4.2	7.23	8.2	3.6:1
4.3	7.22	8.2	3.6:1
4.4	7.20	8.2	3.6:1
4.5	7.18	8.1	3.6:1
4.6	7.16	8.1	3.6:1
4.7	7.14	8.1	3.6:1
4.8	7.12	8.1	3.6:1
4.9	7.11	8.1	3.6:1
5.0	7.09	8.0	3.5:1
5.1	7.07	8.0	3.5:1
5.2	7.05	8.0	3.5:1
5.3	7.03	8.0	3.5:1
5.4	7.01	8.0	3.5:1
5.5	7.00	8.0	3.5:1
5.6	6.98	7.9	3.5:1
5.7	6.96	7.9	3.5:1
5.8	6.94	7.9	3.5:1
5.9	6.92	7.9	3.5:1
6.0	6.91	7.9	3.5:1
6.1	6.89	7.8	3.4:1
6.2	6.87	7.8	3.4:1
6.3	6.85	7.8	3.4:1
6.4	6.83	7.8	3.4:1
6.5	6.81	7.8	3.4:1
6.6	6.80	7.8	3.4:1
6.7	6.78	7.7	3.4:1
6.8	6.76	7.7	3.4:1
6.9	6.74	7.7	3.4:1
7.0	6.72	7.7	3.4:1

APPENDIX B - 1
Continued

94 lb BAG

THICKENING TABLES

From 5:1 to 4:1

2 BAG BATCH

Grout in Line and Agitator (cu ft)	MAKEUP MIX		
	Water (cu ft)	Mix Vol (cu ft)	Mix w:c
7.1	6.70	7.7	3.4:1
7.2	6.69	7.6	3.3:1
7.3	6.67	7.6	3.3:1
7.4	6.65	7.6	3.3:1
7.5	6.63	7.6	3.3:1
7.6	6.61	7.6	3.3:1
7.7	6.59	7.6	3.3:1
7.8	6.58	7.5	3.3:1
7.9	6.56	7.5	3.3:1
8.0	6.54	7.5	3.3:1
8.1	6.52	7.5	3.3:1
8.2	6.50	7.5	3.3:1
8.3	6.49	7.4	3.2:1
8.4	6.47	7.4	3.2:1
8.5	6.45	7.4	3.2:1
8.6	6.43	7.4	3.2:1
8.7	6.41	7.4	3.2:1
8.8	6.39	7.4	3.2:1
8.9	6.38	7.3	3.2:1
9.0	6.36	7.3	3.2:1
9.1	6.34	7.3	3.2:1
9.2	6.32	7.3	3.2:1
9.3	6.30	7.3	3.2:1
9.4	6.28	7.2	3.1:1
9.5	6.27	7.2	3.1:1
9.6	6.25	7.2	3.1:1
9.7	6.23	7.2	3.1:1
9.8	6.21	7.2	3.1:1
9.9	6.19	7.2	3.1:1
10.0	6.18	7.1	3.1:1
10.1	6.16	7.1	3.1:1
10.2	6.14	7.1	3.1:1
10.3	6.12	7.1	3.1:1
10.4	6.10	7.1	3.1:1

Grout in Line and Agitator (cu ft)	MAKEUP MIX		
	Water (cu ft)	Mix Vol (cu ft)	Mix w:c
10.5	6.08	7.0	3.0:1
10.6	6.07	7.0	3.0:1
10.7	6.05	7.0	3.0:1
10.8	6.03	7.0	3.0:1
10.9	6.01	7.0	3.0:1
11.0	5.99	7.0	3.0:1
11.1	5.97	6.9	3.0:1
11.2	5.96	6.9	3.0:1
11.3	5.94	6.9	3.0:1
11.4	5.92	6.9	3.0:1
11.5	5.90	6.9	3.0:1
11.6	5.88	6.8	2.9:1
11.7	5.86	6.8	2.9:1
11.8	5.85	6.8	2.9:1
11.9	5.83	6.8	2.9:1
12.0	5.81	6.8	2.9:1
12.1	5.79	6.8	2.9:1
12.2	5.77	6.7	2.9:1
12.3	5.76	6.7	2.9:1
12.4	5.74	6.7	2.9:1
12.5	5.72	6.7	2.9:1
12.6	5.70	6.7	2.9:1
12.7	5.68	6.6	2.8:1
12.8	5.66	6.6	2.8:1
12.9	5.65	6.6	2.8:1
13.0	5.63	6.6	2.8:1
13.1	5.61	6.6	2.8:1
13.2	5.59	6.6	2.8:1
13.3	5.57	6.5	2.8:1
13.4	5.55	6.5	2.8:1
13.5	5.54	6.5	2.8:1
13.6	5.52	6.5	2.8:1
13.7	5.50	6.5	2.7:1
13.8	5.48	6.4	2.7:1
13.9	5.46	6.4	2.7:1
14.0	5.45	6.4	2.7:1

APPENDIX B - 1

Continued

94 lb BAG

THICKENING TABLES

From 5:1 to 4:1

2 BAG BATCH

Grout in Line and Agitator (cu ft)	MAKEUP MIX Water (cu ft)	MAKEUP MIX Mix Vol (cu ft)	MAKEUP MIX Mix w:c
14.1	5.43	6.4	2.7:1
14.2	5.41	6.4	2.7:1
14.3	5.39	6.4	2.7:1
14.4	5.37	6.3	2.7:1
14.5	5.35	6.3	2.7:1
14.6	5.34	6.3	2.7:1
14.7	5.32	6.3	2.7:1
14.8	5.30	6.3	2.6:1
14.9	5.28	6.2	2.6:1
15.0	5.26	6.2	2.6:1
15.1	5.24	6.2	2.6:1
15.2	5.23	6.2	2.6:1
15.3	5.21	6.2	2.6:1
15.4	5.19	6.1	2.6:1
15.5	5.17	6.1	2.6:1
15.6	5.15	6.1	2.6:1
15.7	5.14	6.1	2.6:1
15.8	5.12	6.1	2.6:1
15.9	5.10	6.1	2.5:1
16.0	5.08	6.0	2.5:1
16.1	5.06	6.0	2.5:1
16.2	5.04	6.0	2.5:1
16.3	5.03	6.0	2.5:1
16.4	5.01	6.0	2.5:1
16.5	4.99	5.9	2.5:1
16.6	4.97	5.9	2.5:1
16.7	4.95	5.9	2.5:1
16.8	4.93	5.9	2.5:1
16.9	4.92	5.9	2.5:1
17.0	4.90	5.9	2.4:1
17.1	4.88	5.8	2.4:1
17.2	4.86	5.8	2.4:1
17.3	4.84	5.8	2.4:1
17.4	4.82	5.8	2.4:1
17.5	4.81	5.8	2.4:1
17.6	4.79	5.7	2.4:1
17.7	4.77	5.7	2.4:1
17.8	4.75	5.7	2.4:1
17.9	4.73	5.7	2.4:1
18.0	4.72	5.7	2.4:1
18.1	4.70	5.7	2.3:1
18.2	4.68	5.6	2.3:1
18.3	4.66	5.6	2.3:1
18.4	4.64	5.6	2.3:1
18.5	4.62	5.6	2.3:1
18.6	4.61	5.6	2.3:1
18.7	4.59	5.5	2.3:1
18.8	4.57	5.5	2.3:1
18.9	4.55	5.5	2.3:1
19.0	4.53	5.5	2.3:1
19.1	4.51	5.5	2.3:1
19.2	4.50	5.5	2.2:1
19.3	4.48	5.4	2.2:1
19.4	4.46	5.4	2.2:1
19.5	4.44	5.4	2.2:1
19.6	4.42	5.4	2.2:1
19.7	4.41	5.4	2.2:1
19.8	4.39	5.3	2.2:1
19.9	4.37	5.3	2.2:1
20.0	4.35	5.3	2.2:1

APPENDIX B - 1

94 lb BAG

THICKENING TABLES

From 4:1 to 3:1
(by loose volume)

See Section 4.1–8 for explanations and methods of use.

Grout in Line and Agitator (cu ft)	MAKEUP MIX			Grout in Line and Agitator (cu ft)	MAKEUP MIX		
	Water (cu ft)	Mix Vol (cu ft)	Mix w:c		Water (cu ft)	Mix Vol (cu ft)	Mix w:c
			1 BAG BATCH				
0.0	3.00	3.5	3.0:1	2.5	2.44	2.9	2.4:1
0.1	2.98	3.5	3.0:1	2.6	2.42	2.9	2.4:1
0.2	2.96	3.4	3.0:1	2.7	2.40	2.9	2.4:1
0.3	2.93	3.4	2.9:1	2.8	2.38	2.9	2.4:1
0.4	2.91	3.4	2.9:1	2.9	2.35	2.8	2.4:1
0.5	2.89	3.4	2.9:1	3.0	2.33	2.8	2.3:1
0.6	2.87	3.3	2.9:1				
0.7	2.84	3.3	2.8:1	3.1	2.31	2.8	2.3:1
0.8	2.82	3.3	2.8:1	3.2	2.29	2.8	2.3:1
0.9	2.80	3.3	2.8:1	3.3	2.26	2.7	2.3:1
1.0	2.78	3.3	2.8:1	3.4	2.24	2.7	2.2:1
				3.5	2.22	2.7	2.2:1
1.1	2.75	3.2	2.8:1	3.6	2.20	2.7	2.2:1
1.2	2.73	3.2	2.7:1	3.7	2.17	2.7	2.2:1
1.3	2.71	3.2	2.7:1	3.8	2.15	2.6	2.2:1
1.4	2.69	3.2	2.7:1	3.9	2.13	2.6	2.1:1
1.5	2.67	3.1	2.7:1	4.0	2.11	2.6	2.1:1
1.6	2.64	3.1	2.6:1				
1.7	2.62	3.1	2.6:1	4.1	2.08	2.6	2.1:1
1.8	2.60	3.1	2.6:1	4.2	2.06	2.5	2.1:1
1.9	2.58	3.1	2.6:1	4.3	2.04	2.5	2.0:1
2.0	2.55	3.0	2.6:1	4.4	2.02	2.5	2.0:1
				4.5	2.00	2.5	2.0:1
2.1	2.53	3.0	2.5:1	4.6	1.97	2.5	2.0:1
2.2	2.51	3.0	2.5:1	4.7	1.95	2.4	2.0:1
2.3	2.49	3.0	2.5:1	4.8	1.93	2.4	1.9:1
2.4	2.46	2.9	2.5:1	4.9	1.91	2.4	1.9:1
				5.0	1.88	2.4	1.9:1

APPENDIX B - 1

Continued

94 lb BAG

THICKENING TABLES

From 4:1 to 3:1

1 BAG BATCH

Grout in Line and Agitator (cu ft)	MAKEUP MIX		
	Water (cu ft)	Mix Vol (cu ft)	Mix w:c
5.1	1.86	2.3	1.9:1
5.2	1.84	2.3	1.8:1
5.3	1.82	2.3	1.8:1
5.4	1.79	2.3	1.8:1
5.5	1.77	2.3	1.8:1
5.6	1.75	2.2	1.7:1
5.7	1.73	2.2	1.7:1
5.8	1.71	2.2	1.7:1
5.9	1.68	2.2	1.7:1
6.0	1.66	2.1	1.7:1
6.1	1.64	2.1	1.6:1
6.2	1.62	2.1	1.6:1
6.3	1.59	2.1	1.6:1
6.4	1.57	2.1	1.6:1
6.5	1.55	2.0	1.5:1
6.6	1.53	2.0	1.5:1
6.7	1.50	2.0	1.5:1
6.8	1.48	2.0	1.5:1
6.9	1.46	1.9	1.5:1
7.0	1.44	1.9	1.4:1
7.1	1.42	1.9	1.4:1
7.2	1.39	1.9	1.4:1
7.3	1.37	1.9	1.4:1
7.4	1.35	1.8	1.3:1
7.5	1.33	1.8	1.3:1
7.6	1.30	1.8	1.3:1
7.7	1.28	1.8	1.3:1
7.8	1.26	1.7	1.3:1
7.9	1.24	1.7	1.2:1
8.0	1.21	1.7	1.2:1
8.1	1.19	1.7	1.2:1
8.2	1.17	1.6	1.2:1
8.3	1.15	1.6	1.1:1
8.4	1.12	1.6	1.1:1

Grout in Line and Agitator (cu ft)	MAKEUP MIX		
	Water (cu ft)	Mix Vol (cu ft)	Mix w:c
8.5	1.10	1.6	1.1:1
8.6	1.08	1.6	1.1:1
8.7	1.06	1.5	1.1:1
8.8	1.04	1.5	1.0:1
8.9	1.01	1.5	1.0:1
9.0	0.99	1.5	1.0:1
9.1	0.97	1.4	1.0:1
9.2	0.95	1.4	0.9:1
9.3	0.92	1.4	0.9:1
9.4	0.90	1.4	0.9:1
9.5	0.88	1.4	0.9:1
9.6	0.86	1.3	0.9:1
9.7	0.83	1.3	0.8:1
9.8	0.81	1.3	0.8:1
9.9	0.79	1.3	0.8:1
10.0	0.77	1.2	0.8:1
10.1	0.75	1.2	0.7:1
10.2	0.72	1.2	0.7:1
10.3	0.70	1.2	0.7:1
10.4	0.68	1.2	0.7:1
10.5	0.66	1.1	0.7:1
10.6	0.63	1.1	0.6:1
10.7	0.61	1.1	0.6:1
10.8	0.59	1.1	0.6:1
10.9	0.57	1.0	0.6:1
11.0	0.54	1.0	0.5:1
11.1	0.52	1.0	0.5:1
11.2	0.50	1.0	0.5:1
11.3	0.48	1.0	0.5:1
11.4	0.46	0.9	0.5:1

APPENDIX B - 1
Continued

94 lb BAG

THICKENING TABLES
From 4:1 to 3:1

Grout in Line and Agitator (cu ft)	MAKEUP MIX		
	Water (cu ft)	Mix Vol (cu ft)	Mix w:c
2 BAG BATCH			
0.0	6.00	7.0	3.0:1
0.1	5.98	6.9	3.0:1
0.2	5.96	6.9	3.0:1
0.3	5.93	6.9	3.0:1
0.4	5.91	6.9	3.0:1
0.5	5.89	6.8	2.9:1
0.6	5.87	6.8	2.9:1
0.7	5.84	6.8	2.9:1
0.8	5.82	6.8	2.9:1
0.9	5.80	6.8	2.9:1
1.0	5.78	6.7	2.9:1
1.1	5.75	6.7	2.9:1
1.2	5.73	6.7	2.9:1
1.3	5.71	6.7	2.9:1
1.4	5.69	6.6	2.8:1
1.5	5.67	6.6	2.8:1
1.6	5.64	6.6	2.8:1
1.7	5.62	6.6	2.8:1
1.8	5.60	6.6	2.8:1
1.9	5.58	6.5	2.8:1
2.0	5.55	6.5	2.8:1
2.1	5.53	6.5	2.8:1
2.2	5.51	6.5	2.8:1
2.3	5.49	6.4	2.7:1
2.4	5.46	6.4	2.7:1
2.5	5.44	6.4	2.7:1
2.6	5.42	6.4	2.7:1
2.7	5.40	6.4	2.7:1
2.8	5.38	6.3	2.7:1
2.9	5.35	6.3	2.7:1
3.0	5.33	6.3	2.7:1
3.1	5.31	6.3	2.7:1
3.2	5.29	6.2	2.6:1
3.3	5.26	6.2	2.6:1
3.4	5.24	6.2	2.6:1
3.5	5.22	6.2	2.6:1
3.6	5.20	6.2	2.6:1
3.7	5.17	6.1	2.6:1
3.8	5.15	6.1	2.6:1
3.9	5.13	6.1	2.6:1
4.0	5.11	6.1	2.6:1
4.1	5.08	6.0	2.5:1
4.2	5.06	6.0	2.5:1
4.3	5.04	6.0	2.5:1
4.4	5.02	6.0	2.5:1
4.5	5.00	6.0	2.5:1
4.6	4.97	5.9	2.5:1
4.7	4.95	5.9	2.5:1
4.8	4.93	5.9	2.5:1
4.9	4.91	5.9	2.5:1
5.0	4.88	5.8	2.4:1
5.1	4.86	5.8	2.4:1
5.2	4.84	5.8	2.4:1
5.3	4.82	5.8	2.4:1
5.4	4.79	5.8	2.4:1
5.5	4.77	5.7	2.4:1
5.6	4.75	5.7	2.4:1
5.7	4.73	5.7	2.4:1
5.8	4.71	5.7	2.4:1
5.9	4.68	5.6	2.3:1
6.0	4.66	5.6	2.3:1
6.1	4.64	5.6	2.3:1
6.2	4.62	5.6	2.3:1
6.3	4.59	5.6	2.3:1
6.4	4.57	5.5	2.3:1
6.5	4.55	5.5	2.3:1
6.6	4.53	5.5	2.3:1
6.7	4.50	5.5	2.3:1
6.8	4.48	5.4	2.2:1
6.9	4.46	5.4	2.2:1
7.0	4.44	5.4	2.2:1

APPENDIX B - 1
Continued

94 lb BAG

THICKENING TABLES

From 4:1 to 3:1

2 BAG BATCH

Grout in Line and Agitator (cu ft)	MAKEUP MIX		
	Water (cu ft)	Mix Vol (cu ft)	Mix w:c
7.1	4.42	5.4	2.2:1
7.2	4.39	5.4	2.2:1
7.3	4.37	5.3	2.2:1
7.4	4.35	5.3	2.2:1
7.5	4.33	5.3	2.2:1
7.6	4.30	5.3	2.2:1
7.7	4.28	5.2	2.1:1
7.8	4.26	5.2	2.1:1
7.9	4.24	5.2	2.1:1
8.0	4.21	5.2	2.1:1
8.1	4.19	5.2	2.1:1
8.2	4.17	5.1	2.1:1
8.3	4.15	5.1	2.1:1
8.4	4.12	5.1	2.1:1
8.5	4.10	5.1	2.1:1
8.6	4.08	5.0	2.0:1
8.7	4.06	5.0	2.0:1
8.8	4.04	5.0	2.0:1
8.9	4.01	5.0	2.0:1
9.0	3.99	5.0	2.0:1
9.1	3.97	4.9	2.0:1
9.2	3.95	4.9	2.0:1
9.3	3.92	4.9	2.0:1
9.4	3.90	4.9	2.0:1
9.5	3.88	4.8	1.9:1
9.6	3.86	4.8	1.9:1
9.7	3.83	4.8	1.9:1
9.8	3.81	4.8	1.9:1
9.9	3.79	4.8	1.9:1
10.0	3.77	4.7	1.9:1
10.1	3.75	4.7	1.9:1
10.2	3.72	4.7	1.9:1
10.3	3.70	4.7	1.9:1
10.4	3.68	4.6	1.8:1

Grout in Line and Agitator (cu ft)	MAKEUP MIX		
	Water (cu ft)	Mix Vol (cu ft)	Mix w:c
10.5	3.66	4.6	1.8:1
10.6	3.63	4.6	1.8:1
10.7	3.61	4.6	1.8:1
10.8	3.59	4.5	1.8:1
10.9	3.57	4.5	1.8:1
11.0	3.54	4.5	1.8:1
11.1	3.52	4.5	1.8:1
11.2	3.50	4.5	1.7:1
11.3	3.48	4.4	1.7:1
11.4	3.46	4.4	1.7:1
11.5	3.43	4.4	1.7:1
11.6	3.41	4.4	1.7:1
11.7	3.39	4.3	1.7:1
11.8	3.37	4.3	1.7:1
11.9	3.34	4.3	1.7:1
12.0	3.32	4.3	1.7:1
12.1	3.30	4.3	1.6:1
12.2	3.28	4.2	1.6:1
12.3	3.25	4.2	1.6:1
12.4	3.23	4.2	1.6:1
12.5	3.21	4.2	1.6:1
12.6	3.19	4.1	1.6:1
12.7	3.17	4.1	1.6:1
12.8	3.14	4.1	1.6:1
12.9	3.12	4.1	1.6:1
13.0	3.10	4.1	1.5:1
13.1	3.08	4.0	1.5:1
13.2	3.05	4.0	1.5:1
13.3	3.03	4.0	1.5:1
13.4	3.01	4.0	1.5:1
13.5	2.99	3.9	1.5:1
13.6	2.96	3.9	1.5:1
13.7	2.94	3.9	1.5:1
13.8	2.92	3.9	1.5:1
13.9	2.90	3.9	1.4:1
14.0	2.87	3.8	1.4:1

APPENDIX B - 1
Continued

94 lb BAG

THICKENING TABLES
From 4:1 to 3:1

Grout in Line and Agitator (cu ft)	MAKEUP MIX			Grout in Line and Agitator (cu ft)	MAKEUP MIX		
	Water (cu ft)	Mix Vol (cu ft)	Mix w:c		Water (cu ft)	Mix Vol (cu ft)	Mix w:c
			2 BAG BATCH				
14.1	2.85	3.8	1.4:1	17.5	2.09	3.1	1.0:1
14.2	2.83	3.8	1.4:1	17.6	2.07	3.0	1.0:1
14.3	2.81	3.8	1.4:1	17.7	2.05	3.0	1.0:1
14.4	2.79	3.7	1.4:1	17.8	2.03	3.0	1.0:1
14.5	2.76	3.7	1.4:1	17.9	2.00	3.0	1.0:1
14.6	2.74	3.7	1.4:1	18.0	1.98	2.9	1.0:1
14.7	2.72	3.7	1.4:1				
14.8	2.70	3.7	1.3:1	18.1	1.96	2.9	1.0:1
14.9	2.67	3.6	1.3:1	18.2	1.94	2.9	1.0:1
15.0	2.65	3.6	1.3:1	18.3	1.92	2.9	1.0:1
				18.4	1.89	2.9	0.9:1
15.1	2.63	3.6	1.3:1	18.5	1.87	2.8	0.9:1
15.2	2.61	3.6	1.3:1	18.6	1.85	2.8	0.9:1
15.3	2.58	3.5	1.3:1	18.7	1.83	2.8	0.9:1
15.4	2.56	3.5	1.3:1	18.8	1.80	2.8	0.9:1
15.5	2.54	3.5	1.3:1	18.9	1.78	2.7	0.9:1
15.6	2.52	3.5	1.3:1	19.0	1.76	2.7	0.9:1
15.7	2.50	3.5	1.2:1				
15.8	2.47	3.4	1.2:1	19.1	1.74	2.7	0.9:1
15.9	2.45	3.4	1.2:1	19.2	1.71	2.7	0.9:1
16.0	2.43	3.4	1.2:1	19.3	1.69	2.7	0.8:1
				19.4	1.67	2.6	0.8:1
16.1	2.41	3.4	1.2:1	19.5	1.65	2.6	0.8:1
16.2	2.38	3.3	1.2:1	19.6	1.62	2.6	0.8:1
16.3	2.36	3.3	1.2:1	19.7	1.60	2.6	0.8:1
16.4	2.34	3.3	1.2:1	19.8	1.58	2.5	0.8:1
16.5	2.32	3.3	1.2:1	19.9	1.56	2.5	0.8:1
16.6	2.29	3.3	1.1:1	20.0	1.54	2.5	0.8:1
16.7	2.27	3.2	1.1:1				
16.8	2.25	3.2	1.1:1				
16.9	2.23	3.2	1.1:1				
17.0	2.21	3.2	1.1:1				
17.1	2.18	3.1	1.1:1				
17.2	2.16	3.1	1.1:1				
17.3	2.14	3.1	1.1:1				
17.4	2.12	3.1	1.1:1				

APPENDIX B - 1 — 94 lb BAG

THICKENING TABLES

From 3:1 to 2:1
(by loose volume)

See Section 4.1–8 for explanations and methods of use.

1 BAG BATCH

Grout in Line and Agitator (cu ft)	MAKEUP MIX Water (cu ft)	MAKEUP MIX Mix Vol (cu ft)	MAKEUP MIX Mix w:c
0.0	2.00	2.5	2.0:1
0.1	1.97	2.5	2.0:1
0.2	1.94	2.4	1.9:1
0.3	1.91	2.4	1.9:1
0.4	1.89	2.4	1.9:1
0.5	1.86	2.3	1.9:1
0.6	1.83	2.3	1.8:1
0.7	1.80	2.3	1.8:1
0.8	1.77	2.3	1.8:1
0.9	1.74	2.2	1.7:1
1.0	1.71	2.2	1.7:1
1.1	1.68	2.2	1.7:1
1.2	1.66	2.1	1.7:1
1.3	1.63	2.1	1.6:1
1.4	1.60	2.1	1.6:1
1.5	1.57	2.0	1.6:1
1.6	1.54	2.0	1.5:1
1.7	1.51	2.0	1.5:1
1.8	1.48	2.0	1.5:1
1.9	1.45	1.9	1.5:1
2.0	1.43	1.9	1.4:1
2.1	1.40	1.9	1.4:1
2.2	1.37	1.8	1.4:1
2.3	1.34	1.8	1.3:1
2.4	1.31	1.8	1.3:1
2.5	1.28	1.8	1.3:1

Grout in Line and Agitator (cu ft)	MAKEUP MIX Water (cu fl)	MAKEUP MIX Mix Vol (cu ft)	MAKEUP MIX Mix w:c
2.6	1.25	1.7	1.3:1
2.7	1.22	1.7	1.2:1
2.8	1.20	1.7	1.2:1
2.9	1.17	1.6	1.2:1
3.0	1.14	1.6	1.1:1
3.1	1.11	1.6	1.1:1
3.2	1.08	1.6	1.1:1
3.3	1.05	1.5	1.1:1
3.4	1.02	1.5	1.0:1
3.5	0.99	1.5	1.0:1
3.6	0.97	1.4	1.0:1
3.7	0.94	1.4	0.9:1
3.8	0.91	1.4	0.9:1
3.9	0.88	1.4	0.9:1
4.0	0.85	1.3	0.9:1
4.1	0.82	1.3	0.8:1
4.2	0.79	1.3	0.8:1
4.3	0.76	1.2	0.8:1
4.4	0.74	1.2	0.7:1
4.5	0.71	1.2	0.7:1
4.6	0.68	1.2	0.7:1
4.7	0.65	1.1	0.6:1
4.8	0.62	1.1	0.6:1
4.9	0.59	1.1	0.6:1
5.0	0.56	1.0	0.6:1
5.1	0.53	1.0	0.5:1

APPENDIX B - 1
Continued

94 lb BAG

THICKENING TABLES
From 3:1 to 2:1

Grout in Line and Agitator (cu ft)	MAKEUP MIX		
	Water (cu ft)	Mix Vol (cu ft)	Mix w:c
2 BAG BATCH			
0.0	4.00	5.0	2.0:1
0.1	3.97	4.9	2.0:1
0.2	3.94	4.9	2.0:1
0.3	3.91	4.9	2.0:1
0.4	3.89	4.8	1.9:1
0.5	3.86	4.8	1.9:1
0.6	3.83	4.8	1.9:1
0.7	3.80	4.8	1.9:1
0.8	3.77	4.7	1.9:1
0.9	3.74	4.7	1.9:1
1.0	3.71	4.7	1.9:1
1.1	3.68	4.6	1.8:1
1.2	3.66	4.6	1.8:1
1.3	3.63	4.6	1.8:1
1.4	3.60	4.6	1.8:1
1.5	3.57	4.5	1.8:1
1.6	3.54	4.5	1.8:1
1.7	3.51	4.5	1.8:1
1.8	3.48	4.4	1.7:1
1.9	3.45	4.4	1.7:1
2.0	3.43	4.4	1.7:1
2.1	3.40	4.4	1.7:1
2.2	3.37	4.3	1.7:1
2.3	3.34	4.3	1.7:1
2.4	3.31	4.3	1.7:1
2.5	3.28	4.2	1.6:1
2.6	3.25	4.2	1.6:1
2.7	3.22	4.2	1.6:1
2.8	3.20	4.2	1.6:1
2.9	3.17	4.1	1.6:1
3.0	3.14	4.1	1.6:1
3.1	3.11	4.1	1.6:1
3.2	3.08	4.0	1.5:1
3.3	3.05	4.0	1.5:1
3.4	3.02	4.0	1.5:1
3.5	2.99	4.0	1.5:1
3.6	2.97	3.9	1.5:1
3.7	2.94	3.9	1.5:1
3.8	2.91	3.9	1.5:1
3.9	2.88	3.8	1.4:1
4.0	2.85	3.8	1.4:1
4.1	2.82	3.8	1.4:1
4.2	2.79	3.8	1.4:1
4.3	2.76	3.7	1.4:1
4.4	2.74	3.7	1.4:1
4.5	2.71	3.7	1.4:1
4.6	2.68	3.6	1.3:1
4.7	2.65	3.6	1.3:1
4.8	2.62	3.6	1.3:1
4.9	2.59	3.6	1.3:1
5.0	2.56	3.5	1.3:1
5.1	2.53	3.5	1.3:1
5.2	2.51	3.5	1.3:1
5.3	2.48	3.4	1.2:1
5.4	2.45	3.4	1.2:1
5.5	2.42	3.4	1.2:1
5.6	2.39	3.4	1.2:1
5.7	2.36	3.3	1.2:1
5.8	2.33	3.3	1.2:1
5.9	2.30	3.3	1.2:1
6.0	2.28	3.2	1.1:1
6.1	2.25	3.2	1.1:1
6.2	2.22	3.2	1.1:1
6.3	2.19	3.1	1.1:1
6.4	2.16	3.1	1.1:1
6.5	2.13	3.1	1.1:1
6.6	2.10	3.1	1.1:1
6.7	2.07	3.0	1.0:1
6.8	2.05	3.0	1.0:1
6.9	2.02	3.0	1.0:1
7.0	1.99	2.9	1.0:1

APPENDIX B - 1 **94 lb BAG**

Continued

THICKENING TABLES

From 3:1 to 2:1

Grout in Line and Agitator (cu ft)	MAKEUP MIX			Grout in Line and Agitator (cu ft)	MAKEUP MIX		
	Water (cu ft)	Mix Vol (cu ft)	Mix w:c		Water (cu ft)	Mix Vol (cu ft)	Mix w:c
			2 BAG BATCH				
7.1	1.96	2.9	1.0:1	9.1	1.39	2.3	0.7:1
7.2	1.93	2.9	1.0:1	9.2	1.36	2.3	0.7:1
7.3	1.90	2.9	1.0:1	9.3	1.33	2.3	0.7:1
7.4	1.87	2.8	0.9:1	9.4	1.30	2.3	0.6:1
7.5	1.84	2.8	0.9:1	9.5	1.27	2.2	0.6:1
7.6	1.82	2.8	0.9:1	9.6	1.24	2.2	0.6:1
7.7	1.79	2.7	0.9:1	9.7	1.21	2.2	0.6:1
7.8	1.76	2.7	0.9:1	9.8	1.18	2.1	0.6:1
7.9	1.73	2.7	0.9:1	9.9	1.16	2.1	0.6:1
8.0	1.70	2.7	0.9:1	10.0	1.13	2.1	0.6:1
8.1	1.67	2.6	0.8:1	10.1	1.10	2.1	0.5:1
8.2	1.64	2.6	0.8:1	10.2	1.07	2.0	0.5:1
8.3	1.61	2.6	0.8:1	10.3	1.04	2.0	0.5:1
8.4	1.59	2.5	0.8:1	10.4	1.01	2.0	0.5:1
8.5	1.56	2.5	0.8:1	10.5	0.98	1.9	0.5:1
8.6	1.53	2.5	0.8:1	10.6	0.95	1.9	0.5:1
8.7	1.50	2.5	0.7:1	10.7	0.93	1.9	0.5:1
8.8	1.47	2.4	0.7:1				
8.9	1.44	2.4	0.7:1				
9.0	1.41	2.4	0.7:1				
			3 BAG BATCH				
0.0	6.00	7.4	2.0:1	1.6	5.54	7.0	1.8:1
0.1	5.97	7.4	2.0:1	1.7	5.51	7.0	1.8:1
0.2	5.94	7.4	2.0:1	1.8	5.48	6.9	1.8:1
0.3	5.91	7.4	2.0:1	1.9	5.45	6.9	1.8:1
0.4	5.89	7.3	2.0:1	2.0	5.43	6.9	1.8:1
0.5	5.86	7.3	2.0:1				
0.6	5.83	7.3	1.9:1	2.1	5.40	6.8	1.8:1
0.7	5.80	7.2	1.9:1	2.2	5.37	6.8	1.8:1
0.8	5.77	7.2	1.9:1	2.3	5.34	6.8	1.8:1
0.9	5.74	7.2	1.9:1	2.4	5.31	6.8	1.8:1
1.0	5.71	7.2	1.9:1	2.5	5.28	6.7	1.8:1
				2.6	5.25	6.7	1.8:1
1.1	5.68	7.1	1.9:1	2.7	5.22	6.7	1.7:1
1.2	5.66	7.1	1.9:1	2.8	5.20	6.6	1.7:1
1.3	5.63	7.1	1.9:1	2.9	5.17	6.6	1.7:1
1.4	5.60	7.0	1.9:1	3.0	5.14	6.6	1.7:1
1.5	5.57	7.0	1.9:1				

APPENDIX B - 1
Continued

94 lb BAG

THICKENING TABLES
From 3:1 to 2:1

Grout in Line and Agitator (cu ft)	MAKEUP MIX Water (cu ft)	Mix Vol (cu ft)	Mix w:c	Grout in Line and Agitator (cu ft)	MAKEUP MIX Water (cu ft)	Mix Vol (cu ft)	Mix w:c
3 BAG BATCH							
3.1	5.11	6.5	1.7:1	6.6	4.10	5.5	1.4:1
3.2	5.08	6.5	1.7:1	6.7	4.07	5.5	1.4:1
3.3	5.05	6.5	1.7:1	6.8	4.05	5.5	1.3:1
3.4	5.02	6.5	1.7:1	6.9	4.02	5.5	1.3:1
3.5	4.99	6.4	1.7:1	7.0	3.99	5.4	1.3:1
3.6	4.97	6.4	1.7:1				
3.7	4.94	6.4	1.6:1	7.1	3.96	5.4	1.3:1
3.8	4.91	6.3	1.6:1	7.2	3.93	5.4	1.3:1
3.9	4.88	6.3	1.6:1	7.3	3.90	5.3	1.3:1
4.0	4.85	6.3	1.6:1	7.4	3.87	5.3	1.3:1
				7.5	3.84	5.3	1.3:1
4.1	4.82	6.3	1.6:1	7.6	3.82	5.3	1.3:1
4.2	4.79	6.2	1.6:1	7.7	3.79	5.2	1.3:1
4.3	4.76	6.2	1.6:1	7.8	3.76	5.2	1.3:1
4.4	4.74	6.2	1.6:1	7.9	3.73	5.2	1.2:1
4.5	4.71	6.1	1.6:1	8.0	3.70	5.1	1.2:1
4.6	4.68	6.1	1.6:1				
4.7	4.65	6.1	1.5:1	8.1	3.67	5.1	1.2:1
4.8	4.62	6.1	1.5:1	8.2	3.64	5.1	1.2:1
4.9	4.59	6.0	1.5:1	8.3	3.61	5.1	1.2:1
5.0	4.56	6.0	1.5:1	8.4	3.59	5.0	1.2:1
				8.5	3.56	5.0	1.2:1
5.1	4.53	6.0	1.5:1	8.6	3.53	5.0	1.2:1
5.2	4.51	5.9	1.5:1	8.7	3.50	4.9	1.2:1
5.3	4.48	5.9	1.5:1	8.8	3.47	4.9	1.2:1
5.4	4.45	5.9	1.5:1	8.9	3.44	4.9	1.1:1
5.5	4.42	5.9	1.5:1	9.0	3.41	4.9	1.1:1
5.6	4.39	5.8	1.5:1				
5.7	4.36	5.8	1.5:1	9.1	3.39	4.8	1.1:1
5.8	4.33	5.8	1.4:1	9.2	3.36	4.8	1.1:1
5.9	4.30	5.7	1.4:1	9.3	3.33	4.8	1.1:1
6.0	4.28	5.7	1.4:1	9.4	3.30	4.7	1.1:1
				9.5	3.27	4.7	1.1:1
6.1	4.25	5.7	1.4:1	9.6	3.24	4.7	1.1:1
6.2	4.22	5.7	1.4:1	9.7	3.21	4.7	1.1:1
6.3	4.19	5.6	1.4:1	9.8	3.18	4.6	1.1:1
6.4	4.16	5.6	1.4:1	9.9	3.16	4.6	1.1:1
6.5	4.13	5.6	1.4:1	10.0	3.13	4.6	1.0:1

APPENDIX B - 1 **94 lb BAG**

Continued

THICKENING TABLES

From 3:1 to 2:1

Grout in Line and Agitator (cu ft)	MAKEUP MIX			Grout in Line and Agitator (cu ft)	MAKEUP MIX		
	Water (cu ft)	Mix Vol (cu ft)	Mix w:c		Water (cu ft)	Mix Vol (cu ft)	Mix w:c
3 BAG BATCH							
10.1	3.10	4.5	1.0:1	13.1	2.24	3.7	0.7:1
10.2	3.07	4.5	1.0:1	13.2	2.21	3.6	0.7:1
10.3	3.04	4.5	1.0:1	13.3	2.18	3.6	0.7:1
10.4	3.01	4.5	1.0:1	13.4	2.15	3.6	0.7:1
10.5	2.98	4.4	1.0:1	13.5	2.12	3.6	0.7:1
10.6	2.95	4.4	1.0:1	13.6	2.09	3.5	0.7:1
10.7	2.93	4.4	1.0:1	13.7	2.06	3.5	0.7:1
10.8	2.90	4.3	1.0:1	13.8	2.03	3.5	0.7:1
10.9	2.87	4.3	1.0:1	13.9	2.01	3.4	0.7:1
11.0	2.84	4.3	0.9:1	14.0	1.98	3.4	0.7:1
11.1	2.81	4.3	0.9:1	14.1	1.95	3.4	0.6:1
11.2	2.78	4.2	0.9:1	14.2	1.92	3.4	0.6:1
11.3	2.75	4.2	0.9:1	14.3	1.89	3.3	0.6:1
11.4	2.72	4.2	0.9:1	14.4	1.86	3.3	0.6:1
11.5	2.70	4.1	0.9:1	14.5	1.83	3.3	0.6:1
11.6	2.67	4.1	0.9:1	14.6	1.80	3.2	0.6:1
11.7	2.64	4.1	0.9:1	14.7	1.78	3.2	0.6:1
11.8	2.61	4.0	0.9:1	14.8	1.75	3.2	0.6:1
11.9	2.58	4.0	0.9:1	14.9	1.72	3.2	0.6:1
12.0	2.55	4.0	0.9:1	15.0	1.69	3.1	0.6:1
12.1	2.52	4.0	0.8:1	15.1	1.66	3.1	0.6:1
12.2	2.49	3.9	0.8:1	15.2	1.63	3.1	0.5:1
12.3	2.47	3.9	0.8:1	15.3	1.60	3.0	0.5:1
12.4	2.44	3.9	0.8:1	15.4	1.57	3.0	0.5:1
12.5	2.41	3.8	0.8:1	15.5	1.55	3.0	0.5:1
12.6	2.38	3.8	0.8:1	15.6	1.52	3.0	0.5:1
12.7	2.35	3.8	0.8:1	15.7	1.49	2.9	0.5:1
12.8	2.32	3.8	0.8:1	15.8	1.46	2.9	0.5:1
12.9	2.29	3.7	0.8:1	15.9	1.43	2.9	0.5:1
13.0	2.26	3.7	0.8:1	16.0	1.40	2.8	0.5:1
4 BAG BATCH							
0.0	8.00	9.9	2.0:1	0.5	7.86	9.8	2.0:1
0.1	7.97	9.9	2.0:1	0.6	7.83	9.7	2.0:1
0.2	7.94	9.9	2.0:1	0.7	7.80	9.7	1.9:1
0.3	7.91	9.8	2.0:1	0.8	7.77	9.7	1.9:1
0.4	7.89	9.8	2.0:1	0.9	7.74	9.7	1.9:1
				1.0	7.71	9.6	1.9:1

APPENDIX B - 1
Continued

94 lb BAG

THICKENING TABLES

From 3:1 to 2:1

Grout in Line and Agitator (cu ft)	MAKEUP MIX Water (cu ft)	MAKEUP MIX Mix Vol (cu ft)	MAKEUP MIX Mix w:c	Grout in Line and Agitator (cu ft)	MAKEUP MIX Water (cu ft)	MAKEUP MIX Mix Vol (cu ft)	MAKEUP MIX Mix w:c
4 BAG BATCH							
1.1	7.68	9.6	1.9:1	5.1	6.53	8.5	1.6:1
1.2	7.66	9.6	1.9:1	5.2	6.51	8.4	1.6:1
1.3	7.63	9.5	1.9:1	5.3	6.48	8.4	1.6:1
1.4	7.60	9.5	1.9:1	5.4	6.45	8.4	1.6:1
1.5	7.57	9.5	1.9:1	5.5	6.42	8.3	1.6:1
1.6	7.54	9.5	1.9:1	5.6	6.39	8.3	1.6:1
1.7	7.51	9.4	1.9:1	5.7	6.36	8.3	1.6:1
1.8	7.48	9.4	1.9:1	5.8	6.33	8.3	1.6:1
1.9	7.45	9.4	1.9:1	5.9	6.30	8.2	1.6:1
2.0	7.43	9.3	1.9:1	6.0	6.28	8.2	1.6:1
2.1	7.40	9.3	1.8:1	6.1	6.25	8.2	1.6:1
2.2	7.37	9.3	1.8:1	6.2	6.22	8.1	1.6:1
2.3	7.34	9.3	1.8:1	6.3	6.19	8.1	1.5:1
2.4	7.31	9.2	1.8:1	6.4	6.16	8.1	1.5:1
2.5	7.28	9.2	1.8:1	6.5	6.13	8.1	1.5:1
2.6	7.25	9.2	1.8:1	6.6	6.10	8.0	1.5:1
2.7	7.22	9.1	1.8:1	6.7	6.07	8.0	1.5:1
2.8	7.20	9.1	1.8:1	6.8	6.05	8.0	1.5:1
2.9	7.17	9.1	1.8:1	6.9	6.02	7.9	1.5:1
3.0	7.14	9.1	1.8:1	7.0	5.99	7.9	1.5:1
3.1	7.11	9.0	1.8:1	7.1	5.96	7.9	1.5:1
3.2	7.08	9.0	1.8:1	7.2	5.93	7.9	1.5:1
3.3	7.05	9.0	1.8:1	7.3	5.90	7.8	1.5:1
3.4	7.02	8.9	1.8:1	7.4	5.87	7.8	1.5:1
3.5	6.99	8.9	1.7:1	7.5	5.84	7.8	1.5:1
3.6	6.97	8.9	1.7:1	7.6	5.82	7.7	1.5:1
3.7	6.94	8.9	1.7:1	7.7	5.79	7.7	1.4:1
3.8	6.91	8.8	1.7:1	7.8	5.76	7.7	1.4:1
3.9	6.88	8.8	1.7:1	7.9	5.73	7.6	1.4:1
4.0	6.85	8.8	1.7:1	8.0	5.70	7.6	1.4:1
4.1	6.82	8.7	1.7:1	8.1	5.67	7.6	1.4:1
4.2	6.79	8.7	1.7:1	8.2	5.64	7.6	1.4:1
4.3	6.76	8.7	1.7:1	8.3	5.61	7.5	1.4:1
4.4	6.74	8.7	1.7:1	8.4	5.59	7.5	1.4:1
4.5	6.71	8.6	1.7:1	8.5	5.56	7.5	1.4:1
4.6	6.68	8.6	1.7:1	8.6	5.53	7.4	1.4:1
4.7	6.65	8.6	1.7:1	8.7	5.50	7.4	1.4:1
4.8	6.62	8.5	1.7:1	8.8	5.47	7.4	1.4:1
4.9	6.59	8.5	1.6:1	8.9	5.44	7.4	1.4:1
5.0	6.56	8.5	1.6:1	9.0	5.41	7.3	1.4:1

APPENDIX B - 1
Continued

94 lb BAG

THICKENING TABLES
From 3:1 to 2:1

Grout in Line and Agitator (cu ft)	MAKEUP MIX			Grout in Line and Agitator (cu ft)	MAKEUP MIX		
	Water (cu ft)	Mix Vol (cu ft)	Mix w:c		Water (cu ft)	Mix Vol (cu ft)	Mix w:c
			4 BAG BATCH				
9.1	5.39	7.3	1.3:1	13.1	4.24	6.2	1.1:1
9.2	5.36	7.3	1.3:1	13.2	4.21	6.1	1.1:1
9.3	5.33	7.2	1.3:1	13.3	4.18	6.1	1.0:1
9.4	5.30	7.2	1.3:1	13.4	4.15	6.1	1.0:1
9.5	5.27	7.2	1.3:1	13.5	4.12	6.0	1.0:1
9.6	5.24	7.2	1.3:1	13.6	4.09	6.0	1.0:1
9.7	5.21	7.1	1.3:1	13.7	4.06	6.0	1.0:1
9.8	5.18	7.1	1.3:1	13.8	4.03	6.0	1.0:1
9.9	5.16	7.1	1.3:1	13.9	4.01	5.9	1.0:1
10.0	5.13	7.0	1.3:1	14.0	3.98	5.9	1.0:1
10.1	5.10	7.0	1.3:1	14.1	3.95	5.9	1.0:1
10.2	5.07	7.0	1.3:1	14.2	3.92	5.8	1.0:1
10.3	5.04	7.0	1.3:1	14.3	3.89	5.8	1.0:1
10.4	5.01	6.9	1.3:1	14.4	3.86	5.8	1.0:1
10.5	4.98	6.9	1.2:1	14.5	3.83	5.8	1.0:1
10.6	4.95	6.9	1.2:1	14.6	3.80	5.7	1.0:1
10.7	4.93	6.8	1.2:1	14.7	3.78	5.7	0.9:1
10.8	4.90	6.8	1.2:1	14.8	3.75	5.7	0.9:1
10.9	4.87	6.8	1.2:1	14.9	3.72	5.6	0.9:1
11.0	4.84	6.8	1.2:1	15.0	3.69	5.6	0.9:1
11.1	4.81	6.7	1.2:1	15.1	3.66	5.6	0.9:1
11.2	4.78	6.7	1.2:1	15.2	3.63	5.6	0.9:1
11.3	4.75	6.7	1.2:1	15.3	3.60	5.5	0.9:1
11.4	4.72	6.6	1.2:1	15.4	3.57	5.5	0.9:1
11.5	4.70	6.6	1.2:1	15.5	3.55	5.5	0.9:1
11.6	4.67	6.6	1.2:1	15.6	3.52	5.4	0.9:1
11.7	4.64	6.6	1.2:1	15.7	3.49	5.4	0.9:1
11.8	4.61	6.5	1.2:1	15.8	3.46	5.4	0.9:1
11.9	4.58	6.5	1.1:1	15.9	3.43	5.4	0.9:1
12.0	4.55	6.5	1.1:1	16.0	3.40	5.3	0.9:1
12.1	4.52	6.4	1.1:1	16.1	3.37	5.3	0.8:1
12.2	4.49	6.4	1.1:1	16.2	3.34	5.3	0.8:1
12.3	4.47	6.4	1.1:1	16.3	3.32	5.2	0.8:1
12.4	4.44	6.4	1.1:1	16.4	3.29	5.2	0.8:1
12.5	4.41	6.3	1.1:1	16.5	3.26	5.2	0.8:1
12.6	4.38	6.3	1.1:1	16.6	3.23	5.1	0.8:1
12.7	4.35	6.3	1.1:1	16.7	3.20	5.1	0.8:1
12.8	4.32	6.2	1.1:1	16.8	3.17	5.1	0.8:1
12.9	4.29	6.2	1.1:1	16.9	3.14	5.1	0.8:1
13.0	4.26	6.2	1.1:1	17.0	3.11	5.0	0.8:1

APPENDIX B - 1
Continued

94 lb BAG

THICKENING TABLES

From 3:1 to 2:1

Grout in Line and Agitator (cu ft)	MAKEUP MIX			Grout in Line and Agitator (cu ft)	MAKEUP MIX		
	Water (cu ft)	Mix Vol (cu ft)	Mix w:c		Water (cu ft)	Mix Vol (cu ft)	Mix w:c
4 BAG BATCH							
17.1	3.09	5.0	0.8:1	19.1	2.51	4.4	0.6:1
17.2	3.06	5.0	0.8:1	19.2	2.48	4.4	0.6:1
17.3	3.03	4.9	0.8:1	19.3	2.45	4.4	0.6:1
17.4	3.00	4.9	0.7:1	19.4	2.43	4.3	0.6:1
17.5	2.97	4.9	0.7:1	19.5	2.40	4.3	0.6:1
17.6	2.94	4.9	0.7:1	19.6	2.37	4.3	0.6:1
17.7	2.91	4.8	0.7:1	19.7	2.34	4.3	0.6:1
17.8	2.89	4.8	0.7:1	19.8	2.31	4.2	0.6:1
17.9	2.86	4.8	0.7:1	19.9	2.28	4.2	0.6:1
18.0	2.83	4.7	0.7:1	20.0	2.25	4.2	0.6:1
18.1	2.80	4.7	0.7:1				
18.2	2.77	4.7	0.7:1				
18.3	2.74	4.7	0.7:1				
18.4	2.71	4.6	0.7:1				
18.5	2.68	4.6	0.7:1				
18.6	2.66	4.6	0.7:1				
18.7	2.63	4.5	0.7:1				
18.8	2.60	4.5	0.6:1				
18.9	2.57	4.5	0.6:1				
19.0	2.54	4.5	0.6:1				

APPENDIX B - 1

94 lb BAG

THICKENING TABLES

From 2:1 to 1:1
(by loose volume)

See Section 4.1–8 for explanations and methods of use.

Grout in Line and Agitator (cu ft)	MAKEUP MIX Water (cu ft)	MAKEUP MIX Mix Vol (cu ft)	MAKEUP MIX Mix w:c
1 BAG BATCH			
0.0	1.00	1.5	1.0:1
0.1	0.96	1.4	1.0:1
0.2	0.92	1.4	0.9:1
0.3	0.88	1.4	0.9:1
0.4	0.84	1.3	0.8:1
0.5	0.80	1.3	0.8:1
0.6	0.76	1.2	0.8:1
0.7	0.72	1.2	0.7:1
0.8	0.68	1.2	0.7:1
0.9	0.64	1.1	0.6:1
1.0	0.60	1.1	0.6:1
1.1	0.56	1.0	0.6:1
1.2	0.52	1.0	0.5:1
1.3	0.48	1.0	0.5:1
2 BAG BATCH			
0.0	2.00	3.0	1.0:1
0.1	1.96	2.9	1.0:1
0.2	1.92	2.9	1.0:1
0.3	1.88	2.8	0.9:1
0.4	1.84	2.8	0.9:1
0.5	1.80	2.8	0.9:1
0.6	1.76	2.7	0.9:1
0.7	1.72	2.7	0.9:1
0.8	1.68	2.6	0.8:1
0.9	1.64	2.6	0.8:1
1.0	1.60	2.6	0.8:1
1.1	1.56	2.5	0.8:1
1.2	1.52	2.5	0.8:1
1.3	1.48	2.4	0.7:1
1.4	1.44	2.4	0.7:1
1.5	1.40	2.4	0.7:1
1.6	1.35	2.3	0.7:1
1.7	1.31	2.3	0.7:1
1.8	1.27	2.2	0.6:1
1.9	1.23	2.2	0.6:1
2.0	1.19	2.2	0.6:1
2.1	1.15	2.1	0.6:1
2.2	1.11	2.1	0.6:1
2.3	1.07	2.0	0.5:1
2.4	1.03	2.0	0.5:1
2.5	0.99	2.0	0.5:1
2.6	0.95	1.9	0.5:1
2.7	0.91	1.9	0.5:1

APPENDIX B - 1
Continued

94 lb BAG

THICKENING TABLES

From 2:1 to 1:1

Grout in Line and Agitator (cu ft)	MAKEUP MIX Water (cu ft)	MAKEUP MIX Mix Vol (cu ft)	MAKEUP MIX Mix w:c
3 BAG BATCH			
0.0	3.00	4.4	1.0:1
0.1	2.96	4.4	1.0:1
0.2	2.92	4.4	1.0:1
0.3	2.88	4.3	1.0:1
0.4	2.84	4.3	0.9:1
0.5	2.80	4.2	0.9:1
0.6	2.76	4.2	0.9:1
0.7	2.72	4.2	0.9:1
0.8	2.68	4.1	0.9:1
0.9	2.64	4.1	0.9:1
1.0	2.60	4.0	0.9:1
1.1	2.56	4.0	0.9:1
1.2	2.52	4.0	0.8:1
1.3	2.48	3.9	0.8:1
1.4	2.44	3.9	0.8:1
1.5	2.40	3.8	0.8:1
1.6	2.35	3.8	0.8:1
1.7	2.31	3.8	0.8:1
1.8	2.27	3.7	0.8:1
1.9	2.23	3.7	0.7:1
2.0	2.19	3.6	0.7:1
2.1	2.15	3.6	0.7:1
2.2	2.11	3.6	0.7:1
2.3	2.07	3.5	0.7:1
2.4	2.03	3.5	0.7:1
2.5	1.99	3.4	0.7:1
2.6	1.95	3.4	0.7:1
2.7	1.91	3.4	0.6:1
2.8	1.87	3.3	0.6:1
2.9	1.83	3.3	0.6:1
3.0	1.79	3.2	0.6:1
3.1	1.75	3.2	0.6:1
3.2	1.71	3.1	0.6:1
3.3	1.67	3.1	0.6:1
3.4	1.63	3.1	0.5:1
3.5	1.59	3.0	0.5:1
3.6	1.55	3.0	0.5:1
3.7	1.51	2.9	0.5:1
3.8	1.47	2.9	0.5:1
3.9	1.43	2.9	0.5:1
4.0	1.39	2.8	0.5:1
4 BAG BATCH			
0.0	4.00	5.9	1.0:1
0.1	3.96	5.9	1.0:1
0.2	3.92	5.8	1.0:1
0.3	3.88	5.8	1.0:1
0.4	3.84	5.8	1.0:1
0.5	3.80	5.7	0.9:1
0.6	3.76	5.7	0.9:1
0.7	3.72	5.6	0.9:1
0.8	3.68	5.6	0.9:1
0.9	3.64	5.6	0.9:1
1.0	3.60	5.5	0.9:1
1.1	3.56	5.5	0.9:1
1.2	3.52	5.4	0.9:1
1.3	3.48	5.4	0.9:1
1.4	3.44	5.4	0.9:1
1.5	3.40	5.3	0.8:1
1.6	3.35	5.3	0.8:1
1.7	3.31	5.2	0.8:1
1.8	3.27	5.2	0.8:1
1.9	3.23	5.2	0.8:1
2.0	3.19	5.1	0.8:1
2.1	3.15	5.1	0.8:1
2.2	3.11	5.0	0.8:1
2.3	3.07	5.0	0.8:1
2.4	3.03	5.0	0.8:1
2.5	2.99	4.9	0.7:1
2.6	2.95	4.9	0.7:1
2.7	2.91	4.8	0.7:1
2.8	2.87	4.8	0.7:1
2.9	2.83	4.8	0.7:1
3.0	2.79	4.7	0.7:1

APPENDIX B - 1

Continued

94 lb BAG

THICKENING TABLES

From 2:1 to 1:1

4 BAG BATCH

Grout in Line and Agitator (cu ft)	MAKEUP MIX Water (cu ft)	MAKEUP MIX Mix Vol (cu ft)	MAKEUP MIX Mix w:c
3.1	2.75	4.7	0.7:1
3.2	2.71	4.6	0.7:1
3.3	2.67	4.6	0.7:1
3.4	2.63	4.5	0.7:1
3.5	2.59	4.5	0.6:1
3.6	2.55	4.5	0.6:1
3.7	2.[illegible]1	4.4	0.6:1
3.8	2.47	4.4	0.6:1
3.9	2.43	4.3	0.6:1
4.0	2.39	4.3	0.6:1
4.1	2.35	4.3	0.6:1
4.2	2.31	4.2	0.6:1
4.3	2.27	4.2	0.6:1
4.4	2.23	4.1	0.6:1

Grout in Line and Agitator (cu ft)	MAKEUP MIX Water (cu ft)	MAKEUP MIX Mix Vol (cu ft)	MAKEUP MIX Mix w:c
4.5	2.19	4.1	0.5:1
4.6	2.15	4.1	0.5:1
4.7	2.10	4.0	0.5:1
4.8	2.06	4.0	0.5:1
4.9	2.02	3.9	0.5:1
5.0	1.98	3.9	0.5:1
5.1	1.94	3.9	0.5:1
5.2	1.90	3.8	0.5:1
5.3	1.86	3.8	0.5:1
5.4	1.82	3.7	0.5:1

5 BAG BATCH

Grout in Line and Agitator (cu ft)	MAKEUP MIX Water (cu ft)	MAKEUP MIX Mix Vol (cu ft)	MAKEUP MIX Mix w:c
0.0	5.00	7.4	1.0:1
0.1	4.96	7.4	1.0:1
0.2	4.92	7.3	1.0:1
0.3	4.88	7.3	1.0:1
0.4	4.84	7.2	1.0:1
0.5	4.80	7.2	1.0:1
0.6	4.76	7.2	1.0:1
0.7	4.72	7.1	0.9:1
0.8	4.68	7.1	0.9:1
0.9	4.64	7.0	0.9:1
1.0	4.60	7.0	0.9:1
1.1	4.56	7.0	0.9:1
1.2	4.52	6.9	0.9:1
1.3	4.48	6.9	0.9:1
1.4	4.44	6.8	0.9:1
1.5	4.40	6.8	0.9:1
1.6	4.35	6.8	0.9:1
1.7	4.31	6.7	0.9:1
1.8	4.27	6.7	0.9:1
1.9	4.23	6.6	0.8:1
2.0	4.19	6.6	0.8:1

Grout in Line and Agitator (cu ft)	MAKEUP MIX Water (cu ft)	MAKEUP MIX Mix Vol (cu ft)	MAKEUP MIX Mix w:c
2.1	4.15	6.6	0.8:1
2.2	4.11	6.5	0.8:1
2.3	4.07	6.5	0.8:1
2.4	4.03	6.4	0.8:1
2.5	3.99	6.4	0.8:1
2.6	3.95	6.4	0.8:1
2.7	3.91	6.3	0.8:1
2.8	3.87	6.3	0.8:1
2.9	3.83	6.2	0.8:1
3.0	3.79	6.2	0.8:1
3.1	3.75	6.1	0.8:1
3.2	3.71	6.1	0.7:1
3.3	3.67	6.1	0.7:1
3.4	3.63	6.0	0.7:1
3.5	3.59	6.0	0.7:1
3.6	3.55	5.9	0.7:1
3.7	3.51	5.9	0.7:1
3.8	3.47	5.9	0.7:1
3.9	3.43	5.8	0.7:1
4.0	3.39	5.8	0.7:1

APPENDIX B - 1
Continued

94 lb BAG

THICKENING TABLES
From 2:1 to 1:1

Grout in Line and Agitator (cu ft)	MAKEUP MIX			Grout in Line and Agitator (cu ft)	MAKEUP MIX		
	Water (cu ft)	Mix Vol (cu ft)	Mix w:c		Water (cu ft)	Mix Vol (cu ft)	Mix w:c
			5 BAG BATCH				
4.1	3.35	5.7	0.7:1	5.5	2.78	5.2	0.6:1
4.2	3.31	5.7	0.7:1	5.6	2.74	5.1	0.5:1
4.3	3.27	5.7	0.7:1	5.7	2.70	5.1	0.5:1
4.4	3.23	5.6	0.6:1	5.8	2.66	5.1	0.5:1
4.5	3.19	5.6	0.6:1	5.9	2.62	5.0	0.5:1
4.6	3.15	5.5	0.6:1	6.0	2.58	5.0	0.5:1
4.7	3.10	5.5	0.6:1				
4.8	3.06	5.5	0.6:1	6.1	2.54	4.9	0.5:1
4.9	3.02	5.4	0.6:1	6.2	2.50	4.9	0.5:1
5.0	2.98	5.4	0.6:1	6.3	2.46	4.9	0.5:1
				6.4	2.42	4.8	0.5:1
5.1	2.94	5.3	0.6:1	6.5	2.38	4.8	0.5:1
5.2	2.90	5.3	0.6:1	6.6	2.34	4.7	0.5:1
5.3	2.86	5.3	0.6:1	6.7	2.30	4.7	0.5:1
5.4	2.82	5.2	0.6:1	6.8	2.26	4.7	0.5:1

APPENDIX B - 1

94 lb BAG

THICKENING TABLES

From 1:1 to 0.8:1
(by loose volume)

See Section 4.1–8 for explanations and methods of use.

Grout in Line and Agitator (cu ft)	MAKEUP MIX		
	Water (cu ft)	Mix Vol (cu ft)	Mix w:c
1 BAG BATCH			
0.0	0.80	1.3	0.8:1
0.1	0.79	1.3	0.8:1
0.2	0.77	1.3	0.8:1
0.3	0.76	1.2	0.8:1
0.4	0.75	1.2	0.7:1
0.5	0.73	1.2	0.7:1
0.6	0.72	1.2	0.7:1
0.7	0.71	1.2	0.7:1
0.8	0.69	1.2	0.7:1
0.9	0.68	1.2	0.7:1
1.0	0.66	1.1	0.7:1
1.1	0.65	1.1	0.7:1
1.2	0.64	1.1	0.6:1
1.3	0.62	1.1	0.6:1
1.4	0.61	1.1	0.6:1
1.5	0.60	1.1	0.6:1
1.6	0.58	1.1	0.6:1
1.7	0.57	1.1	0.6:1
1.8	0.56	1.0	0.6:1
1.9	0.54	1.0	0.5:1
2.0	0.53	1.0	0.5:1
2.1	0.52	1.0	0.5:1
2.2	0.50	1.0	0.5:1
2.3	0.49	1.0	0.5:1
2.4	0.48	1.0	0.5:1
2.5	0.46	0.9	0.5:1
2 BAG BATCH			
0.0	1.60	2.6	0.8:1
0.1	1.59	2.5	0.8:1
0.2	1.57	2.5	0.8:1
0.3	1.56	2.5	0.8:1
0.4	1.55	2.5	0.8:1
0.5	1.53	2.5	0.8:1
0.6	1.52	2.5	0.8:1
0.7	1.51	2.5	0.8:1
0.8	1.49	2.5	0.7:1
0.9	1.48	2.4	0.7:1
1.0	1.46	2.4	0.7:1
1.1	1.45	2.4	0.7:1
1.2	1.44	2.4	0.7:1
1.3	1.42	2.4	0.7:1
1.4	1.41	2.4	0.7:1
1.5	1.40	2.4	0.7:1
1.6	1.38	2.3	0.7:1
1.7	1.37	2.3	0.7:1
1.8	1.36	2.3	0.7:1
1.9	1.34	2.3	0.7:1
2.0	1.33	2.3	0.7:1

APPENDIX B - 1
Continued

94 lb BAG

THICKENING TABLES

From 1:1 to 0.8:1

Grout in Line and Agitator (cu ft)	MAKEUP MIX Water (cu ft)	MAKEUP MIX Mix Vol (cu ft)	MAKEUP MIX Mix w:c
	2 BAG BATCH		
2.1	1.32	2.3	0.7:1
2.2	1.30	2.3	0.7:1
2.3	1.29	2.2	0.6:1
2.4	1.28	2.2	0.6:1
2.5	1.26	2.2	0.6:1
2.6	1.25	2.2	0.6:1
2.7	1.24	2.2	0.6:1
2.8	1.22	2.2	0.6:1
2.9	1.21	2.2	0.6:1
3.0	1.19	2.2	0.6:1
3.1	1.18	2.1	0.6:1
3.2	1.17	2.1	0.6:1
3.3	1.15	2.1	0.6:1
3.4	1.14	2.1	0.6:1
3.5	1.13	2.1	0.6:1
3.6	1.11	2.1	0.6:1
3.7	1.10	2.1	0.5:1
3.8	1.09	2.0	0.5:1
3.9	1.07	2.0	0.5:1
4.0	1.06	2.0	0.5:1
4.1	1.05	2.0	0.5:1
4.2	1.03	2.0	0.5:1
4.3	1.02	2.0	0.5:1
4.4	1.01	2.0	0.5:1
4.5	0.99	2.0	0.5:1
4.6	0.98	1.9	0.5:1
4.7	0.96	1.9	0.5:1
4.8	0.95	1.9	0.5:1
4.9	0.94	1.9	0.5:1
5.0	0.92	1.9	0.5:1
	3 BAG BATCH		
0.0	2.40	3.8	0.8:1
0.1	2.39	3.8	0.8:1
0.2	2.37	3.8	0.8:1
0.3	2.36	3.8	0.8:1
0.4	2.35	3.8	0.8:1
0.5	2.33	3.8	0.8:1
0.6	2.32	3.8	0.8:1
0.7	2.31	3.7	0.8:1
0.8	2.29	3.7	0.8:1
0.9	2.28	3.7	0.8:1
1.0	2.26	3.7	0.8:1
1.1	2.25	3.7	0.8:1
1.2	2.24	3.7	0.7:1
1.3	2.22	3.7	0.7:1
1.4	2.21	3.7	0.7:1
1.5	2.20	3.6	0.7:1
1.6	2.18	3.6	0.7:1
1.7	2.17	3.6	0.7:1
1.8	2.16	3.6	0.7:1
1.9	2.14	3.6	0.7:1
2.0	2.13	3.6	0.7:1
2.1	2.12	3.6	0.7:1
2.2	2.10	3.5	0.7:1
2.3	2.09	3.5	0.7:1
2.4	2.08	3.5	0.7:1
2.5	2.06	3.5	0.7:1
2.6	2.05	3.5	0.7:1
2.7	2.04	3.5	0.7:1
2.8	2.02	3.5	0.7:1
2.9	2.01	3.4	0.7:1
3.0	1.99	3.4	0.7:1
3.1	1.98	3.4	0.7:1
3.2	1.97	3.4	0.7:1
3.3	1.95	3.4	0.7:1
3.4	1.94	3.4	0.6:1
3.5	1.93	3.4	0.6:1
3.6	1.91	3.4	0.6:1
3.7	1.90	3.3	0.6:1
3.8	1.89	3.3	0.6:1
3.9	1.87	3.3	0.6:1
4.0	1.86	3.3	0.6:1

APPENDIX B - 1 **94 lb BAG**

Continued

THICKENING TABLES

From 1:1 to 0.8:1

Grout in Line and Agitator (cu ft)	MAKEUP MIX Water (cu ft)	MAKEUP MIX Mix Vol (cu ft)	MAKEUP MIX Mix w:c	Grout in Line and Agitator (cu ft)	MAKEUP MIX Water (cu ft)	MAKEUP MIX Mix Vol (cu ft)	MAKEUP MIX Mix w:c
3 BAG BATCH							
4.1	1.85	3.3	0.6:1	6.1	1.58	3.0	0.5:1
4.2	1.83	3.3	0.6:1	6.2	1.56	3.0	0.5:1
4.3	1.82	3.3	0.6:1	6.3	1.55	3.0	0.5:1
4.4	1.81	3.2	0.6:1	6.4	1.54	3.0	0.5:1
4.5	1.79	3.2	0.6:1	6.5	1.52	3.0	0.5:1
4.6	1.78	3.2	0.6:1	6.6	1.51	2.9	0.5:1
4.7	1.76	3.2	0.6:1	6.7	1.49	2.9	0.5:1
4.8	1.75	3.2	0.6:1	6.8	1.48	2.9	0.5:1
4.9	1.74	3.2	0.6:1	6.9	1.47	2.9	0.5:1
5.0	1.72	3.2	0.6:1	7.0	1.45	2.9	0.5:1
5.1	1.71	3.2	0.6:1	7.1	1.44	2.9	0.5:1
5.2	1.70	3.1	0.6:1	7.2	1.43	2.9	0.5:1
5.3	1.68	3.1	0.6:1	7.3	1.41	2.9	0.5:1
5.4	1.67	3.1	0.6:1	7.4	1.40	2.8	0.5:1
5.5	1.66	3.1	0.6:1	7.5	1.39	2.8	0.5:1
5.6	1.64	3.1	0.5:1	7.6	1.37	2.8	0.5:1
5.7	1.63	3.1	0.5:1	7.7	1.36	2.8	0.5:1
5.8	1.62	3.1	0.5:1				
5.9	1.60	3.0	0.5:1				
6.0	1.59	3.0	0.5:1				
4 BAG BATCH							
0.0	3.20	5.1	0.8:1	1.5	3.00	4.9	0.7:1
0.1	3.19	5.1	0.8:1	1.6	2.98	4.9	0.7:1
0.2	3.17	5.1	0.8:1	1.7	2.97	4.9	0.7:1
0.3	3.16	5.1	0.8:1	1.8	2.96	4.9	0.7:1
0.4	3.15	5.1	0.8:1	1.9	2.94	4.9	0.7:1
0.5	3.13	5.1	0.8:1	2.0	2.93	4.8	0.7:1
0.6	3.12	5.0	0.8:1				
0.7	3.11	5.0	0.8:1	2.1	2.92	4.8	0.7:1
0.8	3.09	5.0	0.8:1	2.2	2.90	4.8	0.7:1
0.9	3.08	5.0	0.8:1	2.3	2.89	4.8	0.7:1
1.0	3.06	5.0	0.8:1	2.4	2.88	4.8	0.7:1
				2.5	2.86	4.8	0.7:1
1.1	3.05	5.0	0.8:1	2.6	2.85	4.8	0.7:1
1.2	3.04	5.0	0.8:1	2.7	2.84	4.8	0.7:1
1.3	3.02	4.9	0.8:1	2.8	2.82	4.7	0.7:1
1.4	3.01	4.9	0.8:1	2.9	2.81	4.7	0.7:1
				3.0	2.79	4.7	0.7:1

APPENDIX B - 1
Continued

94 lb BAG

THICKENING TABLES
From 1:1 to 0.8:1

4 BAG BATCH

Grout in Line and Agitator (cu ft)	MAKEUP MIX Water (cu ft)	MAKEUP MIX Mix Vol (cu ft)	MAKEUP MIX Mix w:c
3.1	2.78	4.7	0.7:1
3.2	2.77	4.7	0.7:1
3.3	2.75	4.7	0.7:1
3.4	2.74	4.7	0.7:1
3.5	2.73	4.6	0.7:1
3.6	2.71	4.6	0.7:1
3.7	2.70	4.6	0.7:1
3.8	2.69	4.6	0.7:1
3.9	2.67	4.6	0.7:1
4.0	2.66	4.6	0.7:1
4.1	2.65	4.6	0.7:1
4.2	2.63	4.6	0.7:1
4.3	2.62	4.5	0.7:1
4.4	2.61	4.5	0.7:1
4.5	2.59	4.5	0.6:1
4.6	2.58	4.5	0.6:1
4.7	2.56	4.5	0.6:1
4.8	2.55	4.5	0.6:1
4.9	2.54	4.5	0.6:1
5.0	2.52	4.4	0.6:1
5.1	2.51	4.4	0.6:1
5.2	2.50	4.4	0.6:1
5.3	2.48	4.4	0.6:1
5.4	2.47	4.4	0.6:1
5.5	2.46	4.4	0.6:1
5.6	2.44	4.4	0.6:1
5.7	2.43	4.3	0.6:1
5.8	2.42	4.3	0.6:1
5.9	2.40	4.3	0.6:1
6.0	2.39	4.3	0.6:1
6.1	2.38	4.3	0.6:1
6.2	2.36	4.3	0.6:1
6.3	2.35	4.3	0.6:1
6.4	2.34	4.3	0.6:1

Grout in Line and Agitator (cu ft)	MAKEUP MIX Water (cu ft)	MAKEUP MIX Mix Vol (cu ft)	MAKEUP MIX Mix w:c
6.5	2.32	4.2	0.6:1
6.6	2.31	4.2	0.6:1
6.7	2.29	4.2	0.6:1
6.8	2.28	4.2	0.6:1
6.9	2.27	4.2	0.6:1
7.0	2.25	4.2	0.6:1
7.1	2.24	4.2	0.6:1
7.2	2.23	4.1	0.6:1
7.3	2.21	4.1	0.6:1
7.4	2.20	4.1	0.5:1
7.5	2.19	4.1	0.5:1
7.6	2.17	4.1	0.5:1
7.7	2.16	4.1	0.5:1
7.8	2.15	4.1	0.5:1
7.9	2.13	4.1	0.5:1
8.0	2.12	4.0	0.5:1
8.1	2.11	4.0	0.5:1
8.2	2.09	4.0	0.5:1
8.3	2.08	4.0	0.5:1
8.4	2.06	4.0	0.5:1
8.5	2.05	4.0	0.5:1
8.6	2.04	4.0	0.5:1
8.7	2.02	3.9	0.5:1
8.8	2.01	3.9	0.5:1
8.9	2.00	3.9	0.5:1
9.0	1.98	3.9	0.5:1
9.1	1.97	3.9	0.5:1
9.2	1.96	3.9	0.5:1
9.3	1.94	3.9	0.5:1
9.4	1.93	3.8	0.5:1
9.5	1.92	3.8	0.5:1
9.6	1.90	3.8	0.5:1
9.7	1.89	3.8	0.5:1
9.8	1.88	3.8	0.5:1
9.9	1.86	3.8	0.5:1
10.0	1.85	3.8	0.5:1

APPENDIX B - 1

94 lb BAG

THICKENING TABLES

From 0.8:1 to 0.6:1
(by loose volume)

See Section 4.1–8 for explanations and methods of use.

Grout in Line and Agitator (cu ft)	MAKEUP MIX			Grout in Line and Agitator (cu ft)	MAKEUP MIX		
	Water (cu ft)	Mix Vol (cu ft)	Mix w:c		Water (cu ft)	Mix Vol (cu ft)	Mix w:c
1 BAG BATCH							
0.0	0.60	1.1	0.6:1				
0.1	0.58	1.1	0.6:1				
0.2	0.57	1.0	0.6:1				
0.3	0.55	1.0	0.6:1				
0.4	0.54	1.0	0.5:1				
0.5	0.52	1.0	0.5:1				
0.6	0.51	1.0	0.5:1				
0.7	0.49	1.0	0.5:1				
0.8	0.47	1.0	0.5:1				
0.9	0.46	0.9	0.5:1				
2 BAG BATCH							
0.0	1.20	2.2	0.6:1	1.1	1.03	2.0	0.5:1
0.1	1.18	2.1	0.6:1	1.2	1.01	2.0	0.5:1
0.2	1.17	2.1	0.6:1	1.3	1.00	2.0	0.5:1
0.3	1.15	2.1	0.6:1	1.4	0.98	1.9	0.5:1
0.4	1.14	2.1	0.6:1	1.5	0.97	1.9	0.5:1
0.5	1.12	2.1	0.6:1	1.6	0.95	1.9	0.5:1
0.6	1.11	2.1	0.6:1	1.7	0.93	1.9	0.5:1
0.7	1.09	2.1	0.5:1	1.8	0.92	1.9	0.5:1
0.8	1.07	2.0	0.5:1	1.9	0.90	1.9	0.5:1
0.9	1.06	2.0	0.5:1				
1.0	1.04	2.0	0.5:1				
3 BAG BATCH							
0.0	1.80	3.2	0.6:1	0.5	1.72	3.2	0.6:1
0.1	1.78	3.2	0.6:1	0.6	1.71	3.1	0.6:1
0.2	1.77	3.2	0.6:1	0.7	1.69	3.1	0.6:1
0.3	1.75	3.2	0.6:1	0.8	1.67	3.1	0.6:1
0.4	1.74	3.2	0.6:1	0.9	1.66	3.1	0.6:1
				1.0	1.64	3.1	0.5:1

APPENDIX B - 1
Continued

94 lb BAG

THICKENING TABLES
From 0.8:1 to 0.6:1

Grout in Line and Agitator (cu ft)	MAKEUP MIX Water (cu ft)	MAKEUP MIX Mix Vol (cu ft)	MAKEUP MIX Mix w:c	Grout in Line and Agitator (cu ft)	MAKEUP MIX Water (cu ft)	MAKEUP MIX Mix Vol (cu ft)	MAKEUP MIX Mix w:c
			3 BAG BATCH				
1.1	1.63	3.1	0.5:1	2.1	1.47	2.9	0.5:1
1.2	1.61	3.1	0.5:1	2.2	1.46	2.9	0.5:1
1.3	1.60	3.0	0.5:1	2.3	1.44	2.9	0.5:1
1.4	1.58	3.0	0.5:1	2.4	1.42	2.9	0.5:1
1.5	1.57	3.0	0.5:1	2.5	1.41	2.8	0.5:1
1.6	1.55	3.0	0.5:1	2.6	1.39	2.8	0.5:1
1.7	1.53	3.0	0.5:1	2.7	1.38	2.8	0.5:1
1.8	1.52	3.0	0.5:1	2.8	1.36	2.8	0.5:1
1.9	1.50	2.9	0.5:1				
2.0	1.49	2.9	0.5:1				
			4 BAG BATCH				
0.0	2.40	4.3	0.6:1	2.1	2.07	4.0	0.5:1
0.1	2.38	4.3	0.6:1	2.2	2.06	4.0	0.5:1
0.2	2.37	4.3	0.6:1	2.3	2.04	4.0	0.5:1
0.3	2.35	4.3	0.6:1	2.4	2.02	3.9	0.5:1
0.4	2.34	4.3	0.6:1	2.5	2.01	3.9	0.5:1
0.5	2.32	4.2	0.6:1	2.6	1.99	3.9	0.5:1
0.6	2.31	4.2	0.6:1	2.7	1.98	3.9	0.5:1
0.7	2.29	4.2	0.6:1	2.8	1.96	3.9	0.5:1
0.8	2.27	4.2	0.6:1	2.9	1.95	3.9	0.5:1
0.9	2.26	4.2	0.6:1	3.0	1.93	3.9	0.5:1
1.0	2.24	4.2	0.6:1				
				3.1	1.92	3.8	0.5:1
1.1	2.23	4.1	0.6:1	3.2	1.90	3.8	0.5:1
1.2	2.21	4.1	0.6:1	3.3	1.88	3.8	0.5:1
1.3	2.20	4.1	0.5:1	3.4	1.87	3.8	0.5:1
1.4	2.18	4.1	0.5:1	3.5	1.85	3.8	0.5:1
1.5	2.17	4.1	0.5:1	3.6	1.84	3.8	0.5:1
1.6	2.15	4.1	0.5:1	3.7	1.82	3.7	0.5:1
1.7	2.13	4.1	0.5:1	3.8	1.81	3.7	0.5:1
1.8	2.12	4.0	0.5:1				
1.9	2.10	4.0	0.5:1				
2.0	2.09	4.0	0.5:1				
			5 BAG BATCH				
0.0	3.00	5.4	0.6:1	0.5	2.92	5.3	0.6:1
0.1	2.98	5.4	0.6:1	0.6	2.91	5.3	0.6:1
0.2	2.97	5.4	0.6:1	0.7	2.89	5.3	0.6:1
0.3	2.95	5.4	0.6:1	0.8	2.87	5.3	0.6:1
0.4	2.94	5.3	0.6:1	0.9	2.86	5.3	0.6:1
				1.0	2.84	5.2	0.6:1

APPENDIX B - 1
Continued

94 lb BAG

THICKENING TABLES
From 0.8:1 to 0.6:1

Grout in Line and Agitator (cu ft)	MAKEUP MIX Water (cu ft)	MAKEUP MIX Mix Vol (cu ft)	MAKEUP MIX Mix w:c	Grout in Line and Agitator (cu ft)	MAKEUP MIX Water (cu ft)	MAKEUP MIX Mix Vol (cu ft)	MAKEUP MIX Mix w:c
5 BAG BATCH							
1.1	2.83	5.2	0.6:1	3.1	2.52	4.9	0.5:1
1.2	2.81	5.2	0.6:1	3.2	2.50	4.9	0.5:1
1.3	2.80	5.2	0.6:1	3.3	2.48	4.9	0.5:1
1.4	2.78	5.2	0.6:1	3.4	2.47	4.9	0.5:1
1.5	2.77	5.2	0.6:1	3.5	2.45	4.9	0.5:1
1.6	2.75	5.1	0.5:1	3.6	2.44	4.8	0.5:1
1.7	2.73	5.1	0.5:1	3.7	2.42	4.8	0.5:1
1.8	2.72	5.1	0.5:1	3.8	2.41	4.8	0.5:1
1.9	2.70	5.1	0.5:1	3.9	2.39	4.8	0.5:1
2.0	2.69	5.1	0.5:1	4.0	2.37	4.8	0.5:1
2.1	2.67	5.1	0.5:1	4.1	2.36	4.8	0.5:1
2.2	2.66	5.1	0.5:1	4.2	2.34	4.7	0.5:1
2.3	2.64	5.0	0.5:1	4.3	2.33	4.7	0.5:1
2.4	2.62	5.0	0.5:1	4.4	2.31	4.7	0.5:1
2.5	2.61	5.0	0.5:1	4.5	2.30	4.7	0.5:1
2.6	2.59	5.0	0.5:1	4.6	2.28	4.7	0.5:1
2.7	2.58	5.0	0.5:1	4.7	2.27	4.7	0.5:1
2.8	2.56	5.0	0.5:1				
2.9	2.55	4.9	0.5:1				
3.0	2.53	4.9	0.5:1				

APPENDIX B - 1

94 lb BAG

THICKENING TABLES

From 0.6:1 to 0.5:1
(by loose volume)

See Section 4.1–8 for explanations and methods of use.

Grout in Line and Agitator (cu ft)	MAKEUP MIX Water (cu ft)	MAKEUP MIX Mix Vol (cu ft)	MAKEUP MIX Mix w:c	Grout in Line and Agitator (cu ft)	MAKEUP MIX Water (cu ft)	MAKEUP MIX Mix Vol (cu ft)	MAKEUP MIX Mix w:c
1 BAG BATCH							
0.0	0.50	1.0	0.5:1				
0.1	0.49	1.0	0.5:1				
0.2	0.48	1.0	0.5:1				
0.3	0.47	1.0	0.5:1				
0.4	0.46	0.9	0.5:1				
0.5	0.45	0.9	0.5:1				
2 BAG BATCH							
0.0	1.00	2.0	0.5:1	0.5	0.95	1.9	0.5:1
0.1	0.99	2.0	0.5:1	0.6	0.94	1.9	0.5:1
0.2	0.98	1.9	0.5:1	0.7	0.94	1.9	0.5:1
0.3	0.97	1.9	0.5:1	0.8	0.93	1.9	0.5:1
0.4	0.96	1.9	0.5:1	0.9	0.92	1.9	0.5:1
				1.0	0.91	1.9	0.5:1
3 BAG BATCH							
0.0	1.50	2.9	0.5:1	1.1	1.40	2.8	0.5:1
0.1	1.49	2.9	0.5:1	1.2	1.39	2.8	0.5:1
0.2	1.48	2.9	0.5:1	1.3	1.38	2.8	0.5:1
0.3	1.47	2.9	0.5:1	1.4	1.37	2.8	0.5:1
0.4	1.46	2.9	0.5:1	1.5	1.36	2.8	0.5:1
0.5	1.45	2.9	0.5:1	1.6	1.35	2.8	0.5:1
0.6	1.44	2.9	0.5:1				
0.7	1.44	2.9	0.5:1				
0.8	1.43	2.9	0.5:1				
0.9	1.42	2.9	0.5:1				
1.0	1.41	2.8	0.5:1				
4 BAG BATCH							
0.0	2.00	3.9	0.5:1	0.5	1.95	3.9	0.5:1
0.1	1.99	3.9	0.5:1	0.6	1.94	3.9	0.5:1
0.2	1.98	3.9	0.5:1	0.7	1.94	3.9	0.5:1
0.3	1.97	3.9	0.5:1	0.8	1.93	3.8	0.5:1
0.4	1.96	3.9	0.5:1	0.9	1.92	3.8	0.5:1

APPENDIX B - 1 — 94 lb BAG

Continued

THICKENING TABLES
From 0.6:1 to 0.5:1

Grout in Line and Agitator (cu ft)	MAKEUP MIX Water (cu ft)	MAKEUP MIX Mix Vol (cu ft)	MAKEUP MIX Mix w:c	Grout in Line and Agitator (cu ft)	MAKEUP MIX Water (cu ft)	MAKEUP MIX Mix Vol (cu ft)	MAKEUP MIX Mix w:c
4 BAG BATCH							
1.0	1.91	3.8	0.5:1	1.5	1.86	3.8	0.5:1
1.1	1.90	3.8	0.5:1	1.6	1.85	3.8	0.5:1
1.2	1.89	3.8	0.5:1	1.7	1.84	3.8	0.5:1
1.3	1.88	3.8	0.5:1	1.8	1.83	3.8	0.5:1
1.4	1.87	3.8	0.5:1	1.9	1.82	3.7	0.5:1
				2.0	1.81	3.7	0.5:1
5 BAG BATCH							
0.0	2.50	4.9	0.5:1	2.1	2.31	4.7	0.5:1
0.1	2.49	4.9	0.5:1	2.2	2.30	4.7	0.5:1
0.2	2.48	4.9	0.5:1	2.3	2.29	4.7	0.5:1
0.3	2.47	4.9	0.5:1	2.4	2.28	4.7	0.5:1
0.4	2.46	4.9	0.5:1	2.5	2.27	4.7	0.5:1
0.5	2.45	4.9	0.5:1	2.6	2.26	4.7	0.5:1
0.6	2.44	4.8	0.5:1				
0.7	2.44	4.8	0.5:1				
0.8	2.43	4.8	0.5:1				
0.9	2.42	4.8	0.5:1				
1.0	2.41	4.8	0.5:1				
1.1	2.40	4.8	0.5:1				
1.2	2.39	4.8	0.5:1				
1.3	2.38	4.8	0.5:1				
1.4	2.37	4.8	0.5:1				
1.5	2.36	4.8	0.5:1				
1.6	2.35	4.8	0.5:1				
1.7	2.34	4.7	0.5:1				
1.8	2.33	4.7	0.5:1				
1.9	2.32	4.7	0.5:1				
2.0	2.31	4.7	0.5:1				

APPENDIX B - 2

CU FT

THINNING TABLES

From 4:1
(by loose volume)

See Section 4.1–9 for explanations and methods of use.

Grout in Line and Agitator (cu ft)	THIN-NING WATER (cu ft)	New vol (cu ft)	Grout in Line and Agitator (cu ft)	THIN-NING WATER (cu ft)	New vol (cu ft)	Grout in Line and Agitator (cu ft)	THIN-NING WATER (cu ft)	New vol (cu ft)
TO 5:1								
1.0	0.22	1.2	3.5	0.78	4.3	6.1	1.36	7.5
			3.6	0.80	4.4	6.2	1.38	7.6
1.1	0.25	1.3	3.7	0.83	4.5	6.3	1.41	7.7
1.2	0.27	1.5	3.8	0.85	4.6	6.4	1.43	7.8
1.3	0.29	1.6	3.9	0.87	4.8	6.5	1.45	8.0
1.4	0.31	1.7	4.0	0.89	4.9	6.6	1.47	8.1
1.5	0.33	1.8				6.7	1.50	8.2
1.6	0.36	2.0	4.1	0.92	5.0	6.8	1.52	8.3
1.7	0.38	2.1	4.2	0.94	5.1	6.9	1.54	8.4
1.8	0.40	2.2	4.3	0.96	5.3	7.0	1.56	8.6
1.9	0.42	2.3	4.4	0.98	5.4			
2.0	0.45	2.4	4.5	1.00	5.5	7.1	1.58	8.7
			4.6	1.03	5.6	7.2	1.61	8.8
2.1	0.47	2.6	4.7	1.05	5.7	7.3	1.63	8.9
2.2	0.49	2.7	4.8	1.07	5.9	7.4	1.65	9.1
2.3	0.51	2.8	4.9	1.09	6.0	7.5	1.67	9.2
2.4	0.54	2.9	5.0	1.12	6.1	7.6	1.70	9.3
2.5	0.56	3.1				7.7	1.72	9.4
2.6	0.58	3.2	5.1	1.14	6.2	7.8	1.74	9.5
2.7	0.60	3.3	5.2	1.16	6.4	7.9	1.76	9.7
2.8	0.62	3.4	5.3	1.18	6.5	8.0	1.79	9.8
2.9	0.65	3.5	5.4	1.21	6.6			
3.0	0.67	3.7	5.5	1.23	6.7	8.1	1.81	9.9
			5.6	1.25	6.8	8.2	1.83	10.0
3.1	0.69	3.8	5.7	1.27	7.0	8.3	1.85	10.2
3.2	0.71	3.9	5.8	1.29	7.1	8.4	1.87	10.3
3.3	0.74	4.0	5.9	1.32	7.2			
3.4	0.76	4.2	6.0	1.34	7.3			

APPENDIX B - 2 CU FT

Continued

THINNING TABLES

From 4:1

TO 5:1

Grout in Line and Agitator (cu ft)	THINNING WATER (cu ft)	New vol (cu ft)
8.5	1.90	10.4
8.6	1.92	10.5
8.7	1.94	10.6
8.8	1.96	10.8
8.9	1.99	10.9
9.0	2.01	11.0
9.1	2.03	11.1
9.2	2.05	11.3
9.3	2.08	11.4
9.4	2.10	11.5
9.5	2.12	11.6
9.6	2.14	11.7
9.7	2.17	11.9
9.8	2.19	12.0
9.9	2.21	12.1
10.0	2.23	12.2
10.1	2.25	12.4
10.2	2.28	12.5
10.3	2.30	12.6
10.4	2.32	12.7
10.5	2.34	12.8
10.6	2.37	13.0
10.7	2.39	13.1
10.8	2.41	13.2
10.9	2.43	13.3
11.0	2.46	13.5
11.1	2.48	13.6
11.2	2.50	13.7
11.3	2.52	13.8
11.4	2.54	13.9
11.5	2.57	14.1
11.6	2.59	14.2
11.7	2.61	14.3
11.8	2.63	14.4
11.9	2.66	14.6
12.0	2.68	14.7
12.1	2.70	14.8
12.2	2.72	14.9
12.3	2.75	15.0
12.4	2.77	15.2
12.5	2.79	15.3
12.6	2.81	15.4
12.7	2.83	15.5
12.8	2.86	15.7
12.9	2.88	15.8
13.0	2.90	15.9
13.1	2.92	16.0
13.2	2.95	16.1
13.3	2.97	16.3
13.4	2.99	16.4
13.5	3.01	16.5
13.6	3.04	16.6
13.7	3.06	16.8
13.8	3.08	16.9
13.9	3.10	17.0
14.0	3.12	17.1
14.1	3.15	17.2
14.2	3.17	17.4
14.3	3.19	17.5
14.4	3.21	17.6
14.5	3.24	17.7
14.6	3.26	17.9
14.7	3.28	18.0
14.8	3.30	18.1
14.9	3.33	18.2
15.0	3.35	18.3
15.1	3.37	18.5
15.2	3.39	18.6
15.3	3.42	18.7
15.4	3.44	18.8
15.5	3.46	19.0
15.6	3.48	19.1
15.7	3.50	19.2
15.8	3.53	19.3
15.9	3.55	19.4
16.0	3.57	19.6
16.1	3.59	19.7
16.2	3.62	19.8
16.3	3.64	19.9

APPENDIX B - 2 **CU FT**

THINNING TABLES

From 3:1
(by loose volume)

See Section 4.1–9 for explanations and methods of use.

Grout in Line and Agitator (cu ft)	THIN-NING WATER (cu ft)	New vol (cu ft)	Grout in Line and Agitator (cu ft)	THIN-NING WATER (cu ft)	New vol (cu ft)	Grout in Line and Agitator (cu ft)	THIN-NING WATER (cu ft)	New vol (cu ft)
TO 4:1								
1.0	0.29	1.3	3.5	1.01	4.5	6.1	1.75	7.9
			3.6	1.03	4.6	6.2	1.78	8.0
1.1	0.32	1.4	3.7	1.06	4.8	6.3	1.81	8.1
1.2	0.34	1.5	3.8	1.09	4.9	6.4	1.84	8.2
1.3	0.37	1.7	3.9	1.12	5.0	6.5	1.87	8.4
1.4	0.40	1.8	4.0	1.15	5.1	6.6	1.90	8.5
1.5	0.43	1.9				6.7	1.93	8.6
1.6	0.46	2.1	4.1	1.18	5.3	6.8	1.95	8.8
1.7	0.49	2.2	4.2	1.21	5.4	6.9	1.98	8.9
1.8	0.52	2.3	4.3	1.24	5.5	7.0	2.01	9.0
1.9	0.55	2.4	4.4	1.26	5.7			
2.0	0.57	2.6	4.5	1.29	5.8	7.1	2.04	9.1
			4.6	1.32	5.9	7.2	2.07	9.3
2.1	0.60	2.7	4.7	1.35	6.1	7.3	2.10	9.4
2.2	0.63	2.8	4.8	1.38	6.2	7.4	2.13	9.5
2.3	0.66	3.0	4.9	1.41	6.3	7.5	2.16	9.7
2.4	0.69	3.1	5.0	1.44	6.4	7.6	2.18	9.8
2.5	0.72	3.2				7.7	2.21	9.9
2.6	0.75	3.3	5.1	1.47	6.6	7.8	2.24	10.0
2.7	0.78	3.5	5.2	1.49	6.7	7.9	2.27	10.2
2.8	0.80	3.6	5.3	1.52	6.8	8.0	2.30	10.3
2.9	0.83	3.7	5.4	1.55	7.0			
3.0	0.86	3.9	5.5	1.58	7.1	8.1	2.33	10.4
			5.6	1.61	7.2	8.2	2.36	10.6
3.1	0.89	4.0	5.7	1.64	7.3	8.3	2.39	10.7
3.2	0.92	4.1	5.8	1.67	7.5	8.4	2.41	10.8
3.3	0.95	4.2	5.9	1.70	7.6			
3.4	0.98	4.4	6.0	1.72	7.7			

APPENDIX B - 2 CU FT

Continued

THINNING TABLES

From 3:1

Grout in Line and Agitator (cu ft)	THIN-NING WATER (cu ft)	New vol (cu ft)	Grout in Line and Agitator (cu ft)	THIN-NING WATER (cu ft)	New vol (cu ft)	Grout in Line and Agitator (cu ft)	THIN-NING WATER (cu ft)	New vol (cu ft)
TO 4:1								
8.5	2.44	10.9	11.1	3.19	14.3	13.6	3.91	17.5
8.6	2.47	11.1	11.2	3.22	14.4	13.7	3.94	17.6
8.7	2.50	11.2	11.3	3.25	14.5	13.8	3.97	17.8
8.8	2.53	11.3	11.4	3.28	14.7	13.9	3.99	17.9
8.9	2.56	11.5	11.5	3.30	14.8	14.0	4.02	18.0
9.0	2.59	11.6	11.6	3.33	14.9			
			11.7	3.36	15.1	14.1	4.05	18.2
9.1	2.61	11.7	11.8	3.39	15.2	14.2	4.08	18.3
9.2	2.64	11.8	11.9	3.42	15.3	14.3	4.11	18.4
9.3	2.67	12.0	12.0	3.45	15.4	14.4	4.14	18.5
9.4	2.70	12.1				14.5	4.17	18.7
9.5	2.73	12.2	12.1	3.48	15.6	14.6	4.20	18.8
9.6	2.76	12.4	12.2	3.51	15.7	14.7	4.22	18.9
9.7	2.79	12.5	12.3	3.53	15.8	14.8	4.25	19.1
9.8	2.82	12.6	12.4	3.56	16.0	14.9	4.28	19.2
9.9	2.84	12.7	12.5	3.59	16.1	15.0	4.31	19.3
10.0	2.87	12.9	12.6	3.62	16.2			
			12.7	3.65	16.3	15.1	4.34	19.4
10.1	2.90	13.0	12.8	3.68	16.5	15.2	4.37	19.6
10.2	2.93	13.1	12.9	3.71	16.6	15.3	4.40	19.7
10.3	2.96	13.3	13.0	3.74	16.7	15.4	4.43	19.8
10.4	2.99	13.4				15.5	4.45	20.0
10.5	3.02	13.5	13.1	3.76	16.9			
10.6	3.05	13.6	13.2	3.79	17.0			
10.7	3.07	13.8	13.3	3.82	17.1			
10.8	3.10	13.9	13.4	3.85	17.3			
10.9	3.13	14.0	13.5	3.88	17.4			
11.0	3.16	14.2						
TO 5:1								
1.0	0.57	1.6	2.1	1.21	3.3	3.1	1.78	4.9
1.1	0.63	1.7	2.2	1.26	3.5	3.2	1.84	5.0
1.2	0.69	1.9	2.3	1.32	3.6	3.3	1.90	5.2
1.3	0.75	2.0	2.4	1.38	3.8	3.4	1.95	5.4
1.4	0.80	2.2	2.5	1.44	3.9	3.5	2.01	5.5
1.5	0.86	2.4	2.6	1.49	4.1	3.6	2.07	5.7
1.6	0.92	2.5	2.7	1.55	4.3	3.7	2.13	5.8
1.7	0.98	2.7	2.8	1.61	4.4	3.8	2.18	6.0
1.8	1.03	2.8	2.9	1.67	4.6	3.9	2.24	6.1
1.9	1.09	3.0	3.0	1.72	4.7	4.0	2.30	6.3
2.0	1.15	3.1						

APPENDIX B - 2

Continued

CU FT

THINNING TABLES

From 3:1

Grout in Line and Agitator (cu ft)	THIN-NING WATER (cu ft)	New vol (cu ft)
TO 5:1		
4.1	2.36	6.5
4.2	2.41	6.6
4.3	2.47	6.8
4.4	2.53	6.9
4.5	2.59	7.1
4.6	2.64	7.2
4.7	2.70	7.4
4.8	2.76	7.6
4.9	2.82	7.7
5.0	2.87	7.9
5.1	2.93	8.0
5.2	2.99	8.2
5.3	3.05	8.3
5.4	3.10	8.5
5.5	3.16	8.7
5.6	3.22	8.8
5.7	3.28	9.0
5.8	3.33	9.1
5.9	3.39	9.3
6.0	3.45	9.4
6.1	3.51	9.6
6.2	3.56	9.8
6.3	3.62	9.9
6.4	3.68	10.1
6.5	3.74	10.2
6.6	3.79	10.4
6.7	3.85	10.6
6.8	3.91	10.7
6.9	3.97	10.9
7.0	4.02	11.0
7.1	4.08	11.2
7.2	4.14	11.3
7.3	4.20	11.5
7.4	4.25	11.7
7.5	4.31	11.8
7.6	4.37	12.0
7.7	4.43	12.1
7.8	4.48	12.3
7.9	4.54	12.4
8.0	4.60	12.6
8.1	4.66	12.8
8.2	4.71	12.9
8.3	4.77	13.1
8.4	4.83	13.2
8.5	4.89	13.4
8.6	4.94	13.5
8.7	5.00	13.7
8.8	5.06	13.9
8.9	5.11	14.0
9.0	5.17	14.2
9.1	5.23	14.3
9.2	5.29	14.5
9.3	5.34	14.6
9.4	5.40	14.8
9.5	5.46	15.0
9.6	5.52	15.1
9.7	5.57	15.3
9.8	5.63	15.4
9.9	5.69	15.6
10.0	5.75	15.7
10.1	5.80	15.9
10.2	5.86	16.1
10.3	5.92	16.2
10.4	5.98	16.4
10.5	6.03	16.5
10.6	6.09	16.7
10.7	6.15	16.8
10.8	6.21	17.0
10.9	6.26	17.2
11.0	6.32	17.3
11.1	6.38	17.5
11.2	6.44	17.6
11.3	6.49	17.8
11.4	6.55	18.0
11.5	6.61	18.1
11.6	6.67	18.3
11.7	6.72	18.4
11.8	6.78	18.6
11.9	6.84	18.7
12.0	6.90	18.9
12.1	6.95	19.1
12.2	7.01	19.2
12.3	7.07	19.4
12.4	7.13	19.5
12.5	7.18	19.7
12.6	7.24	19.8
12.7	7.30	20.0

APPENDIX B - 2 CU FT

THINNING TABLES

From 2:1
(by loose volume)

See Section 4.1–9 for explanations and methods of use.

Grout in Line and Agitator (cu ft)	THIN-NING WATER (cu ft)	New vol (cu ft)
TO 3:1		
1.0	0.40	1.4
1.1	0.44	1.5
1.2	0.48	1.7
1.3	0.52	1.8
1.4	0.56	2.0
1.5	0.60	2.1
1.6	0.65	2.2
1.7	0.69	2.4
1.8	0.73	2.5
1.9	0.77	2.7
2.0	0.81	2.8
2.1	0.85	2.9
2.2	0.89	3.1
2.3	0.93	3.2
2.4	0.97	3.4
2.5	1.01	3.5
2.6	1.05	3.6
2.7	1.09	3.8
2.8	1.13	3.9
2.9	1.17	4.1
3.0	1.21	4.2
3.1	1.25	4.3
3.2	1.29	4.5
3.3	1.33	4.6
3.4	1.37	4.8
3.5	1.41	4.9
3.6	1.45	5.1
3.7	1.49	5.2
3.8	1.53	5.3
3.9	1.57	5.5
4.0	1.61	5.6
4.1	1.65	5.8
4.2	1.69	5.9
4.3	1.73	6.0
4.4	1.77	6.2
4.5	1.81	6.3
4.6	1.85	6.5
4.7	1.90	6.6
4.8	1.94	6.7
4.9	1.98	6.9
5.0	2.02	7.0
5.1	2.06	7.2
5.2	2.10	7.3
5.3	2.14	7.4
5.4	2.18	7.6
5.5	2.22	7.7
5.6	2.26	7.9
5.7	2.30	8.0
5.8	2.34	8.1
5.9	2.38	8.3
6.0	2.42	8.4
6.1	2.46	8.6
6.2	2.50	8.7
6.3	2.54	8.8
6.4	2.58	9.0
6.5	2.62	9.1
6.6	2.66	9.3
6.7	2.70	9.4
6.8	2.74	9.5
6.9	2.78	9.7
7.0	2.82	9.8
7.1	2.86	10.0
7.2	2.90	10.1
7.3	2.94	10.2
7.4	2.98	10.4
7.5	3.02	10.5
7.6	3.06	10.7
7.7	3.10	10.8
7.8	3.15	10.9
7.9	3.19	11.1
8.0	3.23	11.2
8.1	3.27	11.4
8.2	3.31	11.5
8.3	3.35	11.6
8.4	3.39	11.8
8.5	3.43	11.9

APPENDIX B - 2 **CU FT**

Continued

THINNING TABLES

From 2:1

Grout in Line and Agitator (cu ft)	THINNING WATER (cu ft)	New vol (cu ft)
TO 3:1		
8.6	3.47	12.1
8.7	3.51	12.2
8.8	3.55	12.3
8.9	3.59	12.5
9.0	3.63	12.6
9.1	3.67	12.8
9.2	3.71	12.9
9.3	3.75	13.0
9.4	3.79	13.2
9.5	3.83	13.3
9.6	3.87	13.5
9.7	3.91	13.6
9.8	3.95	13.8
9.9	3.99	13.9
10.0	4.03	14.0
10.1	4.07	14.2
10.2	4.11	14.3
10.3	4.15	14.5
10.4	4.19	14.6
10.5	4.23	14.7
10.6	4.27	14.9
10.7	4.31	15.0
10.8	4.35	15.2
10.9	4.40	15.3
11.0	4.44	15.4
11.1	4.48	15.6
11.2	4.52	15.7
11.3	4.56	15.9
11.4	4.60	16.0
11.5	4.64	16.1
11.6	4.68	16.3
11.7	4.72	16.4
11.8	4.76	16.6
11.9	4.80	16.7
12.0	4.84	16.8
12.1	4.88	17.0
12.2	4.92	17.1
12.3	4.96	17.3
12.4	5.00	17.4
12.5	5.04	17.5
12.6	5.08	17.7
12.7	5.12	17.8
12.8	5.16	18.0
12.9	5.20	18.1
13.0	5.24	18.2
13.1	5.28	18.4
13.2	5.32	18.5
13.3	5.36	18.7
13.4	5.40	18.8
13.5	5.44	18.9
13.6	5.48	19.1
13.7	5.52	19.2
13.8	5.56	19.4
13.9	5.60	19.5
14.0	5.65	19.6
14.1	5.69	19.8
14.2	5.73	19.9
TO 4:1		
1.0	0.81	1.8
1.1	0.89	2.0
1.2	0.97	2.2
1.3	1.05	2.3
1.4	1.13	2.5
1.5	1.21	2.7
1.6	1.29	2.9
1.7	1.37	3.1
1.8	1.45	3.3
1.9	1.53	3.4
2.0	1.61	3.6
2.1	1.69	3.8
2.2	1.77	4.0
2.3	1.85	4.2
2.4	1.94	4.3
2.5	2.02	4.5
2.6	2.10	4.7
2.7	2.18	4.9
2.8	2.26	5.1
2.9	2.34	5.2
3.0	2.42	5.4
3.1	2.50	5.6
3.2	2.58	5.8
3.3	2.66	6.0
3.4	2.74	6.1
3.5	2.82	6.3
3.6	2.90	6.5
3.7	2.98	6.7
3.8	3.06	6.9
3.9	3.15	7.0
4.0	3.23	7.2

APPENDIX B - 2

CU FT

Continued

THINNING TABLES

From 2:1

Grout in Line and Agitator (cu ft)	THIN-NING WATER (cu ft)	New vol (cu ft)	Grout in Line and Agitator (cu ft)	THIN-NING WATER (cu ft)	New vol (cu ft)	Grout in Line and Agitator (cu ft)	THIN-NING WATER (cu ft)	New vol (cu ft)
TO 4:1								
4.1	3.31	7.4	6.1	4.92	11.0	8.1	6.53	14.6
4.2	3.39	7.6	6.2	5.00	11.2	8.2	6.61	14.8
4.3	3.47	7.8	6.3	5.08	11.4	8.3	6.69	15.0
4.4	3.55	7.9	6.4	5.16	11.6	8.4	6.77	15.2
4.5	3.63	8.1	6.5	5.24	11.7	8.5	6.85	15.4
4.6	3.71	8.3	6.6	5.32	11.9	8.6	6.94	15.6
4.7	3.79	8.5	6.7	5.40	12.1	8.7	7.02	15.7
4.8	3.87	8.7	6.8	5.48	12.3	8.8	7.10	15.9
4.9	3.95	8.9	6.9	5.56	12.5	8.9	7.18	16.1
5.0	4.03	9.0	7.0	5.65	12.6	9.0	7.26	16.3
5.1	4.11	9.2	7.1	5.73	12.8	9.1	7.34	16.4
5.2	4.19	9.4	7.2	5.81	13.0	9.2	7.42	16.6
5.3	4.27	9.6	7.3	5.89	13.2	9.3	7.50	16.8
5.4	4.35	9.8	7.4	5.97	13.4	9.4	7.58	17.0
5.5	4.44	9.9	7.5	6.05	13.5	9.5	7.66	17.2
5.6	4.52	10.1	7.6	6.13	13.7	9.6	7.74	17.3
5.7	4.60	10.3	7.7	6.21	13.9	9.7	7.82	17.5
5.8	4.68	10.5	7.8	6.29	14.1	9.8	7.90	17.7
5.9	4.76	10.7	7.9	6.37	14.3	9.9	7.98	17.9
6.0	4.84	10.8	8.0	6.45	14.5	10.0	8.06	18.1
						10.1	8.15	18.2
						10.2	8.23	18.4
						10.3	8.31	18.6
						10.4	8.39	18.8
						10.5	8.47	19.0
TO 5:1								
1.0	1.21	2.2	2.1	2.54	4.6	3.1	3.75	6.8
			2.2	2.66	4.9	3.2	3.87	7.1
1.1	1.33	2.4	2.3	2.78	5.1	3.3	3.99	7.3
1.2	1.45	2.7	2.4	2.90	5.3	3.4	4.11	7.5
1.3	1.57	2.9	2.5	3.02	5.5	3.5	4.23	7.7
1.4	1.69	3.1	2.6	3.15	5.7	3.6	4.35	8.0
1.5	1.81	3.3	2.7	3.27	6.0	3.7	4.48	8.2
1.6	1.94	3.5	2.8	3.39	6.2	3.8	4.60	8.4
1.7	2.06	3.8	2.9	3.51	6.4	3.9	4.72	8.6
1.8	2.18	4.0	3.0	3.63	6.6	4.0	4.84	8.8
1.9	2.30	4.2						
2.0	2.42	4.4						

APPENDIX B - 2

Continued

CU FT

THINNING TABLES

From 2:1

TO 5:1

Grout in Line and Agitator (cu ft)	THIN-NING WATER (cu ft)	New vol (cu ft)	Grout in Line and Agitator (cu ft)	THIN-NING WATER (cu ft)	New vol (cu ft)	Grout in Line and Agitator (cu ft)	THIN-NING WATER (cu ft)	New vol (cu ft)
4.1	4.96	9.1	6.1	7.38	13.5	8.1	9.80	17.9
4.2	5.08	9.3	6.2	7.50	13.7	8.2	9.92	18.1
4.3	5.20	9.5	6.3	7.62	13.9	8.3	10.04	18.3
4.4	5.32	9.7	6.4	7.74	14.1	8.4	10.16	18.6
4.5	5.44	9.9	6.5	7.86	14.4	8.5	10.28	18.8
4.6	5.56	10.2	6.6	7.98	14.6	8.6	10.40	19.0
4.7	5.69	10.4	6.7	8.10	14.8	8.7	10.52	19.2
4.8	5.81	10.6	6.8	8.23	15.0	8.8	10.65	19.4
4.9	5.93	10.8	6.9	8.35	15.2	8.9	10.77	19.7
5.0	6.05	11.0	7.0	8.47	15.5	9.0	10.89	19.9
5.1	6.17	11.3	7.1	8.59	15.7			
5.2	6.29	11.5	7.2	8.71	15.9			
5.3	6.41	11.7	7.3	8.83	16.1			
5.4	6.53	11.9	7.4	8.95	16.4			
5.5	6.65	12.2	7.5	9.07	16.6			
5.6	6.77	12.4	7.6	9.19	16.8			
5.7	6.90	12.6	7.7	9.31	17.0			
5.8	7.02	12.8	7.8	9.44	17.2			
5.9	7.14	13.0	7.9	9.56	17.5			
6.0	7.26	13.3	8.0	9.68	17.7			

APPENDIX B - 2

CU FT

THINNING TABLES

From 1:1
(by loose volume)

See Section 4.1–9 for explanations and methods of use.

Grout in Line and Agitator (cu ft)	THIN-NING WATER (cu ft)	New vol (cu ft)
TO 2:1		
1.0	0.68	1.7
1.1	0.74	1.8
1.2	0.81	2.0
1.3	0.88	2.2
1.4	0.95	2.3
1.5	1.01	2.5
1.6	1.08	2.7
1.7	1.15	2.8
1.8	1.22	3.0
1.9	1.28	3.2
2.0	1.35	3.4
2.1	1.42	3.5
2.2	1.49	3.7
2.3	1.55	3.9
2.4	1.62	4.0
2.5	1.69	4.2
2.6	1.76	4.4
2.7	1.82	4.5
2.8	1.89	4.7
2.9	1.96	4.9
3.0	2.03	5.0
3.1	2.09	5.2
3.2	2.16	5.4
3.3	2.23	5.5
3.4	2.30	5.7

Grout in Line and Agitator (cu ft)	THIN-NING WATER (cu ft)	New vol (cu ft)
3.5	2.36	5.9
3.6	2.43	6.0
3.7	2.50	6.2
3.8	2.57	6.4
3.9	2.64	6.5
4.0	2.70	6.7
4.1	2.77	6.9
4.2	2.84	7.0
4.3	2.91	7.2
4.4	2.97	7.4
4.5	3.04	7.5
4.6	3.11	7.7
4.7	3.18	7.9
4.8	3.24	8.0
4.9	3.31	8.2
5.0	3.38	8.4
5.1	3.45	8.5
5.2	3.51	8.7
5.3	3.58	8.9
5.4	3.65	9.0
5.5	3.72	9.2
5.6	3.78	9.4
5.7	3.85	9.6
5.8	3.92	9.7
5.9	3.99	9.9
6.0	4.05	10.1

Grout in Line and Agitator (cu ft)	THIN-NING WATER (cu ft)	New vol (cu ft)
6.1	4.12	10.2
6.2	4.19	10.4
6.3	4.26	10.6
6.4	4.32	10.7
6.5	4.39	10.9
6.6	4.46	11.1
6.7	4.53	11.2
6.8	4.59	11.4
6.9	4.66	11.6
7.0	4.73	11.7
7.1	4.80	11.9
7.2	4.86	12.1
7.3	4.93	12.2
7.4	5.00	12.4
7.5	5.07	12.6
7.6	5.14	12.7
7.7	5.20	12.9
7.8	5.27	13.1
7.9	5.34	13.2
8.0	5.41	13.4
8.1	5.47	13.6
8.2	5.54	13.7
8.3	5.61	13.9
8.4	5.68	14.1
8.5	5.74	14.2
8.6	5.81	14.4

APPENDIX B - 2 **CU FT**

Continued

THINNING TABLES

From 1:1

Grout in Line and Agitator (cu ft)	THIN-NING WATER (cu ft)	New vol (cu ft)	Grout in Line and Agitator (cu ft)	THIN-NING WATER (cu ft)	New vol (cu ft)	Grout in Line and Agitator (cu ft)	THIN-NING WATER (cu ft)	New vol (cu ft)
TO 2:1								
8.7	5.88	14.6	10.1	6.82	16.9	11.1	7.50	18.6
8.8	5.95	14.7	10.2	6.89	17.1	11.2	7.57	18.8
8.9	6.01	14.9	10.3	6.96	17.3	11.3	7.64	18.9
9.0	6.08	15.1	10.4	7.03	17.4	11.4	7.70	19.1
			10.5	7.09	17.6	11.5	7.77	19.3
9.1	6.15	15.2	10.6	7.16	17.8	11.6	7.84	19.4
9.2	6.22	15.4	10.7	7.23	17.9	11.7	7.91	19.6
9.3	6.28	15.6	10.8	7.30	18.1	11.8	7.97	19.8
9.4	6.35	15.8	10.9	7.36	18.3	11.9	8.04	19.9
9.5	6.42	15.9	11.0	7.43	18.4			
9.6	6.49	16.1						
9.7	6.55	16.3						
9.8	6.62	16.4						
9.9	6.69	16.6						
10.0	6.76	16.8						
TO 3:1								
1.0	1.35	2.4	3.1	4.19	7.3	5.1	6.89	12.0
			3.2	4.32	7.5	5.2	7.03	12.2
1.1	1.49	2.6	3.3	4.46	7.8	5.3	7.16	12.5
1.2	1.62	2.8	3.4	4.59	8.0	5.4	7.30	12.7
1.3	1.76	3.1	3.5	4.73	8.2	5.5	7.43	12.9
1.4	1.89	3.3	3.6	4.86	8.5	5.6	7.57	13.2
1.5	2.03	3.5	3.7	5.00	8.7	5.7	7.70	13.4
1.6	2.16	3.8	3.8	5.14	8.9	5.8	7.84	13.6
1.7	2.30	4.0	3.9	5.27	9.2	5.9	7.97	13.9
1.8	2.43	4.2	4.0	5.41	9.4	6.0	8.11	14.1
1.9	2.57	4.5						
2.0	2.70	4.7	4.1	5.54	9.6	6.1	8.24	14.3
			4.2	5.68	9.9	6.2	8.38	14.6
2.1	2.84	4.9	4.3	5.81	10.1	6.3	8.51	14.8
2.2	2.97	5.2	4.4	5.95	10.3	6.4	8.65	15.0
2.3	3.11	5.4	4.5	6.08	10.6	6.5	8.78	15.3
2.4	3.24	5.6	4.6	6.22	10.8	6.6	8.92	15.5
2.5	3.38	5.9	4.7	6.35	11.1	6.7	9.05	15.8
2.6	3.51	6.1	4.8	6.49	11.3	6.8	9.19	16.0
2.7	3.65	6.3	4.9	6.62	11.5	6.9	9.32	16.2
2.8	3.78	6.6	5.0	6.76	11.8	7.0	9.46	16.5
2.9	3.92	6.8						
3.0	4.05	7.1						

APPENDIX B - 2 **CU FT**

Continued

THINNING TABLES

From 1:1

Grout in Line and Agitator (cu ft)	THINNING WATER (cu ft)	New vol (cu ft)
TO 3:1		
7.1	9.59	16.7
7.2	9.73	16.9
7.3	9.86	17.2
7.4	10.00	17.4
7.5	10.14	17.6
7.6	10.27	17.9
7.7	10.41	18.1
7.8	10.54	18.3
7.9	10.68	18.6
8.0	10.81	18.8
8.1	10.95	19.0
8.2	11.08	19.3
8.3	11.22	19.5
8.4	11.35	19.8
TO 4:1		
1.0	2.03	3.0
1.1	2.23	3.3
1.2	2.43	3.6
1.3	2.64	3.9
1.4	2.84	4.2
1.5	3.04	4.5
1.6	3.24	4.8
1.7	3.45	5.1
1.8	3.65	5.4
1.9	3.85	5.8
2.0	4.05	6.1
2.1	4.26	6.4
2.2	4.46	6.7
2.3	4.66	7.0
2.4	4.86	7.3
2.5	5.07	7.6
2.6	5.27	7.9
2.7	5.47	8.2
2.8	5.68	8.5
2.9	5.88	8.8
3.0	6.08	9.1
3.1	6.28	9.4
3.2	6.49	9.7
3.3	6.69	10.0
3.4	6.89	10.3
3.5	7.09	10.6
3.6	7.30	10.9
3.7	7.50	11.2
3.8	7.70	11.5
3.9	7.91	11.8
4.0	8.11	12.1
4.1	8.31	12.4
4.2	8.51	12.7
4.3	8.72	13.0
4.4	8.92	13.3
4.5	9.12	13.6
4.6	9.32	13.9
4.7	9.53	14.2
4.8	9.73	14.5
4.9	9.93	14.8
5.0	10.14	15.1
5.1	10.34	15.4
5.2	10.54	15.7
5.3	10.74	16.0
5.4	10.95	16.3
5.5	11.15	16.6
5.6	11.35	17.0
5.7	11.55	17.3
5.8	11.76	17.6
5.9	11.96	17.9
6.0	12.16	18.2
6.1	12.36	18.5
6.2	12.57	18.8
6.3	12.77	19.1
6.4	12.97	19.4
6.5	13.18	19.7
TO 5:1		
1.0	2.70	3.7
1.1	2.97	4.1
1.2	3.24	4.4
1.3	3.51	4.8
1.4	3.78	5.2
1.5	4.05	5.6
1.6	4.32	5.9
1.7	4.59	6.3
1.8	4.86	6.7
1.9	5.14	7.0
2.0	5.41	7.4
2.1	5.68	7.8
2.2	5.95	8.1
2.3	6.22	8.5
2.4	6.49	8.9
2.5	6.76	9.3
2.6	7.03	9.6
2.7	7.30	10.0
2.8	7.57	10.4
2.9	7.84	10.7
3.0	8.11	11.1

APPENDIX B - 2
Continued

CU FT

THINNING TABLES

From 1:1

Grout in Line and Agitator (cu ft)	THIN-NING WATER (cu ft)	New vol (cu ft)	Grout in Line and Agitator (cu ft)	THIN-NING WATER (cu ft)	New vol (cu ft)	Grout in Line and Agitator (cu ft)	THIN-NING WATER (cu ft)	New vol (cu ft)
TO 5:1								
3.1	8.38	11.5	4.1	11.08	15.2	5.1	13.78	18.9
3.2	8.65	11.8	4.2	11.35	15.6	5.2	14.05	19.3
3.3	8.92	12.2	4.3	11.62	15.9	5.3	14.32	19.6
3.4	9.19	12.6	4.4	11.89	16.3			
3.5	9.46	13.0	4.5	12.16	16.7			
3.6	9.73	13.3	4.6	12.43	17.0			
3.7	10.00	13.7	4.7	12.70	17.4			
3.8	10.27	14.1	4.8	12.97	17.8			
3.9	10.54	14.4	4.9	13.24	18.1			
4.0	10.81	14.8	5.0	13.51	18.5			

APPENDIX B - 2 CU FT

THINNING TABLES

From 0.8:1
(by loose volume)

See Section 4.1–9 for explanations and methods of use.

Grout in Line and Agitator (cu ft)	THIN-NING WATER (cu ft)	New vol (cu ft)
TO 1:1		
1.0	0.16	1.2
1.1	0.17	1.3
1.2	0.19	1.4
1.3	0.20	1.5
1.4	0.22	1.6
1.5	0.23	1.7
1.6	0.25	1.8
1.7	0.27	2.0
1.8	0.28	2.1
1.9	0.30	2.2
2.0	0.31	2.3
2.1	0.33	2.4
2.2	0.34	2.5
2.3	0.36	2.7
2.4	0.38	2.8
2.5	0.39	2.9
2.6	0.41	3.0
2.7	0.42	3.1
2.8	0.44	3.2
2.9	0.45	3.4
3.0	0.47	3.5
3.1	0.48	3.6
3.2	0.50	3.7
3.3	0.52	3.8
3.4	0.53	3.9

Grout in Line and Agitator (cu ft)	THIN-NING WATER (cu ft)	New vol (cu ft)
3.5	0.55	4.0
3.6	0.56	4.2
3.7	0.58	4.3
3.8	0.59	4.4
3.9	0.61	4.5
4.0	0.62	4.6
4.1	0.64	4.7
4.2	0.66	4.9
4.3	0.67	5.0
4.4	0.69	5.1
4.5	0.70	5.2
4.6	0.72	5.3
4.7	0.73	5.4
4.8	0.75	5.5
4.9	0.77	5.7
5.0	0.78	5.8
5.1	0.80	5.9
5.2	0.81	6.0
5.3	0.83	6.1
5.4	0.84	6.2
5.5	0.86	6.4
5.6	0.87	6.5
5.7	0.89	6.6
5.8	0.91	6.7
5.9	0.92	6.8
6.0	0.94	6.9

Grout in Line and Agitator (cu ft)	THIN-NING WATER (cu ft)	New vol (cu ft)
6.1	0.95	7.1
6.2	0.97	7.2
6.3	0.98	7.3
6.4	1.00	7.4
6.5	1.02	7.5
6.6	1.03	7.6
6.7	1.05	7.7
6.8	1.06	7.9
6.9	1.08	8.0
7.0	1.09	8.1
7.1	1.11	8.2
7.2	1.13	8.3
7.3	1.14	8.4
7.4	1.16	8.6
7.5	1.17	8.7
7.6	1.19	8.8
7.7	1.20	8.9
7.8	1.22	9.0
7.9	1.23	9.1
8.0	1.25	9.2
8.1	1.27	9.4
8.2	1.28	9.5
8.3	1.30	9.6
8.4	1.31	9.7
8.5	1.33	9.8
8.6	1.34	9.9

APPENDIX B - 2

Continued

CU FT

THINNING TABLES

From 0.8:1

Grout in Line and Agitator (cu ft)	THIN-NING WATER (cu ft)	New vol (cu ft)
TO 1:1		
8.7	1.36	10.1
8.8	1.37	10.2
8.9	1.39	10.3
9.0	1.41	10.4
9.1	1.42	10.5
9.2	1.44	10.6
9.3	1.45	10.8
9.4	1.47	10.9
9.5	1.48	11.0
9.6	1.50	11.1
9.7	1.52	11.2
9.8	1.53	11.3
9.9	1.55	11.4
10.0	1.56	11.6
10.1	1.58	11.7
10.2	1.59	11.8
10.3	1.61	11.9
10.4	1.63	12.0
10.5	1.64	12.1
10.6	1.66	12.3
10.7	1.67	12.4
10.8	1.69	12.5
10.9	1.70	12.6
11.0	1.72	12.7
11.1	1.73	12.8
11.2	1.75	12.9
11.3	1.77	13.1
11.4	1.78	13.2
11.5	1.80	13.3
11.6	1.81	13.4
11.7	1.83	13.5
11.8	1.84	13.6
11.9	1.86	13.8
12.0	1.87	13.9
12.1	1.89	14.0
12.2	1.91	14.1
12.3	1.92	14.2
12.4	1.94	14.3
12.5	1.95	14.5
12.6	1.97	14.6
12.7	1.98	14.7
12.8	2.00	14.8
12.9	2.02	14.9
13.0	2.03	15.0
13.1	2.05	15.1
13.2	2.06	15.3
13.3	2.08	15.4
13.4	2.09	15.5
13.5	2.11	15.6
13.6	2.13	15.7
13.7	2.14	15.8
13.8	2.16	16.0
13.9	2.17	16.1
14.0	2.19	16.2
14.1	2.20	16.3
14.2	2.22	16.4
14.3	2.23	16.5
14.4	2.25	16.6
14.5	2.27	16.8
14.6	2.28	16.9
14.7	2.30	17.0
14.8	2.31	17.1
14.9	2.33	17.2
15.0	2.34	17.3
15.1	2.36	17.5
15.2	2.38	17.6
15.3	2.39	17.7
15.4	2.41	17.8
15.5	2.42	17.9
15.6	2.44	18.0
15.7	2.45	18.2
15.8	2.47	18.3
15.9	2.48	18.4
16.0	2.50	18.5
16.1	2.52	18.6
16.2	2.53	18.7
16.3	2.55	18.8
16.4	2.56	19.0
16.5	2.58	19.1
16.6	2.59	19.2
16.7	2.61	19.3
16.8	2.63	19.4
16.9	2.64	19.5
17.0	2.66	19.7
17.1	2.67	19.8
17.2	2.69	19.9

APPENDIX B - 2 **CU FT**

Continued

THINNING TABLES

From 0.8:1

Grout in Line and Agitator (cu ft)	THINNING WATER (cu ft)	New vol (cu ft)	Grout in Line and Agitator (cu ft)	THINNING WATER (cu ft)	New vol (cu ft)	Grout in Line and Agitator (cu ft)	THINNING WATER (cu ft)	New vol (cu ft)
TO 2:1								
1.0	0.94	1.9	4.1	3.84	7.9	7.1	6.66	13.8
			4.2	3.94	8.1	7.2	6.75	13.9
1.1	1.03	2.1	4.3	4.03	8.3	7.3	6.84	14.1
1.2	1.13	2.3	4.4	4.12	8.5	7.4	6.94	14.3
1.3	1.22	2.5	4.5	4.22	8.7	7.5	7.03	14.5
1.4	1.31	2.7	4.6	4.31	8.9	7.6	7.13	14.7
1.5	1.41	2.9	4.7	4.41	9.1	7.7	7.22	14.9
1.6	1.50	3.1	4.8	4.50	9.3	7.8	7.31	15.1
1.7	1.59	3.3	4.9	4.59	9.5	7.9	7.41	15.3
1.8	1.69	3.5	5.0	4.69	9.7	8.0	7.50	15.5
1.9	1.78	3.7						
2.0	1.87	3.9	5.1	4.78	9.9	8.1	7.59	15.7
			5.2	4.88	10.1	8.2	7.69	15.9
2.1	1.97	4.1	5.3	4.97	10.3	8.3	7.78	16.1
2.2	2.06	4.3	5.4	5.06	10.5	8.4	7.88	16.3
2.3	2.16	4.5	5.5	5.16	10.7	8.5	7.97	16.5
2.4	2.25	4.6	5.6	5.25	10.8	8.6	8.06	16.7
2.5	2.34	4.8	5.7	5.34	11.0	8.7	8.16	16.9
2.6	2.44	5.0	5.8	5.44	11.2	8.8	8.25	17.0
2.7	2.53	5.2	5.9	5.53	11.4	8.9	8.34	17.2
2.8	2.63	5.4	6.0	5.62	11.6	9.0	8.44	17.4
2.9	2.72	5.6						
3.0	2.81	5.8	6.1	5.72	11.8	9.1	8.53	17.6
			6.2	5.81	12.0	9.2	8.62	17.8
3.1	2.91	6.0	6.3	5.91	12.2	9.3	8.72	18.0
3.2	3.00	6.2	6.4	6.00	12.4	9.4	8.81	18.2
3.3	3.09	6.4	6.5	6.09	12.6	9.5	8.91	18.4
3.4	3.19	6.6	6.6	6.19	12.8	9.6	9.00	18.6
3.5	3.28	6.8	6.7	6.28	13.0	9.7	9.09	18.8
3.6	3.38	7.0	6.8	6.38	13.2	9.8	9.19	19.0
3.7	3.47	7.2	6.9	6.47	13.4	9.9	9.28	19.2
3.8	3.56	7.4	7.0	6.56	13.6	10.0	9.37	19.4
3.9	3.66	7.6				10.1	9.47	19.6
4.0	3.75	7.7				10.2	9.56	19.8
TO 3:1								
1.0	1.72	2.7	1.6	2.75	4.3	2.1	3.61	5.7
1.1	1.89	3.0	1.7	2.92	4.6	2.2	3.78	6.0
1.2	2.06	3.3	1.8	3.09	4.9	2.3	3.95	6.3
1.3	2.23	3.5	1.9	3.27	5.2	2.4	4.13	6.5
1.4	2.41	3.8	2.0	3.44	5.4	2.5	4.30	6.8
1.5	2.58	4.1						

APPENDIX B - 2

Continued

CU FT

THINNING TABLES

From 0.8:1

Grout in Line and Agitator (cu ft)	THINNING WATER (cu ft)	New vol (cu ft)	Grout in Line and Agitator (cu ft)	THINNING WATER (cu ft)	New vol (cu ft)	Grout in Line and Agitator (cu ft)	THINNING WATER (cu ft)	New vol (cu ft)
TO 3:1								
2.6	4.47	7.1	4.1	7.05	11.1	5.6	9.62	15.2
2.7	4.64	7.3	4.2	7.22	11.4	5.7	9.80	15.5
2.8	4.81	7.6	4.3	7.39	11.7	5.8	9.97	15.8
2.9	4.98	7.9	4.4	7.56	12.0	5.9	10.14	16.0
3.0	5.16	8.2	4.5	7.73	12.2	6.0	10.31	16.3
			4.6	7.91	12.5			
3.1	5.33	8.4	4.7	8.08	12.8	6.1	10.48	16.6
3.2	5.50	8.7	4.8	8.25	13.0	6.2	10.66	16.9
3.3	5.67	9.0	4.9	8.42	13.3	6.3	10.83	17.1
3.4	5.84	9.2	5.0	8.59	13.6	6.4	11.00	17.4
3.5	6.02	9.5				6.5	11.17	17.7
3.6	6.19	9.8	5.1	8.77	13.9	6.6	11.34	17.9
3.7	6.36	10.1	5.2	8.94	14.1	6.7	11.52	18.2
3.8	6.53	10.3	5.3	9.11	14.4	6.8	11.69	18.5
3.9	6.70	10.6	5.4	9.28	14.7	6.9	11.86	18.8
4.0	6.87	10.9	5.5	9.45	15.0	7.0	12.03	19.0
TO 4:1								
1.0	2.50	3.5	3.1	7.75	10.8	5.1	12.75	17.8
			3.2	8.00	11.2	5.2	13.00	18.2
1.1	2.75	3.8	3.3	8.25	11.5	5.3	13.25	18.5
1.2	3.00	4.2	3.4	8.50	11.9	5.4	13.50	18.9
1.3	3.25	4.5	3.5	8.75	12.2	5.5	13.75	19.2
1.4	3.50	4.9	3.6	9.00	12.6	5.6	14.00	19.6
1.5	3.75	5.2	3.7	9.25	12.9	5.7	14.25	19.9
1.6	4.00	5.6	3.8	9.50	13.3			
1.7	4.25	5.9	3.9	9.75	13.6			
1.8	4.50	6.3	4.0	10.00	14.0			
1.9	4.75	6.6						
2.0	5.00	7.0	4.1	10.25	14.3			
			4.2	10.50	14.7			
2.1	5.25	7.3	4.3	10.75	15.0			
2.2	5.50	7.7	4.4	11.00	15.4			
2.3	5.75	8.0	4.5	11.25	15.7			
2.4	6.00	8.4	4.6	11.50	16.1			
2.5	6.25	8.7	4.7	11.75	16.4			
2.6	6.50	9.1	4.8	12.00	16.8			
2.7	6.75	9.4	4.9	12.25	17.1			
2.8	7.00	9.8	5.0	12.50	17.5			
2.9	7.25	10.1						
3.0	7.50	10.5						

APPENDIX B - 2

Continued

CU FT

THINNING TABLES
From 0.8:1

Grout in Line and Agitator (cu ft)	THIN-NING WATER (cu ft)	New vol (cu ft)	Grout in Line and Agitator (cu ft)	THIN-NING WATER (cu ft)	New vol (cu ft)	Grout in Line and Agitator (cu ft)	THIN-NING WATER (cu ft)	New vol (cu ft)
TO 5:1								
1.0	3.28	4.3	2.1	6.89	9.0	3.1	10.17	13.3
			2.2	7.22	9.4	3.2	10.50	13.7
1.1	3.61	4.7	2.3	7.55	9.8	3.3	10.83	14.1
1.2	3.94	5.1	2.4	7.87	10.3	3.4	11.16	14.6
1.3	4.27	5.6	2.5	8.20	10.7	3.5	11.48	15.0
1.4	4.59	6.0	2.6	8.53	11.1	3.6	11.81	15.4
1.5	4.92	6.4	2.7	8.86	11.6	3.7	12.14	15.8
1.6	5.25	6.8	2.8	9.19	12.0	3.8	12.47	16.3
1.7	5.58	7.3	2.9	9.52	12.4	3.9	12.80	16.7
1.8	5.91	7.7	3.0	9.84	12.8	4.0	13.12	17.1
1.9	6.23	8.1						
2.0	6.56	8.6				4.1	13.45	17.6
						4.2	13.78	18.0
						4.3	14.11	18.4
						4.4	14.44	18.8
						4.5	14.77	19.3
						4.6	15.09	19.7

APPENDIX B - 2 **CU FT**

THINNING TABLES

From 0.6:1
(by loose volume)

See Section 4.1–9 for explanations and methods of use.

TO 0.8:1

Grout in Line and Agitator (cu ft)	THINNING WATER (cu ft)	New vol (cu ft)
1.0	0.19	1.2
1.1	0.20	1.3
1.2	0.22	1.4
1.3	0.24	1.5
1.4	0.26	1.7
1.5	0.28	1.8
1.6	0.30	1.9
1.7	0.31	2.0
1.8	0.33	2.1
1.9	0.35	2.3
2.0	0.37	2.4
2.1	0.39	2.5
2.2	0.41	2.6
2.3	0.43	2.7
2.4	0.44	2.8
2.5	0.46	3.0
2.6	0.48	3.1
2.7	0.50	3.2
2.8	0.52	3.3
2.9	0.54	3.4
3.0	0.56	3.6
3.1	0.57	3.7
3.2	0.59	3.8
3.3	0.61	3.9
3.4	0.63	4.0

Grout in Line and Agitator (cu ft)	THINNING WATER (cu ft)	New vol (cu ft)
3.5	0.65	4.1
3.6	0.67	4.3
3.7	0.69	4.4
3.8	0.70	4.5
3.9	0.72	4.6
4.0	0.74	4.7
4.1	0.76	4.9
4.2	0.78	5.0
4.3	0.80	5.1
4.4	0.81	5.2
4.5	0.83	5.3
4.6	0.85	5.5
4.7	0.87	5.6
4.8	0.89	5.7
4.9	0.91	5.8
5.0	0.93	5.9
5.1	0.94	6.0
5.2	0.96	6.2
5.3	0.98	6.3
5.4	1.00	6.4
5.5	1.02	6.5
5.6	1.04	6.6
5.7	1.06	6.8
5.8	1.07	6.9
5.9	1.09	7.0
6.0	1.11	7.1

Grout in Line and Agitator (cu ft)	THINNING WATER (cu ft)	New vol (cu ft)
6.1	1.13	7.2
6.2	1.15	7.3
6.3	1.17	7.5
6.4	1.19	7.6
6.5	1.20	7.7
6.6	1.22	7.8
6.7	1.24	7.9
6.8	1.26	8.1
6.9	1.28	8.2
7.0	1.30	8.3
7.1	1.31	8.4
7.2	1.33	8.5
7.3	1.35	8.7
7.4	1.37	8.8
7.5	1.39	8.9
7.6	1.41	9.0
7.7	1.43	9.1
7.8	1.44	9.2
7.9	1.46	9.4
8.0	1.48	9.5
8.1	1.50	9.6
8.2	1.52	9.7
8.3	1.54	9.8
8.4	1.56	10.0
8.5	1.57	10.1
8.6	1.59	10.2

APPENDIX B - 2
Continued

CU FT

THINNING TABLES
From 0.6:1

Grout in Line and Agitator (cu ft)	THIN-NING WATER (cu ft)	New vol (cu ft)	Grout in Line and Agitator (cu ft)	THIN-NING WATER (cu ft)	New vol (cu ft)	Grout in Line and Agitator (cu ft)	THIN-NING WATER (cu ft)	New vol (cu ft)
TO 0.8:1								
8.7	1.61	10.3	11.6	2.15	13.7	14.6	2.70	17.3
8.8	1.63	10.4	11.7	2.17	13.9	14.7	2.72	17.4
8.9	1.65	10.5	11.8	2.19	14.0	14.8	2.74	17.5
9.0	1.67	10.7	11.9	2.20	14.1	14.9	2.76	17.7
9.1	1.69	10.8	12.0	2.22	14.2	15.0	2.78	17.8
9.2	1.70	10.9	12.1	2.24	14.3	15.1	2.80	17.9
9.3	1.72	11.0	12.2	2.26	14.5	15.2	2.81	18.0
9.4	1.74	11.1	12.3	2.28	14.6	15.3	2.83	18.1
9.5	1.76	11.3	12.4	2.30	14.7	15.4	2.85	18.3
9.6	1.78	11.4	12.5	2.31	14.8	15.5	2.87	18.4
9.7	1.80	11.5	12.6	2.33	14.9	15.6	2.89	18.5
9.8	1.81	11.6	12.7	2.35	15.1	15.7	2.91	18.6
9.9	1.83	11.7	12.8	2.37	15.2	15.8	2.93	18.7
10.0	1.85	11.9	12.9	2.39	15.3	15.9	2.94	18.8
10.1	1.87	12.0	13.0	2.41	15.4	16.0	2.96	19.0
10.2	1.89	12.1	13.1	2.43	15.5	16.1	2.98	19.1
10.3	1.91	12.2	13.2	2.44	15.6	16.2	3.00	19.2
10.4	1.93	12.3	13.3	2.46	15.8	16.3	3.02	19.3
10.5	1.94	12.4	13.4	2.48	15.9	16.4	3.04	19.4
10.6	1.96	12.6	13.5	2.50	16.0	16.5	3.06	19.6
10.7	1.98	12.7	13.6	2.52	16.1	16.6	3.07	19.7
10.8	2.00	12.8	13.7	2.54	16.2	16.7	3.09	19.8
10.9	2.02	12.9	13.8	2.56	16.4	16.8	3.11	19.9
11.0	2.04	13.0	13.9	2.57	16.5	16.9	3.13	20.0
11.1	2.06	13.2	14.0	2.59	16.6			
11.2	2.07	13.3	14.1	2.61	16.7			
11.3	2.09	13.4	14.2	2.63	16.8			
11.4	2.11	13.5	14.3	2.65	16.9			
11.5	2.13	13.6	14.4	2.67	17.1			
			14.5	2.69	17.2			
TO 1:1								
1.0	0.37	1.4	1.6	0.59	2.2	2.1	0.78	2.9
1.1	0.41	1.5	1.7	0.63	2.3	2.2	0.81	3.0
1.2	0.44	1.6	1.8	0.67	2.5	2.3	0.85	3.2
1.3	0.48	1.8	1.9	0.70	2.6	2.4	0.89	3.3
1.4	0.52	1.9	2.0	0.74	2.7	2.5	0.93	3.4
1.5	0.56	2.1						

APPENDIX B - 2 **CU FT**

Continued

THINNING TABLES

From 0.6:1

Grout in Line and Agitator (cu ft)	THIN-NING WATER (cu ft)	New vol (cu ft)	Grout in Line and Agitator (cu ft)	THIN-NING WATER (cu ft)	New vol (cu ft)	Grout in Line and Agitator (cu ft)	THIN-NING WATER (cu ft)	New vol (cu ft)
TO 1:1								
2.6	0.96	3.6	6.1	2.26	8.4	9.6	3.56	13.2
2.7	1.00	3.7	6.2	2.30	8.5	9.7	3.59	13.3
2.8	1.04	3.8	6.3	2.33	8.6	9.8	3.63	13.4
2.9	1.07	4.0	6.4	2.37	8.8	9.9	3.67	13.6
3.0	1.11	4.1	6.5	2.41	8.9	10.0	3.70	13.7
			6.6	2.44	9.0			
3.1	1.15	4.2	6.7	2.48	9.2	10.1	3.74	13.8
3.2	1.19	4.4	6.8	2.52	9.3	10.2	3.78	14.0
3.3	1.22	4.5	6.9	2.56	9.5	10.3	3.81	14.1
3.4	1.26	4.7	7.0	2.59	9.6	10.4	3.85	14.3
3.5	1.30	4.8				10.5	3.89	14.4
3.6	1.33	4.9	7.1	2.63	9.7	10.6	3.93	14.5
3.7	1.37	5.1	7.2	2.67	9.9	10.7	3.96	14.7
3.8	1.41	5.2	7.3	2.70	10.0	10.8	4.00	14.8
3.9	1.44	5.3	7.4	2.74	10.1	10.9	4.04	14.9
4.0	1.48	5.5	7.5	2.78	10.3	11.0	4.07	15.1
			7.6	2.81	10.4			
4.1	1.52	5.6	7.7	2.85	10.6	11.1	4.11	15.2
4.2	1.56	5.8	7.8	2.89	10.7	11.2	4.15	15.3
4.3	1.59	5.9	7.9	2.93	10.8	11.3	4.19	15.5
4.4	1.63	6.0	8.0	2.96	11.0	11.4	4.22	15.6
4.5	1.67	6.2				11.5	4.26	15.8
4.6	1.70	6.3	8.1	3.00	11.1	11.6	4.30	15.9
4.7	1.74	6.4	8.2	3.04	11.2	11.7	4.33	16.0
4.8	1.78	6.6	8.3	3.07	11.4	11.8	4.37	16.2
4.9	1.81	6.7	8.4	3.11	11.5	11.9	4.41	16.3
5.0	1.85	6.9	8.5	3.15	11.6	12.0	4.44	16.4
			8.6	3.19	11.8			
5.1	1.89	7.0	8.7	3.22	11.9	12.1	4.48	16.6
5.2	1.93	7.1	8.8	3.26	12.1	12.2	4.52	16.7
5.3	1.96	7.3	8.9	3.30	12.2	12.3	4.56	16.9
5.4	2.00	7.4	9.0	3.33	12.3	12.4	4.59	17.0
5.5	2.04	7.5				12.5	4.63	17.1
5.6	2.07	7.7	9.1	3.37	12.5	12.6	4.67	17.3
5.7	2.11	7.8	9.2	3.41	12.6	12.7	4.70	17.4
5.8	2.15	7.9	9.3	3.44	12.7	12.8	4.74	17.5
5.9	2.19	8.1	9.4	3.48	12.9	12.9	4.78	17.7
6.0	2.22	8.2	9.5	3.52	13.0	13.0	4.81	17.8

APPENDIX B - 2 **CU FT**

Continued

THINNING TABLES

From 0.6:1

TO 1:1

Grout in Line and Agitator (cu ft)	THINNING WATER (cu ft)	New vol (cu ft)
13.1	4.85	18.0
13.2	4.89	18.1
13.3	4.93	18.2
13.4	4.96	18.4
13.5	5.00	18.5
13.6	5.04	18.6
13.7	5.07	18.8
13.8	5.11	18.9
13.9	5.15	19.0
14.0	5.19	19.2
14.1	5.22	19.3
14.2	5.26	19.5
14.3	5.30	19.6
14.4	5.33	19.7
14.5	5.37	19.9

TO 2:1

Grout in Line and Agitator (cu ft)	THINNING WATER (cu ft)	New vol (cu ft)
1.0	1.30	2.3
1.1	1.43	2.5
1.2	1.56	2.8
1.3	1.69	3.0
1.4	1.81	3.2
1.5	1.94	3.4
1.6	2.07	3.7
1.7	2.20	3.9
1.8	2.33	4.1
1.9	2.46	4.4
2.0	2.59	4.6
2.1	2.72	4.8
2.2	2.85	5.1
2.3	2.98	5.3
2.4	3.11	5.5
2.5	3.24	5.7
2.6	3.37	6.0
2.7	3.50	6.2
2.8	3.63	6.4
2.9	3.76	6.7
3.0	3.89	6.9
3.1	4.02	7.1
3.2	4.15	7.3
3.3	4.28	7.6
3.4	4.41	7.8
3.5	4.54	8.0
3.6	4.67	8.3
3.7	4.80	8.5
3.8	4.93	8.7
3.9	5.06	9.0
4.0	5.19	9.2
4.1	5.31	9.4
4.2	5.44	9.6
4.3	5.57	9.9
4.4	5.70	10.1
4.5	5.83	10.3
4.6	5.96	10.6
4.7	6.09	10.8
4.8	6.22	11.0
4.9	6.35	11.3
5.0	6.48	11.5
5.1	6.61	11.7
5.2	6.74	11.9
5.3	6.87	12.2
5.4	7.00	12.4
5.5	7.13	12.6
5.6	7.26	12.9
5.7	7.39	13.1
5.8	7.52	13.3
5.9	7.65	13.5
6.0	7.78	13.8
6.1	7.91	14.0
6.2	8.04	14.2
6.3	8.17	14.5
6.4	8.30	14.7
6.5	8.43	14.9
6.6	8.56	15.2
6.7	8.69	15.4
6.8	8.81	15.6
6.9	8.94	15.8
7.0	9.07	16.1
7.1	9.20	16.3
7.2	9.33	16.5
7.3	9.46	16.8
7.4	9.59	17.0
7.5	9.72	17.2
7.6	9.85	17.5
7.7	9.98	17.7
7.8	10.11	17.9
7.9	10.24	18.1
8.0	10.37	18.4
8.1	10.50	18.6
8.2	10.63	18.8
8.3	10.76	19.1
8.4	10.89	19.3
8.5	11.02	19.5
8.6	11.15	19.7

APPENDIX B - 2 CU FT
Continued

THINNING TABLES
From 0.6:1

Grout in Line and Agitator (cu ft)	THINNING WATER (cu ft)	New vol (cu ft)	Grout in Line and Agitator (cu ft)	THINNING WATER (cu ft)	New vol (cu ft)	Grout in Line and Agitator (cu ft)	THINNING WATER (cu ft)	New vol (cu ft)
TO 3:1								
1.0	2.22	3.2	3.1	6.89	10.0	5.1	11.33	16.4
			3.2	7.11	10.3	5.2	11.56	16.8
1.1	2.44	3.5	3.3	7.33	10.6	5.3	11.78	17.1
1.2	2.67	3.9	3.4	7.56	11.0	5.4	12.00	17.4
1.3	2.89	4.2	3.5	7.78	11.3	5.5	12.22	17.7
1.4	3.11	4.5	3.6	8.00	11.6	5.6	12.44	18.0
1.5	3.33	4.8	3.7	8.22	11.9	5.7	12.67	18.4
1.6	3.56	5.2	3.8	8.44	12.2	5.8	12.89	18.7
1.7	3.78	5.5	3.9	8.67	12.6	5.9	13.11	19.0
1.8	4.00	5.8	4.0	8.89	12.9	6.0	13.33	19.3
1.9	4.22	6.1						
2.0	4.44	6.4	4.1	9.11	13.2	6.1	13.56	19.7
			4.2	9.33	13.5			
2.1	4.67	6.8	4.3	9.56	13.9			
2.2	4.89	7.1	4.4	9.78	14.2			
2.3	5.11	7.4	4.5	10.00	14.5			
2.4	5.33	7.7	4.6	10.22	14.8			
2.5	5.56	8.1	4.7	10.44	15.1			
2.6	5.78	8.4	4.8	10.67	15.5			
2.7	6.00	8.7	4.9	10.89	15.8			
2.8	6.22	9.0	5.0	11.11	16.1			
2.9	6.44	9.3						
3.0	6.67	9.7						
TO 4:1								
1.0	3.15	4.1	2.5	7.87	10.4	4.1	12.91	17.0
			2.6	8.19	10.8	4.2	13.22	17.4
1.1	3.46	4.6	2.7	8.50	11.2	4.3	13.54	17.8
1.2	3.78	5.0	2.8	8.81	11.6	4.4	13.85	18.3
1.3	4.09	5.4	2.9	9.13	12.0	4.5	14.17	18.7
1.4	4.41	5.8	3.0	9.44	12.4	4.6	14.48	19.1
1.5	4.72	6.2				4.7	14.80	19.5
1.6	5.04	6.6	3.1	9.76	12.9	4.8	15.11	19.9
1.7	5.35	7.1	3.2	10.07	13.3			
1.8	5.67	7.5	3.3	10.39	13.7			
1.9	5.98	7.9	3.4	10.70	14.1			
2.0	6.30	8.3	3.5	11.02	14.5			
			3.6	11.33	14.9			
2.1	6.61	8.7	3.7	11.65	15.3			
2.2	6.93	9.1	3.8	11.96	15.8			
2.3	7.24	9.5	3.9	12.28	16.2			
2.4	7.56	10.0	4.0	12.59	16.6			

APPENDIX B - 2

Continued

CU FT

THINNING TABLES

From 0.6:1

Grout in Line and Agitator (cu ft)	THIN-NING WATER (cu ft)	New vol (cu ft)	Grout in Line and Agitator (cu ft)	THIN-NING WATER (cu ft)	New vol (cu ft)	Grout in Line and Agitator (cu ft)	THIN-NING WATER (cu ft)	New vol (cu ft)
TO 5:1								
1.0	4.07	5.1	2.1	8.56	10.7	3.1	12.63	15.7
			2.2	8.96	11.2	3.2	13.04	16.2
1.1	4.48	5.6	2.3	9.37	11.7	3.3	13.44	16.7
1.2	4.89	6.1	2.4	9.78	12.2	3.4	13.85	17.3
1.3	5.30	6.6	2.5	10.19	12.7	3.5	14.26	17.8
1.4	5.70	7.1	2.6	10.59	13.2	3.6	14.67	18.3
1.5	6.11	7.6	2.7	11.00	13.7	3.7	15.07	18.8
1.6	6.52	8.1	2.8	11.41	14.2	3.8	15.48	19.3
1.7	6.93	8.6	2.9	11.81	14.7	3.9	15.89	19.8
1.8	7.33	9.1	3.0	12.22	15.2			
1.9	7.74	9.6						
2.0	8.15	10.1						

APPENDIX B - 2 **CU FT**

THINNING TABLES

From 0.5:1
(by loose volume)

See Section 4.1–9 for explanations and methods of use.

Grout in Line and Agitator (cu ft)	THIN-NING WATER (cu ft)	New vol (cu ft)
TO 0.6:1		
1.0	0.10	1.1
1.1	0.11	1.2
1.2	0.12	1.3
1.3	0.13	1.4
1.4	0.14	1.5
1.5	0.15	1.7
1.6	0.16	1.8
1.7	0.17	1.9
1.8	0.18	2.0
1.9	0.19	2.1
2.0	0.20	2.2
2.1	0.21	2.3
2.2	0.22	2.4
2.3	0.23	2.5
2.4	0.24	2.6
2.5	0.26	2.8
2.6	0.27	2.9
2.7	0.28	3.0
2.8	0.29	3.1
2.9	0.30	3.2
3.0	0.31	3.3
3.1	0.32	3.4
3.2	0.33	3.5
3.3	0.34	3.6
3.4	0.35	3.7
3.5	0.36	3.9
3.6	0.37	4.0
3.7	0.38	4.1
3.8	0.39	4.2
3.9	0.40	4.3
4.0	0.41	4.4
4.1	0.42	4.5
4.2	0.43	4.6
4.3	0.44	4.7
4.4	0.45	4.8
4.5	0.46	5.0
4.6	0.47	5.1
4.7	0.48	5.2
4.8	0.49	5.3
4.9	0.50	5.4
5.0	0.51	5.5
5.1	0.52	5.6
5.2	0.53	5.7
5.3	0.54	5.8
5.4	0.55	6.0
5.5	0.56	6.1
5.6	0.57	6.2
5.7	0.58	6.3
5.8	0.59	6.4
5.9	0.60	6.5
6.0	0.61	6.6
6.1	0.62	6.7
6.2	0.63	6.8
6.3	0.64	6.9
6.4	0.65	7.1
6.5	0.66	7.2
6.6	0.67	7.3
6.7	0.68	7.4
6.8	0.69	7.5
6.9	0.70	7.6
7.0	0.71	7.7
7.1	0.72	7.8
7.2	0.73	7.9
7.3	0.74	8.0
7.4	0.76	8.2
7.5	0.77	8.3
7.6	0.78	8.4
7.7	0.79	8.5
7.8	0.80	8.6
7.9	0.81	8.7
8.0	0.82	8.8
8.1	0.83	8.9
8.2	0.84	9.0
8.3	0.85	9.1
8.4	0.86	9.3
8.5	0.87	9.4
8.6	0.88	9.5

APPENDIX B - 2 CU FT
Continued

THINNING TABLES

From 0.5:1

TO 0.6:1

Grout in Line and Agitator (cu ft)	THINNING WATER (cu ft)	New vol (cu ft)
8.7	0.89	9.6
8.8	0.90	9.7
8.9	0.91	9.8
9.0	0.92	9.9
9.1	0.93	10.0
9.2	0.94	10.1
9.3	0.95	10.2
9.4	0.96	10.4
9.5	0.97	10.5
9.6	0.98	10.6
9.7	0.99	10.7
9.8	1.00	10.8
9.9	1.01	10.9
10.0	1.02	11.0
10.1	1.03	11.1
10.2	1.04	11.2
10.3	1.05	11.4
10.4	1.06	11.5
10.5	1.07	11.6
10.6	1.08	11.7
10.7	1.09	11.8
10.8	1.10	11.9
10.9	1.11	12.0
11.0	1.12	12.1
11.1	1.13	12.2
11.2	1.14	12.3
11.3	1.15	12.5
11.4	1.16	12.6
11.5	1.17	12.7
11.6	1.18	12.8
11.7	1.19	12.9
11.8	1.20	13.0
11.9	1.21	13.1
12.0	1.22	13.2
12.1	1.23	13.3
12.2	1.24	13.4
12.3	1.26	13.6
12.4	1.27	13.7
12.5	1.28	13.8
12.6	1.29	13.9
12.7	1.30	14.0
12.8	1.31	14.1
12.9	1.32	14.2
13.0	1.33	14.3
13.1	1.34	14.4
13.2	1.35	14.5
13.3	1.36	14.7
13.4	1.37	14.8
13.5	1.38	14.9
13.6	1.39	15.0
13.7	1.40	15.1
13.8	1.41	15.2
13.9	1.42	15.3
14.0	1.43	15.4
14.1	1.44	15.5
14.2	1.45	15.6
14.3	1.46	15.8
14.4	1.47	15.9
14.5	1.48	16.0
14.6	1.49	16.1
14.7	1.50	16.2
14.8	1.51	16.3
14.9	1.52	16.4
15.0	1.53	16.5
15.1	1.54	16.6
15.2	1.55	16.8
15.3	1.56	16.9
15.4	1.57	17.0
15.5	1.58	17.1
15.6	1.59	17.2
15.7	1.60	17.3
15.8	1.61	17.4
15.9	1.62	17.5
16.0	1.63	17.6
16.1	1.64	17.7
16.2	1.65	17.9
16.3	1.66	18.0
16.4	1.67	18.1
16.5	1.68	18.2
16.6	1.69	18.3
16.7	1.70	18.4
16.8	1.71	18.5
16.9	1.72	18.6
17.0	1.73	18.7

APPENDIX B - 2 **CU FT**

Continued

THINNING TABLES

From 0.5:1

TO 0.8:1

Grout in Line and Agitator (cu ft)	THIN-NING WATER (cu ft)	New vol (cu ft)
1.0	0.31	1.3
1.1	0.34	1.4
1.2	0.37	1.6
1.3	0.40	1.7
1.4	0.43	1.8
1.5	0.46	2.0
1.6	0.49	2.1
1.7	0.52	2.2
1.8	0.55	2.4
1.9	0.58	2.5
2.0	0.61	2.6
2.1	0.64	2.7
2.2	0.67	2.9
2.3	0.70	3.0
2.4	0.73	3.1
2.5	0.77	3.3
2.6	0.80	3.4
2.7	0.83	3.5
2.8	0.86	3.7
2.9	0.89	3.8
3.0	0.92	3.9
3.1	0.95	4.0
3.2	0.98	4.2
3.3	1.01	4.3
3.4	1.04	4.4
3.5	1.07	4.6
3.6	1.10	4.7
3.7	1.13	4.8
3.8	1.16	5.0
3.9	1.19	5.1
4.0	1.22	5.2
4.1	1.26	5.4
4.2	1.29	5.5
4.3	1.32	5.6
4.4	1.35	5.7
4.5	1.38	5.9
4.6	1.41	6.0
4.7	1.44	6.1
4.8	1.47	6.3
4.9	1.50	6.4
5.0	1.53	6.5
5.1	1.56	6.7
5.2	1.59	6.8
5.3	1.62	6.9
5.4	1.65	7.1
5.5	1.68	7.2
5.6	1.71	7.3
5.7	1.74	7.4
5.8	1.78	7.6
5.9	1.81	7.7
6.0	1.84	7.8
6.1	1.87	8.0
6.2	1.90	8.1
6.3	1.93	8.2
6.4	1.96	8.4
6.5	1.99	8.5
6.6	2.02	8.6
6.7	2.05	8.8
6.8	2.08	8.9
6.9	2.11	9.0
7.0	2.14	9.1
7.1	2.17	9.3
7.2	2.20	9.4
7.3	2.23	9.5
7.4	2.27	9.7
7.5	2.30	9.8
7.6	2.33	9.9
7.7	2.36	10.1
7.8	2.39	10.2
7.9	2.42	10.3
8.0	2.45	10.4
8.1	2.48	10.6
8.2	2.51	10.7
8.3	2.54	10.8
8.4	2.57	11.0
8.5	2.60	11.1
8.6	2.63	11.2
8.7	2.66	11.4
8.8	2.69	11.5
8.9	2.72	11.6
9.0	2.76	11.8
9.1	2.79	11.9
9.2	2.82	12.0
9.3	2.85	12.1
9.4	2.88	12.3
9.5	2.91	12.4
9.6	2.94	12.5
9.7	2.97	12.7
9.8	3.00	12.8
9.9	3.03	12.9
10.0	3.06	13.1
10.1	3.09	13.2
10.2	3.12	13.3
10.3	3.15	13.5
10.4	3.18	13.6
10.5	3.21	13.7
10.6	3.24	13.8
10.7	3.28	14.0
10.8	3.31	14.1
10.9	3.34	14.2
11.0	3.37	14.4
11.1	3.40	14.5
11.2	3.43	14.6
11.3	3.46	14.8
11.4	3.49	14.9
11.5	3.52	15.0

APPENDIX B - 2 CU FT

Continued

THINNING TABLES

From 0.5:1

Grout in Line and Agitator (cu ft)	THINNING WATER (cu ft)	New vol (cu ft)	Grout in Line and Agitator (cu ft)	THINNING WATER (cu ft)	New vol (cu ft)	Grout in Line and Agitator (cu ft)	THINNING WATER (cu ft)	New vol (cu ft)
TO 0.8:1								
11.6	3.55	15.2	13.1	4.01	17.1	14.1	4.32	18.4
11.7	3.58	15.3	13.2	4.04	17.2	14.2	4.35	18.5
11.8	3.61	15.4	13.3	4.07	17.4	14.3	4.38	18.7
11.9	3.64	15.5	13.4	4.10	17.5	14.4	4.41	18.8
12.0	3.67	15.7	13.5	4.13	17.6	14.5	4.44	18.9
			13.6	4.16	17.8	14.6	4.47	19.1
12.1	3.70	15.8	13.7	4.19	17.9	14.7	4.50	19.2
12.2	3.73	15.9	13.8	4.22	18.0	14.8	4.53	19.3
12.3	3.77	16.1	13.9	4.26	18.2	14.9	4.56	19.5
12.4	3.80	16.2	14.0	4.29	18.3	15.0	4.59	19.6
12.5	3.83	16.3						
12.6	3.86	16.5				15.1	4.62	19.7
12.7	3.89	16.6				15.2	4.65	19.9
12.8	3.92	16.7						
12.9	3.95	16.8						
13.0	3.98	17.0						
TO 1:1								
1.0	0.51	1.5	3.1	1.58	4.7	5.1	2.60	7.7
			3.2	1.63	4.8	5.2	2.65	7.9
1.1	0.56	1.7	3.3	1.68	5.0	5.3	2.70	8.0
1.2	0.61	1.8	3.4	1.73	5.1	5.4	2.76	8.2
1.3	0.66	2.0	3.5	1.79	5.3	5.5	2.81	8.3
1.4	0.71	2.1	3.6	1.84	5.4	5.6	2.86	8.5
1.5	0.77	2.3	3.7	1.89	5.6	5.7	2.91	8.6
1.6	0.82	2.4	3.8	1.94	5.7	5.8	2.96	8.8
1.7	0.87	2.6	3.9	1.99	5.9	5.9	3.01	8.9
1.8	0.92	2.7	4.0	2.04	6.0	6.0	3.06	9.1
1.9	0.97	2.9						
2.0	1.02	3.0	4.1	2.09	6.2	6.1	3.11	9.2
			4.2	2.14	6.3	6.2	3.16	9.4
2.1	1.07	3.2	4.3	2.19	6.5	6.3	3.21	9.5
2.2	1.12	3.3	4.4	2.24	6.6	6.4	3.27	9.7
2.3	1.17	3.5	4.5	2.30	6.8	6.5	3.32	9.8
2.4	1.22	3.6	4.6	2.35	6.9	6.6	3.37	10.0
2.5	1.28	3.8	4.7	2.40	7.1	6.7	3.42	10.1
2.6	1.33	3.9	4.8	2.45	7.2	6.8	3.47	10.3
2.7	1.38	4.1	4.9	2.50	7.4	6.9	3.52	10.4
2.8	1.43	4.2	5.0	2.55	7.6	7.0	3.57	10.6
2.9	1.48	4.4						
3.0	1.53	4.5						

APPENDIX B - 2 — CU FT

Continued

THINNING TABLES

From 0.5:1

Grout in Line and Agitator (cu ft)	THIN-NING WATER (cu ft)	New vol (cu ft)	Grout in Line and Agitator (cu ft)	THIN-NING WATER (cu ft)	New vol (cu ft)	Grout in Line and Agitator (cu ft)	THIN-NING WATER (cu ft)	New vol (cu ft)
TO 1:1								
7.1	3.62	10.7	9.1	4.64	13.7	11.1	5.66	16.8
7.2	3.67	10.9	9.2	4.69	13.9	11.2	5.71	16.9
7.3	3.72	11.0	9.3	4.74	14.0	11.3	5.77	17.1
7.4	3.78	11.2	9.4	4.80	14.2	11.4	5.82	17.2
7.5	3.83	11.3	9.5	4.85	14.3	11.5	5.87	17.4
7.6	3.88	11.5	9.6	4.90	14.5	11.6	5.92	17.5
7.7	3.93	11.6	9.7	4.95	14.6	11.7	5.97	17.7
7.8	3.98	11.8	9.8	5.00	14.8	11.8	6.02	17.8
7.9	4.03	11.9	9.9	5.05	15.0	11.9	6.07	18.0
8.0	4.08	12.1	10.0	5.10	15.1	12.0	6.12	18.1
8.1	4.13	12.2	10.1	5.15	15.3	12.1	6.17	18.3
8.2	4.18	12.4	10.2	5.20	15.4	12.2	6.22	18.4
8.3	4.23	12.5	10.3	5.26	15.6	12.3	6.28	18.6
8.4	4.29	12.7	10.4	5.31	15.7	12.4	6.33	18.7
8.5	4.34	12.8	10.5	5.36	15.9	12.5	6.38	18.9
8.6	4.39	13.0	10.6	5.41	16.0	12.6	6.43	19.0
8.7	4.44	13.1	10.7	5.46	16.2	12.7	6.48	19.2
8.8	4.49	13.3	10.8	5.51	16.3	12.8	6.53	19.3
8.9	4.54	13.4	10.9	5.56	16.5	12.9	6.58	19.5
9.0	4.59	13.6	11.0	5.61	16.6	13.0	6.63	19.6
TO 2:1								
1.0	1.53	2.5	2.6	3.98	6.6	4.1	6.28	10.4
			2.7	4.13	6.8	4.2	6.43	10.6
1.1	1.68	2.8	2.8	4.29	7.1	4.3	6.58	10.9
1.2	1.84	3.0	2.9	4.44	7.3	4.4	6.73	11.1
1.3	1.99	3.3	3.0	4.59	7.6	4.5	6.89	11.4
1.4	2.14	3.5				4.6	7.04	11.6
1.5	2.30	3.8	3.1	4.74	7.8	4.7	7.19	11.9
1.6	2.45	4.0	3.2	4.90	8.1	4.8	7.35	12.1
1.7	2.60	4.3	3.3	5.05	8.4	4.9	7.50	12.4
1.8	2.76	4.6	3.4	5.20	8.6	5.0	7.65	12.7
1.9	2.91	4.8	3.5	5.36	8.9			
2.0	3.06	5.1	3.6	5.51	9.1	5.1	7.81	12.9
			3.7	5.66	9.4	5.2	7.96	13.2
2.1	3.21	5.3	3.8	5.82	9.6	5.3	8.11	13.4
2.2	3.37	5.6	3.9	5.97	9.9	5.4	8.27	13.7
2.3	3.52	5.8	4.0	6.12	10.1	5.5	8.42	13.9
2.4	3.67	6.1						
2.5	3.83	6.3						

APPENDIX B - 2 — CU FT

Continued

THINNING TABLES

From 0.5:1

Grout in Line and Agitator (cu ft)	THINNING WATER (cu ft)	New vol (cu ft)	Grout in Line and Agitator (cu ft)	THINNING WATER (cu ft)	New vol (cu ft)	Grout in Line and Agitator (cu ft)	THINNING WATER (cu ft)	New vol (cu ft)
TO 2:1								
5.6	8.57	14.2	6.1	9.34	15.4	7.1	10.87	18.0
5.7	8.72	14.4	6.2	9.49	15.7	7.2	11.02	18.2
5.8	8.88	14.7	6.3	9.64	15.9	7.3	11.17	18.5
5.9	9.03	14.9	6.4	9.80	16.2	7.4	11.33	18.7
6.0	9.18	15.2	6.5	9.95	16.4	7.5	11.48	19.0
			6.6	10.10	16.7	7.6	11.63	19.2
			6.7	10.26	17.0	7.7	11.79	19.5
			6.8	10.41	17.2	7.8	11.94	19.7
			6.9	10.56	17.5			
			7.0	10.71	17.7			
TO 3:1								
1.0	2.55	3.6	2.6	6.63	9.2	4.1	10.46	14.6
			2.7	6.89	9.6	4.2	10.71	14.9
1.1	2.81	3.9	2.8	7.14	9.9	4.3	10.97	15.3
1.2	3.06	4.3	2.9	7.40	10.3	4.4	11.22	15.6
1.3	3.32	4.6	3.0	7.65	10.7	4.5	11.48	16.0
1.4	3.57	5.0				4.6	11.73	16.3
1.5	3.83	5.3	3.1	7.91	11.0	4.7	11.99	16.7
1.6	4.08	5.7	3.2	8.16	11.4	4.8	12.24	17.0
1.7	4.34	6.0	3.3	8.42	11.7	4.9	12.50	17.4
1.8	4.59	6.4	3.4	8.67	12.1	5.0	12.76	17.8
1.9	4.85	6.7	3.5	8.93	12.4			
2.0	5.10	7.1	3.6	9.18	12.8	5.1	13.01	18.1
			3.7	9.44	13.1	5.2	13.27	18.5
2.1	5.36	7.5	3.8	9.69	13.5	5.3	13.52	18.8
2.2	5.61	7.8	3.9	9.95	13.8	5.4	13.78	19.2
2.3	5.87	8.2	4.0	10.20	14.2	5.5	14.03	19.5
2.4	6.12	8.5				5.6	14.29	19.9
2.5	6.38	8.9						
TO 4:1								
1.0	3.57	4.6	2.1	7.50	9.6	3.1	11.07	14.2
1.1	3.93	5.0	2.2	7.86	10.1	3.2	11.43	14.6
1.2	4.29	5.5	2.3	8.21	10.5	3.3	11.79	15.1
1.3	4.64	5.9	2.4	8.57	11.0	3.4	12.14	15.5
1.4	5.00	6.4	2.5	8.93	11.4	3.5	12.50	16.0
1.5	5.36	6.9	2.6	9.29	11.9	3.6	12.86	16.5
1.6	5.71	7.3	2.7	9.64	12.3	3.7	13.21	16.9
1.7	6.07	7.8	2.8	10.00	12.8	3.8	13.57	17.4
1.8	6.43	8.2	2.9	10.36	13.3	3.9	13.93	17.8
1.9	6.79	8.7	3.0	10.71	13.7	4.0	14.29	18.3
2.0	7.14	9.1						

APPENDIX B - 2
Continued

CU FT

THINNING TABLES

From 0.5:1

TO 5:1

Grout in Line and Agitator (cu ft)	THINNING WATER (cu ft)	New vol (cu ft)
1.0	4.59	5.6
1.1	5.05	6.2
1.2	5.51	6.7
1.3	5.97	7.3
1.4	6.43	7.8
1.5	6.89	8.4
1.6	7.35	8.9
1.7	7.81	9.5
1.8	8.27	10.1
1.9	8.72	10.6
2.0	9.18	11.2
2.1	9.64	11.7
2.2	10.10	12.3
2.3	10.56	12.9
2.4	11.02	13.4
2.5	11.48	14.0
2.6	11.94	14.5
2.7	12.40	15.1
2.8	12.86	15.7
2.9	13.32	16.2
3.0	13.78	16.8
3.1	14.23	17.3
3.2	14.69	17.9
3.3	15.15	18.5
3.4	15.61	19.0
3.5	16.07	19.6

APPENDIX B - 3

94 lb BAG and BULK

QUANTITY OF CEMENT IN GROUT

5 : 1
(by loose volume)

See Section 4.1–11 for explanations and methods of use.

Grout Volume (cu ft)	Cement Content (bags)	Cement Content (lb)
0.1	0.02	2
0.2	0.04	3
0.3	0.05	5
0.4	0.07	7
0.5	0.09	9
0.6	0.11	10
0.7	0.13	12
0.8	0.15	14
0.9	0.16	15
1.0	0.18	17
1.1	0.20	19
1.2	0.22	21
1.3	0.24	22
1.4	0.26	24
1.5	0.27	26
1.6	0.29	27
1.7	0.31	29
1.8	0.33	31
1.9	0.35	33
2.0	0.36	34
2.1	0.38	36
2.2	0.40	38
2.3	0.42	39
2.4	0.44	41
2.5	0.46	43
2.6	0.47	45
2.7	0.49	46
2.8	0.51	48
2.9	0.53	50
3.0	0.55	51
3.1	0.57	53
3.2	0.58	55
3.3	0.60	57
3.4	0.62	58
3.5	0.64	60
3.6	0.66	62
3.7	0.68	63
3.8	0.69	65
3.9	0.71	67
4.0	0.73	69
4.1	0.75	70
4.2	0.77	72
4.3	0.78	74
4.4	0.80	75
4.5	0.82	77
4.6	0.84	79
4.7	0.86	81
4.8	0.88	82
4.9	0.89	84
5.0	0.91	86
5.1	0.93	87
5.2	0.95	89
5.3	0.97	91
5.4	0.99	93
5.5	1.00	94
5.6	1.02	96
5.7	1.04	98
5.8	1.06	99
5.9	1.08	101
6.0	1.09	103
6.1	1.11	105
6.2	1.13	106
6.3	1.15	108
6.4	1.17	110
6.5	1.19	111
6.6	1.20	113
6.7	1.22	115
6.8	1.24	117
6.9	1.26	118
7.0	1.28	120
7.1	1.30	122
7.2	1.31	124
7.3	1.33	125
7.4	1.35	127
7.5	1.37	129
7.6	1.39	130
7.7	1.41	132
7.8	1.42	134
7.9	1.44	136
8.0	1.46	137
8.1	1.48	139
8.2	1.50	141
8.3	1.51	142
8.4	1.53	144
8.5	1.55	146
8.6	1.57	148
8.7	1.59	149
8.8	1.61	151
8.9	1.62	153
9.0	1.64	154

APPENDIX B - 3
Continued

94 lb BAG and BULK

QUANTITY OF CEMENT IN GROUT

5 : 1

Grout Volume	Cement Content		Grout Volume	Cement Content		Grout Volume	Cement Content	
(cu ft)	(bags)	(lb)	(cu ft)	(bags)	(lb)	(cu ft)	(bags)	(lb)
9.1	1.66	156	13.1	2.39	225	17.1	3.12	293
9.2	1.68	158	13.2	2.41	226	17.2	3.14	295
9.3	1.70	160	13.3	2.43	228	17.3	3.16	297
9.4	1.72	161	13.4	2.45	230	17.4	3.18	298
9.5	1.73	163	13.5	2.46	232	17.5	3.19	300
9.6	1.75	165	13.6	2.48	233	17.6	3.21	302
9.7	1.77	166	13.7	2.50	235	17.7	3.23	304
9.8	1.79	168	13.8	2.52	237	17.8	3.25	305
9.9	1.81	170	13.9	2.54	238	17.9	3.27	307
10.0	1.82	172	14.0	2.55	240	18.0	3.28	309
10.1	1.84	173	14.1	2.57	242	18.1	3.30	310
10.2	1.86	175	14.2	2.59	244	18.2	3.32	312
10.3	1.88	177	14.3	2.61	245	18.3	3.34	314
10.4	1.90	178	14.4	2.63	247	18.4	3.36	316
10.5	1.92	180	14.5	2.65	249	18.5	3.38	317
10.6	1.93	182	14.6	2.66	250	18.6	3.39	319
10.7	1.95	184	14.7	2.68	252	18.7	3.41	321
10.8	1.97	185	14.8	2.70	254	18.8	3.43	322
10.9	1.99	187	14.9	2.72	256	18.9	3.45	324
11.0	2.01	189	15.0	2.74	257	19.0	3.47	326
11.1	2.03	190	15.1	2.76	259	19.1	3.49	328
11.2	2.04	192	15.2	2.77	261	19.2	3.50	329
11.3	2.06	194	15.3	2.79	262	19.3	3.52	331
11.4	2.08	196	15.4	2.81	264	19.4	3.54	333
11.5	2.10	197	15.5	2.83	266	19.5	3.56	334
11.6	2.12	199	15.6	2.85	268	19.6	3.58	336
11.7	2.14	201	15.7	2.86	269	19.7	3.59	338
11.8	2.15	202	15.8	2.88	271	19.8	3.61	340
11.9	2.17	204	15.9	2.90	273	19.9	3.63	341
12.0	2.19	206	16.0	2.92	274	20.0	3.65	343
12.1	2.21	208	16.1	2.94	276			
12.2	2.23	209	16.2	2.96	278			
12.3	2.24	211	16.3	2.97	280			
12.4	2.26	213	16.4	2.99	281			
12.5	2.28	214	16.5	3.01	283			
12.6	2.30	216	16.6	3.03	285			
12.7	2.32	218	16.7	3.05	286			
12.8	2.34	220	16.8	3.07	288			
12.9	2.35	221	16.9	3.08	290			
13.0	2.37	223	17.0	3.10	292			

APPENDIX B - 3

94 lb BAG and BULK

QUANTITY OF CEMENT IN GROUT

4 : 1
(by loose volume)

See Section 4.1–11 for explanations and methods of use.

Grout Volume	Cement Content		Grout Volume	Cement Content		Grout Volume	Cement Content	
(cu ft)	(bags)	(lb)	(cu ft)	(bags)	(lb)	(cu ft)	(bags)	(lb)
0.1	0.02	2	3.1	0.69	65	6.1	1.36	128
0.2	0.04	4	3.2	0.71	67	6.2	1.38	130
0.3	0.07	6	3.3	0.74	69	6.3	1.41	132
0.4	0.09	8	3.4	0.76	71	6.4	1.43	134
0.5	0.11	10	3.5	0.78	73	6.5	1.45	136
0.6	0.13	13	3.6	0.80	76	6.6	1.47	138
0.7	0.16	15	3.7	0.83	78	6.7	1.50	141
0.8	0.18	17	3.8	0.85	80	6.8	1.52	143
0.9	0.20	19	3.9	0.87	82	6.9	1.54	145
1.0	0.22	21	4.0	0.89	84	7.0	1.56	147
1.1	0.25	23	4.1	0.92	86	7.1	1.58	149
1.2	0.27	25	4.2	0.94	88	7.2	1.61	151
1.3	0.29	27	4.3	0.96	90	7.3	1.63	153
1.4	0.31	29	4.4	0.98	92	7.4	1.65	155
1.5	0.33	31	4.5	1.00	94	7.5	1.67	157
1.6	0.36	34	4.6	1.03	97	7.6	1.70	159
1.7	0.38	36	4.7	1.05	99	7.7	1.72	162
1.8	0.40	38	4.8	1.07	101	7.8	1.74	164
1.9	0.42	40	4.9	1.09	103	7.9	1.76	166
2.0	0.45	42	5.0	1.12	105	8.0	1.79	168
2.1	0.47	44	5.1	1.14	107	8.1	1.81	170
2.2	0.49	46	5.2	1.16	109	8.2	1.83	172
2.3	0.51	48	5.3	1.18	111	8.3	1.85	174
2.4	0.54	50	5.4	1.21	113	8.4	1.88	176
2.5	0.56	52	5.5	1.23	115	8.5	1.90	178
2.6	0.58	55	5.6	1.25	117	8.6	1.92	180
2.7	0.60	57	5.7	1.27	120	8.7	1.94	183
2.8	0.63	59	5.8	1.29	122	8.8	1.96	185
2.9	0.65	61	5.9	1.32	124	8.9	1.99	187
3.0	0.67	63	6.0	1.34	126	9.0	2.01	189

APPENDIX B - 3
Continued

94 lb BAG and BULK

QUANTITY OF CEMENT IN GROUT

4 : 1

Grout Volume (cu ft)	Cement Content (bags)	Cement Content (lb)	Grout Volume (cu ft)	Cement Content (bags)	Cement Content (lb)	Grout Volume (cu ft)	Cement Content (bags)	Cement Content (lb)
9.1	2.03	191	13.1	2.92	275	17.1	3.82	359
9.2	2.05	193	13.2	2.95	277	17.2	3.84	361
9.3	2.08	195	13.3	2.97	279	17.3	3.86	363
9.4	2.10	197	13.4	2.99	281	17.4	3.88	365
9.5	2.12	199	13.5	3.01	283	17.5	3.91	367
9.6	2.14	201	13.6	3.04	285	17.6	3.93	369
9.7	2.17	204	13.7	3.06	287	17.7	3.95	371
9.8	2.19	206	13.8	3.08	290	17.8	3.97	373
9.9	2.21	208	13.9	3.10	292	17.9	4.00	376
10.0	2.23	210	14.0	3.13	294	18.0	4.02	378
10.1	2.25	212	14.1	3.15	296	18.1	4.04	380
10.2	2.28	214	14.2	3.17	298	18.2	4.06	382
10.3	2.30	216	14.3	3.19	300	18.3	4.08	384
10.4	2.32	218	14.4	3.21	302	18.4	4.11	386
10.5	2.34	220	14.5	3.24	304	18.5	4.13	388
10.6	2.37	222	14.6	3.26	306	18.6	4.15	390
10.7	2.39	225	14.7	3.28	308	18.7	4.17	392
10.8	2.41	227	14.8	3.30	311	18.8	4.20	394
10.9	2.43	229	14.9	3.33	313	18.9	4.22	397
11.0	2.46	231	15.0	3.35	315	19.0	4.24	399
11.1	2.48	233	15.1	3.37	317	19.1	4.26	401
11.2	2.50	235	15.2	3.39	319	19.2	4.29	403
11.3	2.52	237	15.3	3.42	321	19.3	4.31	405
11.4	2.54	239	15.4	3.44	323	19.4	4.33	407
11.5	2.57	241	15.5	3.46	325	19.5	4.35	409
11.6	2.59	243	15.6	3.48	327	19.6	4.37	411
11.7	2.61	245	15.7	3.50	329	19.7	4.40	413
11.8	2.63	248	15.8	3.53	332	19.8	4.42	415
11.9	2.66	250	15.9	3.55	334	19.9	4.44	418
12.0	2.68	252	16.0	3.57	336	20.0	4.46	420
12.1	2.70	254	16.1	3.59	338			
12.2	2.72	256	16.2	3.62	340			
12.3	2.75	258	16.3	3.64	342			
12.4	2.77	260	16.4	3.66	344			
12.5	2.79	262	16.5	3.68	346			
12.6	2.81	264	16.6	3.71	348			
12.7	2.83	266	16.7	3.73	350			
12.8	2.86	269	16.8	3.75	352			
12.9	2.88	271	16.9	3.77	355			
13.0	2.90	273	17.0	3.79	357			

APPENDIX B - 3

94 lb BAG and BULK

QUANTITY OF CEMENT IN GROUT

3 : 1
(by loose volume)

See Section 4.1–11 for explanations and methods of use.

Grout Volume	Cement Content		Grout Volume	Cement Content		Grout Volume	Cement Content	
(cu ft)	(bags)	(lb)	(cu ft)	(bags)	(lb)	(cu ft)	(bags)	(lb)
0.1	0.03	3	3.1	0.89	84	6.1	1.75	165
0.2	0.06	5	3.2	0.92	86	6.2	1.78	167
0.3	0.09	8	3.3	0.95	89	6.3	1.81	170
0.4	0.11	11	3.4	0.98	92	6.4	1.84	173
0.5	0.14	14	3.5	1.01	95	6.5	1.87	176
0.6	0.17	16	3.6	1.03	97	6.6	1.90	178
0.7	0.20	19	3.7	1.06	100	6.7	1.93	181
0.8	0.23	22	3.8	1.09	103	6.8	1.95	184
0.9	0.26	24	3.9	1.12	105	6.9	1.98	186
1.0	0.29	27	4.0	1.15	108	7.0	2.01	189
1.1	0.32	30	4.1	1.18	111	7.1	2.04	192
1.2	0.34	32	4.2	1.21	113	7.2	2.07	194
1.3	0.37	35	4.3	1.24	116	7.3	2.10	197
1.4	0.40	38	4.4	1.26	119	7.4	2.13	200
1.5	0.43	41	4.5	1.29	122	7.5	2.16	203
1.6	0.46	43	4.6	1.32	124	7.6	2.18	205
1.7	0.49	46	4.7	1.35	127	7.7	2.21	208
1.8	0.52	49	4.8	1.38	130	7.8	2.24	211
1.9	0.55	51	4.9	1.41	132	7.9	2.27	213
2.0	0.57	54	5.0	1.44	135	8.0	2.30	216
2.1	0.60	57	5.1	1.47	138	8.1	2.33	219
2.2	0.63	59	5.2	1.49	140	8.2	2.36	221
2.3	0.66	62	5.3	1.52	143	8.3	2.39	224
2.4	0.69	65	5.4	1.55	146	8.4	2.41	227
2.5	0.72	68	5.5	1.58	149	8.5	2.44	230
2.6	0.75	70	5.6	1.61	151	8.6	2.47	232
2.7	0.78	73	5.7	1.64	154	8.7	2.50	235
2.8	0.80	76	5.8	1.67	157	8.8	2.53	238
2.9	0.83	78	5.9	1.70	159	8.9	2.56	240
3.0	0.86	81	6.0	1.72	162	9.0	2.59	243

APPENDIX B - 3
Continued

94 lb BAG and BULK

QUANTITY OF CEMENT IN GROUT

3 : 1

Grout Volume (cu ft)	Cement Content (bags)	Cement Content (lb)
9.1	2.61	246
9.2	2.64	249
9.3	2.67	251
9.4	2.70	254
9.5	2.73	257
9.6	2.76	259
9.7	2.79	262
9.8	2.82	265
9.9	2.84	267
10.0	2.87	270
10.1	2.90	273
10.2	2.93	276
10.3	2.96	278
10.4	2.99	281
10.5	3.02	284
10.6	3.05	286
10.7	3.07	289
10.8	3.10	292
10.9	3.13	294
11.0	3.16	297
11.1	3.19	300
11.2	3.22	303
11.3	3.25	305
11.4	3.28	308
11.5	3.30	311
11.6	3.33	313
11.7	3.36	316
11.8	3.39	319
11.9	3.42	321
12.0	3.45	324
12.1	3.48	327
12.2	3.51	330
12.3	3.53	332
12.4	3.56	335
12.5	3.59	338
12.6	3.62	340
12.7	3.65	343
12.8	3.68	346
12.9	3.71	348
13.0	3.74	351
13.1	3.76	354
13.2	3.79	357
13.3	3.82	359
13.4	3.85	362
13.5	3.88	365
13.6	3.91	367
13.7	3.94	370
13.8	3.97	373
13.9	3.99	375
14.0	4.02	378
14.1	4.05	381
14.2	4.08	384
14.3	4.11	386
14.4	4.14	389
14.5	4.17	392
14.6	4.20	394
14.7	4.22	397
14.8	4.25	400
14.9	4.28	402
15.0	4.31	405
15.1	4.34	408
15.2	4.37	411
15.3	4.40	413
15.4	4.43	416
15.5	4.45	419
15.6	4.48	421
15.7	4.51	424
15.8	4.54	427
15.9	4.57	429
16.0	4.60	432
16.1	4.63	435
16.2	4.66	438
16.3	4.68	440
16.4	4.71	443
16.5	4.74	446
16.6	4.77	448
16.7	4.80	451
16.8	4.83	454
16.9	4.86	456
17.0	4.89	459
17.1	4.91	462
17.2	4.94	465
17.3	4.97	467
17.4	5.00	470
17.5	5.03	473
17.6	5.06	475
17.7	5.09	478
17.8	5.11	481
17.9	5.14	484
18.0	5.17	486
18.1	5.20	489
18.2	5.23	492
18.3	5.26	494
18.4	5.29	497
18.5	5.32	500
18.6	5.34	502
18.7	5.37	505
18.8	5.40	508
18.9	5.43	511
19.0	5.46	513
19.1	5.49	516
19.2	5.52	519
19.3	5.55	521
19.4	5.57	524
19.5	5.60	527
19.6	5.63	529
19.7	5.66	532
19.8	5.69	535
19.9	5.72	538
20.0	5.75	540

APPENDIX B - 3

94 lb BAG and BULK

QUANTITY OF CEMENT IN GROUT

2 : 1
(by loose volume)

See Section 4.1–11 for explanations and methods of use.

Grout Volume (cu ft)	Cement Content (bags)	Cement Content (lb)	Grout Volume (cu ft)	Cement Content (bags)	Cement Content (lb)	Grout Volume (cu ft)	Cement Content (bags)	Cement Content (lb)
0.1	0.04	4	3.1	1.25	117	6.1	2.46	231
0.2	0.08	8	3.2	1.29	121	6.2	2.50	235
0.3	0.12	11	3.3	1.33	125	6.3	2.54	239
0.4	0.16	15	3.4	1.37	129	6.4	2.58	243
0.5	0.20	19	3.5	1.41	133	6.5	2.62	246
0.6	0.24	23	3.6	1.45	136	6.6	2.66	250
0.7	0.28	27	3.7	1.49	140	6.7	2.70	254
0.8	0.32	30	3.8	1.53	144	6.8	2.74	258
0.9	0.36	34	3.9	1.57	148	6.9	2.78	262
1.0	0.40	38	4.0	1.61	152	7.0	2.82	265
1.1	0.44	42	4.1	1.65	155	7.1	2.86	269
1.2	0.48	45	4.2	1.69	159	7.2	2.90	273
1.3	0.52	49	4.3	1.73	163	7.3	2.94	277
1.4	0.56	53	4.4	1.77	167	7.4	2.98	280
1.5	0.60	57	4.5	1.81	171	7.5	3.02	284
1.6	0.65	61	4.6	1.85	174	7.6	3.06	288
1.7	0.69	64	4.7	1.90	178	7.7	3.10	292
1.8	0.73	68	4.8	1.94	182	7.8	3.15	296
1.9	0.77	72	4.9	1.98	186	7.9	3.19	299
2.0	0.81	76	5.0	2.02	190	8.0	3.23	303
2.1	0.85	80	5.1	2.06	193	8.1	3.27	307
2.2	0.89	83	5.2	2.10	197	8.2	3.31	311
2.3	0.93	87	5.3	2.14	201	8.3	3.35	315
2.4	0.97	91	5.4	2.18	205	8.4	3.39	318
2.5	1.01	95	5.5	2.22	208	8.5	3.43	322
2.6	1.05	99	5.6	2.26	212	8.6	3.47	326
2.7	1.09	102	5.7	2.30	216	8.7	3.51	330
2.8	1.13	106	5.8	2.34	220	8.8	3.55	334
2.9	1.17	110	5.9	2.38	224	8.9	3.59	337
3.0	1.21	114	6.0	2.42	227	9.0	3.63	341

APPENDIX B - 3
Continued

94 lb BAG and BULK

QUANTITY OF CEMENT IN GROUT

2 : 1

Grout Volume (cu ft)	Cement Content (bags)	Cement Content (lb)	Grout Volume (cu ft)	Cement Content (bags)	Cement Content (lb)	Grout Volume (cu ft)	Cement Content (bags)	Cement Content (lb)
9.1	3.67	345	13.1	5.28	497	17.1	6.90	648
9.2	3.71	349	13.2	5.32	500	17.2	6.94	652
9.3	3.75	352	13.3	5.36	504	17.3	6.98	656
9.4	3.79	356	13.4	5.40	508	17.4	7.02	660
9.5	3.83	360	13.5	5.44	512	17.5	7.06	663
9.6	3.87	364	13.6	5.48	515	17.6	7.10	667
9.7	3.91	368	13.7	5.52	519	17.7	7.14	671
9.8	3.95	371	13.8	5.56	523	17.8	7.18	675
9.9	3.99	375	13.9	5.60	527	17.9	7.22	678
10.0	4.03	379	14.0	5.65	531	18.0	7.26	682
10.1	4.07	383	14.1	5.69	534	18.1	7.30	686
10.2	4.11	387	14.2	5.73	538	18.2	7.34	690
10.3	4.15	390	14.3	5.77	542	18.3	7.38	694
10.4	4.19	394	14.4	5.81	546	18.4	7.42	697
10.5	4.23	398	14.5	5.85	550	18.5	7.46	701
10.6	4.27	402	14.6	5.89	553	18.6	7.50	705
10.7	4.31	406	14.7	5.93	557	18.7	7.54	709
10.8	4.35	409	14.8	5.97	561	18.8	7.58	713
10.9	4.40	413	14.9	6.01	565	18.9	7.62	716
11.0	4.44	417	15.0	6.05	569	19.0	7.66	720
11.1	4.48	421	15.1	6.09	572	19.1	7.70	724
11.2	4.52	425	15.2	6.13	576	19.2	7.74	728
11.3	4.56	428	15.3	6.17	580	19.3	7.78	732
11.4	4.60	432	15.4	6.21	584	19.4	7.82	735
11.5	4.64	436	15.5	6.25	587	19.5	7.86	739
11.6	4.68	440	15.6	6.29	591	19.6	7.90	743
11.7	4.72	443	15.7	6.33	595	19.7	7.94	747
11.8	4.76	447	15.8	6.37	599	19.8	7.98	750
11.9	4.80	451	15.9	6.41	603	19.9	8.02	754
12.0	4.84	455	16.0	6.45	606	20.0	8.06	758
12.1	4.88	459	16.1	6.49	610			
12.2	4.92	462	16.2	6.53	614			
12.3	4.96	466	16.3	6.57	618			
12.4	5.00	470	16.4	6.61	622			
12.5	5.04	474	16.5	6.65	625			
12.6	5.08	478	16.6	6.69	629			
12.7	5.12	481	16.7	6.73	633			
12.8	5.16	485	16.8	6.77	637			
12.9	5.20	489	16.9	6.81	641			
13.0	5.24	493	17.0	6.85	644			

APPENDIX B - 3

94 lb BAG and BULK

QUANTITY OF CEMENT IN GROUT

1 : 1
(by loose volume)

See Section 4.1–11 for explanations and methods of use.

Grout Volume	Cement Content		Grout Volume	Cement Content		Grout Volume	Cement Content	
(cu ft)	(bags)	(lb)	(cu ft)	(bags)	(lb)	(cu ft)	(bags)	(lb)
0.1	0.07	6	3.1	2.09	197	6.1	4.12	387
0.2	0.14	13	3.2	2.16	203	6.2	4.19	394
0.3	0.20	19	3.3	2.23	210	6.3	4.26	400
0.4	0.27	25	3.4	2.30	216	6.4	4.32	406
0.5	0.34	32	3.5	2.36	222	6.5	4.39	413
0.6	0.41	38	3.6	2.43	229	6.6	4.46	419
0.7	0.47	44	3.7	2.50	235	6.7	4.53	426
0.8	0.54	51	3.8	2.57	241	6.8	4.59	432
0.9	0.61	57	3.9	2.64	248	6.9	4.66	438
1.0	0.68	64	4.0	2.70	254	7.0	4.73	445
1.1	0.74	70	4.1	2.77	260	7.1	4.80	451
1.2	0.81	76	4.2	2.84	267	7.2	4.86	457
1.3	0.88	83	4.3	2.91	273	7.3	4.93	464
1.4	0.95	89	4.4	2.97	279	7.4	5.00	470
1.5	1.01	95	4.5	3.04	286	7.5	5.07	476
1.6	1.08	102	4.6	3.11	292	7.6	5.14	483
1.7	1.15	108	4.7	3.18	299	7.7	5.20	489
1.8	1.22	114	4.8	3.24	305	7.8	5.27	495
1.9	1.28	121	4.9	3.31	311	7.9	5.34	502
2.0	1.35	127	5.0	3.38	318	8.0	5.41	508
2.1	1.42	133	5.1	3.45	324	8.1	5.47	514
2.2	1.49	140	5.2	3.51	330	8.2	5.54	521
2.3	1.55	146	5.3	3.58	337	8.3	5.61	527
2.4	1.62	152	5.4	3.65	343	8.4	5.68	534
2.5	1.69	159	5.5	3.72	349	8.5	5.74	540
2.6	1.76	165	5.6	3.78	356	8.6	5.81	546
2.7	1.82	171	5.7	3.85	362	8.7	5.88	553
2.8	1.89	178	5.8	3.92	368	8.8	5.95	559
2.9	1.96	184	5.9	3.99	375	8.9	6.01	565
3.0	2.03	191	6.0	4.05	381	9.0	6.08	572

APPENDIX B - 3
Continued

94 lb BAG and BULK

QUANTITY OF CEMENT IN GROUT

1 : 1

Grout Volume (cu ft)	Cement Content (bags)	Cement Content (lb)	Grout Volume (cu ft)	Cement Content (bags)	Cement Content (lb)	Grout Volume (cu ft)	Cement Content (bags)	Cement Content (lb)
9.1	6.15	578	13.1	8.85	832	17.1	11.55	1086
9.2	6.22	584	13.2	8.92	838	17.2	11.62	1092
9.3	6.28	591	13.3	8.99	845	17.3	11.69	1099
9.4	6.35	597	13.4	9.05	851	17.4	11.76	1105
9.5	6.42	603	13.5	9.12	857	17.5	11.82	1111
9.6	6.49	610	13.6	9.19	864	17.6	11.89	1118
9.7	6.55	616	13.7	9.26	870	17.7	11.96	1124
9.8	6.62	622	13.8	9.32	876	17.8	12.03	1131
9.9	6.69	629	13.9	9.39	883	17.9	12.09	1137
10.0	6.76	635	14.0	9.46	889	18.0	12.16	1143
10.1	6.82	641	14.1	9.53	896	18.1	12.23	1150
10.2	6.89	648	14.2	9.59	902	18.2	12.30	1156
10.3	6.96	654	14.3	9.66	908	18.3	12.36	1162
10.4	7.03	661	14.4	9.73	915	18.4	12.43	1169
10.5	7.09	667	14.5	9.80	921	18.5	12.50	1175
10.6	7.16	673	14.6	9.86	927	18.6	12.57	1181
10.7	7.23	680	14.7	9.93	934	18.7	12.64	1188
10.8	7.30	686	14.8	10.00	940	18.8	12.70	1194
10.9	7.36	692	14.9	10.07	946	18.9	12.77	1200
11.0	7.43	699	15.0	10.14	953	19.0	12.84	1207
11.1	7.50	705	15.1	10.20	959	19.1	12.91	1213
11.2	7.57	711	15.2	10.27	965	19.2	12.97	1219
11.3	7.64	718	15.3	10.34	972	19.3	13.04	1226
11.4	7.70	724	15.4	10.41	978	19.4	13.11	1232
11.5	7.77	730	15.5	10.47	984	19.5	13.18	1239
11.6	7.84	737	15.6	10.54	991	19.6	13.24	1245
11.7	7.91	743	15.7	10.61	997	19.7	13.31	1251
11.8	7.97	749	15.8	10.68	1004	19.8	13.38	1258
11.9	8.04	756	15.9	10.74	1010	19.9	13.45	1264
12.0	8.11	762	16.0	10.81	1016	20.0	13.51	1270
12.1	8.18	769	16.1	10.88	1023			
12.2	8.24	775	16.2	10.95	1029			
12.3	8.31	781	16.3	11.01	1035			
12.4	8.38	788	16.4	11.08	1042			
12.5	8.45	794	16.5	11.15	1048			
12.6	8.51	800	16.6	11.22	1054			
12.7	8.58	807	16.7	11.28	1061			
12.8	8.65	813	16.8	11.35	1067			
12.9	8.72	819	16.9	11.42	1073			
13.0	8.78	826	17.0	11.49	1080			

APPENDIX B - 3

94 lb BAG and BULK

QUANTITY OF CEMENT IN GROUT

0.8 : 1
(by loose volume)

See Section 4.1–11 for explanations and methods of use.

Grout Volume	Cement Content		Grout Volume	Cement Content		Grout Volume	Cement Content	
(cu ft)	(bags)	(lb)	(cu ft)	(bags)	(lb)	(cu ft)	(bags)	(lb)
0.1	0.08	7	3.1	2.42	228	6.1	4.77	448
0.2	0.16	15	3.2	2.50	235	6.2	4.84	455
0.3	0.23	22	3.3	2.58	242	6.3	4.92	463
0.4	0.31	29	3.4	2.66	250	6.4	5.00	470
0.5	0.39	37	3.5	2.73	257	6.5	5.08	477
0.6	0.47	44	3.6	2.81	264	6.6	5.16	485
0.7	0.55	51	3.7	2.89	272	6.7	5.23	492
0.8	0.63	59	3.8	2.97	279	6.8	5.31	499
0.9	0.70	66	3.9	3.05	286	6.9	5.39	507
1.0	0.78	73	4.0	3.13	294	7.0	5.47	514
1.1	0.86	81	4.1	3.20	301	7.1	5.55	521
1.2	0.94	88	4.2	3.28	308	7.2	5.63	529
1.3	1.02	95	4.3	3.36	316	7.3	5.70	536
1.4	1.09	103	4.4	3.44	323	7.4	5.78	543
1.5	1.17	110	4.5	3.52	330	7.5	5.86	551
1.6	1.25	117	4.6	3.59	338	7.6	5.94	558
1.7	1.33	125	4.7	3.67	345	7.7	6.02	565
1.8	1.41	132	4.8	3.75	352	7.8	6.09	573
1.9	1.48	140	4.9	3.83	360	7.9	6.17	580
2.0	1.56	147	5.0	3.91	367	8.0	6.25	587
2.1	1.64	154	5.1	3.98	375	8.1	6.33	595
2.2	1.72	162	5.2	4.06	382	8.2	6.41	602
2.3	1.80	169	5.3	4.14	389	8.3	6.48	610
2.4	1.88	176	5.4	4.22	397	8.4	6.56	617
2.5	1.95	184	5.5	4.30	404	8.5	6.64	624
2.6	2.03	191	5.6	4.38	411	8.6	6.72	632
2.7	2.11	198	5.7	4.45	419	8.7	6.80	639
2.8	2.19	206	5.8	4.53	426	8.8	6.87	646
2.9	2.27	213	5.9	4.61	433	8.9	6.95	654
3.0	2.34	220	6.0	4.69	441	9.0	7.03	661

APPENDIX B - 3
Continued

94 lb BAG and BULK

QUANTITY OF CEMENT IN GROUT

0.8 : 1

Grout Volume (cu ft)	Cement Content (bags)	Cement Content (lb)	Grout Volume (cu ft)	Cement Content (bags)	Cement Content (lb)	Grout Volume (cu ft)	Cement Content (bags)	Cement Content (lb)
9.1	7.11	668	13.1	10.23	962	17.1	13.36	1256
9.2	7.19	676	13.2	10.31	969	17.2	13.44	1263
9.3	7.27	683	13.3	10.39	977	17.3	13.52	1270
9.4	7.34	690	13.4	10.47	984	17.4	13.59	1278
9.5	7.42	698	13.5	10.55	991	17.5	13.67	1285
9.6	7.50	705	13.6	10.63	999	17.6	13.75	1292
9.7	7.58	712	13.7	10.70	1006	17.7	13.83	1300
9.8	7.66	720	13.8	10.78	1013	17.8	13.91	1307
9.9	7.73	727	13.9	10.86	1021	17.9	13.98	1315
10.0	7.81	734	14.0	10.94	1028	18.0	14.06	1322
10.1	7.89	742	14.1	11.02	1035	18.1	14.14	1329
10.2	7.97	749	14.2	11.09	1043	18.2	14.22	1337
10.3	8.05	756	14.3	11.17	1050	18.3	14.30	1344
10.4	8.13	764	14.4	11.25	1057	18.4	14.38	1351
10.5	8.20	771	14.5	11.33	1065	18.5	14.45	1359
10.6	8.28	778	14.6	11.41	1072	18.6	14.53	1366
10.7	8.36	786	14.7	11.48	1080	18.7	14.61	1373
10.8	8.44	793	14.8	11.56	1087	18.8	14.69	1381
10.9	8.52	800	14.9	11.64	1094	18.9	14.77	1388
11.0	8.59	808	15.0	11.72	1102	19.0	14.84	1395
11.1	8.67	815	15.1	11.80	1109	19.1	14.92	1403
11.2	8.75	822	15.2	11.88	1116	19.2	15.00	1410
11.3	8.83	830	15.3	11.95	1124	19.3	15.08	1417
11.4	8.91	837	15.4	12.03	1131	19.4	15.16	1425
11.5	8.98	845	15.5	12.11	1138	19.5	15.23	1432
11.6	9.06	852	15.6	12.19	1146	19.6	15.31	1439
11.7	9.14	859	15.7	12.27	1153	19.7	15.39	1447
11.8	9.22	867	15.8	12.34	1160	19.8	15.47	1454
11.9	9.30	874	15.9	12.42	1168	19.9	15.55	1461
12.0	9.38	881	16.0	12.50	1175	20.0	15.63	1469
12.1	9.45	889	16.1	12.58	1182			
12.2	9.53	896	16.2	12.66	1190			
12.3	9.61	903	16.3	12.73	1197			
12.4	9.69	911	16.4	12.81	1204			
12.5	9.77	918	16.5	12.89	1212			
12.6	9.84	925	16.6	12.97	1219			
12.7	9.92	933	16.7	13.05	1226			
12.8	10.00	940	16.8	13.13	1234			
12.9	10.08	947	16.9	13.20	1241			
13.0	10.16	955	17.0	13.28	1248			

APPENDIX B - 3

94 lb BAG and BULK

QUANTITY OF CEMENT IN GROUT

0.6 : 1
(by loose volume)

See Section 4.1–11 for explanations and methods of use.

Grout Volume	Cement Content		Grout Volume	Cement Content		Grout Volume	Cement Content	
(cu ft)	(bags)	(lb)	(cu fl)	(bags)	(lb)	(cu ft)	(bags)	(lb)
0.1	0.09	9	3.1	2.87	270	6.1	5.65	531
0.2	0.19	17	3.2	2.96	279	6.2	5.74	540
0.3	0.28	26	3.3	3.06	287	6.3	5.83	548
0.4	0.37	35	3.4	3.15	296	6.4	5.93	557
0.5	0.46	44	3.5	3.24	305	6.5	6.02	566
0.6	0.56	52	3.6	3.33	313	6.6	6.11	574
0.7	0.65	61	3.7	3.43	322	6.7	6.20	583
0.8	0.74	70	3.8	3.52	331	6.8	6.30	592
0.9	0.83	78	3.9	3.61	339	6.9	6.39	601
1.0	0.93	87	4.0	3.70	348	7.0	6.48	609
1.1	1.02	96	4.1	3.80	357	7.1	6.57	618
1.2	1.11	104	4.2	3.89	366	7.2	6.67	627
1.3	1.20	113	4.3	3.98	374	7.3	6.76	635
1.4	1.30	122	4.4	4.07	383	7.4	6.85	644
1.5	1.39	131	4.5	4.17	392	7.5	6.94	653
1.6	1.48	139	4.6	4.26	400	7.6	7.04	661
1.7	1.57	148	4.7	4.35	409	7.7	7.13	670
1.8	1.67	157	4.8	4.44	418	7.8	7.22	679
1.9	1.76	165	4.9	4.54	426	7.9	7.31	688
2.0	1.85	174	5.0	4.63	435	8.0	7.41	696
2.1	1.94	183	5.1	4.72	444	8.1	7.50	705
2.2	2.04	191	5.2	4.81	453	8.2	7.59	714
2.3	2.13	200	5.3	4.91	461	8.3	7.69	722
2.4	2.22	209	5.4	5.00	470	8.4	7.78	731
2.5	2.31	218	5.5	5.09	479	8.5	7.87	740
2.6	2.41	226	5.6	5.19	487	8.6	7.96	749
2.7	2.50	235	5.7	5.28	496	8.7	8.06	757
2.8	2.59	244	5.8	5.37	505	8.8	8.15	766
2.9	2.69	252	5.9	5.46	514	8.9	8.24	775
3.0	2.78	261	6.0	5.56	522	9.0	8.33	783

APPENDIX B - 3
Continued

94 lb BAG and BULK

QUANTITY OF CEMENT IN GROUT

0.6 : 1

Grout Volume (cu ft)	Cement Content (bags)	Cement Content (lb)	Grout Volume (cu ft)	Cement Content (bags)	Cement Content (lb)	Grout Volume (cu ft)	Cement Content (bags)	Cement Content (lb)
9.1	8.43	792	13.1	12.13	1140	17.1	15.83	1488
9.2	8.52	801	13.2	12.22	1149	17.2	15.93	1497
9.3	8.61	809	13.3	12.31	1158	17.3	16.02	1506
9.4	8.70	818	13.4	12.41	1166	17.4	16.11	1514
9.5	8.80	827	13.5	12.50	1175	17.5	16.20	1523
9.6	8.89	836	13.6	12.59	1184	17.6	16.30	1532
9.7	8.98	844	13.7	12.69	1192	17.7	16.39	1541
9.8	9.07	853	13.8	12.78	1201	17.8	16.48	1549
9.9	9.17	862	13.9	12.87	1210	17.9	16.57	1558
10.0	9.26	870	14.0	12.96	1219	18.0	16.67	1567
10.1	9.35	879	14.1	13.06	1227	18.1	16.76	1575
10.2	9.44	888	14.2	13.15	1236	18.2	16.85	1584
10.3	9.54	896	14.3	13.24	1245	18.3	16.94	1593
10.4	9.63	905	14.4	13.33	1253	18.4	17.04	1601
10.5	9.72	914	14.5	13.43	1262	18.5	17.13	1610
10.6	9.81	923	14.6	13.52	1271	18.6	17.22	1619
10.7	9.91	931	14.7	13.61	1279	18.7	17.31	1628
10.8	10.00	940	14.8	13.70	1288	18.8	17.41	1636
10.9	10.09	949	14.9	13.80	1297	18.9	17.50	1645
11.0	10.19	957	15.0	13.89	1306	19.0	17.59	1654
11.1	10.28	966	15.1	13.98	1314	19.1	17.69	1662
11.2	10.37	975	15.2	14.07	1323	19.2	17.78	1671
11.3	10.46	984	15.3	14.17	1332	19.3	17.87	1680
11.4	10.56	992	15.4	14.26	1340	19.4	17.96	1689
11.5	10.65	1001	15.5	14.35	1349	19.5	18.06	1697
11.6	10.74	1010	15.6	14.44	1358	19.6	18.15	1706
11.7	10.83	1018	15.7	14.54	1366	19.7	18.24	1715
11.8	10.93	1027	15.8	14.63	1375	19.8	18.33	1723
11.9	11.02	1036	15.9	14.72	1384	19.9	18.43	1732
12.0	11.11	1044	16.0	14.81	1393	20.0	18.52	1741
12.1	11.20	1053	16.1	14.91	1401			
12.2	11.30	1062	16.2	15.00	1410			
12.3	11.39	1071	16.3	15.09	1419			
12.4	11.48	1079	16.4	15.19	1427			
12.5	11.57	1088	16.5	15.28	1436			
12.6	11.67	1097	16.6	15.37	1445			
12.7	11.76	1105	16.7	15.46	1454			
12.8	11.85	1114	16.8	15.56	1462			
12.9	11.94	1123	16.9	15.65	1471			
13.0	12.04	1131	17.0	15.74	1480			

APPENDIX B - 3

94 lb BAG and BULK

QUANTITY OF CEMENT IN GROUT

0.5 : 1
(by loose volume)

See Section 4.1–11 for explanations and methods of use.

Grout Volume (cu ft)	Cement Content (bags)	Cement Content (lb)	Grout Volume (cu ft)	Cement Content (bags)	Cement Content (lb)	Grout Volume (cu ft)	Cement Content (bags)	Cement Content (lb)
0.1	0.10	10	3.1	3.16	297	6.1	6.22	585
0.2	0.20	19	3.2	3.27	307	6.2	6.33	595
0.3	0.31	29	3.3	3.37	317	6.3	6.43	604
0.4	0.41	38	3.4	3.47	326	6.4	6.53	614
0.5	0.51	48	3.5	3.57	336	6.5	6.63	623
0.6	0.61	58	3.6	3.67	345	6.6	6.73	633
0.7	0.71	67	3.7	3.78	355	6.7	6.84	643
0.8	0.82	77	3.8	3.88	364	6.8	6.94	652
0.9	0.92	86	3.9	3.98	374	6.9	7.04	662
1.0	1.02	96	4.0	4.08	384	7.0	7.14	671
1.1	1.12	106	4.1	4.18	393	7.1	7.24	681
1.2	1.22	115	4.2	4.29	403	7.2	7.35	691
1.3	1.33	125	4.3	4.39	412	7.3	7.45	700
1.4	1.43	134	4.4	4.49	422	7.4	7.55	710
1.5	1.53	144	4.5	4.59	432	7.5	7.65	719
1.6	1.63	153	4.6	4.69	441	7.6	7.76	729
1.7	1.73	163	4.7	4.80	451	7.7	7.86	739
1.8	1.84	173	4.8	4.90	460	7.8	7.96	748
1.9	1.94	182	4.9	5.00	470	7.9	8.06	758
2.0	2.04	192	5.0	5.10	480	8.0	8.16	767
2.1	2.14	201	5.1	5.20	489	8.1	8.27	777
2.2	2.24	211	5.2	5.31	499	8.2	8.37	787
2.3	2.35	221	5.3	5.41	508	8.3	8.47	796
2.4	2.45	230	5.4	5.51	518	8.4	8.57	806
2.5	2.55	240	5.5	5.61	528	8.5	8.67	815
2.6	2.65	249	5.6	5.71	537	8.6	8.78	825
2.7	2.76	259	5.7	5.82	547	8.7	8.88	834
2.8	2.86	269	5.8	5.92	556	8.8	8.98	844
2.9	2.96	278	5.9	6.02	566	8.9	9.08	854
3.0	3.06	288	6.0	6.12	576	9.0	9.18	863

APPENDIX B - 3
Continued

94 lb BAG and BULK

QUANTITY OF CEMENT IN GROUT

0.5 : 1

Grout Volume	Cement Content		Grout Volume	Cement Content		Grout Volume	Cement Content	
(cu ft)	(bags)	(lb)	(cu ft)	(bags)	(lb)	(cu ft)	(bags)	(lb)
9.1	9.29	873	13.1	13.37	1257	17.1	17.45	1640
9.2	9.39	882	13.2	13.47	1266	17.2	17.55	1650
9.3	9.49	892	13.3	13.57	1276	17.3	17.65	1659
9.4	9.59	902	13.4	13.67	1285	17.4	17.76	1669
9.5	9.69	911	13.5	13.78	1295	17.5	17.86	1679
9.6	9.80	921	13.6	13.88	1304	17.6	17.96	1688
9.7	9.90	930	13.7	13.98	1314	17.7	18.06	1698
9.8	10.00	940	13.8	14.08	1324	17.8	18.16	1707
9.9	10.10	950	13.9	14.18	1333	17.9	18.27	1717
10.0	10.20	959	14.0	14.29	1343	18.0	18.37	1727
10.1	10.31	969	14.1	14.39	1352	18.1	18.47	1736
10.2	10.41	978	14.2	14.49	1362	18.2	18.57	1746
10.3	10.51	988	14.3	14.59	1372	18.3	18.67	1755
10.4	10.61	998	14.4	14.69	1381	18.4	18.78	1765
10.5	10.71	1007	14.5	14.80	1391	18.5	18.88	1774
10.6	10.82	1017	14.6	14.90	1400	18.6	18.98	1784
10.7	10.92	1026	14.7	15.00	1410	18.7	19.08	1794
10.8	11.02	1036	14.8	15.10	1420	18.8	19.18	1803
10.9	11.12	1046	14.9	15.20	1429	18.9	19.29	1813
11.0	11.22	1055	15.0	15.31	1439	19.0	19.39	1822
11.1	11.33	1065	15.1	15.41	1448	19.1	19.49	1832
11.2	11.43	1074	15.2	15.51	1458	19.2	19.59	1842
11.3	11.53	1084	15.3	15.61	1468	19.3	19.69	1851
11.4	11.63	1093	15.4	15.71	1477	19.4	19.80	1861
11.5	11.73	1103	15.5	15.82	1487	19.5	19.90	1870
11.6	11.84	1113	15.6	15.92	1496	19.6	20.00	1880
11.7	11.94	1122	15.7	16.02	1506	19.7	20.10	1890
11.8	12.04	1132	15.8	16.12	1516	19.8	20.20	1899
11.9	12.14	1141	15.9	16.22	1525	19.9	20.31	1909
12.0	12.24	1151	16.0	16.33	1535	20.0	20.41	1918
12.1	12.35	1161	16.1	16.43	1544			
12.2	12.45	1170	16.2	16.53	1554			
12.3	12.55	1180	16.3	16.63	1563			
12.4	12.65	1189	16.4	16.73	1573			
12.5	12.76	1199	16.5	16.84	1583			
12.6	12.86	1209	16.6	16.94	1592			
12.7	12.96	1218	16.7	17.04	1602			
12.8	13.06	1228	16.8	17.14	1611			
12.9	13.16	1237	16.9	17.24	1621			
13.0	13.27	1247	17.0	17.35	1631			

Batching Tables for Grout: Metric Units

See Appendix B for traditional units

The tables in this Appendix are calculated on the basis of

a 40 kg bag = 27 liters (loose), or
a 50 kg bag = 35 liters (loose).

APPENDIX C - 1

40 kg BAG

THICKENING TABLES

From 5:1 to 4:1
(by loose volume)

See Section 4.1–8 for explanations and methods of use.

Grout in Line and Agitator (liters)	MAKEUP MIX			Grout in Line and Agitator (liters)	MAKEUP MIX		
	Water (liters)	Mix Vol (liters)	Mix w:c		Water (liters)	Mix Vol (liters)	Mix w:c
1 BAG BATCH							
0	108	121	4.0:1	155	80	92	3.0:1
5	107	120	4.0:1	160	79	91	2.9:1
10	106	119	3.9:1	165	78	91	2.9:1
15	105	118	3.9:1	170	77	90	2.8:1
20	104	117	3.9:1	175	76	89	2.8:1
25	103	116	3.8:1	180	75	88	2.8:1
30	103	115	3.8:1	185	74	87	2.7:1
35	102	114	3.8:1	190	73	86	2.7:1
40	101	113	3.7:1	195	72	85	2.7:1
45	100	112	3.7:1	200	71	84	2.6:1
50	99	112	3.7:1	205	71	83	2.6:1
55	98	111	3.6:1	210	70	82	2.6:1
60	97	110	3.6:1	215	69	81	2.5:1
65	96	109	3.6:1	220	68	80	2.5:1
70	95	108	3.5:1	225	67	80	2.5:1
75	94	107	3.5:1	230	66	79	2.4:1
80	93	106	3.5:1	235	65	78	2.4:1
85	92	105	3.4:1	240	64	77	2.4:1
90	92	104	3.4:1	245	63	76	2.3:1
95	91	103	3.4:1	250	62	75	2.3:1
100	90	102	3.3:1	255	61	74	2.3:1
105	89	102	3.3:1	260	60	73	2.2:1
110	88	101	3.3:1	265	60	72	2.2:1
115	87	100	3.2:1	270	59	71	2.2:1
120	86	99	3.2:1	275	58	70	2.1:1
125	85	98	3.2:1	280	57	70	2.1:1
130	84	97	3.1:1	285	56	69	2.1:1
135	83	96	3.1:1	290	55	68	2.0:1
140	82	95	3.1:1	295	54	67	2.0:1
145	81	94	3.0:1	300	53	66	2.0:1
150	81	93	3.0:1				

APPENDIX C - 1

Continued

40 kg BAG

THICKENING TABLES

From 5:1 to 4:1

Grout in Line and Agitator (liters)	MAKEUP MIX			Grout in Line and Agitator (liters)	MAKEUP MIX		
	Water (liters)	Mix Vol (liters)	Mix w:c		Water (liters)	Mix Vol (liters)	Mix w:c
			1 BAG BATCH				
305	52	65	1.9:1	355	43	56	1.6:1
310	51	64	1.9:1	360	42	55	1.6:1
315	50	63	1.9:1	365	41	54	1.5:1
320	50	62	1.8:1	370	40	53	1.5:1
325	49	61	1.8:1	375	39	52	1.5:1
330	48	60	1.8:1	380	39	51	1.4:1
335	47	59	1.7:1	385	38	50	1.4:1
340	46	59	1.7:1	390	37	49	1.4:1
345	45	58	1.7:1	395	36	48	1.3:1
350	44	57	1.6:1	400	35	48	1.3:1
			2 BAG BATCH				
0	216	241	4.0:1	150	189	214	3.5:1
5	215	240	4.0:1	155	188	213	3.5:1
10	214	240	4.0:1	160	187	212	3.5:1
15	213	239	3.9:1	165	186	211	3.4:1
20	212	238	3.9:1	170	185	210	3.4:1
25	211	237	3.9:1	175	184	209	3.4:1
30	211	236	3.9:1	180	183	208	3.4:1
35	210	235	3.9:1	185	182	208	3.4:1
40	209	234	3.9:1	190	181	207	3.4:1
45	208	233	3.8:1	195	180	206	3.3:1
50	207	232	3.8:1	200	179	205	3.3:1
55	206	231	3.8:1	205	179	204	3.3:1
60	205	230	3.8:1	210	178	203	3.3:1
65	204	230	3.8:1	215	177	202	3.3:1
70	203	229	3.8:1	220	176	201	3.3:1
75	202	228	3.7:1	225	175	200	3.2:1
80	201	227	3.7:1	230	174	199	3.2:1
85	200	226	3.7:1	235	173	198	3.2:1
90	200	225	3.7:1	240	172	198	3.2:1
95	199	224	3.7:1	245	171	197	3.2:1
100	198	223	3.7:1	250	170	196	3.2:1
105	197	222	3.6:1	255	169	195	3.1:1
110	196	221	3.6:1	260	168	194	3.1:1
115	195	220	3.6:1	265	168	193	3.1:1
120	194	219	3.6:1	270	167	192	3.1:1
125	193	219	3.6:1	275	166	191	3.1:1
130	192	218	3.6:1	280	165	190	3.1:1
135	191	217	3.5:1	285	164	189	3.0:1
140	190	216	3.5:1	290	163	188	3.0:1
145	189	215	3.5:1	295	162	187	3.0:1

APPENDIX C - 1

40 kg BAG

THICKENING TABLES

From 4:1 to 3:1
(by loose volume)

See Section 4.1–8 for explanations and methods of use.

Grout in Line and Agitator (liters)	MAKEUP MIX		
	Water (liters)	Mix Vol (liters)	Mix w:c
1 BAG BATCH			
0	81	94	3.0:1
5	80	93	3.0:1
10	79	91	2.9:1
15	78	90	2.9:1
20	77	89	2.8:1
25	75	88	2.8:1
30	74	87	2.8:1
35	73	86	2.7:1
40	72	85	2.7:1
45	71	84	2.6:1
50	70	83	2.6:1
55	69	81	2.5:1
60	68	80	2.5:1
65	66	79	2.5:1
70	65	78	2.4:1
75	64	77	2.4:1
80	63	76	2.3:1
85	62	75	2.3:1
90	61	74	2.3:1
95	60	72	2.2:1
100	59	71	2.2:1
105	58	70	2.1:1
110	56	69	2.1:1
115	55	68	2.0:1
120	54	67	2.0:1
125	53	66	2.0:1
130	52	65	1.9:1
135	51	63	1.9:1
140	50	62	1.8:1
145	49	61	1.8:1
150	47	60	1.8:1
155	46	59	1.7:1
160	45	58	1.7:1
165	44	57	1.6:1
170	43	56	1.6:1
175	42	55	1.6:1
180	41	53	1.5:1
185	40	52	1.5:1
190	38	51	1.4:1
195	37	50	1.4:1
200	36	49	1.3:1
205	35	48	1.3:1
210	34	47	1.3:1
215	33	46	1.2:1
220	32	44	1.2:1
225	31	43	1.1:1
230	30	42	1.1:1
235	28	41	1.1:1
240	27	40	1.0:1
245	26	39	1.0:1
250	25	38	0.9:1
255	24	37	0.9:1
260	23	36	0.8:1
265	22	34	0.8:1
270	21	33	0.8:1
275	19	32	0.7:1
280	18	31	0.7:1
285	17	30	0.6:1
290	16	29	0.6:1
295	15	28	0.6:1
300	14	27	0.5:1
305	13	25	0.5:1

APPENDIX C - 1
Continued

40 kg BAG

THICKENING TABLES
From 4:1 to 3:1

Grout in Line and Agitator (liters)	MAKEUP MIX			Grout in Line and Agitator (liters)	MAKEUP MIX		
	Water (liters)	Mix Vol (liters)	Mix w:c		Water (liters)	Mix Vol (liters)	Mix w:c
2 BAG BATCH							
0	162	187	3.0:1	205	116	142	2.2:1
5	161	186	3.0:1	210	115	140	2.1:1
10	160	185	3.0:1	215	114	139	2.1:1
15	159	184	2.9:1	220	113	138	2.1:1
20	158	183	2.9:1	225	112	137	2.1:1
25	156	182	2.9:1	230	111	136	2.0:1
30	155	181	2.9:1	235	109	135	2.0:1
35	154	180	2.9:1	240	108	134	2.0:1
40	153	178	2.8:1	245	107	133	2.0:1
45	152	177	2.8:1	250	106	131	2.0:1
50	151	176	2.8:1	255	105	130	1.9:1
55	150	175	2.8:1	260	104	129	1.9:1
60	149	174	2.8:1	265	103	128	1.9:1
65	147	173	2.7:1	270	102	127	1.9:1
70	146	172	2.7:1	275	100	126	1.9:1
75	145	171	2.7:1	280	99	125	1.8:1
80	144	170	2.7:1	285	98	124	1.8:1
85	143	168	2.6:1	290	97	123	1.8:1
90	142	167	2.6:1	295	96	121	1.8:1
95	141	166	2.6:1	300	95	120	1.8:1
100	140	165	2.6:1	305	94	119	1.7:1
105	139	164	2.6:1	310	93	118	1.7:1
110	137	163	2.5:1	315	92	117	1.7:1
115	136	162	2.5:1	320	90	116	1.7:1
120	135	161	2.5:1	325	89	115	1.7:1
125	134	159	2.5:1	330	88	114	1.6:1
130	133	158	2.5:1	335	87	112	1.6:1
135	132	157	2.4:1	340	86	111	1.6:1
140	131	156	2.4:1	345	85	110	1.6:1
145	130	155	2.4:1	350	84	109	1.6:1
150	128	154	2.4:1	355	83	108	1.5:1
155	127	153	2.4:1	360	81	107	1.5:1
160	126	152	2.3:1	365	80	106	1.5:1
165	125	150	2.3:1	370	79	105	1.5:1
170	124	149	2.3:1	375	78	104	1.4:1
175	123	148	2.3:1	380	77	102	1.4:1
180	122	147	2.3:1	385	76	101	1.4:1
185	121	146	2.2:1	390	75	100	1.4:1
190	119	145	2.2:1	395	74	99	1.4:1
195	118	144	2.2:1	400	73	98	1.3:1
200	117	143	2.2:1				

APPENDIX C - 1

40 kg BAG

THICKENING TABLES

From 3:1 to 2:1
(by loose volume)

See Section 4.1–8 for explanations and methods of use.

Grout in Line and Agitator (liters)	MAKEUP MIX		
	Water (liters)	Mix Vol (liters)	Mix w:c
	1 BAG BATCH		
0	54	67	2.0:1
5	53	65	1.9:1
10	51	64	1.9:1
15	50	62	1.8:1
20	48	61	1.8:1
25	47	59	1.7:1
30	45	58	1.7:1
35	44	57	1.6:1
40	42	55	1.6:1
45	41	54	1.5:1
50	40	52	1.5:1
55	38	51	1.4:1
60	37	49	1.4:1
65	35	48	1.3:1
70	34	47	1.3:1
75	32	45	1.2:1
80	31	44	1.1:1
85	30	42	1.1:1
90	28	41	1.0:1
95	27	39	1.0:1
100	25	38	0.9:1
105	24	36	0.9:1
110	22	35	0.8:1
115	21	34	0.8:1
120	19	32	0.7:1
125	18	31	0.7:1
130	17	29	0.6:1
135	15	28	0.6:1
140	14	26	0.5:1
145	12	25	0.5:1

Grout in Line and Agitator (liters)	MAKEUP MIX		
	Water (liters)	Mix Vol (liters)	Mix w:c
	2 BAG BATCH		
0	108	133	2.0:1
5	107	132	2.0:1
10	105	131	1.9:1
15	104	129	1.9:1
20	102	128	1.9:1
25	101	126	1.9:1
30	99	125	1.8:1
35	98	123	1.8:1
40	96	122	1.8:1
45	95	120	1.8:1
50	94	119	1.7:1
55	92	118	1.7:1
60	91	116	1.7:1
65	89	115	1.7:1
70	88	113	1.6:1
75	86	112	1.6:1
80	85	110	1.6:1
85	84	109	1.5:1
90	82	107	1.5:1
95	81	106	1.5:1
100	79	105	1.5:1
105	78	103	1.4:1
110	76	102	1.4:1
115	75	100	1.4:1
120	73	99	1.4:1
125	72	97	1.3:1
130	71	96	1.3:1
135	69	94	1.3:1
140	68	93	1.3:1
145	66	92	1.2:1
150	65	90	1.2:1

APPENDIX C - 1
Continued

40 kg BAG

THICKENING TABLES
From 3:1 to 2:1

Grout in Line and Agitator (liters)	MAKEUP MIX Water (liters)	MAKEUP MIX Mix Vol (liters)	MAKEUP MIX Mix w:c	Grout in Line and Agitator (liters)	MAKEUP MIX Water (liters)	MAKEUP MIX Mix Vol (liters)	MAKEUP MIX Mix w:c
2 BAG BATCH							
155	63	89	1.2:1	225	43	69	0.8:1
160	62	87	1.1:1	230	42	67	0.8:1
165	60	86	1.1:1	235	40	66	0.7:1
170	59	84	1.1:1	240	39	64	0.7:1
175	58	83	1.1:1	245	37	63	0.7:1
180	56	82	1.0:1	250	36	61	0.7:1
185	55	80	1.0:1	255	35	60	0.6:1
190	53	79	1.0:1	260	33	58	0.6:1
195	52	77	1.0:1	265	32	57	0.6:1
200	50	76	0.9:1	270	30	56	0.6:1
205	49	74	0.9:1	275	29	54	0.5:1
210	47	73	0.9:1	280	27	53	0.5:1
215	46	71	0.9:1	285	26	51	0.5:1
220	45	70	0.8:1	290	24	50	0.5:1
3 BAG BATCH							
0	162	200	2.0:1	130	125	163	1.5:1
5	161	199	2.0:1	135	123	161	1.5:1
10	159	197	2.0:1	140	122	160	1.5:1
15	158	196	1.9:1	145	120	158	1.5:1
20	156	194	1.9:1	150	119	157	1.5:1
25	155	193	1.9:1	155	117	155	1.4:1
30	153	191	1.9:1	160	116	154	1.4:1
35	152	190	1.9:1	165	114	153	1.4:1
40	150	189	1.9:1	170	113	151	1.4:1
45	149	187	1.8:1	175	112	150	1.4:1
50	148	186	1.8:1	180	110	148	1.4:1
55	146	184	1.8:1	185	109	147	1.3:1
60	145	183	1.8:1	190	107	145	1.3:1
65	143	181	1.8:1	195	106	144	1.3:1
70	142	180	1.8:1	200	104	142	1.3:1
75	140	178	1.7:1	205	103	141	1.3:1
80	139	177	1.7:1	210	101	140	1.3:1
85	138	176	1.7:1	215	100	138	1.2:1
90	136	174	1.7:1	220	99	137	1.2:1
95	135	173	1.7:1	225	97	135	1.2:1
100	133	171	1.6:1	230	96	134	1.2:1
105	132	170	1.6:1	235	94	132	1.2:1
110	130	168	1.6:1	240	93	131	1.1:1
115	129	167	1.6:1	245	91	129	1.1:1
120	127	166	1.6:1	250	90	128	1.1:1
125	126	164	1.6:1	255	89	127	1.1:1

APPENDIX C - 1

Continued

40 kg BAG

THICKENING TABLES
From 3:1 to 2:1

Grout in Line and Agitator (liters)	MAKEUP MIX Water (liters)	MAKEUP MIX Mix Vol (liters)	MAKEUP MIX Mix w:c	Grout in Line and Agitator (liters)	MAKEUP MIX Water (liters)	MAKEUP MIX Mix Vol (liters)	MAKEUP MIX Mix w:c
3 BAG BATCH							
260	87	125	1.1:1	335	65	104	0.8:1
265	86	124	1.1:1	340	64	102	0.8:1
270	84	122	1.0:1	345	63	101	0.8:1
275	83	121	1.0:1	350	61	99	0.8:1
280	81	119	1.0:1	355	60	98	0.7:1
285	80	118	1.0:1	360	58	96	0.7:1
290	78	117	1.0:1	365	57	95	0.7:1
295	77	115	1.0:1	370	55	93	0.7:1
300	76	114	0.9:1	375	54	92	0.7:1
305	74	112	0.9:1	380	52	91	0.6:1
310	73	111	0.9:1	385	51	89	0.6:1
315	71	109	0.9:1	390	50	88	0.6:1
320	70	108	0.9:1	395	48	86	0.6:1
325	68	106	0.8:1	400	47	85	0.6:1
330	67	105	0.8:1				

APPENDIX C - 1

40 kg BAG

THICKENING TABLES

From 2:1 to 1:1
(by loose volume)

See Section 4.1–8 for explanations and methods of use.

Grout in Line and Agitator (liters)	MAKEUP MIX			Grout in Line and Agitator (liters)	MAKEUP MIX		
	Water (liters)	Mix Vol (liters)	Mix w:c		Water (liters)	Mix Vol (liters)	Mix w:c
1 BAG BATCH							
0	27	40	1.0:1	20	19	32	0.7:1
5	25	38	0.9:1	25	17	30	0.6:1
10	23	36	0.9:1	30	15	28	0.6:1
15	21	34	0.8:1	35	13	26	0.5:1
2 BAG BATCH							
0	54	79	1.0:1	40	38	63	0.7:1
5	52	77	1.0:1	45	36	61	0.7:1
10	50	75	0.9:1	50	34	59	0.6:1
15	48	73	0.9:1	55	32	57	0.6:1
20	46	71	0.9:1	60	30	55	0.6:1
25	44	69	0.8:1	65	28	53	0.5:1
30	42	67	0.8:1	70	26	51	0.5:1
35	40	65	0.7:1				
3 BAG BATCH							
0	81	119	1.0:1	55	59	97	0.7:1
5	79	117	1.0:1	60	57	95	0.7:1
10	77	115	1.0:1	65	55	93	0.7:1
15	75	113	0.9:1	70	53	91	0.7:1
20	73	111	0.9:1	75	51	89	0.6:1
25	71	109	0.9:1	80	49	87	0.6:1
30	69	107	0.9:1	85	47	85	0.6:1
35	67	105	0.8:1	90	45	83	0.6:1
40	65	103	0.8:1	95	43	81	0.5:1
45	63	101	0.8:1	100	41	79	0.5:1
50	61	99	0.8:1	105	38	77	0.5:1
				110	36	75	0.5:1
4 BAG BATCH							
0	108	159	1.0:1	20	100	151	0.9:1
5	106	157	1.0:1	25	98	149	0.9:1
10	104	155	1.0:1	30	96	147	0.9:1
15	102	153	0.9:1	35	94	145	0.9:1

APPENDIX C - 1

Continued

40 kg BAG

THICKENING TABLES

From 2:1 to 1:1

Grout in Line and Agitator (liters)	MAKEUP MIX			Grout in Line and Agitator (liters)	MAKEUP MIX		
	Water (liters)	Mix Vol (liters)	Mix w:c		Water (liters)	Mix Vol (liters)	Mix w:c
			4 BAG BATCH				
40	92	143	0.9:1	95	70	120	0.6:1
45	90	141	0.8:1	100	68	118	0.6:1
50	88	139	0.8:1	105	65	116	0.6:1
55	86	137	0.8:1	110	63	114	0.6:1
60	84	135	0.8:1	115	61	112	0.6:1
65	82	132	0.8:1	120	59	110	0.6:1
70	80	130	0.7:1	125	57	108	0.5:1
75	78	128	0.7:1	130	55	106	0.5:1
80	76	126	0.7:1	135	53	104	0.5:1
85	74	124	0.7:1	140	51	102	0.5:1
90	72	122	0.7:1	145	49	100	0.5:1
			5 BAG BATCH				
0	135	198	1.0:1	95	97	160	0.7:1
5	133	196	1.0:1	100	95	158	0.7:1
10	131	194	1.0:1	105	92	156	0.7:1
15	129	192	1.0:1	110	90	154	0.7:1
20	127	190	0.9:1	115	88	152	0.7:1
25	125	188	0.9:1	120	86	150	0.6:1
30	123	186	0.9:1	125	84	148	0.6:1
35	121	184	0.9:1	130	82	146	0.6:1
40	119	182	0.9:1	135	80	144	0.6:1
45	117	180	0.9:1	140	78	142	0.6:1
50	115	178	0.9:1	145	76	140	0.6:1
55	113	176	0.8:1	150	74	138	0.6:1
60	111	174	0.8:1	155	72	136	0.5:1
65	109	172	0.8:1	160	70	134	0.5:1
70	107	170	0.8:1	165	68	132	0.5:1
75	105	168	0.8:1	170	66	130	0.5:1
80	103	166	0.8:1	175	64	128	0.5:1
85	101	164	0.7:1	180	62	126	0.5:1
90	99	162	0.7:1				

APPENDIX C - 1 **40 kg BAG**

THICKENING TABLES

From 1:1 to 0.8:1
(by loose volume)

See Section 4.1–8 for explanations and methods of use.

Grout in Line and Agitator (liters)	MAKEUP MIX			Grout in Line and Agitator (liters)	MAKEUP MIX		
	Water (liters)	Mix Vol (liters)	Mix w:c		Water (liters)	Mix Vol (liters)	Mix w:c
			1 BAG BATCH				
0	22	34	0.8:1	35	17	30	0.6:1
5	21	34	0.8:1	40	16	29	0.6:1
10	20	33	0.7:1	45	15	28	0.6:1
15	20	32	0.7:1	50	15	27	0.5:1
20	19	32	0.7:1	55	14	27	0.5:1
25	18	31	0.7:1	60	13	26	0.5:1
30	18	30	0.6:1	65	13	25	0.5:1
			2 BAG BATCH				
0	43	69	0.8:1	70	34	59	0.6:1
5	43	68	0.8:1	75	33	58	0.6:1
10	42	67	0.8:1	80	32	58	0.6:1
15	41	67	0.8:1	85	32	57	0.6:1
20	40	66	0.7:1	90	31	56	0.6:1
25	40	65	0.7:1	95	30	56	0.6:1
30	39	65	0.7:1	100	30	55	0.5:1
35	38	64	0.7:1	105	29	54	0.5:1
40	38	63	0.7:1	110	28	54	0.5:1
45	37	62	0.7:1	115	28	53	0.5:1
50	36	62	0.7:1	120	27	52	0.5:1
55	36	61	0.7:1	125	26	52	0.5:1
60	35	60	0.6:1	130	26	51	0.5:1
65	34	60	0.6:1	135	25	50	0.5:1
			3 BAG BATCH				
0	65	103	0.8:1	40	59	97	0.7:1
5	64	102	0.8:1	45	59	97	0.7:1
10	63	102	0.8:1	50	58	96	0.7:1
15	63	101	0.8:1	55	57	95	0.7:1
20	62	100	0.8:1	60	57	95	0.7:1
25	61	99	0.8:1	65	56	94	0.7:1
30	61	99	0.7:1	70	55	93	0.7:1
35	60	98	0.7:1	75	55	93	0.7:1

APPENDIX C - 1

Continued

40 kg BAG

THICKENING TABLES

From 1:1 to 0.8:1

Grout in Line and Agitator (liters)	MAKEUP MIX Water (liters)	MAKEUP MIX Mix Vol (liters)	MAKEUP MIX Mix w:c	Grout in Line and Agitator (liters)	MAKEUP MIX Water (liters)	MAKEUP MIX Mix Vol (liters)	MAKEUP MIX Mix w:c
3 BAG BATCH							
80	54	92	0.7:1	145	45	83	0.6:1
85	53	91	0.7:1	150	44	82	0.5:1
90	53	91	0.6:1	155	44	82	0.5:1
95	52	90	0.6:1	160	43	81	0.5:1
100	51	89	0.6:1	165	42	80	0.5:1
105	51	89	0.6:1	170	42	80	0.5:1
110	50	88	0.6:1	175	41	79	0.5:1
115	49	87	0.6:1	180	40	78	0.5:1
120	48	87	0.6:1	185	40	78	0.5:1
125	48	86	0.6:1	190	39	77	0.5:1
130	47	85	0.6:1	195	38	76	0.5:1
135	46	85	0.6:1	200	38	76	0.5:1
140	46	84	0.6:1	205	37	75	0.5:1
4 BAG BATCH							
0	86	137	0.8:1	135	68	119	0.6:1
5	86	137	0.8:1	140	67	118	0.6:1
10	85	136	0.8:1	145	67	117	0.6:1
15	84	135	0.8:1	150	66	117	0.6:1
20	84	134	0.8:1	155	65	116	0.6:1
25	83	134	0.8:1	160	65	115	0.6:1
30	82	133	0.8:1	165	64	115	0.6:1
35	82	132	0.8:1	170	63	114	0.6:1
40	81	132	0.7:1	175	63	113	0.6:1
45	80	131	0.7:1	180	62	113	0.6:1
50	80	130	0.7:1	185	61	112	0.6:1
55	79	130	0.7:1	190	61	111	0.6:1
60	78	129	0.7:1	195	60	111	0.6:1
65	78	128	0.7:1	200	59	110	0.5:1
70	77	128	0.7:1	205	59	109	0.5:1
75	76	127	0.7:1	210	58	109	0.5:1
80	76	126	0.7:1	215	57	108	0.5:1
85	75	126	0.7:1	220	56	107	0.5:1
90	74	125	0.7:1	225	56	107	0.5:1
95	73	124	0.7:1	230	55	106	0.5:1
100	73	124	0.7:1	235	54	105	0.5:1
105	72	123	0.7:1	240	54	105	0.5:1
110	71	122	0.7:1	245	53	104	0.5:1
115	71	122	0.7:1	250	52	103	0.5:1
120	70	121	0.6:1	255	52	103	0.5:1
125	69	120	0.6:1	260	51	102	0.5:1
130	69	120	0.6:1	265	50	101	0.5:1

APPENDIX C - 1
Continued

40 kg BAG

THICKENING TABLES
From 1:1 to 0.8:1

Grout in Line and Agitator (liters)	MAKEUP MIX			Grout in Line and Agitator (liters)	MAKEUP MIX		
	Water (liters)	Mix Vol (liters)	Mix w:c		Water (liters)	Mix Vol (liters)	Mix w:c
			4 BAG BATCH				
270	50	100	0.5:1	275	49	100	0.5:1
			5 BAG BATCH				
0	108	171	0.8:1	155	87	150	0.6:1
5	107	171	0.8:1	160	86	150	0.6:1
10	107	170	0.8:1	165	86	149	0.6:1
15	106	169	0.8:1	170	85	148	0.6:1
20	105	169	0.8:1	175	84	148	0.6:1
25	105	168	0.8:1	180	84	147	0.6:1
30	104	167	0.8:1	185	83	146	0.6:1
35	103	167	0.8:1	190	82	146	0.6:1
40	103	166	0.8:1	195	81	145	0.6:1
45	102	165	0.8:1	200	81	144	0.6:1
50	101	165	0.7:1	205	80	144	0.6:1
55	101	164	0.7:1	210	79	143	0.6:1
60	100	163	0.7:1	215	79	142	0.6:1
65	99	163	0.7:1	220	78	142	0.6:1
70	98	162	0.7:1	225	77	141	0.6:1
75	98	161	0.7:1	230	77	140	0.6:1
80	97	161	0.7:1	235	76	140	0.6:1
85	96	160	0.7:1	240	75	139	0.6:1
90	96	159	0.7:1	245	75	138	0.6:1
95	95	159	0.7:1	250	74	137	0.5:1
100	94	158	0.7:1	255	73	137	0.5:1
105	94	157	0.7:1	260	73	136	0.5:1
110	93	157	0.7:1	265	72	135	0.5:1
115	92	156	0.7:1	270	71	135	0.5:1
120	92	155	0.7:1	275	71	134	0.5:1
125	91	154	0.7:1	280	70	133	0.5:1
130	90	154	0.7:1	285	69	133	0.5:1
135	90	153	0.7:1	290	69	132	0.5:1
140	89	152	0.7:1	295	68	131	0.5:1
145	88	152	0.7:1	300	67	131	0.5:1
150	88	151	0.6:1				

APPENDIX C - 1 40 kg BAG

THICKENING TABLES

From 0.8:1 to 0.6:1
(by loose volume)

See Section 4.1–8 for explanations and methods of use.

Grout in Line and Agitator (liters)	MAKEUP MIX			Grout in Line and Agitator (liters)	MAKEUP MIX		
	Water (liters)	Mix Vol (liters)	Mix w:c		Water (liters)	Mix Vol (liters)	Mix w:c
			1 BAG BATCH				
0	16	29	0.6:1	15	14	27	0.5:1
5	15	28	0.6:1	20	13	26	0.5:1
10	15	27	0.5:1	25	12	25	0.5:1
			2 BAG BATCH				
0	32	58	0.6:1	30	28	53	0.5:1
5	32	57	0.6:1	35	27	52	0.5:1
10	31	56	0.6:1	40	26	51	0.5:1
15	30	55	0.6:1	45	25	51	0.5:1
20	29	55	0.5:1	50	25	50	0.5:1
25	28	54	0.5:1				
			3 BAG BATCH				
0	49	87	0.6:1	40	42	80	0.5:1
5	48	86	0.6:1	45	42	80	0.5:1
10	47	85	0.6:1	50	41	79	0.5:1
15	46	84	0.6:1	55	40	78	0.5:1
20	45	84	0.6:1	60	39	77	0.5:1
25	45	83	0.6:1	65	38	76	0.5:1
30	44	82	0.5:1	70	38	76	0.5:1
35	43	81	0.5:1	75	37	75	0.5:1
			4 BAG BATCH				
0	65	116	0.6:1	55	56	107	0.5:1
5	64	115	0.6:1	60	55	106	0.5:1
10	63	114	0.6:1	65	55	105	0.5:1
15	62	113	0.6:1	70	54	105	0.5:1
20	62	112	0.6:1	75	53	104	0.5:1
25	61	112	0.6:1	80	52	103	0.5:1
30	60	111	0.6:1	85	51	102	0.5:1
35	59	110	0.5:1	90	51	101	0.5:1
40	59	109	0.5:1	95	50	101	0.5:1
45	58	109	0.5:1	100	49	100	0.5:1
50	57	108	0.5:1				

APPENDIX C - 1
Continued

40 kg BAG

THICKENING TABLES

From 0.8:1 to 0.6:1

Grout in Line and Agitator (liters)	MAKEUP MIX			Grout in Line and Agitator (liters)	MAKEUP MIX		
	Water (liters)	Mix Vol (liters)	Mix w:c		Water (liters)	Mix Vol (liters)	Mix w:c
5 BAG BATCH							
0	81	144	0.6:1	65	71	134	0.5:1
5	80	144	0.6:1	70	70	133	0.5:1
10	79	143	0.6:1	75	69	133	0.5:1
15	79	142	0.6:1	80	68	132	0.5:1
20	78	141	0.6:1	85	68	131	0.5:1
25	77	141	0.6:1	90	67	130	0.5:1
30	76	140	0.6:1	95	66	130	0.5:1
35	75	139	0.6:1	100	65	129	0.5:1
40	75	138	0.6:1	105	64	128	0.5:1
45	74	137	0.5:1	110	64	127	0.5:1
50	73	137	0.5:1	115	63	126	0.5:1
55	72	136	0.5:1	120	62	126	0.5:1
60	72	135	0.5:1	125	61	125	0.5:1

APPENDIX C - 1 — 40 kg BAG

THICKENING TABLES

From 0.6:1 to 0.5:1
(by loose volume)

See Section 4.1–8 for explanations and methods of use.

Grout in Line and Agitator (liters)	MAKEUP MIX			Grout in Line and Agitator (liters)	MAKEUP MIX		
	Water (liters)	Mix Vol (liters)	Mix w:c		Water (liters)	Mix Vol (liters)	Mix w:c
1 BAG BATCH							
0	13	26	0.5:1	10	13	25	0.5:1
5	13	26	0.5:1				
2 BAG BATCH							
0	27	52	0.5:1	15	26	51	0.5:1
5	27	52	0.5:1	20	25	51	0.5:1
10	26	51	0.5:1	25	25	50	0.5:1
3 BAG BATCH							
0	40	79	0.5:1	25	38	76	0.5:1
5	40	78	0.5:1	30	38	76	0.5:1
10	40	78	0.5:1	35	37	75	0.5:1
15	39	77	0.5:1	40	37	75	0.5:1
20	39	77	0.5:1				
4 BAG BATCH							
0	54	105	0.5:1	30	51	102	0.5:1
5	54	104	0.5:1	35	51	102	0.5:1
10	53	104	0.5:1	40	50	101	0.5:1
15	53	103	0.5:1	45	50	101	0.5:1
20	52	103	0.5:1	50	49	100	0.5:1
25	52	102	0.5:1	55	49	100	0.5:1
5 BAG BATCH							
0	67	131	0.5:1	40	64	127	0.5:1
5	67	131	0.5:1	45	63	127	0.5:1
10	67	130	0.5:1	50	63	126	0.5:1
15	66	130	0.5:1	55	62	126	0.5:1
20	66	129	0.5:1	60	62	125	0.5:1
25	65	129	0.5:1	65	61	125	0.5:1
30	65	128	0.5:1	70	61	124	0.5:1
35	64	128	0.5:1				

APPENDIX C - 2

50 kg BAG and BULK

THICKENING TABLES

From 5:1 to 4:1
(by loose volume)

See Section 4.1–8 for explanations and methods of use.

Grout in Line and Agitator (liters)	MAKEUP MIX			Grout in Line and Agitator (liters)	MAKEUP MIX		
	Water (liters)	Mix Vol (liters)	Mix w:c		Water (liters)	Mix Vol (liters)	Mix w:c
			1 BAG BATCH				
0	140	156	4.0:1	155	112	127	3.2:1
5	139	155	4.0:1	160	111	127	3.2:1
10	138	154	3.9:1	165	110	126	3.1:1
15	137	153	3.9:1	170	109	125	3.1:1
20	136	152	3.9:1	175	108	124	3.1:1
25	135	151	3.9:1	180	107	123	3.1:1
30	134	150	3.8:1	185	106	122	3.0:1
35	134	149	3.8:1	190	105	121	3.0:1
40	133	149	3.8:1	195	104	120	3.0:1
45	132	148	3.8:1	200	103	119	3.0:1
50	131	147	3.7:1	205	102	118	2.9:1
55	130	146	3.7:1	210	101	117	2.9:1
60	129	145	3.7:1	215	101	116	2.9:1
65	128	144	3.7:1	220	100	116	2.8:1
70	127	143	3.6:1	225	99	115	2.8:1
75	126	142	3.6:1	230	98	114	2.8:1
80	125	141	3.6:1	235	97	113	2.8:1
85	124	140	3.6:1	240	96	112	2.7:1
90	123	139	3.5:1	245	95	111	2.7:1
95	123	138	3.5:1	250	94	110	2.7:1
100	122	138	3.5:1	255	93	109	2.7:1
105	121	137	3.4:1	260	92	108	2.6:1
110	120	136	3.4:1	265	91	107	2.6:1
115	119	135	3.4:1	270	90	106	2.6:1
120	118	134	3.4:1	275	90	105	2.6:1
125	117	133	3.3:1	280	89	105	2.5:1
130	116	132	3.3:1	285	88	104	2.5:1
135	115	131	3.3:1	290	87	103	2.5:1
140	114	130	3.3:1	295	86	102	2.5:1
145	113	129	3.2:1	300	85	101	2.4:1
150	112	128	3.2:1	305	84	100	2.4:1

APPENDIX C - 2

Continued

50 kg BAG and BULK

THICKENING TABLES

From 5:1 to 4:1

Grout in Line and Agitator (liters)	MAKEUP MIX Water (liters)	MAKEUP MIX Mix Vol (liters)	MAKEUP MIX Mix w:c	Grout in Line and Agitator (liters)	MAKEUP MIX Water (liters)	MAKEUP MIX Mix Vol (liters)	MAKEUP MIX Mix w:c
1 BAG BATCH							
310	83	99	2.4:1	360	74	90	2.1:1
315	82	98	2.3:1	365	73	89	2.1:1
320	81	97	2.3:1	370	72	88	2.1:1
325	80	96	2.3:1	375	71	87	2.0:1
330	79	95	2.3:1	380	70	86	2.0:1
335	79	94	2.2:1	385	69	85	2.0:1
340	78	94	2.2:1	390	68	84	2.0:1
345	77	93	2.2:1	395	68	83	1.9:1
350	76	92	2.2:1	400	67	83	1.9:1
355	75	91	2.1:1				
2 BAG BATCH							
0	280	312	4.0:1	150	252	284	3.6:1
5	279	311	4.0:1	155	252	283	3.6:1
10	278	310	4.0:1	160	251	282	3.6:1
15	277	309	4.0:1	165	250	281	3.6:1
20	276	308	3.9:1	170	249	281	3.6:1
25	275	307	3.9:1	175	248	280	3.5:1
30	274	306	3.9:1	180	247	279	3.5:1
35	274	305	3.9:1	185	246	278	3.5:1
40	273	304	3.9:1	190	245	277	3.5:1
45	272	303	3.9:1	195	244	276	3.5:1
50	271	303	3.9:1	200	243	275	3.5:1
55	270	302	3.9:1	205	242	274	3.5:1
60	269	301	3.8:1	210	241	273	3.4:1
65	268	300	3.8:1	215	241	272	3.4:1
70	267	299	3.8:1	220	240	271	3.4:1
75	266	298	3.8:1	225	239	270	3.4:1
80	265	297	3.8:1	230	238	270	3.4:1
85	264	296	3.8:1	235	237	269	3.4:1
90	263	295	3.8:1	240	236	268	3.4:1
95	263	294	3.8:1	245	235	267	3.4:1
100	262	293	3.7:1	250	234	266	3.3:1
105	261	292	3.7:1	255	233	265	3.3:1
110	260	292	3.7:1	260	232	264	3.3:1
115	259	291	3.7:1	265	231	263	3.3:1
120	258	290	3.7:1	270	230	262	3.3:1
125	257	289	3.7:1	275	230	261	3.3:1
130	256	288	3.7:1	280	229	260	3.3:1
135	255	287	3.6:1	285	228	259	3.3:1
140	254	286	3.6:1	290	227	259	3.2:1
145	253	285	3.6:1	295	226	258	3.2:1

APPENDIX C - 2

50 kg BAG and BULK

THICKENING TABLES

From 4:1 to 3:1
(by loose volume)

See Section 4.1–8 for explanations and methods of use

1 BAG BATCH

Grout in Line and Agitator (liters)	MAKEUP MIX Water (liters)	MAKEUP MIX Mix Vol (liters)	MAKEUP MIX Mix w:c	Grout in Line and Agitator (liters)	MAKEUP MIX Water (liters)	MAKEUP MIX Mix Vol (liters)	MAKEUP MIX Mix w:c
0	105	121	3.0:1	155	70	86	2.0:1
5	104	120	3.0:1	160	69	85	2.0:1
10	103	119	2.9:1	165	68	84	1.9:1
15	102	118	2.9:1	170	67	83	1.9:1
20	101	116	2.9:1	175	66	82	1.9:1
25	99	115	2.8:1	180	65	80	1.8:1
30	98	114	2.8:1	185	63	79	1.8:1
35	97	113	2.8:1	190	62	78	1.8:1
40	96	112	2.7:1	195	61	77	1.7:1
45	95	111	2.7:1	200	60	76	1.7:1
50	94	110	2.7:1	205	59	75	1.7:1
55	93	109	2.6:1	210	58	74	1.7:1
60	92	107	2.6:1	215	57	73	1.6:1
65	90	106	2.6:1	220	56	71	1.6:1
70	89	105	2.6:1	225	54	70	1.6:1
75	88	104	2.5:1	230	53	69	1.5:1
80	87	103	2.5:1	235	52	68	1.5:1
85	86	102	2.5:1	240	51	67	1.5:1
90	85	101	2.4:1	245	50	66	1.4:1
95	84	100	2.4:1	250	49	65	1.4:1
100	83	98	2.4:1	255	48	64	1.4:1
105	81	97	2.3:1	260	47	62	1.3:1
110	80	96	2.3:1	265	45	61	1.3:1
115	79	95	2.3:1	270	44	60	1.3:1
120	78	94	2.2:1	275	43	59	1.2:1
125	77	93	2.2:1	280	42	58	1.2:1
130	76	92	2.2:1	285	41	57	1.2:1
135	75	91	2.1:1	290	40	56	1.1:1
140	74	89	2.1:1	295	39	55	1.1:1
145	72	88	2.1:1	300	38	54	1.1:1
150	71	87	2.0:1	305	37	52	1.0:1

APPENDIX C - 2
Continued

50 kg BAG and BULK

THICKENING TABLES
From 4:1 to 3:1

Grout in Line and Agitator (liters)	MAKEUP MIX			Grout in Line and Agitator (liters)	MAKEUP MIX		
	Water (liters)	Mix Vol (liters)	Mix w:c		Water (liters)	Mix Vol (liters)	Mix w:c
1 BAG BATCH							
310	35	51	1.0:1	355	25	41	0.7:1
315	34	50	1.0:1	360	24	40	0.7:1
320	33	49	0.9:1	365	23	39	0.7:1
325	32	48	0.9:1	370	22	38	0.6:1
330	31	47	0.9:1	375	21	37	0.6:1
335	30	46	0.9:1	380	20	36	0.6:1
340	29	45	0.8:1	385	19	34	0.5:1
345	28	43	0.8:1	390	17	33	0.5:1
350	26	42	0.8:1	395	16	32	0.5:1
2 BAG BATCH							
0	210	242	3.0:1	155	175	207	2.5:1
5	209	241	3.0:1	160	174	206	2.5:1
10	208	240	3.0:1	165	173	205	2.5:1
15	207	238	3.0:1	170	172	204	2.5:1
20	206	237	2.9:1	175	171	202	2.4:1
25	204	236	2.9:1	180	170	201	2.4:1
30	203	235	2.9:1	185	168	200	2.4:1
35	202	234	2.9:1	190	167	199	2.4:1
40	201	233	2.9:1	195	166	198	2.4:1
45	200	232	2.9:1	200	165	197	2.4:1
50	199	231	2.8:1	205	164	196	2.3:1
55	198	229	2.8:1	210	163	195	2.3:1
60	197	228	2.8:1	215	162	193	2.3:1
65	195	227	2.8:1	220	161	192	2.3:1
70	194	226	2.8:1	225	159	191	2.3:1
75	193	225	2.8:1	230	158	190	2.3:1
80	192	224	2.7:1	235	157	189	2.2:1
85	191	223	2.7:1	240	156	188	2.2:1
90	190	222	2.7:1	245	155	187	2.2:1
95	189	220	2.7:1	250	154	186	2.2:1
100	188	219	2.7:1	255	153	184	2.2:1
105	186	218	2.7:1	260	152	183	2.2:1
110	185	217	2.6:1	265	150	182	2.1:1
115	184	216	2.6:1	270	149	181	2.1:1
120	183	215	2.6:1	275	148	180	2.1:1
125	182	214	2.6:1	280	147	179	2.1:1
130	181	213	2.6:1	285	146	178	2.1:1
135	180	211	2.6:1	290	145	177	2.1:1
140	179	210	2.6:1	295	144	176	2.1:1
145	177	209	2.5:1	300	143	174	2.0:1
150	176	208	2.5:1	305	142	173	2.0:1

APPENDIX C - 2

Continued

50 kg BAG and BULK

THICKENING TABLES

From 4:1 to 3:1

Grout in Line and Agitator (liters)	MAKEUP MIX			Grout in Line and Agitator (liters)	MAKEUP MIX		
	Water (liters)	Mix Vol (liters)	Mix w:c		Water (liters)	Mix Vol (liters)	Mix w:c
2 BAG BATCH							
310	140	172	2.0:1	360	129	161	1.8:1
315	139	171	2.0:1	365	128	160	1.8:1
320	138	170	2.0:1	370	127	159	1.8:1
325	137	169	2.0:1	375	126	158	1.8:1
330	136	168	1.9:1	380	125	156	1.8:1
335	135	167	1.9:1	385	124	155	1.8:1
340	134	165	1.9:1	390	122	154	1.7:1
345	133	164	1.9:1	395	121	153	1.7:1
350	131	163	1.9:1	400	120	152	1.7:1
355	130	162	1.9:1				

APPENDIX C - 2

50 kg BAG and BULK

THICKENING TABLES

From 3:1 to 2:1
(by loose volume)

See Section 4.1–8 for explanations and methods of use.

Grout in Line and Agitator (liters)	MAKEUP MIX			Grout in Line and Agitator (liters)	MAKEUP MIX		
	Water (liters)	Mix Vol (liters)	Mix w:c		Water (liters)	Mix Vol (liters)	Mix w:c
			1 BAG BATCH				
0	70	86	2.0:1	95	42	58	1.2:1
5	69	84	2.0:1	100	41	57	1.2:1
10	67	83	1.9:1	105	40	55	1.1:1
15	66	82	1.9:1	110	38	54	1.1:1
20	64	80	1.8:1	115	37	53	1.0:1
25	63	79	1.8:1	120	35	51	1.0:1
30	61	77	1.8:1	125	34	50	1.0:1
35	60	76	1.7:1	130	32	48	0.9:1
40	58	74	1.7:1	135	31	47	0.9:1
45	57	73	1.6:1	140	29	45	0.8:1
50	56	71	1.6:1	145	28	44	0.8:1
55	54	70	1.5:1	150	27	42	0.8:1
60	53	68	1.5:1	155	25	41	0.7:1
65	51	67	1.5:1	160	24	40	0.7:1
70	50	66	1.4:1	165	22	38	0.6:1
75	48	64	1.4:1	170	21	37	0.6:1
80	47	63	1.3:1	175	19	35	0.6:1
85	45	61	1.3:1	180	18	34	0.5:1
90	44	60	1.3:1	185	16	32	0.5:1
			2 BAG BATCH				
0	140	172	2.0:1	55	124	156	1.8:1
5	139	170	2.0:1	60	123	154	1.8:1
10	137	169	2.0:1	65	121	153	1.7:1
15	136	167	1.9:1	70	120	151	1.7:1
20	134	166	1.9:1	75	118	150	1.7:1
25	133	165	1.9:1	80	117	149	1.7:1
30	131	163	1.9:1	85	115	147	1.6:1
35	130	162	1.9:1	90	114	146	1.6:1
40	128	160	1.8:1	95	112	144	1.6:1
45	127	159	1.8:1	100	111	143	1.6:1
50	126	157	1.8:1	105	110	141	1.6:1

APPENDIX C - 2
Continued

50 kg BAG and BULK

THICKENING TABLES
From 3:1 to 2:1

Grout in Line and Agitator (liters)	MAKEUP MIX			Grout in Line and Agitator (liters)	MAKEUP MIX		
	Water (liters)	Mix Vol (liters)	Mix w:c		Water (liters)	Mix Vol (liters)	Mix w:c
2 BAG BATCH							
110	108	140	1.5:1	245	69	101	1.0:1
115	107	138	1.5:1	250	68	99	1.0:1
120	105	137	1.5:1	255	66	98	0.9:1
125	104	136	1.5:1	260	65	96	0.9:1
130	102	134	1.5:1	265	63	95	0.9:1
135	101	133	1.4:1	270	62	94	0.9:1
140	99	131	1.4:1	275	60	92	0.9:1
145	98	130	1.4:1	280	59	91	0.8:1
150	97	128	1.4:1	285	57	89	0.8:1
155	95	127	1.4:1	290	56	88	0.8:1
160	94	125	1.3:1	295	55	86	0.8:1
165	92	124	1.3:1	300	53	85	0.8:1
170	91	123	1.3:1	305	52	83	0.7:1
175	89	121	1.3:1	310	50	82	0.7:1
180	88	120	1.3:1	315	49	81	0.7:1
185	86	118	1.2:1	320	47	79	0.7:1
190	85	117	1.2:1	325	46	78	0.7:1
195	84	115	1.2:1	330	44	76	0.6:1
200	82	114	1.2:1	335	43	75	0.6:1
205	81	112	1.2:1	340	42	73	0.6:1
210	79	111	1.1:1	345	40	72	0.6:1
215	78	109	1.1:1	350	39	70	0.6:1
220	76	108	1.1:1	355	37	69	0.5:1
225	75	107	1.1:1	360	36	68	0.5:1
230	73	105	1.0:1	365	34	66	0.5:1
235	72	104	1.0:1	370	33	65	0.5:1
240	71	102	1.0:1				
3 BAG BATCH							
0	210	258	2.0:1	65	191	239	1.8:1
5	209	256	2.0:1	70	190	237	1.8:1
10	207	255	2.0:1	75	188	236	1.8:1
15	206	253	2.0:1	80	187	234	1.8:1
20	204	252	1.9:1	85	185	233	1.8:1
25	203	250	1.9:1	90	184	232	1.8:1
30	201	249	1.9:1	95	182	230	1.7:1
35	200	247	1.9:1	100	181	229	1.7:1
40	198	246	1.9:1	105	180	227	1.7:1
45	197	245	1.9:1	110	178	226	1.7:1
50	196	243	1.9:1	115	177	224	1.7:1
55	194	242	1.8:1	120	175	223	1.7:1
60	193	240	1.8:1	125	174	221	1.7:1

APPENDIX C - 2

Continued

50 kg BAG and BULK

THICKENING TABLES

From 3:1 to 2:1

3 BAG BATCH

Grout in Line and Agitator (liters)	MAKEUP MIX Water (liters)	MAKEUP MIX Mix Vol (liters)	MAKEUP MIX Mix w:c
130	172	220	1.6:1
135	171	219	1.6:1
140	169	217	1.6:1
145	168	216	1.6:1
150	167	214	1.6:1
155	165	213	1.6:1
160	164	211	1.6:1
165	162	210	1.5:1
170	161	208	1.5:1
175	159	207	1.5:1
180	158	205	1.5:1
185	156	204	1.5:1
190	155	203	1.5:1
195	154	201	1.5:1
200	152	200	1.4:1
205	151	198	1.4:1
210	149	197	1.4:1
215	148	195	1.4:1
220	146	194	1.4:1
225	145	192	1.4:1
230	143	191	1.4:1
235	142	190	1.4:1
240	141	188	1.3:1
245	139	187	1.3:1
250	138	185	1.3:1
255	136	184	1.3:1
260	135	182	1.3:1
265	133	181	1.3:1
270	132	179	1.3:1

Grout in Line and Agitator (liters)	MAKEUP MIX Water (liters)	MAKEUP MIX Mix Vol (liters)	MAKEUP MIX Mix w:c
275	130	178	1.2:1
280	129	177	1.2:1
285	127	175	1.2:1
290	126	174	1.2:1
295	125	172	1.2:1
300	123	171	1.2:1
305	122	169	1.2:1
310	120	168	1.1:1
315	119	166	1.1:1
320	117	165	1.1:1
325	116	164	1.1:1
330	114	162	1.1:1
335	113	161	1.1:1
340	112	159	1.1:1
345	110	158	1.0:1
350	109	156	1.0:1
355	107	155	1.0:1
360	106	153	1.0:1
365	104	152	1.0:1
370	103	150	1.0:1
375	101	149	1.0:1
380	100	148	1.0:1
385	99	146	0.9:1
390	97	145	0.9:1
395	96	143	0.9:1
400	94	142	0.9:1

APPENDIX C - 2

50 kg BAG and BULK

THICKENING TABLES

From 2:1 to 1:1
(by loose volume)

See Section 4.1–8 for explanations and methods of use.

Grout in Line and Agitator (liters)	MAKEUP MIX			Grout in Line and Agitator (liters)	MAKEUP MIX		
	Water (liters)	Mix Vol (liters)	Mix w:c		Water (liters)	Mix Vol (liters)	Mix w:c
			1 BAG BATCH				
0	35	51	1.0:1	25	25	41	0.7:1
5	33	49	0.9:1	30	23	39	0.7:1
10	31	47	0.9:1	35	21	37	0.6:1
15	29	45	0.8:1	40	19	35	0.5:1
20	27	43	0.8:1	45	17	33	0.5:1
			2 BAG BATCH				
0	70	102	1.0:1	50	50	81	0.7:1
5	68	100	1.0:1	55	48	79	0.7:1
10	66	98	0.9:1	60	46	77	0.7:1
15	64	96	0.9:1	65	44	75	0.6:1
20	62	94	0.9:1	70	41	73	0.6:1
25	60	92	0.9:1	75	39	71	0.6:1
30	58	90	0.8:1	80	37	69	0.5:1
35	56	87	0.8:1	85	35	67	0.5:1
40	54	85	0.8:1	90	33	65	0.5:1
45	52	83	0.7:1				
			3 BAG BATCH				
0	105	153	1.0:1	70	76	124	0.7:1
5	103	151	1.0:1	75	74	122	0.7:1
10	101	149	1.0:1	80	72	120	0.7:1
15	99	147	0.9:1	85	70	118	0.7:1
20	97	144	0.9:1	90	68	116	0.7:1
25	95	142	0.9:1	95	66	114	0.6:1
30	93	140	0.9:1	100	64	112	0.6:1
35	91	138	0.9:1	105	62	110	0.6:1
40	89	136	0.8:1	110	60	108	0.6:1
45	87	134	0.8:1	115	58	106	0.6:1
50	85	132	0.8:1	120	56	104	0.5:1
55	83	130	0.8:1	125	54	102	0.5:1
60	81	128	0.8:1	130	52	100	0.5:1
65	79	126	0.7:1	135	50	98	0.5:1

APPENDIX C - 2
Continued

50 kg BAG and BULK

THICKENING TABLES
From 2:1 to 1:1

Grout in Line and Agitator (liters)	MAKEUP MIX			Grout in Line and Agitator (liters)	MAKEUP MIX		
	Water (liters)	Mix Vol (liters)	Mix w:c		Water (liters)	Mix Vol (liters)	Mix w:c
4 BAG BATCH							
0	140	203	1.0:1	100	99	163	0.7:1
5	138	201	1.0:1	105	97	161	0.7:1
10	136	199	1.0:1	110	95	159	0.7:1
15	134	197	1.0:1	115	93	157	0.7:1
20	132	195	0.9:1	120	91	155	0.7:1
25	130	193	0.9:1	125	89	153	0.6:1
30	128	191	0.9:1	130	87	151	0.6:1
35	126	189	0.9:1	135	85	148	0.6:1
40	124	187	0.9:1	140	83	146	0.6:1
45	122	185	0.9:1	145	81	144	0.6:1
50	120	183	0.9:1	150	79	142	0.6:1
55	118	181	0.8:1	155	77	140	0.5:1
60	116	179	0.8:1	160	75	138	0.5:1
65	114	177	0.8:1	165	73	136	0.5:1
70	111	175	0.8:1	170	71	134	0.5:1
75	109	173	0.8:1	175	69	132	0.5:1
80	107	171	0.8:1	180	67	130	0.5:1
85	105	169	0.8:1	185	65	128	0.5:1
90	103	167	0.7:1				
95	101	165	0.7:1				
5 BAG BATCH							
0	175	254	1.0:1	100	134	214	0.8:1
5	173	252	1.0:1	105	132	212	0.8:1
10	171	250	1.0:1	110	130	210	0.7:1
15	169	248	1.0:1	115	128	207	0.7:1
20	167	246	1.0:1	120	126	205	0.7:1
25	165	244	0.9:1	125	124	203	0.7:1
30	163	242	0.9:1	130	122	201	0.7:1
35	161	240	0.9:1	135	120	199	0.7:1
40	159	238	0.9:1	140	118	197	0.7:1
45	157	236	0.9:1	145	116	195	0.7:1
50	155	234	0.9:1	150	114	193	0.7:1
55	153	232	0.9:1	155	112	191	0.6:1
60	151	230	0.9:1	160	110	189	0.6:1
65	149	228	0.8:1	165	108	187	0.6:1
70	146	226	0.8:1	170	106	185	0.6:1
75	144	224	0.8:1	175	104	183	0.6:1
80	142	222	0.8:1	180	102	181	0.6:1
85	140	220	0.8:1	185	100	179	0.6:1
90	138	218	0.8:1	190	98	177	0.6:1
95	136	216	0.8:1	195	96	175	0.5:1

APPENDIX C - 2

50 kg BAG and BULK

THICKENING TABLES

From 1:1 to 0.8:1
(by loose volume)

See Section 4.1–8 for explanations and methods of use.

Grout in Line and Agitator (liters)	MAKEUP MIX			Grout in Line and Agitator (liters)	MAKEUP MIX		
	Water (liters)	Mix Vol (liters)	Mix w:c		Water (liters)	Mix Vol (liters)	Mix w:c
1 BAG BATCH							
0	28	44	0.8:1	45	22	38	0.6:1
5	27	43	0.8:1	50	21	37	0.6:1
10	27	42	0.8:1	55	20	36	0.6:1
15	26	42	0.7:1	60	20	36	0.6:1
20	25	41	0.7:1	65	19	35	0.5:1
25	25	40	0.7:1	70	18	34	0.5:1
30	24	40	0.7:1	75	18	34	0.5:1
35	23	39	0.7:1	80	17	33	0.5:1
40	22	38	0.6:1	85	16	32	0.5:1
2 BAG BATCH							
0	56	88	0.8:1	90	44	75	0.6:1
5	55	87	0.8:1	95	43	75	0.6:1
10	55	86	0.8:1	100	42	74	0.6:1
15	54	86	0.8:1	105	42	73	0.6:1
20	53	85	0.8:1	110	41	73	0.6:1
25	53	84	0.8:1	115	40	72	0.6:1
30	52	84	0.7:1	120	39	71	0.6:1
35	51	83	0.7:1	125	39	71	0.6:1
40	50	82	0.7:1	130	38	70	0.5:1
45	50	82	0.7:1	135	37	69	0.5:1
50	49	81	0.7:1	140	37	68	0.5:1
55	48	80	0.7:1	145	36	68	0.5:1
60	48	79	0.7:1	150	35	67	0.5:1
65	47	79	0.7:1	155	35	66	0.5:1
70	46	78	0.7:1	160	34	66	0.5:1
75	46	77	0.7:1	165	33	65	0.5:1
80	45	77	0.6:1	170	33	64	0.5:1
85	44	76	0.6:1	175	32	64	0.5:1
3 BAG BATCH							
0	84	132	0.8:1	10	83	130	0.8:1
5	83	131	0.8:1	15	82	130	0.8:1

APPENDIX C - 2

Continued

50 kg BAG and BULK

THICKENING TABLES

From 1:1 to 0.8:1

Grout in Line and Agitator (liters)	MAKEUP MIX			Grout in Line and Agitator (liters)	MAKEUP MIX		
	Water (liters)	Mix Vol (liters)	Mix w:c		Water (liters)	Mix Vol (liters)	Mix w:c
3 BAG BATCH							
20	81	129	0.8:1	145	64	112	0.6:1
25	81	128	0.8:1	150	63	111	0.6:1
30	80	127	0.8:1	155	63	110	0.6:1
35	79	127	0.8:1	160	62	110	0.6:1
40	78	126	0.7:1	165	61	109	0.6:1
45	78	125	0.7:1	170	61	108	0.6:1
50	77	125	0.7:1	175	60	108	0.6:1
55	76	124	0.7:1	180	59	107	0.6:1
60	76	123	0.7:1	185	59	106	0.6:1
65	75	123	0.7:1	190	58	105	0.6:1
70	74	122	0.7:1	195	57	105	0.5:1
75	74	121	0.7:1	200	56	104	0.5:1
80	73	121	0.7:1	205	56	103	0.5:1
85	72	120	0.7:1	210	55	103	0.5:1
90	72	119	0.7:1	215	54	102	0.5:1
95	71	119	0.7:1	220	54	101	0.5:1
100	70	118	0.7:1	225	53	101	0.5:1
105	70	117	0.7:1	230	52	100	0.5:1
110	69	116	0.7:1	235	52	99	0.5:1
115	68	116	0.6:1	240	51	99	0.5:1
120	67	115	0.6:1	245	50	98	0.5:1
125	67	114	0.6:1	250	50	97	0.5:1
130	66	114	0.6:1	255	49	97	0.5:1
135	65	113	0.6:1	260	48	96	0.5:1
140	65	112	0.6:1	265	48	95	0.5:1
4 BAG BATCH							
0	112	175	0.8:1	75	102	165	0.7:1
5	111	175	0.8:1	80	101	164	0.7:1
10	111	174	0.8:1	85	100	164	0.7:1
15	110	173	0.8:1	90	100	163	0.7:1
20	109	173	0.8:1	95	99	162	0.7:1
25	109	172	0.8:1	100	98	162	0.7:1
30	108	171	0.8:1	105	98	161	0.7:1
35	107	171	0.8:1	110	97	160	0.7:1
40	106	170	0.8:1	115	96	160	0.7:1
45	106	170	0.8:1	120	95	159	0.7:1
50	105	169	0.8:1	125	95	158	0.7:1
55	104	128	0.7:1	130	94	158	0.7:1
60	104	127	0.7:1	135	93	157	0.7:1
65	103	167	0.7:1	140	93	156	0.7:1
70	102	166	0.7:1	145	92	156	0.7:1

APPENDIX C - 2
Continued

50 kg BAG and BULK

THICKENING TABLES
From 1:1 to 0.8:1

Grout in Line and Agitator (liters)	MAKEUP MIX Water (liters)	MAKEUP MIX Mix Vol (liters)	MAKEUP MIX Mix w:c	Grout in Line and Agitator (liters)	MAKEUP MIX Water (liters)	MAKEUP MIX Mix Vol (liters)	MAKEUP MIX Mix w:c
4 BAG BATCH							
150	91	155	0.7:1	230	80	144	0.6:1
155	91	154	0.6:1	235	80	143	0.6:1
160	90	153	0.6:1	240	79	142	0.6:1
165	89	153	0.6:1	245	78	142	0.6:1
170	89	152	0.6:1	250	78	141	0.6:1
175	88	151	0.6:1	255	77	140	0.5:1
180	87	151	0.6:1	260	76	140	0.5:1
185	87	150	0.6:1	265	76	139	0.5:1
190	86	149	0.6:1	270	75	138	0.5:1
195	85	149	0.6:1	275	74	138	0.5:1
200	84	148	0.6:1	280	73	137	0.5:1
205	84	147	0.6:1	285	73	136	0.5:1
210	83	147	0.6:1	290	72	136	0.5:1
215	82	146	0.6:1	295	71	135	0.5:1
220	82	145	0.6:1	300	71	134	0.5:1
225	81	145	0.6:1				

APPENDIX C - 2

50 kg BAG and BULK

THICKENING TABLES

From 0.8:1 to 0.6:1
(by loose volume)

See Section 4.1–8 for explanations and methods of use.

Grout in Line and Agitator (liters)	MAKEUP MIX Water (liters)	MAKEUP MIX Mix Vol (liters)	MAKEUP MIX Mix w:c	Grout in Line and Agitator (liters)	MAKEUP MIX Water (liters)	MAKEUP MIX Mix Vol (liters)	MAKEUP MIX Mix w:c
1 BAG BATCH							
0	21	37	0.6:1	20	18	34	0.5:1
5	20	36	0.6:1	25	17	33	0.5:1
10	19	35	0.6:1	30	16	32	0.5:1
15	19	34	0.5:1				
2 BAG BATCH							
0	42	74	0.6:1	35	36	68	0.5:1
5	41	73	0.6:1	40	36	67	0.5:1
10	40	72	0.6:1	45	35	67	0.5:1
15	40	71	0.6:1	50	34	66	0.5:1
20	39	71	0.6:1	55	33	65	0.5:1
25	38	70	0.5:1	60	32	64	0.5:1
30	37	69	0.5:1	65	32	63	0.5:1
3 BAG BATCH							
0	63	111	0.6:1	50	55	103	0.5:1
5	62	110	0.6:1	55	54	102	0.5:1
10	61	109	0.6:1	60	53	101	0.5:1
15	61	108	0.6:1	65	53	100	0.5:1
20	60	107	0.6:1	70	52	99	0.5:1
25	59	107	0.6:1	75	51	99	0.5:1
30	58	106	0.6:1	80	50	98	0.5:1
35	57	105	0.5:1	85	49	97	0.5:1
40	57	104	0.5:1	90	49	96	0.5:1
45	56	103	0.5:1	95	48	95	0.5:1
4 BAG BATCH							
0	84	147	0.6:1	35	78	142	0.6:1
5	83	147	0.6:1	40	78	141	0.6:1
10	82	146	0.6:1	45	77	140	0.5:1
15	82	145	0.6:1	50	76	140	0.5:1
20	81	144	0.6:1	55	75	139	0.5:1
25	80	144	0.6:1	60	74	138	0.5:1
30	79	143	0.6:1	65	74	137	0.5:1

APPENDIX C - 2
Continued

50 kg BAG and BULK

THICKENING TABLES
From 0.8:1 to 0.6:1

Grout in Line and Agitator (liters)	MAKEUP MIX			Grout in Line and Agitator (liters)	MAKEUP MIX		
	Water (liters)	Mix Vol (liters)	Mix w:c		Water (liters)	Mix Vol (liters)	Mix w:c
			4 BAG BATCH				
70	73	136	0.5:1	105	67	131	0.5:1
75	72	136	0.5:1	110	66	130	0.5:1
80	71	135	0.5:1	115	66	129	0.5:1
85	70	134	0.5:1	120	65	128	0.5:1
90	70	133	0.5:1	125	64	128	0.5:1
95	69	132	0.5:1	130	63	127	0.5:1
100	68	132	0.5:1				
			5 BAG BATCH				
0	105	184	0.6:1	85	91	171	0.5:1
5	104	184	0.6:1	90	91	170	0.5:1
10	103	183	0.6:1	95	90	169	0.5:1
15	103	182	0.6:1	100	89	168	0.5:1
20	102	181	0.6:1	105	88	168	0.5:1
25	101	180	0.6:1	110	87	167	0.5:1
30	100	180	0.6:1	115	87	166	0.5:1
35	99	179	0.6:1	120	86	165	0.5:1
40	99	178	0.6:1	125	85	164	0.5:1
45	98	177	0.6:1	130	84	164	0.5:1
50	97	176	0.6:1	135	83	163	0.5:1
55	96	176	0.5:1	140	83	162	0.5:1
60	95	175	0.5:1	145	82	161	0.5:1
65	95	174	0.5:1	150	81	160	0.5:1
70	94	173	0.5:1	155	80	160	0.5:1
75	93	172	0.5:1	160	79	159	0.5:1
80	92	172	0.5:1				

APPENDIX C - 2

50 kg BAG and BULK

THICKENING TABLES

From 0.6:1 to 0.5:1
(by loose volume)

See Section 4.1–8 for explanations and methods of use.

Grout in Line and Agitator (liters)	MAKEUP MIX			Grout in Line and Agitator (liters)	MAKEUP MIX		
	Water (liters)	Mix Vol (liters)	Mix w:c		Water (liters)	Mix Vol (liters)	Mix w:c
			1 BAG BATCH				
0	18	33	0.5:1	10	17	32	0.5:1
5	17	33	0.5:1	15	16	32	0.5:1
			2 BAG BATCH				
0	35	67	0.5:1	20	33	65	0.5:1
5	35	66	0.5:1	25	33	64	0.5:1
10	34	66	0.5:1	30	32	64	0.5:1
15	34	65	0.5:1	35	32	63	0.5:1
			3 BAG BATCH				
0	53	100	0.5:1	30	50	97	0.5:1
5	52	100	0.5:1	35	49	97	0.5:1
10	52	99	0.5:1	40	49	96	0.5:1
15	51	99	0.5:1	45	48	96	0.5:1
20	51	98	0.5:1	50	48	95	0.5:1
25	50	98	0.5:1	55	47	95	0.5:1
			4 BAG BATCH				
0	70	133	0.5:1	40	66	130	0.5:1
5	70	133	0.5:1	45	66	129	0.5:1
10	69	133	0.5:1	50	65	129	0.5:1
15	69	132	0.5:1	55	65	128	0.5:1
20	68	132	0.5:1	60	64	128	0.5:1
25	68	131	0.5:1	65	64	127	0.5:1
30	67	131	0.5:1	70	63	127	0.5:1
35	67	130	0.5:1				
			5 BAG BATCH				
0	88	167	0.5:1	35	84	164	0.5:1
5	87	166	0.5:1	40	84	163	0.5:1
10	87	166	0.5:1	45	83	163	0.5:1
15	86	165	0.5:1	50	83	162	0.5:1
20	86	165	0.5:1	55	82	162	0.5:1
25	85	164	0.5:1	60	82	161	0.5:1
30	85	164	0.5:1	65	81	161	0.5:1

APPENDIX C - 3 LITERS

THINNING TABLES

From 4:1
(by loose volume)

See Section 4.1–9 for explanations and methods of use.

Grout in Line and Agitator (liters)	THINNING WATER (liters)	New vol (liters)	Grout in Line and Agitator (liters)	THINNING WATER (liters)	New vol (liters)	Grout in Line and Agitator (liters)	THINNING WATER (liters)	New vol (liters)
TO 5:1								
20	4	24	175	39	214			
25	6	31	180	40	220			
30	7	37	185	41	226			
35	8	43	190	43	233			
40	9	49	195	44	239			
45	10	55	200	45	245			
50	11	61	205	46	251			
55	12	67	210	47	257			
60	13	73	215	48	263			
65	15	80	220	49	269			
70	16	86	225	50	275			
75	17	92	230	51	281			
80	18	98	235	53	288			
85	19	104	240	54	294			
90	20	110	245	55	300			
95	21	116	250	56	306			
100	22	122	255	57	312			
105	23	128	260	58	318			
110	25	135	265	59	324			
115	26	141	270	60	330			
120	27	147	275	62	337			
125	28	153	280	63	343			
130	29	159	285	64	349			
135	30	165	290	65	355			
140	31	171	295	66	361			
145	32	177	300	67	367			
150	34	184	305	68	373			
155	35	190	310	69	379			
160	36	196	315	70	385			
165	37	202	320	72	392			
170	38	208	325	73	398			

APPENDIX C - 3

LITERS

THINNING TABLES

From 3:1
(by loose volume)

See Section 4.1–9 for explanations and methods of use.

Grout in Line and Agitator (liters)	THIN-NING WATER (liters)	New vol (liters)	Grout in Line and Agitator (liters)	THIN-NING WATER (liters)	New vol (liters)	Grout in Line and Agitator (liters)	THIN-NING WATER (liters)	New vol (liters)
TO 4:1								
20	6	26	175	50	225			
25	7	32	180	52	232			
30	9	39	185	53	238			
35	10	45	190	55	245			
40	12	52	195	56	251			
45	13	58	200	58	258			
50	14	64	205	59	264			
55	16	71	210	61	271			
60	17	77	215	62	277			
65	19	84	220	63	283			
70	20	90	225	65	290			
75	22	97	230	66	296			
80	23	103	235	68	303			
85	24	109	240	69	309			
90	26	116	245	71	316			
95	27	122	250	72	322			
100	29	129	255	73	328			
105	30	135	260	75	335			
110	32	142	265	76	341			
115	33	148	270	78	348			
120	35	155	275	79	354			
125	36	161	280	81	361			
130	37	167	285	82	367			
135	39	174	290	84	374			
140	40	180	295	85	380			
145	42	187	300	86	386			
150	43	193	305	88	393			
155	45	200	310	89	399			
160	46	206						
165	48	213						
170	49	219						

APPENDIX C - 3 **LITERS**
Continued

THINNING TABLES

From 3:1

Grout in Line and Agitator (liters)	THIN-NING WATER (liters)	New vol (liters)	Grout in Line and Agitator (liters)	THIN-NING WATER (liters)	New vol (liters)	Grout in Line and Agitator (liters)	THIN-NING WATER (liters)	New vol (liters)
TO 5:1								
20	12	32	155	89	244			
25	14	39	160	92	252			
30	17	47	165	95	260			
35	20	55	170	98	268			
40	23	63	175	101	276			
45	26	71	180	104	284			
50	29	79	185	107	292			
55	32	87	190	110	300			
60	35	95	195	112	307			
65	37	102	200	115	315			
70	40	110	205	118	323			
75	43	118	210	121	331			
80	46	126	215	124	339			
85	49	134	220	127	347			
90	52	142	225	130	355			
95	55	150	230	133	363			
100	58	158	235	135	370			
105	61	166	240	138	378			
110	63	173	245	141	386			
115	66	181	250	144	394			
120	69	189	255	147	402			
125	72	197						
130	75	205						
135	78	213						
140	81	221						
145	84	229						
150	86	236						

APPENDIX C - 3 LITERS

THINNING TABLES

From 2:1
(by loose volume)

See Section 4.1–9 for explanations and methods of use.

Grout in Line and Agitator (liters)	THIN-NING WATER (liters)	New vol (liters)	Grout in Line and Agitator (liters)	THIN-NING WATER (liters)	New vol (liters)	Grout in Line and Agitator (liters)	THIN-NING WATER (liters)	New vol (liters)
TO 3:1								
20	8	28	110	45	155	200	81	281
25	10	35	115	47	162	205	83	288
30	12	42	120	49	169	210	85	295
35	14	49	125	51	176	215	87	302
40	16	56	130	53	183	220	89	309
45	18	63	135	55	190	225	91	316
50	20	70	140	57	197	230	93	323
55	22	77	145	59	204	235	95	330
60	24	84	150	61	211	240	97	337
65	26	91	155	63	218	245	99	344
70	28	98	160	65	225	250	101	351
75	30	105	165	67	232	255	103	358
80	32	112	170	69	239	260	105	365
85	34	119	175	71	246	265	107	372
90	36	126	180	73	253	270	109	379
95	38	133	185	75	260	275	111	386
100	40	140	190	77	267	280	113	393
105	43	148	195	79	274	285	115	400
TO 4:1								
20	16	36	75	61	136	130	105	235
25	20	45	80	65	145	135	109	244
30	24	54	85	69	154	140	113	253
35	28	63	90	73	163	145	117	262
40	32	72	95	77	172	150	121	271
45	36	81	100	81	181	155	125	280
50	40	90	105	85	190	160	130	290
55	45	100	110	89	199	165	134	299
60	49	109	115	93	208	170	138	308
65	53	118	120	97	217	175	142	317
70	57	127	125	101	226	180	146	326

APPENDIX C - 3 **LITERS**
Continued

THINNING TABLES

From 2:1

Grout in Line and Agitator (liters)	THIN-NING WATER (liters)	New vol (liters)	Grout in Line and Agitator (liters)	THIN-NING WATER (liters)	New vol (liters)	Grout in Line and Agitator (liters)	THIN-NING WATER (liters)	New vol (liters)
TO 4:1								
185	150	335	200	162	362	215	174	389
190	154	344	205	166	371	220	178	398
195	158	353	210	170	380			
TO 5:1								
20	24	44	80	97	177	140	170	310
25	30	55	85	103	188	145	176	321
30	36	66	90	109	199	150	182	332
35	43	78	95	115	210	155	188	343
40	49	89	100	121	221	160	194	354
45	55	100	105	128	233	165	200	365
50	61	111	110	134	244	170	206	376
55	67	122	115	140	255	175	213	388
60	73	133	120	146	266	180	219	399
65	79	144	125	152	277	185	225	410
70	85	155	130	158	288			
75	91	166	135	164	299			

APPENDIX C - 3 LITERS

THINNING TABLES

From 1:1
(by loose volume)

See Section 4.1–9 for explanations and methods of use.

Grout in Line and Agitator (liters)	THIN-NING WATER (liters)	New vol (liters)	Grout in Line and Agitator (liters)	THIN-NING WATER (liters)	New vol (liters)	Grout in Line and Agitator (liters)	THIN-NING WATER (liters)	New vol (liters)
TO 2:1								
20	14	34	100	68	168	175	119	294
25	17	42	105	71	176	180	122	302
30	20	50	110	75	185	185	126	311
35	24	59	115	78	193	190	129	319
40	27	67	120	82	202	195	133	328
45	31	76	125	85	210	200	136	336
50	34	84	130	88	218	205	139	344
55	37	92	135	92	227	210	143	353
60	41	101	140	95	235	215	146	361
65	44	109	145	99	244	220	150	370
70	48	118	150	102	252	225	153	378
75	51	126	155	105	260	230	156	386
80	54	134	160	109	269	235	160	395
85	58	143	165	112	277	240	163	403
90	61	151	170	116	286			
95	65	160						
TO 3:1								
20	27	47	85	116	201	150	204	354
25	34	59	90	122	212	155	211	366
30	41	71	95	129	224	160	218	378
35	48	83	100	136	236			
40	54	94	105	143	248			
45	61	106	110	150	260			
50	68	118	115	156	271			
55	75	130	120	163	283			
60	82	142	125	170	295			
65	88	153	130	177	307			
70	95	165	135	184	319			
75	102	177	140	190	330			
80	109	189	145	197	342			

APPENDIX C - 3 **LITERS**

Continued

THINNING TABLES

From 1:1

Grout in Line and Agitator (liters)	THIN-NING WATER (liters)	New vol (liters)	Grout in Line and Agitator (liters)	THIN-NING WATER (liters)	New vol (liters)	Grout in Line and Agitator (liters)	THIN-NING WATER (liters)	New vol (liters)
TO 4:1								
20	41	61	60	122	182	100	204	304
25	51	76	65	133	198	105	214	319
30	61	91	70	143	213	110	224	334
35	71	106	75	153	228	115	235	350
40	82	122	80	163	243	120	245	365
45	92	137	85	173	258	125	255	380
50	102	152	90	184	274	130	265	395
55	112	167	95	194	289			
TO 5:1								
20	54	74						
25	68	93						
30	82	112						
35	95	130						
40	109	149						
45	122	167						
50	136	186						
55	150	205						
60	163	223						
65	177	242						
70	190	260						
75	204	279						
80	218	298						
85	231	316						
90	245	335						
95	258	353						
100	272	372						

APPENDIX C - 3 — LITERS

THINNING TABLES

From 0.8:1
(by loose volume)

See Section 4.1–9 for explanations and methods of use.

Grout in Line and Agitator (liters)	THIN-NING WATER (liters)	New vol (liters)	Grout in Line and Agitator (liters)	THIN-NING WATER (liters)	New vol (liters)	Grout in Line and Agitator (liters)	THIN-NING WATER (liters)	New vol (liters)
TO 1:1								
20	3	23	175	28	203	325	51	376
25	4	29	180	28	208	330	52	382
30	5	35	185	29	214	335	53	388
35	6	41	190	30	220	340	54	394
40	6	46	195	31	226	345	54	399
45	7	52	200	31	231			
50	8	58	205	32	237			
55	9	64	210	33	243			
60	9	69	215	34	249			
65	10	75	220	35	255			
70	11	81	225	35	260			
75	12	87	230	36	266			
80	13	93	235	37	272			
85	13	98	240	38	278			
90	14	104	245	39	284			
95	15	110	250	39	289			
100	16	116	255	40	295			
105	17	122	260	41	301			
110	17	127	265	42	307			
115	18	133	270	43	313			
120	19	139	275	43	318			
125	20	145	280	44	324			
130	20	150	285	45	330			
135	21	156	290	46	336			
140	22	162	295	46	341			
145	23	168	300	47	347			
150	24	174	305	48	353			
155	24	179	310	49	359			
160	25	185	315	50	365			
165	26	191	320	50	370			
170	27	197						

APPENDIX C - 3
Continued

LITERS

THINNING TABLES
From 0.8:1

Grout in Line and Agitator (liters)	THIN-NING WATER (liters)	New vol (liters)	Grout in Line and Agitator (liters)	THIN-NING WATER (liters)	New vol (liters)	Grout in Line and Agitator (liters)	THIN-NING WATER (liters)	New vol (liters)
TO 2:1								
20	19	39	85	80	165	150	142	292
25	24	49	90	85	175	155	146	301
30	28	58	95	90	185	160	151	311
35	33	68	100	94	194	165	156	321
40	38	78	105	99	204	170	161	331
45	43	88	110	104	214	175	165	340
50	47	97	115	109	224	180	170	350
55	52	107	120	113	233	185	175	360
60	57	117	125	118	243	190	179	369
65	61	126	130	123	253	195	184	379
70	66	136	135	128	263	200	189	389
75	71	146	140	132	272	205	194	399
80	76	156	145	137	282			
TO 3:1								
20	35	55	65	113	178	110	191	301
25	43	68	70	121	191	115	199	314
30	52	82	75	130	205	120	208	328
35	61	96	80	139	219	125	216	341
40	69	109	85	147	232	130	225	355
45	78	123	90	156	246	135	234	369
50	87	137	95	165	260	140	242	382
55	95	150	100	173	273	145	251	396
60	104	164	105	182	287			
TO 4:1								
20	50	70	55	139	194	90	227	317
25	63	88	60	151	211	95	239	334
30	76	106	65	164	229	100	252	352
35	88	123	70	176	246	105	265	370
40	101	141	75	189	264	110	277	387
45	113	158	80	202	282			
50	126	176	85	214	299			
TO 5:1								
20	66	86	50	165	215	80	265	345
25	83	108	55	182	237	85	281	366
30	99	129	60	198	258	90	298	388
35	116	151	65	215	280			
40	132	172	70	231	301			
45	149	194	75	248	323			

APPENDIX C - 3 — LITERS

THINNING TABLES

From 0.6:1
(by loose volume)

See Section 4.1–9 for explanations and methods of use.

Grout in Line and Agitator (liters)	THINNING WATER (liters)	New vol (liters)
TO 0.8:1		
20	4	24
25	5	30
30	6	36
35	7	42
40	7	47
45	8	53
50	9	59
55	10	65
60	11	71
65	12	77
70	13	83
75	14	89
80	15	95
85	16	101
90	17	107
95	18	113
100	19	119
105	20	125
110	21	131
115	21	136
120	22	142
125	23	148
130	24	154
135	25	160
140	26	166
145	27	172
150	28	178
155	29	184
160	30	190
165	31	196
170	32	202
175	33	208
180	34	214
185	35	220
190	36	226
195	36	231
200	37	237
205	38	243
210	39	249
215	40	255
220	41	261
225	42	267
230	43	273
235	44	279
240	45	285
245	46	291
250	47	297
255	48	303
260	49	309
265	50	315
270	50	320
275	51	326
280	52	332
285	53	338
290	54	344
295	55	350
300	56	356
305	57	362
310	58	368
315	59	374
320	60	380
325	61	386
330	62	392
335	63	398

APPENDIX C - 3 **LITERS**

Continued

THINNING TABLES

From 0.6:1

Grout in Line and Agitator (liters)	THIN-NING WATER (liters)	New vol (liters)	Grout in Line and Agitator (liters)	THIN-NING WATER (liters)	New vol (liters)	Grout in Line and Agitator (liters)	THIN-NING WATER (liters)	New vol (liters)
TO 1:1								
20	7	27	115	43	158	210	78	288
25	9	34	120	45	165	215	80	295
30	11	41	125	47	172	220	82	302
35	13	48	130	49	179	225	84	309
40	15	55	135	50	185	230	86	316
45	17	62	140	52	192	235	88	323
50	19	69	145	54	199	240	90	330
55	21	76	150	56	206	245	92	337
60	22	82	155	58	213	250	93	343
65	24	89	160	60	220	255	95	350
70	26	96	165	62	227	260	97	357
75	28	103	170	64	234	265	99	364
80	30	110	175	65	240	270	101	371
85	32	117	180	67	247	275	103	378
90	34	124	185	69	254	280	105	385
95	36	131	190	71	261	285	107	392
100	37	137	195	73	268	290	108	398
105	39	144	200	75	275			
110	41	151	205	77	282			
TO 2:1								
20	26	46	70	92	162	120	157	277
25	33	58	75	98	173	125	164	289
30	39	69	80	105	185	130	170	300
35	46	81	85	111	196	135	177	312
40	52	92	90	118	208	140	183	323
45	59	104	95	124	219	145	190	335
50	65	115	100	131	231	150	196	346
55	72	127	105	137	242	155	203	358
60	78	138	110	144	254	160	209	369
65	85	150	115	150	265	165	216	381
						170	222	392
TO 3:1								
20	45	65	55	123	178	90	202	292
25	56	81	60	135	195	95	213	308
30	67	97	65	146	211	100	224	324
35	78	113	70	157	227	105	235	340
40	90	130	75	168	243	110	247	357
45	101	146	80	179	259	115	258	373
50	112	162	85	191	276	120	269	389

APPENDIX C - 3

Continued

LITERS

THINNING TABLES

From 0.6:1

Grout in Line and Agitator (liters)	THIN-NING WATER (liters)	New vol (liters)	Grout in Line and Agitator (liters)	THIN-NING WATER (liters)	New vol (liters)	Grout in Line and Agitator (liters)	THIN-NING WATER (liters)	New vol (liters)
TO 4:1								
20	64	84	50	159	209	80	254	334
25	79	104	55	175	230	85	270	355
30	95	125	60	191	251	90	286	376
35	111	146	65	206	271	95	302	397
40	127	167	70	222	292			
45	143	188	75	238	313			
TO 5:1								
20	82	102	40	164	204	60	247	307
25	103	128	45	185	230	65	267	332
30	123	153	50	206	256	70	288	358
35	144	179	55	226	281	75	308	383

APPENDIX C - 3

LITERS

THINNING TABLES

From 0.5:1
(by loose volume)

See Section 4.1–9 for explanations and methods of use.

TO 0.6:1

Grout in Line and Agitator (liters)	THINNING WATER (liters)	New vol (liters)
20	2	22
25	3	28
30	3	33
35	4	39
40	4	44
45	5	50
50	5	55
55	6	61
60	6	66
65	7	72
70	7	77
75	8	83
80	8	88
85	9	94
90	9	99
95	10	105
100	10	110
105	11	116
110	11	121
115	12	127
120	12	132
125	13	138
130	13	143
135	14	149
140	14	154
145	15	160
150	15	165
155	16	171
160	16	176
165	17	182
170	18	188
175	18	193
180	19	199
185	19	204
190	20	210
195	20	215
200	21	221
205	21	226
210	22	232
215	22	237
220	23	243
225	23	248
230	24	254
235	24	259
240	25	265
245	25	270
250	26	276
255	26	281
260	27	287
265	27	292
270	28	298
275	28	303
280	29	309
285	29	314
290	30	320
295	30	325
300	31	331
305	31	336
310	32	342
315	32	347
320	33	353
325	33	358
330	34	364
335	35	370
340	35	375
345	36	381
350	36	386
355	37	392
360	37	397

APPENDIX C - 3 — LITERS

Continued

THINNING TABLES

From 0.5:1

Grout in Line and Agitator (liters)	THINNING WATER (liters)	New vol (liters)	Grout in Line and Agitator (liters)	THINNING WATER (liters)	New vol (liters)	Grout in Line and Agitator (liters)	THINNING WATER (liters)	New vol (liters)
TO 0.8:1								
20	6	26	120	37	157	220	68	288
25	8	33	125	39	164	225	70	295
30	9	39	130	40	170	230	71	301
35	11	46	135	42	177	235	73	308
40	12	52	140	43	183	240	74	314
45	14	59	145	45	190	245	76	321
50	15	65	150	46	196	250	77	327
55	17	72	155	48	203	255	79	334
60	19	79	160	49	209	260	80	340
65	20	85	165	51	216	265	82	347
70	22	92	170	53	223	270	83	353
75	23	98	175	54	229	275	85	360
80	25	105	180	56	236	280	87	367
85	26	111	185	57	242	285	88	373
90	28	118	190	59	249	290	90	380
95	29	124	195	60	255	295	91	386
100	31	131	200	62	262	300	93	393
105	32	137	205	63	268	305	94	399
110	34	144	210	65	275			
115	36	151	215	66	281			
TO 1:1								
20	10	30	115	59	174	210	108	318
25	13	38	120	62	182	215	111	326
30	15	45	125	64	189	220	113	333
35	18	53	130	67	197	225	116	341
40	21	61	135	70	205	230	119	349
45	23	68	140	72	212	235	121	356
50	26	76	145	75	220	240	124	364
55	28	83	150	77	227	245	126	371
60	31	91	155	80	235	250	129	379
65	33	98	160	82	242	255	131	386
70	36	106	165	85	250	260	134	394
75	39	114	170	88	258			
80	41	121	175	90	265			
85	44	129	180	93	273			
90	46	136	185	95	280			
95	49	144	190	98	288			
100	52	152	195	100	295			
105	54	159	200	103	303			
110	57	167	205	106	311			

APPENDIX C - 3
Continued

LITERS

THINNING TABLES
From 0.5:1

Grout in Line and Agitator (liters)	THIN-NING WATER (liters)	New vol (liters)	Grout in Line and Agitator (liters)	THIN-NING WATER (liters)	New vol (liters)	Grout in Line and Agitator (liters)	THIN-NING WATER (liters)	New vol (liters)
TO 2:1								
20	31	51	70	108	178	120	186	306
25	39	64	75	116	191	125	193	318
30	46	76	80	124	204	130	201	331
35	54	89	85	131	216	135	209	344
40	62	102	90	139	229	140	216	356
45	70	115	95	147	242	145	224	369
50	77	127	100	155	255	150	232	382
55	85	140	105	162	267	155	240	395
60	93	153	110	170	280			
65	100	165	115	178	293			
TO 3:1								
20	52	72	55	142	197	90	232	322
25	64	89	60	155	215	95	245	340
30	77	107	65	167	232	100	258	358
35	90	125	70	180	250	105	271	376
40	103	143	75	193	268	110	283	393
45	116	161	80	206	286			
50	129	179	85	219	304			
TO 4:1								
20	72	92	45	162	207	70	252	322
25	90	115	50	180	230	75	271	346
30	108	138	55	198	253	80	289	369
35	126	161	60	216	276	85	307	392
40	144	184	65	234	299			
TO 5:1								
20	93	113	40	186	226	60	278	338
25	116	141	45	209	254	65	301	366
30	139	169	50	232	282	70	325	395
35	162	197	55	255	310			

APPENDIX C - 4 — 40 kg BAG

QUANTITY OF CEMENT IN GROUT

5 : 1
(by loose volume)

See Section 4.1–11 for explanations and methods of use.

Grout Volume (liters)	Cement Content (bags)	Cement Content (kg)	Grout Volume (liters)	Cement Content (bags)	Cement Content (kg)	Grout Volume (liters)	Cement Content (bags)	Cement Content (kg)
0	0.00	0.0	150	1.02	40.6	300	2.03	81.2
5	0.03	1.4	155	1.05	42.0	305	2.07	82.6
10	0.07	2.7	160	1.08	43.3	310	2.10	84.0
15	0.10	4.1	165	1.12	44.7	315	2.13	85.3
20	0.14	5.4	170	1.15	46.0	320	2.17	86.7
25	0.17	6.8	175	1.18	47.4	325	2.20	88.0
30	0.20	8.1	180	1.22	48.7	330	2.23	89.4
35	0.24	9.5	185	1.25	50.1	335	2.27	90.7
40	0.27	10.8	190	1.29	51.5	340	2.30	92.1
45	0.30	12.2	195	1.32	52.8	345	2.34	93.4
50	0.34	13.5	200	1.35	54.2	350	2.37	94.8
55	0.37	14.9	205	1.39	55.5	355	2.40	96.1
60	0.41	16.2	210	1.42	56.9	360	2.44	97.5
65	0.44	17.6	215	1.46	58.2	365	2.47	98.9
70	0.47	19.0	220	1.49	59.6	370	2.51	100.2
75	0.51	20.3	225	1.52	60.9	375	2.54	101.6
80	0.54	21.7	230	1.56	62.3	380	2.57	102.9
85	0.58	23.0	235	1.59	63.6	385	2.61	104.3
90	0.61	24.4	240	1.62	65.0	390	2.64	105.6
95	0.64	25.7	245	1.66	66.4	395	2.67	107.0
100	0.68	27.1	250	1.69	67.7	400	2.71	108.3
105	0.71	28.4	255	1.73	69.1			
110	0.74	29.8	260	1.76	70.4			
115	0.78	31.1	265	1.79	71.8			
120	0.81	32.5	270	1.83	73.1			
125	0.85	33.9	275	1.86	74.5			
130	0.88	35.2	280	1.90	75.8			
135	0.91	36.6	285	1.93	77.2			
140	0.95	37.9	290	1.96	78.5			
145	0.98	39.3	295	2.00	79.9			

APPENDIX C - 4

40 kg BAG

QUANTITY OF CEMENT IN GROUT

4 : 1
(by loose volume)

See Section 4.1–11 for explanations and methods of use.

Grout Volume	Cement Content		Grout Volume	Cement Content		Grout Volume	Cement Content	
(liters)	(bags)	(kg)	(liters)	(bags)	(kg)	(liters)	(bags)	(kg)
0	0.00	0.0	150	1.24	49.7	300	2.49	99.4
5	0.04	1.7	155	1.28	51.4	305	2.53	101.1
10	0.08	3.3	160	1.33	53.0	310	2.57	102.7
15	0.12	5.0	165	1.37	54.7	315	2.61	104.4
20	0.17	6.6	170	1.41	56.3	320	2.65	106.0
25	0.21	8.3	175	1.45	58.0	325	2.69	107.7
30	0.25	9.9	180	1.49	59.7	330	2.73	109.4
35	0.29	11.6	185	1.53	61.3	335	2.78	111.0
40	0.33	13.3	190	1.57	63.0	340	2.82	112.7
45	0.37	14.9	195	1.62	64.6	345	2.86	114.3
50	0.41	16.6	200	1.66	66.3	350	2.90	116.0
55	0.46	18.2	205	1.70	67.9	355	2.94	117.6
60	0.50	19.9	210	1.74	69.6	360	2.98	119.3
65	0.54	21.5	215	1.78	71.3	365	3.02	121.0
70	0.58	23.2	220	1.82	72.9	370	3.07	122.6
75	0.62	24.9	225	1.86	74.6	375	3.11	124.3
80	0.66	26.5	230	1.91	76.2	380	3.15	125.9
85	0.70	28.2	235	1.95	77.9	385	3.19	127.6
90	0.75	29.8	240	1.99	79.5	390	3.23	129.2
95	0.79	31.5	245	2.03	81.2	395	3.27	130.9
100	0.83	33.1	250	2.07	82.9	400	3.31	132.6
105	0.87	34.8	255	2.11	84.5			
110	0.91	36.5	260	2.15	86.2			
115	0.95	38.1	265	2.20	87.8			
120	0.99	39.8	270	2.24	89.5			
125	1.04	41.4	275	2.28	91.1			
130	1.08	43.1	280	2.32	92.8			
135	1.12	44.7	285	2.36	94.5			
140	1.16	46.4	290	2.40	96.1			
145	1.20	48.1	295	2.44	97.8			

APPENDIX C - 4

40 kg BAG

QUANTITY OF CEMENT IN GROUT

3 : 1
(by loose volume)

See Section 4.1–11 for explanations and methods of use.

Grout Volume	Cement Content		Grout Volume	Cement Content		Grout Volume	Cement Content	
(liters)	(bags)	(kg)	(liters)	(bags)	(kg)	(liters)	(bags)	(kg)
0	0.00	0.0	150	1.60	64.0	300	3.20	128.1
5	0.05	2.1	155	1.65	66.2	305	3.26	130.2
10	0.11	4.3	160	1.71	68.3	310	3.31	132.3
15	0.16	6.4	165	1.76	70.4	315	3.36	134.5
20	0.21	8.5	170	1.81	72.6	320	3.42	136.6
25	0.27	10.7	175	1.87	74.7	325	3.47	138.7
30	0.32	12.8	180	1.92	76.8	330	3.52	140.9
35	0.37	14.9	185	1.97	79.0	335	3.58	143.0
40	0.43	17.1	190	2.03	81.1	340	3.63	145.1
45	0.48	19.2	195	2.08	83.2	345	3.68	147.3
50	0.53	21.3	200	2.13	85.4	350	3.74	149.4
55	0.59	23.5	205	2.19	87.5	355	3.79	151.6
60	0.64	25.6	210	2.24	89.6	360	3.84	153.7
65	0.69	27.7	215	2.29	91.8	365	3.90	155.8
70	0.75	29.9	220	2.35	93.9	370	3.95	158.0
75	0.80	32.0	225	2.40	96.1	375	4.00	160.1
80	0.85	34.2	230	2.45	98.2	380	4.06	162.2
85	0.91	36.3	235	2.51	100.3	385	4.11	164.4
90	0.96	38.4	240	2.56	102.5	390	4.16	166.5
95	1.01	40.6	245	2.61	104.6	395	4.22	168.6
100	1.07	42.7	250	2.67	106.7	400	4.27	170.8
105	1.12	44.8	255	2.72	108.9			
110	1.17	47.0	260	2.77	111.0			
115	1.23	49.1	265	2.83	113.1			
120	1.28	51.2	270	2.88	115.3			
125	1.33	53.4	275	2.93	117.4			
130	1.39	55.5	280	2.99	119.5			
135	1.44	57.6	285	3.04	121.7			
140	1.49	59.8	290	3.10	123.8			
145	1.55	61.9	295	3.15	125.9			

APPENDIX C - 4 **40 kg BAG**

QUANTITY OF CEMENT IN GROUT

2 : 1
(by loose volume)

See Section 4.1–11 for explanations and methods of use.

Grout Volume	Cement Content		Grout Volume	Cement Content		Grout Volume	Cement Content	
(liters)	(bags)	(kg)	(liters)	(bags)	(kg)	(liters)	(bags)	(kg)
0	0.00	0.0	150	2.25	90.0	300	4.50	179.9
5	0.07	3.0	155	2.32	93.0	305	4.57	182.9
10	0.15	6.0	160	2.40	96.0	310	4.65	185.9
15	0.22	9.0	165	2.47	99.0	315	4.72	188.9
20	0.30	12.0	170	2.55	102.0	320	4.80	191.9
25	0.37	15.0	175	2.62	105.0	325	4.87	194.9
30	0.45	18.0	180	2.70	107.9	330	4.95	197.9
35	0.52	21.0	185	2.77	110.9	335	5.02	200.9
40	0.60	24.0	190	2.85	113.9	340	5.10	203.9
45	0.67	27.0	195	2.92	116.9	345	5.17	206.9
50	0.75	30.0	200	3.00	119.9	350	5.25	209.9
55	0.82	33.0	205	3.07	122.9	355	5.32	212.9
60	0.90	36.0	210	3.15	125.9	360	5.40	215.9
65	0.97	39.0	215	3.22	128.9	365	5.47	218.9
70	1.05	42.0	220	3.30	131.9	370	5.55	221.9
75	1.12	45.0	225	3.37	134.9	375	5.62	224.9
80	1.20	48.0	230	3.45	137.9	380	5.70	227.9
85	1.27	51.0	235	3.52	140.9	385	5.77	230.9
90	1.35	54.0	240	3.60	143.9	390	5.85	233.9
95	1.42	57.0	245	3.67	146.9	395	5.92	236.9
100	1.50	60.0	250	3.75	149.9	400	6.00	239.9
105	1.57	63.0	255	3.82	152.9			
110	1.65	66.0	260	3.90	155.9			
115	1.72	69.0	265	3.97	158.9			
120	1.80	72.0	270	4.05	161.9			
125	1.87	75.0	275	4.12	164.9			
130	1.95	78.0	280	4.20	167.9			
135	2.02	81.0	285	4.27	170.9			
140	2.10	84.0	290	4.35	173.9			
145	2.17	87.0	295	4.42	176.9			

APPENDIX C - 4

40 kg BAG

QUANTITY OF CEMENT IN GROUT

1 : 1

(by loose volume)

See Section 4.1–11 for explanations and methods of use.

Grout Volume	Cement Content		Grout Volume	Cement Content		Grout Volume	Cement Content	
(liters)	(bags)	(kg)	(liters)	(bags)	(kg)	(liters)	(bags)	(kg)
0	0.00	0.0	150	3.78	151.1	300	7.56	302.3
5	0.13	5.0	155	3.90	156.2	305	7.68	307.3
10	0.25	10.1	160	4.03	161.2	310	7.81	312.4
15	0.38	15.1	165	4.16	166.3	315	7.93	317.4
20	0.50	20.2	170	4.28	171.3	320	8.06	322.4
25	0.63	25.2	175	4.41	176.3	325	8.19	327.5
30	0.76	30.2	180	4.53	181.4	330	8.31	332.5
35	0.88	35.3	185	4.66	186.4	335	8.44	337.5
40	1.01	40.3	190	4.79	191.4	340	8.56	342.6
45	1.13	45.3	195	4.91	196.5	345	8.69	347.6
50	1.26	50.4	200	5.04	201.5	350	8.82	352.7
55	1.39	55.4	205	5.16	206.6	355	8.94	357.7
60	1.51	60.5	210	5.29	211.6	360	9.07	362.7
65	1.64	65.5	215	5.42	216.6	365	9.19	367.8
70	1.76	70.5	220	5.54	221.7	370	9.32	372.8
75	1.89	75.6	225	5.67	226.7	375	9.45	377.8
80	2.02	80.6	230	5.79	231.7	380	9.57	382.9
85	2.14	85.6	235	5.92	236.8	385	9.70	387.9
90	2.27	90.7	240	6.05	241.8	390	9.82	393.0
95	2.39	95.7	245	6.17	246.9	395	9.95	398.0
100	2.52	100.8	250	6.30	251.9	400	10.08	403.0
105	2.64	105.8	255	6.42	256.9			
110	2.77	110.8	260	6.55	262.0			
115	2.90	115.9	265	6.68	267.0			
120	3.02	120.9	270	6.80	272.1			
125	3.15	125.9	275	6.93	277.1			
130	3.27	131.0	280	7.05	282.1			
135	3.40	136.0	285	7.18	287.2			
140	3.53	141.1	290	7.31	292.2			
145	3.65	146.1	295	7.43	297.2			

APPENDIX C - 4

40 kg BAG

QUANTITY OF CEMENT IN GROUT

0.8 : 1
(by loose volume)

See Section 4.1–11 for explanations and methods of use.

Grout Volume	Cement Content		Grout Volume	Cement Content		Grout Volume	Cement Content	
(liters)	(bags)	(kg)	(liters)	(bags)	(kg)	(liters)	(bags)	(kg)
0	0.00	0.0	150	4.37	174.9	300	8.75	349.9
5	0.15	5.8	155	4.52	180.8	305	8.89	355.7
10	0.29	11.7	160	4.66	186.6	310	9.04	361.5
15	0.44	17.5	165	4.81	192.4	315	9.18	367.4
20	0.58	23.3	170	4.96	198.3	320	9.33	373.2
25	0.73	29.2	175	5.10	204.1	325	9.48	379.0
30	0.87	35.0	180	5.25	209.9	330	9.62	384.9
35	1.02	40.8	185	5.39	215.8	335	9.77	390.7
40	1.17	46.6	190	5.54	221.6	340	9.91	396.5
45	1.31	52.5	195	5.69	227.4	345	10.06	402.4
50	1.46	58.3	200	5.83	233.2	350	10.20	408.2
55	1.60	64.1	205	5.98	239.1	355	10.35	414.0
60	1.75	70.0	210	6.12	244.9	360	10.50	419.8
65	1.90	75.8	215	6.27	250.7	365	10.64	425.7
70	2.04	81.6	220	6.41	256.6	370	10.79	431.5
75	2.19	87.5	225	6.56	262.4	375	10.93	437.3
80	2.33	93.3	230	6.71	268.2	380	11.08	443.2
85	2.48	99.1	235	6.85	274.1	385	11.23	449.0
90	2.62	105.0	240	7.00	279.9	390	11.37	454.8
95	2.77	110.8	245	7.14	285.7	395	11.52	460.7
100	2.92	116.6	250	7.29	291.6	400	11.66	466.5
105	3.06	122.5	255	7.43	297.4			
110	3.21	128.3	260	7.58	303.2			
115	3.35	134.1	265	7.73	309.1			
120	3.50	139.9	270	7.87	314.9			
125	3.64	145.8	275	8.02	320.7			
130	3.79	151.6	280	8.16	326.5			
135	3.94	157.4	285	8.31	332.4			
140	4.08	163.3	290	8.46	338.2			
145	4.23	169.1	295	8.60	344.0			

APPENDIX C - 4

40 kg BAG

QUANTITY OF CEMENT IN GROUT

0.6 : 1
(by loose volume)

See Section 4.1–11 for explanations and methods of use.

Grout Volume	Cement Content	
(liters)	(bags)	(kg)
0	0.00	0.0
5	0.17	6.9
10	0.35	13.8
15	0.52	20.8
20	0.69	27.7
25	0.87	34.6
30	1.04	41.5
35	1.21	48.4
40	1.38	55.4
45	1.56	62.3
50	1.73	69.2
55	1.90	76.1
60	2.08	83.0
65	2.25	90.0
70	2.42	96.9
75	2.60	103.8
80	2.77	110.7
85	2.94	117.7
90	3.11	124.6
95	3.29	131.5
100	3.46	138.4
105	3.63	145.3
110	3.81	152.3
115	3.98	159.2
120	4.15	166.1
125	4.33	173.0
130	4.50	179.9
135	4.67	186.9
140	4.84	193.8
145	5.02	200.7
150	5.19	207.6
155	5.36	214.5
160	5.54	221.5
165	5.71	228.4
170	5.88	235.3
175	6.06	242.2
180	6.23	249.1
185	6.40	256.1
190	6.57	263.0
195	6.75	269.9
200	6.92	276.8
205	7.09	283.8
210	7.27	290.7
215	7.44	297.6
220	7.61	304.5
225	7.79	311.4
230	7.96	318.4
235	8.13	325.3
240	8.30	332.2
245	8.48	339.1
250	8.65	346.0
255	8.82	353.0
260	9.00	359.9
265	9.17	366.8
270	9.34	373.7
275	9.52	380.6
280	9.69	387.6
285	9.86	394.5
290	10.04	401.4
295	10.21	408.3
300	10.38	415.2
305	10.55	422.2
310	10.73	429.1
315	10.90	436.0
320	11.07	442.9
325	11.25	449.9
330	11.42	456.8
335	11.59	463.7
340	11.77	470.6
345	11.94	477.5
350	12.11	484.5
355	12.28	491.4
360	12.46	498.3
365	12.63	505.2
370	12.80	512.1
375	12.98	519.1
380	13.15	526.0
385	13.32	532.9
390	13.50	539.8
395	13.67	546.7
400	13.84	553.7

APPENDIX C - 4

40 kg BAG

QUANTITY OF CEMENT IN GROUT

0.5 : 1
(by loose volume)

See Section 4.1–11 for explanations and methods of use.

Grout Volume (liters)	Cement Content (bags)	Cement Content (kg)
0	0.00	0.0
5	0.19	7.6
10	0.38	15.3
15	0.57	22.9
20	0.76	30.5
25	0.95	38.2
30	1.15	45.8
35	1.34	53.4
40	1.53	61.1
45	1.72	68.7
50	1.91	76.3
55	2.10	84.0
60	2.29	91.6
65	2.48	99.2
70	2.67	106.9
75	2.86	114.5
80	3.05	122.1
85	3.24	129.8
90	3.44	137.4
95	3.63	145.0
100	3.82	152.7
105	4.01	160.3
110	4.20	167.9
115	4.39	175.6
120	4.58	183.2
125	4.77	190.9
130	4.96	198.5
135	5.15	206.1
140	5.34	213.8
145	5.53	221.4
150	5.73	229.0
155	5.92	236.7
160	6.11	244.3
165	6.30	251.9
170	6.49	259.6
175	6.68	267.2
180	6.87	274.8
185	7.06	282.5
190	7.25	290.1
195	7.44	297.7
200	7.63	305.4
205	7.82	313.0
210	8.02	320.6
215	8.21	328.3
220	8.40	335.9
225	8.59	343.5
230	8.78	351.2
235	8.97	358.8
240	9.16	366.4
245	9.35	374.1
250	9.54	381.7
255	9.73	389.3
260	9.92	397.0
265	10.12	404.6
270	10.31	412.2
275	10.50	419.9
280	10.69	427.5
285	10.88	435.1
290	11.07	442.8
295	11.26	450.4
300	11.45	458.0
305	11.64	465.7
310	11.83	473.3
315	12.02	480.9
320	12.21	488.6
325	12.41	496.2
330	12.60	503.8
335	12.79	511.5
340	12.98	519.1
345	13.17	526.7
350	13.36	534.4
355	13.55	542.0
360	13.74	549.7
365	13.93	557.3
370	14.12	564.9
375	14.31	572.6
380	14.50	580.2
385	14.70	587.8
390	14.89	595.5
395	15.08	603.1
400	15.27	610.7

APPENDIX C - 5

50 kg BAG and BULK

QUANTITY OF CEMENT IN GROUT

5 : 1
(by loose volume)

See Section 4.1–11 for explanations and methods of use.

Grout Volume	Cement Content		Grout Volume	Cement Content		Grout Volume	Cement Content	
(liters)	(bags)	(kg)	(liters)	(bags)	(kg)	(liters)	(bags)	(kg)
0	0.00	0.0	150	0.79	39.3	300	1.57	78.6
5	0.03	1.3	155	0.81	40.6	305	1.60	79.9
10	0.05	2.6	160	0.84	41.9	310	1.62	81.2
15	0.08	3.9	165	0.86	43.2	315	1.65	82.5
20	0.10	5.2	170	0.89	44.5	320	1.68	83.8
25	0.13	6.5	175	0.92	45.8	325	1.70	85.1
30	0.16	7.9	180	0.94	47.2	330	1.73	86.4
35	0.18	9.2	185	0.97	48.5	335	1.76	87.8
40	0.21	10.5	190	1.00	49.8	340	1.78	89.1
45	0.24	11.8	195	1.02	51.1	345	1.81	90.4
50	0.26	13.1	200	1.05	52.4	350	1.83	91.7
55	0.29	14.4	205	1.07	53.7	355	1.86	93.0
60	0.31	15.7	210	1.10	55.0	360	1.89	94.3
65	0.34	17.0	215	1.13	56.3	365	1.91	95.6
70	0.37	18.3	220	1.15	57.6	370	1.94	96.9
75	0.39	19.6	225	1.18	58.9	375	1.96	98.2
80	0.42	21.0	230	1.20	60.2	380	1.99	99.5
85	0.45	22.3	235	1.23	61.6	385	2.02	100.9
90	0.47	23.6	240	1.26	62.9	390	2.04	102.2
95	0.50	24.9	245	1.28	64.2	395	2.07	103.5
100	0.52	26.2	250	1.31	65.5	400	2.10	104.8
105	0.55	27.5	255	1.34	66.8			
110	0.58	28.8	260	1.36	68.1			
115	0.60	30.1	265	1.39	69.4			
120	0.63	31.4	270	1.41	70.7			
125	0.65	32.7	275	1.44	72.0			
130	0.68	34.1	280	1.47	73.3			
135	0.71	35.4	285	1.49	74.7			
140	0.73	36.7	290	1.52	76.0			
145	0.76	38.0	295	1.55	77.3			

APPENDIX C - 5

50 kg BAG and BULK

QUANTITY OF CEMENT IN GROUT

4 : 1
(by loose volume)

See Section 4.1–11 for explanations and methods of use.

Grout Volume (liters)	Cement Content (bags)	Cement Content (kg)
0	0.00	0.0
5	0.03	1.6
10	0.06	3.2
15	0.10	4.8
20	0.13	6.4
25	0.16	8.0
30	0.19	9.6
35	0.22	11.2
40	0.26	12.8
45	0.29	14.4
50	0.32	16.0
55	0.35	17.6
60	0.38	19.2
65	0.42	20.9
70	0.45	22.5
75	0.48	24.1
80	0.51	25.7
85	0.55	27.3
90	0.58	28.9
95	0.61	30.5
100	0.64	32.1
105	0.67	33.7
110	0.71	35.3
115	0.74	36.9
120	0.77	38.5
125	0.80	40.1
130	0.83	41.7
135	0.87	43.3
140	0.90	44.9
145	0.93	46.5
150	0.96	48.1
155	0.99	49.7
160	1.03	51.3
165	1.06	52.9
170	1.09	54.5
175	1.12	56.1
180	1.15	57.7
185	1.19	59.3
190	1.22	60.9
195	1.25	62.6
200	1.28	64.2
205	1.32	65.8
210	1.35	67.4
215	1.38	69.0
220	1.41	70.6
225	1.44	72.2
230	1.48	73.8
235	1.51	75.4
240	1.54	77.0
245	1.57	78.6
250	1.60	80.2
255	1.64	81.8
260	1.67	83.4
265	1.70	85.0
270	1.73	86.6
275	1.76	88.2
280	1.80	89.8
285	1.83	91.4
290	1.86	93.0
295	1.89	94.6
300	1.92	96.2
305	1.96	97.8
310	1.99	99.4
315	2.02	101.0
320	2.05	102.6
325	2.09	104.3
330	2.12	105.9
335	2.15	107.5
340	2.18	109.1
345	2.21	110.7
350	2.25	112.3
355	2.28	113.9
360	2.31	115.5
365	2.34	117.1
370	2.37	118.7
375	2.41	120.3
380	2.44	121.9
385	2.47	123.5
390	2.50	125.1
395	2.53	126.7
400	2.57	128.3

APPENDIX C - 5

50 kg BAG and BULK

QUANTITY OF CEMENT IN GROUT

3 : 1
(by loose volume)

See Section 4.1–11 for explanations and methods of use.

Grout Volume (liters)	Cement Content (bags)	Cement Content (kg)	Grout Volume (liters)	Cement Content (bags)	Cement Content (kg)	Grout Volume (liters)	Cement Content (bags)	Cement Content (kg)
0	0.00	0.0	150	1.24	62.0	300	2.48	124.1
5	0.04	2.1	155	1.28	64.1	305	2.52	126.2
10	0.08	4.1	160	1.32	66.2	310	2.56	128.2
15	0.12	6.2	165	1.37	68.3	315	2.61	130.3
20	0.17	8.3	170	1.41	70.3	320	2.65	132.4
25	0.21	10.3	175	1.45	72.4	325	2.69	134.4
30	0.25	12.4	180	1.49	74.5	330	2.73	136.5
35	0.29	14.5	185	1.53	76.5	335	2.77	138.6
40	0.33	16.5	190	1.57	78.6	340	2.81	140.6
45	0.37	18.6	195	1.61	80.7	345	2.85	142.7
50	0.41	20.7	200	1.65	82.7	350	2.90	144.8
55	0.46	22.8	205	1.70	84.8	355	2.94	146.8
60	0.50	24.8	210	1.74	86.9	360	2.98	148.9
65	0.54	26.9	215	1.78	88.9	365	3.02	151.0
70	0.58	29.0	220	1.82	91.0	370	3.06	153.1
75	0.62	31.0	225	1.86	93.1	375	3.10	155.1
80	0.66	33.1	230	1.90	95.1	380	3.14	157.2
85	0.70	35.2	235	1.94	97.2	385	3.19	159.3
90	0.74	37.2	240	1.99	99.3	390	3.23	161.3
95	0.79	39.3	245	2.03	101.3	395	3.27	163.4
100	0.83	41.4	250	2.07	103.4	400	3.31	165.5
105	0.87	43.4	255	2.11	105.5			
110	0.91	45.5	260	2.15	107.6			
115	0.95	47.6	265	2.19	109.6			
120	0.99	49.6	270	2.23	111.7			
125	1.03	51.7	275	2.28	113.8			
130	1.08	53.8	280	2.32	115.8			
135	1.12	55.8	285	2.36	117.9			
140	1.16	57.9	290	2.40	120.0			
145	1.20	60.0	295	2.44	122.0			

APPENDIX C - 5

50 kg BAG and BULK

QUANTITY OF CEMENT IN GROUT

2 : 1
(by loose volume)

See Section 4.1–11 for explanations and methods of use.

Grout Volume	Cement Content	
(liters)	(bags)	(kg)
0	0.00	0.0
5	0.06	2.9
10	0.12	5.8
15	0.17	8.7
20	0.23	11.6
25	0.29	14.6
30	0.35	17.5
35	0.41	20.4
40	0.47	23.3
45	0.52	26.2
50	0.58	29.1
55	0.64	32.0
60	0.70	34.9
65	0.76	37.8
70	0.82	40.8
75	0.87	43.7
80	0.93	46.6
85	0.99	49.5
90	1.05	52.4
95	1.11	55.3
100	1.16	58.2
105	1.22	61.1
110	1.28	64.0
115	1.34	67.0
120	1.40	69.9
125	1.46	72.8
130	1.51	75.7
135	1.57	78.6
140	1.63	81.5
145	1.69	84.4
150	1.75	87.3
155	1.80	90.2
160	1.86	93.2
165	1.92	96.1
170	1.98	99.0
175	2.04	101.9
180	2.10	104.8
185	2.15	107.7
190	2.21	110.6
195	2.27	113.5
200	2.33	116.5
205	2.39	119.4
210	2.45	122.3
215	2.50	125.2
220	2.56	128.1
225	2.62	131.0
230	2.68	133.9
235	2.74	136.8
240	2.79	139.7
245	2.85	142.7
250	2.91	145.6
255	2.97	148.5
260	3.03	151.4
265	3.09	154.3
270	3.14	157.2
275	3.20	160.1
280	3.26	163.0
285	3.32	165.9
290	3.38	168.9
295	3.44	171.8
300	3.49	174.7
305	3.55	177.6
310	3.61	180.5
315	3.67	183.4
320	3.73	186.3
325	3.78	189.2
330	3.84	192.1
335	3.90	195.1
340	3.96	198.0
345	4.02	200.9
350	4.08	203.8
355	4.13	206.7
360	4.19	209.6
365	4.25	212.5
370	4.31	215.4
375	4.37	218.3
380	4.43	221.3
385	4.48	224.2
390	4.54	227.1
395	4.60	230.0
400	4.66	232.9

APPENDIX C - 5

50 kg BAG and BULK

QUANTITY OF CEMENT IN GROUT

1 : 1
(by loose volume)

See Section 4.1–11 for explanations and methods of use.

Grout Volume (liters)	Cement Content (bags)	Cement Content (kg)	Grout Volume (liters)	Cement Content (bags)	Cement Content (kg)	Grout Volume (liters)	Cement Content (bags)	Cement Content (kg)
0	0.00	0.0	150	2.95	147.4	300	5.90	294.9
5	0.10	4.9	155	3.05	152.3	305	6.00	299.8
10	0.20	9.8	160	3.15	157.3	310	6.09	304.7
15	0.29	14.7	165	3.24	162.2	315	6.19	309.6
20	0.39	19.7	170	3.34	167.1	320	6.29	314.5
25	0.49	24.6	175	3.44	172.0	325	6.39	319.4
30	0.59	29.5	180	3.54	176.9	330	6.49	324.3
35	0.69	34.4	185	3.64	181.8	335	6.59	329.3
40	0.79	39.3	190	3.73	186.7	340	6.68	334.2
45	0.88	44.2	195	3.83	191.7	345	6.78	339.1
50	0.98	49.1	200	3.93	196.6	350	6.88	344.0
55	1.08	54.1	205	4.03	201.5	355	6.98	348.9
60	1.18	59.0	210	4.13	206.4	360	7.08	353.8
65	1.28	63.9	215	4.23	211.3	365	7.17	358.7
70	1.38	68.8	220	4.32	216.2	370	7.27	363.7
75	1.47	73.7	225	4.42	221.1	375	7.37	368.6
80	1.57	78.6	230	4.52	226.1	380	7.47	373.5
85	1.67	83.5	235	4.62	231.0	385	7.57	378.4
90	1.77	88.5	240	4.72	235.9	390	7.67	383.3
95	1.87	93.4	245	4.82	240.8	395	7.76	388.2
100	1.97	98.3	250	4.91	245.7	400	7.86	393.1
105	2.06	103.2	255	5.01	250.6			
110	2.16	108.1	260	5.11	255.5			
115	2.26	113.0	265	5.21	260.5			
120	2.36	117.9	270	5.31	265.4			
125	2.46	122.9	275	5.41	270.3			
130	2.56	127.8	280	5.50	275.2			
135	2.65	132.7	285	5.60	280.1			
140	2.75	137.6	290	5.70	285.0			
145	2.85	142.5	295	5.80	289.9			

APPENDIX C - 5

50 kg BAG and BULK

QUANTITY OF CEMENT IN GROUT

0.8 : 1
(by loose volume)

See Section 4.1–11 for explanations and methods of use.

Grout Volume (liters)	Cement Content (bags)	Cement Content (kg)	Grout Volume (liters)	Cement Content (bags)	Cement Content (kg)	Grout Volume (liters)	Cement Content (bags)	Cement Content (kg)
0	0.00	0.0	150	3.42	170.9	300	6.84	341.9
5	0.11	5.7	155	3.53	176.6	305	6.95	347.6
10	0.23	11.4	160	3.65	182.3	310	7.07	353.3
15	0.34	17.1	165	3.76	188.0	315	7.18	359.0
20	0.46	22.8	170	3.87	193.7	320	7.29	364.7
25	0.57	28.5	175	3.99	199.4	325	7.41	370.4
30	0.68	34.2	180	4.10	205.1	330	7.52	376.1
35	0.80	39.9	185	4.22	210.8	335	7.64	381.8
40	0.91	45.6	190	4.33	216.5	340	7.75	387.5
45	1.03	51.3	195	4.44	222.2	345	7.86	393.2
50	1.14	57.0	200	4.56	227.9	350	7.98	398.9
55	1.25	62.7	205	4.67	233.6	355	8.09	404.6
60	1.37	68.4	210	4.79	239.3	360	8.21	410.3
65	1.48	74.1	215	4.90	245.0	365	8.32	416.0
70	1.60	79.8	220	5.01	250.7	370	8.43	421.7
75	1.71	85.5	225	5.13	256.4	375	8.55	427.4
80	1.82	91.2	230	5.24	262.1	380	8.66	433.1
85	1.94	96.9	235	5.36	267.8	385	8.78	438.8
90	2.05	102.6	240	5.47	273.5	390	8.89	444.5
95	2.17	108.3	245	5.58	279.2	395	9.00	450.2
100	2.28	114.0	250	5.70	284.9	400	9.12	455.9
105	2.39	119.7	255	5.81	290.6			
110	2.51	125.4	260	5.93	296.3			
115	2.62	131.1	265	6.04	302.0			
120	2.74	136.8	270	6.15	307.7			
125	2.85	142.5	275	6.27	313.4			
130	2.96	148.2	280	6.38	319.1			
135	3.08	153.9	285	6.50	324.8			
140	3.19	159.6	290	6.61	330.5			
145	3.30	165.2	295	6.72	336.2			

APPENDIX C - 5

50 kg BAG and BULK

QUANTITY OF CEMENT IN GROUT

0.6 : 1
(by loose volume)

See Section 4.1–11 for explanations and methods of use.

Grout Volume	Cement Content		Grout Volume	Cement Content		Grout Volume	Cement Content	
(liters)	(bags)	(kg)	(liters)	(bags)	(kg)	(liters)	(bags)	(kg)
0	0.00	0.0	150	4.07	203.4	300	8.14	406.8
5	0.14	6.8	155	4.20	210.2	305	8.27	413.6
10	0.27	13.6	160	4.34	217.0	310	8.41	420.4
15	0.41	20.3	165	4.47	223.7	315	8.54	427.1
20	0.54	27.1	170	4.61	230.5	320	8.68	433.9
25	0.68	33.9	175	4.75	237.3	325	8.81	440.7
30	0.81	40.7	180	4.88	244.1	330	8.95	447.5
35	0.95	47.5	185	5.02	250.9	335	9.09	454.3
40	1.08	54.2	190	5.15	257.6	340	9.22	461.0
45	1.22	61.0	195	5.29	264.4	345	9.36	467.8
50	1.36	67.8	200	5.42	271.2	350	9.49	474.6
55	1.49	74.6	205	5.56	278.0	355	9.63	481.4
60	1.63	81.4	210	5.70	284.8	360	9.76	488.2
65	1.76	88.1	215	5.83	291.5	365	9.90	494.9
70	1.90	94.9	220	5.97	298.3	370	10.03	501.7
75	2.03	101.7	225	6.10	305.1	375	10.17	508.5
80	2.17	108.5	230	6.24	311.9	380	10.31	515.3
85	2.31	115.3	235	6.37	318.7	385	10.44	522.1
90	2.44	122.0	240	6.51	325.4	390	10.58	528.8
95	2.58	128.8	245	6.64	332.2	395	10.71	535.6
100	2.71	135.6	250	6.78	339.0	400	10.85	542.4
105	2.85	142.4	255	6.92	345.8			
110	2.98	149.2	260	7.05	352.6			
115	3.12	155.9	265	7.19	359.3			
120	3.25	162.7	270	7.32	366.1			
125	3.39	169.5	275	7.46	372.9			
130	3.53	176.3	280	7.59	379.7			
135	3.66	183.1	285	7.73	386.5			
140	3.80	189.8	290	7.86	393.2			
145	3.93	196.6	295	8.00	400.0			

APPENDIX C - 5

50 kg BAG and BULK

QUANTITY OF CEMENT IN GROUT

0.5 : 1
(by loose volume)

See Section 4.1–11 for explanations and methods of use.

Grout Volume (liters)	Cement Content (bags)	Cement Content (kg)	Grout Volume (liters)	Cement Content (bags)	Cement Content (kg)	Grout Volume (liters)	Cement Content (bags)	Cement Content (kg)
0	0.00	0.0	150	4.49	224.7	300	8.99	449.5
5	0.15	7.5	155	4.64	232.2	305	9.14	457.0
10	0.30	15.0	160	4.79	239.7	310	9.29	464.4
15	0.45	22.5	165	4.94	247.2	315	9.44	471.9
20	0.60	30.0	170	5.09	254.7	320	9.59	479.4
25	0.75	37.5	175	5.24	262.2	325	9.74	486.9
30	0.90	44.9	180	5.39	269.7	330	9.89	494.4
35	1.05	52.4	185	5.54	277.2	335	10.04	501.9
40	1.20	59.9	190	5.69	284.7	340	10.19	509.4
45	1.35	67.4	195	5.84	292.2	345	10.34	516.9
50	1.50	74.9	200	5.99	299.6	350	10.49	524.4
55	1.65	82.4	205	6.14	307.1	355	10.64	531.9
60	1.80	89.9	210	6.29	314.6	360	10.79	539.4
65	1.95	97.4	215	6.44	322.1	365	10.94	546.8
70	2.10	104.9	220	6.59	329.6	370	11.09	554.3
75	2.25	112.4	225	6.74	337.1	375	11.24	561.8
80	2.40	119.9	230	6.89	344.6	380	11.39	569.3
85	2.55	127.3	235	7.04	352.1	385	11.54	576.8
90	2.70	134.8	240	7.19	359.6	390	11.69	584.3
95	2.85	142.3	245	7.34	367.1	395	11.84	591.8
100	3.00	149.8	250	7.49	374.6	400	11.99	599.3
105	3.15	157.3	255	7.64	382.0			
110	3.30	164.8	260	7.79	389.5			
115	3.45	172.3	265	7.94	397.0			
120	3.60	179.8	270	8.09	404.5			
125	3.75	187.3	275	8.24	412.0			
130	3.90	194.8	280	8.39	419.5			
135	4.05	202.3	285	8.54	427.0			
140	4.20	209.8	290	8.69	434.5			
145	4.34	217.2	295	8.84	442.0			

Comparison of Water:Cement Ratios by Weight and Volume

Water:cement ratios can be specified in various ways. They are compared here.

As explained in Section 4.1–2, "volume" is the loose volume of a bag of cement. The conversion factors used here are:

1.5 when changing from weight to volume, and

0.66 when doing the reverse.

Tables comparing water:cement ratios follow.

Water:Cement Ratios

By Volume (As used in this book)		By Weight
5:1	=	3.3:1
4:1	=	2.6:1
3:1	=	2:1
2:1	=	1.3:1
1:1	=	0.7:1
0.8:1	=	0.5:1
0.6:1	=	0.4:1
0.5:1	=	0.3:1

Appendix D
Continued

Water:Cement Ratios

By Weight		By Volume (As used in this book)
4:1	=	6:1
3:1	=	4.5:1
2:1	=	3:1
1:1	=	1.5:1
0.8:1	=	1.2:1
0.6:1	=	0.9:1
0.4:1	=	0.6:1
0.3:1	=	0.5:1

Reverse Ratios

- By Weight ONLY

W : C		C : W
4:1	=	0.25
3.3:1	=	0.30
3:1	=	0.33
2.6:1	=	0.38
2:1	=	0.50
1.3:1	=	0.77
1:1	=	1.00
0.8:1	=	1.25
0.7:1	=	1.43
0.6:1	=	1.67
0.5:1	=	2.00
0.4:1	=	2.50
0.3:1	=	3.33

APPENDIX E

Pressures for Water Testing in Exploratory Holes

This Appendix shows test pressures for water testing in exploratory holes in pounds per square inch and in bars, for normal and for weak rock. See Section 8.2–8 for method of use and comments.

APPENDIX E - 1

PSI

NORMAL ROCK

Depth to Bottom of Test Stage. (feet)	Test Pressure				
	First Run (psi)	Second Run (psi)	Third Run (psi)	Fourth Run (psi)	Fifth Run (psi)
5	5	10	10	10	5
10	5	10	10	10	5
15	5	10	15	10	5
20	10	15	20	15	10
25	10	15	25	15	10
30	10	20	30	20	10
35	10	20	35	20	10
40	10	25	40	25	10
45	15	30	45	30	15
50	20	35	50	35	20
55	20	40	55	40	20
60	25	40	60	40	25
65	25	45	65	45	25
70	30	45	70	45	30
75	30	50	75	50	30
80	35	55	80	55	35
85	35	55	85	55	35
90	40	60	90	60	40
95	40	60	95	60	40
100	40	65	100	65	40
105	40	65	105	65	40
110	40	70	110	70	40
115	40	75	115	75	40
120	40	80	120	80	40
125	40	85	125	85	40
130	40	90	130	90	40
135	40	95	135	95	40
140	40	100	140	100	40
145	40	100	145	100	40
150	40	100	150	100	40
>150	40	100	150	100	40

APPENDIX E - 1

PSI

WEAK ROCK

Depth to Bottom of Test Stage. (feet)	Test Pressure				
	First Run (psi)	Second Run (psi)	Third Run (psi)	Fourth Run (psi)	Fifth Run (psi)
5	5	5	5	5	5
10	5	5	5	5	5
15	5	5	5	5	5
20	5	5	10	5	5
25	5	5	10	5	5
30	5	10	15	10	5
35	5	10	15	10	5
40	10	15	20	15	10
45	10	15	25	15	10
50	10	15	25	15	10
55	10	20	30	20	10
60	10	20	30	20	10
65	10	20	35	20	10
70	10	20	35	20	10
75	10	25	40	25	10
80	10	25	40	25	10
85	15	30	45	30	15
90	15	35	50	35	15
95	15	35	50	35	15
100	15	35	55	35	15
105	20	40	55	40	20
110	20	40	60	40	20
115	20	40	60	40	20
120	20	40	65	40	20
125	20	40	65	40	20
130	25	45	70	45	25
135	25	45	70	45	25
140	25	50	75	50	25
145	25	55	80	55	25
150	25	55	80	55	25
>150	25	55	85	55	25

APPENDIX E - 2

BARS

NORMAL ROCK

Depth to Bottom of Test Stage. (meters)	Test Pressure				
	First Run (bars)	Second Run (bars)	Third Run (bars)	Fourth Run (bars)	Fifth Run (bars)
5	0.5	0.8	1.0	0.8	0.5
6	0.8	1.0	1.4	1.0	0.8
7	0.8	1.0	1.6	1.0	0.8
8	0.8	1.0	1.6	1.0	0.8
9	0.8	1.4	2.0	1.4	0.8
10	0.8	1.4	2.5	1.4	0.8
12	0.8	1.6	2.8	1.6	0.8
14	1.0	2.0	3.0	2.0	1.0
16	1.4	2.5	3.5	2.5	1.4
18	1.6	2.8	4.0	2.8	1.6
20	1.6	3.0	4.5	3.0	1.6
22	2.0	3.5	5.0	3.5	2.0
24	2.5	4.0	5.5	4.0	2.5
26	2.5	4.0	5.8	4.0	2.5
28	2.8	4.0	6.3	4.0	2.8
30	2.8	4.5	6.8	4.5	2.8
32	2.8	4.5	7.3	4.5	2.8
34	2.8	5.0	7.8	5.0	2.8
36	2.8	5.5	8.3	5.5	2.8
38	2.8	6.0	8.8	6.0	2.8
40	2.8	6.3	9.0	6.3	2.8
42	2.8	6.5	9.5	6.5	2.8
44	2.8	7.0	10.0	7.0	2.8
46	2.8	7.0	10.0	7.0	2.8
>46	2.8	7.0	10.0	7.0	2.8

APPENDIX E - 2

BARS

NORMAL ROCK

Depth to Bottom of Test Stage. (meters)	Test Pressure				
	First Run (bars)	Second Run (bars)	Third Run (bars)	Fourth Run (bars)	Fifth Run (bars)
5	0.4	0.4	0.4	0.4	0.4
6	0.4	0.4	0.7	0.4	0.4
7	0.4	0.4	0.7	0.4	0.4
8	0.4	0.4	0.7	0.4	0.4
9	0.4	0.7	1.0	0.7	0.4
10	0.4	0.7	1.0	0.7	0.4
12	0.7	1.0	1.4	1.0	0.7
14	0.7	1.0	1.8	1.0	0.7
16	0.7	1.0	1.8	1.0	0.7
18	0.7	1.4	2.0	1.4	0.7
20	0.7	1.4	2.5	1.4	0.7
22	0.7	1.8	2.5	1.8	0.7
24	0.7	1.8	3.0	1.8	0.7
26	1.0	2.0	3.0	2.0	1.0
28	1.0	2.5	3.5	2.5	1.0
30	1.0	2.5	3.8	2.5	1.0
32	1.4	3.0	3.8	3.0	1.4
34	1.4	3.0	4.0	3.0	1.4
36	1.4	3.0	4.5	3.0	1.4
38	1.4	3.0	4.5	3.0	1.4
40	1.8	3.8	5.0	3.8	1.8
42	1.8	3.8	5.0	3.8	1.8
44	1.8	3.8	5.5	3.8	1.8
46	1.8	3.8	5.5	3.8	1.8
48	1.8	3.8	6.0	3.8	1.8
50	1.8	4.0	6.0	4.0	1.8
52	1.8	4.0	6.5	4.0	1.8
54	1.8	4.0	7.0	4.0	1.8
>54	1.8	4.0	7.0	4.0	1.8

About the Author

A native of Sydney, Australia, I was born in 1929. Involvement with grouting started in 1954. Prior to that, my civil engineering qualifications were obtained at a part-time course while working in an engineering office.

The first five years of working with grouting were ones that could be regarded as an apprenticeship of sorts, with crews who had a grouting tradition going back 30 years or more on a series of dams. The job at that time was the construction of Warragamba Dam, the largest concrete gravity dam in this part of the world and a large one by any standards. My involvement included all forms of rock grouting, contraction joint grouting, tunnel grouting, penstock grouting, and various other forms of grouting. There were general construction duties as well.

When the Warragamba dam was finished, I changed to another government department that was engaged in building a series of dams and there stayed until retirement. During those 27 years, I took part in every aspect of rock grouting as well as other engineering work, including design. Involvement in grouting included exploratory work on new sites, design of the grouting, oversight of the construction, daily assessments, and overall evaluations. And after that, in my other hat as leader of the department's dam safety and monitoring group, I was able to keep a detailed watch on the long-term performance of the grouting. This was an exceptional opportunity to evaluate grouting methods and decide which were good and which weren't. This book benefits, hopefully, from these lessons.

In due course I was lent out to other government departments in various parts of Australia to help with their grouting problems after things had gone wrong or to give advice on tricky sites. This activity later extended overseas.

In 1980, I was brought from antipodean remoteness to the U.S. scene by Professor Norbert Schmidt, of the University of Missouri at Rolla. He had heard rumors about a distant fellow and wanted someone to give the cement grouting section of his annual short course on grouting. After getting a person he called a "spy" to hear me lecture in Australia, off I went to the States—for the first time. Each year thereafter I've enjoyed being a speaker at the Course; even more I have liked the informal chats about grouting experiences with the people there, both experienced and otherwise. As mentioned in Chapter 10, I suspect that in retrospect this course will be regarded as a significant event in the history of grouting. It has become a forum on the subject and is very well attended, and the know-how resulting from the course has spread far and wide.

Over the years I have lost count of the number of dam grouting jobs worked on to various extents—it must be over fifty by now. It became apparent that written guidance on the construction side of grouting was needed, so I wrote (and illustrated, edited, and supervised the printing of) the *Grouting Manual* in 1972 for the Water Resources Commission. This has proved popular and useful. It has been translated into other languages and printed in other countries. However, it is not quite up to date. Being a manual, it did not give explanations or comment, and its brevity precluded mention of much ancillary information. Hence the need for this present book, which includes the whole of the old manual.

This book has been a retirement project over the last three years. I hope it proves interesting and helpful.

INDEX

9 780471 516293

Made in the USA
Monee, IL
14 September 2026

63198321R00260